T0342248

Advances in Time-Domain Computational Electromagnetic Methods

Advances in Time-Domain Computational Electromagnetic Methods

Edited by Qiang Ren, Su Yan, and Atef Z. Elsherbeni

IEEE Antennas and Propagation Society

IEEE Press Series on Electromagnetic Wave Theory

Published by John Wiley & Sons, Inc., Hoboken, New Jersey.
Published simultaneously in Canada.

For general information on our other products and services or for technical support, please contact our Customer Care Department within the United States at (800) 762-2974, outside the United States at (317) 572-3993 or fax (317) 572-4002.

Wiley also publishes its books in a variety of electronic formats. Some content that appears in print may not be available in electronic formats. For more information about Wiley products, visit our web site at www.wiley.com.

Library of Congress Cataloging-in-Publication Data applied for:
ISBN: 9781119808374

Cover Image and Design: Wiley

Set in 9.5/12.5pt STIXTwoText by Straive, Chennai, India

Contents

About the Editors

Qiang Ren (S'12–M'16–SM'22) received both the B.S. and M.S. degrees in electrical engineering from Beihang University, Beijing, China, and Institute of Acoustics, Chinese Academy of Sciences, Beijing, China, in 2008 and 2011, respectively, and the PhD degree in Electrical Engineering from Duke University, Durham, NC, in 2015. From 2016 to 2017, he was a postdoctoral researcher with the Computational Electromagnetics and Antennas Research Laboratory (CEARL) of the Pennsylvania State University, University Park, PA. In Sept 2017, he joined the School of Electronics and Information Engineering, Beihang University, Beijing, China, as an "Excellent Hundred" Associate Professor.

Dr. Ren is the recipient of The Applied Computational Electromagnetics Society (ACES) Early Career Award in 2021, the Young Scientist Award of APEMC 2022, and the Young Scientist Award of 2018 International Applied Computational Electromagnetics Society (ACES) Symposium in China. He serves as the Associate Editor of *IEEE Journal of Multiscale and Multiphysics Simulation Techniques*, *ACES Journal*, and *Microwave and Optical Technology Letters* (MOTL), and also serves as a reviewer for more than 30 journals. He has published more than 90 papers in peer-reviewed journals and conferences. The students he advised have received multiple awards including Best Student Paper 3rd Prize of PIERS 2021 and Best Paper Shortlist of CSRSWTC 2021, and Honorable Mention Award of AP-S URSI 2022. His current research interests include numerical modeling methods for complex media, multiscale and multiphysics problems, inverse scattering, deep learning, and parallel computing.

Dr. Su Yan (S'08-M'12-SM'17) received the B.S. degree in electromagnetics and microwave technology from the University of Electronic Science and Technology of China, Chengdu, China, in 2005, and the M.S. and Ph.D. degrees in electrical and computer engineering from the University of Illinois at Urbana–Champaign (UIUC), Urbana, IL, USA, in 2012 and 2016, respectively. He joined the Howard University, Washington, DC, USA, as an Assistant Professor in 2018, and serves as the Director of Graduate Studies in the Department of Electrical Engineering and Computer Science since 2020. He has authored or coauthored over 110 papers in refereed journals. His current research interests include nonlinear electromagnetic and multiphysics problems, electromagnetic scattering and radiation, numerical methods in computational electromagnetics, especially continuous and discontinuous Galerkin finite element methods, integral-equation-based methods, domain decomposition methods, fast algorithms, and preconditioning techniques.

Dr. Yan is a Senior Member of the Institute of Electrical and Electronics Engineers (IEEE) and a Life Member of the Applied Computational Electromagnetics Society (ACES). He was a recipient of the ACES Early Career Award "for contributions to linear and nonlinear electromagnetic and multiphysics modeling and simulation methods" in 2020, the P. D. Coleman Outstanding Research Award, and the Yuen T. Lo Outstanding Research Award by the Department of Electrical and Computer Engineering, UIUC, in 2015 and 2014, respectively. He was also a recipient of the Edward E. Altschuler AP-S Magazine Prize Paper Award by IEEE Antennas and Propagation Society in 2020, the Best Student Paper Award (the first place winner) at the IEEE ICWITS/ACES 2016 Conference, Honolulu, HI, USA, in 2016, the USNC/URSI Travel Fellowship Grant Award by the National Academies in 2015, the Best Student Paper Award (the first place winner) at the 27th International Review of Progress in ACES, Williamsburg, VA, USA, in 2011, and the Best Student Paper Award by the IEEE Chengdu Section in 2010. He serves as an Associate Editor for *IEEE Access*, an Associate Editor for the *International Journal of Numerical Modelling: Electronic Networks, Devices and Fields*, and a reviewer for multiple journals and conferences.

Atef Z. Elsherbeni received an honor's B.Sc. degree in Electronics and Communications, an honor's B.Sc. degree in Applied Physics, and a M.Eng. degree in Electrical Engineering, all from Cairo University, Cairo, Egypt, in 1976, 1979, and 1982, respectively, and a Ph.D. degree in Electrical Engineering from Manitoba University, Winnipeg, Manitoba, Canada, in 1987. He started his engineering career as a part-time Software and System Design Engineer from March 1980 to December 1982 at the Automated Data System Center, Cairo, Egypt. From January to August 1987, he was a Post-Doctoral Fellow at Manitoba University. Dr. Elsherbeni joined the faculty at the University of Mississippi in August 1987 as an Assistant Professor of Electrical Engineering. He advanced to the rank of Associate Professor in July 1991 and to the rank of Professor in July 1997. He was the Associate Dean of the College of Engineering for Research and Graduate Programs from July 2009 to July 2013 at the University of Mississippi. He then joined the Electrical Engineering and Computer Science (EECS) Department at Colorado School of Mines in August 2013 as the Dobelman Distinguished Chair Professor. He was appointed the Interim Department Head for (EECS) from 2015 to 2016 and from 2016 to 2018 he was the Electrical Engineering Department Head. He spent a sabbatical term in 1996 at the Electrical Engineering Department, University of California at Los Angeles (UCLA), and was a Visiting Professor at Magdeburg University during the summer of 2005 and at Tampere University of Technology in Finland during the summer of 2007. In 2009, he was selected as Finland Distinguished Professor by the Academy of Finland and TEKES.

Over the years, Dr. Elsherbeni participated in acquiring millions of dollars to support his research group activities dealing with scattering and diffraction of EM waves by dielectric and metal objects, finite difference time domain analysis of antennas and microwave devices, field visualization and software development for EM education, interactions of electromagnetic waves with human body, RFID and sensor integrated FRID systems, reflector and printed antennas and antenna arrays for radars, UAV, and personal communication systems, antennas for wideband applications, and measurements of antenna characteristics and material properties. Dr. Elsherbeni is an IEEE life fellow and ACES fellow. He is the Editor-in-Chief for *ACES Journal* and a past Associate Editor of the *Radio Science Journal*. He was the Chair of the Engineering and Physics Division of the

Mississippi Academy of Science, the Chair of the Educational Activity Committee for IEEE Region 3 Section, and the general Chair for the 2014 APS-URSI Symposium and the president of ACES Society from 2013 to 2015. Dr. Elsherbeni is selected as Distinguished Lecturer for IEEE Antennas and Propagation Society for 2020-2023.

List of Contributors

David S. Abraham
Department of Electrical and
Computer Engineering
McGill University
Montréal, Québec
Canada

Amir Akbari
Department of Electrical & Computer
Engineering
McGill University
Montréal, Québec
Canada

Ali Akbarzadeh-Sharbaf
Department of Electrical and
Computer Engineering
McGill University
Montréal, Québec
Canada

Abdullah Algarni
Department of Electrical Engineering
King Fahd University of Petroleum
and Minerals, Dhahran
Saudi Arabia

Shirook Ali
Faculty of Applied Science and
Technology
Sheridan College
Brampton, Ontario
Canada

Francesco P. Andriulli
Department of Electronics
Politecnico di Torino
Torino
Italy

Hakan Bagci
Electrical and Computer Engineering
(ECE) Program
Division of Computer, Electrical, and
Mathematical
Science and Engineering (CEMSE)
King Abdullah University of Science
and Technology (KAUST)
Thuwal
Saudi Arabia

Mohamed H. Bakr
Department of Electrical and
Computer Engineering
McMaster University
Hamilton, Ontario
Canada

Jiefu Chen
Department of Electrical and
Computer Engineering
University of Houston
Houston
United States

Liang Chen
Electrical and Computer Engineering
(ECE) Program, Division of Computer,
Electrical, and Mathematical Science
and Engineering (CEMSE)
King Abdullah University of Science
and Technology (KAUST)
Thuwal
Saudi Arabia

Rui Chen
Division of Computer, Electrical, and
Mathematical Science and
Engineering (CEMSE)
King Abdullah University of Science
and Technology (KAUST)
Thuwal
Saudi Arabia

Xibi Chen
Tsinghua University
Beijing
China

Kristof Cools
Department of Information Tech.,
Ghent University
Belgium

Alexandre Dély
Department of Electronics
Politecnico di Torino
Torino
Italy

Veysel Demir
Electrical Engineering Department
Northern Illinois University
Chicago, IL
USA

Ming Dong
Electrical and Computer Engineering
(ECE) Program, Division of Computer,
Electrical, and Mathematical Science
and Engineering (CEMSE)
King Abdullah University of Science
and Technology (KAUST)
Thuwal
Saudi Arabia

Atef Z. Elsherbeni
Electrical Engineering Department
Colorado School of Mines
Golden, CO
USA

Dennis D. Giannacopoulos
Department of Electrical and
Computer Engineering
McGill University
Montréal, Québec
Canada

Mohammed Hadi
Electrical Engineering Department
Colorado School of Mines
Golden, CO
USA

Yanyan Hu
Department of Electrical and
Computer Engineering
University of Houston
Houston
USA

Yuyang Hu
Department of Electronic Engineering
Shanghai Jiao Tong University
Shanghai
P.R. China

Shifeng Huang
Department of Electronic Engineering
Shanghai Jiao Tong University
Shanghai
P.R. China

Zhixiang Huang
Key Laboratory of Intelligent
Computing and Signal Processing
Ministry of Education
Anhui University
Hefei
China

Yunfeng Jia
Research Institute of Frontier Science
Beihang University
Beijing
China

Lijun Jiang
Department of Electrical and
Electronic Engineering
The University of Hongkong
Hongkong
China

Yuchen Jin
Department of Electrical and
Computer Engineering
University of Houston
Houston
USA

Joshua M. Kast
Colorado School of Mines
Electrical Engineering Department
Golden, CO
USA

Chao Li
School of Electronics and Information
Engineering
Beihang University
Beijing
China

Ping Li
Department of Electrical Engineering
Shanghai Jiao Tong University
Shanghai
China

Rui Liu
Department of Electronic Engineering
Shanghai Jiao Tong University
Shanghai
P.R. China

Adrien Merlini
Microwave department
IMT Atlantique
Brest
France

Jiamei Mi
School of Electronics and Information
Engineering
Beihang University
Beijing
China

Qiang Ren
School of Electronics and Information
Engineering
Beihang University
Beijing
China

Xingang Ren
Key Laboratory of Intelligent
Computing and Signal Processing,
Ministry of Education
Anhui University
Hefei
China

Sadeed B. Sayed
School of Electrical and Electronic
Engineering
Nanyang Technological University
Singapore

Wei E.I. Sha
Key Laboratory of Micro-Nano
Electronic Devices and Smart Systems
of Zhejiang Province
College of Information Science and
Electronic Engineering
Zhejiang University
Hangzhou
China

Xuezhe Tian
Department of Electronic Engineering
Shanghai Jiao Tong University
Shanghai
P.R. China

Huseyin A. Ulku
Department of Electronics
Engineering
Gebze Technical University
Kocaeli
Turkey

Alec Weiss
Electrical Engineering Department
Colorado School of Mines
Golden, CO
USA

Pengfei Wen
School of Electronics and Information
Engineering
Beihang University
Beijing
China

Kaiming Wu
School of Electronics and Information
Engineering
Beihang University
Beijing
China

Xuqing Wu
Department of Information and
Logistics Technology
University of Houston
Houston
USA

Gaobiao Xiao
Department of Electronic Engineering
Shanghai Jiao Tong University
Shanghai
P.R. China

Guoda Xie
Key Laboratory of Intelligent
Computing and Signal Processing
Ministry of Education
Anhui University
Hefei
China

Su Yan
Department of Electrical Engineering
and Computer Science
Howard University
Washington, DC
USA

Fan Yang
Tsinghua University
Beijing
China

Wei Zhang
Sino French Engineering School
Beihang University
Beijing
China

Preface

Computational electromagnetics (CEM) research aims at the modeling and simulation of scientific and engineering problems based on the solution of Maxwell's equations or their variations through the development of numerical algorithms and computer programs for the evaluation, prediction, and optimization purposes. Since their early developments in 1960s, CEM methods have been used in widespread areas, such as radar scattering evaluation, antennas and array design, microwave circuit analysis, electronics and nanodevice development, magnetic and electric machine modeling, high-power microwave system simulation, bioelectromagnetic effect and biomedical device modeling, and electromagnetic imaging and inversion, just to name a few.

Through the past 70 some years, CEM methods have evolved into two major categories, asymptotic and full-wave methods. The asymptotic methods are based on an optical description of electromagnetic waves at high frequencies when the problem is electrically large. This category includes the basic geometrical optics (GO) and physical optics (PO) methods, which are later extended with the theory of diffraction to obtain the methods based on the geometrical theory of diffraction (GTD) and the physical theory of diffraction (PTD). Later developments include the unified theory of diffraction (UTD) and the shooting and bouncing ray (SBR) methods. These asymptotic methods, in general, are very efficient in the simulation of electromagnetic problems due to their simplified description of electromagnetic waves and are accurate only in an asymptotic sense, which means their accuracy is good only when the operating frequency is very high, or the problem size is very large compared to the wavelength. As a result, the asymptotic methods are usually regarded as "high-frequency methods."

Another type of methods, known as the full-wave methods, are based on the rigorous numerical solution of Maxwell's equations, the Helmholtz equation, or various integral equations, in either the frequency or the time domain. Two types of numerical methods have been developed based on the mathematical nature of the governing equations, including the ones that solve partial differential equations (PDEs) and those that solve integral equations (IEs). Well-known PDE solvers are the frequency-domain finite difference method (FDM), finite-element method (FEM), and their time-domain counterparts, the finite-difference time-domain (FDTD) method, and the finite-element time-domain (FETD) method. On the IE side, the method of moments (MoM) has been widely used to solve surface integral equations (SIEs), volume integral equations (VIEs), and their combination, the volume–surface integral equations (VSIEs). The FDM and FDTD are very popular in electromagnetic and optical modeling and simulations due to their simplicity in formulation and implementation. But they generally suffer from the low geometrical modeling accuracy due to the use of structured meshes, the low spatial and temporal interpolation accuracy due to the use of finite differencing that is usually second-order accurate, and a large number of time steps due to the use of conditionally stable time integration schemes. Compared to FDM and FDTD, the FEM and FETD are very flexible and accurate in describing complex geometrical structures with unstructured conformal meshes, convenient in achieving higher-order accuracy with high-order interpolatory or hierarchical basis functions, and efficient in time integration with unconditionally stabile time marching schemes in the temporal discretization. However, although FEM and FETD convert PDEs into sparse matrix equations that have a linear storage complexity, the dimension of the matrix equations is usually very large since the simulation domains need to be discretized into volumetric meshes and consequently, the numerical solution of the matrix equations can be very time consuming. Based on the solution of IEs, the MoM only requires the discretization of either the surface or the volume of the objects without the need of modeling their surrounding background or the truncation boundary. As a result, the overall dimension of the matrix equations is much smaller compared to that of the same problem modeled by FEM. Unfortunately, due to the use of Green's function that depicts the global coupling of fields between every two points in the objects, the resulting system matrix is a fully populated dense matrix that requires an $\mathcal{O}(N^2)$ storage with N being the number of degrees of freedoms (DoFs) and either $\mathcal{O}(N^3)$ or $\mathcal{O}(N^2)$ solution cost with a direct or an iterative solver, respectively. This greatly limits the size of the problem that a full-wave method can handle. Since 1990s, various fast algorithms have been developed to reduce the storage and computational complexities, and domain decomposition methods based on the philosophy of "divide and conquer" have been developed. The further application of large-scale parallel computation boosted the modeling and simulation capabilities of the modern CEM methods significantly. Many EM problems that cannot be tackled by the full-wave methods in the past can now be solved with high accuracy and good efficiency.

Full-wave simulations can be performed in both the frequency and the time domains. The numerical solutions obtained from these two types of methods are related by Fourier/inverse Fourier transform. When performing wideband or transient simulations, the frequency- and time-domain methods are equivalent. On the one hand, frequency-domain simulations can be performed on multiple frequencies of interest and their solutions can be inverse Fourier transformed to the time domain to construct time-domain results. On the other hand, time-domain simulations can be performed using a transient excitation to calculate the time-domain solutions, which can then be Fourier transformed to the frequency domain and sampled at the frequencies of interest. Apparently, in solving ultra-wideband problems, the time-domain simulation will outperform its frequency-domain counterpart since the excitation in these applications only lasts for a very short period but will lead to a very large number of sampling frequencies and a very long simulation time when employing frequency-domain solvers. There are, in fact, many other scenarios where time-domain simulations are not only preferred but, many times, required. Typical examples include the simulation of time-modulated materials, nonlinear problems, and multiphysics problems.

In problems involving time-varying/time-modulated materials, such as those encountered in the research of metamaterials and metasurfaces, many material parameters or properties can be tuned using electrical/magnetic/optical/thermal modulation. For instance, an important application of tenability is to break the time invariance, thus achieving the magnetless Lorentz nonreciprocity, which may lead to many exotic phenomena. To simulate such problems, the only approach viable is the time-domain method. In multiphysics problems, multiple physical phenomena, such as EM, thermal, mechanical, and even chemical phenomena, are coupled together in a sense that the variation of the physical quantities in one physical process affects those in other physical processes and vice versa. To simulate such problems, the time-domain methods are also necessary because many of these multiphysics problems involve transient processes, and the mutual couplings usually take place in the temporal domain. More importantly, many multiphysics problems involve the nonlinear coupling of physics, which is naturally suitable and more convenient to simulate in the time domain than in the frequency domain that usually requires a time-harmonic assumption that is not always satisfied.

In this edited book, many important aspects and latest developments of time-domain methods have been covered and the interesting progress of time-domain simulation applications has been reported. These methods and applications have been divided into the following six themes.

A. Time-domain methods for analyzing nonlinear phenomena
B. Time-domain methods for multiphysics and multiscale modeling
C. Time-domain integral equation methods for scattering analysis
D. Applications of deep learning in time-domain methods

E. Parallel computation schemes for time-domain methods

F. Multidisciplinary explorations of time-domain methods

Specifically, Chapters 1–3 focus on the recent developments of time-domain methods for nonlinear phenomena. In Chapter 1, nonlinear circuit elements are integrated into the FDTD modeling of circuit problems, where three categories of methods are discussed in detail, including the use of specialized updating equations, the development of co-simulation approaches, and the employment of data-based models. It is shown that the FDTD technique is extremely well suited to the simulation of nonlinear devices due to its time-domain nature. Chapter 2 employs Maxwell-hydrodynamic model and the finite difference time domain perturbation method to analyze the electromagnetic response of the nonlinear metasurfaces. Several optical components are designed using the proposed algorithm. In Chapter 3, the FETD modeling of dispersive and nonlinear media is reported where very general modeling techniques based on the FETD method are described. These techniques are capable of accurately simulating complex materials exhibiting any combination of electric dispersion, magnetic dispersion, and electric nonlinearity.

Chapters 4–6 cover the time-domain methods for multiphysics and multiscale problems. In Chapter 4, a discontinuous Galerkin time-domain (DGTD) method for the modeling and simulation of a variety of EM problems is described. In this chapter, the DGTD formulation and implementation are explained, followed by the demonstration of its modeling capabilities using several real-life practical problems, from nanostructure simulation to multiphysics implementations. Chapter 5 describes an adaptive DGTD method for the modeling and simulation of both pure EM and multiphysics problems. Through the development of both the dynamic h- and p-adaptation algorithms, the strong nonlinear and multiphysics interactions between EM fields and plasma fluids have been successfully simulated. A DGTD method for periodic and quasi-periodic structures has been reported in Chapter 6, where a memory-efficient DGTD method is developed for the simulation of finite large periodic/quasi-periodic arrays. This memory-efficient DGTD method is an advanced version of the subdomain-level DGTD method that treats each cell in the array as a subdomain to reduce the overall memory consumption.

The third theme of this book, time-domain IE (TDIE) based methods for scattering analysis, includes Chapters 7–9. An explicit marching-on-in-time (MOT) solver is presented in Chapter 7 to solve second-kind TDIEs. In this chapter, a class of explicit MOT schemes recently developed is described. This new class of explicit MOT solvers casts the second-kind TDIE in the form of ordinary differential equations (ODEs) that relate the variable of interest to its temporal derivative, and is shown to be very efficient and accurate. Convolution quadrature TDIE methods for EM scattering are presented in Chapter 8, which aims to provide an overview of the convolution quadrature methods and their applications in CEM. The key ingredient in these convolution quadrature methods is evaluation of frequency-domain integral operators at matrix-valued frequencies. The integration of these methods with the existing frequency-domain boundary element method implementations has been sketched. Chapter 9 discusses the solution of EM scattering problems using impulse responses. In this chapter, impulse responses are defined for TDIEs. The stability properties of TDIEs are investigated from the properties of their impulse responses. The spurious solutions related to the time-domain electric-field integral equation, the time-domain magnetic-field integral equation, and the time-domain combined-field integral equation are discussed.

Chapters 10 and 11 discuss the applications of deep learning techniques in solving time-domain problems. In Chapter 10, time-domain forward and inverse modeling methods are reported using a differentiable programming platform, where a trainable theory-guided recurrent neural network (RNN) equivalent to the FDTD method is exploited to formulate EM propagation and solve Maxwell's equations on differentiable programming platform PyTorch. Due to the specific performance-focused design of PyTorch, the computation efficiency is substantially improved compared to the conventional FDTD implemented on Matlab. Chapter 11 discusses machine learning applications for the modeling and design optimization of high-frequency structures. Focused on this chapter are the applications of artificial neural networks (ANNs) for the efficient EM modeling and design optimization. Different types of DNN layers are discussed and their applications in several EM problems are exploited.

Parallel computation schemes for time-domain methods are discussed in Chapters 12–14. In Chapter 12, an acceleration of FDTD code using Matlab's parallel computing toolbox is described. While the FDTD method has been successfully parallelized in many other programming languages, their implementations are verbose and require an in-depth knowledge of the hardware and the libraries. It is shown in this chapter that using MATLAB's parallel computing toolbox circumvents this problem by providing an abstraction layer between the user and the complex libraries for parallel computing, which provides a flexible and easy way of FDTD acceleration. In Chapter 13, an automatic parallelization scheme of the subdomain-level DGTD method using message passing interface (MPI) technique is presented. A parallel algorithm and automatic load balancing strategy of the subdomain-level DGTD method have been discussed and demonstrated by several large examples. Chapter 14 introduces several parallelization strategies for FETD formulation, including simple and straightforward parallelization strategies and a relatively recent highly parallelizable approach known as the finite-element Gaussian belief propagation (FGaBP) method. Their combination and applications are also discussed.

As the last theme covered in this book, multidisciplinary topics of time-domain methods have been explored in Chapters 15 and 16. A symplectic FDTD method for solving the coupled Maxwell–Schrödinger (M–S) equations is introduced in Chapter 15. In this chapter, the basic theory of the symplectic FDTD is introduced, different types of coupled M–S equations without and with the symplectic structure are discussed, and the related symplectic FDTD technique is presented. Chapter 16 presents a detailed FDTD method formulation based on cylindrical coordinates (CFDTD) with emphases on low-frequency applications, the formulation of the proper absorbing boundary conditions, and the integration of linear circuit elements models for practical simulations.

This book covers many exciting topics in the recent advancements of time-domain modeling and simulation techniques for EM problems. Nonetheless, there are still some difficulties requiring further investigation. Summarized below are some of the editors' thoughts to promote interests in future research.

1. Enhancement of the computational efficiency for multiphysics problems
   Besides the coupling between different governing equations, multiphysics problems are also usually characterized with multiscale nature, i.e. the wavelengths of different waves may vary by several orders. A uniform mesh and time integration will lead to unnecessary DoFs and unaffordable simulation time for the wave with a large wavelength. Therefore, independent mesh and time integration scheme for respective fields or adaptive mesh and adaptive time step interval are highly desired.

2. Fast and efficient simulation of dispersive media
   Different from the frequency-domain methods, which are straightforward in dealing with dispersive media, additional procedure is required in time-domain approaches for dispersion modeling, such as additional differential equations, recursive convolution, and Z transformation. The time-domain simulation methods for the classical dispersive models, such as Drude, Lorentz, and Debye models are mature. However, for more complex dispersive models, the time-domain simulation approach is still not well developed. For example, the Cole–Cole model and Havriliak–Negami model have fractional differential order, which are relatively more difficult to handle. In addition, chiral materials and metasurfaces may not only be dispersive but also be anisotropic, which raises further challenges to time-domain simulation methods.

3. Dimension reduction techniques for efficient geometrical modeling
   The widely employed time-domain EM simulation methods, including the FDTD, FETD, and DGTD methods, are all based on volumetric discretization of the computational domain. This will lead to a very high modeling and computational burden, especially for cases with thin sheets or tiny objects. As a result, appropriate dimension reduction technique is an effective approach to

enhance the modeling efficiency. For example, surface impedance boundary condition, impedance transition boundary condition, or surface current boundary condition can be applied to the thin sheets, such as graphene, black phosphorus, or fractures of formation. The 2.5D approximation can be employed for the IC modeling and geophysics exploration through the exploration of their layered structures.

Qiang Ren
Beihang University, Beijing, China

Su Yan
Howard University, Washington, DC, USA

Atef Z. Elsherbeni
Colorado School of Mines, Golden, CO, USA

Part I

Time-Domain Methods for Analyzing Nonlinear Phenomena

1

Integration of Nonlinear Circuit Elements into FDTD Method Formulation

Joshua M. Kast and Atef Z. Elsherbeni

Colorado School of Mines, Electrical Engineering Department, Golden, CO, USA

1.1 Introduction

Over the past three decades, finite-difference time-domain (FDTD) simulations have been applied dramatically, with three key factors: innovation in simulation boundary conditions, improved capabilities of computing hardware, and novel updating equations to model special conditions within the FDTD. While FDTD simulations initially enabled the exploration of interactions between electromagnetic fields and arbitrary material geometries, a host of additional processes may now be explored by embedding lumped-circuit elements and even entire sub-circuits into FDTD simulations. With this approach, also called "lumped-element finite-difference time-domain" (LE-FDTD) or "hybrid FDTD," complex circuits may be simulated with simultaneous simulation of circuit theory and Maxwell's equations.

There are two major application areas for these evolving techniques: the design of analog microwave circuits and the design of high-speed digital circuits. In both cases, there are demands for increased frequency, bandwidth, and complexity, in addition to low energy consumption and footprint. In microwave circuits, nonlinearity is used in the design of mixers and energy-efficient amplifiers. Digital circuits, comprised of logic gates, are fundamentally nonlinear in their operation. In both cases, effects of harmonics, crosstalk, and distortion are key factors in the implementation of a design.

An important work in this area was performed by Sui et al. [1, 2], who demonstrated the extension of 2D FDTD simulations by incorporating lumped-element circuit components: resistor, capacitor, inductor, and diode, as well as a small-signal model for a transistor. This work shed light on the transmission-line method (TLM), its similarities to FDTD [3], and previous successful implementations of nonlinear components into TLM simulations [4]. In their 2D simulations,

Advances in Time-Domain Computational Electromagnetic Methods, First Edition.
Edited by Qiang Ren, Su Yan, and Atef Z. Elsherbeni.

Sui et al. demonstrated a circuit comprising a linear resistor and a nonlinear diode, which was in favorable agreement with circuit-theory SPICE-type simulations.

This work in 2D FDTD was extended by Piket-May et al. [5, 6]. In these works, several important contributions were made to the technique. First, 3D formulations were described for lumped elements including resistor, capacitor, inductor, and diode. Second, the updating formulation of diodes was modified based on [2], by replacing the explicit implementation of the Shockley diode equation by an implicit form, which was solved using Newton–Raphson iteration. Also, a bipolar junction transistor (BJT) was implemented based on the Ebers–Moll model [7], which accounts for both small- and large-signal effects. The Ebers–Moll model was implemented directly in the form of FDTD updating equations, which were solved using Newton–Raphson iteration in each FDTD time step.

An additional novel technique was presented in [5]: a software link between an FDTD simulator and a circuit-theory SPICE simulator [8]. The possibility of integrating a time-domain circuit simulator into the FDTD is extremely appealing because it enables the integration of lumped elements, or even sub-circuits into an FDTD simulation, without the need to create bespoke updating equations for each sub-circuit. A wealth of semiconductor formulations within SPICE [9] may be readily incorporated into the FDTD. While the FDTD–SPICE connection was successfully demonstrated, accuracy and stability problems were encountered with microstrip geometries commonly used for microwave circuits [5]. Nonetheless, this work laid important foundations for future developments in the field.

Based on these foundational works, numerous approaches have been described for simulating FDTD with integrated lumped elements, sub-circuits, and active semiconductor devices. We broadly divide these into three categories for further discussion. The first category is specialized updating equations used to simulate a specific type of component within the FDTD. The second category is co-simulation approaches, such as the one described in [5], where an offboard circuit simulator is used to model the activity of a sub-circuit or circuit element. The third category includes data-based models, which emulate the behavior of a device based on stored measurement or simulation data (often S-parameters).

1.2 FDTD Updating Equations for Nonlinear Elements

Initial efforts to incorporate lumped-element devices into the FDTD focused primarily on developing specialized updating formulations for Yee cells occupied by circuit components [2, 6]. This is an advantageous approach because circuit components may be updated using the same central difference (second-order accurate) differentiation scheme used for updating the electric and magnetic fields. Also, the computational overhead is minimized when separate simulation engines are not invoked.

As a basis for further discussion, we assume an FDTD simulation with an electric-field updating equation of the form:

$$
\begin{aligned}
E_z\big|_{i,j,k}^{n+1} = {} & \frac{2\varepsilon_z\big|_{i,j,k} - \Delta t\,\sigma_z^e\big|_{i,j,k}}{2\varepsilon_z\big|_{i,j,k} + \Delta t\,\sigma_z^e\big|_{i,j,k}}\, E_z\big|_{i,j,k}^{n} \\
& + \frac{2\Delta t}{\left(2\varepsilon_z\big|_{i,j,k} + \Delta t\,\sigma_z^e\big|_{i,j,k}\right)\Delta x}\left(H_y\big|_{i,j,k}^{n+\frac{1}{2}} - H_y\big|_{i-1,j,k}^{n+\frac{1}{2}}\right) \\
& - \frac{2\Delta t}{\left(2\varepsilon_z\big|_{i,j,k} + \Delta t\,\sigma_z^e\big|_{i,j,k}\right)\Delta y}\left(H_x\big|_{i,j,k}^{n+\frac{1}{2}} - H_x\big|_{i,j-1,k}^{n+\frac{1}{2}}\right) \\
& - \frac{2\Delta t}{2\varepsilon_z\big|_{i,j,k} + \Delta t\,\sigma_z^e\big|_{i,j,k}}\, J_{iz}\big|_{i,j,k}^{n+\frac{1}{2}}
\end{aligned}
\tag{1.1}
$$

and with magnetic fields updated by:

$$
\begin{aligned}
H_z\big|_{i,j,k}^{n+\frac{1}{2}} = {} & \frac{2\mu_z\big|_{i,j,k} - \Delta t\,\sigma_z^m\big|_{i,j,k}}{2\mu_z\big|_{i,j,k} + \Delta t\,\sigma_z^m\big|_{i,j,k}}\, H_z\big|_{i,j,k}^{n-\frac{1}{2}} \\
& + \frac{2\Delta t}{\left(2\mu_z\big|_{i,j,k} + \Delta t\,\sigma_z^m\big|_{i,j,k}\right)\Delta y}\left(E_x\big|_{i,j+1,k}^{n} - E_x\big|_{i,j,k}^{n}\right) \\
& - \frac{2\Delta t}{\left(2\mu_z\big|_{i,j,k} + \Delta t\,\sigma_z^m\big|_{i,j,k}\right)\Delta x}\left(E_y\big|_{i+1,j,k}^{n} - E_y\big|_{i,j,k}^{n}\right) \\
& - \frac{2\Delta t}{2\mu_z\big|_{i,j,k} + \Delta t\,\sigma_z^m\big|_{i,j,k}}\, M_{iz}\big|_{i,j,k}^{n}
\end{aligned}
\tag{1.2}
$$

These equations are similar to those listed in [6] and are based on the notation and parameter definitions in [10].

1.2.1 Junction Diode

The diode serves as an excellent model for the integration of a nonlinear device into FDTD: it is passive, has only two ports, and can occupy one FDTD cell. The diode was integrated into 2D FDTD by Sui et al. [2], and this approach was improved and extended to 3D FDTD by Piket-May et al. [6]. Both methods use the "ideal" diode equation proposed by Shockley [11]:

$$
I_D = I_S\left(e^{\frac{qV_D}{\eta kT}} - 1\right)
\tag{1.3}
$$

In this equation, the current through the diode I_D is a function of the voltage across the diode V_D, as well as several physical properties of the diode: temperature T and the ideality factor η. These are related by q and K: the charge of an electron and Boltzmann's constant, respectively.

We may adapt (1.1) for a diode pointing in the negative z direction by incorporating (1.3) and understanding that the voltage across the diode V_D at time step $n + \frac{1}{2}$ will be $V_D\big|^{n+\frac{1}{2}} = \Delta z \frac{E_z|^n_{i,j,k} + E_z|^{n+1}_{i,j,k}}{2}$. The current density through the cell $J_{iz}\big|_{i,j,k}$ is replaced by the diode current and the cross-sectional area of the cell $\left(\frac{I_D}{\Delta x \Delta y} \right)$, giving:

$$
\begin{aligned}
E_z\big|^{n+1}_{i,j,k} =\ & \frac{2\varepsilon_z\big|_{i,j,k} - \Delta t\, \sigma^e_z\big|_{i,j,k}}{2\varepsilon_z\big|_{i,j,k} + \Delta t\, \sigma^e_z\big|_{i,j,k}} E_z\big|^n_{i,j,k} \\[2mm]
& + \frac{2\Delta t}{\left(2\varepsilon_z\big|_{i,j,k} + \Delta t\, \sigma^e_z\big|_{i,j,k} \right)\Delta x} \left(H_y\big|^{n+\frac{1}{2}}_{i,j,k} - H_y\big|^{n+\frac{1}{2}}_{i-1,j,k} \right) \\[2mm]
& - \frac{2\Delta t}{\left(2\varepsilon_z\big|_{i,j,k} + \Delta t\, \sigma^e_z\big|_{i,j,k} \right)\Delta y} \left(H_x\big|^{n+\frac{1}{2}}_{i,j,k} - H_x\big|^{n+\frac{1}{2}}_{i,j-1,k} \right) \\[2mm]
& - \frac{2\Delta t}{2\varepsilon_z\big|_{i,j,k} + \Delta t\, \sigma^e_z\big|_{i,j,k}} \frac{I_S\left(e^{\frac{q\Delta z\left(E_z|^n_{i,j,k} + E_z|^{n+1}_{i,j,k} \right)}{2\eta kT}} - 1 \right)}{\Delta x \Delta y}
\end{aligned} \tag{1.4}
$$

The transcendental Eq. (1.4) cannot be solved analytically. In the work of Sui et al. this problem was resolved by calculating V_D based only on the value of $E_z\big|^n_{i,j,k}$ [2], which limited the approach to only very small time steps and low voltages across the diode. A more capable approach, proposed by Piket-May, uses an iterative process to resolve (1.4). In this method, we rearrange (1.4) as follows:

$$
\begin{aligned}
f\left(E_z\big|^n_{i,j,k} \right) =\ & \frac{2\Delta t}{2\varepsilon_z\big|_{i,j,k} + \Delta t\, \sigma^e_z\big|_{i,j,k}} \frac{I_S}{\Delta x \Delta y} e^{\frac{q\,\Delta z E_z|^n_{i,j,k}}{2\eta kT}} e^{\frac{q\, E_z|^{n+1}_{i,j,k}}{2\eta kT}} + E_z\big|^{n+1}_{i,j,k} \\[2mm]
& - \frac{2\Delta t}{2\varepsilon_z\big|_{i,j,k} + \Delta t\, \sigma^e_z\big|_{i,j,k}} \frac{I_S}{\Delta x \Delta y} - \frac{2\varepsilon_z\big|_{i,j,k} - \Delta t\, \sigma^e_z\big|_{i,j,k}}{2\varepsilon_z\big|_{i,j,k} + \Delta t\, \sigma^e_z\big|_{i,j,k}} E_z\big|^n_{i,j,k} \\[2mm]
& - \frac{2\Delta t}{\left(2\varepsilon_z\big|_{i,j,k} + \Delta t\, \sigma^e_z\big|_{i,j,k} \right)\Delta x} \left(H_y\big|^{n+\frac{1}{2}}_{i,j,k} - H_y\big|^{n+\frac{1}{2}}_{i-1,j,k} \right) \\[2mm]
& + \frac{2\Delta t}{\left(2\varepsilon_z\big|_{i,j,k} + \Delta t\, \sigma^e_z\big|_{i,j,k} \right)\Delta y} \left(H_x\big|^{n+\frac{1}{2}}_{i,j,k} - H_x\big|^{n+\frac{1}{2}}_{i,j-1,k} \right)
\end{aligned} \tag{1.5}
$$

with the goal of creating an equation with the format of $f(x) = Ae^{Bx} + x + C$ and finding the root where $f(x) = 0$. In this case:

$$A = \frac{2\Delta t}{2\varepsilon_z|_{i,j,k} + \Delta t \sigma^e_z|_{i,j,k}} \frac{I_S}{\Delta x \Delta y} e^{\frac{q\Delta z\, E_z|^n_{i,j,k}}{2\eta kT}}$$

$$B = \frac{q\Delta z}{2\eta kT}$$

$$C = -\frac{2\Delta t}{2\varepsilon_z|_{i,j,k} + \Delta t \sigma^e_z|_{i,j,k}} \frac{I_S}{\Delta x \Delta y} - \frac{2\varepsilon_z|_{i,j,k} - \Delta t \sigma^e_z|_{i,j,k}}{2\varepsilon_z|_{i,j,k} + \Delta t \sigma^e_z|_{i,j,k}} E_z|^n_{i,j,k}$$
$$- \frac{2\Delta t}{\left(2\varepsilon_z|_{i,j,k} + \Delta t \sigma^e_z|_{i,j,k}\right)\Delta x}\left(H_y\Big|^{n+\frac{1}{2}}_{i,j,k} - H_y\Big|^{n+\frac{1}{2}}_{i-1,j,k}\right)$$
$$+ \frac{2\Delta t}{\left(2\varepsilon_z|_{i,j,k} + \Delta t \sigma^e_z|_{i,j,k}\right)\Delta y}\left(H_x\Big|^{n+\frac{1}{2}}_{i,j,k} - H_x\Big|^{n+\frac{1}{2}}_{i,j-1,k}\right) \qquad (1.6)$$

Given an initial guess for the quantity x, we may improve upon it by solving the following one:

$$x_{\text{updated}} = x - \frac{f(x)}{f'(x)} = x - \frac{Ae^{Bx} + x + C}{ABe^{Bx} + 1} \qquad (1.7)$$

By iteratively solving (1.7), with the constants found in (1.6), the updated value of the electric field across the diode $E_z|^{n+1}_{i,j,k}$ may be calculated, while preserving the second-order accuracy and central difference nature of the FDTD formulation.

A further improvement of this technique was proposed by Ciampolini et al. [12], who provided a more sophisticated model of the junction diode, which includes its nonlinear capacitance, as illustrated in Figure 1.1. In their model, the total

Figure 1.1 Improved model of junction diode incorporating junction and diffusion capacitance.

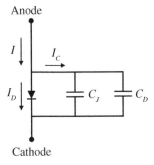

capacitance of a diode C_D is the sum of the junction capacitance C_J and the diffusion capacitance C_D. Both capacitance values are functions of the current through the diode I_D.

The total current through the diode I is given by:

$$I|^{n+\frac{1}{2}} = I_D|^{n+\frac{1}{2}} + \left[C_J|^{n+\frac{1}{2}} + C_D|^{n+\frac{1}{2}} \right] \frac{V_D|^{n+1} - V_D|^n}{\Delta t} \tag{1.8}$$

In this equation, capacitance values are time dependent because they depend on the voltage across the diode V_D. The diffusion capacitance is calculated as:

$$C_D(V_D) = \frac{q}{KT} \tau_D I_S \left(e^{\frac{qV_D}{KT}} - 1 \right) \tag{1.9}$$

where τ_D is the diode's transit time. The junction capacitance is a piecewise function of V_D:

$$C_J(V_D) = \begin{cases} C_{J0} \left(1 - \frac{V_D}{\Phi_0} \right)^{-m} & \text{if } V_D < F_C \cdot \Phi_0 \\ \dfrac{C_{J0}}{(1 - F_C)^{1+m}} \left((1 - F_C(1 + m)) - \dfrac{m \cdot V_D}{\Phi_0} \right) & \text{if } V_D \geq F_C \cdot \Phi_0 \end{cases} \tag{1.10}$$

The values m and F_C are specific to each model of diode and relate to its composition. The value Φ_0 is the "built-in voltage" of the diode, with a default value in SPICE2 of $1\,V$ (but this value may change depending on the specific diode used) [9].

This addition of nonlinear capacitance allows for a more nuanced model of the diode to be incorporated into an FDTD simulation. Ciampolini et al. noted that it is still possible to use Newton–Raphson iteration to resolve self-referential updating equations. However, the increased complexity of the equations and the piecewise nature of (1.10) impose additional demands on the numerical method. To maintain stability in the updating solutions, very short time steps (below the Courant-Friedrichs-Lewy [CFL] limit) and a "damped" Newton–Raphson method were required [12].

1.2.2 Bipolar Junction Transistors: Small-Signal Model

To simulate active devices such as amplifiers, it is often necessary to incorporate a transistor model into a circuit or full-wave simulation. BJTs are commonly employed in analog and radio frequency (RF) electronic devices, and they are capable of both linear and nonlinear modes of operation. During linear operation in the forward-active mode, a BJT may be modeled as a current-controlled current source, as described in [13] and shown in Figure 1.2. The values of the parameters α, R_e, and R_o depend on the type of the transistor, and the selected direct current (DC) operating point, as determined by supply voltage and bias resistors.

Figure 1.2 Schematic symbol and small-signal model of bipolar junction transistor.

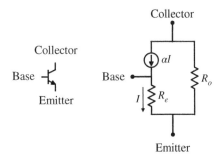

ElMahgoub and Elsherbeni described the introduction of multiple types of dependent sources into FDTD [14], including voltage-controlled voltage sources, voltage-controlled current sources, current-controlled voltage sources, and current-controlled current sources. This selection of dependent sources enables the implementation of small-signal amplifier and transistor models in FDTD.

In [14], the BJT was simulated by first designing a BJT common-collector amplifier circuit and then determining the amplifier's DC operating condition. Next, the updating Eq. (1.1) was modified to function as a current-controlled current source:

$$
\begin{aligned}
E_z\big|_{i,j,k}^{n+1} = {} & \frac{2\varepsilon_z\big|_{i,j,k} - \Delta t\, \sigma_z^e\big|_{i,j,k}}{2\varepsilon_z\big|_{i,j,k} + \Delta t\, \sigma_z^e\big|_{i,j,k}}\, E_z\big|_{i,j,k}^{n} \\
& + \frac{2\Delta t}{\left(2\varepsilon_z\big|_{i,j,k} + \Delta t\, \sigma_z^e\big|_{i,j,k}\right)\Delta x}\left(H_y\big|_{i,j,k}^{n+\frac{1}{2}} - H_y\big|_{i-1,j,k}^{n+\frac{1}{2}}\right) \\
& - \frac{2\Delta t}{\left(2\varepsilon_z\big|_{i,j,k} + \Delta t\, \sigma_z^e\big|_{i,j,k}\right)\Delta y}\left(H_x\big|_{i,j,k}^{n+\frac{1}{2}} - H_x\big|_{i,j-1,k}^{n+\frac{1}{2}}\right) \\
& + \frac{2\Delta t}{\Delta x \Delta y\left(2\varepsilon_z\big|_{i,j,k} + \Delta t\, \sigma_z^e\big|_{i,j,k}\right)}\, \alpha I_{z\,\text{sampled}}\big|_{i,j,k}^{n+\frac{1}{2}}
\end{aligned}
\tag{1.11}
$$

In this modified updating equation, the variable $I_{z\,\text{sampled}}\big|_{i,j,k}^{n+\frac{1}{2}}$ is the current at another location in the circuit, sampled by integration of the magnetic field following Ampere's law. The constant α is the gain of the dependent source and is chosen depending on the DC analysis of the transistor [13].

In the work of [14], the active component of the transistor was simulated by the current-controlled current source updating equation of (1.11). The bias resistors were modeled using lumped-element FDTD components, following the formulations of [14]. When the amplifier circuit was excited by a sinusoidal waveform, the circuit was seen to function as a voltage amplifier with a high gain. The output voltage observed in the FDTD simulation was comparable favorably with analytical predictions, demonstrating the feasibility of this small-signal approach.

1.2.3 Bipolar Junction Transistors: Ebers–Moll Model

The formulation for simulation of a nonlinear BJT may be viewed as an extension of the formulation for the junction diode. The BJT is a three-terminal device wherein the flow of current from the collector terminal to the emitter is controlled by the current applied at the base. In the Ebers–Moll model [7], the two PN semiconductor junctions of the BJT are modeled as junction diodes and the current flow is modeled using dependent current sources, as shown in Figure 1.3.

Following the formulation of [5], we will develop an FDTD updating formulation for the BJT. This three-terminal device spans two cells, with two distinct voltages, the base–collector voltage V_{BC} and the base–emitter voltage V_{BE}. These voltages are related to the electric fields within the Yee cells at time step $n + \frac{1}{2}$ by:

$$V_{BC}\Big|^{n+\frac{1}{2}} = \Delta z \frac{E_z\big|_{BC}^{n} + E_z\big|_{BC}^{n+1}}{2} \tag{1.12}$$

and

$$V_{BE}\Big|^{n+\frac{1}{2}} = -\Delta z \frac{E_z\big|_{EB}^{n} + E_z\big|_{EB}^{n+1}}{2} \tag{1.13}$$

We adapt (1.4) to solve for $E_z\big|_{BC}^{n+1}$. We assume that $\sigma_z^e = 0$ within this region:

$$
\begin{aligned}
E_z\big|_{BC}^{n+1} = E_z\big|_{BC}^{n} &+ \frac{\Delta t}{\varepsilon_z\big|_{BC}\Delta x}\left(H_y\big|_{BC}^{n+\frac{1}{2}} - H_y\big|_{BC+(-1,0,0)}^{n+\frac{1}{2}} \right) \\
&- \frac{\Delta t}{\varepsilon_z\big|_{BC}\Delta y}\left(H_x\big|_{BC}^{n+\frac{1}{2}} - H_x\big|_{BC+(0,-1,0)}^{n+\frac{1}{2}} \right) \\
&- \frac{\Delta t}{\varepsilon_z\big|_{BC}} \frac{I_S\left(e^{\frac{q\Delta z\left(E_z\big|_{BC}^{n} + E_z\big|_{BC}^{n+1} \right)}{2\eta kT}} - 1 \right) - \alpha_F I_S\left(e^{\frac{-q\Delta z\left(E_z\big|_{EB}^{n} + E_z\big|_{EB}^{n+1} \right)}{2\eta kT}} - 1 \right)}{\Delta x \Delta y}
\end{aligned}
\tag{1.14}
$$

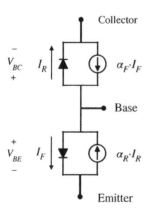

Collector

Base

Emitter

Figure 1.3 Ebers–Moll model of a bipolar junction transistor.

For $E_z|_{EB}^{n+1}$, we find a similar expression:

$$
E_z|_{EB}^{n+1} = E_z|_{EB}^n + \frac{\Delta t}{\varepsilon_z|_{EB}\Delta x} \left(H_y\Big|_{EB}^{n+\frac{1}{2}} - H_y\Big|_{EB+(-1,0,0)}^{n+\frac{1}{2}} \right)
$$

$$
- \frac{\Delta t}{\varepsilon_z|_{EB}\Delta y} \left(H_x\Big|_{EB}^{n+\frac{1}{2}} - H_x\Big|_{EB+(0,-1,0)}^{n+\frac{1}{2}} \right)
$$

$$
- \frac{\Delta t}{\varepsilon_z|_{EB}} \frac{\alpha_R I_S \left(e^{\frac{q\Delta z\left(E_z|_{BC}^n + E_z|_{BC}^{n+1} \right)}{2\eta kT}} - 1 \right) - I_S \left(e^{\frac{-q\Delta z\left(E_z|_{EB}^n + E_z|_{EB}^{n+1} \right)}{2\eta kT}} - 1 \right)}{\Delta x \Delta y}
$$

$$
\text{(1.15)}
$$

These expressions are rearranged to form a pair of expressions in the following format:

$$
f_1(\vec{x}) = A_{1,1}e^{B_{1,1}\vec{x}_1} + A_{1,2}e^{B_{1,2}\vec{x}_2} + \vec{x}_1 + \vec{C}_1
$$
$$
f_2(\vec{x}) = A_{2,1}e^{B_{2,1}\vec{x}_1} + A_{2,2}e^{B_{2,2}\vec{x}_2} + \vec{x}_2 + \vec{C}_2
$$

$$
\text{(1.16)}
$$

as:

$$
f_1\left(\left[E_z|_{EB}^{n+1}, E_z|_{BC}^{n+1} \right]^T \right) = \frac{\Delta t I_S e^{\frac{q\Delta z\, E_z|_{BC}^n}{2\eta kT}}}{\varepsilon_z|_{BC}\Delta x \Delta y} e^{\frac{q\,\Delta z E_z|_{BC}^{n+1}}{2\eta kT}}
$$

$$
- \frac{\Delta t \alpha_F I_S e^{\frac{-q\,\Delta z E_z|_{EB}^n}{2\eta kT}}}{\varepsilon_z|_{BC}\Delta x \Delta y} e^{\frac{-q\Delta z\, E_z|_{EB}^{n+1}}{2\eta kT}} + E_z|_{BC}^{n+1}
$$

$$
- E_z|_{BC}^n - \frac{\Delta t}{\varepsilon_z|_{BC}\Delta x} \left(H_y\Big|_{BC}^{n+\frac{1}{2}} - H_y\Big|_{BC+(-1,0,0)}^{n+\frac{1}{2}} \right)
$$

$$
+ \frac{\Delta t}{\varepsilon_z|_{BC}\Delta y} \left(H_x\Big|_{BC}^{n+\frac{1}{2}} - H_x\Big|_{BC+(0,-1,0)}^{n+\frac{1}{2}} \right)
$$

$$
+ (\alpha_F - 1)\frac{\Delta t I_S}{\varepsilon_z|_{BC}\Delta x \Delta y}
$$

$$
\text{(1.17)}
$$

and

$$
f_2\left(\left[E_z|_{EB}^{n+1}, E_z|_{BC}^{n+1} \right]^T \right) = \frac{\Delta t \alpha_R I_S e^{\frac{q\Delta z\, E_z|_{BC}^n}{2\eta kT}}}{\varepsilon_z|_{EB}\Delta x \Delta y} e^{\frac{q\,\Delta z E_z|_{BC}^{n+1}}{2\eta kT}}
$$

$$
- \frac{\Delta t I_S e^{\frac{-q\Delta z\, E_z|_{EB}^n}{2\eta kT}}}{\varepsilon_z|_{EB}\Delta x \Delta y} e^{\frac{-q\,\Delta z E_z|_{EB}^{n+1}}{2\eta kT}} + E_\varepsilon|_{ED}^{n+1}
$$

$$
- E_z|_{EB}^n - \frac{\Delta t}{\varepsilon_z|_{EB}\Delta x} \left(H_y\Big|_{EB}^{n+\frac{1}{2}} - H_y\Big|_{EB+(-1,0,0)}^{n+\frac{1}{2}} \right)
$$

$$
+ \frac{\Delta t}{\varepsilon_z|_{EB}\Delta y} \left(H_x\Big|_{EB}^{n+\frac{1}{2}} - H_x\Big|_{EB+(0,-1,0)}^{n+\frac{1}{2}} \right)
$$

$$
+ (1 - \alpha_R)\frac{\Delta t I_S}{\varepsilon_z|_{EB}\Delta x \Delta y}
$$

$$
\text{(1.18)}
$$

The values of $E_z|_{EB}^{n+1}$ and $E_z|_{BC}^{n+1}$ are numerically approximated by iterating:

$$\vec{x}^{\text{updated}} = \vec{x} - \boldsymbol{F}^{-1}(\vec{x})f(\vec{x}) \tag{1.19}$$

where

$$\boldsymbol{F}(\vec{x}) = \begin{bmatrix} \dfrac{\partial f_1(\vec{x})}{\partial \vec{x}_1} & \dfrac{\partial f_1(\vec{x})}{\partial \vec{x}_2} \\[2ex] \dfrac{\partial f_2(\vec{x})}{\partial \vec{x}_1} & \dfrac{\partial f_2(\vec{x})}{\partial \vec{x}_2} \end{bmatrix} \tag{1.20}$$

Iteration of (1.19) is carried out until the solution converges satisfactorily, by ensuring that $|\vec{x}^{\text{updated}} - \vec{x}|$ reaches a sufficiently low value. When this technique is applied in circuits with realistic biasing configurations (source voltages and resistances), functional amplifiers may be modeled with the transistor capable of operating in the cut-off, active, and saturation modes [5].

The work of Ciampolini et al. on an improved diode model suggests similar possibilities for the simulation of BJTs [12]. Indeed, in a follow-up work on this subject, the authors demonstrated the extension of the Ebers–Moll model by simulating a BJT having nonlinear capacitances on its emitter–base and base–collector junctions [15]. This is achieved quite simply as extensions of (1.14) and (1.15), which incorporate the nonlinear capacitances of (1.9) and (1.10). We write these modified updating expressions as follows:

$$E_z|_{BC}^{n+1} = E_z|_{BC}^{n} + \frac{\Delta t}{\varepsilon_z|_{BC}\Delta x}\left(H_y\Big|_{BC}^{n+\frac{1}{2}} - H_y\Big|_{BC+(-1,0,0)}^{n+\frac{1}{2}} \right)$$

$$- \frac{\Delta t}{\varepsilon_z|_{BC}\Delta y}\left(H_x\Big|_{BC}^{n+\frac{1}{2}} - H_x\Big|_{BC+(0,-1,0)}^{n+\frac{1}{2}} \right) - \frac{\Delta t}{\varepsilon_z|_{BC}\Delta x\Delta y}$$

$$\times \left(\begin{aligned} & I_S\left(e^{\frac{q\Delta z\left(E_z|_{BC}^{n} + E_z|_{BC}^{n+1} \right)}{2\eta kT}} - 1 \right) - \alpha_F I_S\left(e^{\frac{-q\Delta z\left(E_z|_{EB}^{n} + E_z|_{EB}^{n+1} \right)}{2\eta kT}} - 1 \right) \\ & + \left(C_{JBC}\left(-\Delta z\frac{E_z|_{BC}^{n} + E_z|_{BC}^{n+1}}{2} \right) \right. \\ & \left. + C_{DBC}\left(-\Delta z\frac{E_z|_{BC}^{n} + E_z|_{BC}^{n+1}}{2} \right) \right)\Delta z\frac{E_z|_{BC}^{n+1} - E_z|_{BC}^{n}}{\Delta t} \end{aligned} \right)$$

$$\tag{1.21}$$

and

$$E_z|_{EB}^{n+1} = E_z|_{EB}^{n} + \frac{\Delta t}{\varepsilon_z|_{EB}\Delta x}\left(H_y\Big|_{EB}^{n+\frac{1}{2}} - H_y\Big|_{EB+(-1,0,0)}^{n+\frac{1}{2}} \right)$$

$$- \frac{\Delta t}{\varepsilon_z|_{EB}\Delta y}\left(H_x\Big|_{EB}^{n+\frac{1}{2}} - H_x\Big|_{EB+(0,-1,0)}^{n+\frac{1}{2}} \right) - \frac{\Delta t}{\varepsilon_z|_{EB}\Delta x\Delta y}$$

$$\times \left(\begin{array}{l} \alpha_R I_S \left(e^{\frac{q\Delta z\left(E_z|_{BC}^n + E_z|_{BC}^{n+1} \right)}{2\eta kT}} - 1 \right) - I_S \left(e^{\frac{-q\Delta z\left(E_z|_{EB}^n + E_z|_{EB}^{n+1} \right)}{2\eta kT}} - 1 \right) \\ \\ + \left(C_{JEB} \left(\Delta z \dfrac{E_z|_{EB}^n + E_z|_{EB}^{n+1}}{2} \right) \right. \\ \\ \left. + C_{DEB} \left(\Delta z \dfrac{E_z|_{EB}^n + E_z|_{EB}^{n+1}}{2} \right) \right) \Delta z \dfrac{E_z|_{EB}^{n+1} - E_z|_{EB}^n}{\Delta t} \end{array} \right)$$

$$(1.22)$$

As noted in [12], the additional complexity of these updating equations, with their multiple (and piecewise) nonlinear elements, imposes increased demands on the numerical computations required for these updating equations. In [15], a time-adaptive technique has been proposed, where failure of Newton–Raphson iteration to converge signals a reduction in Δt of the FDTD simulation. This feature is of particular necessity for signals with abrupt transitions.

1.2.4   Bipolar Junction Transistors: Gummel–Poon Model

Improvements upon the Ebers–Moll BJT model were proposed by Gummel and Poon [16]. This model for the BJT is widely used in simulation programs including SPICE, and when certain device constants are omitted, it becomes equivalent to the Ebers–Moll model [9]. The schematic representation of the Gummel–Poon model is shown in Figure 1.4.

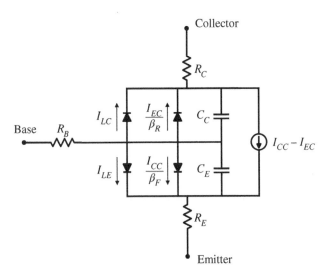

Figure 1.4   Gummel–Poon model of a bipolar junction transistor.

We write a subset of this model, with the current at the collector after [17] as follows:

$$I_C = I_{LC} + \left(1 + \frac{1}{\beta_R}\right) I_{EC} - I_{CC} + C_C \frac{\partial V_{BC}}{\partial t} \tag{1.23}$$

and at the emitter:

$$I_E = -\left(I_{LE} + \left(1 + \frac{1}{\beta_F}\right) I_{CC} - I_{EC} + C_E \frac{\partial V_{BE}}{\partial t}\right) \tag{1.24}$$

where

$$I_{CC} = \frac{I_S}{q_b} \left(e^{\frac{qV_{BE}}{n_F kT}} - 1\right)$$

$$I_{LE} = I_{SE} \left(e^{\frac{qV_{BE}}{n_E kT}} - 1\right)$$

$$I_{EC} = \frac{I_S}{q_b} \left(e^{\frac{qV_{BC}}{n_R kT}} - 1\right)$$

and

$$I_{LC} = I_{SC} \left(e^{\frac{qV_{BC}}{n_C kT}} - 1\right)$$

where q_b is given by:

$$q_b = \frac{1}{2} + \frac{V_{BE}}{2V_B} + \frac{V_{BC}}{2V_A}$$

$$+ \sqrt{\left(\frac{1}{2} + \frac{V_{BE}}{2V_B} + \frac{V_{BC}}{2V_A}\right)^2 + \frac{I_{SS}}{I_{KF}} \left(e^{\frac{qV_{BE}}{n_F kT}} - 1\right) + \frac{I_{SS}}{I_{KR}} \left(e^{\frac{qV_{BC}}{n_R kT}} - 1\right)} \tag{1.25}$$

Nonlinear capacitances, resulting from junction and diffusion effects, can be found by the piecewise expressions:

$$C_E = \begin{cases} \tau_F \dfrac{\partial I_{CC}}{\partial V_{BE}} + C_{JE} \left(1 - \dfrac{V_{BE}}{\Phi_E}\right)^{-m_E} & \text{if } V_{BE} < (FC \cdot \Phi_E) \\[4mm] \tau_F \dfrac{\partial I_{CC}}{\partial V_{BE}} + \dfrac{C_{JE}}{F_{2E}} \left(F_{3E} - \dfrac{m_E V_{BE}}{\Phi_E}\right) & \text{if } V_{BE} \geq (FC \cdot \Phi_E) \end{cases} \tag{1.26}$$

and

$$C_C = \begin{cases} \tau_R \dfrac{\partial I_{EC}}{\partial V_{BC}} + C_{JC} \left(1 - \dfrac{V_{BC}}{\Phi_C}\right)^{-m_C} & \text{if } V_{BC} < (FC \cdot \Phi_C) \\[4mm] \tau_R \dfrac{\partial I_{EC}}{\partial V_{BC}} + \dfrac{C_{JC}}{F_{2C}} \left(F_{3C} - \dfrac{m_C V_{BC}}{\Phi_C}\right) & \text{if } V_{BC} \geq (FC \cdot \Phi_C) \end{cases} \tag{1.27}$$

The terms F_{2E} and F_{2C} are calculated as follows:

$$F_{2\_} = (1 - FC)^{1+m\_}$$

and F_{3C} and F_{3C}:

$$F_{3\_} = 1 - FC \cdot (1 + m\_)$$

Definitions for constant terms are listed in Table 1.1 after [9, 17]. Where applicable, default values used in the SPICE2 program are listed. Note that when these default values are applied, the additional effects of the Gummel–Poon model are negated, and the model reduces to the Ebers–Moll formulation.

Implementation of the Gummel–Poon model for FDTD simulation poses many of the challenges similar to that of the modified Ebers–Moll model in [15]. Namely, the updating equations become significantly more complex, and the nonlinear equations defining the capacitance can exceed the capabilities of the Newton–Raphson method for small time step values near the CFL limit. This challenge is particularly acute for the simulation of the BJT, where a two-dimensional implementation of Newton–Raphson iteration is required.

Table 1.1 Description of numerical constants for the Gummel–Poon BJT model.

Symbol	Description	Default value	Unit
β_F, β_R	Ideal maximum beta (forward, reverse)	100, 1	
n_F, n_R	Current emission coefficient (forward, reverse)	1	
n_C	Base–collector leakage current emission coefficient	2	
V_A, V_B	Early voltage (forward, reverse)	∞	V
I_S	Saturation current	10^{-16}	A
I_{SE}, I_{SC}	Leakage saturation current (base–emitter, base–collector)	0	A
I_{KF}, I_{KR}	Corner for β high-current roll-off (forward, reverse)	∞	A
τ_F, τ_R	Ideal transit time (forward, reverse)	0	s
C_{JE}, C_{JC}	Zero-bias capacitance (base–emitter, base–collector)	0	F
Φ_E, Φ_C	Junction built-in potential (base–emitter, base–collector)	0.75	V
m_E, m_C	Junction grading coefficient	0.33	
FC	Forward bias depletion capacitance coefficient	0.5	

One example of a BJT simulated using FDTD is reported by Kung and Chuah [17]. In their work, the process for updating the emitter and collector current at time step $n + \frac{1}{2}$ begins by creating Taylor expansions for (1.23) and (1.24) at time n. In this approach, the need for numerical approximation of nonlinear equations is avoided, and reasonable accuracy can be achieved when sufficient terms are used to describe the Taylor series. Good accuracy and stability are provided by this approach, but the authors note that harmonics produced by nonlinear devices may exceed the capability of absorbing boundaries, leading to high-frequency modes accumulating in the simulation.

The current state of the art in BJT simulation using FDTD appears to remain a Taylor approximation of the Gummel–Poon model. This is likely due to the great complexity of FDTD updating equations based on the full Gummel–Poon formulation and small benefits provided beyond the Taylor approximation. An extension of the work in [17] includes the estimation of thermal self-heating effects during each simulation step, but retains the Taylor approximation for the transistor model.

1.2.5 Field-Effect Transistors: Small-Signal Modeling

Field-effect transistors are commonly employed in RF circuits, owing to higher frequencies of operation, lower noise figure, and capability for high-power operation [18]. Because small-signal models only address the transistor's linear region of operation, they apply primarily to the design of class A amplifiers [19]. Nonetheless, the small-signal approach forms the basis for introductions to amplifier design, including those of [20, 21], as it effectively addresses input matching, output matching, and stability issues under linear operation.

Lumped-circuit models may be embedded into FDTD simulations by using the approach described in [22]: at each time step, current at a port within the FDTD simulation is updated as a function of voltage, where the capacitance of the FDTD cell is accounted for in the updating formulation. For complex circuit models, an external simulation program such as SPICE may be used to update the state equations of the circuit model during each time step, with resulting current values passed back to the FDTD simulation during the electric-field-updating step.

A widely used 15-element field-effect transistor (FET) model [23] is illustrated in Figure 1.5. In this type of model, the active effect of the transistor is emulated by means of a controlled current source. Non-ideality of the transistor is emulated by capacitors, resistors, and inductors. Additionally, the effect of the transistor packaging is modeled by two capacitors C_{pd} and C_{pg}. In more advanced formulations, additional components are introduced to increase the accuracy of the model, and resistance or capacitance values are varied nonlinearly as a function of the current through the model.

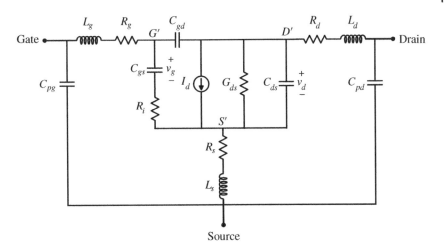

Figure 1.5 Field-effect transistor model.

Kuo et al. introduced a model similar to that in Figure 1.5 into an FDTD simulation at the junction of two transmission lines, with provisions for DC-bias current to be applied [24, 25]. The circuit was simulated in FDTD using two separate approaches: an FDTD–SPICE linked simulation as described in [22] and a direct incorporation of a circuit state-space model into the FDTD updating formulation using a differential equation.

1.2.6 Field-Effect Transistors: Large-Signal Modeling

While the integration of small-signal FET models into the FDTD simulation provides an important simulation tool for active microwave circuits, modern applications of the FET routinely use nonlinear regions of operation. Amplifiers operating outside of class A conditions are significantly more energy efficient but are more complex to design and optimize. Indeed, an accurate full-wave simulation is an important tool in the design of nonlinear amplifiers.

The nonlinear properties of a transistor may be incorporated into a circuit model using a combination of approaches. First, controlled current sources, such as I_d in Figure 1.5, may be controlled by nonlinear or transcendental functions of voltage or current elsewhere in the circuit: this approach was used to model the junction diode in Section 1.2.1. Second, values of circuit elements such as resistors and capacitors may depend upon voltages or currents in the circuit, possibly by a nonlinear function. Finally, nonlinear effects may be emulated by the introduction of junction diodes into the circuit model.

In one approach [26], the metal semiconductor field effect transistor (MESFET) model similar to that of Curtice [27] was used to simulate a nonlinear amplifier

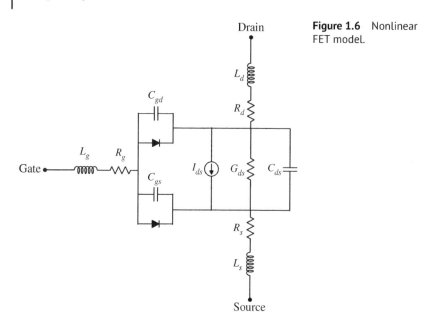

Figure 1.6 Nonlinear FET model.

circuit. The transistor model used is illustrated in schematic form in Figure 1.6: the nonlinear current source I_{ds} is represented in terms of the voltages at the source, gate, and drain terminals as follows:

$$I_{ds} = \begin{cases} \beta(1 + \lambda V_{ds})(V_{gs} - V_{to})^2 \tanh(\alpha V_{ds}) & \text{if } V_{gs} > V_{to} \\ 0 & \text{if } V_{gs} < V_{to} \end{cases} \quad (1.28)$$

where the parameters α, β, and λ are specific to the construction of the transistor, and V_{to} is the threshold voltage of the transistor.

In the approach of [26], the lumped components L_d, R_d, L_g, R_g, L_s, and R_s were each placed into their own FDTD cell and simulated using the approach of [6]. Similarly, each pairing of diode and parallel capacitance occupied its own cell, and these were simulated according to the approach of [12]. Finally, the parallel combination of $I_{ds}\|R_{ds}\|C_{ds}$ was addressed using a nonlinear FDTD updating formulation, which was solved numerically at each time step using Newton–Raphson iteration [28].

A similar work was performed by Kuo et al., wherein the model of Figure 1.5 was extended through a nonlinear capacitance C_{gs} and a nonlinear source for I_{ds}. The nonlinear capacitance is controlled by the following formula:

$$C_{gs}(v_g) = \frac{C_{gs0}}{\sqrt{1 - \frac{v_g}{\phi_{bi}}}} \quad (1.29)$$

and the drain current is a function of both external and internal node voltages:

$$I_{ds}\left(v_d, v_{G'S'}\right) = \left(A_0 + A_1 v_{G'S'} + A_2 v_{G'S'}^2 + A_3 v_{G'S'}^3\right) \tanh(\alpha v_d) \quad (1.30)$$

Table 1.2 Description of numerical constants for the MESFET model.

Symbol	Description	Value	Unit
R_g, R_d	Gate, drain resistance	0.5	Ω
R_s	Source resistance	0.7	Ω
R_i	Intrinsic resistance	1.0	Ω
L_g, L_d	Gate, drain inductance	0.05	nH
L_S	Source inductance	0.1	nH
C_{gd}	Gate–drain capacitance	0.2	pF
C_{ds}	Drain–source capacitance	0.6	pF
C_{gs0}	Zero-bias gate capacitance	3	pF
A_0	Model coefficients	0.5304	
A_1		0.2595	
A_2		−0.0542	
A_3		−0.0305	
α	Hyperbolic tangent parameter	1.0	

The numerical constants used for the MESFET in [29] are listed in Table 1.2.

Unlike the approaches of [24, 25], where only the linear operation of the MES-FET was addressed, and also unlike the approach of [26], where lumped components were modeled separately within the FDTD simulation, the simulation approach of [29] encapsulates the entire transistor model into a single FDTD cell. The five nonlinear state equations of the MESFET model, including the internal node voltages G', S', and D', as well as the total voltage across the FDTD cell are converted into a matrix form, and solved iteratively at each time step using the Newton–Raphson approach. Nonlinear behavior of the device was shown in the form of compression in the amplifier's output power and in the appearance of 2nd and 3rd harmonic components, as the input signal power was increased.

Simulation of active devices within FDTD poses an additional challenge in the form of coupling between device leads. In an active device with a high gain, there is a strong possibility for coupling between the device leads leading to unwanted self-oscillation of the circuit. In a work similar to [29], Chen and Fusco placed a section of simulated absorbing boundary between the gate and drain terminals of an FET [30].

The approach of [29] was further extended by modifying the interface region between the simulated active device and the conducting transmission line [31]. By modeling the device as a pair of "active sheets," with the capability for varying the cross-sectional voltage, it is possible to effectively match the simulated nonlinear device to modes on the transmission line. Mode matching and launching is an important area of consideration for the FDTD simulation of planar circuits and in particular for the simulation of circuit elements. This topic is further discussed in [32–34].

Many FDTD simulation packages incorporate simulation elements such as voltage probes, current probes, and voltage sources as preprogrammed software modules. Where possible, it is advantageous to reuse such modules instead of writing and testing specialized updating equations for each new type of component. Mix et al. used a modified Curtice MESFET model, where incoming currents are determined by magnetic field integration around the conducting terminals of the MESFET device [35]. In turn, the MESFET's response was emulated in the FDTD simulation by means of a controlled voltage source. In this simplified model, an explicit formulation was used for the Curtice MESFET equations, so that an iterative solution was not required. This approach reduces the computational complexity of the simulation, but at the possible cost of accuracy and stability.

1.3 FDTD–SPICE

In many of the examples discussed earlier in the chapter, simulation of a new type of circuit element required the development of an entirely new FDTD updating formulation, specialized to that device type. It is advantageous to connect an FDTD simulation to an external solver, which is already equipped with a variety of device models. The well-known SPICE circuit simulator is one such software, and is considered the industry standard for analog circuit simulation. In this program, arbitrary arrangements of passive, active, and nonlinear components can be constructed and simulated in the time or frequency domain [36].

A simulator combining the capabilities of FDTD and SPICE was demonstrated by Piket-May [5, 6]. In this approach, Ampere's law is written as follows:

$$\varepsilon \frac{\partial \vec{E}}{\partial t} + \vec{J}(\vec{E}) = \nabla \times \vec{H} \tag{1.31}$$

where $\vec{J}(\vec{E})$ represents the current through a circuit component as a function of the electric field across it. In terms of an electrical circuit (1.31) may be rewritten as follows:

$$C \frac{dV}{dt} + I_{\text{SPICE}}(V) = I_{\text{total}} \tag{1.32}$$

where I_{total} is the total current through the FDTD cell and C is the capacitance of the cell. For a Z-directed lumped element, C can be found as follows:

$$C = \frac{\varepsilon \Delta x \Delta y}{\Delta z} \tag{1.33}$$

Figure 1.7 Schematic illustration of FDTD–SPICE integration.

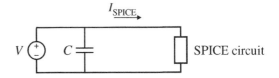

From the perspective of the SPICE simulator, a circuit similar to that in Figure 1.7 is constructed. In this illustration, a one-port (two-terminal) device is illustrated, though the technique is extensible to networks of many ports. At each time step, voltage V is updated based on the electric field within the FDTD cell. Current I is a function of voltage V and the response of the circuit to be modeled (labeled "SPICE Circuit").

At each simulation time step, the FDTD–SPICE interaction occurs one time. Voltage V is updated in the SPICE simulation, and then current I_{SPICE} is retrieved and updated in the FDTD simulation by means of current density J. One disadvantage to this approach occurs because FDTD and SPICE address the updating of voltage and current in different ways. In SPICE, both voltages and currents are updated simultaneously at each time step. Conversely, in FDTD, the electric and magnetic fields are updated in a "leapfrog" manner.

In the ideal case, the value of voltage V supplied to the SPICE simulation will be based on an average of $E_z|_{i,j,k}^{n}$ and $E_z|_{i,j,k}^{n+1}$, because current I_{SPICE} must be valid at time $n + \frac{1}{2}$. Inserting (1.32) into (1.1) leads to an electric-field updating formulation of:

$$
E_z|_{i,j,k}^{n+1} = \frac{2\varepsilon_z|_{i,j,k} - \Delta t\,\sigma_z^e|_{i,j,k}}{2\varepsilon_z|_{i,j,k} + \Delta t\,\sigma_z^e|_{i,j,k}}\, E_z|_{i,j,k}^{n}
$$

$$
+ \frac{2\Delta t}{\left(2\varepsilon_z|_{i,j,k} + \Delta t\,\sigma_z^e|_{i,j,k}\right)\Delta x}\left(H_y|_{i,j,k}^{n+\frac{1}{2}} - H_y|_{i-1,j,k}^{n+\frac{1}{2}} \right)
$$

$$
- \frac{2\Delta t}{\left(2\varepsilon_z|_{i,j,k} + \Delta t\,\sigma_z^e|_{i,j,k}\right)\Delta y}\left(H_x|_{i,j,k}^{n+\frac{1}{2}} - H_x|_{i,j-1,k}^{n+\frac{1}{2}} \right)
$$

$$
- \frac{2\Delta t}{2\varepsilon_z|_{i,j,k} + \Delta t\,\sigma_z^e|_{i,j,k}}\frac{I_{\text{SPICE}}\left(\dfrac{\Delta z\left(E_z|_{i,j,k}^{n} + E_z|_{i,j,k}^{n+1} \right)}{2} \right)}{\Delta x\Delta y} \tag{1.34}
$$

While this equation is numerically implicit, it is not possible to solve using a traditional SPICE simulator, because the value of $E_z|_{i,j,k}^{n+1}$ is not known at the time when the SPICE function is executed. Thus, an explicit form is required:

$$
E_z|_{i,j,k}^{n+1} = \frac{2\varepsilon_z|_{i,j,k} - \Delta t\, \sigma_z^e|_{i,j,k}}{2\varepsilon_z|_{i,j,k} + \Delta t\, \sigma_z^e|_{i,j,k}} E_z|_{i,j,k}^{n}
$$

$$
+ \frac{2\Delta t}{\left(2\varepsilon_z|_{i,j,k} + \Delta t\, \sigma_z^e|_{i,j,k}\right)\Delta x} \left(H_y\Big|_{i,j,k}^{n+\frac{1}{2}} - H_y\Big|_{i-1,j,k}^{n+\frac{1}{2}} \right)
$$

$$
- \frac{2\Delta t}{\left(2\varepsilon_z|_{i,j,k} + \Delta t\, \sigma_z^e|_{i,j,k}\right)\Delta y} \left(H_x\Big|_{i,j,k}^{n+\frac{1}{2}} - H_x\Big|_{i,j-1,k}^{n+\frac{1}{2}} \right)
$$

$$
- \frac{2\Delta t}{2\varepsilon_z|_{i,j,k} + \Delta t\, \sigma_z^e|_{i,j,k}} \frac{I_{\text{SPICE}}\left(E_z|_{i,j,k}^{n}\right)}{\Delta x \Delta y} \tag{1.35}
$$

This explicit formulation is practical to solve on an unmodified SPICE software, but using an explicit formulation may lead to accuracy or stability problems. Indeed, Piket-May noted that certain configurations of an FDTD–SPICE amplifier simulation had diverging solutions [5]. Similarly, Thomas et al. noted that an FDTD–SPICE simulation of a digital circuit suffered from low accuracy at high frequencies [22]. Also, it was noted that certain circuits may be impossible to simulate using this configuration and the use of the poisson and continuity equation solver (PISCES) simulator [37] was suggested in place of SPICE [38].

The explicit nature of the FDTD–SPICE updating formulation was observed by Li and Sui, who proposed an alternative technique based on solving for I_{SPICE} at time step $n + 1$ instead of $n + \frac{1}{2}$ [39]. In this approach, the SPICE simulation is supplied with a current value from the FDTD simulation, taken at time $n + \frac{1}{2}$. Within the SPICE netlist, this current is represented as a Norton-equivalent current source with a parallel resistance R_{grid}. The resulting current into the SPICE-simulated circuit is updated back into the FDTD simulation at time $n + 1$. This method allowed to simulate both passive and active circuits, including a mixer and pulse generator, demonstrating accuracy at even high frequencies and nonlinear operating conditions [39, 40].

In many cases, time steps employed in SPICE simulators are varied depending on the convergence of nonlinear updating equations. In situations that render this convergence difficult, such as rapid changes in voltage or highly nonlinear device operation, the time step is reduced [41]. Sui noted that an FDTD–SPICE simulator can operate with variable Δt. During normal operation, the time step is chosen according to the CFL condition – the maximum possible for an FDTD simulation. When poor convergence is detected within the SPICE simulator, the time step may be reduced in both FDTD and SPICE to account for this. Such a method allows for improved accuracy and stability while maintaining computational efficiency [42].

1.4 Data-Based Models

In the previously described approaches for the incorporation of lumped-element circuits into FDTD simulations, the circuit components were described in the form of physical models. Such models may be represented by a single schematic symbol, as in the case of the junction diode, or they may be represented by complex sub-circuits, such as the MESFET model illustrated in Figure 1.6. In all cases, the physics of the sub-circuits are modeled by the values of circuit components and by the mathematical formulas used to update dependent voltage sources and nonlinear circuit components.

In many cases, it may be impractical or fully impossible to acquire an exact physical model of a circuit component. If the physical details of a component (material composition, geometry) are available, then a physical model may be generated by numerical simulations of the semiconductor's operation. If the device can be acquired and measured in the laboratory, then the device parameters may be extracted [23]. However, exact internal details of a device may be unavailable due to intellectual property restrictions, and transistor parameter extraction in the laboratory requires specialized equipment and skills.

It is desirable to be able to simulate a device based on characteristics that have been measured in the laboratory or provided by the manufacturer. For linear devices, the common language of device description is S-parameters: they may be measured in most microwave laboratories using a vector network analyzer. Also, many component manufacturers provide S-parameter data for the devices they sell. For nonlinear components, an emerging characterization is X-parameters, which describe a component in terms of its harmonic response to an input signal [43].

1.4.1 Linear Lumped Elements: S-Parameter Approaches

While FDTD is a time-domain technique, it lends itself well to extraction of S-parameters from simulation results through a Fourier transform of electric and magnetic field results. The reverse case has also been demonstrated: inserting lumped elements into FDTD simulations, which are described by their S-parameters. This is a useful approach for characterizing a device whose parameters have already been measured in the context of a larger microwave device such as an antenna or transmission line. Because the S-parameters only address the linear operation of circuit components, only a superficial review will be presented for this topic.

In the approach of Zhang and Wang [44], thse scattering **S** matrix is converted into an admittance **Y** matrix, using a method such as that described by Pozar [20]. For a two-port network, the current at each port can be calculated in the frequency domain as the matrix product of the admittance matrix and the port voltages:

$$\begin{bmatrix} I_1(\omega) \\ I_2(\omega) \end{bmatrix} = \begin{bmatrix} Y_{11}(\omega) & Y_{12}(\omega) \\ Y_{22}(\omega) & Y_{22}(\omega) \end{bmatrix} \begin{bmatrix} V_1(\omega) \\ V_2(\omega) \end{bmatrix} \tag{1.36}$$

For a time-domain simulation, voltages must be updated at discrete time steps. The inverse Fourier transforms of the admittance terms are taken to provide time-domain admittance parameters:

$$\widehat{Y}_{i,j}(t) = \mathcal{F}^{-1}(Y_{i,j}(\omega)) \tag{1.37}$$

At each time step, the currents at two ports of the network can be found by convolving the time-domain admittance with the port voltages:

$$I_1(t) = \widehat{Y}_{11}(t) * V_1(t) + \widehat{Y}_{12}(t) * V_2(t)$$
$$I_2(t) = \widehat{Y}_{21}(t) * V_1(t) + \widehat{Y}_{22}(t) * V_2(t) \tag{1.38}$$

For example, the current in the z direction at port 1 of the network, as a function of all previous voltages, can be written as [44]:

$$I_z\big|_{\text{port 1}}^{n+\frac{1}{2}} = \Delta z \sum_{k=1}^{n} \left[\widehat{Y}_{11}(k)\, E_z\big|_{\text{port 1}}^{n-k} + \widehat{Y}_{12}(k)\, E_z\big|_{\text{port 2}}^{n-k} \right] \tag{1.39}$$

We note that while the required convolution is performed in this formulation, it is an explicit formulation because the value of $E_z|^{n+1}$ is not considered, which may limit accuracy at large time steps. An extension of this technique was described by Ye and Drewniak, wherein the generalized pencil-of-function technique was used to convert the time-domain Y-parameters into a series of exponential functions [45]. This approach reduces memory consumption and computational time for the simulation.

Pereda et al. described an extension to the embedding of arbitrary lumped elements into the FDTD grid [46]. In their approach, the impedance of the network is transformed to the Laplace domain and approximated by means of a bilinear transformation. This bilinear expression forms the basis for a set of updating equations that have implicit formulation and, therefore, match the central difference scheme of the FDTD method.

Numerous additional schemes have been described for the incorporation of arbitrary linear lumped elements into FDTD simulations. For example, the work of Pereda et al. was extended by Gonzalez and Pereda to two-port [47] and active [48] networks, using a similar bilinear approach to [46]. Other techniques have been developed to reduce computational load, such as the use of recursive techniques [49, 50]. Also, methods are available for the incorporation of linear lumped elements based on their circuit models, such as [51–54].

1.4.2 Nonlinear Lumped Elements: X-Parameters

As noted in the previous section, S-parameters can be used to model active devices such as amplifiers. However, the use of S-parameters for active-device modeling is limited to linear cases. Two factors limit the applicability of S-parameters in the modeling of nonlinear devices. First, S-parameters are considered on a frequency-by-frequency basis. A signal at frequency ω_1 will not affect a device's response to a signal being transmitted simultaneously at frequency ω_2. Second, S-parameters do not consider the capability for a device to generate harmonics: for example, when a signal at frequency ω_1 enters a nonlinear device, we may see signals of $2\omega_1$, $3\omega_1$, etc. propagating from the device's ports. To properly address these effects, an alternative model is needed.

One such solution is offered by X-parameters [55]. In this approach, nonlinear responses of a device are linearized in the frequency domain around a specific operating frequency. For example, an amplifier can be excited with a frequency ω_1 at port 1. Then, small additional signals of frequencies $[\omega_1, 2\omega_1..., K\omega_1]$ are applied to the network, and changes in the outgoing waves are measured, again at the frequencies $[\omega_1, 2\omega_1..., K\omega_1]$.

In the nomenclature of X-parameters, waves incident on a network are written as $A_{q,l}$, where q is the index of the port and l is the index of the harmonic ($l = 1$ is the fundamental frequency of the excitation signal). Waves propagating outward from the network are written as $B_{p,k}$, again with p referring to the port index and k the harmonic. The outgoing wave $B_{p,k}$ can be approximated as [43]:

$$B_{p,k} \cong X_{p,k}^{(FB)} P^k + \sum_{\substack{q=1 \\ l=1 \\ (q,l)\neq(1,1)}}^{\substack{q=N \\ l=K}} \left[X_{p,k;q,l}^{(S)} A_{q,l} P^{k-l} + X_{p,k;q,l}^{(T)} A_{q,l}^* P^{k+l} \right] \tag{1.40}$$

In place of the $N \times N$ matrix of the S-parameters, there are three matrices of X-parameters required for this formulation. The parameter $X_{p,k}^{(FB)}$ is a matrix of dimension $N \times K$ (for a system with N ports and where K harmonics are considered) describing the device's large-signal response. The parameters $X_{p,k;q,l}^{(S)}$ and $X_{p,k;q,l}^{(T)}$ both have dimension $N \times K \times N \times K$, and describe the small changes in the outgoing waves $B_{p,k}$, based on small changes in the incoming signals $A_{q,l}$ at any port and harmonic. The constant P serves the purpose of phase normalization and has the value $P = e^{j \angle A_{11}}$.

Kast and Elsherbeni described the extraction of large-signal $\left(X_{p,k}^{(FB)} \right)$ X-parameters from the simulation of a one-port circuit containing a junction diode [55]. In this approach, a discrete Fourier transform was used to convert voltages and currents at the network ports into the frequency domain for each of the harmonics. This approach was extended to a two-port case, where both large-signal and small-signal $\left(X_{p,k;q,l}^{(S)}, X_{p,k;q,l}^{(T)}\right)$ terms were extracted from the simulation results [56]. While these techniques do not yet provide a method for simulating a device based on its X-parameters, they create a necessary foundation for the further study of nonlinear devices within FDTD simulations.

1.5 Conclusions

Myriad applications of RF and microwave technologies depend on nonlinear components to achieve their desired function. Indeed, at a fundamental level, all real-world RF active devices and amplifiers are nonlinear components, and RF oscillators require the presence of a nonlinear component to operate [57]. In this context, we find that the capability to simulate nonlinear circuit components is a valuable asset to any microwave simulation technique. Indeed, the FDTD technique is extremely well suited to simulation of nonlinear devices due to its time-domain nature: nonlinear components may be directly embedded into the FDTD updating formulation, and results extracted in the time domain, or converted into the frequency domain allowing evaluation of harmonics and other effects.

References

1 Sui, W. (1991). *An Extended Finite Difference Time Domain Method for Hybrid Electromagnetic Systems with Active and Passive Lumped Elements*. University of Utah.

2 W. Sui, D. A. Christensen, and C. H. Durney, "Extending the two-dimensional FDTD method to hybrid electromagnetic systems with active and passive lumped elements," *IEEE Transactions on Microwave Theory and Techniques*, vol. 40, no. 4, pp. 724–730, Apr. 1992, doi: https://doi.org/10.1109/22.127522.

3 P. B. Johns, "On the relationship between TLM and finite-difference methods for Maxwell's equations (short paper)," *IEEE Transactions on Microwave Theory and Techniques*, vol. 35, no. 1, pp. 60–61, Jan. 1987, doi: https://doi.org/10.1109/TMTT.1987.1133595.

4 R. H. Voelker and R. J. Lomax, "A finite-difference transmission line matrix method incorporating a nonlinear device model," *IEEE Transactions on Microwave Theory and Techniques*, vol. 38, no. 3, pp. 302–312, Mar. 1990, doi: https://doi.org/10.1109/22.45349.

5 Piket-May, M.J. (2021). Three-dimensional time-domain numerical studies of pulse behavior in digital interconnect circuits with passive and active loads. PhD Northwestern University, United States – Illinois. https://www.proquest .com/docview/304057910/abstract/B04340F3C8194D1APQ/1 (accessed 20 June 2022).

6 Piket-May, M., Taflove, A., and Baron, J. (1994). FD-TD modeling of digital signal propagation in 3-D circuits with passive and active loads. *IEEE Transactions on Microwave Theory and Techniques* 42 (8): 1514–1523. https://doi.org/10 .1109/22.297814.

7 J. J. Ebers and J. L. Moll, "Large-signal behavior of junction transistors," *Proceedings of the IRE*, vol. 42, no. 12, pp. 1761–1772, Dec. 1954, doi: https:// doi.org/10.1109/JRPROC.1954.274797.

8 Nagel, L.W. (1970). SPICE2: a computer program to simulate semiconductor circuits. PhD Dissertations University of California, Berkeley. https://ci.nii.ac .jp/naid/10014667404/ (accessed 30 June 2022).

9 Massobrio, G. and Antognetti, P. (1998). *Semiconductor Device Modeling*, 1e. New York: McGraw-Hill Education.

10 Elsherbeni, A.Z. and Demir, V. (2016). *The finite-difference time-domain method for electromagnetics with MATLAB simulations*, ACES Series on Computational Electromagnetics and Engineering, 2e. Edison, NJ: SciTech Publishing, an Imprint of IET.

11 W. Shockley, "The theory of p-n junctions in semiconductors and p–n junction transistors," *Bell System Technical Journal*, vol. 28, no. 3, pp. 435–489, Jul. 1949, doi: https://doi.org/10.1002/j.1538-7305.1949.tb03645.x.

12 P. Ciampolini, P. Mezzanotte, L. Roselli, D. Sereni, R. Sorrentino, and P. Torti, "Simulation of HF circuits with FDTD technique including non-ideal lumped elements," in *Proceedings of 1995 IEEE MTT-S International Microwave Symposium*, May 1995, pp. 361–364 vol. 2. doi: https://doi.org/10.1109/MWSYM.1995 .405969.

13 Sedra, A.S. and Smith, K.C. (2014). *Microelectronic Circuits*, The Oxford Series in Electrical and Computer Engineering, 7e. New York, NY: Oxford University Press.

14 ElMahgoub, K. and Elsherbeni, A.Z. (2014). FDTD implementations of integrated dependent sources in full-wave electromagnetic simulations. *Applied Computational Electromagnetics Society (ACES) Journal* 29 (12): 10.

15 P. Ciampolini, P. Mezzanotte, L. Roselli, and R. Sorrentino, "Accurate and efficient circuit simulation with lumped-element FDTD technique," *IEEE Transactions on Microwave Theory and Techniques*, vol. 44, no. 12, pp. 2207–2215, Dec. 1996, doi: https://doi.org/10.1109/22.556448.

16 H. K. Gummel and H. C. Poon, "An integral charge control model of bipolar transistors," *Bell System Technical Journal*, vol. 49, no. 5, pp. 827–852, May 1970, doi: https://doi.org/10.1002/j.1538-7305.1970.tb01803.x.

17 F. Kung and H. T. Chuah, "Modeling of bipolar junction transistor in FDTD simulation of printed circuit board," *Progress In Electromagnetics Research*, vol. 36, pp. 179–192, 2002, doi: https://doi.org/10.2528/PIER02013001.

18 Bahl, I. (2009). *Fundamentals of RF and Microwave Transistor Amplifiers*, 1e. Hoboken, N.J: Wiley-Interscience.

19 Razavi, B. (2011). *RF Microelectronics*, 2e. Upper Saddle River, NJ: Prentice Hall.

20 Pozar, D.M. (2011). *Microwave Engineering*, 4e. Hoboken, NJ: Wiley.

21 Ludwig, R. and Bogdanov, G. (2008). *RF Circuit Design: Theory & Applications*, 2e. Upper Saddle River, NJ: Pearson.

22 V. A. Thomas, M. E. Jones, M. Piket-May, A. Taflove, and E. Harrigan, "The use of SPICE lumped circuits as sub-grid models for FDTD analysis," *IEEE Microwave and Guided Wave Letters*, vol. 4, no. 5, pp. 141–143, May 1994, doi: https://doi.org/10.1109/75.289516.

23 Kompa, G. (2019). *Parameter Extraction and Complex Nonlinear Transistor Models*. Artech House.

24 Chien-Nan Kuo, B.H, and T. Itoh, "FDTD analysis of active circuits with equivalent current source approach," in *IEEE Antennas and Propagation Society International Symposium. 1995 Digest*, Jun. 1995, vol. 3, pp. 1510–1513. doi: https://doi.org/10.1109/APS.1995.530863.

25 Chien-Nan Kuo, V. A. Thomas, Siou Teck Chew, B. Houshmand, and T. Itoh, "Small signal analysis of active circuits using FDTD algorithm," *IEEE Microwave and Guided Wave Letters*, vol. 5, no. 7, pp. 216–218, 1995, doi: https://doi.org/10.1109/75.392279.

26 M. Matteucci, P. Mezzanotte, L. Roselli, and P. Ciampolini, "Numerical analysis of electronic circuits with FDTD-LE technique," *WIT Transactions on Engineering Sciences*, vol. 11, 1996, doi: https://doi.org/10.2495/ES960211.

27 W. R. Curtice, "A MESFET model for use in the design of GaAs integrated circuits," *IEEE Transactions on Microwave Theory and Techniques*, vol. 28, no. 5, pp. 448–456, May 1980, doi: https://doi.org/10.1109/TMTT.1980.1130099.

28 P. Ciampolini, P. Mezzanotte, L. Roselli, and R. Sorrentino, "Efficient simulation of high speed digital circuits using time adaptive FD-TD technique," in *1995 25th European Microwave Conference*, IEEE, 1995, vol. 2, pp. 636–640. doi: https://doi.org/10.1109/EUMA.1995.337038.

29 Chien-Nan Kuo, B. Houshmand, and T. Itoh, "Full-wave analysis of packaged microwave circuits with active and nonlinear devices: an FDTD approach," *IEEE Transactions on Microwave Theory and Techniques*, vol. 45, no. 5, pp. 819–826, 1997, doi: https://doi.org/10.1109/22.575606.

30 Q. Chen and V. F. Fusco, "Hybrid FDTD large-signal modeling of three-terminal active devices," *IEEE Transactions on Microwave Theory and Techniques*, vol. 45, no. 8, pp. 1267–1270, Aug. 1997, doi: https://doi.org/10.1109/22.618419.

31 V. S. Reddy and R. Garg, "An improved extended FDTD formulation for active microwave circuits," *IEEE Transactions on Microwave Theory and Techniques*, vol. 47, no. 9, pp. 1603–1608, Sep. 1999, doi: https://doi.org/10.1109/22.788599.

32 I. Wolff, "Finite difference time-domain simulation of electromagnetic fields and microwave circuits," *International Journal of Numerical Modelling: Electronic Networks, Devices and Fields*, vol. 5, no. 3, pp. 163–182, 1992, doi: https://doi.org/10.1002/jnm.1660050306.

33 C. H. Durney, Wenquan Sui, D. A. Christensen, and Jingyi Zhu, "A general formulation for connecting sources and passive lumped-circuit elements across multiple 3-D FDTD cells," *Microwave and Guided Wave Letters*, vol. 6, no. 2, pp. 85-, 1996, doi: https://doi.org/10.1109/75.481997.

34 W. K. Gwarek and M. Celuch-Marcysiak, "Wide-band *S*-parameter extraction from FD-TD simulations for propagating and evanescent modes in inhomogeneous guides," *IEEE Transactions on Microwave Theory and Techniques*, vol. 51, no. 8, pp. 1920–1928, Aug. 2003, doi: https://doi.org/10.1109/TMTT.2003.815265.

35 Mix, J., Dixon, J., Popovic, Z., and Piket-May, M. (1999). Incorporating non-linear lumped elements in FDTD: the equivalent source method. *International Journal of Numerical Modelling: Electronic Networks, Devices and Fields* 12 (1–2): 157–170.

36 Vladimirescu, A. (1994). *The SPICE Book*. New York: John Wiley & Sons.

37 Pinto, M.R., Rafferty, C.S., and Dutton, R.W. (1984). *PISCES II: Poisson and Continuity Solver*. Stanford, CA: Stanford Electronics Laboratories, Stanford University.

38 V. A. Thomas, M. E. Jones, and R. J. Mason, "Coupling of the PISCES device modeler to a 3-D Maxwell FDTD solver," *IEEE Transactions on Microwave Theory and Techniques*, vol. 43, no. 9, pp. 2170–2172, Sep. 1995, doi: https://doi.org/10.1109/22.414557.

39 Li, T. (1999). *FDTD-Based Full Wave Co-simulation Model for Hybrid Electromagnetic Systems*. Newark, NJ: New Jersey Institute of Technology.

40 T. Li and W. Sui, "Extending Spice-like analog simulator with a time-domain full-wave field solver," in *2001 IEEE MTT-S International Microwave Symposium Digest (Cat. No.01CH37157)*, 2001, vol. 2, pp. 1023–1026doi: https://doi.org/10.1109/MWSYM.2001.967066.

41 Najm, F.N. (2010). *Circuit Simulation*. Hoboken, NJ: John Wiley & Sons.

42 Sui, W. (2001). *Time-Domain Computer Analysis of Nonlinear Hybrid Systems*. CRC Press.

43 Root, D., Verspecht, J., Horn, J., and Marcu, M. (2013). *X-Parameters: Characterization, Modeling, and Design of Nonlinear RF and Microwave Components*. Cambridge University Press.

44 Jiazong Zhang and Yunyi Wang, "FDTD analysis of active circuits based on the *S*-parameters [microwave hybrid ICs]," in *Proceedings of 1997 Asia-Pacific Microwave Conference*, IEEE, Dec. 1997, vol. 3, pp. 1049–1052. doi: https://doi .org/10.1109/APMC.1997.656395.

45 X. Ye and J. L. Drewniak, "Incorporating two-port networks with S-parameters into FDTD," *IEEE Microwave and Wireless Components Letters*, vol. 11, no. 2, pp. 77–79, Feb. 2001, doi: https://doi.org/10.1109/7260.914308.

46 J. A. Pereda, F. Alimenti, P. Mezzanotte, L. Roselli, and R. Sorrentino, "A new algorithm for the incorporation of arbitrary linear lumped networks into FDTD simulators," *IEEE Transactions on Microwave Theory and Techniques*, vol. 47, no. 6, pp. 943–949, Jun. 1999, doi: https://doi.org/10.1109/22.769330.

47 O. Gonzalez, J. A. Pereda, A. Herrera, and A. Vegas, "An extension of the lumped-network FDTD method to linear two-port lumped circuits," *IEEE Transactions on Microwave Theory and Techniques*, vol. 54, no. 7, pp. 3045–3051, Jul. 2006, doi: https://doi.org/10.1109/TMTT.2006.877058.

48 O. Gonzalez, J. A. Pereda, A. Herrera, A. Grande, and A. Vegas, "Combining the FDTD method and rational-fitting techniques for modeling active devices characterized by measured S -parameters," *IEEE Microwave and Wireless Components Letters*, vol. 17, no. 7, pp. 477–479, 2007, doi: https://doi.org/10 .1109/LMWC.2007.899292.

49 Jung-Yub Lee, Jeong-Hae Lee, and Hyun-Kyo Jung, "Linear lumped loads in the FDTD method using piecewise linear recursive convolution method," *IEEE Microwave and Wireless Components Letters*, vol. 16, no. 4, pp. 158–160, Apr. 2006, doi: https://doi.org/10.1109/LMWC.2006.872148.

50 C. Wang and C. Kuo, "An efficient scheme for processing arbitrary lumped multiport devices in the finite-difference time-domain method," *IEEE Transactions on Microwave Theory and Techniques*, vol. 55, no. 5, pp. 958–965, May 2007, doi: https://doi.org/10.1109/TMTT.2007.895652.

51 H. E. A. El-Raouf, W. Yu, and R. Mittra, "Application of the Z-transform technique to modelling linear lumped loads in the FDTD," *IEE Proceedings - Microwaves, Antennas and Propagation*, vol. 151, no. 1, pp. 67–70, Feb. 2004, doi: https://doi.org/10.1049/ip-map:20040067.

52 Tzong-Lin Wu, Sin-Ting Chen, and Yi-Shang Huang, "A novel approach for the incorporation of arbitrary linear lumped network into FDTD method," *IEEE Microwave and Wireless Components Letters*, vol. 14, no. 2, pp. 74–76, Feb. 2004, doi: https://doi.org/10.1109/LMWC.2003.822567.

53 Zhenhai Shao and M. Fujise, "An improved FDTD formulation for general linear lumped microwave circuits based on matrix theory," *IEEE Transactions* *on Microwave Theory and Techniques,* vol. 53, no. 7, pp. 2261–2266, 2005, doi: https://doi.org/10.1109/TMTT.2005.850450.

54 Demir, V. (2016). Formulations for modeling voltage sources with RLC impedances in the FDTD method. *The Applied Computational Electromagnetics Society Journal (ACES)* 31 (9): 1020–1027.

55 J. M. Kast and A. Z. Elsherbeni, "Extraction of nonlinear X-parameters from FDTD simulation of a one-port device," in *2021 United States National Committee of URSI National Radio Science Meeting (USNC-URSI NRSM),* Jan. 2021, pp. 89–90. doi: https://doi.org/10.23919/USNC-URSINRSM51531.2021.9336444.

56 J. M. Kast and A. Z. Elsherbeni, "Extraction of X-parameters from FDTD simulation of a two-port nonlinear circuit," *2021 International Applied Computational Electromagnetics Society Symposium (ACES),* , Aug. 2021.

57 Pedro, J.C., Root, D.E., Xu, J., and Nunes, L.C. (2018). *Nonlinear Circuit Simulation and Modeling: Fundamentals for Microwave Design,* 1e, 1–44. Cambridge, Cambridge University Press.

2

FDTD Method for Nonlinear Metasurface Analysis

Xibi Chen and Fan Yang

Tsinghua University, Beijing, China

2.1 Introduction to Nonlinear Metasurface

2.1.1 What is Nonlinear Metasurface?

Metasurface, which originates from the rapid development of modern electromagnetics (EM), has become an extremely well-known concept in advanced EM researches. Dating back to 1960s or even earlier, when the academic terminology "metasurface" was even not invented, people had already shown great interests in using engineered 2D/planar EM structures for various practical applications. One typical example is the frequency selective surface (FSS) [1], which is still an important type of metasurface nowadays. Starting from FSS, the increasingly demanding nature of EM and antenna engineers further pushed a wealth of creations in 2D/planar EM structures, including electromagnetic band gap (EBG) structure [2, 3], high impedance surface (HIS) [4], reflectarray/transmitarray antennas [5] and so on. Ranging from microwaves to photonics, these EM structures generated extraordinary contributions to many EM related technologies in the past 60 years.

Since the great values of those artificial 2D/planar EM structures have been proven as years went by, researchers in academies decided to discover some common senses in all those structures, and finally named them all as "metasurfaces." In spite of the rather long history for the inventions of different types of metasurfaces, a general theory for all kinds of metasurfaces is always lacked and is now welcomed. To establish that general theory, recent studies about metasurfaces start to change their approach from a "technology-based" scheme to a "science-based" scheme. As illustrated in Figure 2.1, the "technology-based" scheme focuses more on the applications and engineering at the very beginning, and only cares about the hidden common rules after the laboring. In contrast, the "science-based" scheme thinks in a reverse way, and tends to well manage

Advances in Time-Domain Computational Electromagnetic Methods, First Edition.
Edited by Qiang Ren, Su Yan, and Atef Z. Elsherbeni.

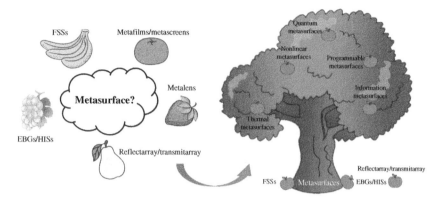

Figure 2.1 The trend for metasurface research: from "technology-based" scheme (left) to "science-based" scheme (right).

the theories first. Once the fundamental theory for the general case has been set up, it behaves like the stem of a knowledge tree, and all the other metasurface subjects are just like those branches grown from the stem.

Nonlinear metasurface (NMS) is a consequence of this "science-based" scheme shown in Figure 2.1. It is a specialized kind of metasurface, which combines nonlinearities with the concept of metasurface, exhibiting certain nonlinear responses when interacting with electromagnetic waves. Thanks to these nonlinear effects, NMSs have a wide range of applications in microwave [6–9], terahertz [10] and optical [11–17] regions, including harmonic generation, wave mixing, switching, parametric amplifying, phase conjugation, etc.

Although NMS can be seen as a universal concept among the entire frequency spectrum, the origin of each NMS's nonlinearities is quite diverse. Optical NMSs are usually considered as a hybrid of nonlinear photonics and metasurfaces. Their nonlinearities come from certain optical nonlinear materials in most cases [11–17]. While at the lower end, microwave NMSs usually utilize nonlinear lumped electronic devices (e.g. transistors, varactors, PIN diodes) as their nonlinearity sources [6–8]. Moreover, terahertz NMSs are seen as a transition from microwave NMSs to optical NMSs. Both nonlinear materials and lumped electronic devices can be used in terahertz NMSs.

Intriguingly, nonlinearities in lumped electronic devices actually originate from the materials as well, especially semiconductor materials. So in fact, the nonlinearities in all kinds of NMSs share the same source. Based on this point, no matter what kind of NMS is being considered, it is always important to analyze the interactions between electromagnetic fields and nonlinear materials. However, modeling or computing nonlinear materials in different electromagnetic structures is generally a challenging problem, since different nonlinear materials hold various nonlinear mechanisms. Thus, giving a general solution in these situations is a difficult task. A practical solution, then, is to analyze different nonlinearities on a case-by-case basis. Keeping this difficulty in mind, this chapter will mainly focus on the analysis of a certain type of NMS, which its materials satisfy with several restrictions. This will be clarified in a clearer way in the following sections of this chapter.

2.1.2 Material Modeling

The aforementioned difficulty in NMS analysis leads to a critical question: how to model the materials used inside an NMS? Answering this question will be helpful for the readers to easily and correctly understand the rest of the chapter. Nevertheless, a physically rigorous answer to this question is somehow logically tricky and beyond the scope of this chapter. Thus, a simplified description is provided here.

Generally speaking, there are three categories of material modeling in EM, namely, classical approach [18], semi-classical (semi-quantum) approach [13, 19, 20], and full-quantum approach [21]. Each approach has its unique physics and mathematical expressions, and in most engineering cases, especially linear cases, different approach will give similar/same results. Nonetheless, for nonlinear EM problems, choosing a suitable approach for material modeling becomes extremely critical. The figure of merit in here is to find the most handy theory while keeping an acceptable accuracy for the measuring of those generally small nonlinear effects. Again, finding a general solution for all NMSs is nearly impossible, and one has to make compromise for certain problems. The bottomline here in this chapter is, the chosen approach should be elegant enough in mathematics so that it can be easily implemented into numerical algorithms, while at the same time, it is still suitable for quite some NMSs.

Now let's take a closer look at the three approaches. Basically, the three approaches are distinguished by a simple criterion: whether quantum theory has been applied to electrons and/or photons? If neither the electrons nor the photons are analyzed by quantum theory, it is called classical approach. If the electrons are modeled by quantum physics while the photons (EM fields/waves) are still described in classical Maxwell equations, it is called semi-classical (semi-quantum) approach. No need to say, if both electrons and photons are interpreted as quantum particles, full-quantum approach pops out. The relations of the three approaches can be seen in Figure 2.2, and the corresponding details are explained as follows.

2.1.2.1 Classical Approach

The classical approach is probably the most popular way to model materials in EM research field. Greatly benefits from the well-ordered mathematics and clear

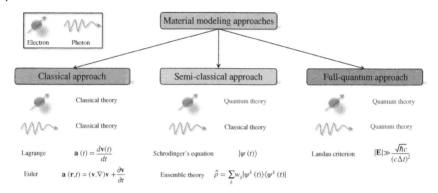

Figure 2.2 Different approaches for material modeling.

physical meanings, implementing classical theories into numerical algorithms occupies the majority of computational electromagnetics (CEM). Aiming to solve the electrons' motions inside materials, as well as their interactions with EM fields, two kinds of descriptions are further developed based on the classical approach. One is called as "Lagrangian description," and the other is called as "Eulerian description." While rigorous discussions about these two descriptions can be found in [22], some general elaborations are included here.

Lagrangian description is sometimes known as a "particle tracking method," which tracks the motion of one particular particle (e.g. electron) inside the material during the whole physical process. Once the dynamics for this particular electron has been well understood, the whole electron system of the material can be simply treated as multiple replicas of this "one electron analysis." To reasonably form this "one electron analysis," all the external interactions and the possible mutual couplings between electrons and lattice/electrons are modeled as equivalent forces to one single electron. The detailed expressions for these equivalent forces then determine the performances of that electron, and eventually, the entire material. Since an electron only occupies an electrically small volume in the material, the local reaction of that electron usually satisfies the "quasi-static" condition. Thus, a simple harmonic oscillator is often applied as the physical model for the electron dynamics. As can be seen in Figure 2.3, the external EM fields will provide an equivalent driving force to the oscillator, while the internal coupling between the electron and the lattice can be modeled as the restoring force. If the high-order effects (like the nonlinear effects) are being considered, extra high-order terms can be added to the left-hand side of that oscillator equation as well. Meanwhile, the radiation damping effect caused by the electron acceleration can be described by another damping coefficient, which contributes to the material loss eventually.

In contrast, Eulerian description is more like a "field approach." It can also be called as "hydrodynamic method." Different from the Lagrangian description,

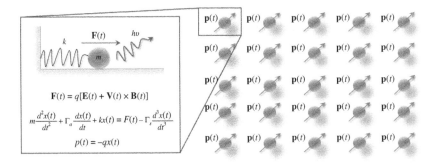

Figure 2.3 Lagrangian description for classical approach. The oscillator equation is expressed in a one-dimensional form for simplicity, but it is easy to further extend it to three-dimensional form. Detailed discussion about these equations can be found in [23]. Source: Adapted from Yu et al. [23].

Eulerian description treats the multi-electron system as a whole piece, and serve the system as an equivalent "electron field," just like what people do to waters in hydrodynamics [18]. According to this idea, the exact motions for individual electrons are not that important, whereas the "global effects" of a whole bunch of electrons deserve specific cares. Since the "quasi-static" condition is usually broken in the scale of the whole material, original Maxwell's equations are normally required for solving problems based on Eulerian description. The field coupling between the EM field and the electron field needs special treatment as well. Thanks to the brilliance of physicists/mathematicians in thermodynamics and statistics, a super compact, elegant, classical, and mesoscopic model had been established for the purpose of Eulerian description, namely, Boltzmann transport equation [18], which can be regarded as a field equation for electron field. The Boltzmann transport equation naturally contains all the high-order effects, including nonlinearities, in its initial expression. Thus, generally speaking, little modification of the equation is needed for most of the practical cases. This is a huge advantage compared with Lagrangian description, where different theoretical equations are required for different cases in order to get accurate enough results. For numerical algorithms, the universality of the theoretical equations leads to great convenience for the coding. Nevertheless, Boltzmann transport equation suffers from its high standard mathematics and can hardly get analytical results. That is why people still value the importance of having Lagrangian description in classical physics.

From a computational view, both Lagrange and Eulerian descriptions have their own pros and cons. Having Lagrangian description in a completely linear problem is usually more welcomed than having an equivalent Eulerian description. Researchers in CEM field tend to first extract the phenomenological macroscopic model (e.g. dielectric constants) for the materials, which is way more mathematical friendly than the original physical model, then insert those phenomenological model into detailed EM algorithms. However, when nonlinear effects are involved into the problem, even a phenomenological model could become too annoying in mathematics to be easily implemented into numerical algorithms. In that case, the biggest advantage of Lagrangian description disappears, while the compactness and universality of the original physical model in Eulerian description turns to be more competitive and preferred.

One of the strongest attacks to classical approach might be, how come a fully classical theory based approach can correctly solve the behaviors of electrons, which are supposed to be governed by quantum laws? As a matter of fact, in most of the macroscopic problems, classical approach can always generate satisfying results with experimental data. The reason behind it is that, for macroscopic problems, the quantum effects of a single electron will be easily overwhelmed or degraded by the messy interference from large amount of particles and high system energy, so that eventually, the lost information caused by using classical approach won't show any observable contributions to the macroscopic responses. This is the main reason for classical approach still playing an important role, regardless of the fever for quantum theories nowadays.

2.1.2.2 Semi-Classical (Semi-Quantum) Approach

Semi-classical (semi-quantum) approach is the most common way for material modeling in photonics, especially nonlinear photonics [13, 24]. Applying quantum theories into solid-state materials had created remarkable achievement in modern physics and technologies, where even several subjects were invented to change the whole world's civilization, including solid-state physics, microelectronics, and integrated circuits (IC). Semi-classical (semi-quantum) approach is all about applying non-relativity quantum theories into describing the electron behaviors, while still keep the Maxwell's equations in their original forms (without second quantization).

Compares to the classical approach, semi-classical (semi-quantum) approach is way more interesting and closer to the real physics inside the microscopic world of materials. The unique consequences generated from quantum theories, like energy levels and band structures, significantly broadened people's visions in material physics, and perfectly explained tons of inconceivable natures of electrons. Just like the classical approach, there are mainly two branches for semi-classical (semi-quantum) approach.

In the most common case, photonics people would like to model the material by first analyzing one single electron in it, with the simplest isolated Schrodinger's equation [20]. The analytical solution of Schrodinger's equation basically tells everything about the eigen-characters for that electron. One thus further assume that these eigen-characters, like energy levels and wave functions, can be applied to all the electrons in the material. The phenomenological macroscopic model for the material (e.g. dielectric constants) can be efficiently calculated or extracted afterwards. It is not hard to see that, the above process is just like a quantum version of Lagrangian description in classical approach. High-order effects like quantum transitions here generally produce nonlinearities inside materials.

With the same logic, it is reasonable to expect the existence of a quantum version for Eulerian description. Indeed, there is something like that in semi-classical (semi-quantum) approach as well, which is officially named as ensemble theory [20]. No need to say, ensemble theory treats the entire multi-electron stuff as a large quantum system, and focuses on the global effects. The unique property here is that, in quantum theory, ensemble will have peculiar quantum phenomena that can never be repeated by simply adding many isolated particles together. In other words, unlike the classical approach, certain cases in semi-classical (semi-quantum) approach will make the implementation of ensemble theory become a must. In most cases, nonlinear equations are implemented in ensemble theory and thus cause nonlinearities.

Semi-classical (semi-quantum) approach is generally a super useful tool for simulating quantum effects, along with the macroscopic EM interfaces. Typical examples would be quantum computers [25], photonic resonators [24], and microwave-photonic devices [26]. Many inspiring and fancy concepts/terminologies in modern physics today, like Bloch sphere [25] and Rabi oscillation [25], are also generated from this approach. Some intuitive pictures about these applications are given in Figure 2.4. The drawback here is that, implementing semi-classical (semi-quantum) approach into numerical algorithms is usually more complicated than classical approach.

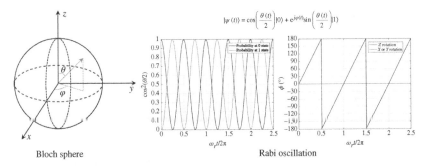

Bloch sphere   Rabi oscillation

Figure 2.4 Intuitive illustrations for two most popular quantum concepts nowadays, namely, Bloch sphere and Rabi oscillation [25]. Source: Adapted from Bardin et al. [25].

2.1.2.3 Full-Quantum Approach

Full-quantum approach is kind of like an ultimate dream of material modeling, which not only applies quantum theories to electrons, but also rewrite the Maxwell's equations with relativity quantum mechanics. A general theory for this approach is quantum electrodynamics (QED) [21], or in some recent studies [27, 28], quantum electromagnetics (QEM). Second quantization for Maxwell's equations, and complicated dynamic equations like Dirac's equation [21, 29], are required for correctly analyzing the electron motions with photons. Until now, this approach is still something under developing, and applying it into numerical algorithms is apparently challenging to many extents. The good news is, because of the Landau criterion [21], full-quantum approach is rarely necessary for most of the practical engineering situations. Thus, as long as stuffs like quantum computers, quantum information, or quantum communications are not being considered, learning full-quantum approach really makes no difference to the analysis, and it is certainly beyond the scope of this chapter.

In this chapter, the Eulerian description in classical approach is chosen for the material modeling in NMSs. To be more specific, the actual model is called "Maxwell-hydrodynamic" model, named by Fang et al. [10], as will be derived in Section 2.2. This model shows outstanding compatibility with numerical methods, and is suitable for quite some NMSs in microwave and optical societies. No doubt that it suffers from certain restrictions as well, which will be discussed in Section 2.2 along with the derivations. Any NMS analysis that breaks those restrictions will exceed the scope of this chapter.

2.1.3 Computational Methods for NMS Analysis

With the help of Maxwell-hydrodynamic model, the main target of computational methods for NMS analysis is to simultaneously solve the hydrodynamic model with Maxwell equations, and develop an efficient numerical algorithm for computing the model. Several related works have been published in previous research with the most common methods for solving Maxwell-hydrodynamic model being divided into two branches: one based on frequency-domain equations [30] and the other based on time-domain equations [10, 31].

Using frequency-domain equations to solve Maxwell-hydrodynamic model is more closely resembles a physical method rather than a purely numerical one. Frequency-domain perturbation method (FDPM) [30] is often implemented so that the complex nonlinear partial differential equations can be divided into series of orders, the greatest advantage of FDPM being that in the most common cases, only the zeroth and the first order equations are needed to give accurate enough results [30]. However, this method only shows great convenience in monotone excitation problems. Once the excitation becomes multi-frequency or wide-band, it must derive significantly more equations, further complicating the entire process.

To give full-wave simulations for wide-band nonlinear problems, directly solving the time-domain equations is better. One of the earliest works about this is proposed by Jinjie Liu et al. [31]. By implementing the finite difference time domain (FDTD) method [32] into the time-domain Maxwell-hydrodynamic model, they present a three-step splitting method to calculate second-harmonic generations (SHGs) in periodic split-ring resonator (SRR) and other metallic structures. Ming Fang et al. further demonstrate another two-step splitting method to solve the time-domain Maxwell-hydrodynamic model with FDTD algorithm [10]. However, because of the Yee's grid topologies [32], it is rather difficult for these FDTD algorithms to compute the strong discontinuities on the materials' surfaces in an accurate way. Furthermore, their algorithms' stabilities are difficult to determine as nonlinearities may unexpectedly cause instability.

Another approach for solving Maxwell-hydrodynamic model, named the time-domain perturbation method (TDPM), is proposed by the authors of this chapter in [33]. By using the TDPM, the nonlinear equations of Maxwell-hydrodynamic model are greatly simplified while keeping their major nonlinear effects and time-domain expressions. An FDTD algorithm can be then implemented to numerically solve the theoretical equations. Certain boundary conditions for calculating the charge accumulations on material surfaces can be derived and well arranged in the Yee's scheme. Furthermore, quantitative stable conditions of the proposed algorithm are able to be presented and analytically proven. Most of the discussions in this chapter will be based on [33].

2.2 Fundamentals of Classical Models

To prepare necessary knowledge of Maxwell-hydrodynamic model, which will be used in this chapter, some fundamentals in microelectronics [34] are presented here. It is interesting to note that, although the majority of this chapter only targeted on EM and photonics, physics in microelectronics still valid very well and turn to be tremendously useful. The seemingly separated two subjects, electronics and photonics, reach to the same point in this topic. This is a common fact of NMS analysis.

2.2.1 Carrier Transport Equations

As mentioned in Section 2.1, Eulerian description in classical approach models the material as an equivalent "electron field," or an "electron sea," and monitors the flow of electrons as a kind of liquid. To mathematically describe the motion of electrons by Eulerian description, several theories in microelectronics [34] had been developed accordingly, including drift-diffusion model, Shockley equations and so on. In the most general case, these theories/models/equations are all named as "carrier transport equations."[1]

Since different alternatives of carrier transport equations are designated with specific physical backgrounds and problems, a proper choice of equation as the start point will be critical. In this chapter, the Boltzmann transport equation [18], which treats the electron sea in a statistical way, is chosen to be the origin of material modeling.

Inherited from the tradition of classical mechanics and statistics, one cares about the probability density function (PDF), $f(\mathbf{r}, \mathbf{p}, t)$, of carriers with respect to the generalized coordinates $(\mathbf{r}, \mathbf{p}, t)$ [22]. Here, $\mathbf{r}$ is the three-dimensional spatial coordinates, $\mathbf{p}$ is the three-dimensional momentum coordinates, and t is the time. $f(\mathbf{r}, \mathbf{p}, t)$ basically describes the probability of an electron that exists at point $(\mathbf{r}, \mathbf{p}, t)$, with the normalization condition given as:

$$\iiint d\mathbf{r} \iiint d\mathbf{p} f(\mathbf{r}, \mathbf{p}, t) = N(t) \tag{2.1}$$

where $N(t)$ is the total carrier number in the material system, which is usually a time invariant (e.g. $N(t) = N$). Most of the times, carrier density $n(\mathbf{r}, t)$ deserves more attentions, which can be calculated as:

$$\iiint d\mathbf{p} f(\mathbf{r}, \mathbf{p}, t) = n(\mathbf{r}, t) \tag{2.2}$$

In fact, $f(\mathbf{r}, \mathbf{p}, t)$ implicitly tells that, the probability of carrier existence is fully determined by the position and momentum of carrier. Moreover, it is a time varying process. In macroscopic EM problems, once the probability of carrier existence is known, everything of that material should be able to solve in theory. Therefore, a fundamental law for this PDF $f(\mathbf{r}, \mathbf{p}, t)$ is needed, which should be able to demonstrate the general mechanism of carrier behaviors inside any material. This is the Boltzmann transport equation [18]:

$$\mathbf{v} \cdot \nabla_{\mathbf{r}} f(\mathbf{r}, \mathbf{p}, t) + \mathbf{F}(\mathbf{r}, t) \cdot \nabla_{\mathbf{p}} f(\mathbf{r}, \mathbf{p}, t) + \frac{\partial f(\mathbf{r}, \mathbf{p}, t)}{\partial t} = \left. \frac{\partial f(\mathbf{r}, \mathbf{p}, t)}{\partial t} \right|_{\text{Collision}} \tag{2.3}$$

where $\mathbf{v}$ is the carrier velocity, $\mathbf{F}$ is the external force that applies to the material (carriers), $\nabla_{\mathbf{r}}$ and $\nabla_{\mathbf{p}}$ are the nabla operators with $\mathbf{r}$ and $\mathbf{p}$ coordinates, respectively. Note that the time derivative of PDF $f(\mathbf{r}, \mathbf{p}, t)$ has two parts. One is based on the continuous motion of carriers driven by the external force or thermal energy, which is written on the left side of (2.3). While the other one is caused

---

1 In this chapter, mostly, the term "carrier" simply means "electron."

by the sudden collisions between carriers and lattice,[2] which is written on the right-hand side of (2.3).

One interesting thing to note here is, the Boltzmann transport equation (2.3) is a mixture of macroscopic and microscopic parameters. The PDF $f(\mathbf{r}, \mathbf{p}, t)$, carrier velocity $\mathbf{v}$, and carrier momentum $\mathbf{p}$ are all microscopic parameters.[3] At the meantime, the external force $\mathbf{F}$ is a macroscopic parameter.[4] For that reason, Boltzmann transport equation (2.3) is also known as a "mesoscopic" model. Multiple advantages will come from this mesoscopic property, as will be notified in the latter contents.

2.2.2 Momentum Equations

The Boltzmann transport equation (2.3) turns out to be super powerful in many practical problems, but the complicated (though compact) mathematics makes it become challenging for numerical methods. In that case, some simplifications of (2.3) need to be developed so that the mathematics become more friendly.

As mentioned before, Boltzmann transport equation itself is a mesoscopic model, while in a general EM problem, only the macroscopic responses matter. This inspires us to extract macroscopic descriptions from the original Boltzmann transport equation. That is where "momentum equations" [35] come into play.

Note that the momentum coordinates $\mathbf{p}$ won't appear in the EM equations (e.g. Maxwell's equations), eliminating them by integrating (2.3) in $\mathbf{p}$ coordinates should be a wise choice. Thus, in order to do that, we first define some generalized quantities for the convenience of future narrations.

Let $\phi(\mathbf{p})$ be any microscopic parameter in the material system, which only varies with $\mathbf{p}$. To eliminate $\mathbf{p}$, consider the following integral:

$$n_\phi(\mathbf{r}, t) = \iiint d\mathbf{p}\phi(\mathbf{p})f(\mathbf{r}, \mathbf{p}, t) \tag{2.4}$$

where $n_\phi(\mathbf{r}, t)$ should be some kind of macroscopic parameter related to $\phi(\mathbf{p})$ then. $n_\phi(\mathbf{r}, t)$ is intentionally named as "generalized carrier density."[5] If we normalize this $n_\phi(\mathbf{r}, t)$ with the real carrier density $n(\mathbf{r}, t)$, we can further define a "macroscopic average" of $\phi(\mathbf{p})$, given as:

$$\langle \phi(\mathbf{p}) \rangle = \frac{n_\phi(\mathbf{r}, t)}{n(\mathbf{r}, t)} \tag{2.5}$$

---

2 Strictly speaking, this kind of sudden change is not a continuous function at all, so writing it into a time derivative is non-sense mathematically. However, we write it like that anyway, since there is no better general expression for it. This is also consistent with mainstream textbooks, like [34].

3 Think them as parameters for a single carrier.

4 Think it as a global parameter applies to the entire material system.

5 The reason for this naming will be discovered later.

where the bracket "$\langle \rangle$" denotes for "averaging." Thus,

$$n_\phi(\mathbf{r}, t) = \iiint d\mathbf{p}\phi(\mathbf{p})f(\mathbf{r}, \mathbf{p}, t) = n(\mathbf{r}, t)\langle\phi(\mathbf{p})\rangle \tag{2.6}$$

meaning that the generalized carrier density can be obtained by the product of carrier density and $\phi(\mathbf{p})$ average. Note that $n_\phi(\mathbf{r}, t) = n(\mathbf{r}, t)$ if $\phi(\mathbf{p}) = 1$, which is obviously consistent with the original definition of PDF.

Repeat the operation in (2.4) for every term in (2.3), we have:

$$\iiint d\mathbf{p}\phi(\mathbf{p})\mathbf{v} \cdot \nabla_r f(\mathbf{r}, \mathbf{p}, t) + \iiint d\mathbf{p}\phi(\mathbf{p})\mathbf{F}(\mathbf{r}, t) \cdot \nabla_\mathbf{p} f(\mathbf{r}, \mathbf{p}, t)$$

$$+ \iiint d\mathbf{p}\phi(\mathbf{p})\frac{\partial f(\mathbf{r}, \mathbf{p}, t)}{\partial t} = \iiint d\mathbf{p}\phi(\mathbf{p})\frac{\partial f(\mathbf{r}, \mathbf{p}, t)}{\partial t}\bigg|_{\text{Collision}} \tag{2.7}$$

Now assume the right-hand side term can be simply replaced by a relaxation process,[6]

$$\iiint d\mathbf{p}\phi(\mathbf{p})\frac{\partial f(\mathbf{r}, \mathbf{p}, t)}{\partial t}\bigg|_{\text{Collision}} = \frac{n_\phi(\mathbf{r}, t) - n_\phi^0(\mathbf{r}, t)}{\tau_\phi} \tag{2.8}$$

where τ_ϕ is the equivalent relaxation time, and $n_\phi^0(\mathbf{r}, t)$ (the "0" index) denotes for the generalized carrier density under thermal equilibrium [34]. Further, define the following quantities:

$$\mathbf{F}_\phi(\mathbf{r}, t) = \iiint d\mathbf{p}\phi(\mathbf{p})\mathbf{v}f(\mathbf{r}, \mathbf{p}, t) \tag{2.9}$$

$$\mathbf{G}_\phi(\mathbf{r}, t) = \mathbf{F} \cdot \iiint d\mathbf{p}f(\mathbf{r}, \mathbf{p}, t)\nabla_\mathbf{p}\phi(\mathbf{p}) \tag{2.10}$$

$$\mathbf{R}_\phi(\mathbf{r}, t) = \frac{n_\phi(\mathbf{r}, t) - n_\phi^0(\mathbf{r}, t)}{\tau_\phi} \tag{2.11}$$

where $\mathbf{F}_\phi(\mathbf{r}, t)$, $\mathbf{G}_\phi(\mathbf{r}, t)$ and $\mathbf{R}_\phi(\mathbf{r}, t)$ are intentionally named as "generalized carrier flow," "generalized carrier generation rate," and "generalized carrier recombination rate," respectively. Inserting (2.6) and (2.9)–(2.11) into (2.7), with the help of some fundamental vector field mathematics, and be careful of the two coordinates used in those nabla operators, we can equivalently write (2.7) into:

$$\frac{\partial n_\phi(\mathbf{r}, t)}{\partial t} + \nabla \cdot \mathbf{F}_\phi(\mathbf{r}, t) = \mathbf{G}_\phi(\mathbf{r}, t) - \mathbf{R}_\phi(\mathbf{r}, t) \tag{2.12}$$

Note that here, the lower index of $\mathbf{r}$ is already omitted for the nabla operator, since there is no $\mathbf{p}$ variable in the equation any more.

Equation (2.12) looks exactly like the "carrier conservation law" in Shockley equations [34]. Indeed, this is the major reason for defining those generalized $n_\phi(\mathbf{r}, t)$, $\mathbf{F}_\phi(\mathbf{r}, t)$, $\mathbf{G}_\phi(\mathbf{r}, t)$ and $\mathbf{R}_\phi(\mathbf{r}, t)$ terms, and (2.12) is called "generalized carrier conservation law." Having this interesting coincidence between Boltzmann

---

6 This is a brave but smart assumption, detail physics about it can be found in [19].

transport equation and Shockley equations is not surprising though, as long as one can notice the identical deep physics behind them. Unfortunately, the beautiful physics in behind is not the task for this chapter, thus we need to stop here and move on to the next step.

To get the "momentum equations," we just need to substitute specific expressions of $\phi(\mathbf{p})$ into the generalized carrier conservation law (2.12). In general, we care about three momentums, zeroth order momentum (carrier density), first-order momentum (carrier momentum), and second-order momentum (carrier kinetic energy), defined as:

$$\phi(\mathbf{p}) = 1, \quad n_\phi(\mathbf{r}, t) = n(\mathbf{r}, t) \tag{2.13}$$

$$\phi(\mathbf{p}) = \mathbf{p}, \quad n_\phi(\mathbf{r}, t) = \mathbf{P}(\mathbf{r}, t) \tag{2.14}$$

$$\phi(\mathbf{p}) = \frac{p^2}{2m^*}, \quad n_\phi(\mathbf{r}, t) = E_k(\mathbf{r}, t) \tag{2.15}$$

where capital letter "P" is used to denote the macroscopic momentum distribution $\mathbf{P}(\mathbf{r}, t)$, while $E_k(\mathbf{r}, t)$ is the kinetic energy distribution, m^* is the effective mass [34] of carrier. Substituting (2.13)–(2.15) into (2.12), and utilizing (2.6) yields the following three equations:

$$\frac{\partial n}{\partial t} + \nabla \cdot (n \langle \mathbf{v} \rangle) = \frac{n - n_0}{\tau_n} \tag{2.16}$$

$$\frac{\partial (n\mathbf{P})}{\partial t} + \nabla \cdot \overline{\mathbf{F}}_\mathbf{P} = n\mathbf{F} - \frac{n\mathbf{P} - n_0\mathbf{P}_0}{\tau_m} \tag{2.17}$$

$$\frac{\partial (nE_k)}{\partial t} + \nabla \cdot \mathbf{F}_W = \mathbf{F} \cdot n \langle \mathbf{v} \rangle - \frac{nE_k - n_0E_{k0}}{\tau_E} \tag{2.18}$$

where $\langle \mathbf{v} \rangle$ is the average electron velocity, $\overline{\mathbf{F}}_\mathbf{P}$ and $\mathbf{F}_W$ are the generalized carrier flows for carrier momentum and kinetic energy, respectively. Note that, since the momentum $\mathbf{P}$ itself is already a vector, from the definition of generalized carrier flow, $\overline{\mathbf{F}}_\mathbf{P}$ should be a second-order tensor.[7]

Up to this point, we have not applied any pre-knowledge to the PDF yet. With several reasonable assumptions to the PDF and carrier velocity distribution, (2.16)–(2.18) can be further simplified.

In thermal equilibrium, the carrier velocity is often described by the classical Maxwellian distribution [34], where the average velocity $\langle \mathbf{v} \rangle$ is simply zero. Thus, it is easy to note that $\mathbf{P}_0 = 0$ as well, since $\mathbf{P}_0 = m^* \langle \mathbf{v} \rangle_0$. When external force is induced on the material system, the condition of thermal equilibrium will definitely be broken, so that $\langle \mathbf{v} \rangle$ usually has a finite value. However, in most cases, the internal interactions (e.g. collisions) between the carriers and the lattice will cause the carrier velocity distribution to stabilize after an external force has been added

---

7 Basically like a square matrix.

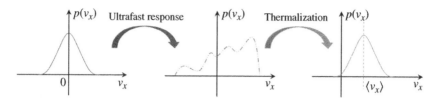

Figure 2.5 A simple but straightforward explanation for probability variation in an ultra-fast thermalization process. Here, the carrier velocity is only shown in a one-dimensional manner, but there is no limitation to extend it into three-dimensional problems.

for a while. This stabilization process is called "thermalization," and thermalization usually can be considered as a ultra-fast process [34]. The fast response for thermalization causes the stabilized carrier velocity intends to maintain the original Maxwellian distribution but with a constant shift[8] in $\langle \mathbf{v} \rangle$ [34], like illustrated in Figure 2.5. Based on this assumption, we are allowed to express the microscopic velocity for each carrier as:

$$\mathbf{v} = \langle \mathbf{v} \rangle + \mathbf{c} \tag{2.19}$$

where $\mathbf{c}$ is a stochastical process that follows normal Maxwellian distribution. This condition is sometimes called as "quasi-equilibrium" [34].

From Figure 2.5, it seems that quasi-equilibrium will make the integral of PDF in momentum coordinates become exactly the same as the one in thermal equilibrium, since the exact Maxwellian distribution is kept identical. In that case, the carrier density $n = n_0$ everywhere in the material. However, it should be clarified that this declaration is only correct when there is no extra carrier generation/recombination caused by external sources (e.g. light, heat, ultrasound). Otherwise, there will be excess carriers [34] that breaks the condition of $n = n_0$. Nevertheless, in many practical situations, another assumption called "low level injection" (LLI) can always be valid, which basically says, the excess carrier density (no matter you really have it or not) is always negligible compared to the majority carrier density. Thus, if only majority carriers[9] are taken into consideration, we will always have $n \approx n_0$.

By taking into account of all the above assumptions, we can further simplify (2.16)–(2.18) into the following forms:

$$\frac{\partial n}{\partial t} + \nabla \cdot (n \langle \mathbf{v} \rangle) = 0 \tag{2.20}$$

$$\frac{\partial (n\mathbf{P})}{\partial t} + \nabla \cdot \overline{\mathbf{F}}_\mathbf{P} = n\mathbf{F} - \frac{n\mathbf{P}}{\tau_m} \tag{2.21}$$

---

8 This is pretty much like adding a constant DC term to an AC signal.
9 Like electrons in n-type semiconductors and metals, or holes in p-type semiconductors.

$$\frac{\partial(nE_k)}{\partial t} + \nabla \cdot \mathbf{F}_W = \mathbf{F} \cdot n \langle \mathbf{v} \rangle - \frac{nE_k - n_0 E_{k0}}{\tau_E} \tag{2.22}$$

where some terms can be deleted without bringing too much inaccuracy.

The next step is to look at the detail expression for $\mathbf{F}_W$. From (2.9),

$$\mathbf{F}_W = \iiint d\mathbf{p} v f \frac{1}{2} m^* v^2 = \frac{1}{2} m^* \iiint d\mathbf{p} v^2 \mathbf{v} f \tag{2.23}$$

Inserting (2.19) into it,

$$\mathbf{F}_W = \left(\frac{1}{2} m^* \iiint d\mathbf{p} v^2 f \right) \langle \mathbf{v} \rangle + \frac{1}{2} m^* \iiint d\mathbf{p} v^2 \mathbf{c} f \tag{2.24}$$

Further expanding v^2 by (2.19) and note (2.6),

$$\mathbf{F}_W = nE_k \langle \mathbf{v} \rangle + \frac{1}{2} m^* \iiint d\mathbf{p} \left(|\langle \mathbf{v} \rangle|^2 + 2 \langle \mathbf{v} \rangle \cdot \mathbf{c} + c^2 \right) \mathbf{c} f \tag{2.25}$$

The Maxwellian distribution yields,

$$\iiint d\mathbf{p} \mathbf{c} f = 0 \tag{2.26}$$

so that,

$$\mathbf{F}_W = nE_k \langle \mathbf{v} \rangle + 0 + nm^* \langle \mathbf{v} \rangle \cdot \langle \mathbf{cc} \rangle + \frac{1}{2} nm^* \langle c^2 \mathbf{c} \rangle \tag{2.27}$$

Define the temperature tensor $\overline{\mathbf{T}}$ and thermal flow density $\mathbf{Q}$ as [18]:

$$nm^* \langle \mathbf{cc} \rangle = nk\overline{\mathbf{T}}, \quad \frac{1}{2} nm^* \langle c^2 \mathbf{c} \rangle = \mathbf{Q} \tag{2.28}$$

where k is the Boltzmann constant. Then,

$$\mathbf{F}_W = nE_k \langle \mathbf{v} \rangle + nk \langle \mathbf{v} \rangle \cdot \overline{\mathbf{T}} + \mathbf{Q} \tag{2.29}$$

With similar logic, $\overline{\mathbf{F}}_P$ can be rewritten as:

$$\overline{\mathbf{F}}_P = nm^* \langle \mathbf{vv} \rangle = nm^* \langle \mathbf{v} \rangle \langle \mathbf{v} \rangle + nm^* \langle \mathbf{cc} \rangle = n\mathbf{P} \langle \mathbf{v} \rangle + nk\overline{\mathbf{T}} \tag{2.30}$$

Now applying the Maxwell's approximations from thermodynamics [18],

$$\overline{\mathbf{T}} = T\overline{\mathbf{I}}, \quad \mathbf{Q} = 0 \tag{2.31}$$

where $\overline{\mathbf{I}}$ is the identity tensor (matrix), T has the unit of Kelvin, which is basically the normal temperature. Finally, we get:

$$\frac{\partial n}{\partial t} + \nabla \cdot (n \langle \mathbf{v} \rangle) = 0 \tag{2.32}$$

$$\frac{\partial(n\mathbf{P})}{\partial t} + \nabla \cdot (n\mathbf{P} \langle \mathbf{v} \rangle) = n\mathbf{F} - \nabla(nkT) - \frac{n\mathbf{P}}{\tau_m} \tag{2.33}$$

$$\frac{\partial(nE_k)}{\partial t} + \nabla \cdot \left(nE_k \langle \mathbf{v} \rangle \right) = \mathbf{F} \cdot n \langle \mathbf{v} \rangle - \nabla \cdot (nkT \langle \mathbf{v} \rangle) - \frac{nE_k - n_0 E_{k0}}{\tau_E} \tag{2.34}$$

These are the so-called "momentum equations" [35]. Since no more microscopic parameters will be discussed in the rest of this chapter, for the sake of convenience, replace $\langle \mathbf{v} \rangle$ by just $\mathbf{v}$, rewrite the momentum equations as follows:

$$\frac{\partial n}{\partial t} + \nabla \cdot (n\mathbf{v}) = 0 \tag{2.35}$$

$$\frac{\partial (n\mathbf{P})}{\partial t} + \nabla \cdot (n\mathbf{P}\mathbf{v}) = n\mathbf{F} - \nabla(nkT) - \frac{n\mathbf{P}}{\tau_m} \tag{2.36}$$

$$\frac{\partial (nE_k)}{\partial t} + \nabla \cdot \left(nE_k\mathbf{v}\right) = \mathbf{F} \cdot n\mathbf{v} - \nabla \cdot (nkT\mathbf{v}) - \frac{nE_k - n_0 E_{k0}}{\tau_E} \tag{2.37}$$

We will reference (2.35)–(2.37) as the momentum equations in the rest of this chapter.

2.2.3 Maxwell-Hydrodynamic Model

For the ease of narration, from now on, the material modeling in this chapter will only deal with electrons. Use m_e to represent the effective mass m^* of electron, omit the lower index of momentum relaxation time τ_m, and insert (2.35) into (2.36), we can get a new compact equation from the momentum equations. Given as,[10]

$$\frac{\partial \mathbf{v}}{\partial t} + (\mathbf{v} \cdot \nabla)\mathbf{v} = \frac{\mathbf{F}}{m_e} - \frac{kT}{m_e n}\nabla n - \frac{\mathbf{v}}{\tau} \tag{2.38}$$

which is called as "drifting-plasma model" in some literatures [36, 37]. Note that here, (2.38) contains the information from both (2.35) and (2.36), while the information in (2.37) has been intentionally abandoned. This means that (2.38) will only generate reasonable results for the carrier density and momentum, while the calculated kinetic energy of carriers might be inaccurate. Usually, however, solving macroscopic EM problems won't require any ability for kinetic energy calculation in material modeling. Thus, the drifting-plasma model is often good enough for EM. As a matter of fact, precisely solving kinetic energy for electrons requires the complete picture of band structures, where quantum theories are definitely needed. Since we only focus on classical approach here, strictly speaking, the computation of kinetic energy will be incorrect anyway. Nevertheless, classical approach used in here is still valid for the computation of carrier density and momentum, because solid-state physics [19] proves consistent results between classical and quantum interpretations for these two quantities.[11] Moreover, throwing away (2.37) doesn't lead to any violence of energy conservation law, since both the kinetic energy and the energy bands (potentials) need to be taken into account,

---

10 Here, we also assumed the homogeneity of temperature T, which is quite reasonable in most practical cases. Besides, apparently $\mathbf{P} = m_e\mathbf{v}$.
11 Referring to the term "pseudo-momentum" in [19].

and they will always compensate for the energy deficit automatically.[12] This is something that one should always keep in mind.

To make (2.38) be more compatible with Maxwell's equations, substitute the external force $\mathbf{F}$ with Lorentz force $\mathbf{F} = -q\,(\mathbf{E} + \mathbf{v} \times \mathbf{B})$, where q is the absolute value of electron charge. By further representing the carrier velocity $\mathbf{v}$ into the form of current density $\mathbf{J} = -nq\mathbf{v}$, we have:

$$\frac{\partial \mathbf{J}}{\partial t} = -\frac{\mathbf{J}}{\tau} - \frac{q}{m_e}(-nq\mathbf{E} + \mathbf{J} \times \mathbf{B}) - \frac{kT}{m_e}\nabla(-nq) + \nabla \cdot \left(\frac{\mathbf{J}\mathbf{J}}{nq}\right) \tag{2.39}$$

This is the equation that can be easily implemented into most CEM algorithms. In [31], (2.39) is named as "nonlinear Drude model," in analog with the traditional Drude model [38]. Solving (2.39) together with Maxwell's equations gives the so-called "Maxwell-hydrodynamic model,"[13] which is the major material model used in this chapter.

The physical meaning of (2.39) is quite straightforward. The left-hand side of (2.39) gives the time varying situation of the current density, and the right-hand side of (2.39) indicates the reasons for that variation. The first term on the right-hand side of (2.39) represents the relaxation process of the electrons, which guarantees that the electrons intend to calm down their motions when there are no external forces. The second term is the Lorentz force added to the electrons by the electromagnetic fields. This is how the electromagnetic fields couple to the electrons. The third term is a thermal term that illustrates the diffusion effect of the electrons, arising due to how the electrons are viewed as liquid in the hydrodynamic model. The last term on the right-hand side of (2.39) shows the convection of electrons inside the materials, which is a nonlinear process of the electron motions.

2.2.4 Simplified Models at Low Frequencies

Some even simpler models can be further derived based on (2.39). Deriving these even simpler models has nothing to do with the task of this chapter. However, for the completeness of fundamentals of classical models, which is the theme of this section, it is worthy to note them here as well.

Check the expression of (2.39), if we already know that the external EM forces will change much slower than the relaxation time τ, which basically indicates a

---

12 The logic here is tricky. Since we don't add energy restrictions to the carriers, that part of energy will always be computed at the end. As long as we still believe in the correctness of energy conservation law, we will just compute that energy by the subtraction of other energies from the total energy. Then apparently, carrier energy will always compensate for the energy deficit of the system.

13 First named in [10].

"low frequency problem," then the time derivative on the left-hand side of (2.39) can be deleted. Moreover, if the effect of magnetic field is much weaker than electric field, which is usually the case at low frequencies, then the **B** term in (2.39) can be omitted as well. Finally, if we do not care about nonlinear effects, which are usually small second-order effects [34], then the convection term (the last term on the right-hand side) can be removed. Combining all these simplifications, a much simpler model can be obtained for low frequency problems, where,

$$\mathbf{J} = \frac{nq^2\tau}{m_e}\mathbf{E} + \frac{kTq\tau}{m_e}\nabla n \tag{2.40}$$

Now if we define two important parameters in microelectronics [34], electron mobility μ_n, and electron diffusion coefficient D_n as:

$$\mu_n = \frac{q\tau}{m_e}, \quad D_n = \frac{kT\tau}{m_e} \tag{2.41}$$

then (2.40) becomes,

$$\mathbf{J} = nq\mu_n\mathbf{E} + qD_n\nabla n \tag{2.42}$$

which is exactly the well-known drift-diffusion model in microelectronics [34]. Note that, we also have the relation,

$$\frac{D_n}{\mu_n} = \frac{kT}{q} \tag{2.43}$$

This is the Einstein relation [34], which we prove it by the way.

Last but not the least, if the diffusion term in (2.42) is negligible, a simplest expression can be obtained as:

$$\mathbf{J} = nq\mu_n\mathbf{E} = \sigma\mathbf{E} \tag{2.44}$$

where the $\sigma = nq\mu_n$ is the definition of conductivity. No need to clarify the importance of this concept.

As mentioned before, these simplifications only valid at low frequencies, where circuits problems often locate. Thus, in microelectronics and ICs, (2.42) and (2.44) are way more commonly used than (2.39).

2.2.5 Review and Restrictions

At the end of this section, let's take a brief review on how did we start from the Boltzmann transport equation, all the way to the Maxwell-hydrodynamic model, as well as the major restrictions for implementing the model.

The basic steps of deriving Maxwell-hydrodynamic model from Boltzmann transport equation is illustrated in Figure 2.6. Recall that many assumptions have been made for this derivation, and note that each assumption will lead

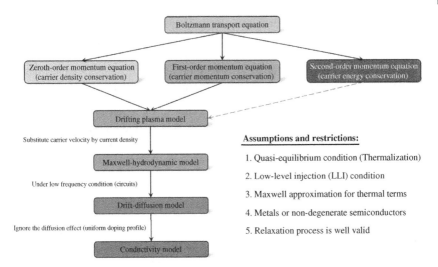

Figure 2.6 The basic steps of deriving Maxwell-hydrodynamic model from Boltzmann transport equation, together with related interesting models, as well as the assumptions and restrictions.

to a corresponding restriction for the model. The major restrictions for this Maxwell-hydrodynamic model can be concluded as follows:

(1) Quasi-equilibrium condition must be well satisfied. This indicates that the EM fields must demonstrate a much slower reaction than the thermalization process inside the material, which is usually the case in microwave, terahertz, and optical regions.

(2) There is no excess carriers inside the material, or LLI condition has been satisfied for majority carriers. For semiconductor materials, this is something that needs to be carefully checked. However, in metallic materials, LLI assumption is usually fine for electrons. This restriction also implicitly indicates that only majority carriers can be correctly computed by this model.

(3) The Maxwell approximation in thermodynamics must always be valid. Detail thermal restrictions for this approximation can be found in [18], which can usually be well satisfied in normal EM problems though.

(4) The effective mass of carrier is considered to be a constant in a certain problem. This condition is implicitly included by the derivation of momentum equations. For metals, the parabolic shape of their conduction band often induces a near constant effective mass for electrons. Nevertheless, for semiconductors, the band structures could be messy, so that this condition actually requires a tiny filling amount of electrons (holes) in the conduction band (valence band), which means the semiconductor needs to be non-degenerate [34].

(5) The relaxation process has been assumed to explain some mechanisms inside the material. This is somehow hard to rigorously explain by theoretical physics. However, it is extremely important for the material modeling, and it is the reason why we have the term "relaxation time." Fortunately, experiments show that using the relaxation process for material modeling rarely fails.

With the above restrictions, we conclude the introduction of necessary fundamentals here, and come up with the Maxwell-hydrodynamic model for later sections. Only by satisfying the aforementioned restrictions, can Maxwell-hydrodynamic model work efficiently in CEM numerical algorithms.

2.3 FDTD Analysis

To make things clear, the following analysis will be based on metallic materials, where nonlinearity mainly comes from the massive electron couplings, and mainly from the authors' previous publication [33]. However, it should be kept in mind that same analysis may be applied to semiconductor materials as well, as long as the material satisfies the restrictions listed in Section 2.2.

2.3.1 Time-Domain Perturbation Method (TDPM)

It is interesting to note that, although all the terms on the right-hand side of (2.39) affects the current density's variance, the actual effects are not on the same level for different terms. This fact inspires the use of separate operations for different levels of the effects and some relatively non-essential factors in (2.39) to be ignored. To achieve that, several assumptions must first be clarified with appropriate physical explanations.

First, it is apparent that the nonlinear effects are much weaker than the linear effects when metallic materials are interacting with electromagnetic fields. Second, the magnetic field's influences (forces) are much smaller than the electric field, as illustrated in the expression of electromagnetic force. When the magnetic field adds forces to the electrons, the direct effects are given by the magnetic induction **B** rather than the magnetic field **H**. Because the constant μ_0 is small in numeral value, the magnetic induction should be much smaller than the magnetic field. Considering that the numeric value of the magnetic field is at the same level with the electric field, it is reasonable to assume that the magnetic forces are significantly smaller than the electric forces.[14] Furthermore, since

---

14 Strictly speaking, the comparison here is not physically correct, since the units are different. The intrinsical reason for treating magnetic force as small quantity is the rather low speed of electron motion inside the material. However, rigorously proving this low speed is complicated

there is no intentional doping profile in the materials, the spatial variance of the current density should be regarded as minor effects as well.

Based on the assumptions above, the TDPM equations are then derived in the following manner. Suppose the current density $\mathbf{J}$ can be divided into two parts, $\mathbf{J}_0$ and $\mathbf{J}_{NL}$, respectively, corresponding to the linear and nonlinear responses. Then, based on the Maxwell equations, one has $\epsilon_0 \nabla \cdot \mathbf{E} = \rho = q(N - n)$, where N is the static value of electron density in the materials. Thereafter, (2.39) is rewritten into:

$$\frac{\partial \mathbf{J}_0}{\partial t} + \frac{\partial \mathbf{J}_{NL}}{\partial t} = -\frac{\mathbf{J}_0}{\tau} - \frac{\mathbf{J}_{NL}}{\tau} + \frac{Nq^2}{m_e}\mathbf{E} - \frac{\epsilon_0 q}{m_e}(\nabla \cdot \mathbf{E})\mathbf{E}$$

$$-\frac{q}{m_e}\mathbf{J} \times \mathbf{B} - \frac{kT}{m_e}\nabla(-nq) + \nabla \cdot \left(\frac{\mathbf{JJ}}{nq}\right) \quad (2.45)$$

Consider that the $\mathbf{J}_{NL}$, $\mathbf{B}$, and the operator ∇ are the first-order small quantities, and each term contains one of them can be also seen as the first-order small quantity. In that case, if only the zeroth- and first-order small quantities are preserved in (2.45), one can get two equations as shown in (2.46) and (2.47):

$$\frac{\partial \mathbf{J}_0}{\partial t} = -\frac{\mathbf{J}_0}{\tau} + \frac{Nq^2}{m_e}\mathbf{E}_0 \quad (2.46)$$

$$\frac{\partial \mathbf{J}_{NL}}{\partial t} = -\frac{\mathbf{J}_{NL}}{\tau} + \frac{Nq^2}{m_e}\mathbf{E}_1$$

$$-\frac{\epsilon_0 q}{m_e}(\nabla \cdot \mathbf{E}_0)\mathbf{E}_0 - \frac{q}{m_e}\mathbf{J}_0 \times \mathbf{B}_0 + \nabla \cdot \left(\frac{\mathbf{J}_0\mathbf{J}_0}{Nq}\right) \quad (2.47)$$

Notice that the lower indices 0 and 1 of EM fields in (2.46) and (2.47) represent linear and nonlinear responses respectively, as the two curl equations in Maxwell equations get the similar perturbation forms:

$$\nabla \times \mathbf{E}_0 = -\mu_0 \frac{\partial \mathbf{H}_0}{\partial t} \quad (2.48)$$

$$\nabla \times \mathbf{E}_1 = -\mu_0 \frac{\partial \mathbf{H}_1}{\partial t} \quad (2.49)$$

$$\nabla \times \mathbf{H}_0 = \mathbf{J}_0 + \epsilon_0 \frac{\partial \mathbf{E}_0}{\partial t} \quad (2.50)$$

$$\nabla \times \mathbf{H}_1 = \mathbf{J}_{NL} + \epsilon_0 \frac{\partial \mathbf{E}_1}{\partial t} \quad (2.51)$$

The diffusion term in (2.45) disappears in (2.46) and (2.47) because it contains $\nabla n = \nabla(N - (\epsilon_0/q)\nabla \cdot \mathbf{E})$, which equals 0 plus a second order small quantity.

---

and beyond the scope of this chapter. Thus, we made the assumption here based on a fudge but straightforward explanation. Nevertheless, numerical results do satisfy the assumption made here once the units are unified.

The reason is similar for the transformation from $\mathbf{J}$ to $\mathbf{J}_0$, n to N between (2.45) and (2.47).

Respectively, Eqs. (2.46) and (2.47) are the zeroth- and the first-order TDPM equations and it is important to note that these two equations are totally independent with the numerical method or the excitation frequencies. Thus, the TDPM equations (2.46) and (2.47) can be universally used in any kinds of numerical method under any frequency situations.

2.3.2 Numerical Algorithm: FDTD-TDPM

Aiming to solve the Maxwell-hydrodynamic model in numerical ways, the FDTD method [32] is implemented into (2.46)–(2.51) simultaneously. In order to make the algorithm be even more feasible, Eqs. (2.47) and (2.51) are further split into the following forms:

$$\frac{\partial \mathbf{J}_1}{\partial t} = -\frac{\mathbf{J}_1}{\tau} + \frac{Nq^2}{m_e}\mathbf{E}_1 \tag{2.52}$$

$$\frac{\partial \mathbf{J}_e}{\partial t} = -\frac{\mathbf{J}_e}{\tau} - \frac{\epsilon_0 q}{m_e}(\nabla \cdot \mathbf{E}_0)\mathbf{E}_0 - \frac{q}{m_e}\mathbf{J}_0 \times \mathbf{B}_0 + \nabla \cdot \left(\frac{\mathbf{J}_0 \mathbf{J}_0}{Nq}\right) \tag{2.53}$$

$$\nabla \times \mathbf{H}_1 = \mathbf{J}_e + \mathbf{J}_1 + \epsilon_0 \frac{\partial \mathbf{E}_1}{\partial t} \tag{2.54}$$

where the $\mathbf{J}_{NL}$ in (2.47) and (2.51) are divided into $\mathbf{J}_1$ and $\mathbf{J}_e$, which are intentionally corresponded to different parts of the right-hand side terms in (2.47). This can be seen as a strategy for the numerical algorithm with the following reason for doing so. Since (2.52) has the exactly same form as (2.46) (except for the change of lower indices), and the right-hand side terms in (2.53) (except the $\mathbf{J}_e$ term) can be explicitly calculated once (2.48), (2.50), and (2.46) are solved, $\mathbf{J}_e$ can be regarded as an extra nonlinear source which is explicitly determined by the linear responses, and $\mathbf{J}_1$ becomes the passive nonlinear current response excited by $\mathbf{J}_e$. Thus, once (2.48), (2.50), and (2.46) are solved in the algorithm, $\mathbf{J}_e$ could be explicitly computed by solving (2.53). And after $\mathbf{J}_e$ is computed, one can just recall exactly the same linear solver, which solves (2.48), (2.50), and (2.46), to compute (2.49), (2.54), and (2.47) by simply changing all the lower indices from 0 to 1 and adding an extra current source $\mathbf{J}_e$. In summary, the system diagram of the FDTD algorithm is demonstrated in Figure 2.7, and its details are described as follows.

2.3.2.1 Computational Grids

Before the specific updating equations are derived, one must first determine proper computational grids so that the discretization and the colocation for all the fields and the current densities can be well arranged. The computational grids in this work are generated from the original Yee's grid [32, 39]. Figure 2.8a shows the

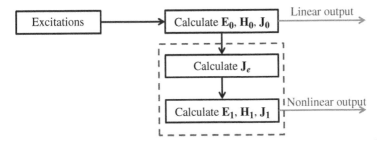

Figure 2.7 System diagram of the FDTD-TDPM algorithm.

Figure 2.8 Computational grids
in FDTD algorithm. (a) Standard
Yee's grid. (b) A denser grid
generated from the standard Yee's
grid, which is named as the node
grid for nonlinear current
calculation.

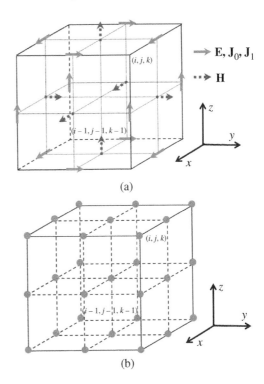

traditional Yee's grid for the electromagnetic fields, where the current densities
are sampled at the same positions with the electric fields.

Further considerations is necessary if (2.53) is going to be discretized in the
standard Yee's grid. The sample points at the standard Yee's grid are not convenient to give a simple solution for those complex spatial differentials in (2.53). In
that case, another grid is generated from the standard Yee's grid to circumvent
this problem, as shown in Figure 2.8b. The sample points of the electromagnetic
fields and the current densities are all set at the positions of each nodes in

Figure 2.8b. Notice that the grid in Figure 2.8b are eight times' denser than the grid in Figure 2.8a. For narrative convenience, the grid shown in Figure 2.8b is referred to the node grid.

The two sets of grids in Figure 2.8 are used simultaneously in the iteration process of the algorithm. Figure 2.8a is used for the updating equations of (2.48), (2.49), (2.50), (2.54), (2.46) and (2.52), and Figure 2.8b is used for the updating equation of (2.53).

2.3.2.2 Linear FDTD Solver

Based on the system diagram demonstrated in Figure 2.7 and previous discussions, to realize the FDTD algorithm, one needs to develop a linear FDTD solver to solve (2.48), (2.49), (2.50), (2.54), (2.46), and (2.52), together with the calculation of J_e. This subsection presents updating equations for the linear FDTD solver.

By using the standard Yee's grid as shown in Figure 2.8a and following the standard iteration manner in FDTD method, all the updating equations for the linear FDTD solver could be written in a unified form, as shown in (2.55)–(2.61).

$$
\begin{aligned}
H_{ax,i,j-1/2,k-1/2}^{m+1/2} = {} & H_{ax,i,j-1/2,k-1/2}^{m-1/2} \\
& - \frac{\Delta t}{\mu_0} \left[\frac{E_{az,i,j,k-1/2}^m - E_{az,i,j-1,k-1/2}^m}{\Delta y} \right. \\
& \left. - \frac{E_{ay,i,j-1/2,k}^m - E_{ay,i,j-1/2,k-1}^m}{\Delta z} \right]
\end{aligned}
\tag{2.55}
$$

$$
\begin{aligned}
H_{ay,i-1/2,j,k-1/2}^{m+1/2} = {} & H_{ay,i-1/2,j,k-1/2}^{m-1/2} \\
& - \frac{\Delta t}{\mu_0} \left[\frac{E_{ax,i-1/2,j,k}^m - E_{ax,i-1/2,j,k-1}^m}{\Delta z} \right. \\
& \left. - \frac{E_{az,i,j,k-1/2}^m - E_{az,i-1,j,k-1/2}^m}{\Delta x} \right]
\end{aligned}
\tag{2.56}
$$

$$
\begin{aligned}
H_{az,i-1/2,j-1/2,k}^{m+1/2} = {} & H_{az,i-1/2,j-1/2,k}^{m-1/2} \\
& - \frac{\Delta t}{\mu_0} \left[\frac{E_{ay,i,j-1/2,k}^m - E_{ay,i-1,j-1/2,k}^m}{\Delta x} \right. \\
& \left. - \frac{E_{ax,i-1/2,j,k}^m - E_{ax,i-1/2,j-1,k}^m}{\Delta y} \right]
\end{aligned}
\tag{2.57}
$$

$$
J_\alpha^{m+1/2} = \frac{1 - \Delta t/2\tau}{1 + \Delta t/2\tau} J_\alpha^{m-1/2} + \frac{\Delta t}{1 + \Delta t/2\tau} \frac{Nq^2}{m_e} E_\alpha^m
\tag{2.58}
$$

$$
E_{ax,i-1/2,j,k}^{m+1} = E_{ax,i-1/2,j,k}^{m} - \frac{\Delta t}{\epsilon_0} J_{ax,i-1/2,j,k}^{m+1/2} - \frac{\alpha \Delta t}{\epsilon_0} J_{ex,i-1/2,j,k}^{m+1/2}
$$

$$
+ \frac{\Delta t}{\epsilon_0} \left[\frac{H_{az,i-1/2,j+1/2,k}^{m+1/2} - H_{az,i-1/2,j-1/2,k}^{m+1/2}}{\Delta y} \right.
$$

$$
\left. - \frac{H_{ay,i-1/2,j,k+1/2}^{m+1/2} - H_{ay,i-1/2,j,k-1/2}^{m+1/2}}{\Delta z} \right] \tag{2.59}
$$

$$
E_{ay,i,j-1/2,k}^{m+1} = E_{ay,i,j-1/2,k}^{m} - \frac{\Delta t}{\epsilon_0} J_{ay,i,j-1/2,k}^{m+1/2} - \frac{\alpha \Delta t}{\epsilon_0} J_{ey,i,j-1/2,k}^{m+1/2}
$$

$$
+ \frac{\Delta t}{\epsilon_0} \left[\frac{H_{ax,i,j-1/2,k+1/2}^{m+1/2} - H_{ax,i,j-1/2,k-1/2}^{m+1/2}}{\Delta z} \right.
$$

$$
\left. - \frac{H_{az,i+1/2,j-1/2,k}^{m+1/2} - H_{az,i-1/2,j-1/2,k}^{m+1/2}}{\Delta x} \right] \tag{2.60}
$$

$$
E_{az,i,j,k-1/2}^{m+1} = E_{az,i,j,k-1/2}^{m} - \frac{\Delta t}{\epsilon_0} J_{az,i,j,k-1/2}^{m+1/2} - \frac{\alpha \Delta t}{\epsilon_0} J_{ez,i,j,k-1/2}^{m+1/2}
$$

$$
+ \frac{\Delta t}{\epsilon_0} \left[\frac{H_{ay,i+1/2,j,k-1/2}^{m+1/2} - H_{ay,i-1/2,j,k-1/2}^{m+1/2}}{\Delta x} \right.
$$

$$
\left. - \frac{H_{ax,i,j+1/2,k-1/2}^{m+1/2} - H_{ax,i,j-1/2,k-1/2}^{m+1/2}}{\Delta y} \right] \tag{2.61}
$$

Here, $\alpha = 0,1$. The upper index m is the time step of the iteration and as seen above, the electric fields are sampled at integral time steps while the magnetic fields and the current densities are sampled at half-integral time steps. The lower indices i, j, and k represent the positions of the fields in Yee's grid. Δx, Δy, Δz, and Δt are, respectively, the spatial lengths for each cell in the three-dimensional Yee's grid and the time interval in the iteration. The spatial indices in (2.58) are omitted by directly using the vector form of expressions as there is no spatial difference in that equation.

It is obvious that, by letting $\alpha = 0$, (2.55)–(2.61) form a closed loop for the time iteration of linear responses in the FDTD algorithm. If it is considered that the J_e is an extra current source which has already been known, then the updating equations of (2.49), (2.54), and (2.52) can be immediately derived by changing α from 0 to 1 in (2.55)–(2.61).

In that case, the updating processes of the linear responses and the nonlinear responses can actually share the same linear FDTD solver, just as the illustration in Figure 2.7. The general flow chart of this linear FDTD solver is shown in

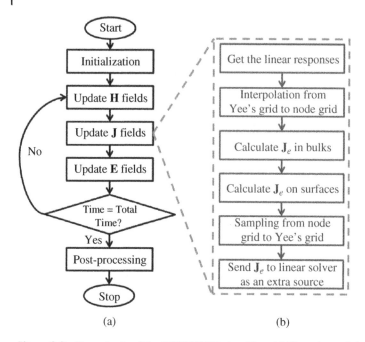

Figure 2.9 Flow charts of the FDTD-TDPM algorithm. (a) Flow chart of the general linear FDTD solver. (b) Flow chart of the calculation process for J_e.

Figure 2.9a. The algorithm will first call the iteration process of (2.55)–(2.61) with $\alpha = 0$, then while the linear responses are being solved, J_e will be calculated as well through the same time iteration. Finally, the algorithm will recall the same linear solver for the iteration process of (2.55)–(2.61) with $\alpha = 1$, as well as by regarding J_e as an extra nonlinear current source, and thereby acquiring the nonlinear responses.

Now the only thing left is that the extra nonlinear current source J_e, which contains all the generations of nonlinearities, needs to be calculated carefully. This problem is discussed in the following subsection.

2.3.2.3 Extra Nonlinear Current Source

As can be seen from the previous discussions, all the nonlinearities are generated from (2.53) since all the other theoretical equations in the TDPM are linear equations. Thus, the computation of J_e would significantly contribute to the accuracy of nonlinear responses in the algorithm. However, even if the calculation of J_e has a rather explicit form once the linear responses are solved, the numerical computation of (2.53) should still be considered carefully. Due to both nonlinear terms and complex spatial differentials, the original Yee's grid shown in Figure 2.8a is not an ideal candidate for discretizing (2.53). Therefore, the node grid shown in Figure 2.8b is chosen as the computational grid for (2.53). The concrete operations are shown as follows:

$$
\begin{aligned}
\mathbf{J}_e^{m+1/2} &= \frac{1-\Delta t/2\tau}{1+\Delta t/2\tau}\mathbf{J}_e^{m-1/2} \\
&\quad - \frac{\Delta t}{1+\Delta t/2\tau}\left[\frac{\epsilon_0 q}{m_e}(\nabla\cdot\mathbf{E}_0^m)\mathbf{E}_0^m + \frac{\mu_0 q}{m_e}\mathbf{J}_0^m\times\mathbf{H}_0^m \right. \\
&\qquad\qquad\qquad\qquad \left. -\nabla\cdot\left(\frac{\mathbf{J}_0^m\mathbf{J}_0^m}{Nq}\right)\right]
\end{aligned}
\tag{2.62}
$$

First, all the data points of the linear responses in the Yee's grid are interpolated to the nodes of the node grid. Then $\mathbf{J}_e$ is computed by (2.62), which is derived from (2.53), where the spatial indices are omitted again. This calculation of (2.62) should be done after every time step in the iteration process of linear responses. Thus, it is also an updating equation. Since (2.55)–(2.61) with $\alpha = 0$ should have been updated before the calculation of $\mathbf{J}_e$ at each time step, the approximations $\mathbf{J}_0^m = (\mathbf{J}_0^{m+1/2} + \mathbf{J}_0^{m-1/2})/2$ and $\mathbf{H}_0^m = (\mathbf{H}_0^{m+1/2} + \mathbf{H}_0^{m-1/2})/2$ are applied into (2.62). Furthermore, all the spatial differentials in (2.62) should be replaced by ordinary spatial center differences in the numerical algorithm. Notice that these spatial center differences in (2.62) have the same accuracies as the differences in Yee's grid because the length of the cell in node grid is half of the length in Yee's grid.

Unfortunately, only (2.62) is not enough for a complete updating process of (2.53). Unlike the updating equations for electromagnetic fields in Yee's grid, where the continuous boundary conditions at the interfaces of the grid cells are automatically satisfied [39], additional attentions for the surfaces of the metallic materials are needed when updating (2.62).

Figure 2.10 illustrates the surface model for metallic materials. The surfaces of the materials can be regarded as two kinds of boundaries, external boundary and internal boundary which are marked by surface A and surface B respectively. The region between surfaces A and B is the charge accumulation layer. This model has a clear physical meaning, as the electrons will accumulate at the surfaces of the materials when interacting with electromagnetic fields. The thickness of the charge accumulating layer is denoted by a parameter l, which should be much smaller than the cell size in computational grids. In that case, the surface effects of the materials cannot be adequately sampled in the computational grids

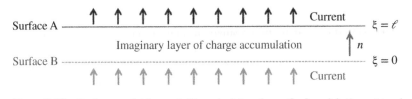

Figure 2.10 Surface model for metallic material surfaces. Surface A is the external boundary and surface B is the internal boundary.

shown in Figure 2.8, while they still have significant influences to the nonlinear effects. Thus, the integral of the current densities on the material boundaries is implemented and the concept of surface currents is applied. With the notations in Figure 2.10, one derives:

$$
\int_{\xi=0}^{\xi=l} \mathbf{J}_e^{m+1/2} \, d\xi = \frac{1 - \Delta t/2\tau}{1 + \Delta t/2\tau} \int_{\xi=0}^{\xi=l} \mathbf{J}_e^{m-1/2} \, d\xi
$$

$$
- \frac{\Delta t}{1 + \Delta t/2\tau} \left[\int_{\xi=0}^{\xi=l} \frac{\epsilon_0 q}{m_e} (\nabla \cdot \mathbf{E}_0^m) \mathbf{E}_0^m \, d\xi \right.
$$

$$
\left. - \int_{\xi=0}^{\xi=l} \nabla \cdot \left(\frac{\mathbf{J}_0^m \mathbf{J}_0^m}{Nq} \right) d\xi \right] \tag{2.63}
$$

The magnetic field term in (2.62) is omitted in (2.63) because it does not contain any spatial differentials, thus it does not contain sudden changes, or in other words, extremely large numeral values, at the material surfaces. Thus, when the parameter l tends to 0, that term disappears. By utilizing the "delta-function technique," let l goes to 0, which is a reasonable operation since the numeral value of l is far beyond the resolution of the numerical algorithm, (2.63) would then become into an updating equation for surface currents, given as:

$$
\mathbf{J}_{es}^{m+1/2} = \frac{1 - \Delta t/2\tau}{1 + \Delta t/2\tau} \mathbf{J}_{es}^{m-1/2}
$$

$$
+ \frac{\Delta t}{1 + \Delta t/2\tau} \left[\frac{-\epsilon_0 q}{m_e} \mathbf{E}_{0t0}^m (E_{0nl}^m - E_{0n0}^m) - \frac{1}{Nq} \mathbf{J}_{0t0}^m J_{0n0}^m \right]
$$

$$
+ \frac{\Delta t}{1 + \Delta t/2\tau} \left[\frac{-\epsilon_0 q}{2m_e} \left((E_{0nl}^m)^2 - (E_{0n0}^m)^2 \right) - \frac{1}{Nq} (J_{0n0}^m)^2 \right] \tag{2.64}
$$

where the indices t and n represent the tangential and normal components, respectively, at the material surfaces. After the update of (2.64), the surface currents in (2.64) will be transformed into current densities again by using the approximation $\mathbf{J}_e^{m+1/2} = \mathbf{J}_{es}^{m+1/2}/\Delta s$, where Δs (e.g. Δx, Δy, Δz) represents the cell length in Yee's grid.

Equations (2.62) and (2.64) form the updating equations of $\mathbf{J}_e$ together, where (2.62) is for bulky currents and (2.64) is for surface currents. To recall the linear FDTD solver as discussed before, all the data points from the node grid are sampled back to the Yee's grid after the results of $\mathbf{J}_e$ are obtained in each time step. The whole process of the $\mathbf{J}_e$ calculation is shown in Figure 2.9b.

In a brief summary, during each time step of the FDTD iteration, the algorithm first updates (2.55)–(2.61) with $\alpha = 0$, then updates (2.62) inside the metallic materials and (2.64) on the material surfaces, and finally, the algorithm updates

(2.55)–(2.61) with $\alpha = 1$. Linear responses and nonlinear responses could be obtained simultaneously in the algorithm.

2.3.3 Stability Issues

The stability of the algorithm is vital in time-domain methods because the unstable factors in the numerical algorithm may easily make the iteration results diverge. Based on the updating equations shown before, the algorithm stability is carefully analyzed, and two conditions for numerical parameters to satisfy in order to guarantee absolutely stable iterations are presented as:

$$\Delta t \leq \frac{1}{c\sqrt{(1/\Delta x)^2 + (1/\Delta y)^2 + (1/\Delta z)^2}} \tag{2.65}$$

$$\Delta t < \sqrt{\frac{4\epsilon_0 m_e}{Nq^2} - \frac{\epsilon_0^2 m_e^2}{N^2 q^4 \tau^2}} \tag{2.66}$$

Equation (2.65) is actually the CFL condition in traditional FDTD algorithms [39], where c is the speed of light in free space, and the detail proof of (2.66) is shown as follows.

For clarity and ease of understanding, the stability for (2.55)–(2.61) with $\alpha = 0$ is analyzed first.

There are basically two sets of updating equations in (2.55)–(2.61) with $\alpha = 0$. One set is the updating equations for electromagnetic fields, (2.55)–(2.57) and (2.59)–(2.61), and the other is the updating equations for linear current densities, (2.58). As for (2.55)–(2.57) and (2.59)–(2.61) with $\alpha = 0$, they are just the standard updating equations in traditional FDTD methods except for the additional current density source $\mathbf{J}_0$. In that case, if $\mathbf{J}_0$ does not diverge and maintains in a finite numeral value, then the stability of (2.55)–(2.57) and (2.59)–(2.61) with $\alpha = 0$ would be guaranteed by the CFL condition (2.65).

But to guarantee that $\mathbf{J}_0$ is stable, one needs to consider (2.58) together with (2.59)–(2.61) because the currents are strongly coupled to the electric fields. The x direction equations could be used to discuss their stabilities and the other directions can be operated in the same way. From (2.58) and (2.59) with $\alpha = 0$, one has (2.67) and (2.68):

$$J_{0,x}^{m+1/2} = \frac{1 - \Delta t/2\tau}{1 + \Delta t/2\tau} J_{0,x}^{m-1/2} + \frac{\Delta t}{1 + \Delta t/2\tau} \frac{Nq^2}{m_e} E_{0,x}^m \tag{2.67}$$

$$E_{0,x}^m = E_{0,x}^{m-1} - \frac{\Delta t}{\epsilon_0} J_{0,x}^{m-1/2} + S_{H_0,x}^{m-1/2} \tag{2.68}$$

Here, the effects of magnetic fields can be seen as an extra source $S_{H_0,x}^{m-1/2}$ to (2.68), thus only the stability of the electric field and current density in (2.67) and (2.68) needs to be discussed once the magnetic fields are stable, which can be preserved by using the CFL condition discussed before. For further clarity, (2.67) and (2.68) are rewritten into matrix forms, by substituting (2.68) into (2.67):

$$\begin{bmatrix} J \\ E \end{bmatrix}_m \equiv \begin{bmatrix} J_{0,x}^{m+1/2} \\ E_{0,x}^m \end{bmatrix}, \quad \begin{bmatrix} S_1 \\ S_2 \end{bmatrix}_m \equiv \begin{bmatrix} \frac{\Delta t}{1+\Delta t/2\tau} \frac{Nq^2}{m_e} S_{H_0,x}^{m+1/2} \\ S_{H_0,x}^{m+1/2} \end{bmatrix} \tag{2.69}$$

$$\begin{bmatrix} J \\ E \end{bmatrix}_m = \begin{bmatrix} A & B \\ C & D \end{bmatrix} \begin{bmatrix} J \\ E \end{bmatrix}_{m-1} + \begin{bmatrix} S_1 \\ S_2 \end{bmatrix}_{m-1} \tag{2.70}$$

where the coefficients of the updating matrix are:

$$A = \frac{1-\Delta t/2\tau}{1+\Delta t/2\tau} - \frac{\Delta t}{1+\Delta t/2\tau} \frac{Nq^2}{m_e} \frac{\Delta t}{\epsilon_0}$$

$$B = \frac{\Delta t}{1+\Delta t/2\tau} \frac{Nq^2}{m_e}, \quad C = -\frac{\Delta t}{\epsilon_0}, \quad D = 1 \tag{2.71}$$

To make the iteration of (2.70) stable, both of the two eigenvalues of the updating matrix should have less than unit length. The characteristic-equation of the matrix is given by:

$$\lambda^2 - (A + D)\lambda + AD - BC = 0 \tag{2.72}$$

Suppose λ_1 and λ_2 are the two solutions of (2.72), then by using the Vieta theorem, two relations are realized in (2.73):

$$\lambda_1 + \lambda_2 = A + D = \left(2 - \frac{\Delta t^2}{\epsilon_0} \frac{Nq^2}{m_e}\right) \Big/ \left(1 + \frac{\Delta t}{2\tau}\right)$$

$$\lambda_1 \lambda_2 = AD - BC = \frac{1-\Delta t/2\tau}{1+\Delta t/2\tau} \tag{2.73}$$

Since the metallic NMSs discussed here are often operated in hundreds of THz and the collision frequency (the reciprocal of the relaxation time) is at tens or hundreds of THz, the time interval in FDTD algorithm Δt should be much smaller than the relaxation time τ of the materials once (2.65) is satisfied.[15] Thus, the product of λ_1 and λ_2 should be very close to one but still less than one. In which case, it is obvious that an absolutely stable condition is when the discriminant of (2.72) is less than zero. The two eigenvalues will then become a pair of conjugated complex numbers which share the same length less than unit. Otherwise, one of the eigenvalue may have[16] length larger than unit since the two eigenvalues would both be real numbers and have a product near one. Therefore by calculating the

---

15 If semiconductor material is being considered, both the collision frequency and the operating frequency will drop with the same orders. Thus, the condition here will still be valid.
16 But not necessary.

discriminant of (2.72) and letting it be negative, the absolutely stable condition for $\mathbf{J}_0$ could be demonstrated as:

$$\Delta t < \sqrt{\frac{4\epsilon_0 m_e}{Nq^2} - \frac{\epsilon_0^2 m_e^2}{N^2 q^4 \tau^2}} \tag{2.74}$$

This is exactly the stable condition presented in (2.66).

It is obvious that once (2.55)–(2.61) with $\alpha = 0$ are stable, (2.55)–(2.61) with $\alpha = 1$ are automatically stable as they follow the same updating manners. Now the only problem is the stability of (2.62) and (2.64). As shown in (2.62) and (2.64), no matter which kind of currents is being updated, bulky currents or surface currents, the updating equations for them can be always seen as:

$$\mathbf{J}_e^{m+1/2} = \frac{1 - \Delta t/2\tau}{1 + \Delta t/2\tau} \mathbf{J}_e^{m-1/2} + \mathbf{L}^m \tag{2.75}$$

where $\mathbf{L}^m$ can be regarded as a nonlinear source given from $\mathbf{E}_0$, $\mathbf{H}_0$, and $\mathbf{J}_0$. As per the discussion above, the coefficient for updating (2.75) is close to one but still less than one, meaning that (2.75) would be completely stable if $\mathbf{L}^m$ keeps stable and the value of $\mathbf{J}_e^{m+1/2}$ would closely follow the variance of $\mathbf{L}^m$, indicating that once the linear responses are stable, the extra nonlinear current source $\mathbf{J}_e$ will stabilize as well.

So in summary, the CFL condition (2.65) guarantees the stability of electromagnetic fields when current densities are stable; Eq. (2.66) guarantees the stabilities of linear currents when magnetic fields are stable and the stabilities of linear currents automatically lead the nonlinear currents to stabilize. Essentially, the whole iteration process would be guaranteed stable once conditions (2.65) and (2.66) are both satisfied at the same time.

Both (2.65) and (2.66) limit the length of time interval in the algorithm. Equation (2.65) presents the spatial limitation for the time interval, which is restricted by the grid size. Equation (2.66) shows the material limitation for the time interval, which is restricted by the material parameters. To make sure that (2.65) and (2.66) are satisfied at the same time, one need to set small enough time intervals in the FDTD algorithm.

2.3.4 Numerical Results and Validations

Using the FDTD-TDPM algorithm developed before, several NMSs with different element structures are calculated to verify the accuracy of the algorithm. Figure 2.11 illustrates the simulation setups for the unit cells of NMSs, together with the element geometries.

The metallic material of the NMSs is chosen to be gold and the parameters of gold are set as: $\tau = 9.3$ fs, $m_e = 0.9 \times 10^{-30}$ kg, $N = 5.7 \times 10^{28}$ m^{-3} in (2.39), which are typical numeral values for real golden materials. Detailed geometries of the

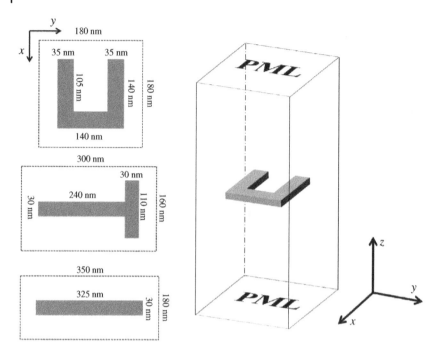

Figure 2.11 The element geometries and unit cell setup of nonlinear metasurfaces. The element structures are named as (from top to bottom): Split-ring resonator (SRR), T-shape resonator and dipole. Periodic boundary conditions (PBCs) are applied to the four vertical boundaries surrounding the element, and perfect matching layers (PMLs) are set both on the top and bottom of the computational domain. All materials of the elements are chosen to be gold and the thickness of all the elements is set to 20 nm.

elements are also illustrated in Figure 2.11, and the thickness of all the elements is set to 20 nm.

Periodic boundary conditions (PBCs) [2, 40] are set at the four vertical boundaries of the unit cell so that one can easily get the electromagnetic responses of the whole metasurface by only calculating this unit cell. Besides, perfect matched layers (PMLs) [2, 40] are implemented as the absorbing boundary conditions to truncate the computational domain from infinity to a finite size along the z direction. The y-polarized normally incident plane wave is used as the excitation of the NMSs, and the time-domain excitation signal is the standard modulated Gaussian wave [2, 40].

Both the linear and nonlinear responses are calculated based on the above setups, and the numerical results are demonstrated in the following discussions. The computational results are compared with both the commercial software and the results in [30], which show reasonable good agreement with the experimental data. Since the commercial software could not simulate the nonlinear responses, the nonlinear numerical results are verified only using the results in [30]. It is also worthwhile to note that the element geometries shown in Figure 2.11 have almost the same structures as the ones shown in [30] except for a minor adjustment of the depth in the SRR element (from 103 to 105 nm) so that it would be convenient to mesh them in the algorithm. In that case, it should be quite reasonable comparing the computational results with [30].

2.3.4.1 Linear Responses

In order to verify the accuracy for the linear responses of the metasurfaces, a wide-band excitation signal is set in the algorithm and the linear transmittance, reflectance and absorbance of the metasurfaces are computed under a wide frequency range. Then the linear responses in [30] are extracted for comparisons. Figure 2.12 shows the numerical results.

As seen from Figure 2.12, the linear responses calculated by the algorithm agree well with the results shown in [30] except for minor blue shifts near the resonances. This may occur due to the mesh differences between the algorithm and the references. Same situations are also simulated in *CST Studio Suites* [41], where the simulated results concur with the results in Figure 2.12 and are thus omitted for brevity. The results indicate that the algorithm can give out acceptable accuracy on calculating linear effects of the NMSs.

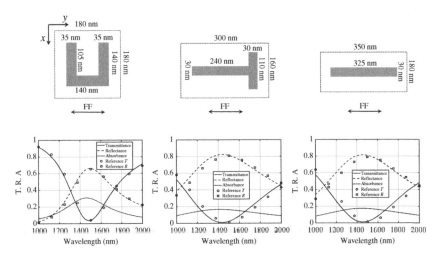

Figure 2.12 Linear responses for different kinds of element structures. The first row: element structures. The second row: FDTD results (proposed), simulation results in [30] are presented by the circular and square dots as references. Polarizations of fundamental frequency (FF) observation fields are demonstrated below each element structures. Polarizations of the excitation fields are all aligned with y axes.

2.3.4.2 Nonlinear Responses

As for the nonlinear responses, since the nonlinearities would be strongest near the resonant frequencies, a narrow-band excitation signal with its center frequency near the fundamental resonant frequency of the NMSs is set, and the SHGs in the transmitting waves are monitored. Here, as can be seen from Figure 2.12, the fundamental resonant frequency for the three kinds of NMSs is about 200 THz (1500 nm). Thus, the second harmonics (SHs) should be observed at 400 THz (750 nm). Since the nonlinear responses would vary from the intensity of the exciting electric field, to make reasonable comparisons with reference [30], the peak value for electric field intensity of the excitation in the algorithm is set to be the same as the intensity in [30], which is 2×10^7 V/m.

It is worthwhile to argue that, although the SHG is chosen to verify the nonlinear responses so that the results could be compared with the FDPM in [30], the ability of calculating nonlinearities in the algorithm is far beyond this monotone excited situation.

Figure 2.13 illustrates the SHG performances of the NMSs. Based on the intrinsic symmetries of the elements' geometries and the relevant discussions in [30], the SHGs are observed in the x-polarized transmitting wave for the SRR structure and in the y-polarized transmitting waves for the T-shape and dipole structures.

It can be found in Figure 2.13 that different element structures exhibit different intensities of SHG. The SHGs of the NMSs under both the condition that includes and excludes the bulky nonlinear currents are calculated as well. The one that

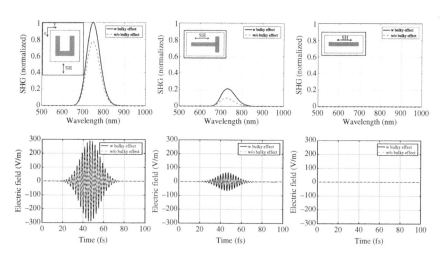

Figure 2.13 Nonlinear responses for different kinds of element structures. The first row: SHG in spectrum-domain (normalized), the corresponding element structures are shown in the insets. The second row: SHG in time-domain. Polarizations of second-harmonic (SH) observation fields are demonstrated near each element structures. Polarizations of the excitation fields are all aligned with y axes.

includes bulky nonlinear currents is just the algorithm, while the one without the bulky effects is obtained by intentionally setting all the bulky nonlinear currents in the algorithm to be 0. The solid black lines in Figure 2.13 present the SHG results with bulky effects, and the dash lines demonstrate the SHG results without bulky effects. It is obvious that the bulky nonlinear effects in the NMSs, which are always neglected in FDPM [30], actually generate certain contributions to the SHGs. To clarify the relative intensities of different nonlinear effects, all the SHG results in the frequency domain are normalized to the SHG result of SRR structure with bulky effects. The absolute values of SHG intensities could be observed in the time-domain results.

To verify the accuracy of the nonlinear responses obtained by the algorithm. The conversion efficiencies of SHGs are calculated and compared with the results in [30]. Table 2.1 summarizes these conversion efficiencies of SHGs.

A tricky thing appears for the conversion efficiency definition of SHGs. Unlike the FDPM discussed in [30], where the excitation is a single-frequency signal so that the definition of conversion efficiencies is straightforward. The algorithm operates in time domain, and thus, only a wave packet with limited time length rather than an infinite continuous wave could be obtained. In that case, the absolute peak value of the SH wave packet E_p is chosen to represent the conversion efficiency. Furthermore, considering the energy differences between real signals and complex signals, the conversion efficiency should be redefined as[17]:

$$\eta_{eff} = \frac{4E_p^2}{(2 \times 10^7 \text{ V/m})^2} \tag{2.76}$$

Results in Table 2.1 are obtained by this definition.

Table 2.1 SHG conversion efficiencies.

Elements	Methods	Conversion efficiencies
	FDTD-TDPM (*with bulky effects*)	8.67×10^{-10}
SRR	FDTD-TDPM (*without bulky effects*)	5.16×10^{-10}
	Reference [30]	6.90×10^{-10}
	FDTD-TDPM (*with bulky effects*)	4.27×10^{-11}
T-Shape	FDTD-TDPM (*without bulky effects*)	0.92×10^{-11}
	Reference [30]	1.10×10^{-11}
	FDTD-TDPM (*with bulky effects*)	$< 10^{-29}$
Dipole	FDTD-TDPM (*without bulky effects*)	$< 10^{-29}$
	Reference [30]	0

---

17 Note the extra factor of 4.

Based on the results shown in Table 2.1, the conversion efficiencies calculated in the proposed algorithm without bulky effects agree quite well with the reference results, while the results involving bulky effects are higher than the reference results. The result of the T-shape element also illustrates that the bulky effects sometimes even contribute significantly to the nonlinear responses.

With all these numerical results, it can be seen that the FDTD-TDPM algorithm is accurate enough for certain type of NMS analysis. Implementing this algorithm into some NMS designs will lead to a few interesting applications for the algorithm, which will be discussed in Section 2.4.

2.4 Applications

Further stemming from the aforementioned FDTD-TDPM algorithm, multiple functional optical components can be efficiently designed based on the NMS scenario. Having those NMS-based components will greatly benefit the mainstream studies in photonics nowadays, which can be seen as the major application for the algorithm. In this section, we try to somehow reveal the power of the algorithm by showing some rather simple but interesting designs, with implementation of the algorithm itself, as well as some extended calculations. Most of the discussions in this section can be found in the authors' previous publications [17] and [42].

2.4.1 Nonlinear Surface Susceptibility Extraction

In practical cases, computing a finite size NMS array with an electrically large aperture is actively required, where only implementing the PBCs for a unit cell simulation is far from enough. Full-wave simulation for the entire aperture will be needed then, which is not only time consuming, but also, mostly in optical studies, even impossible. In that case, a proper solution for computing finite, electrically large apertured NMS array needs to be developed, which takes into account the trade-offs between computational costs and accuracies. One possible idea is explained as follows.

Take the SRR element in Section 2.3 as an example. By applying the FDTD-TDPM algorithm, accurate results for both the linear and nonlinear responses of a single unit cell (under PBCs) can be obtained efficiently. Then, inspired by one of the common operations in photonics studies [13, 24], we can further extract some equivalent macroscopic parameters for the SRR element structure. Those parameters are, namely, equivalent surface susceptibilities [43]. The overall EM response of a large SRR array can be approximately computed with the help of those surface susceptibilities afterwards.

Specifically, for the SRR element, since the nonlinearity is rather small, only the first two orders of surface susceptibilities $\chi_s^{(1)}$ and $\chi_s^{(2)}$ need to be extracted, where $\chi_s^{(1)}$ contributes to the linear responses while $\chi_s^{(2)}$ contributes to the nonlinear effects (up to the second order). Generally, both $\chi_s^{(1)}$ and $\chi_s^{(2)}$ are tensors, which contains multiple scalar components. By noting the intrinsical symmetry of SRR geometry, however, most of the tensor elements are out of our considerations. Precisely, suppose the incident light is polarized only at y direction in Figure 2.11, then by symmetry, the linear reflection/transmission will only happen at y direction, while the nonlinear reflection/transmission will only happen at x direction. Detailed reasons for this critical observation can be found in [30]. Thus, only two scalar elements $\chi_{s,yy}^{(1)}$ and $\chi_{s,xyy}^{(2)}$ are needed in our calculations. Inheriting the fundamental concepts in nonlinear photonics [13, 24], we immediately have,

$$P_{s,y}^{(1)}(\omega_k) = \epsilon_0 \chi_{s,yy}^{(1)}(\omega_k) E_y(\omega_k) \tag{2.77}$$

$$P_{s,x}^{(2)}(\omega_k) = \frac{1}{2}\epsilon_0 \sum_n \sum_m \chi_{s,xyy}^{(2)}(-\omega_k; \omega_n, \omega_m) E_y(\omega_m) E_y(\omega_n) \tag{2.78}$$

where $P_{s,y}^{(1)}(\omega_k)$ and $P_{s,x}^{(2)}(\omega_k)$ are the equivalent first- and second-order electric polarizations at a certain frequency ω_k. Here, we can see a wave mixing property in (2.78), where new frequencies can be generated. This is, apparently, a typical reveal of nonlinearity.

Two remarks need to be mentioned here. First, the above extraction strategy implicitly assumed that the equivalent surface susceptibilities are uniform among the whole aperture.[18] This, without a doubt, is not necessarily correct. As a matter of fact, it only valid when all unit cells in the array share exactly the same geometry, and each unit cell has a sub-wavelength periodicity as well. Though a large NMS array with identical elements are common to see in optical society, one should keep in mind that once different unit cells are being used, these equivalent surface susceptibilities are also location dependent.

Second, the parameter extractions in (2.77) and (2.78) are all done by a unit cell simulation (PBCs and FDTD-TDPM applied) with normal incident wave only. This definitely limits the fidelity of the model at large oblique incident angles. What we are actually doing here, is to assume that the unit cell of the NMS array is not spatially dispersive. In other words, to implement these equivalent surface susceptibilities, the unit cells shouldn't deviate their EM responses (either linear or nonlinear part) a lot at oblique incidences, from normal incidence. Again, no way to prove this will be the general case. Nevertheless, it turns out, in practice, making

---

18 This is actually a popular trick used in EM studies as well, such as [43] and [44].

this normal incidence approximation often gives us acceptable results. Taking its simplicity into consideration,[19] it is reasonable to merit this strategy eventually.

Utilizing these equivalent surface susceptibilities, one can then easily, approximately though, calculate both the linear and nonlinear responses of a finite-sized NMS array, just by straightforward theoretical equations [13, 24]. We will go back to this issue with more detail once we encounter some particular design problems later in this section.

2.4.2 All-Optical Switch (AOS)

Even if we just stick on the concrete FDTD-TDPM algorithm, some interesting designs of NMS can already be realized. Here, utilizing the frequency mixing function of the NMS element, an NMS-based all-optical switch (AOS) [17] is designed, which can be treated as a fundamental optical component.

As shown in Figure 2.14, suppose we have an infinitely large periodic NMS array with identical unit cells of SRR, which has the same geometrical parameters as Figure 2.14 showed. Now if two y-polarized EM waves normally incident onto the NMS, the nonlinear effect of SRR will couple the frequencies of incident waves together (wave mixing), and generate reflection/transmission waves with new frequencies at x polarization. By intentionally set the two excitation (incident) frequencies as $f_1 = 200\,\text{THz}$ and $f_2 = 2f_1 = 400\,\text{THz}$, a difference frequency of $f_d = f_2 - f_1 = 200\,\text{THz}$ will be generated with an x polarization. As can be seen, $f_d = f_1$. This frequency relationship indicates that, if one y-polarized wave with f_1 and one y-polarized wave with f_2 are injected into the NMS, another

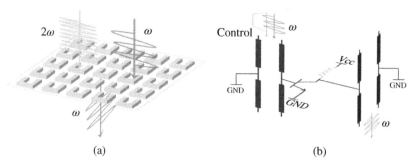

(a)   (b)

Figure 2.14 All-optical switch (AOS) configuration: (a) AOS geometry, and (b) analogy to a transistor-based switch circuit.

---

19 Calculating oblique incidences in FDTD for a unit cell under PBCs could be a really hard job. The time domain character of FDTD is in fact unfriendly for periodic structures when the Bloch or Floquet boundaries involve non-zero phase delays. Avoiding something like non-causal factors will require the actual algorithm be more complicated by an enormous amount. Some related discussion can be found in [2], referring to the "constant-k" method.

x-polarized wave with f_1 will be transmitted through. On the other hand, if only one *y*-polarized wave with f_1 is injected (without the other f_2 wave), no *x*-polarized wave with f_1 can be seen at transmission (or reflection).[20] Thus, the whole scenario in Figure 2.14 forms an equivalent AOS [16], where the transmission of f_1 wave can be "switched" by the existence of f_2 wave. Note that here, the equivalent AOS treats the *y*-polarized f_1 wave as input, but regards the *x*-polarized f_1 wave as output. This polarization transformation essentially isolates the real output from other spurious scatterings.[21] The working principle of this AOS is pretty much like a transistor-based switch circuit in normal electronics, like shown in Figure 2.14. The intrinsic difference here is just that, in circuit, the control signal for switching is a dynamic DC signal, while in AOS, the control signal is literally a "DC light" [13] with a frequency of f_2.[22]

Implementing the FDTD-TDPM algorithm introduced in Section 2.3, the EM responses of the AOS can be computed accordingly, as shown in Figure 2.15. From the simulation results in Figure 2.15, it is obvious that, when the f_2 light (pump light, second harmonic, SH) is turned off, there is no transmitted f_1 light (signal light, fundamental frequency, FF) at *x* polarization. Once the f_2 light

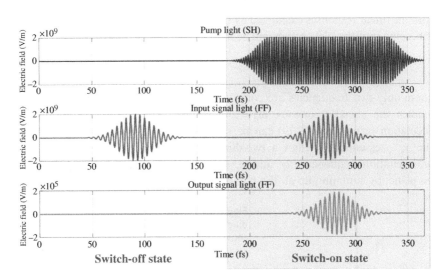

Figure 2.15 Simulation results for the AOS by FDTD-TDPM algorithm.

---

20 Only high-order harmonics of f_1 will be there.
21 The linear response will also send out f_1 transmission waves even without the f_2 excitation, but in *y* polarization.
22 Interestingly, from the signal perspective, DC light has exactly the same function as normal DC signal in electronics. However, since real DC signal can never turn to be lights, the function of DC can only be realized by the concept of DC light in optical region.

turns on, however, we immediately see the transmitted signal light at x polarization. Note that the f_1 excitation light is always there during the whole simulation process. Clearly, Figure 2.15 proves the functionality of this NMS-based AOS.

2.4.3 Harmonic-Modulated NMS (HM-NMS)

Analogy of frequency and wave vectors indicates that, the nonlinearity of NMS is not only able to mix frequencies, but also mix wave vectors (**k** vectors). Taking the coupling effect of **k** vectors into consideration, even more intriguing phenomena can be observed by the same NMS structure.

Still originates from the SRR structure shown in Figure 2.11, we now consider a finite-size array with oblique incident lights. As will be revealed in this subsection, a so-called "harmonic modulated NMS" (HM-NMS) can be designed with versatile interesting functions, based on a generalized phase conjugation principle [42]. Since oblique incidence and finite size are being considered here, only do the element simulation under PBCs is far from enough. The aforementioned surface susceptibility extraction strategy should be involved now.

Following the traditional steps in [43], it is rather straightforward to extract $\chi_{s,yy}^{(1)}$ from the linear responses calculated by FDTD-TDPM algorithm. For $\chi_{s,xy}^{(2)}$ extraction, some intuitive elaboration will be needed. Rewrite (2.78) here,

$$P_{s,x}^{(2)}(\omega_3) = \frac{1}{2}\epsilon_0 \sum_{\pm\omega_s,\pm\omega_p} \sum_{\pm\omega_s,\pm\omega_p} \chi_{s,xyy}^{(2)}(-\omega_3;\omega_1,\omega_2)E_y(\omega_1)E_y(\omega_2) \tag{2.79}$$

where only two frequencies are considered in this case. Specifically, $\omega_s = 2\pi f_s$ denotes the signal frequency, $\omega_p = 2\pi f_p$ denotes the pump frequency. The calculation principle of (2.79) can be intuitively demonstrated in Figure 2.16. The $\chi_{s,xyy}^{(2)}$ essentially creates a two-dimensional plane for wave mixing. The two excitation frequencies, ω_s and ω_p generates four corresponding points on the plane, by summing up all the points on the black dash line, we can get the

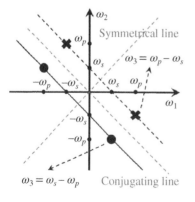

Figure 2.16 Two-dimensional dispersion diagram for understanding the behaviors of second-order nonlinear surface susceptibility.

second-order nonlinear response with a new frequency $\omega_3 = \omega_p - \omega_s$, which is basically the same as ω_s since $\omega_p = 2\omega_s$ are intentionally set. By noticing the intrinsic symmetrical and conjugation properties of $\chi_{s,xy}^{(2)}$, we further know that the two points on the black dash line give identical contributions for the nonlinear polarization. Moreover, the black solid line is conjugated to the black dash line. Thus, the four points in Figure 2.16 represent exactly the same information about $\chi_{s,xyy}^{(2)}(-\omega_s, \omega_p)$. Besides, since the linear responses are in a different polarization, they really has no influence to the $\chi_{s,xyy}^{(2)}(-\omega_s, \omega_p)$ extraction. All the above characters here guarantee the possibility of efficiently extracting $\chi_{s,xyy}^{(2)}(-\omega_s, \omega_p)$, from the calculation results of FDTD-TDPM algorithm.

By simulating a single element of SRR with FDTD-TDPM algorithm, we can have the normalized nonlinear responses at x polarization as shown in Figure 2.17. Substituting the numerical values in Figure 2.17 to (2.79) and regarding the facts in Figure 2.16, $\chi_{s,xyy}^{(2)}(-\omega_s, \omega_p)$ can be well extracted.

Once both $\chi_{s,yy}^{(1)}$ and $\chi_{s,xy}^{(2)}$ are obtained, we can easily calculate the EM responses of a finite-size SRR array, with any incident angle, by book keeping theoretical equations [3, 45–47]. In this example, we choose to calculate an SRR array with 500 by 500 elements.

As shown in Figure 2.18, once two waves with ω_s and ω_p frequencies are obliquely incident to the NMS array, something called "phase conjugation" (PC) [13] may happen because of the nonlinear effects. The traditional PC principle claims that the transverse wave vector of the nonlinear scattering wave could

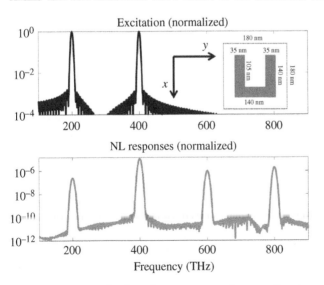

Figure 2.17 Normalized nonlinear responses for extracting the equivalent second-order nonlinear surface susceptibility.

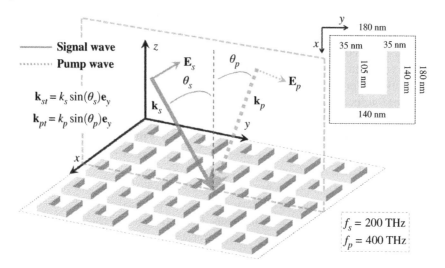

Figure 2.18 Harmonic-modulated NMS (HM-NMS) configuration.

have opposite directions with the incident wave. The spatial nonlinearities of the NMS, however, actually possess abilities far beyond this level. Thus, a generalized PC principle can be derived in the following manner.

According to (2.79), it is clear to see that when two plane waves interact with the NMS simultaneously, the NMS would behave like a multiplier for the two waves. As shown in (2.80) and (2.81):

$$2A\cos(\omega_s t - \mathbf{k}_{st} \cdot \mathbf{s})B\cos(\omega_p t - \mathbf{k}_{pt} \cdot \mathbf{s})$$
$$\sim AB\cos[(\omega_p - \omega_s)t - (\mathbf{k}_{pt} - \mathbf{k}_{st}) \cdot \mathbf{s}] \tag{2.80}$$

$$\omega_{NL} = \omega_p - \omega_s, \qquad \mathbf{k}_{NL,t} = \mathbf{k}_{pt} - \mathbf{k}_{st} \tag{2.81}$$

Here, $\mathbf{s}$ is the position vector inside the *xoy* plane in Figure 2.18. The multiplying process in (2.80) generates a new plane wave with the angular frequency of ω_{NL} and the transverse wave vector $\mathbf{k}_{NL,t}$, as demonstrated in (2.81). This indicates that $\mathbf{k}_{NL,t}$ could be adjusted by $\mathbf{k}_{pt}$ and $\mathbf{k}_{st}$. Equation (2.81) forms the generalized PC principle under NMS situation. When $\mathbf{k}_{pt} = 0$ in (2.81), it degenerates to the traditional PC principle.

The generalized PC principle could also be explained by using the collisions of photons, as illustrated in Figure 2.19. Based on the energy and momentum conservation laws in quantum theory, the photons which interacts with the NMS must satisfy the relationship demonstrated in Figure 2.19. Notice that the equations on the right bottom of Figure 2.19 are actually the same as (2.81) (the Planck constant $\hbar$ could be cancelled).

Based on this generalized PC principle, the NMS array shown in Figure 2.18 can be treated as a kind of "harmonic modulated surface," since the pump light

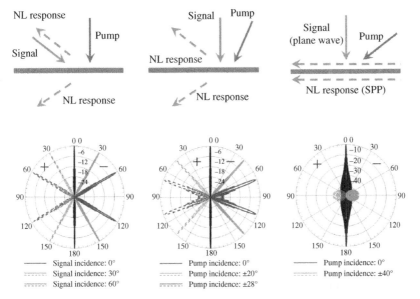

$\hbar\omega_p, \hbar\mathbf{k}_{pt}$ $\hbar\omega_s, \hbar\mathbf{k}_{st}$ **Linear case (Collisions → Loss)**

$-\hbar\omega_p, -\hbar\mathbf{k}_{pt}$ $-\hbar\omega_s, -\hbar\mathbf{k}_{st}$ $\hbar\omega_L = \hbar\omega_{p,s} - \hbar\omega_{p,s} = 0$ $\hbar\mathbf{k}_{L,t} = \hbar\mathbf{k}_{p,st} - \hbar\mathbf{k}_{p,st} = 0$

$\hbar\omega_p, \hbar\mathbf{k}_{pt}$ **Nonlinear case (Collisions → PC)**

$-\hbar\omega_s, -\hbar\mathbf{k}_{st}$ $\hbar\omega_{NL} = \hbar\omega_p - \hbar\omega_s$ $\hbar\mathbf{k}_{NL,t} = \hbar\mathbf{k}_{pt} - \hbar\mathbf{k}_{st}$

Figure 2.19 Quantum interpretations for the generalized phase conjugation (PC) principle. The nonlinearity will initiate the mutual coupling between photons with different energies or momentums.

Figure 2.20 Simulation results for HM-NMS with versatile functions. Left column: phase conjugation mirror (PCM); middle column: optically controlled beam steering (OCBS) array; right column: spoof surface plasmon polaritons (SSPPs) generator.

(ω_p) is actually the second harmonic of the signal light (ω_s), and $\omega_{NL} = \omega_s$. The transverse wave vector of the signal light can be tuned by the pump light, as if it is being modulated. We thus name it as HM-NMS, as declared previously. Versatile functions can be realized by this scenario. Also note that,

$$\mathbf{k}_{NL,t} = [2k_{s0}\sin(\theta_p) - k_{s0}\sin(\theta_s)]\mathbf{e}_y \tag{2.82}$$

By setting $\theta_p = 0$ in (2.82), $\mathbf{k}_{NL,t} = -k_{s0}\sin(\theta_s)\mathbf{e}_y$, which indicates that, by setting the pump light into normal incidence, the nonlinear response of the signal wave would reverse its transverse wave vector. As can be seen in the left column of

Figure 2.20, the nonlinear scatterings of the NMS (dash lines) occupy exactly the opposite beam directions with the linear scattering waves (solid lines). The whole NMS performs like a phase conjugation mirror (PCM) [14] under this condition. This PCM property may lead to some exciting applications like the perfect lens [14].

By setting $\theta_s = 0$ in (2.82), $\mathbf{k}_{NL,t} = 2k_{s0}\sin(\theta_p)\mathbf{e}_y$. This illustrates that if the signal wave is kept being normally incident on the NMS, then the beam direction of the nonlinear scattering waves could actually be adjusted by the incident angle of the pump wave, as shown in the middle column of Figure 2.20. The NMS achieves an optically-controlled beam steering (OCBS) array in this situation. In addition, the major scattering angle of the nonlinear scattering wave θ_{NL} satisfies: $\theta_{NL} = \arcsin(2\sin(\theta_p))$. By scanning θ_p from $0°$ to $30°$, the NMS could achieve a $0°$ to $90°$ beam steering accordingly. Hence, the NMS could achieve a wide-angle beam steering by using narrow-angle incident waves.

Finally, if the θ_p discussed above is over $30°$, then the θ_{NL} would become a complex angle which means that $\mathbf{k}_{NL,t}$ is larger than $\mathbf{k}_{s0} = \omega_s\sqrt{\epsilon_0\mu_0}$. In that case, the nonlinear scattering waves would become surface waves which are sometimes named as spoof surface plasmon polaritons (SSPPs) [48] in optical societies, as demonstrated in the right column of Figure 2.20. The results show that the NMS could transform the plane wave excitations into SSPPs under this configuration, which evolves a new method for exciting SSPPs.

2.5 Summary

In this chapter, we first introduced some basic concepts of NMS, as well as the backgrounds of its computational methods. Material modeling techniques and a brief review of microelectronic fundamentals are then given. Based on the Maxwell-hydrodynamic model, an FDTD-TDPM algorithm is presented with numerical validations. Implementing the FDTD-TDPM algorithm, together with some extended calculations, several NMS-based optical components can be efficiently designed, including AOS and HM-NMS with versatile functions like PCM, OCBS, and SSPPs generation, which are regarded as typical applications for the proposed algorithm.

References

1 Munk, B.A. (2005). *Frequency Selective Surfaces: Theory and Design*. Wiley.

2 Yang, F. and Rahmat-Samii, Y. (2009). *Electromagnetic Band Gap Structures in Antenna Engineering*. Cambridge: Cambridge University Press.

3 Yablonovitch, E. (1993). Photonic band-gap structures. *JOSA B* 10 (2): 283–295.

4 Sievenpiper, D. (1999). High-impedance electromagnetic surfaces. PhD dissertation. Los Angeles, Los Angeles, CA: Electrical Engineering Department University of California, p. 33.

5 Yang, F. and Rahmat-Samii, Y. (eds.) (2019). *Surface Electromagnetics: With Applications in Antenna, Microwave, and Optical Engineering.* Cambridge University Press.

6 Sievenpiper, D.F. (2011). Nonlinear grounded metasurfaces for suppression of high-power pulsed RF currents. *IEEE Antennas and Wireless Propagation Letters* 10: 1516–1519.

7 Li, A., Luo, Z., Wakatsuchi, H. et al. (2017). Nonlinear, active, and tunable metasurfaces for advanced electromagnetics applications. *IEEE Access* 5: 27439–27452.

8 Malyuskin, O., Fusco, V., and Schuchinsky, A.G. (2006). Microwave phase conjugation using nonlinearly loaded wire arrays. *IEEE Transactions on Antennas and Propagation* 54 (1): 192–203.

9 Wen, Y. and Zhou, J. (2017). Artificial nonlinearity generated from electromagnetic coupling metamolecule. *Physical Review Letters* 118 (16): 167401.

10 Fang, M., Huang, Z., Sha, W.E., and Wu, X. (2017). Maxwell-hydrodynamic model for simulating nonlinear terahertz generation from plasmonic metasurfaces. *IEEE Journal on Multiscale and Multiphysics Computational Techniques (J-MMCT)* 2: 194–201.

11 Lee, J., Tymchenko, M., Argyropoulos, C. et al. (2014). Giant nonlinear response from plasmonic metasurfaces coupled to intersubband transitions. *Nature* 511 (7507): 65.

12 Ye, W., Zeuner, F., Li, X. et al. (2016). Spin and wavelength multiplexed nonlinear metasurface holography. *Nature Communications* 7: 11930.

13 Guo, Y., Chiang, K.S., and Li H. (2002). *Nonlinear Photonics: Nonlinearities in Optics, Optoelectronics, and Fiber Communications*, vol. 8. Chinese University Press.

14 Pendry, J.B. (2000). Negative refraction makes a perfect lens. *Physical Review Letters* 85 (18): 3966.

15 Maslovski, S. and Tretyakov, S. (2003). Phase conjugation and perfect lensing. *Journal of Applied Physics* 94 (7): 4241–4243.

16 Krasnok, A., Li, S., Lepeshov, S. et al. (2018). All-optical switching and unidirectional plasmon launching with nonlinear dielectric nanoantennas. *Physical Review Applied* 9 (1): 014015.

17 Chen, X., Yang, F., Li, M., and Xu, S. (2018). FDTD solver with time-domain perturbation method for simulating an all-optical switch realized by nonlinear metasurface. *2018 International Applied Computational Electromagnetics Society Symposium - China (ACES)*, pp. 1–2.

18 Lundstrom, M. (2009). *Fundamentals of Carrier Transport*. Cambridge University Press.

19 Kittel, C. (1976). *Introduction to Solid State Physics*, vol. 8. New York: Wiley.

20 Lifshitz, E.M. and Landau (JB), L.D. (1965). *Quantum Mechanics; Non-relativistic Theory*. Pergamon Press.

21 Berestetskii, V.B., Lifshitz, E.M., and Pitaevskii, L.P. (1982). *Quantum Electrodynamics*, vol. 4. Butterworth-Heinemann.

22 Landau, L. and Lifshitz, E. (1960). *Course of Theoretical Physics: Mechanics*, vol. 1. Oxford.

23 Yu, N., Genevet, P., Aieta, F. et al. (2013). Flat optics: controlling wavefronts with optical antenna metasurfaces. *IEEE Journal of Selected Topics in Quantum Electronics* 19 (3): 4700423.

24 Boyd, R.W. (2019). *Nonlinear Optics*. Academic Press.

25 Bardin, J.C., Sank, D., Naaman, O., and Jeffrey, E. (2020). Quantum computing: an introduction for microwave engineers. *IEEE Microwave Magazine* 21 (8): 24–44.

26 Razavi, B. (2003). *Design of Integrated Circuits for Optical Communications*. McGraw Hill: New York, NY.

27 Chew, W.C., Boag, A., and Hanson, G.W. (2020). Comparing classical and quantum electromagnetics [Special Series: Guest Editorial]. *IEEE Antennas and Propagation Magazine* 62 (4): 14.

28 Sha, W.E.I., Liu, A.Y., and Chew, W.C. (2018). Dissipative quantum electromagnetics. *IEEE Journal on Multiscale and Multiphysics Computational Techniques* 3: 198–213.

29 Feynman, R.P., Leighton, R.B., and Sands, M.L. (1998). *[Lectures on Physics]; The Feynman Lectures on Physics. 3. Quantum Mechanics*. Narosa Publishing House.

30 Ciracì, C., Poutrina, E., Scalora, M., and Smith, D.R. (2012). Origin of second-harmonic generation enhancement in optical split-ring resonators. *Physical Review B* 85 (20): 201403.

31 Liu, J., Brio, M., Zeng, Y. et al. (2010). Generalization of the FDTD algorithm for simulations of hydrodynamic nonlinear Drude model. *Journal of Computational Physics* 229 (17): 5921–5932.

32 Yee, K. (1966). Numerical solution of initial boundary value problems involving Maxwell's equations in isotropic media. *IEEE Transactions on Antennas and Propagation* 14 (3): 302–307.

33 Chen, X., Yang, F., Li, M., and Xu, S. (2020). Analysis of nonlinear metallic metasurface elements using maxwell-hydrodynamic model with time-domain perturbation method. *IEEE Transactions on Antennas and Propagation*. vol. 68, no. 3, pp. 2213–2223.

34 Del Alamo, J.A., Tuller, H., and Scholvin, J. (2009). *Integrated Microelectronic Devices*. Pearson.

35 Alsunaidi, M.A., Imtiaz, S.S., and El-Ghazaly, S.M. (1996). Electromagnetic wave effects on microwave transistors using a full-wave time-domain model. *IEEE Transactions on Microwave Theory and Techniques* 44 (6): 799–808.

36 Mustafa, F. and Hashim, A.M. (2010). Generalized 3D transverse magnetic mode method for analysis of interaction between drifting plasma waves in 2DEG-structured semiconductors and electromagnetic space harmonic waves. *Progress in Electromagnetics Research* 102: 315–335.

37 Mustafa, F. and Hashim, A.M. (2010). Properties of electromagnetic fields and effective permittivity excited by drifting plasma waves in semiconductor-insulator interface structure and equivalent transmission line technique for multi-layered structure. *Progress in Electromagnetics Research* 104: 403–425.

38 Yu, N., Genevet, P., Kats, M.A. et al. (2011). Light propagation with phase discontinuities: generalized laws of reflection and refraction. *Science* 334 (6054): 333–337.

39 Taflove, A. and Hagness, S.C. (2005). *Computational Electrodynamics: The Finite-Difference Time-Domain Method*. Artech House.

40 Elsherbeni, A.Z. and Demir, V. (2016). *The Finite-Difference Time-Domain Method for Electromagnetics with MATLAB Simulations*. The Institution of Engineering and Technology.

41 CST Studio Suite (2013).

42 Chen, X. and Yang, F. (2019). Harmonic-modulated nonlinear metasurface based on generalized phase conjugation principle. *2019 IEEE International Symposium on Antennas and Propagation and USNC-URSI Radio Science Meeting*, pp. 241–242.

43 Holloway, C.L., Kuester, E.F., and Haddab, A.H. (2019). Retrieval approach for determining surface susceptibilities and surface porosities of a symmetric metascreen from reflection and transmission coefficients. arXiv preprint arXiv:190208703.

44 Idemen, M.M. (2011). *Discontinuities in the Electromagnetic Field*, vol. 40. Wiley.

45 Kong, J.A. (1986). *Electromagnetic Wave Theory*. Wiley. New York.

46 Jin, J.M. (2011). *Theory and Computation of Electromagnetic Fields*. Wiley.

47 Harrington, R.F. (1993). *Field Computation by Moment Methods*. Wiley-IEEE Press.

48 Chen, H., Ma, H., Li, Y. et al. (2018). Wideband frequency scanning spoof surface plasmon polariton planar antenna based on transmissive phase gradient metasurface. *IEEE Antennas and Wireless Propagation Letters* 17 (3): 463–467.

3

The Finite-Element Time-Domain Method for Dispersive and Nonlinear Media

David S. Abraham, Ali Akbarzadeh-Sharbaf, and Dennis D. Giannacopoulos

Department of Electrical and Computer Engineering, McGill University, Montréal, Québec, Canada

3.1 Background and Motivation

While electromagnetic problems in general can prove to be quite nuanced and complex, simplifying assumptions may often be made about the systems and materials being studied, to ease their analysis. In particular, assuming symmetry, time-independence, the presence of only a single frequency component (time-harmonic form), or that materials respond linearly and instantaneously to any applied fields all significantly reduce the complexity of the underlying equations. It is no surprise, therefore, that the most common numerical methods for Maxwell's equations operate in these idealistic regimes.

Unfortunately, for many problems such simplifications may be invalid or result in unacceptable losses in accuracy. For instance, material dispersion, in which a medium's response to electromagnetic waves is not instantaneous but rather depends on the frequency, is a ubiquitous phenomenon in nature [1]. While negligible under certain circumstances, dispersion can significantly impact the shape and structure of electromagnetic signals in others. As a result, being able to effectively model these frequency-dependent interactions can have far-reaching impacts, such as in the study of dispersive biological tissue samples in medical imaging [2, 3], dispersive environmental elements in radar applications [4, 5], and pulse broadening and distortion in fiber optic cables [6, 7].

Moreover, nonlinear dielectric effects, in which the medium's response can be a complicated function of the applied field strength, can also significantly increase complexity. In particular, the nonlinear response can be immediate, known as an instantaneous nonlinearity, or depend on the frequency of the applied field, known as a dispersive nonlinearity. This nonlinear behavior is most commonly encountered when materials are subjected to very large field strengths, inducing a

Advances in Time-Domain Computational Electromagnetic Methods, First Edition.
Edited by Qiang Ren, Su Yan, and Atef Z. Elsherbeni.

myriad of complex behaviors. Modeling these nonlinear interactions can thus have important implications for areas of study such as nonlinear optics, in which very high intensity laser light is routinely encountered. Such techniques could therefore be used, for example, in the design and investigation of photonic crystals [8, 9] and soliton formation and interaction [10, 11]. Moreover, it has been theoretically and empirically demonstrated that biological tissues, in addition to being dispersive, can also produce nonlinear effects [12–14]. Modeling such behavior thus has the potential not only to increase accuracy in medical imaging but also to detect malignancies. For instance, it has been shown that mitochondria with metabolic disorders generally have suppressed harmonic generation as compared to their healthy counterparts when excited by a single frequency source [15].

In fact, the need for more general numerical methods to model complex material interactions such as these will only continue to grow as our ability improves in precisely designing and engineering materials. Indeed, new materials and structures with properties unseen in nature can now be fabricated, such as so-called negative and zero-index metamaterials [16, 17]. Moreover, as technology advances, existing materials can be further exposed to extreme operating conditions, causing effects usually deemed negligible to manifest more often. As a consequence of these existing and emerging media and their applications, there is a clear and growing need for numerical methods that are able to accurately simulate very general and complex material interactions with electromagnetic fields, and that are free of inaccurate simplifying assumptions.

In response to these current and developing challenges, this chapter will present very general modeling techniques based on the finite-element time-domain (FETD) method, which are capable of accurately simulating complex materials exhibiting any combination of electric dispersion, magnetic dispersion, and electric nonlinearity. While other techniques do currently exist for simulating these materials, the FETD method offers several unique advantages. For instance, arguably the most popular and successful alternative method to FETD is the finite-difference time-domain (FDTD) method. Indeed, FDTD has been successfully adapted to model a wide array of complex materials (including dispersion and nonlinearity). However, despite the numerous successes of FDTD-based methods in capturing the full nonlinear, dispersive, and wide-band response of complex materials within a single computation, there are nevertheless certain drawbacks associated with these methods. Difficulties in modeling curved interfaces, boundaries, and complex geometries have traditionally caused issues for FDTD methods, which generally operate on fixed regular grids, resulting in staircasing errors [18, 19]. Moreover, sharp material discontinuities or abrupt changes in material parameters can significantly weaken the accuracy of the underlying Taylor approximations, if not render them useless. Lastly, the most popular FDTD algorithms tend to be explicit in time and as a result may incur reduced stability [20, 21]. While mitigating strategies are available for each of these issues (such as parameter smoothing [22], correction function methods [23], and implicit temporal discretizations [24]), they inevitably increase the complexity of the underlying method and may not be able to fully counterbalance these weaknesses. In contrast, the FETD methods presented in this chapter do not suffer from any of these difficulties. Of course, this increased versatility comes at a cost: FETD methods in general are far more computationally demanding than FDTD methods. However, often the benefits of FETD outweigh these concerns, particularly when techniques can be adopted (such as those presented in Chapter 6[1]) to increase performance.

It should also be noted that the nonlinearities discussed in this chapter are *electric* in nature, meaning that they are produced by a nonlinear interaction between the electric field and the material permittivity. Of course, magnetic non-linearity and hysteresis are not only possible but pervasive in many fields of study, such as in the design of magnetic machines and electric motors. While these types of nonlinearity will not be discussed in this chapter (where the focus is instead on electric nonlinearities and their applications to optics), the reader is encouraged to consult Refs. [25, 26] for an overview of models for nonlinear magnetic media within the FETD method. Furthermore, while the dielectric models introduced in this chapter are quite general, there remain many phenomena which will not be discussed. For instance, anisotropy, ferroelectricity, and electric hysteresis can all add additional complexity to the polarization models discussed in this chapter. For an in depth comprehensive overview of complex material modeling within FDTD and FETD, including some of these phenomena as well as those covered in this chapter, the reader is encouraged to consult Ref. [27].

Lastly, a brief note on notation. Throughout this chapter vector fields will be denoted by over-arrows, $\vec{E}$, vector quantities by braces, $\{b\}$, and matrices by square brackets, $[A]$. Moreover, it should be noted that while a quantity may depend on multiple variables, e.g. $\vec{E}(\vec{r}, t)$, to increase legibility only the variables of interest in a particular context will be explicitly shown, with the rest being implicitly present, e.g. $\vec{E}(t) = \vec{E}(\vec{r}, t)$.

3.2 Dispersive and Nonlinear Media

When the permittivity (ϵ) and/or permeability (μ) of a material are functions of frequency, the material is termed *dispersive*. Depending on which parameter is frequency-dependent, the material is called either electrically, magnetically, or doubly dispersive. Recall that for linear nondispersive material, the constitutive

---

1 Alternate parallelization strategies for the FETD formulations.

equation relating the electric field intensity $\vec{E}$ and the electric flux density $\vec{D}$ can be written as follows:

$$\vec{D} = \epsilon_0 \vec{E} + \vec{P}$$
$$= \epsilon_0 \vec{E} + \epsilon_0 \chi_e \vec{E}$$
$$= \epsilon_0 (1 + \chi_e) \vec{E}$$
$$= \epsilon_0 \epsilon_r \vec{E}$$
$$= \epsilon \vec{E} \tag{3.1}$$

where χ_e is the electric susceptibility and $\vec{P}$ is the polarization density vector. The polarization density vector accounts for the electric dipole moment induced per unit volume of the dielectric when an external electric field is applied. Note that in Eq. (3.1) a change in the electric field is reflected instantaneously in the polarization field and thus in $\vec{D}$. However, in reality, materials do not respond instantaneously to applied fields. Their ability to react to an incident field depends on a few different parameters, one of which, in general, is the frequency of the field. Therefore, Eq. (3.1) can be expressed in the Fourier domain as

$$\vec{D}(\omega) = \epsilon(\omega) \vec{E}(\omega) \tag{3.2}$$

which can also be expressed in the time-domain by the inverse Fourier transform of the above, resulting in the introduction of a *convolution*:

$$\vec{D}(t) = \epsilon(t) * \vec{E}(t) \tag{3.3}$$

where $*$ denotes the convolution. Of course, similar equations hold for magnetically dispersive materials, in which:

$$\vec{H}(\omega) = \mu^{-1}(\omega) \vec{B}(\omega) \tag{3.4}$$

$$\vec{H}(t) = \mu^{-1}(t) * \vec{B}(t) \tag{3.5}$$

While Eqs. (3.3) and (3.5) are manifestly more complex than in the case of nondispersive media, it is important to note that they remain *linear* in $\vec{E}$ and $\vec{B}$, due to the linearity property of the convolution operator [28].

In contrast, when exposed to high field strengths, a given material may no longer respond linearly to the applied field. Such a material is deemed *nonlinear*, and in the most general case, rather than being directly proportional, the polarization density can instead be an arbitrary function of the electric field, such that:

$$\vec{D} = \epsilon_0 \vec{E} + \vec{P}(\vec{E}) \tag{3.6}$$

Lastly, as noted earlier, all magnetic materials considered in this chapter will be strictly linear in nature, with all sources of nonlinearity being introduced via Eq. (3.6).

3.2.1 Dispersive Material Models

In order to simulate dispersive materials using numerical techniques, they should first be described in terms of mathematical models. In consequence, a variety of material models have been developed for different materials over the years. The most well known of which are the Debye, Lorentz, and Drude models. Since these models will be used in the following sections, they are briefly reviewed here.

Debye materials are characterized by the following frequency-dependent permittivity:

$$\epsilon(\omega) = \epsilon_\infty + \frac{\epsilon_s - \epsilon_\infty}{1 + j\omega\tau_e} \tag{3.7}$$

where ϵ_s and ϵ_∞ are the permittivity at zero and infinite frequencies and τ_e is the relaxation time. Debye materials are commonly used to describe biological tissues.

Lorentz materials, in turn, are characterized by:

$$\epsilon(\omega) = \epsilon_\infty + \frac{(\epsilon_s - \epsilon_\infty)\omega_e^2}{-\omega^2 + 2j\omega\delta_e + \omega_e^2} \tag{3.8}$$

where δ_e is the damping coefficient and ω_e is the frequency of the pole pair. The frequency behavior of some metamaterials can be approximated by Lorentz models.

Lastly, **Drude materials** are characterized by:

$$\epsilon(\omega) = \epsilon_\infty + \frac{\omega_e^2}{j\omega(j\omega + \gamma_e)} \tag{3.9}$$

where γ_e is the inverse of the relaxation time. Metals at optical frequencies are widely described by Drude models.

It is sometimes required to include multiple terms of a specific model or to combine different models together to achieve more accurate models. Therefore, a general rational polynomial embracing different combinations of these models will be assumed for the permittivity and permeability in the FETD formulations derived in this chapter:

$$\epsilon(j\omega), \mu(j\omega) = \frac{a_n(j\omega)^n + a_{n-1}(j\omega)^{n-1} + \cdots + a_1(j\omega) + a_0}{b_n(j\omega)^n + b_{n-1}(j\omega)^{n-1} + \cdots + b_1(j\omega) + b_0} \tag{3.10}$$

In fact, Eq. (3.10) is capable of describing a wide array of material models that, as with the Debye, Lorentz, and Drude formulations, are consequences of polarization and magnetization models formulated in terms of constant-coefficient differential equations. Moreover, since the focus here is on time-invariant dispersive media, the coefficients a_n and b_n are real-valued constants and depend only on the material parameters, while n is an integer. Therefore, the inverse Fourier transform of (3.10) is a sum of exponential functions plus a Dirac delta function in time. The Dirac delta function is the inverse Fourier transform of the constant part that comes from the partial fraction expansion of (3.10). Additionally, it is worth noting that (3.10) cannot be an arbitrary function and must obey the Kramers–Kronig relations, as a result of the causality principle [1].

It should be noted that there do exist some material models that do not fit into this general model. For example, the Cole–Cole model has fractional exponents, whereas the graphene surface conductivity model involves logarithmic terms. Alternatively, in some scenarios, only measured data is available, for which one common approach is to approximate these material's permittivity by rational polynomial interpolation, resulting in an expression similar to Eq. (3.10). In these cases, experimental data can thus be incorporated into the FETD simulations to come without special treatment.

3.2.2 Dispersive Media Modeling Techniques

In order to perform an FETD simulation, the constitutive equations must be solved in synchronism with Maxwell's equations. As a result, several methods to discretize these equations have been proposed, of which the piecewise linear recursive convolution (PLRC) [29], the auxiliary differential equation (ADE) [30, 31], the z-transform [32], and the bilinear transform (BT) [33, 34] are the most widely used. These techniques were initially proposed for the FDTD method; however, due to the fundamental similarities between the FETD and FDTD methods, they are applicable to FETD as well. Here, these methods will be described for the electric constitutive equation, however they can be similarly applied to the magnetic constitutive equation, if desired.

Since the materials in question are assumed to be linear and time-invariant, the electric constitutive equation can be expressed in time by a convolution integral as

$$\vec{D}(t) = \epsilon(t) * \vec{E}(t) = \int_{-\infty}^{t} \vec{E}(t - \tau)\epsilon(\tau)d\tau \tag{3.11}$$

where $\epsilon(t)$ is the inverse Fourier transform of $\epsilon(\omega)$. In the PLRC method, a piecewise-linear approximation of the time-discrete values of $\vec{E}(t)$ are obtained from the FETD simulation and are used to evaluate the integral over each time-step Δt. The convolution can then be computed efficiently in a recursive manner, due to the exponential nature of the permittivity. To that end, recursive accumulators are often introduced, as in [29], which depend only on a few previous values of the electric field intensity.

Alternatively, in the ADE approach, the constitutive equation can be directly written in the form of an ordinary differential equation as

$$b_n \frac{\partial^n \vec{D}(t)}{\partial t^n} + \cdots + b_1 \frac{\partial \vec{D}(t)}{\partial t} + b_0 \vec{D}(t) = a_n \frac{\partial^n \vec{E}(t)}{\partial t^n} + \cdots + a_1 \frac{\partial \vec{E}(t)}{\partial t} + a_0 \vec{E}(t)$$
$$\tag{3.12}$$

This ordinary differential equation can then be discretized in time using various finite-difference schemes, with each choice yielding a different discretization having its own accuracy and stability conditions.

One can equally interpret the constitutive equation as a filter, in which $\epsilon(\omega)$ is the filter transfer function, $\vec{E}(\omega)$ is the input, and $\vec{D}(\omega)$ is the filter response or output. The z-transform and BT are two common approaches in the digital signal processing community to convert a continuous transfer function to a discrete form. They have thus also been adopted by the electromagnetics community to model dispersive media [32–34]. In the z-transform method, $\epsilon(t)$ is first obtained from $\epsilon(\omega)$ using the inverse Fourier transform. Then, it is discretized in time by the sampling rate Δt. Finally, the z-transform is applied to compute $\epsilon(z)$ from the discrete-time form. The bilinear transformation, however, proceeds directly from the Fourier transform to the z-transform via the following substitution:

$$j\omega = \frac{2}{\Delta t}\frac{1 - z^{-1}}{1 + z^{-1}} \tag{3.13}$$

This transformation maps the left-hand side and the entire $j\omega$ axis of the complex s-plane to the interior and boundary of the unit circle in the z-plane, respectively. As a result, the stability of a transfer function is preserved under the bilinear transformation. Stability analyses performed in [34, 35] have also shown that the BT does not restrict the stability condition of FDTD in the case of Debye and Lorentz media.

In fact, (3.13) is nothing but the trapezoidal rule in disguise. If (3.12) is rewritten as a system of first-order differential equations and discretized using the trapezoidal rule, the ADE approach will yield the same discrete-time form as the BT. Therefore, the BT can be seen as a systematic implementation of the ADE method for dispersive materials of arbitrary order. The BT method will thus be used in this chapter to discretize dispersive materials.

3.2.3 Nonlinear Dielectric Models

Earlier it was mentioned that under certain conditions the response of a dielectric to an applied field may no longer be linear, and that in general the polarization density can be an arbitrary function of the electric field:

$$\vec{D} = \epsilon_0 \vec{E} + \vec{P}(\vec{E}) \tag{3.14}$$

However, to simplify the analysis of such nonlinear interactions, it is usually customary to expand the polarization density as a power series in the electric field, in which only the first few dominant terms are kept. While such an expansion would in the most general case involve susceptibility tensors, for many isotropic material models an adequate expansion is instead given by [36]:

$$\vec{P} = \epsilon_0 \chi^{(1)}\vec{E} + \epsilon_0 \chi^{(2)}E\vec{E} + \epsilon_0 \chi^{(3)}E^2\vec{E} \tag{3.15}$$

where $\chi^{(n)}$ is the nth-order susceptibility, and E is the magnitude of the electric field. While a third-order approximation is often sufficient, in principle higher order terms could equally be included in the above expansion. Moreover, for many materials, symmetry within the crystal lattice structure of the medium precludes the existence of the $\chi^{(2)}$ term, a phenomenon known as inversion symmetry [36]. This nonlinear polarization can result in numerous additional and unique phenomena, such as the Kerr effect, frequency doubling, self-phase modulation, solitons, four-wave mixing, and supercontinuum generation [36–38].

The nonlinear polarization in (3.15) represents an *instantaneous* nonlinearity, as changes in the applied field are reflected immediately in the polarization. However, just as in the linear case, this is not necessarily a physical result and the nonlinear response can also be a function of frequency, thereby representing a *dispersive nonlinearity*. Much as in the linear case, this nonlinear dispersion corresponds to the introduction of additional convolutions into the nonlinear model of (3.15) and gives rise to additional phenomenon such as stimulated Raman scattering [36, 39]. As a result, incorporating linear dispersion, instantaneous nonlinearity, as well as dispersive nonlinearity into a single model for the polarization density can be accomplished quite accurately via the following expression for $\vec{P}$ [40, 41]:

$$\vec{P} = \epsilon_0 \chi^{(1)}(t) * \vec{E} + \epsilon_0 \chi^{(3)} \left(\alpha E^2 + (1-\alpha)g(t) * E^2 \right) \vec{E} \tag{3.16}$$

in which the susceptibility $\chi^{(1)}$ models the linear dispersion, the susceptibility $\chi^{(3)}$ an instantaneous (Kerr) and/or dispersive (Raman) nonlinearity (with the α term controlling their relative strengths), and $g(t)$ being a causal response function characterizing the nonlinear dispersion. The polarization in (3.16) represents a very general model capable of including a wide variety of material interactions and phenomena, and will form the basis for the inclusion of these effects within Maxwell's equations in this chapter.

3.3 Finite-Element Time-Domain Formulations

FETD formulations broadly fall into two main categories: the vector wave equation (VWE) formulation and the mixed formulation. The VWE formulation, as the name suggests, is constructed based on the second-order VWE expressed in terms of the electric field intensity $\vec{E}$ [42, 43]. In contrast, the mixed formulation solves the coupled first-order Maxwell curl equations directly [44]. In the following sections both formulations will be reviewed for linear nondispersive media and their inherent properties will be discussed and compared.

3.3.1 Vector Wave Equation Formulation

The VWE can be written in the time-domain as

$$\nabla \times \left(\mu^{-1} \nabla \times \vec{E}(t) \right) + \epsilon \frac{\partial^2 \vec{E}(t)}{\partial t^2} = -\frac{\partial \vec{J}(t)}{\partial t} \tag{3.17}$$

where $\vec{J}(t)$ is the impressed electric current density (excitation) in the continuum case. To discretize this equation, the entire computation domain can be subdivided into small parts called *finite-elements*. The electric field intensity within each finite element (e) can then be approximated as

$$\vec{E}^{(e)}(t) = \sum_{i=1}^{l_e} \vec{W}_i^{(1)(e)} e_i^{(e)}(t) \tag{3.18}$$

where $e_i^{(e)}$ are time-dependent expansion weights, $\vec{W}_i^{(1)(e)}$ is the Whitney 1-form (also known as Whitney *edge* element) corresponding to the ith unknown, and l_e is the total number of unknowns in element (e). Whitney edge elements guarantee continuity of the tangential component of $\vec{E}$ across element boundaries. Following a standard variational or Galerkin finite-element procedure for discretization of the spatial component of Eq. (3.17), the VWE formulation becomes [45]

$$\mu^{-1}[\tilde{S}]\{e\} + \epsilon[\tilde{M}] \frac{\partial^2 \{e\}}{\partial t^2} + \{f\} = 0 \tag{3.19}$$

where $\{e\}$ is the vector of unknowns and the quantities $[\tilde{M}]$, $[\tilde{S}]$, and $\{f\}$ have all been assembled from their local counterparts, which are themselves given by

$$[M^{(e)}]_{ij} = \epsilon[\tilde{M}^{(e)}]_{ij} = \epsilon \int_{\Omega^{(e)}} \vec{W}_i^{(1)(e)} \cdot \vec{W}_j^{(1)(e)} \, d\Omega \tag{3.20}$$

$$[S^{(e)}]_{ij} = \mu^{-1}[\tilde{S}^{(e)}]_{ij} = \mu^{-1} \int_{\Omega^{(e)}} \nabla \times \vec{W}_i^{(1)(e)} \cdot \nabla \times \vec{W}_j^{(1)(e)} \, d\Omega \tag{3.21}$$

$$\{f^{(e)}\}_i = \int_{\Omega^{(e)}} \vec{W}_i^{(1)(e)} \cdot \frac{\partial \vec{J}}{\partial t} \, d\Omega \tag{3.22}$$

It should be noted that constant material properties within each element are assumed throughout this chapter.

The last remaining step is to discretize (3.19) in time. While many temporal discretizations are possible, one of the most popular is the Newmark-β method, which is based on the following approximations [46]

$$u^{n+1} - u^n = \Delta t \frac{\partial u^n}{\partial t} + \beta \Delta t^2 \frac{\partial^2 u^{n+1}}{\partial t^2} + \Delta t^2 \left(\frac{1}{2} - \beta \right) \frac{\partial^2 u^n}{\partial t^2} \tag{3.23}$$

$$\frac{1}{\Delta t} \left(\frac{\partial u^{n+1}}{\partial t} - \frac{\partial u^n}{\partial t} \right) = \frac{1}{2} \left(\frac{\partial^2 u^{n+1}}{\partial t^2} + \frac{\partial^2 u^n}{\partial t^2} \right) \tag{3.24}$$

where $0 \leq 2\beta \leq 1$, and $\frac{\partial u}{\partial t}^n$ and $\frac{\partial^2 u}{\partial t^2}^n$ denote the values of the first and second time derivatives of u at time $t = n\Delta t$, respectively. Applying the Newmark-β method to the VWE thus gives

$$([M] + \beta\Delta t^2[S])\{e\}^{n+1} = (2[M] - (1 - 2\beta)\Delta t^2[S])\{e\}^n$$
$$- ([M] + \beta\Delta t^2[S])\{e\}^{n-1}$$
$$+ \Delta t^2(\beta\{f\}^{n+1} + (1 - 2\beta)\{f\}^n + \beta\{f\}^{n-1}) \quad (3.25)$$

The parameter β allows for adjusting the properties of the scheme. Two choices are of special interest: $\beta = 0$ and $\beta = 1/4$. These values amount to the central difference method and the trapezoidal rule, respectively. It can also be shown that the first choice yields *conditional stability* (CS), whereas the second one guarantees *unconditional stability* (US) [47]. In this context, US means that the method remains stable during time-stepping for arbitrary large time-step sizes, meaning accuracy considerations determine the time-step size Δt used rather than stability. Both schemes are second order accurate in time. For additional details on the notion of numerical stability, see Section 3.5.

3.3.2 Mixed Formulation

In contrast to the VWE method, the mixed formulation solves Maxwell's equations directly. Correspondingly, Faraday and Ampère's laws for linear nondispersive media are expressed as follows:

$$\nabla \times \vec{E}(t) = -\frac{\partial \vec{B}(t)}{\partial t} \quad (3.26)$$

$$\nabla \times \mu^{-1}\vec{B}(t) = \vec{J}(t) + \epsilon\frac{\partial \vec{E}(t)}{\partial t} \quad (3.27)$$

As a result, both the electric field intensity and the magnetic flux density need to be approximated within each finite-element. Whitney 1-form and 2-form basis functions are most often used for this purpose [48], which results in

$$\vec{E}^{(e)}(t) = \sum_{i=1}^{l_e} \vec{W}_i^{(1)(e)} e_i^{(e)}(t) \quad (3.28)$$

$$\vec{B}^{(e)}(t) = \sum_{j=1}^{m_e} \vec{W}_j^{(2)(e)} b_j^{(e)}(t) \quad (3.29)$$

where $b_j^{(e)}(t)$ are time-dependent magnetic expansion weights, $\vec{W}_j^{(2)(e)}$ is the Whitney 2-form (also known as Whitney *face* element) corresponding to the jth unknown, and m_e is the total number of magnetic unknowns per element (e). Whitney face elements guarantee continuity of the normal component of $\vec{B}$ across element boundaries. Following a finite-element Galerkin procedure, the spatially discretized semi-discrete equations can be obtain as

$$\frac{\partial \{d\}}{\partial t} = [C]^{\mathrm{T}}\{h\} + \{f\} \tag{3.30}$$

$$\frac{\partial \{b\}}{\partial t} = -[C]\{e\} \tag{3.31}$$

$$\{d\} = \epsilon[\tilde{M}]\{e\} \tag{3.32}$$

$$\{h\} = \mu^{-1}[\tilde{M}_f]\{b\} \tag{3.33}$$

where the $[C]$ matrix represents the discrete curl operator and is a sparse rectangular matrix consisting of $\{-1,1\}$ nonzero entries. Furthermore, due to the relationship between the 1-forms and 2-forms, it can be shown that $[S] = [C]^{\mathrm{T}}[M_f][C]$ [49]. The quantities $[\tilde{M}]$, $[\tilde{M}_f]$, and $\{f\}$ have again been assembled from their local elemental counterparts, which are given by

$$[M^{(e)}]_{ij} = \epsilon[\tilde{M}^{(e)}]_{ij} = \epsilon \int_{\Omega^{(e)}} \vec{W}_i^{(1)(e)} \cdot \vec{W}_j^{(1)(e)} \, d\Omega \tag{3.34}$$

$$[M_f^{(e)}]_{ij} = \mu^{-1}[\tilde{M}_f^{(e)}]_{ij} = \mu^{-1} \int_{\Omega^{(e)}} \vec{W}_i^{(2)(e)} \cdot \vec{W}_j^{(2)(e)} \, d\Omega \tag{3.35}$$

$$\{f^{(e)}\}_i = -\int_{\Omega^{(e)}} \vec{W}_i^{(1)} \cdot \vec{J} \, d\Omega \tag{3.36}$$

There are two schemes that are widely used to discretize the mixed formulation in time. The first is the *leap-frog* scheme that temporally interweaves the electric and magnetic fields, yielding:

$$[M]\frac{\{e\}^n - \{e\}^{n-1}}{\Delta t} = [C]^{\mathrm{T}}[M_f]\{b\}^{n-1/2} + \{f\}^{n-1/2} \tag{3.37}$$

$$\frac{\{b\}^{n+1/2} - \{b\}^{n-1/2}}{\Delta t} = -[C]\{e\}^n \tag{3.38}$$

and is conditionally stable. By eliminating $\{b\}$, it can be easily shown that this formulation is equivalent to the VWE formulation (3.25) with $\beta = 0$.

The other popular scheme is the *Crank–Nicolson* method, which is based on the following approximation:

$$\frac{\partial g(t, \mathbf{r})}{\partial t} = f(t, \mathbf{r}) \xrightarrow{\text{CN}} \frac{g^n(\mathbf{r}) - g^{n-1}(\mathbf{r})}{\Delta t} = \frac{f^n(\mathbf{r}) + f^{n-1}(\mathbf{r})}{2} \tag{3.39}$$

Applying this scheme to the spatially discretized Maxwell's equation yields [50]

$$[M]\frac{\{e\}^n - \{e\}^{n-1}}{\Delta t} = [C]^{\mathrm{T}}[M_f]\frac{\{b\}^n + \{b\}^{n-1}}{2} + \frac{\{f\}^n + \{f\}^{n-1}}{2} \tag{3.40}$$

$$\frac{\{b\}^n - \{b\}^{n-1}}{\Delta t} = -[C]\frac{\{e\}^n + \{e\}^{n-1}}{2} \tag{3.41}$$

Eliminating $\{b\}^n$ in (3.40) with (3.41) then yields the final set of update equations:

$$\left([M] + \frac{(\Delta t)^2}{4}[C]^{\mathrm{T}}[M_f][C]\right)\{e\}^{n+1} = \left([M] - \frac{(\Delta t)^2}{4}[C]^{\mathrm{T}}[M_f][C]\right)\{e\}^n$$

$$+ \Delta t[C]^{\mathrm{T}}[M_f]\{b\}^n$$

$$+ \frac{\Delta t}{2}(\{f\}^n + \{f\}^{n-1}) \qquad (3.42)$$

$$\{b\}^{n+1} = \{b\}^n - \frac{\Delta t}{2}[C]\left(\{e\}^{n+1} + \{e\}^n\right) \qquad (3.43)$$

Numerical studies have demonstrated that this formulation is US, however no theoretical proof has yet been reported. Similarly to the leap-frog method, by eliminating $\{b\}$ entirely, it can be shown that this technique is equivalent to the VWE formulation (3.25) with $\beta = 1/4$ [51], and may help justify its posited unconditional stability.

Note that eliminating $\{e\}^n$ in (3.41) with (3.40) gives an alternative formulation in terms of only $\{b\}$. However, it is much less efficient as it involves the inversion of the $[M]$ matrix, and although $[M]$ is sparse, $[M]^{-1}$ is not generally sparse.

3.3.3 Remarks on FETD Formulations

The VWE formulation is more popular than the mixed formulation as it requires only one type of basis function. Moreover, frequency-domain finite-element solvers are mainly developed based on the VWE, meaning it is straightforward to extend an existing frequency domain code base to the time-domain using this formulation. However, the VWE supports the nontrivial solution $\vec{E}(t) = (a + bt)\nabla\phi$ in a lossless source-free region, where a and b are constants and ϕ is a time-independent scalar potential. As b is not necessarily zero, the solution of the VWE can drift linearly in time and may cause late-time instabilities when dispersive materials are included in the problem. A notable example of this is in the VWE FETD implementation of anisotropic-medium perfectly matched layers (PMLs), for which late-time instability issues have been reported in several studies [52, 53].

As discussed before, both VWE and mixed FETD formulations have CS and US variants. However, US is not a clear-cut advantage. Choosing a CS or a US formulation is problem-dependent and involves trade-offs between several parameters, some of which are listed here:

- The left-hand side matrix of the FETD formulation can be written in the general form $[M] + \alpha\Delta t^2[S]$ where $\alpha > 0$ for US and $\alpha = 0$ for CS methods. The condition number of $[S]$ increases without bound as the mesh is refined. However, the condition number of $[M]$ remains bounded during refinement as long as the quality of the mesh is preserved. Therefore, the convergence of a US method degrades significantly compared to a CS method during refinement when an iterative solver is employed [54].

- Since the left-hand side matrix is always positive definite, there exists some effective preconditioners to improve the convergence rate of iterative solvers. On the other hand, the spectral properties of a matrix do not affect the performance of direct solvers, which these days may be used even for relative large problems on personal computers.

- Practical meshes often make use of adaptive mesh refinement, in which some regions are more highly refined than others. These regions can severely limit the time-step size in a CS method; however, selecting a time-step beyond the stability condition in a US method not only decreases the accuracy, particularly in refined regions, but can also make the left-hand side matrix more ill conditioned.

- There are other time scales that can also play a role in modeling dispersive media, which are determined by, e.g. relaxation times and resonant frequencies. Properly capturing them places additional constraints on the time-step size, which can be stricter than the stability limit of CS methods.

- The stability condition of a CS method is $\mathcal{O}(\ell_{min}/c)$ in which ℓ_{min} is the minimum edge length in the mesh and c is the speed of light. In some applications, such as low-frequency problems, the frequency corresponding to ℓ_{min} can be thousands of times, or even more, larger than the operating frequency. In other words, a time-step thousands of times greater than the stability limit can provide accurate results.

- Although $[M]$ and $[S]$ are equally sparse and the inverse of a sparse matrix is not necessarily sparse, it has been shown that the interactions between nonlocal degrees of freedom in $[S]^{-1}$ is much stronger than in $[M]^{-1}$. Hence, $[M]^{-1}$ can be well approximated by a sparse matrix [55]. Moreover, some techniques have been developed to dramatically decrease the number of nonzeros or even diagonalize the mass matrix, such as orthogonal basis functions [56] and mass-lumping techniques for edge elements [57], which can greatly speedup the time-stepping in a CS method.

It follows from the discussion above that each method has its own advantages and disadvantages. US techniques can decrease the number of time steps, but they do not necessarily reduce the total computation time as compared to a CS method. Depending on many factors, e.g. size and scale of the problem, accuracy, importance of refined regions, and user preferences, one may rightly prefer either a US or a CS formulation.

3.4 FETD for Dispersive and Nonlinear Media

In the following sections, both the VWE and mixed FETD formulations for linear dispersive, instantaneous nonlinear, and dispersive nonlinear media will be presented along with several numerical studies.

3.4.1 Vector Wave Equation (VWE) Formulation

To begin, the VWE FETD formulation for dispersive and nonlinear media are considered and validated with representative numerical studies.

3.4.1.1 Linear Dispersive Media

As discussed previously, dispersive media have material properties which are functions of time, resulting in the multiplications of Eq. (3.19) being converted into convolutions:

$$\mu^{-1}(t) * ([\tilde{S}]\{e(t)\}) + \epsilon(t) * \left([\tilde{M}]\frac{\partial^2 \{e(t)\}}{\partial t^2}\right) + \{f(t)\} = 0 \tag{3.44}$$

To begin addressing this issue, start by defining two *auxiliary variables* as follows [58]:

$$\{\mathcal{L}_\epsilon(t)\} = \epsilon(t) * ([\tilde{M}]\{e(t)\}) \tag{3.45}$$

$$\{\mathcal{L}_{\mu^{-1}}(t)\} = \mu^{-1}(t) * ([\tilde{S}]\{e(t)\}) \tag{3.46}$$

Applying the bilinear transformation of Eq. (3.13) to the general material model (3.10) yields the general discretized form in the z-domain as

$$\epsilon(z) = \frac{c_0 + c_1 z^{-1} + \cdots + c_p z^{-p}}{1 + d_1 z^{-1} + \cdots + d_p z^{-p}}, \tag{3.47}$$

$$\mu^{-1}(z) = \frac{q_0 + q_1 z^{-1} + \cdots + q_p z^{-p}}{1 + r_1 z^{-1} + \cdots + r_p z^{-p}} \tag{3.48}$$

Substituting the above Eqs. into (3.45) and (3.46), and making use of the time shifting and convolution properties of the z-transform, then results in a discretized equation for the convolution at the next time step in terms of previous convolution and electric field values. However, to implement the update equations more efficiently, an approach known as *Transposed Direct Form II* can be utilized, whereby the convolutions are instead expressed as follows:

$$\{\mathcal{L}_\epsilon\}^n = c_0[\tilde{M}]\{e\}^n + \{\mathcal{W}_1\}^{n-1} \tag{3.49}$$

$$\{\mathcal{L}_{\mu^{-1}}\}^n = q_0[\tilde{S}]\{e\}^n + \{\mathcal{G}_1\}^{n-1} \tag{3.50}$$

in which $\{\mathcal{W}_\alpha\}$ and $\{\mathcal{G}_\alpha\}$ are *auxiliary variables* defined by:

$$\{\mathcal{W}_\alpha\}^n = c_\alpha[\tilde{M}]\{e\}^n - d_\alpha\{\mathcal{L}_\varepsilon\}^n + \{\mathcal{W}_{\alpha+1}\}^{n-1} \quad ; \quad \alpha = 1, 2, \ldots, p-1$$
$$\{\mathcal{W}_\alpha\}^n = c_\alpha[\tilde{M}]\{e\}^n - d_\alpha\{\mathcal{L}_\varepsilon\}^n \quad\quad\quad ; \quad \alpha = p$$
$$(3.51)$$

$$\{\mathcal{G}_\alpha\}^n = q_\alpha[\tilde{S}]\{e\}^n - r_\alpha\{\mathcal{L}_{\mu^{-1}}\}^n + \{\mathcal{G}_{\alpha+1}\}^{n-1} \quad ; \quad \alpha = 1, 2, \ldots, p-1$$
$$\{\mathcal{G}_\alpha\}^n = q_\alpha[\tilde{S}]\{e\}^n - r_\alpha\{\mathcal{L}_{\mu^{-1}}\}^n \quad\quad\quad ; \quad \alpha = p$$
$$(3.52)$$

These auxiliary variables improve the overall efficiency of the convolution update procedure by gradually accumulating past values over time, rather than explicitly storing each past value individually.

Lastly, by substituting (3.49) and (3.50) into (3.44) and applying the Newmark-β method with $\beta = 1/4$, the fully discretized form can finally be obtained as follows:

$$\left([M] + \frac{(\Delta t)^2}{4}[S]\right)\{e\}^{n+1} = 2\left([M] - \frac{(\Delta t)^2}{4}[S]\right)\{e\}^n$$
$$- \left([M] + \frac{(\Delta t)^2}{4}[S]\right)\{e\}^{n-1}$$
$$- \left(\{\mathcal{W}_1\}^n - 2\{\mathcal{W}_1\}^{n-1} + \{\mathcal{W}_1\}^{n-2}\right)$$
$$- \frac{(\Delta t)^2}{4}\left(\{\mathcal{G}_1\}^n + 2\{\mathcal{G}_1\}^{n-1} + \{\mathcal{G}_1\}^{n-2}\right)$$
$$+ \frac{(\Delta t)^2}{4}\left(\{f\}^{n+1} + 2\{f\}^n + \{f\}^{n-1}\right) \quad (3.53)$$

in which $[M] = c_0[\tilde{M}]$ and $[S] = q_0[\tilde{S}]$.

3.4.1.2 Instantaneous Nonlinearity

In this section the linear VWE formulation presented earlier will be adapted for use with nondispersive nonlinear media, i.e. media whose response to an applied field is both nonlinear and *instantaneous*, as originally derived in [59]. For such a material the polarization density vector does not contain any convolutions and can be expanded in an approximate power series in $\vec{E}$ as detailed previously in Section 3.2.3:

$$\vec{P} \approx \epsilon_0 \chi^{(1)}\vec{E} + \epsilon_0 \chi^{(3)}E^2\vec{E} \tag{3.54}$$

where $E = |\vec{E}|$. Substituting this into the constitutive relation for the $\vec{D}$ field thus results in the definition of a nonlinear nondispersive permittivity:

$$\vec{D} = \epsilon_0(1 + \chi^{(1)} + \chi^{(3)}E^2)\vec{E} = \epsilon\vec{E} \tag{3.55}$$

Due to the dependence of ϵ on E and thus on time, when including nonlinearity in Maxwell's equations the permittivity must remain within the temporal derivatives of both Ampère's law:

$$\nabla \times \frac{\vec{B}}{\mu} = \vec{J} + \frac{\partial}{\partial t}\left(\epsilon\vec{E}\right) \tag{3.56}$$

and the resulting VWE:

$$\nabla \times \frac{1}{\mu} \nabla \times \vec{E} + \frac{\partial^2}{\partial t^2} \left(\epsilon \vec{E} \right) = -\frac{\partial \vec{J}}{\partial t} \tag{3.57}$$

To spatially discretize Eq. (3.57) with the finite-element method, a Galerkin procedure can be applied in the exact same manner as it was in the linear case. After assembling or summing the contributions of each individual element into a global system, the following semi-discrete matrix equation results:

$$\frac{d^2}{dt^2} \left([M]\{e\} \right) + [S]\{e\} + \{f\} = 0, \tag{3.58}$$

where the matrix and vector quantities within each individual element before assembly are given by:

$$[M^{(e)}]_{ij} = \int_{\Omega^e} \epsilon \, \vec{W}_i^{(1)(e)} \cdot \vec{W}_j^{(1)(e)} \, d\Omega$$

$$= \int_{\Omega^e} \epsilon_0 \left(1 + \chi^{(1)} + \chi^{(3)} E^2 \right) \vec{W}_i^{(1)(e)} \cdot \vec{W}_j^{(1)(e)} \, d\Omega \tag{3.59}$$

$$[S^{(e)}]_{ij} = \int_{\Omega^e} \frac{1}{\mu} \left(\nabla \times \vec{W}_i^{(1)(e)} \right) \cdot \left(\nabla \times \vec{W}_j^{(1)(e)} \right) \, d\Omega \tag{3.60}$$

$$\{f^{(e)}\}_j = \int_{\Omega^e} \frac{\partial \vec{J}}{\partial t} \cdot \vec{W}_j^{(1)(e)} \, d\Omega + \int_{\partial\Omega^e} \frac{1}{\mu} \left(\nabla \times \vec{E} \right) \times \vec{W}_j^{(1)(e)} \cdot d\vec{S} \tag{3.61}$$

Up until now the formulation has not differed much from that presented for linear media, except for the subtle yet crucial difference of ϵ, and consequently $[M]$, being kept inside the temporal derivatives in (3.57) and (3.58). At this point in the process, however, it becomes necessary to discretize the temporal derivatives, for which additional care must be taken. As was the case in Section 3.3.1 for linear media, here the Newmark-β method will be applied with $\beta = 1/4$. While other temporal discretizations are possible (for example, the well-known central difference method), Newmark-β is preferred due to its implicit formulation lending it increased stability. However, in contrast to the linear case, the Newmark-β equations must now be applied to the matrix-vector product $[M]\{e\}$ as a *whole* rather than solely to the vector term $\{e\}$:

$$[M]^{n+1}\{e\}^{n+1} = [M]^n\{e\}^n + \Delta t \left(\frac{d}{dt}[M]\{e\} \right)^n$$

$$+ \frac{\Delta t^2}{4} \left(\frac{d^2}{dt^2}[M]\{e\} \right)^n + \frac{\Delta t^2}{4} \left(\frac{d^2}{dt^2}[M]\{e\} \right)^{n+1} \tag{3.62}$$

$$\left(\frac{d}{dt}[M]\{e\} \right)^{n+1} = \left(\frac{d}{dt}[M]\{e\} \right)^n + \frac{\Delta t}{2} \left(\frac{d^2}{dt^2}[M]\{e\} \right)^n$$

$$+ \frac{\Delta t}{2} \left(\frac{d^2}{dt^2}[M]\{e\} \right)^{n+1} \tag{3.63}$$

where again it should be emphasized that both $[M]$ and $\{e\}$ are functions of time, owing to the nonlinearity.

Similarly to the linear case, the semi-discrete equation (3.58) can be used to obtain an expression for the second temporal derivatives in Eqs. (3.62) and (3.63), whereas Eq. (3.63) can be used to eliminate the first temporal derivative in Eq. (3.62). The final fully discretized update equation for the electric field in instantaneous nonlinear media is thus obtained as:

$$
\left([M]^{n+1} + \frac{\Delta t^2}{4}[S] \right) \{e\}^{n+1} = 2 \left([M]^n - \frac{\Delta t^2}{4}[S] \right) \{e\}^n
$$
$$
- \left([M]^{n-1} + \frac{\Delta t^2}{4}[S] \right) \{e\}^{n-1}
$$
$$
- \frac{\Delta t^2}{4} \left(\{f\}^{n+1} + 2\{f\}^n + \{f\}^{n-1} \right) \tag{3.64}
$$

The update equation in (3.64) bears a striking similarity to that for linear media obtained previously in Eq. (3.25). Despite this resemblance, however, Eq. (3.64) is of course much more difficult to solve due to the dependence of the $[M]$ matrix on the unknown solution vector $\{e\}$. As a result, straightforward matrix solvers are no longer of any use by themselves and instead methods must be applied which are capable of solving systems of nonlinear equations. While many such methods exist with varying advantages and disadvantages, here the focus will be on the widely popular *Newton–Raphson method* (also known simply as *Newton's method*).

In the Newton–Raphson method, a system of nonlinear equations written in the form $\{F\} = \{0\}$, dependent on a set of unknown variables $\{x\}$, may be solved by an iterative procedure of the form:

$$
\{x\}_{(k+1)} = \{x\}_{(k)} - [J]^{-1}_{(k)}\{F\}_{(k)} \tag{3.65}
$$

where the bracketed subscript (k) indicates the iteration number and $[J]$ is the *Jacobian matrix* defined as:

$$
[J]_{ij} = \frac{\partial \{F\}_i}{\partial \{x\}_j} \tag{3.66}
$$

The nonlinear update equation for the electric field can easily be made to fit this format by moving all terms to the left-hand side. Thus, in order to apply the Newton–Raphson method to the current problem, all that is required is an expression for the Jacobian matrix.

In Eq. (3.64) the unknown variable of interest is the electric field expansion weights $\{e\}^{n+1}$, and so the Jacobian expression in Eq. (3.66) will only take derivatives with respect to this quantity. Since none of the terms on the right-hand side of Eq. (3.64) depend on $\{e\}^{n+1}$, only the left-hand side will have a contribution to the Jacobian:

$$
[J]_{ij} = \frac{\partial}{\partial \{e\}_j^{(n+1)}} \left\{ \left([M]^{n+1} + \frac{\Delta t^2}{4}[S] \right) \{e\}^{n+1} \right\}_i \tag{3.67}
$$

The goal is to now obtain a more useful closed-form expression for the Jacobian matrix in Eq. (3.67), which will allow for the solution of Eq. (3.64) using Newton's method. To this end, the first step is to rewrite the matrix-vector product in Eq. (3.67) more explicitly as a sum:

$$[J]_{ij}^{n+1} = \frac{\partial \{F\}_i}{\partial \{e\}_j^{n+1}} = \frac{\partial}{\partial \{e\}_j^{n+1}} \sum_{k=1}^{l} \left([M]_{ik}^{n+1} + \frac{\Delta t^2}{4} [S]_{ik}^T \right) \{e\}_k^{n+1} \tag{3.68}$$

From here, it is relatively straightforward to distribute the derivative into the sum and apply the product rule to the matrix-vector product:

$$[J]_{ij}^{n+1} = \sum_{k=1}^{l} \left([M]_{ik}^{n+1} + \frac{\Delta t^2}{4} [S]_{ik} \right) \frac{\partial \{e\}_k^{n+1}}{\partial \{e\}_j^{n+1}}$$

$$+ \sum_{k=1}^{l} \frac{\partial}{\partial \{e\}_j^{n+1}} \left([M]_{ik}^{n+1} + \frac{\Delta t^2}{4} [S]_{ik} \right) \{e\}_k^{n+1} \tag{3.69}$$

While the expression in Eq. (3.69) may seem intimidating at first glance, it can be considerably simplified. First, it is noted that since the [S] matrix is a constant and thus independent of $\{e\}^{n+1}$, its derivative is zero. Second, because each of the unknown variables is independent of the others, it can be shown that:

$$\frac{\partial \{e\}_k^{n+1}}{\partial \{e\}_j^{n+1}} = \begin{cases} 1, & \text{for } j = k \\ 0, & \text{for } j \neq k \end{cases} \tag{3.70}$$

meaning that the first term in Eq. (3.69) will only be nonzero when $j = k$. Combining these results, Eq. (3.69) can be recast into the following simplified form:

$$[J]_{ij}^{n+1} = [M]_{ij}^{n+1} + \frac{\Delta t^2}{4} [S]_{ij} + \sum_{k=1}^{l} \frac{\partial [M]_{ik}^{n+1}}{\partial \{e\}_j^{n+1}} \{e\}_k^{n+1} \tag{3.71}$$

To deal with the last unsimplified summation term remaining in Eq. (3.71), the derivative, sum, and unknown can all be brought into the definition of the elemental $[M^{(e)}]$ matrix:

$$\sum_{k=1}^{l_e} \frac{\partial [M^{(e)}]_{ik}^{n+1}}{\partial \{e^{(e)}\}_j^{n+1}} \{e^{(e)}\}_k^{n+1} = \int_{\Omega^e} \frac{\partial \epsilon^{n+1}}{\partial \{e^{(e)}\}_j^{n+1}} \vec{W}_i^{(1)(e)} \cdot \sum_{k=1}^{l_e} \vec{W}_k^{(1)(e)} \{e^{(e)}\}_k^{n+1} \, d\Omega \tag{3.72}$$

Looking more closely at the summation term in Eq. (3.72), it can be recognized as the interpolated form of the electric field within in each element from (3.18), meaning Eq. (3.72) can be recast as:

$$\sum_{k=1}^{l_e} \frac{\partial [M^{(e)}]_{ik}^{n+1}}{\partial \{e^{(e)}\}_j^{n+1}} \{e^{(e)}\}_k^{n+1} = \int_{\Omega^e} \frac{\partial \epsilon^{n+1}}{\partial \{e^{(e)}\}_j^{n+1}} \vec{W}_i^{(1)(e)} \cdot \vec{E}^{n+1} \, d\Omega \tag{3.73}$$

Furthermore, since it is more natural to discuss the dependence of ϵ on the electric field strength E than it is on the unknown variables $\{e\}_j$, the remaining derivative term in Eq. (3.73) can be rewritten using the chain rule:

$$\int_{\Omega^e} \frac{\partial \epsilon^{n+1}}{\partial \{e^{(e)}\}_j^{n+1}} \vec{W}_i^{(1)(e)} \cdot \vec{E}^{n+1} \, d\Omega = \int_{\Omega^e} \frac{\partial \epsilon^{n+1}}{\partial E^{n+1}} \frac{\partial E^{n+1}}{\partial \{e^{(e)}\}_j^{n+1}} \vec{W}_i^{(1)(e)} \cdot \vec{E}^{n+1} \, d\Omega$$

$$(3.74)$$

Lastly, the dependence of E on $\{e\}$ can be shown through straightforward analysis to be equivalent to:

$$\frac{\partial E^{n+1}}{\partial \{e^{(e)}\}_j^{n+1}} = \frac{\vec{E}^{n+1}}{E^{n+1}} \cdot \vec{W}_j^{(1)(e)}$$

$$(3.75)$$

Making this final substitution into Eq. (3.74) and returning to the full expression in (3.71) at last results in the desired simplified Jacobian:

$$[J^{(e)}]_{ij}^{n+1} = [M^{(e)}]_{ij}^{n+1} + \frac{\Delta t^2}{4}[S^{(e)}]_{ij}^{T}$$
$$+ \int_{\Omega^e} \frac{1}{E^{n+1}} \frac{\partial \epsilon^{n+1}}{\partial E^{n+1}} \left(\vec{W}_i^{(1)(e)} \cdot \vec{E}^{n+1} \right) \left(\vec{W}_j^{(1)(e)} \cdot \vec{E}^{n+1} \right) d\Omega \qquad (3.76)$$

where in the case of an instantaneous nonlinearity as in (3.54), the derivative reduces to:

$$\frac{\partial \epsilon^{n+1}}{\partial E^{n+1}} = 2\epsilon_0 \chi^{(3)} E^{n+1}$$

$$(3.77)$$

and the global Jacobian is assembled from the elemental Jacobians in the same fashion as the $[S]$ and $[M]$ matrices. Importantly, it should be noted that the Jacobian expression in Eq. (3.76) results in a symmetric matrix and in the limiting case of a linear problem ($\chi^{(3)} = 0$) reduces to the left-hand side matrix of (3.25).

With this result, the solution may now be found in a straightforward, albeit computationally intensive, manner. In general, the Jacobian in (3.76) will change not only during each time step, but also each iteration of (3.65), as will the $[M]$ matrix in (3.63). They will therefore need to be locally recomputed within each nonlinear element and reassembled into their global counterparts multiple times within each time step. Depending on the amount of nonlinear materials present, this can represent a substantial computational burden.

The general solution procedure for an *instantaneous nonlinearity* is thus as follows:

1. Iterate (3.64) using (3.65) and (3.76), recomputing and assembling $[J]$ and $[M]$ each time, until $\{e\}^{n+1}$ converges to the desired tolerance.
2. Repeat the process until the desired end time.

3.4.1.3 Dispersive Nonlinearity

Having seen how to modify the FETD VWE formulation to accommodate instantaneous nonlinearities, the focus of the present section is to now further expand the generality of these techniques to include not only dispersive nonlinearities, but also all electric phenomena discussed so far.

To that end, recall from Section 3.2.3 that the more general form of the permittivity can be written as:

$$\vec{D} = \epsilon_L * \vec{E} + \epsilon_0 \chi^{(3)} \left(\alpha E^2 + (1 - \alpha)g(t) * E^2 \right) \vec{E} \tag{3.78}$$

and is capable of modeling linear dispersion, instantaneous nonlinearity, and dispersive nonlinearity. Note here that the term ϵ_L is related to $\chi^{(1)}(t)$.

Inserting this new permittivity expression into Ampère's law, deriving the VWE, and applying the finite-element method to the spatial part of the equation all proceeds in very similar fashion as it did in Section 3.4.1.2, resulting in the following semi-discrete system:

$$[\tilde{M}] \frac{d^2}{dt^2} \left(\epsilon_L * \{e\} \right) + \frac{d^2}{dt^2} \left([\hat{M}]\{e\} \right) + [S]\{e\} + \{f\} = 0 \tag{3.79}$$

where $[\tilde{M}]$ is the same matrix seen earlier in (3.20), and $[\hat{M}]$ contains the nonlinear contribution defined by:

$$[\hat{M}^{(e)}]_{ij} = \int_{\Omega^e} \epsilon_0 \chi^{(3)} \left(\alpha E^2 + (1 - \alpha)g(t) * E^2 \right) \vec{W}_i^{(1)(e)} \cdot \vec{W}_j^{(1)(e)} \, d\Omega \tag{3.80}$$

The first term of Eq. (3.79) is identical to that seen previously for linear dispersion. Likewise, the second term is clearly of the same form as that studied for instantaneous nonlinearities in Section 3.4.1.2. The required update equations are thus readily derived by combining both methods with an application of Newmark-β. In fact, substituting the linear convolution update equation from (3.49), $\{\mathcal{L}_e\} = a_0[\tilde{M}] + \{W_1\}$, as well as observing the time-dependence of the $[\hat{M}]$ matrix, yields the full set of update equations:

$$\left([K]^{n+1} + \frac{\Delta t^2}{4}[S] \right) \{e\}^{n+1} = 2 \left([K]^n - \frac{\Delta t^2}{4}[S] \right) \{e\}^n$$
$$- \left([K]^{n-1} + \frac{\Delta t^2}{4}[S] \right) \{e\}^{n-1}$$
$$- \left(\{W_1\}^n - 2\{W_1\}^{n-1} + \{W_1\}^{n-2} \right)$$
$$- \frac{\Delta t^2}{4} \left(\{f\}^{n+1} + 2\{f\}^n + \{f\}^{n-1} \right) \tag{3.81}$$

where the $[K]$ matrix is defined as:

$$[K] = a_0[\tilde{M}] + [\hat{M}] \tag{3.82}$$

or equivalently:

$$[K^{(e)}]_{ij} = \int_{\Omega^e} \left(a_0 + \epsilon_0 \chi^{(3)} E^2 + (1 - \alpha)\epsilon_0 \chi^{(3)} g(t) * E^2 \right) \vec{W}_i^{(1)(e)} \cdot \vec{W}_j^{(1)(e)} \, d\Omega$$

(3.83)

Of course, the effects of linear magnetic dispersion could also be incorporated here by the addition of magnetic auxiliary variables $\{G\}$, as was done previously in Section 3.4.1.1.

Equation (3.81) bears a conspicuous similarity to that derived for instantaneous nonlinearities in the previous section. However, there is now a key difference that has yet to be addressed: the convolution *within* the $[K^{(e)}]$ matrices. Returning to the definition of $[K^{(e)}]$ in (3.83) this means an update equation for the nonlinear convolution term is now required:

$$B(t) \triangleq g(t) * E^2(t)$$

(3.84)

Fortunately, following the procedure originally outlined in [59], the same bilinear transform theory implemented for the treatment of linear dispersion can also be applied in this case without too much additional complications. Indeed, since the nonlinear convolution kernel $g(t)$ is often likewise the result of a constant-coefficient differential equation model, its Laplace transform can also commonly be expressed as a rational function in frequency space:

$$g(s) = \frac{r_p s^p + \cdots + r_0}{w_p s^p + \cdots + w_0}$$

(3.85)

such that, in the Laplace domain, the convolution becomes:

$$B(s) = \frac{r_p s^p + \cdots + r_0}{w_p s^p + \cdots + w_0} E^2(s)$$

(3.86)

Once again applying the bilinear transform method to the above Laplace transform results in a set of auxiliary variables and update equations:

$$G_\alpha^n = h_\alpha (E^2)^n - q_\alpha B^n + G_{\alpha+1}^{n-1} \qquad \alpha < p$$

(3.87)

$$G_\alpha^n = h_\alpha (E^2)^n - q_\alpha B^n \qquad \alpha = p$$

(3.88)

$$B = h_0 (E^2)^n + Q_1^{n-1}$$

(3.89)

where Q_α are the associated new auxiliary variables.

Making the substitution of these new update equations and auxiliary variables into the definition of the $[K^{(e)}]$ matrix yields its final required form:

$$[K^{(e)}]_{ij}^{n+1} = \int_{\Omega^e} \left(a_0 + \epsilon_0 \alpha \chi^{(3)} (E^2)^{n+1} + (1-\alpha)\epsilon_0 \chi^{(3)} \left(h_0 (E^2)^{n+1} + Q_1^n \right) \right)$$

$$\times \vec{W}_i^{(1)(e)} \cdot \vec{W}_j^{(1)(e)} \, d\Omega \tag{3.90}$$

With this result, the only requirement remaining in order to be able to solve the nonlinear update equation in (3.81) is to once again derive the Jacobian matrix. Luckily, due to the similarity between the instantaneous and dispersive update equations in (3.64) and (3.81), the derivation of the Jacobian proceeds in much the same way and results in effectively the same expression:

$$[J^{(e)}]_{ij}^{n+1} = [K^{(e)}]_{ij}^{n+1} + \frac{\Delta t^2}{4}[S^{(e)}]_{ij}$$

$$+ \int_{\Omega^e} \frac{1}{E^{n+1}} \frac{\partial \epsilon^{n+1}}{\partial E^{n+1}} \left(\vec{W}_i^{(1)(e)} \cdot \vec{E}^{n+1} \right) \left(\vec{W}_j^{(1)(e)} \cdot \vec{E}^{n+1} \right) d\Omega \tag{3.91}$$

with the exception that the derivative of the permittivity is now given by:

$$\frac{\partial \epsilon^{n+1}}{\partial E^{n+1}} = 2\epsilon_0 \chi^{(3)} E^{n+1} \left(\alpha + (1-\alpha)h_0 \right) \tag{3.92}$$

The general solution procedure for a medium exhibiting linear dispersion, instantaneous nonlinearity, and dispersive nonlinearity is now as follows:

1. Solve Eq. (3.81) iteratively using (3.65) and (3.91), recomputing and assembling the $[J]$ and $[K]$ matrices each time, until $\{e\}^{n+1}$ converges to the desired tolerance.
2. Update each of the linear auxiliary variables and the linear convolution, using (3.51), to time $n+1$.
3. Update the nonlinear auxiliary variables and the nonlinear convolution, using (3.87)–(3.89), to time $n+1$.
4. Repeat the process until the desired end time.

3.4.1.4 Numerical Studies

Two numerical examples are now provided to validate and demonstrate the formulations discussed so far.

Reflection and Transmission Coefficients of a Dielectric Slab

The first example involves calculation of reflection (Γ) and transmission (T) coefficients of a 5-cm wide dispersive dielectric slab for a normally incident plane wave. Figure 3.1 shows the problem in which the slab is confined inside a 3-D parallel-plate waveguide excited with a TEM mode. The perfect electric and magnetic conductor boundary conditions are imposed on the appropriate walls of the waveguide to support the TEM mode. A first-order absorbing boundary condition

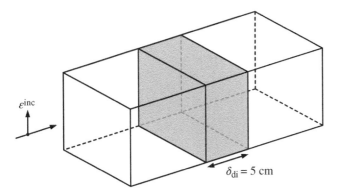

Figure 3.1 Plane wave normally incident on an infinitely large dispersive dielectric slab.

is utilized to terminate the waveguide end and the time step is set to $\Delta t = 0.34$ ps. The slab is doubly dispersive with both the permittivity and permeability being defined by two pairs of Lorentz poles. The permittivity is thus characterized by

$$\epsilon(j\omega) = \epsilon_\infty + G_{e_1} \frac{(\epsilon_s - \epsilon_\infty)\omega_{e_1}^2}{-\omega^2 + 2j\omega\delta_{e_1} + \omega_{e_1}^2} + G_{e_2} \frac{(\epsilon_s - \epsilon_\infty)\omega_{e_2}^2}{-\omega^2 + 2j\omega\delta_{e_2} + \omega_{e_2}^2} \tag{3.93}$$

in which $G_{e_1} = 0.2$, $G_{e_2} = 0.4$, $\delta_{e_1} = 0.025\omega_{e_1}$, $\delta_{e_2} = 0.01\omega_{e_2}$, $\epsilon_s = 5.2\epsilon_0$, $\epsilon_\infty = 3.1\epsilon_0$, $\omega_{e_1} = 3.1\pi \times 10^9$, $\omega_{e_2} = 2.2\pi \times 10^9$. The permeability is similarly characterized by the same model but with: $G_{m_1} = 0.9$, $G_{m_2} = 0.5$, $\delta_{m_1} = 0.03\omega_{m_1}$, $\delta_{m_2} = 0.015\omega_{m_2}$, $\mu_s = 3.7\mu_0$, $\mu_\infty = 1.8\mu_0$, $\omega_{m_1} = 3.3\pi \times 10^9$, $\omega_{m_2} = 4.2\pi \times 10^9$.

The results are plotted within the frequency band 30 MHz to 5 GHz in Figure 3.2. Excellent agreement between the numerical results and the exact analytic solutions can be observed. It is worth mentioning that the simulation was continued up to 8 million time-steps, with no sign of instability being observed.

The Temporal Soliton

To validate and demonstrate the effectiveness of the nonlinear dispersive formulation, the physically significant phenomena of a *Temporal Soliton* will be simulated in this section, in two spatial dimensions. Contrary to propagation in bulk media, a propagating guided wave (such as those found in fiber-optic cables) do not diffract, maintaining in general a fixed transverse profile as it travels. However, depending on the operating frequency and the materials from which the guide is made, the signal can still become distorted. Indeed, the most common source of such distortion is linear material or *chromatic* dispersion. As mentioned in Section 3.2, materials exhibiting linear dispersion in essence have frequency-dependent permittivities, meaning that each spectral component of a pulse or signal will propagate at a slightly different speed, a phenomenon often characterized in terms of the *group velocity dispersion* (GVD). Due to these differences in propagation

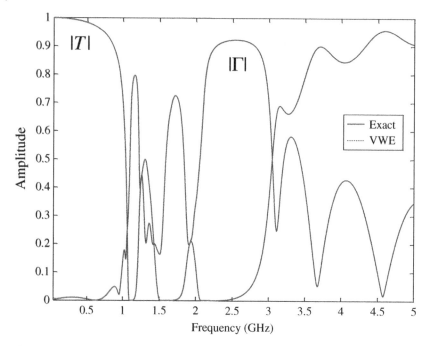

Figure 3.2 Numerical and exact values of the reflection and transmission coefficient amplitudes of a doubly dispersive dielectric slab.

speed, an initially well-formed pulse will gradually broaden, becoming increasingly distorted as its constituent spectral components separate during propagation. Generally speaking such pulse broadening is an unwanted and problematic phenomena, as any signal sent down the guide may not be intelligible upon reception at the other end. Dispersion is thus one main complication limiting bandwidth and transmission rates in optical fibers [60].

One possible way to mitigate the effects of GVD, however, is by leveraging nonlinearity within the optical fiber. Since within a nonlinear medium the permittivity is a function of field strength, different parts of the signal will experience varying propagation velocities according to the wave intensity at that point. In consequence, the resulting self-phase modulation leads to a change in frequency or "chirp" over the length of the pulse, with the leading edge decreasing in frequency and the trailing edge increasing in frequency [36].

In this manner, if a medium exhibits *anomalous* linear dispersion (that is, dispersion in which the high frequency components travel faster than the low frequency ones) as well as precisely the right amount of nonlinearity (tuned via the pulse shape and intensity) the two phenomena can be made to effectively cancel each other out. The resulting pulse thus propagates through the material without significant distortion or broadening and is known as a *temporal soliton*.

To demonstrate this, a problem domain was selected to recreate this phenomenon and simulated using the VWE FETD framework presented in this section (note that the mixed formulation, presented in Section 3.4.2, could also have been used). A dielectric slab waveguide was chosen to mimic the operation of an optical fiber in 2D. The resulting domain was rectangular in shape, measuring 10 μm wide by 100 μm long, and was composed of three dielectric layers. The two exterior cladding regions were selected to be free space, while the middle dielectric measured 2 μm in width and was made to exhibit the full range of electric phenomena discussed so far, including linear dispersion, instantaneous nonlinearity, and dispersive nonlinearity. More specifically, the linear dispersion was selected to be Lorentz, with $\omega_e = 1.885 \times 10^{14}$ (rad/s), $\delta_e = 2 \times 10^{11}$ (rad/s), $\epsilon_s = 6.1$ (F/m), and $\epsilon_\infty = 4.7$ (F/m). Additionally, the parameter α of Eq. (3.16) was selected to be 0.8, meaning both instantaneous and dispersive nonlinearities were present, with $\chi^{(3)} = 1.1 \times 10^{-18}$ (m^2/V^2). In the dispersive case, the nonlinear response function $g(t)$ was also selected to be of the Lorentzian-type, with $\omega_e = 3.114 \times 10^{12}$ (rad/s), $\delta_e = 9.091 \times 10^{12}$ (rad/s), $\epsilon_\infty = 0$, and $\epsilon_s = 1$. Lastly, the incident pulse was excited on the leftmost boundary of the waveguide, in the fundamental transverse magnetic (TM) mode with the temporal profile of a modulated hyperbolic secant function with a full width at half maximum (FWHM) of approximately 52.7 fs and a fundamental frequency of 50 THz.

To illustrate the negative effects of linear dispersion and GVD on their own, an initial simulation with this setup was performed in which all of the nonlinearity was turned off. The results of this simulation (performed with $h = 0.2$ μm, $\Delta t = h/c$, and triangular first-order elements) where h is the average edge length in the mesh, are shown in Figure 3.3. This level of spatial discretization amounts to roughly 15 points per wavelength within the medium and should therefore result in negligible numerical versus chromatic dispersion. While the pulse is initially well localized and compact within the guide, as it propagates the effects of anomalous linear dispersion rapidly take hold, broadening it significantly over time. Indeed, the anomalous nature of the dispersion is partially visible in the pulse as higher frequency components have collected toward the front, leaving the lower frequency components to the rear, resulting in a negative chirp. On the other hand, Figure 3.4 reveals the effects of that same linear dispersion, but with the added influence of the above nonlinearities in effect. This time, the linear dispersion has been largely counterbalanced by the self-phase modulation of the nonlinearity, yielding a temporal soliton. As the pulse propagates, therefore, its initial shape and size remain roughly intact, with no significant alteration in frequency or chirp detected.

3.4.2 Mixed Formulation

Having derived dispersive and nonlinear formulations for the VWE formulation, attention is now turned toward adapting the mixed formulation to also support these types of materials. As will be shown, this is greatly facilitated by the wealth

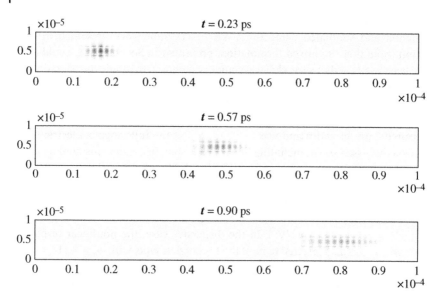

Figure 3.3 Demonstration of the effects of anomalous linear dispersion in a dielectric slab waveguide.

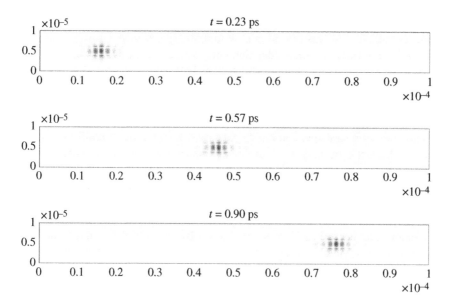

Figure 3.4 Creation of a temporal soliton via the introduction of dielectric nonlinearity.

of similarity between the VWE formulation discretized via Newmark-β and the mixed formulation discretized via Crank–Nicolson. Indeed, in some cases, the required modifications can be taken from the VWE formulations and directly inserted into the mixed formulation, with the new mixed update equations being written down with little extra effort.

3.4.2.1 Linear Dispersive Media

The constitutive equations (3.32) and (3.33) in the mixed formulation are very similar to the auxiliary variables defined in (3.45) and (3.46) for the VWE formulation. Thus, following a similar approach for the mixed formulation as was adopted in the VWE formulation, one can obtain [61]

$$\{d\}^n = c_0[\tilde{M}]\{e\}^n + \{\mathcal{W}_1\}^{n-1} \tag{3.94}$$

$$\{h\}^n = q_0[\tilde{M}_f]\{b\}^n + \{\mathcal{G}_1\}^{n-1} \tag{3.95}$$

in which the update equations for $\{\mathcal{W}_1\}$ and $\{\mathcal{G}_1\}$ are identical to those given in Section 3.4.1.1. Substituting Eqs. (3.94) and (3.95) into (3.30) and (3.31) and applying the Crank–Nicolson scheme yields

$$\left([M] + \frac{(\Delta t)^2}{4}[C]^T[M_f][C]\right)\{e\}^{n+1} = \left([M] - \frac{(\Delta t)^2}{4}[C]^T[M_f][C]\right)\{e\}^n$$
$$+ \Delta t[C]^T[M_f]\{b\}^n$$
$$+ \left(\{\mathcal{W}_1\}^n - \{\mathcal{W}_1\}^{n-1}\right)$$
$$+ \frac{\Delta t}{2}[C]^T\left(\{\mathcal{G}_1\}^n + \{\mathcal{G}_1\}^{n-1}\right)$$
$$+ \frac{\Delta t}{2}(\{f\}^n + \{f\}^{n-1}) \tag{3.96}$$

$$\{b\}^{n+1} = \{b\}^n - \frac{\Delta t}{2}[C]\left(\{e\}^{n+1} + \{e\}^n\right) \tag{3.97}$$

where $[M] = c_0[\tilde{M}]$ and $[M_f] = q_0[\tilde{M}_f]$. An alternative VWE-like formulation can also be obtained by eliminating $\{b\}$ in (3.96) with the aid of (3.97), as follows:

$$\left([M] + \frac{(\Delta t)^2}{4}[S]\right)\{e\}^{n+1} = 2\left([M] - \frac{(\Delta t)^2}{4}[S]\right)\{e\}^n$$
$$- \left([M] + \frac{(\Delta t)^2}{4}[S]\right)\{e\}^{n-1}$$
$$- \left(\{\mathcal{W}_1\}^n - 2\{\mathcal{W}_1\}^{n-1} + \{\mathcal{W}_1\}^{n-2}\right)$$
$$+ \frac{\Delta t}{2}[C]^T\left(\{\mathcal{G}_1\}^n - \{\mathcal{G}_1\}^{n-2}\right)$$
$$+ \frac{\Delta t}{2}(\{f\}^n - \{f\}^{n-2}) \tag{3.98}$$

$$\{b\}^{n+1} = \{b\}^n - \frac{\Delta t}{2}[C]\left(\{e\}^{n+1} + \{e\}^n\right) \tag{3.99}$$

where the equality $[S] = [C]^T[M_f][C]$ has been used. Unlike Eqs. (3.96) and (3.97), Eqs. (3.98) and (3.99) are only coupled through variable $\{G_1\}$. In other words, the formulation exactly reduces to the VWE formulation in a nondispersive or electrically dispersive media ($\{G_1\} = 0$) and (3.99) will be redundant. This formulation will be used in the first two examples in Section 3.4.2.4.

3.4.2.2 Instantaneous Nonlinearity

As seen previously, the introduction of nonlinearity requires that the permittivity remain inside the temporal derivative of Ampère's law. In the case of a mixed method, an application of the exact same finite-element discretization procedure outlined earlier for linear media produces the following semi-discrete system of equations for Faraday and Ampère's laws:

$$[C]\{e\} = -\frac{\partial\{b\}}{\partial t} \tag{3.100}$$

$$[C]^T[M_f]\{b\} = \frac{\partial}{\partial t}([M]\{e\}) + \{f\} \tag{3.101}$$

where it should be emphasized that $[M]$ now depends on the electric field via ϵ and is thus kept inside the temporal derivative as was the case in the VWE formulation.

As a result, once again due to the nonlinearity, the Crank–Nicolson temporal discretization must be applied to the entire matrix-vector product within Ampère's law, such that the temporal derivative in (3.101) is approximated as:

$$\frac{\partial}{\partial t}([M]\{e\}) \approx \frac{[M]^{n+1}\{e\}^{n+1} - [M]^n\{e\}^n}{\Delta t} \tag{3.102}$$

Keeping this in mind, the derivation of the required update equations for the $\vec{E}$ and $\vec{B}$ fields follows a similar procedure to that presented previously, yielding the following [62]:

$$\left([M]^{n+1} + \frac{\Delta t^2}{4}[C]^T[M_f][C]\right)\{e\}^{n+1} = \left([M]^n - \frac{\Delta t^2}{4}[C]^T[M_f][C]\right)\{e\}^n$$
$$+ \Delta t[C]^T[M_f]\{b\}^n$$
$$+ \frac{\Delta t}{2}\left(\{f\}^n + \{f\}^{n+1}\right) \tag{3.103}$$

$$\{b\}^{n+1} = \{b\}^n - \frac{\Delta t}{2}[C]\left(\{e\}^n + \{e\}^{n+1}\right) \tag{3.104}$$

Notice, however, that the left-hand side of Eq. (3.103) is exactly equal to the left-hand side of the VWE update equation in (3.64) when Whitney 1-forms are used, since, as noted earlier:

$$[S] = [C]^T[M_f][C] \tag{3.105}$$

The result is that solving the nonlinear system in Eq. (3.103) is functionally the same as solving that for the VWE formulation. Indeed, since the left-hand side matrix is the only contribution to the Jacobian, it follows that the expression for the mixed method's Jacobian is precisely the same as that obtained earlier for the VWE:

$$[J^{(e)}]_{ij}^{n+1} = [M^{(e)}]_{ij}^{n+1} + \frac{\Delta t^2}{4}[C^{(e)}]^T[M_f^{(e)}][C^{(e)}]_{ij}$$
$$+ \int_{\Omega^e} \frac{1}{E^{n+1}} \frac{\partial \epsilon^{n+1}}{\partial E^{n+1}} \left(\vec{W}_i^{(1)(e)} \cdot \vec{E}^{n+1} \right) \left(\vec{W}_j^{(1)(e)} \cdot \vec{E}^{n+1} \right) d\Omega \qquad (3.106)$$

where once again, for an instantaneous nonlinearity:

$$\frac{\partial \epsilon^{n+1}}{\partial E^{n+1}} = 2\epsilon_0 \chi^{(3)} E^{n+1} \qquad (3.107)$$

The solution procedure is thus the same as that for the VWE formulation, with the exception of also needing to update the magnetic flux density once per time step with Eq. (3.104).

3.4.2.3 Dispersive Nonlinearity
Lastly, for the most general case of linear dispersion, instantaneous nonlinearity, and dispersive nonlinearity, the procedure is, unsurprisingly, very similar to that derived previously for the VWE method. Indeed, applying the spatial discretization procedure detailed earlier produces the following semi-discrete system:

$$[C]\{e\} = -\frac{\partial\{b\}}{\partial t} \qquad (3.108)$$

$$[C]^T[M_f]\{b\} = [\tilde{M}]\frac{\partial}{\partial t}\left(\epsilon_L * \{e\}\right) + \frac{\partial}{\partial t}\left([\hat{M}]\{e\}\right) + \{f\} \qquad (3.109)$$

where $[\tilde{M}]$ and $[\hat{M}]$ have the same definitions as in the VWE formulation.

Applying the Crank–Nicolson method to the semi-discrete equations in (3.108) and (3.109), and borrowing the same implementation for the linear and nonlinear convolutions as used previously, results in the following update equation [62]:

$$\left([K]^{n+1} + \frac{\Delta t^2}{4}[C]^T[M_f][C]\right)\{e\}^{n+1} = \left([K]^n - \frac{\Delta t^2}{4}[C]^T[M_f][C]\right)\{e\}^n$$
$$+ \Delta t[C]^T[M_f]\{b\}^n$$
$$- \{\mathcal{W}_1\}^n + \{\mathcal{W}_1\}^{n-1}$$
$$- \frac{\Delta t}{2}\left(\{f\}^n + \{f\}^{n+1}\right) \qquad (3.110)$$

where the $[K]$ matrix is precisely the same as that seen previously in (3.90):

$$[K^{(e)}]_{ij}^{n+1} = \int_{\Omega^e} \left(a_0 + \epsilon_0\alpha\chi^{(3)}(E^2)^{n+1} + (1-\alpha)\epsilon_0\chi^{(3)}\left(h_0(E^2)^{n+1} + Q_1^n\right)\right)$$
$$\times \vec{W}_i^{(1)(e)} \cdot \vec{W}_j^{(1)(e)} d\Omega \qquad (3.111)$$

Once again it is noted that the left-hand side of Eq. (3.110) is precisely the same as that in the VWE formulation, meaning that the Jacobian is identical to that found previously in (3.91):

$$[J^{(e)}]_{ij}^{n+1} = [K^{(e)}]_{ij}^{n+1} + \frac{\Delta t^2}{4}[C^{(e)}]^T[M_f^{(e)}][C^{(e)}]_{ij}$$

$$+ \int_{\Omega^e} \frac{1}{E^{n+1}} \frac{\partial \epsilon^{n+1}}{\partial E^{n+1}} \left(\vec{W}_i^{(1)(e)} \cdot \vec{E}^{n+1} \right) \left(\vec{W}_j^{(1)(e)} \cdot \vec{E}^{n+1} \right) d\Omega \qquad (3.112)$$

where, as before, for a dispersive nonlinearity:

$$\frac{\partial \epsilon^{n+1}}{\partial E^{n+1}} = 2\epsilon_0 \chi^{(3)} E^{n+1} \left(\alpha + (1-\alpha)h_0 \right) \qquad (3.113)$$

Finally, once more the update procedure for the electric field mirrors that of the VWE method, with the exception that the magnetic flux density must be updated with (3.104) once per time step.

3.4.2.4 Numerical Studies

Several numerical examples will now be provided to validate, demonstrate, and study the performance of the mixed FETD formulations expounded in this chapter.

Reflection and Transmission Coefficients of a Dielectric Slab

This example is identical to the one studied in Section 3.4.1.4, but with the following parameters for the permittivity and permeability of the slab: $G_1 = 2$, $G_2 = 4$, $\delta_{e_1} = 0.1\omega_{e_1}$, $\delta_{e_2} = 0.1\omega_{e_2}$, $\epsilon_s = 6\epsilon_0$, $\epsilon_\infty = 4.3\epsilon_0$, $\omega_{e_1} = 1.2\pi \times 10^9$, $\omega_{e_2} = 3\pi \times 10^9$, $G_{m_1} = 0.2$, $G_{m_2} = 0.4$, $\delta_{m_1} = 2 \times 10^{-8}\omega_{m_1}$, $\delta_{m_2} = 2 \times 10^{-4}\omega_{m_2}$, $\mu_s = 2.5\mu_0$, $\mu_\infty = 1.4\mu_0$, $\omega_{m_1} = 2\pi \times 10^9$, $\omega_{m_2} = 4\pi \times 10^9$.

The reflection and transmission coefficients are depicted in Figure 3.5. Very good agreement can again be observed between the numerical and analytical results. The worst accuracy is obtained in the low-frequency region and near sharp resonances, however more accurate results can be obtained during a longer simulation. It is worth noting that once again the simulation was continued up to 8 million time-steps with the results remaining completely stable.

Perfectly Matched Layers

One of the most effective methods for truncating a computational domain in an open-region problem is the PML. Among the different variants of PML that have been devised over the years, the anisotropic-medium PML formulation [63] is well-suited for FETD implementation, as it can be formulated in terms of a dispersive and anisotropic medium. Indeed, within the PML medium, the modified permittivity and permeability can be expressed as:

$$\epsilon_{PML} = \epsilon \bar{\bar{\Lambda}}, \quad \mu_{PML} = \mu \bar{\bar{\Lambda}} \qquad (3.114)$$

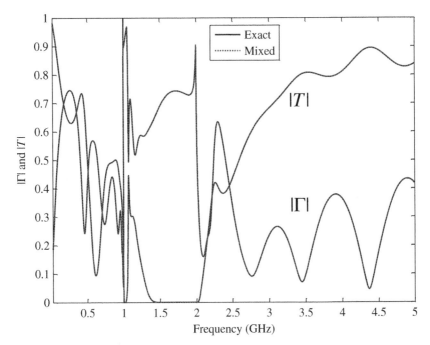

Figure 3.5 Numerical and exact values of the reflection and transmission coefficient amplitudes of a doubly dispersive dielectric slab.

where

$$\overline{\overline{\Lambda}} = \hat{x}\frac{\gamma_y\gamma_z}{\gamma_x}\hat{x} + \hat{y}\frac{\gamma_x\gamma_z}{\gamma_y}\hat{y} + \hat{z}\frac{\gamma_x\gamma_y}{\gamma_z}\hat{z} \tag{3.115}$$

and the parameter $\gamma_\zeta = 1 + \frac{\sigma_\zeta}{j\omega}$ specifies the attenuation along each direction $\zeta = \{x, y, z\}$, with σ_ζ representing the conductivity. σ_ζ is assumed to be already normalized to ε_0 and equal to zero outside the PML.

To demonstrate the effectiveness of this PML formulation when implemented with FETD, a monopole antenna is simulated in free space. As shown in Figure 3.6, the monopole antenna consists of a $h = 32.8$ mm long metallic cylinder with a radius of 1 mm placed on an infinite ground plane. The antenna is fed by an air-filled coaxial cable whose outer radius is 2.3 mm. The coaxial cable is excited by a TEM mode with a Gaussian waveform. A 1-cm thick PML covering a 20 mm × 20 mm × 40 mm box is enclosing the monopole antenna.

Figure 3.7 shows the reflected voltage wave recorded at the excitation port in the time-domain computed using both the VWE and mixed formulations. The horizontal axis represents time normalized to $\tau_a = h/c_0$ in which c_0 is the speed of light in vacuum. The FETD results match the measured results very well, which corroborates the validity and accuracy of both FETD formulations.

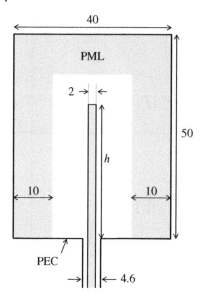

Figure 3.6 A coaxial-fed cylindrical monopole antenna placed on a ground plane (all dimensions are in mm).

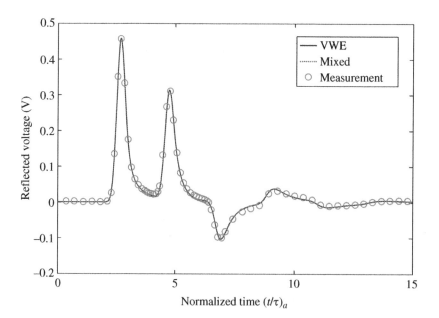

Figure 3.7 Reflected voltage at the terminal of the cylindrical monopole antenna.

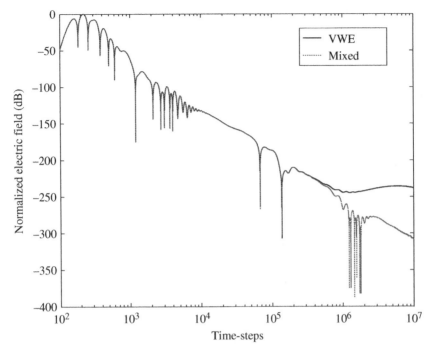

Figure 3.8 Normalized electric field recorded inside the PML over 10 million time steps.

Additionally, to study the long-time stability of the FETD formulations, the simulation was continued up to 10 million time-steps. The size of the problem was reduced and the mesh was coarsened with the goal of decreasing the total number of unknowns. Figure 3.8 shows the resulting normalized electric field recorded *inside* the PML region. Both methods give identical results at the beginning; however, after approximately 250,000 time-steps they begin to diverge. The mixed formulation behaves normally as to what is expected, however the VWE formulation shows unusual fluctuations.

For more numerical examples and a more detailed derivation of the FETD formulations for linear dispersive media, the reader is encouraged to consult reference [64].

The Spatial Soliton

One fundamental aspect of nonguided wave propagation in bulk linear media is *diffraction*, in which an initially confined beam gradually spreads out in space. In consequence, any initially focused beam will tend to rapidly widen and spread out as it propagates, becoming less focused and more diffuse [1]. In a nonlinear medium, however, the dependence of the permittivity or refractive index on the field strength can give rise to a lensing or self-focusing effect. Essentially, the transverse profile of a beam's intensity can yield a gradient in the material's refractive index such that it mimics a convex lens. The result is that if a beam of the correct profile and intensity is emitted into a bulk nonlinear medium, the focusing effect of the nonlinearity can be made to exactly counterbalance the diffraction, yielding the stable propagation of a confined beam over large distances without a guiding structure, known as a spatial soliton [36].

To demonstrate the occurrence of this phenomenon, the mixed formulation was used to simulate a rectangular slab of bulk dielectric media in two dimensions measuring 0.3 m wide by 1 m long (note that the VWE formulation could equally have been used). An initially confined beam with a hyperbolic secant transverse profile is injected on the left side of the domain with a FWHM of approximately 1.46 cm.

Figure 3.9 depicts the result of the simulation for a purely linear medium in which $\epsilon = 4.2\epsilon_0$ (or equivalently for which $\chi^{(1)} = 3.2$), and $\chi^{(2)} = \chi^{(3)} = 0$. The initial bright localized beam can easily be seen on the leftmost boundary, however

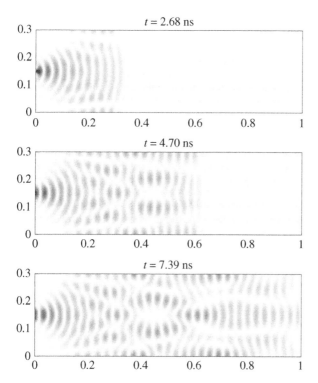

Figure 3.9 Diffuse interference pattern resulting from diffraction in a bulk linear medium.

Figure 3.10 Creation of a focused spatial soliton due to the presence of an instantaneous nonlinearity.

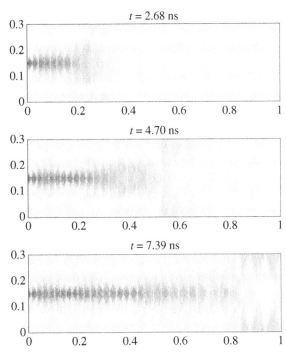

with the lack of nonlinearity the beam rapidly diffracts and spreads out, creating a diffuse interference pattern as the wave rebounds off the PEC walls.

In contrast, Figure 3.10 shows the results of the exact same simulation except where the linear medium has been replaced with one exhibiting an instantaneous nonlinearity of the form presented in Eq. (3.54) with $\chi^{(1)} = 3.2$, and $\chi^{(3)} = 1.5 \times 10^{-19}$ (m^2/V^2). This time the focusing effects of the nonlinearity are abundantly clear, as the beam remains roughly confined to its original focused shape during propagation. Note that in this particular simulation, since the beam intensity is ramped up from zero, an initial nonfocused wavefront occurs until such a time as the beam has reached sufficient strength for the nonlinearity to effectively mitigate diffraction.

Supercontinuum Generation

In this section, a final nonlinear example of a real-world application of the simulation tools developed in this chapter will be demonstrated. More specifically, the phenomenon of *supercontinuum generation* will be exhibited, in which laser light undergoes extreme spectral broadening under the influences of dispersion and nonlinearity, resulting in a very wide continuous optical spectrum. This phenomenon is significant not only for its spectacular visually observable effects, but also because it represents an important source of high power density ultra-broadband radiation [65].

The underlying physical processes behind supercontinuum generation in nonlinear optical fibers can be quite varied and depends on different parameters such as chromatic dispersion, laser pulse intensity, width, and duration, as well as the wavelength or frequency at which the system is operated. However, one common regime of operation in which supercontinuum generation is known to occur is when working with ultrafast femtosecond pulses excited in the vicinity of a waveguide's zero dispersion wavelength (ZDW), the point where the overall fiber transitions from being anomalously dispersive to normally dispersive (or vice versa).

As detailed extensively in [66], when femtosecond pulses are created or "pumped" at a wavelength or frequency in the anomalous dispersion regime, but close to the ZDW, several mechanisms come into effect. Initially, the high intensity of the beam causes the creation of a high-order temporal soliton, similar in nature to that described in Section 3.4.1.4 but with a more complex temporal evolution. After a short distance, however, perturbations cause this high-order soliton to be unstable, resulting in a process called soliton fission. By fissioning, the high-order soliton ejects energy into many smaller distinct low-order or fundamental soliton components. However, due to the proximity of the ZDW, some of this energy is shed into the normally dispersive region, creating dispersive waves. The energy that remains in the anomalous dispersion region, meanwhile, can itself continue to shift to longer wavelengths due to nonlinear soliton phenomena such as self-frequency shift and cross-phase modulation, among others. The result is that, due to the interactions of nonlinearity, anomalous dispersion, and normal dispersion, an initially spectrally confined pulse experiences dramatic spectral broadening, with most of the initial energy now spread out over a wide range of frequencies.

To demonstrate the occurrence of this phenomenon, a numerical simulation was devised to which the mixed FETD method could be applied (again, note that the VWE formulation could equally have been used). This time, a parallel plate waveguide was created and filled with a material exhibiting both linear dispersion and an instantaneous nonlinearity ($\alpha = 1$). The waveguide measured 1 μm in width, 1 mm in length, and was equipped with PEC boundaries. The pulse itself was excited on the leftmost boundary and had a constant transverse component with a modulated hyperbolic secant envelope in time with an approximate FWHM of 74.8 fs. The linear dispersive and instantaneous nonlinear parameters were selected to be similar to those of fused silica, though tweaked in order that supercontinuum generation could be observed over a shorter distance and thus require less computational resources.

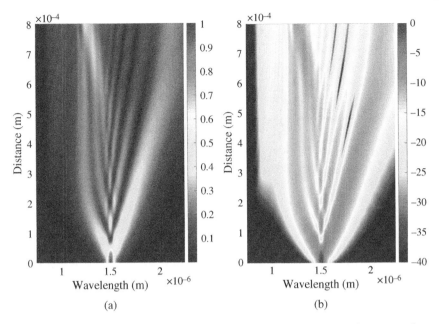

Figure 3.11 Linear (a) and logarithmic (b) visual depiction of the spectral contents of the pulse as it propagates down the guide. The intensity has been normalized to the highest value.

Figure 3.11 depicts a visual representation of the evolution of the pulse's spectral contents during propagation, as measured in the center of the guide at $y = 0.5$ μm. The horizontal axis represents the equivalent free-space wavelength of each frequency, the vertical axis the propagation distance within the guide, and the color the spectral intensity at each point. While most of the spectral energy is initially clustered around the pulse's fundamental frequency of 200 THz (1.5 μm), the pulse rapidly fissions and decomposes, spreading its energy out into a broad range of adjacent wavelengths/frequencies. This is especially evident in Figure 3.12 where the spectral contents of the pulse have been individually plotted for three separate locations in the guide. Again, it can be clearly seen that while the initial pulse is nicely localized around the 200 THz (1.5 μm) fundamental frequency (wavelength), after propagating through the guide the energy has uniformly spread out over a much wider range. The resulting signal has roughly equal spectral power over a 1.4-μm range, whereas the original pulse's energy was contained within a 3-dB bandwidth of only roughly 50 nm.

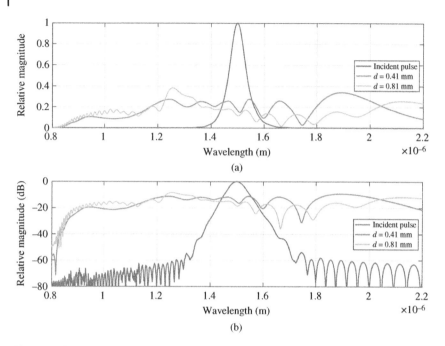

Figure 3.12 Linear (top) and logarithmic (bottom) plots of the pulse's normalized spectral composition after traveling 0.41 mm and 0.81 mm in the guide.

3.4.3 Implementation Issues

3.4.3.1 Newton–Raphson Iteration

A key part of the nonlinear algorithms presented in this chapter was the iterative solution of a nonlinear system of equations via the Newton–Raphson method:

$$\{F\}(\{x\}) = 0 \tag{3.116}$$

$$\{x\}_{(k+1)} = \{x\}_{(k)} - [J]_{(k)}^{-1}\{F\}_{(k)} \tag{3.117}$$

One of the major attractive benefits of the Newton–Raphson method is its fast convergence. More specifically, it can be shown that the Newton–Raphson method exhibits *quadratic convergence*, which is to say:

$$\left|\{x\}_{(k+1)} - \{\hat{x}\}\right| \leq C \left|\{x\}_{(k)} - \{\hat{x}\}\right|^2 \tag{3.118}$$

where $\{x\}_{(k)}$ is the solution after k iterations of the algorithm, $\{\hat{x}\}$ is the exact solution, C is some positive constant, and $|\cdot|$ denotes the Euclidean norm [67].

Of course, the quadratic convergence of Eq. (3.118) is only valid if the solution actually converges. Unfortunately, this is not always guaranteed, and generally requires a starting value $\{x\}_{(0)}$ sufficiently close to the true solution $\{\hat{x}\}$ to ensure convergence. Moreover, should the nonlinear system being solved have multiple possible solutions, the one to which the method converges can exhibit extreme sensitivity to the starting guess, leading to so-called basins of attraction [68]. As a result of these considerations it is imperative that an appropriate starting guess be selected not only to ensure convergence to the correct solution, but also convergence as quickly and efficiently as possible.

Luckily, since the FETD algorithms derived previously all involve time-stepping, a natural choice for the initial guess is available. To illustrate, consider a time-dependent function $u(t)$ whose values at time n and $n + 1$ may be related via a Taylor series expansion:

$$u^{n+1} = u^n + \Delta t \left(\frac{\partial u}{\partial t} \right)^n + \frac{\Delta t^2}{2} \left(\frac{\partial^2 u}{\partial t^2} \right)^n + \cdots \tag{3.119}$$

Rearranging these terms it is thus fairly easy to conclude that

$$\left| u^{n+1} - u^n \right| \approx \mathcal{O}(\Delta t) \tag{3.120}$$

or, in other words, for sufficiently small time steps Δt, the solution at time $n + 1$ will be relatively close to that at time n. This naturally suggests using the previous time step's solution as the initial guess for the current time step [26]:

$$\{e\}_{(0)}^{n+1} = \{e\}^n \tag{3.121}$$

Such a choice should not only result in the convergence of the Newton–Raphson algorithm, but also convergence to the correct solution in an efficient manner.

Despite the availability of a good initial guess, it should be noted however that this is not in and of itself a guarantee of convergence to the correct solution. Owing to this, when performing the iterations it is important to not only monitor the relative change in the solution as a stopping criteria, but also the residual of the original nonlinear system $\{F\}$. Doing so may thus help permit the detection of convergence to an incorrect solution, such as convergence to a local minimum rather than a global one.

3.4.3.2 Evaluation of Elemental Matrices

For a *linear* FETD method, the $[M]$ matrix does not change with time, as the permittivity is a simple scalar constant:

$$[M^{(e)}]_{ij} = \int_{\Omega^e} \epsilon \, \vec{W}_i^{(1)(e)} \cdot \vec{W}_j^{(1)(e)} \, d\Omega \tag{3.122}$$

Since the permittivity in the above equation is constant, the only spatially varying part of the integrand in (3.122) are the 1-form basis functions. At first it may seem a complex task to integrate these basis functions over a triangular or tetrahedral element, however, due to the properties of these basis functions the result is often surprisingly obtainable in exact closed form [45].

Unfortunately, in the nonlinear case the dependence of the permittivity on the electric field, and thus indirectly on the spatial coordinates, means that such closed form expressions are in general no longer obtainable. As a result, the integral in Eq. (3.122) must be approximated numerically.

The easiest and most common way to accomplish this is to use Gaussian Quadrature, in which integrals over the triangle, for example, may be approximated as [45]:

$$\iint_{\Omega^e} f(n_1, n_2, n_3) \, d\Omega \approx \sum_i w_i f(n_{1i}, n_{2i}, n_{3i}) \tag{3.123}$$

where (n_1, n_2, n_3) are *simplex coordinates* within the triangle, (n_{1i}, n_{2i}, n_{3i}) are the corresponding sampling points, and w_i their respective weights. The number of sampling points and weights used depends on the desired level of accuracy, with tabulated values of abscissae and weights available to quite high order [69]. The numerical integration scheme provided by Eq. (3.123) has the benefit of being straightforward to implement, and can achieve, in principle, any desired level of accuracy.

3.4.3.3 Nonlinear Auxiliary Variable Updating

Throughout Sections 3.4.1 and 3.4.2, the inclusion of dispersion (either linear or nonlinear) into the numerical methods generally involved an application of the z-transform that resulted in a series of auxiliary variables and update equations.

While the update procedure for the convolution variables in the linear case is relatively straightforward, a subtlety arises in the update equations for the nonlinear case. Recall that the nonlinear convolution:

$$B = g(t) * E^2(t) \tag{3.124}$$

is approximated via the z-transform method as:

$$B^n = h_0 (E^2)^n + Q_1^{n-1} \tag{3.125}$$

for which the $[K]$ matrix becomes:

$$[K^{(e)}]_{ij}^{n+1} = \int_{\Omega^e} \left(a_0 + \epsilon_0 \chi^{(3)} (E^2)^{n+1} + (1 - \alpha)\epsilon_0 \chi^{(3)} (h_0 (E^2)^{n+1} + Q_1^n) \right)$$
$$\times \vec{W}_i^{(1)(e)} \cdot \vec{W}_j^{(1)(e)} \, d\Omega \tag{3.126}$$

The complicating factor within Eq. (3.126), which differentiates it from the linear case, is that here the auxiliary variable Q_1 is found *within* the elemental integral. The main consequence of this is that, when performing the Gaussian quadrature evaluation of the integral in (3.123), the value of Q_1 will be required at each quadrature or sample point within each element, for as many past time steps are required for the medium's dispersive order. Hence, at every time step, each of the auxiliary variables at each quadrature point must be advanced in time.

3.5 Stability Analysis

When formulating any kind of numerical method or algorithm an important consideration is that of stability. Indeed, in Sections 3.4.1 and 3.4.2 the notion of numerical stability was mentioned several times to distinguish the methods derived in this chapter from existing methods. As a result, in the following sections, the general idea behind numerical stability will be discussed, followed by a stability analysis of the FETD formulations for linear and nonlinear dispersive media which have been presented in this chapter.

3.5.1 Numerical Stability

Depending on context, the notion of stability can have different meanings. However, roughly speaking, a numerical method being *stable* results in the numerical solution to a homogeneous (source-free) problem being bounded by some limit for all time steps. For example, the most common choice usually defines stability as [70]:

$$\left|\{u\}^{n+1}\right| \leq C\left|\{u\}^0\right| \tag{3.127}$$

where $\{u\}$ is the numerical solution, C is some non-negative constant, and $|\cdot|$ is the magnitude or Euclidean norm. In essence, this definition of stability forbids the numerical solution from exhibiting exponential growth, and for linear problems equates to the numerical problem being *well-posed*. To demonstrate this, suppose some numerical algorithm obtains $\{u\}^{n+1}$ from $\{u\}^n$ via the application of some operator $A(\cdot)$, such that:

$$\begin{aligned} \{u\}^{n+1} &= A(\{u\}^n) \\ &= A^n(\{u\}^0) \end{aligned} \tag{3.128}$$

For the method represented by $A(\cdot)$ to be termed well-posed, small perturbations in the initial condition $\{u\}^0$ should result in a comparably small change in the solution at time $\{u\}^{n+1}$ [71]:

$$\left|\{\tilde{u}\}^{n+1} - \{u\}^{n+1}\right| \leq C\left|v^0\right| \tag{3.129}$$

where $\{\tilde{u}\}^{n+1}$ is the solution obtained from the perturbed initial condition

$$\{\tilde{u}\}^0 = \{u\}^0 + \{v\}^0 \tag{3.130}$$

This definition is the same as that used for well-posedness in partial differential equations.

If the operator $A(\cdot)$ is *linear*, the criteria in (3.129) can also be expressed as:

$$\left| \{\tilde{u}\}^{n+1} - \{u\}^n \right| \leq C \left| \{v\}^0 \right| \tag{3.131}$$

$$\left| A^n(\{\tilde{u}\}^0) - A^n(\{u\}^0) \right| \leq C \left| \{v\}^0 \right| \tag{3.132}$$

$$\left| A^n(\{u\}^0) + A^n(\{v\}^0) - A^n(\{u\}^0) \right| \leq C \left| \{v\}^0 \right| \tag{3.133}$$

$$\left| A^n(\{v\}^0) \right| \leq C \left| \{v\}^0 \right| \tag{3.134}$$

$$\left| \{v\}^{n+1} \right| \leq C \left| \{v\}^0 \right| \tag{3.135}$$

which is precisely the original stability criteria stipulated in (3.127). Hence, in the case of a *linear* method, the stability criteria in (3.127) directly implies that condition (3.129) is satisfied.

Such a consistent definition of stability is important, as perturbations to the original differential equation, either due to discretization or finite floating point machine precision, could otherwise result in solutions which deviate substantially from the true solutions and non-physically "explode" or "blow-up," meaning they become unbounded. Having a numerical method that is stable at a given operating point thereby guarantees that such perturbations do not grow, but rather remain bounded. In fact, for a consistent linear finite-difference scheme, due to the Lax equivalence theorem, stability is a sufficient condition to guarantee the solution converges, meaning it approaches the true solution as the grid is refined [72].

As alluded to throughout this chapter, the stability of a given numerical algorithm can generally be classified as either *conditionally* stable or *unconditionally* stable. As the name implies, conditional stability dictates that there is a certain set of constraints or criteria that must be satisfied for the method to be stable. For finite-element analysis, these constraints are generally related to the temporal discretization parameter Δt, as well as the global matrices (themselves dependent on the spatial discretization, i.e. on the size of the elements in the mesh). For instance, when discretized via the central difference method, the linear VWE FETD method is conditionally stable with stability criterion given by [45]:

$$\Delta t \leq \frac{2}{\sqrt{\rho([M]^{-1}[S])}} \tag{3.136}$$

where $\rho(\cdot)$ denotes the *spectral radius* of a matrix, which is its largest eigenvalue. In contrast, an unconditionally stable method is one in which there is no restriction on the temporal step size Δt in order to ensure the stability criteria of Eq. (3.127) is satisfied. For example, when the Newmark-β method is applied to linear FETD problems, it is unconditionally stable for $\beta \geq 1/4$ [45, 47].

Determining whether a given numerical method is stable thus amounts to adopting various strategies to show that the properties of Eqs. (3.127) and (3.129) are satisfied. To that effect, many different techniques have been devised to demonstrate these properties, including energy methods and methods based on discrete system analysis.

3.5.2 Linear Dispersive Media

In this section, a stability analysis of the FETD formulations for linear dispersive media will be presented.

A single update step of FETD can be rewritten as $\{u\}^{n+1} = [A]\{u\}^n$ in which $[A]$ is known as the amplification matrix. In order to have a stable formulation it can be shown that the following two conditions must be satisfied [21]:

1. All eigenvalues of the amplification matrix have to reside on or within the unit circle (in the z-plane). This is a necessary condition for stability, but not sufficient.
2. If there exist some eigenvalues on the unit circle, they have to be either simple or nondefective, meaning their algebraic multiplicity does not exceed their geometric multiplicity. If the amplification matrix is *diagonalizable*, this condition is satisfied.

The first condition (von Neumann criterion) is demonstrated by investigating roots of the characteristic polynomial of $[A]$ using the Routh–Hurwitz criterion [20]. The source term is removed because it has no effect on the stability. Moreover, we restrict ourselves to a homogeneous dispersive medium for the sake of simplicity.

In a source-free and homogeneous dispersive media, the semi-discrete VWE becomes:

$$[S]\mu^{-1}(t) * \{e(t)\} + [M]\epsilon(t) * \frac{\partial^2\{e(t)\}}{\partial t^2} = 0 \tag{3.137}$$

Applying the Newmark-β scheme and taking the z-transform of the resulting update equation gives:

$$(\beta z^2 + (1 - 2\beta)z + \beta)\mu^{-1}(z)[S]\{e(z)\} + \frac{z^2 - 2z + 1}{(\Delta t)^2}\epsilon(z)[M]\{e(z)\} = 0 \tag{3.138}$$

Since matrix $[M]$ is symmetric and positive-definite, all terms can be multiplied by $[M]^{-1}$. The resulting equation can then be considered as an eigenvalue problem [73], in which the amplification matrix's eigenvalues must satisfy the following equation:

$$(\beta z^2 + (1 - 2\beta)z + \beta)\mu^{-1}(z)\Delta t^2 \lambda_i\{e(z)\} + (z^2 - 2z + 1)\epsilon(z)\{e(z)\} = 0 \tag{3.139}$$

where $\lambda_i = \text{eig}([M]^{-1}[S])$ denotes the eigenvalues of the matrix $[M]^{-1}[S]$. Since $[S]$ is symmetric and positive semi-definite, it must follow that $\lambda_i \geq 0$. A stability analysis of the FETD formulations for different types of dispersive media can then be performed by substituting the corresponding permittivity and permeability (transformed to the z-domain by the BT) in (3.139) and verifying whether the zeros of the resultant polynomial lie on or within the unit circle. In order to facilitate this analysis, the eigenvalue polynomials can be mapped to the s-plane using the transformation $z = (s + 1)/(s - 1)$ [20], which yields a general s-dependent polynomial of the form:

$$a_n s^n + a_{n-1} s^{n-1} + \cdots + a_1 s + a_0 = 0 \tag{3.140}$$

This changes the stability criteria from being inside the unit circle to being in the left half plane. The coefficients of the polynomials for $\beta = 1/4$ and $\beta = 0$ are tabulated in Tables 3.1 and 3.2, respectively, for various material models. These models were described previously and are repeated here for convenience:

1. **Lossless medium**: Constant ϵ and μ.
2. **Lossy medium**:

$$\epsilon(\omega) = \epsilon_\infty + \frac{\sigma_e}{j\omega} \tag{3.141}$$

$$\mu(\omega) = \mu_\infty + \frac{\sigma_m}{j\omega} \tag{3.142}$$

Table 3.1 Coefficients of the polynomials for the VWE FETD with $\beta = 1/4$.

Medium	Coefficients
Lossless (ϵ, μ)	$a_2 = \lambda_i \Delta t^2$, $a_1 = 0$, $a_0 = 4\epsilon\mu$
Lossy $(\epsilon(\omega), \mu(\omega))$	$a_2 = \Delta t^2(\lambda_i + \sigma_e\sigma_m)$, $a_1 = 2\Delta t(\epsilon_\infty\sigma_m + \mu_\infty\sigma_e)$ $a_0 = 4\epsilon_\infty\mu_\infty$
Debye $(\epsilon(\omega), \mu(\omega))$	$a_4 = \lambda_i \Delta t^4$, $a_3 = 2\lambda_i \Delta t^3(\tau_e + \tau_m)$ $a_2 = 4\Delta t^2(\epsilon_s\mu_s + \lambda_i\tau_e\tau_m)$, $a_1 = 8\Delta t(\epsilon_\infty\mu_s\tau_e + \epsilon_s\mu_\infty\tau_m)$ $a_0 = 16\epsilon_\infty\mu_\infty\tau_e\tau_m$
Lorentz $(\epsilon(\omega), \mu)$	$a_4 = \lambda_i \Delta t^4\omega_e^2$, $a_3 = 4\lambda_i \Delta t^3\delta_e$, $a_2 = 4\Delta t^2(\epsilon_s\mu\omega_e^2 + \lambda_i)$ $a_1 = 16\Delta t\delta_e\epsilon_\infty\mu$, $a_0 = 16\epsilon_\infty\mu$
Lorentz $(\epsilon, \mu(\omega))$	$a_4 = \lambda_i \Delta t^4\omega_m^2$, $a_3 = 4\lambda_i \Delta t^3\delta_m$, $a_2 = 4\Delta t^2(\epsilon\mu_s\omega_m^2 + \lambda_i)$ $a_1 = 16\Delta t\delta_m\epsilon\mu_\infty$, $a_0 = 16\epsilon\mu_\infty$

Table 3.2 Coefficients of the polynomials for the VWE FETD with $\beta = 0$.

Medium	Coefficients
Lossless (ϵ, μ)	$a_2 = \lambda_i \Delta t^2,\ a_1 = 0,\ a_0 = 4\epsilon\mu - \lambda_i \Delta t^2$
Lossy $(\epsilon(\omega), \mu(\omega))$	$a_2 = \Delta t^2(\lambda_i + \sigma_e\sigma_m),\ a_1 = 2\Delta t(\epsilon_\infty\sigma_m + \mu_\infty\sigma_e)$ $a_0 = 4\epsilon_\infty\mu_\infty - \lambda_i \Delta t^2$
Debye $(\epsilon(\omega), \mu)$	$a_3 = \lambda_i \Delta t^3,\ a_2 = 2\lambda_i \Delta t^2 \tau_e,\ a_1 = 4\epsilon_s\mu\Delta t - \lambda_i \Delta t^3$ $a_0 = 8\epsilon_\infty\mu\tau_e - 2\lambda_i \tau_e \Delta t^2$
Debye $(\epsilon, \mu(\omega))$	$a_3 = \lambda_i \Delta t^3,\ a_2 = 2\lambda_i \Delta t^2 \tau_m,\ a_1 = 4\epsilon\mu_s\Delta t - \lambda_i \Delta t^3$ $a_0 = 8\epsilon\mu_\infty\tau_m - 2\lambda_i \tau_m \Delta t^2$
Lorentz $(\epsilon(\omega), \mu)$	$a_4 = \lambda_i \Delta t^4 \omega_e^2,\ a_3 = 4\lambda_i \Delta t^3 \delta_e$ $a_2 = \Delta t^2(4\epsilon_s\mu\omega_e^2 + 4\lambda_i - \lambda_i \Delta t^2 \omega_e^2)$ $a_1 = 4\delta_e\Delta t(4\epsilon_\infty\mu - \lambda_i \Delta t^2),\ a_0 = 16\epsilon_\infty\mu - 4\lambda_i \Delta t^2$
Lorentz $(\epsilon, \mu(\omega))$	$a_4 = \lambda_i \Delta t^4 \omega_m^2,\ a_3 = 4\lambda_i \Delta t^3 \delta_m$ $a_2 = \Delta t^2(4\epsilon\mu_s\omega_m^2 + 4\lambda_i - \lambda_i \Delta t^2 \omega_m^2)$ $a_1 = 4\delta_m\Delta t(4\epsilon\mu_\infty - \lambda_i \Delta t^2),\ a_0 = 16\epsilon\mu_\infty - 4\lambda_i \Delta t^2$

3. **Debye medium:**

$$\epsilon(\omega) = \epsilon_\infty + \frac{\epsilon_s - \epsilon_\infty}{1 + j\omega\tau_e} \tag{3.143}$$

$$\mu(\omega) = \mu_\infty + \frac{\mu_s - \mu_\infty}{1 + j\omega\tau_m} \tag{3.144}$$

4. **Lorentz medium:**

$$\epsilon(\omega) = \epsilon_\infty + \frac{(\epsilon_s - \epsilon_\infty)\omega_e^2}{-\omega^2 + 2j\omega\delta_e + \omega_e^2} \tag{3.145}$$

$$\mu(\omega) = \mu_\infty + \frac{(\mu_s - \mu_\infty)\omega_m^2}{-\omega^2 + 2j\omega\delta_m + \omega_m^2} \tag{3.146}$$

It should be noted that the stability analysis for some cases can become very complicated and is thus omitted here. For example, a doubly dispersive Debye medium is included here for $\beta = 1/4$ but *not* for $\beta = 0$; only an electrically *or* a magnetically dispersive Debye medium is included for $\beta = 0$. The distinction between electrically, magnetically, or doubly dispersive materials is indicated in Tables 3.1 and 3.2 by the associated material parameters being a function of angular frequency, i.e. $\epsilon(\omega)$ and/or $\mu(\omega)$.

The eigenvalues of Eq. (3.140) can be analyzed via the Routh–Hurwitz criteria, which stipulates that a quadratic polynomial is stable if $a_2, a_1, a_0 > 0$, a cubic polynomial is stable if $a_3, a_2, a_0 > 0$ and $a_1 a_2 > a_0 a_3$, and finally, a fourth degree polynomial is stable if $a_4, a_3, a_0 > 0$, $c_3 = (a_2 a_3 - a_1 a_4)/a_3 > 0$, and $c_3 a_1 > a_0 a_3$.

Having obtained the coefficients for different polynomials, the locations of the eigenvalues can be determined using the Routh–Hurwitz criteria. For $\beta = 1/4$, it can be shown that the VWE FETD eigenvalues remain bounded to the unit circle for any positive value of Δt in lossy and lossless media. In addition to $\Delta t > 0$, it also requires $\epsilon_s > \epsilon_\infty$ and $\mu_s > \mu_\infty$ in Debye and Lorentz media. Since all of these conditions are always satisfied in a practical simulation, the eigenvalues of the VWE FETD with $\beta = 1/4$ never leave the unit circle for the given materials.

For $\beta = 0$, the unit circle boundedness condition becomes: $0 < \Delta t < 2\sqrt{\frac{\epsilon \mu}{\lambda_i}}$ for lossless media, $0 < \Delta t < 2\sqrt{\frac{\epsilon_\infty \mu_\infty}{\lambda_i}}$ for lossy media, $0 < \Delta t < 2\sqrt{\frac{\epsilon_\infty \mu}{\lambda_i}}$ and $\epsilon_s > \epsilon_\infty$ for the electrically dispersive Debye or Lorentz media, and $0 < \Delta t < 2\sqrt{\frac{\epsilon \mu_\infty}{\lambda_i}}$ and $\mu_s > \mu_\infty$ for the magnetically dispersive Debye or Lorentz media. The maximum value of λ_i is denoted by $\rho([M]^{-1}[S])$. It yields the smallest Δt and hence the most restrictive stability condition.

When the amplification matrix has eigenvalues outside the unit circle, the solution vector grows exponentially during the time-marching process. However, if it has eigenvalues on the unit circle and is nondiagonalizable, the method may still exhibit polynomial growth. For example, the coefficients of the polynomials show that all eigenvalues for lossless media lie on the imaginary axis in the s-plane and thus the unit circle in the z-plane. It can be shown that the amplification matrix in this case is nondiagonalizable and thus the method exhibits linear growth for certain field configurations [74]. For the lossy case, there are no eigenvalues on the imaginary axis of the s-plane as long as $\Delta t > 0$ for $\beta = 1/4$. For $\beta = 0$, Δt must meet the stability condition given earlier. Hence, the method is presumed to be stable in lossy media. However, it can be difficult to demonstrate this condition for Debye and Lorentz media.

Although the mixed FETD formulation was shown to be equivalent to the VWE FETD formulation earlier, it has a different amplification matrix and spectral properties. For example, all eigenvalues of the mixed leap-frog FETD formulation can be shown to be on the unit circle for $0 < \Delta t < 2/\sqrt{\rho([M]^{-1}[S])}$ in lossless media; however, unlike the VWE FETD, it remains stable as the amplification matrix is diagonalizable [21]. In the case of lossy media, the mixed FETD formulation is stable under a similar stability condition, because the eigenvalues lie within the unit circle. In other words, the discrete electromagnetic energy of the system decreases as the simulation marches forward in the time-domain [75]. Unfortunately, the stability analysis becomes increasingly difficult for Debye and Lorentz media, as it did for the VWE FETD formulation.

3.5.3 Nonlinear Media

As has been mentioned previously, while unconditionally stable algorithms tend to be more computationally expensive, this is usually somewhat mitigated by their ability to use larger time intervals and therefore fewer overall steps for the same amount of simulated time. Additionally, unconditionally stable methods are also particularly attractive for nonlinear algorithms, due to the constantly changing and evolving nature of the $[M]$ and $[J]$ matrices (and thus their spectral radii and conditional stability criteria). However, a significant caveat to this is that the properties of a given linear algorithm generally do *not* carry over when generalized to more complex and even nonlinear problems. Indeed, this was the case in Section 3.5.2 for linear dispersive materials, in which additional constraints needed to be satisfied to guarantee the unconditionally stability of the method was maintained.

Unfortunately, while a few methods have been devised to analyze the stability of linear FETD methods, the same cannot generally be said for their nonlinear counterparts. In fact, analyzing the stability of nonlinear methods can pose a significant challenge. This difficulty stems, in part, from the fact that for a *nonlinear* operator $A(\cdot)$, the stability condition in (3.127) no longer directly implies (3.129) is satisfied. In other words, a bounded nonlinear operator is not necessarily continuous as it is in the linear case, meaning that additional care must be taken to avoid ambiguity in what is meant by a nonlinear method being termed *stable*. Moreover, there is currently no nonlinear analog of the Lax equivalence theorem, so that even if a nonlinear method is stable in the sense of (3.127) or (3.129) (or both), the solution cannot be guaranteed to converge to and correctly approximate the true solution [76]. Hence, while it is hoped that the unconditional stability of a linear method may be maintained when generalized to nonlinear problems, there is little guarantee that this will occur.

While attempts to prove the unconditional stability of the nonlinear algorithms derived in this chapter have so far been unsuccessful, two promising pieces of evidence do currently support the notion that these algorithms may in fact be numerically stable. The first is the result of empirical numerical studies, which demonstrate bounded solutions for large values of Δt. The second, which is of more theoretical interest, is related to the evolution of the discretized electromagnetic energy stored in the fields during the solution process. More specifically, it can be shown that the nonlinear algorithms derived in this chapter conserve energy exactly, a criteria likely necessary for their unconditional stability, but which may not be sufficient (more generally speaking, it is likely necessary for the energy to be non-increasing). For a more detailed investigation of the stability of these methods, including a proof of energy conservation, the reader is encouraged to consult reference [77].

3.6 Conclusion

In this chapter an overview of FETD methods has been presented for the numerical simulation of electromagnetic problems containing complex material properties. More specifically, the two main variants of the FETD method, the mixed formulation and the VWE formulation, have both been adapted to permit the simulation of arbitrary combinations of linear dispersion, instantaneous nonlinearities, and dispersive nonlinearities.

In the case of linear dispersive media, the frequency dependence of the permittivity required the introduction of convolutions into the formulation, which was accomplished via the bilinear transform method. This technique was shown to have many benefits over other methods, including preserved stability, ease of implementation, and the ability to handle materials of arbitrarily high dispersive order in a systematic and accurate way.

As for nonlinear media, the dependence of the permittivity on the electric field was shown to significantly complicate the FETD formulation, requiring the repeated calculation and assembly of the system matrices and the solving of a nonlinear system of equations at each time step. These challenges were addressed by the use of the Newton–Raphson method as well as the bilinear transform in the case of dispersive nonlinearities. The resulting formulations are among the first to incorporate these effects within implicit FETD, and were shown to be accurate, stable, and easily generalizable.

Lastly, while the permittivity model used in this chapter was quite general and capable of modeling most of the dominant effects routinely encountered in fields such as nonlinear optics, there remain a variety of phenomena which were excluded. Further generalizing the methods discussed in this chapter to scenarios in which dielectric anisotropy, ferroelectricity, conductive losses, coupled problems, or even magnetic nonlinearity are all present in conjunction with the phenomena studied here is undoubtedly possible.

References

1 Jackson, J.D. (1999). *Classical Electrodynamics*, 3e. Hoboken, NJ: Wiley.

2 Gabriel, C. (2007). *Dielectric Properties of Biological Materials, Handbook of Biological Effects of Electromagnetic Fields: Bioengineering and Biophysical Aspects of Electromagnetic Fields*, vol. 1. Boca Raton, FL: Taylor and Francis.

3 Rappaport, C. (2007). A dispersive microwave model for human breast tissue suitable for FDTD computation. *IEEE Antennas and Wireless Propagation Letters* 6: 179–181.

4 Teixeira, F.L., Chew, W.C., Straka, M. et al. (1998). Finite-difference time-domain simulation of ground penetrating radar on dispersive, inhomogeneous, and conductive soils. *IEEE Transactions on Geoscience and Remote Sensing* 36 (6): 1928–1937.

5 Sato, T., Wakayama, T., Takemura, K. (2000). An imaging algorithm of objects embedded in a lossy dispersive medium for subsurface radar data processing. *IEEE Transactions on Geoscience and Remote Sensing* 38 (1): 296–303.

6 DiDomenico, M. (1972). Material dispersion in optical fiber waveguides. *Applied Optics* 11 (3): 652–654.

7 Walmsley, I. (2001). The role of dispersion in ultrafast optics. *Review of Scientific Instruments* 72 (1): 1–29.

8 Kinjo, T., Namihira, Y., Arakaki, K. et al. (2010). Design of highly nonlinear dispersion-flattened square photonic crystal fiber for medical applications. *Optical Review* 17 (2): 61–65.

9 Kim, S., Kim, K., Rotermund, F., and Lim, H. (2009). Computational design of one-dimensional nonlinear photonic crystals with material dispersion for efficient second-harmonic generation. *Optics Express* 17 (21): 19075–19084.

10 Joseph, R.M. and Taflove, A. (1994). Spatial soliton deflection mechanism indicated by FD-TD Maxwell's equations modeling. *IEEE Photonics Technology Letters* 6 (10): 1251–1254.

11 Lubin, Z., Greene, J.H., and Taflove, A. (2011). FDTD computational study of ultra-narrow TM non-paraxial spatial soliton interactions. *IEEE Microwave and Wireless Components Letters* 21 (5): 228–230.

12 Woodward, A.M. and Kell, D.B. (1990). On the nonlinear dielectric properties of biological systems. *Bioelectrochemistry and Bioenergetics* 24: 83–100.

13 Woodward, A.M. and Kell, D.B. (1991). Confirmation by using mutant strains that the membrane-bound H^+-ATPase is the major source of nonlinear dielectricity in *Saccharomyces cerevisiae*. *FEMS Microbiology Letters* 84 (1): 91–96.

14 Woodward, A.M. and Kell, D.B. (1991). On the relationship between the nonlinear dielectric properties and respiratory activity of the obligately aerobic bacterium *Micrococcus luteus*. *Bioelectrochemistry and Bioenergetics* 26 (3): 423–439.

15 Mercier, G.T.S., Palanisami, A., and Miller, J.H. Jr. (2010). Nonlinear dielectric spectroscopy for label-free detection of respiratory activity in whole cells. *Biosensors and Bioelectronics* 25: 2107–2114.

16 Li, J. and Chen, H. (2010). Negative optical phenomena. In: *Proceedings of Advances in Optoelectronics and Micro/Nano-Optics*. Guangzhou, China: IEEE, 13–16.

17 Lin, J., Li, D., and Yu, W. (2017). A review of zero index metamaterials. *Proceedings of the 2017 IEEE Electrical Design of Advanced Packaging and Systems Symposium (EDAPS)*. Haining, China: IEEE.

18 Cangellaris, A.C. and Wright, D.B. (1991). Analysis of the numerical error caused by the stair-stepped approximation of a conducting boundary in FDTD simulations of electromagnetic phenomena. *IEEE Transactions on Antennas and Propagation* 39 (10): 1518–1525.

19 Akyurtlu, A., Werner, D.H., Veremey, V. et al. (1999). Staircasing errors in FDTD at an air-dielectric interface. *IEEE Microwave and Guided Wave Letters* 9 (11): 444–446.

20 Pereda, A., Vielva, L.A., Vegas, A., and Prieto, A. (2001). Analyzing the stability of the FDTD technique by combining the von Neumann method with the Routh-Hurwitz criterion. *IEEE Transactions on Microwave Theory and Techniques* 49 (2): 377–381.

21 Wang, S. and Teixeira, F.L. (2004). Some remarks on the stability of time-domain electromagnetic simulations. *IEEE Transactions on Antennas and Propagation* 52 (3): 895–898.

22 Hwang, K.P. and Cangellaris, A.C. (2001). Effective permittivities for second-order accurate FDTD equations at dielectric interfaces. *IEEE Microwave and Wireless Components Letters* 11 (4): 158–160.

23 Abraham, D.S., Marques, A.N., and Nave, J.C. (2018). A correction function method for the wave equation with interface jump conditions. *Journal of Computational Physics* 353: 281–299.

24 Shi, S.B., Shao, W., Wei, X.K. et al. (2016). A new unconditionally stable FDTD method based on the Newmark-beta algorithm. *IEEE Transactions on Microwave Theory and Techniques* 64 (12): 4082–4090.

25 Yan, S. and Jin, J.M. (2015). Theoretical formulation of a time-domain finite element method for nonlinear magnetic problems in three dimensions. *Progress in Electromagnetics Research* 153: 33–55.

26 Yan, S. and Jin, J.M. (2013). Analysis of nonlinear electromagnetic problems using time-domain finite element method. *Radio Science Meeting (Joint with AP-S Symposium)*, USNC-URSI, p. 99.

27 Teixeira, F.L. (2008). Time-domain finite-difference and finite-element methods for Maxwell equations in complex media. *IEEE Transactions on Antennas and Propagation* 56 (8): 2150–2166.

28 Oppenheim, A.V., Willsky, A.S., and Nawab, S.H. (1996). *Signals and Systems*, 2e. Upper Saddle River, NJ: Prentice Hall.

29 Kelley, D.F. and Luebbers, R.J. (1996). Piecewise linear recursive convolution for dispersive media using FDTD. *IEEE Transactions on Antennas and Propagation* 44 (6): 792–797.

30 Kashiwa, T. and Fukai, I. (1990). A treatment by the FD-TD method of the dispersive characteristics associated with electronic polarization. *Microwave and Optical Technology Letters* 3 (6): 203–205.

31 Joseph, R.M., Hagness, S.C., and Taflove, A. (1991). Direct time integration of Maxwell's equations in linear dispersive media with absorption for scattering and propagation of femtosecond electromagnetic pulses. *Optics Letters* 16 (18): 1412–1414.

32 Sullivan, D.M. (1992). Frequency-dependent FDTD methods using Z transforms. *IEEE Transactions on Antennas and Propagation* 40 (10): 1223–1230.

33 Hulse, C. and Knoesen, A. (1994). Dispersive models for the finite-difference time-domain method: design, analysis, and implementation. *Journal of the Optical Society of America A* 11 (6): 1802–1811.

34 Pereda, J.A., Vegas, A., and Prieto, A. (2002). FDTD modeling of wave propagation in dispersive media by using the Möbius transformation technique. *IEEE Transactions on Microwave Theory and Techniques* 50 (7): 1689–1695.

35 Lin, Z. and Thylén, L. (2009). On the accuracy and stability of several widely used FDTD approaches for modeling Lorentz dielectrics. *IEEE Transactions on Antennas and Propagation* 57 (10): 3378–3381.

36 Boyd, R. (2008). *Nonlinear Optics*, 3e. Burlington, MA: Academic Press.

37 Garrett, I. (1990). Nonlinear effects in fibre communications: an overview. In: *Proceedings of the IEE Colloquium on Non-Linear Effects in Fibre Communications*. London: IET, 1–6.

38 Dudley, J.M., Genty, G., and Coen, S. (2006). Supercontinuum generation in photonic crystal fiber. *Reviews of Modern Physics* 78: 1135–1184.

39 Sirleto, L., Ferrara, M.A., and Vergara, A. (2018). Towards applications of stimulated Raman scattering in nanophotonics. *Proceedings of the 20th International Conference on Transparent Optical Networks*, Bucharest, Romania.

40 Joseph, R.M. and Taflove, A. (1997). FDTD Maxwell's equations models for nonlinear electrodynamics and optics. *IEEE Transactions on Antennas and Propagation* 45 (3): 364–374.

41 Blow, K.J. and Wood, D. (1989). Theoretical description of transient stimulated Raman scattering in optical fibers. *IEEE Journal of Quantum Electronics* 25 (12): 2665–2673.

42 Lynch, D.R. and Pauslen, K.D. (1990). Time-domain integration of the Maxwell equations on finite elements. *IEEE Transactions on Antennas and Propagation* 38 (12): 1933–1942.

43 Lee, J.F. (1994). WETD: A finite-element time-domain approach for solving Maxwell's equations. *IEEE Microwave and Guided Wave Letters* 4 (1): 11–13.

44 Wong, M.F., Picon, O., and Hanna, V.F. (1995). A finite element method based on Whitney forms to solve Maxwell equations in the time domain. *IEEE Transactions on Magnetics* 31 (3): 1618–1621.

45 Jin, J.M. (2014). *The Finite Element Method in Electromagnetics*, 3e. Hoboken, NJ: Wiley-IEEE Press.

46 Newmark, N. (1959). A method of computation for structural dynamics. *Journal of the Engineering Mechanics Division* 85: 67–94.

47 Gedney, S.D. and Navsariwala, U. (1995). An unconditionally stable finite element time-domain solution of the vector wave equation. *IEEE Microwave and Guided Wave Letters* 5 (10): 332–334.

48 Bossavit, A. (1988). Whitney forms: a new class of finite elements for three dimensional computations in electromagnetics. *Proceedings of the Institution of Electrical Engineers* 135: 493–500.

49 Bossavit, A. and Kettunen, L. (1999). Yee-like schemes on a tetrahedral mesh, with diagonal lumping. *International Journal of Numerical Modelling* 12: 129–142.

50 Movahhedi, M., Abdipour, A., Nentchev, A. et al. (2007). Alternating-direction implicit formulation of the finite-element time-domain method. *IEEE Transactions on Microwave Theory and Techniques* 55 (6): 1322–1331.

51 Akbarzadeh-Sharbaf, A. and Giannacopoulos, D. (2013). Implementation of a first-order ABC in mixed finite-element time-domain formulations using equivalent currents. *IEEE Microwave and Wireless Components Letters* 23 (6): 276–278.

52 Rylander, T. and Jin, J.M. (2005). Perfectly matched layer in three dimensions for the time-domain finite element method applied to radiation problems. *IEEE Transactions on Antennas and Propagation* 53 (4): 1489–1499.

53 Wang, S., Lee, R., and Teixeira, F.L. (2006). Anisotropic-medium PML for vector FETD with modified basis functions. *IEEE Transactions on Antennas and Propagation* 54 (1): 20–27.

54 Moon, H., Teixeira, F.L., Kim, J., Omelchenko, Y.A. (2014). Trade-offs for unconditional stability in the finite-element time-domain method. *IEEE Microwave and Wireless Components Letters* 24 (6): 361–363.

55 He, B. and Teixeira, F.L. (2007). Differential forms, Galerkin duality, and sparse inverse approximations in finite element solutions of Maxwell equations. *IEEE Transactions on Antennas and Propagation* 55 (5): 1359–1368.

56 White, D.A. (1999). Orthogonal basis functions for time domain finite element solution of the vector wave equation. *Communications in Numerical Methods in Engineering* 35 (3): 1458–1461.

57 Fisher, A., Rieben, R.N., Rodrigue, G.H., White, D.A. (2005). A generalized mass lumping technique for vector finite-element solutions of the time-dependent Maxwell equations. *IEEE Transactions on Antennas and Propagation* 53 (9): 2900–2910.

58 Akbarzadeh-Sharbaf, A. and Giannacopoulos, D.D. (2015). A stable and efficient direct time integration of the vector wave equation in the finite-element time-domain method for dispersive media. *IEEE Transactions on Antennas and Propagation* 63 (1): 314–321.

59 Abraham, D.S. and Giannacopoulos, D.D. (2019). A convolution-free finite-element time-domain method for the nonlinear dispersive vector wave equation. *IEEE Transactions on Magnetics* 55 (12): 1–4.

60 Liu, J.M. (2005). *Photonic Devices*. Cambridge, UK: Cambridge University Press.

61 Akbarzadeh-Sharbaf, A. and Giannacopoulos, D.D. (2013). Finite-element time-domain solution of the vector wave equation in doubly dispersive media using Möbius transformation technique. *IEEE Transactions on Antennas and Propagation* 61 (8): 4158–4166.

62 Abraham, D.S. and Giannacopoulos, D.D. (2019). A convolution-free mixed finite-element time-domain method for general nonlinear dispersive media. *IEEE Transactions on Antennas and Propagation* 67 (1): 324–334.

63 Gedney, S.D. (1996). An anisotropic perfectly matched layer-absorbing medium for the truncation of FDTD lattices. *IEEE Transactions on Antennas and Propagation* 44 (12): 1630–1639.

64 Akbarzadeh-Sharbaf, A. (2016). Finite-element time-domain modelling of dispersive media and layered structures. PhD thesis. Montréal, Canada: McGill University.

65 Dubietis, A., Couairon, A., and Genty, G. (2019). Supercontinuum generation: introduction. *Journal of the Optical Society of America B* 36 (2): SG1–SG3.

66 Dudley, J.M. and Taylor, J.R. (2010). *Supercontinuum Generation in Optical Fibers*. Cambridge: Cambridge University Press.

67 Ryanben'kii, V.S. and Tsynkov, S.V. (2007). *A Theoretical Introduction to Numerical Analysis*. Chapman & Hall/CRC.

68 Epureanu, B.I. and Greenside, H.S. (1998). Fractal basins of attraction associated with a damped Newton's method. *SIAM Review* 40 (1): 102–109.

69 Dunavant, D.A. (1985). High degree efficient symmetrical Gaussian quadrature rules for the triangle. *International Journal for Numerical Methods in Engineering* 21: 1129–1148.

70 Thomas, J.W. (1995). *Numerical Partial Differential Equations: Finite Difference Methods*. Berlin: Springer Science and Business Media.

71 Strikwerda, J.C. (2004). *Finite Difference Schemes and Partial Differential Equations*, 2e. Philadelphia, PA: Society for Industrial and Applied Mathematics.

72 Lax, P.D. and Richtmyer, R.D. (1956). Survey of the stability of linear finite difference equations. *Communications on Pure and Applied Mathematics* 9: 267–293.

73 Jiao, D. and Jin, J.M. (2002). A general approach for the stability analysis of the time-domain finite-element method for electromagnetic simulations. *IEEE Transactions on Antennas and Propagation* 50 (11): 1624–1632.

74 Chilton, R.A. and Lee, R. (2007). The discrete origin of FETD-Newmark late time instability, and a correction scheme. *Journal of Computational Physics* 224 (2): 1293–1306.

75 Edelvik, F., Schuhmann, R., and Weiland, T. (2004). A general stability analysis of FIT/FDTD applied to lossy dielectrics and lumped elements. *International Journal of Numerical Modelling* 17: 407–419.

76 Ortiz, M. (1986). A note on energy conservation and stability of nonlinear time-stepping algorithms. *Computers and Structures* 24 (1): 167–168.

77 Abraham, D.S. (2020). Finite-element time-domain methods for nonlinear dispersive media. PhD thesis. Montréal, Québec, Canada: McGill University.

Part II

Time-Domain Methods for Multiphysics and Multiscale Modeling

4

Discontinuous Galerkin Time-Domain Method in Electromagnetics: From Nanostructure Simulations to Multiphysics Implementations

Ming Dong[1], Liang Chen[1], Ping Li[2], Lijun Jiang[3], and Hakan Bagci[1]

[1]*Electrical and Computer Engineering (ECE) Program, Division of Computer, Electrical, and Mathematical Science and Engineering (CEMSE), King Abdullah University of Science and Technology (KAUST), Thuwal, Saudi Arabia*
[2]*Department of Electrical Engineering, Shanghai Jiao Tong University, Shanghai, China*
[3]*Department of Electrical and Electronic Engineering, The University of Hongkong, Hongkong, China*

4.1 Introduction to the Discontinuous Galerkin Time-Domain Method

Since the discontinuous Galerkin (DG) method was first applied to solve the neutron transport equation by Reed and Hill in 1973 [1], it has been reformulated and extended to numerous applications including but not limited to gas dynamics, semiconductor physics, magneto-hydrodynamics, fluid dynamics, and electromagnetics [2]. In the field of electromagnetics, one of the very first works on a DG method formulated and implemented to solve the time-domain Maxwell equations is from Bourdel et al. [3]. This DG method uses a zeroth-order (piecewise constant) discretization scheme, but rather quickly after that several high-order discontinuous Galerkin time-domain (DGTD) methods have been developed. These methods can account for unstructured meshes and use various Runge–Kutta schemes to carry out the time integration [4–7]. Since then, other methods have been developed, increasing the popularity of DGTD in the field of computational electromagnetics [8–19]. Finally, in more recent years, the DGTD method has successfully become a powerful numerical technique to tackle a wide variety of challenging, large-scale electromagnetic problems including those that require multiphysics modeling [20–25].

The DGTD method combines advantages of both the finite volume time-domain (FVTD) method and the finite element time-domain (FETD) method [8]. Like the FVTD method, the information exchange between neighboring elements is realized through the use of numerical flux. This approach localizes all spatial

Advances in Time-Domain Computational Electromagnetic Methods, First Edition.
Edited by Qiang Ren, Su Yan, and Atef Z. Elsherbeni.
© 2023 The Institute of Electrical and Electronics Engineers, Inc. Published 2023 by John Wiley & Sons, Inc.

operations to individual discretization elements. The resulting local element-level mass matrix blocks are factorized and stored very efficiently before time integration/marching starts. Indeed, if an explicit time integration scheme is used, the DGTD method becomes an efficient solver with a very small memory footprint (compared to classical time-domain finite element method [FEM]). In addition, the block diagonal mass matrix makes the DGTD method very suitable for parallelization. It can easily be ported on high-performance distributed memory systems or multi-threaded GPUs to solve large-scale electromagnetic problems [26, 27]. Finally, locality of the spatial operations makes it easier to incorporate *h*- and/or *p*-adaptive refinement techniques [7, 8] as well as non-conformal meshing strategies [4].

The DGTD method also comes with the flexibility of choosing different time stepping schemes in different subdomains of the computation domain [12, 14]. For example, an efficient explicit scheme can be used in subdomains with coarser meshes while an unconditionally stable implicit solver can be used in subdomains with dense meshes. This approach allows for the same large time step to be used throughout the whole computation domain potentially reducing the overall computation time.

As a final word, just like other time-domain methods, the DGTD method offers several benefits in simulating various electromagnetic problems: It can provide broadband data with a single execution of the simulation and it allows for hybridization with circuit solvers that can account for nonlinear components [11, 22, 23].

In this chapter, we first describe the basic formulation underlying the DGTD method developed for solving the Maxwell's curl equations using vector basis functions and explain how different boundary conditions can be incorporated within the solver using numerical flux. This is followed by the description of a nodal DGTD scheme, where higher-order polynomials are used as basis functions to expand the three components of the electric and magnetic fields. Then, the chapter continues with the applications of the DGTD method to several real-life practical problems. More specifically, novel techniques developed to enable the application of the DGTD method to electromagnetic analysis of nanostructures and multiphysics simulation of photoconductive devices/antennas are described. For each application, we present several numerical examples to demonstrate the accuracy, efficiency, and robustness of the developed techniques.

4.1.1 The DGTD Formulation for Maxwell's Equations

The analysis of transient electromagnetic field interactions with arbitrarily shaped objects calls for a numerical solution of Maxwell's equations for electric field $\mathbf{E}(\mathbf{r}, t)$ and magnetic field $\mathbf{H}(\mathbf{r}, t)$ [9] or the wave equation for $\mathbf{E}(\mathbf{r}, t)$ or $\mathbf{H}(\mathbf{r}, t)$ [28].

Here, we focus on the formulation of the DGTD schemes for the former and refer the readers to [28] and references therein for the details for the DGTD schemes developed for solving the wave equation. Let $\mathbf{J}(\mathbf{r}, t)$ represent the volumetric current density located in the domain of interest; then Maxwell's curl equations can be written as [21]

$$\varepsilon(\mathbf{r})\frac{\partial \mathbf{E}(\mathbf{r},t)}{\partial t} - \nabla \times \mathbf{H}(\mathbf{r}, t) + \mathbf{J}(\mathbf{r}, t) = 0$$

$$\mu(\mathbf{r})\frac{\partial \mathbf{H}(\mathbf{r},t)}{\partial t} + \nabla \times \mathbf{E}(\mathbf{r}, t) = 0 \tag{4.1}$$

where $\varepsilon(\mathbf{r})$ and $\mu(\mathbf{r})$ denote the permittivity and permeability, respectively. Note that for the sake of simplicity in the presentation, in the rest of the chapter, dependence of the variables on location $\mathbf{r}$ and time t is not explicitly stated (unless it is needed to clarify the formulation/notation).

To develop the DGTD formulation, Maxwell's equations above should be re-expressed in the conservation form as [8]

$$\frac{\partial \mathbf{U}}{\partial t} + \nabla \cdot \mathcal{F}(\mathbf{U}) + \mathbf{S} = 0 \tag{4.2}$$

This can be achieved by defining field vector $\mathbf{U} = [\mathbf{E}, \mathbf{H}]^{\mathrm{T}}$, excitation vector $\mathbf{S} = [\mathbf{J}/\varepsilon, \mathbf{0}]^{\mathrm{T}}$, and the flux vector

$$\mathcal{F}(\mathbf{U}) = \begin{bmatrix} -\hat{x} \times \mathbf{H}/\varepsilon \\ \hat{x} \times \mathbf{E}/\mu \end{bmatrix} \hat{x} + \begin{bmatrix} -\hat{y} \times \mathbf{H}/\varepsilon \\ \hat{y} \times \mathbf{E}/\mu \end{bmatrix} \hat{y} + \begin{bmatrix} -\hat{z} \times \mathbf{H}/\varepsilon \\ \hat{z} \times \mathbf{E}/\mu \end{bmatrix} \hat{z} \tag{4.3}$$

The flux vector $\mathcal{F}(\mathbf{U})$ can be further expressed as a multiplication of the purely material- dependent coefficient matrix $\mathcal{A}$ and the field vector $\mathbf{U}$:

$$\mathcal{F}(\mathbf{U}) = \mathcal{A}\mathbf{U} \tag{4.4}$$

where

$$\mathcal{A} = \begin{bmatrix} 0 & -\frac{1}{\varepsilon}\mathbf{C}_x \\ \frac{1}{\mu}\mathbf{C}_x & 0 \end{bmatrix} \hat{x} + \begin{bmatrix} 0 & -\frac{1}{\varepsilon}\mathbf{C}_y \\ \frac{1}{\mu}\mathbf{C}_y & 0 \end{bmatrix} \hat{y} + \begin{bmatrix} 0 & -\frac{1}{\varepsilon}\mathbf{C}_z \\ \frac{1}{\mu}\mathbf{C}_z & 0 \end{bmatrix} \hat{z}$$

$$\mathbf{C}_x = \begin{bmatrix} 0 & 0 & 0 \\ 0 & 0 & -1 \\ 0 & 1 & 0 \end{bmatrix}, \quad \mathbf{C}_y = \begin{bmatrix} 0 & 0 & 1 \\ 0 & 0 & 0 \\ -1 & 0 & 0 \end{bmatrix}, \quad \mathbf{C}_z = \begin{bmatrix} 0 & -1 & 0 \\ 1 & 0 & 0 \\ 0 & 0 & 0 \end{bmatrix} \tag{4.5}$$

The coefficient matrix $\mathcal{A}$ determines the characteristics of the Maxwell system. It is shown later in this section that the eigenvalues of this flux matrix $\mathcal{A}\mathbf{U}$ are real (a necessary condition for the hyperbolic system) and they are the characteristic speed associated with the Maxwell system.

Let Ω and $\partial\Omega$ denote the computation domain of interest and its boundary, respectively. Ω can be divided into a set of nonoverlapping tetrahedrons Ω_i with boundary $\partial\Omega_i$, i.e. $\Omega = \cup\Omega_i$. In each mesh element Ω_i, $i = 1, \ldots, N$, after applying Galerkin testing [8] to Eq. (4.1) in space with basis function $\Lambda_i = [\Psi_i^e, \Psi_i^h]$ (Ψ_i^e and

Ψ_i^h are used to expand **E** and **H** in element i, respectively, as explained later) and using the divergence (Gauss) theorem twice, we obtain

$$\int_{\Omega_i} \Lambda_i \cdot \left[\frac{\partial \mathbf{U}}{\partial t} + \nabla \cdot \mathcal{F}(\mathbf{U}) + \mathbf{S} \right] d\mathbf{r}$$
$$= -\int_{\partial \Omega_i} \left\{ \hat{\mathbf{n}}_i \cdot [\mathcal{F}^*(\mathbf{U}) - \mathcal{F}(\mathbf{U})] \right\} \cdot \Lambda_i \, d\mathbf{r} \tag{4.6}$$

where $\hat{\mathbf{n}}_i$ is the outward pointing unit normal vector of $\partial \Omega_i$ and $\hat{\mathbf{n}}_i \cdot \mathcal{F}^*(\mathbf{U})$ is the numerical flux. Since the solutions in adjacent elements are allowed to be discontinuous, $\mathcal{F}^*$ must be carefully defined to ensure the uniqueness of the solution.

The definition of the numerical flux starts by solving the corresponding Riemann problem with the Rankine–Hugoniot jump condition. To obtain the Rankine–Hugoniot jump condition [29, 30], the characteristic plot of the Maxwell system has to be obtained in advance from the flux matrix defined as

$$\check{A} = \hat{\mathbf{n}}_i \cdot \mathcal{A} = \begin{bmatrix} \overline{\overline{\varepsilon}}_i & 0 \\ 0 & \overline{\overline{\mu}}_i \end{bmatrix} \cdot \begin{bmatrix} 0 & \overline{\overline{C}}_i \\ -\overline{\overline{C}}_i & 0 \end{bmatrix} \tag{4.7}$$

where $\overline{\overline{\varepsilon}}_i = \mathrm{diag}(\varepsilon_i, \varepsilon_i, \varepsilon_i)$, $\overline{\overline{\mu}}_i = \mathrm{diag}(\mu_i, \mu_i, \mu_i)$, and

$$\overline{\overline{C}}_i = \begin{bmatrix} 0 & n_{i,z} & -n_{i,y} \\ -n_{i,z} & 0 & n_{i,x} \\ n_{i,y} & -n_{i,x} & 0 \end{bmatrix} \tag{4.8}$$

The flux matrix $\check{A}$ has six eigenvalues given by [29, 30]

$$\begin{aligned} \lambda_{1,4} &= 0 \\ \lambda_{2,5} &= \frac{1}{\sqrt{\varepsilon_i \mu_i}} = +c_i \\ \lambda_{3,6} &= -\frac{1}{\sqrt{\varepsilon_i \mu_i}} = -c_i \end{aligned} \tag{4.9}$$

and they contain important information about the solution of the Riemann problem. For a hyperbolic system, the solutions remain unchanged while propagating along the characteristic plot $x_n = \lambda_i t$. Across each solution, Rankine–Hugoniot jump condition holds, that is [8, 29, 30]

$$\check{A}\mathbf{U}^* - \check{A}\mathbf{U}_i = -c_i(\mathbf{U}^* - \mathbf{U}_i) \tag{4.10}$$

$$\check{A}\mathbf{U}^{**} - \check{A}\mathbf{U}^* = 0 \tag{4.11}$$

$$\check{A}\mathbf{U}^{**} - \check{A}\mathbf{U}_j = c_j(\mathbf{U}^{**} - \mathbf{U}^*) \tag{4.12}$$

where $j \in N(i)$, $N(i)$ is the set of elements that are neighboring element i, and $\mathbf{U}^*$ and $\mathbf{U}^{**}$ represent two intermediate states. The explicit jump relations for fields can be simplified as

1. Jump across the characteristic plot $x_n = -c_i t$

$$\frac{1}{\mu_i} \hat{\mathbf{n}}_i \times (\mathbf{E}^* - \mathbf{E}_i) = -c_i(\mathbf{H}^* - \mathbf{H}_i) \tag{4.13}$$

$$\frac{1}{\varepsilon_i} \hat{\mathbf{n}}_i \times (\mathbf{H}_i - \mathbf{H}^*) = -c_i(\mathbf{E}^* - \mathbf{E}_i) \tag{4.14}$$

2. Jump across the characteristic plot $x_n = 0$

$$\hat{\mathbf{n}}_i \times (\mathbf{E}^{**} - \mathbf{E}^*) = 0 \tag{4.15}$$

$$\hat{\mathbf{n}}_i \times (\mathbf{H}^{**} - \mathbf{H}^*) = 0 \tag{4.16}$$

3. Jump across the characteristic plot $x_n = c_j t$

$$\frac{1}{\mu_j} \hat{\mathbf{n}}_i \times (\mathbf{E}_j - \mathbf{E}^{**}) = c_j(\mathbf{H}_j - \mathbf{H}^{**}) \tag{4.17}$$

$$\frac{1}{\varepsilon_j} \hat{\mathbf{n}}_i \times (\mathbf{H}^{**} - \mathbf{H}_j) = c_j(\mathbf{E}_j - \mathbf{E}^{**}) \tag{4.18}$$

With the above jump conditions, we obtain the explicit expression of the numerical flux $\hat{\mathbf{n}}_i \cdot \mathcal{F}^*(\mathbf{U}) = \left[-\hat{\mathbf{n}}_i \times \mathbf{H}^*, \hat{\mathbf{n}}_i \times \mathbf{E}^* \right]^{\mathrm{T}}$ as

$$\hat{\mathbf{n}}_i \times \mathbf{H}^* = \hat{\mathbf{n}}_i \times \frac{(Z_i \mathbf{H}_i + Z_j \mathbf{H}_j) + \hat{\mathbf{n}}_i \times (\mathbf{E}_i - \mathbf{E}_j)}{Z_i + Z_j} \tag{4.19}$$

$$\hat{\mathbf{n}}_i \times \mathbf{E}^* = \hat{\mathbf{n}}_i \times \frac{(Y_i \mathbf{E}_i + Y_j \mathbf{E}_j) - \hat{\mathbf{n}}_i \times (\mathbf{H}_i - \mathbf{H}_j)}{Y_i + Y_j} \tag{4.20}$$

where $Z_{i,j} = \sqrt{\mu_{i,j}/\varepsilon_{i,j}}$ and $Y_{i,j} = 1/Z_{i,j}$ denote the characteristic impedance and admittance, respectively. This numerical flux defined by Eqs. (4.19) and (4.20) is termed "upwind flux," which is the most commonly used type of numerical flux in the DGTD methods for electromagnetic simulation. In the rest of this chapter, it is assumed that upwind numerical is used unless otherwise is stated.

Substituting (4.19) and (4.20) into (4.6), we get two separate equations as

$$\int_{\Omega_i} \left(\varepsilon_i \frac{\partial \mathbf{E}_i}{\partial t} - \nabla \times \mathbf{H}_i + \mathbf{J}_i \right) \cdot \mathbf{\Psi}_{i,k}^e \, d\mathbf{r} = \int_{\partial \Omega_i} \mathbf{\Psi}_{i,k}^e \cdot \left[\hat{\mathbf{n}}_i \times (\mathbf{H}^* - \mathbf{H}_i) \right] d\mathbf{r} \tag{4.21}$$

$$\int_{\Omega_i} \left(\mu_i \frac{\partial \mathbf{H}_i}{\partial t} + \nabla \times \mathbf{E}_i \right) \cdot \mathbf{\Psi}_{i,k}^h \, d\mathbf{r} - \int_{\partial \Omega_i} \mathbf{\Psi}_{i,k}^h \cdot \left[\hat{\mathbf{n}}_i \times (\mathbf{E}_i - \mathbf{E}^*) \right] d\mathbf{r} \tag{4.22}$$

Next, the (local) electric and magnetic fields in element i, $\mathbf{E}_i$ and $\mathbf{H}_i$, are expanded using vector basis functions $\mathbf{\Psi}_{i,l}^e$ and $\mathbf{\Psi}_{i,l}^h$, respectively as [21]

$$\mathbf{E}_i(\mathbf{r}, t) = \sum_{l=1}^{K} e_{i,l}(t) \mathbf{\Psi}_{i,l}^e(\mathbf{r}), \qquad \mathbf{H}_i(\mathbf{r}, t) = \sum_{l=1}^{K} h_{i,l}(t) \mathbf{\Psi}_{i,l}^h(\mathbf{r}) \tag{4.23}$$

where $e_{i,l}$ and $h_{i,l}$ are time-dependent unknown coefficients of $\Psi_{i,l}^e$ and $\Psi_{i,l}^h$, respectively. Note that the numbers of basis functions used for expanding $\mathbf{E}_i$ and $\mathbf{H}_i$ do not have to be the same. Similarly, the number of basis functions can be different in different elements. But for the sake of simplicity in the notation, we assume that all elements use the same number of basis functions for both $\mathbf{E}_i$ and $\mathbf{H}_i$, and that number is K.

Using (4.19)–(4.23), a semi-discrete matrix system can be obtained [24, 31]:

$$\mathbf{M}_i^e \frac{\partial \mathbf{e}_i}{\partial t} = \mathbf{S}_i^e \mathbf{h}_i - \mathbf{j}_i + \mathbf{F}_{ii}^{ee} \mathbf{e}_i - \mathbf{F}_{ij}^{ee} \mathbf{e}_j - \mathbf{F}_{ii}^{eh} \mathbf{h}_i + \mathbf{F}_{ij}^{eh} \mathbf{h}_j \tag{4.24}$$

$$\mathbf{M}_i^h \frac{\partial \mathbf{h}_i}{\partial t} = -\mathbf{S}_i^h \mathbf{e}_i + \mathbf{F}_{ii}^{hh} \mathbf{h}_i - \mathbf{F}_{ij}^{hh} \mathbf{h}_j + \mathbf{F}_{ii}^{he} \mathbf{e}_i - \mathbf{F}_{ij}^{he} \mathbf{e}_j \tag{4.25}$$

where $\mathbf{e}_i$ and $\mathbf{h}_i$ are time-dependent vectors that store unknown basis function coefficients, $\mathbf{M}_i^e$ and $\mathbf{M}_i^h$ are the mass matrices, $\mathbf{S}_i^e$ and $\mathbf{S}_i^h$ are the stiffness matrices, $\mathbf{j}_i$ is a vector that stores the tested current source, and $\mathbf{F}_{ii/ij}^{ee/eh}$ and $\mathbf{F}_{ii/ij}^{hh/he}$ are the flux matrices. The entries of these matrices and vectors are given by [24, 31]:

$$[\mathbf{e}_i]_l = e_{i,l}, \quad [\mathbf{h}_i]_l = h_{i,l}$$

$$[\mathbf{M}_i^e]_{kl} = \int_{\Omega_i} \Psi_{i,k}^e \cdot \varepsilon_i \Psi_{i,l}^e \, d\mathbf{r}, \quad [\mathbf{M}_i^h]_{kl} = \int_{\Omega_i} \Psi_{i,k}^h \cdot \mu_i \Psi_{i,l}^h \, d\mathbf{r}$$

$$[\mathbf{S}_i^e]_{kl} = \int_{\Omega_i} \Psi_{i,k}^e \cdot \nabla \times \Psi_{i,l}^h \, d\mathbf{r}, \quad [\mathbf{S}_i^h]_{kl} = \int_{\Omega_i} \Psi_{i,k}^h \cdot \nabla \times \Psi_{i,l}^e \, d\mathbf{r}$$

$$[\mathbf{j}_i]_{(k)} = \int_{\Omega_i} \Psi_{i,k}^e \cdot \mathbf{J}_i \, d\mathbf{r}$$

$$[\mathbf{F}_{ii}^{ee}]_{kl} = \frac{1}{Z_i + Z_j} \int_{\partial\Omega_{i,j}} \Psi_{i,k}^e \cdot \hat{\mathbf{n}}_i \times (\hat{\mathbf{n}}_i \times \Psi_{i,l}^e) \, d\mathbf{r}$$

$$[\mathbf{F}_{ij}^{ee}]_{kl} = \frac{1}{Z_i + Z_j} \int_{\partial\Omega_{i,j}} \Psi_{i,k}^e \cdot \hat{\mathbf{n}}_i \times (\hat{\mathbf{n}}_i \times \Psi_{j,l}^e) \, d\mathbf{r}$$

$$[\mathbf{F}_{ii}^{eh}]_{kl} = \frac{Z_j}{Z_i + Z_j} \int_{\partial\Omega_{i,j}} \Psi_{i,k}^e \cdot \hat{\mathbf{n}}_i \times \Psi_{i,l}^h \, d\mathbf{r}$$

$$[\mathbf{F}_{ij}^{eh}]_{kl} = \frac{Z_j}{Z_i + Z_j} \int_{\partial\Omega_{i,j}} \Psi_{i,k}^e \cdot \hat{\mathbf{n}}_i \times \Psi_{j,l}^h \, d\mathbf{r}$$

$$[\mathbf{F}_{ii}^{hh}]_{kl} = \frac{1}{Y_i + Y_j} \int_{\partial\Omega_{i,j}} \Psi_{i,k}^h \cdot \hat{\mathbf{n}}_i \times (\hat{\mathbf{n}}_i \times \Psi_{i,l}^h) \, d\mathbf{r}$$

$$[\mathbf{F}_{ij}^{hh}]_{kl} = \frac{1}{Y_i + Y_j} \int_{\partial\Omega_{i,j}} \Psi_{i,k}^h \cdot \hat{\mathbf{n}}_i \times (\hat{\mathbf{n}}_i \times \Psi_{j,l}^h) \, d\mathbf{r}$$

$$[\mathbf{F}_{ii}^{he}]_{kl} = \frac{Y_j}{Y_i + Y_j} \int_{\partial\Omega_{i,j}} \Psi_{i,k}^h \cdot \hat{\mathbf{n}}_i \times \Psi_{i,l}^e \, d\mathbf{r}$$

$$[\mathbf{F}_{ij}^{he}]_{kl} = \frac{Y_j}{Y_i + Y_j} \int_{\partial\Omega_{i,j}} \Psi_{i,k}^h \cdot \hat{\mathbf{n}}_i \times \Psi_{j,l}^e \, d\mathbf{r} \tag{4.26}$$

Note that the semi-discrete Eqs. (4.24) and (4.25) are integrated in time for all $i = 1, \ldots, N$. Before the time integration/time marching starts, $\mathbf{M}_i^e$ and $\mathbf{M}_i^h$, $i = 1, \ldots, N$, each of which is only of dimension $K \times K$, are inverted and stored. Consequently, if an explicit scheme is used for time integration (as described below), the resulting DGTD solver is very efficient with a small memory footprint (compared to traditional time-domain FEM).

Next, we describe how the semi-discrete system in Eqs. (4.24) and (4.25) can be integrated in time to obtain the unknown coefficients $\mathbf{e}_i$ and $\mathbf{h}_i$. We use two examples: Classical Runge–Kutta method [8] and the leap-frog scheme that relies on central finite difference approximation [24].

The fourth-order Runge–Kutta: The system in Eqs. (4.24) and (4.25) is in the form of a first-order ordinary differential equation (ODE) that can be expressed as $\frac{\partial u}{\partial t} = f(u)$. Using the fourth-order Runge–Kutta method, the solution of this ODE at time step $t^{n+1} = (n+1)\Delta t = t^n + \Delta t$ can be expressed as [8]

$$u^{n+1} = u^n + (a^n + 2b^n + 2c^n + d^n)\frac{\Delta t}{6} \tag{4.27}$$

where $a^n = f(u^n)$, $b^n = f(u^n + a^n \Delta t/2)$, $c^n = f(u^n + b^n \Delta t/2)$, and $d^n = f(u^n + c^n \Delta t)$.

Leap-frog scheme: One can use the central finite difference method to approximate the two time derivatives in Eqs. (4.24) and (4.25) at staggered time points [24]:

$$\frac{\partial \mathbf{e}_i}{\partial t}\Big|_{n+\frac{1}{2}} \approx \frac{\mathbf{e}_i^{n+1} - \mathbf{e}_i^n}{\Delta t} + O(\Delta t^2) \tag{4.28}$$

$$\frac{\partial \mathbf{h}_i}{\partial t}\Big|_{n+1} \approx \frac{\mathbf{h}_i^{n+\frac{3}{2}} - \mathbf{h}_i^{n+\frac{1}{2}}}{\Delta t} + O(\Delta t^2) \tag{4.29}$$

Inserting Eqs. (4.28) and (4.29) into Eqs. (4.24) and (4.25) yields the update equations, resulting in the so-called leap frog time marching scheme:

$$\mathbf{M}_i^e \mathbf{e}_i^{n+1} = \mathbf{M}_i^e \mathbf{e}_i^n$$
$$\Delta t \left[\mathbf{S}_i^e \mathbf{h}_i^{n+\frac{1}{2}} - \mathbf{j}_i^{n+\frac{1}{2}} + \mathbf{F}_{ii}^{ee}\mathbf{e}_i^n - \mathbf{F}_{ij}^{ee}\mathbf{e}_j^n - \mathbf{F}_{ii}^{eh}\mathbf{h}_i^{n+\frac{1}{2}} + \mathbf{F}_{ij}^{eh}\mathbf{h}_j^{n+\frac{1}{2}} \right] \tag{4.30}$$

$$\mathbf{M}_i^h \mathbf{h}_i^{n+\frac{3}{2}} = \mathbf{M}_i^h \mathbf{h}_i^{n+\frac{1}{2}}$$
$$+\Delta t \left[\mathbf{S}_i^h \mathbf{e}_i^{n+1} - \mathbf{F}_{ii}^{hh}\mathbf{h}_i^{n+\frac{1}{2}} + \mathbf{F}_{ij}^{hh}\mathbf{h}_j^{n+\frac{1}{2}} - \mathbf{F}_{ii}^{he}\mathbf{e}_i^{n+1} + \mathbf{F}_{ij}^{he}\mathbf{e}_j^{n+1} \right] \tag{4.31}$$

We should emphasize here that the Runge–Kutta method and leap-frog scheme are used as examples because of their popularity. Other explicit schemes can also be used to integrate Eqs. (4.24) and (4.25) in time [14, 32].

4.1.2 Boundary Conditions

Boundary conditions are incorporated within the DGTD method by modifying the numerical flux. The expressions of the upwind flux in Eqs. (4.19) and (4.20) show that the computation of numerical flux for a given element requires the field values on that element and its neighbors (this can be thought of as a decomposition of numerical flux into incoming flux and outgoing flux [30]). However, if an element is "touching" the boundary, the corresponding neighbor element does not exist and the incoming flux is not immediately available in the form of field values. In this case, the relevant boundary condition is used to modify the incoming flux expression and update the numerical flux of the element touching the boundary. Next, we describe how absorbing boundary conditions (ABCs) and the boundary conditions for perfect electrically conducting (PEC) and perfect magnetically conducting (PMC) surfaces can be incorporated within the DGTD method using the numerical flux.

4.1.2.1 Absorbing Boundary Conditions (ABCs)

In simulations of electromagnetic fields in open region problems, the unbounded background domain has to be truncated into a finite computation domain [21]. On the truncation surface, a boundary condition that imitates the radiation relation has to be enforced. The first-order Silver-Muller absorbing boundary condition (SM-ABC) [33, 34]

$$
\begin{aligned}
\hat{\mathbf{n}}_i \times \mathbf{E}_j &= -c_j \mu_j \hat{\mathbf{n}}_i \times (\hat{\mathbf{n}}_i \times \mathbf{H}_j) \\
\hat{\mathbf{n}}_i \times \mathbf{H}_j &= c_j \varepsilon_j \hat{\mathbf{n}}_i \times (\hat{\mathbf{n}}_i \times \mathbf{E}_j)
\end{aligned}
\tag{4.32}
$$

is widely used due to its simplicity and ease of implementation even though its performance is limited. The implementation of SM-ABC in the DGTD method is done by simply setting the field values $\mathbf{E}_j$ and $\mathbf{H}_j$ in the numerical flux expression to zero since all outgoing waves are assumed to be absorbed without any reflection. Then, the numerical flux can be expressed as

$$
\begin{aligned}
\hat{\mathbf{n}}_i \times \mathbf{H}^* &= \hat{\mathbf{n}}_i \times \frac{Z_i \mathbf{H}_i + \hat{\mathbf{n}}_i \times \mathbf{E}_i}{2Z_i} \\
\hat{\mathbf{n}}_i \times \mathbf{E}^* &= \hat{\mathbf{n}}_i \times \frac{Y_i \mathbf{E}_i - \hat{\mathbf{n}}_i \times \mathbf{H}_i}{2Y_i}
\end{aligned}
\tag{4.33}
$$

Note that in Section 4.1.3, a more accurate approach to truncating computation domains is described. This approach hybridizes the DGTD scheme with a time-domain boundary integral (TDBI) method, but just like ABCs, the fields from the TDBI method are coupled to the DGTD method via the use of the numerical flux. Please read Section 4.1.3 for details on the implementation of this hybrid method.

4.1.2.2 Boundary Condition on Perfect Electrically Conducting (PEC) Surfaces

On a PEC surface, the tangential component of $\mathbf{E}_j$ is equal to zero and we set the characteristic impedance $Z_j = 0$, which simplify the expression of the numerical flux to

$$\hat{\mathbf{n}}_i \times \mathbf{H}^* = \hat{\mathbf{n}}_i \times \left(\mathbf{H}_i + \frac{1}{Z_i} \hat{\mathbf{n}}_i \times \mathbf{E}_i \right)$$

$$\hat{\mathbf{n}}_i \times \mathbf{E}^* = 0$$

(4.34)

Alternatively, this expression can be obtained by choosing $Z_j = Z_i$, $\mathbf{E}_j = -\mathbf{E}_i$, and $\mathbf{H}_j = \mathbf{H}_i$.

4.1.2.3 Boundary Condition on Perfect Magnetically Conducting (PMC) Surfaces

On a PMC surface, tangential component of $\mathbf{H}_j$ is equal to zero and we set the characteristic admittance $Y_j = 0$, which simplify the expression of the numerical flux to

$$\hat{\mathbf{n}}_i \times \mathbf{E}^* = \hat{\mathbf{n}}_i \times \left(\mathbf{E}_i - \frac{1}{Y_i} \hat{\mathbf{n}}_i \times \mathbf{H}_i \right)$$

$$\hat{\mathbf{n}}_i \times \mathbf{H}^* = 0$$

(4.35)

Alternatively, this expression can be obtained by choosing $Y_j = Y_i$, $\mathbf{E}_j = \mathbf{E}_i$, and $\mathbf{H}_j = -\mathbf{H}_i$.

4.1.3 Hybridization with Time-Domain Boundary Integral (TDBI) Method

As briefly mentioned before, when analyzing electromagnetic fields in open region problems (e.g. computation of fields scattered from an object) using the DGTD method, the unbounded background medium has to be truncated into a finite computation domain [21]. On the truncation surface, a boundary condition that imitates/approximates the radiation relation has to be enforced. To this end, one can use mathematically exact absorbing boundary conditions (EACs) [35, 36] and their localized but approximate versions (ABCs) [33, 37]. EACs are constructed using the outgoing wave modes defined on the truncation surface and therefore they can only be enforced on planar and/or spherical boundaries. Their accuracy depends on the number of these modes, which cannot be easily estimated for a pre-scribed accuracy of the solution [35, 36]. Localized ABCs are computationally less expensive but their accuracy significantly deteriorates for waves obliquely incident on the truncation surface [33, 37]. Another approach to imitate the radiation relation is to "wrap" the computation domain with perfectly matched layers (PMLs) [38–40]. The attenuation in PML does not depend on the incidence angle [6], but it decreases at low frequencies [41, 42] making the PML less accurate and/or more computationally expensive. Additionally, PML-based truncation methods might introduce late time instabilities in the solution [43] and reduce its accuracy due to spurious reflections at the PML interface [38, 39].

In this section, we describe a scheme [24, 44, 45] that combines the DGTD method with the TDBI method to truncate computation domains without suffering from the shortcomings of the other methods briefly described above. The TDBI method represents the fields on the truncation surface in the form of a retarded time boundary integral defined over a Huygens surface enclosing the scatterer. This results in a mathematically exact replacement for the radiation condition and ensures that the accuracy of the solution is limited only by the discretization error. In addition, since the truncation surface is allowed to conform to the scatterer's shape and can be located very close to its surface, the resulting computation domain can be as small as possible. In the rest of this section, the formulation underlying this hybrid method, which is abbreviated as the DGTDBI method, is described.

The scattering problem under consideration is illustrated in Figure 4.1. The permittivity and permeability inside the scatterer and in the background medium are represented by $\varepsilon(\mathbf{r})$ and $\mu(\mathbf{r})$ and ε_0 and μ_0, respectively. The scatterer is excited by the electromagnetic field $\{\mathbf{E}^{inc}(\mathbf{r}, t), \mathbf{H}^{inc}(\mathbf{r}, t)\}$, and the total field (TF) and scattered field (SF) technique is used to introduce the excitation in the computation domain. In Figure 4.1, Ω is the computation domain, $\partial\Omega$ is the boundary of the computation domain (truncation surface), $\partial\Omega^T$ is the boundary of the TF region, and $\partial\Gamma$ is the Huygens surface that is introduced to facilitate the TDBI method. The spatial discretization used in Ω is exactly same as the one described in Section 4.1.1. The DGTD method solves for the scattered field in the region between $\partial\Omega$ and $\partial\Omega^T$, and solves for the total field in the region enclosed by $\partial\Omega^T$.

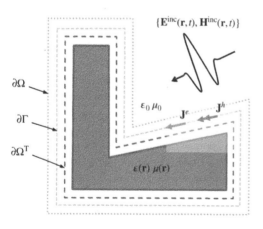

$\{\mathbf{E}^{inc}(\mathbf{r}, t), \mathbf{H}^{inc}(\mathbf{r}, t)\}$

$\partial\Omega$

$\partial\Gamma$

$\partial\Omega^T$

$\varepsilon_0 \mu_0$

J^e J^h

$\varepsilon(\mathbf{r}) \mu(\mathbf{r})$

Figure 4.1 Illustration of the scatterer, computation domain boundary $\partial\Omega$, Huygens surface $\partial\Gamma$, TF/SF surface $\partial\Omega^T$, and the excitation. Source: Reproduced with permission from Li et al. [24]/IEEE.

In this set up, let the upwind numerical flux [6, 8, 9, 34, 46] associated with element i be expressed as

$$\hat{\mathbf{n}}_{i,m} \times \mathbf{H}_m^* = \hat{\mathbf{n}}_{i,m} \times \frac{Z_i \mathbf{H}_i + \hat{\mathbf{n}}_{i,m} \times \mathbf{E}_i + \tilde{Z}\tilde{\mathbf{H}} - \hat{\mathbf{n}}_{i,m} \times \tilde{\mathbf{E}}}{Z_i + \tilde{Z}}$$

$$\hat{\mathbf{n}}_{i,m} \times \mathbf{E}_m^* = \hat{\mathbf{n}}_{i,m} \times \frac{Y_i \mathbf{E}_i - \hat{\mathbf{n}}_{i,m} \times \mathbf{H}_i + \tilde{Y}\tilde{\mathbf{E}} + \hat{\mathbf{n}}_{i,m} \times \tilde{\mathbf{H}}}{Y_i + \tilde{Y}}$$

$$(4.36)$$

where $\hat{\mathbf{n}}_{i,m}$ is the outward pointing unit normal vector on $\partial\Omega_{i,m}$, which is the mth facet of the boundary $\partial\Omega_i$ of element i. In Eq. (4.36), the outgoing flux involves $\mathbf{E}_i$ and $\mathbf{H}_i$ (fields in Ω_i) and the incoming flux involves $\tilde{\mathbf{E}}$ and $\tilde{\mathbf{H}}$ (fields "external" to Ω_i). The incoming flux is used to (i) establish the "connectivity" between fields in Ω_i and other elements "touching" $\partial\Omega_{i,m}$, (ii) enforce the boundary condition on $\partial\Omega$, and (iii) introduce $\{\mathbf{E}^{inc}, \mathbf{H}^{inc}\}$ on $\partial\Omega_T$ [46]. Accordingly, $\tilde{\mathbf{E}}$ and $\tilde{\mathbf{H}}$ in Eq. (4.36) are selected as

$$\tilde{\mathbf{E}} = \begin{cases} \mathbf{E}_{i,m}^{\partial\Omega} & \partial\Omega_{i,m} \in \partial\Omega \\ \mathbf{E}_j \pm \mathbf{E}_{i,m}^{inc} & \partial\Omega_{i,m} \in \partial\Omega^T \\ \mathbf{E}_j & \text{else} \end{cases}$$

$$(4.37)$$

$$\tilde{\mathbf{H}} = \begin{cases} \mathbf{H}_{i,m}^{\partial\Omega} & \partial\Omega_{i,m} \in \partial\Omega \\ \mathbf{H}_j \pm \mathbf{H}_{i,m}^{inc} & \partial\Omega_{i,m} \in \partial\Omega^T \\ \mathbf{H}_j & \text{else} \end{cases}$$

$$(4.38)$$

Here, j is the index of the element that neighbors element i via $\partial\Omega_{i,m}$, $\mathbf{E}_{i,m}^{\partial\Omega}$ and $\mathbf{H}_{i,m}^{\partial\Omega}$ are the fields enforced on $\partial\Omega_{i,m} \in \partial\Omega$, and $\mathbf{E}_{i,m}^{inc}$ and $\mathbf{H}_{i,m}^{inc}$ are the incident fields computed on $\partial\Omega_{i,m} \in \partial\Omega^T$, and finally, "+"/"−" sign should be selected if Ω_i is inside the TF/SF region. Similarly, $\tilde{Z}$ and $\tilde{Y}$ are selected as

$$\tilde{Z} = \begin{cases} Z_0 & \partial\Omega_{i,m} \in \partial\Omega \\ Z_j & \text{else} \end{cases} \quad , \quad \tilde{Y} = \begin{cases} Y_0 & \partial\Omega_{i,m} \in \partial\Omega \\ Y_j & \text{else} \end{cases}$$

$$(4.39)$$

where $Z_0 = \sqrt{\mu_0/\varepsilon_0}$ and $Y_0 = 1/Z_0$ are the wave impedance and admittance in the background medium.

$\mathbf{E}_{i,m}^{\partial\Omega}$ and $\mathbf{H}_{i,m}^{\partial\Omega}$ (as they appear in Eqs. (4.37) and (4.38)) on $\partial\Omega_{i,m} \in \partial\Omega$ are computed using the TDBI method as described next. Let $\partial\Omega_{i',m'}$ represent the triangular facets that discretize $\partial\Gamma$. Here, i' runs over the indices of elements that are outside the volume enclosed by $\partial\Gamma$ and have at least three nodes residing on $\partial\Gamma$, and m' runs over the indices of each element's facets that are described by these nodes. On facet $\partial\Omega_{i',m'}$, the equivalent electric and magnetic surface currents, $\mathbf{J}_{i',m'}^h$

and $\mathbf{J}^e_{i',m'}$ [47, 48], respectively, can be expressed as

$$
\mathbf{J}^h_{i',m'}(\mathbf{r}, t) = \sum_{l'=1}^{K} f^h_{i',l'}(t)\hat{\mathbf{n}}_{i',m'}(\mathbf{r}) \times \Psi^h_{i',l'}(\mathbf{r})
$$

$$
\mathbf{J}^e_{i',m'}(\mathbf{r}, t) = -\sum_{l'=1}^{K} f^e_{i',l'}(t)\hat{\mathbf{n}}_{i',m'}(\mathbf{r}) \times \Psi^e_{i',l'}(\mathbf{r})
\tag{4.40}
$$

Then, $\mathbf{E}^{\partial\Omega}_{i,m}$ and $\mathbf{H}^{\partial\Omega}_{i,m}$ on $\partial\Omega_{i,m} \in \partial\Omega$ are constructed using currents $\mathbf{J}^h_{i',m'}$ and $\mathbf{J}^e_{i',m'}$ [47, 48]:

$$
\mathbf{E}^{\partial\Omega}_{i,m}(\mathbf{r}, t) = \sum_{i'}\sum_{m'} \left[\mu_0 \mathcal{L}_{i',m'}(\mathbf{J}^h_{i',m'}) - \mathcal{K}_{i',m'}(\mathbf{J}^e_{i',m'})\right]
$$

$$
\mathbf{H}^{\partial\Omega}_{i,m}(\mathbf{r}, t) = \sum_{i'}\sum_{m'} \left[\varepsilon_0 \mathcal{L}_{i',m'}(\mathbf{J}^e_{i',m'}) + \mathcal{K}_{i',m'}(\mathbf{J}^h_{i',m'})\right]
\tag{4.41}
$$

Here, the operators $\mathcal{L}_{i',m'}$ and $\mathcal{K}_{i',m'}$ are given by [21, 44]

$$
\mathcal{L}_{i',m'}(\mathbf{J}) = -\frac{1}{4\pi}\int_{\partial\Omega_{i',m'}}\frac{\partial_t\mathbf{J}(\mathbf{r}', t - R/c_0)}{R}d\mathbf{r}'
$$

$$
+ \frac{c_0^2}{4\pi}\nabla\int_{\partial\Omega_{i',m'}}\int_0^{t-R/c_0}\frac{\nabla'\cdot\mathbf{J}(\mathbf{r}', t')}{R}dt'\,d\mathbf{r}'
\tag{4.42}
$$

$$
\mathcal{K}_{i',m'}(\mathbf{J}) = \frac{1}{4\pi}\nabla\times\int_{\partial\Omega_{i',m'}}\frac{\mathbf{J}(\mathbf{r}', t - R/c_0)}{R}d\mathbf{r}'
\tag{4.43}
$$

where $R = |\mathbf{r} - \mathbf{r}'|$ is the distance between the integration point $\mathbf{r}' \in \partial\Omega_{i',m'}$ and the observation point $\mathbf{r}$ and $c_0 = 1/\sqrt{\varepsilon_0\mu_0}$ is the speed of light in the background medium.

To account for the TDBI method, the semi-discrete system in Eqs. (4.24) and (4.25) (Section 4.1.3) is updated using Eqs. (4.36)–(4.43):

$$
\varepsilon_i\mathbf{M}^e_i\partial_t\mathbf{e}_i = [\mathbf{S}^{eh}_i - \mathbf{F}^{eh}_{ii}]\mathbf{h}_i + \mathbf{F}^{ee}_{ii}\mathbf{e}_i + \sum_j[\mathbf{F}^{eh}_{ij}\mathbf{h}_j - \mathbf{F}^{ee}_{ij}\mathbf{e}_j]
$$

$$
- \mathbf{t}^{ee}_i + \mathbf{t}^{eh}_i \mp \mathbf{b}^{ee}_i \pm \mathbf{b}^{eh}_i
\tag{4.44}
$$

$$
\mu_i\mathbf{M}^h_i\partial_t\mathbf{h}_i = [-\mathbf{S}^{he}_i + \mathbf{F}^{he}_{ii}]\mathbf{e}_i + \mathbf{F}^{hh}_{ii}\mathbf{h}_i - \sum_j[\mathbf{F}^{he}_{ij}\mathbf{e}_j + \mathbf{F}^{hh}_{ij}\mathbf{h}_j]
$$

$$
- \mathbf{t}^{hh}_i - \mathbf{t}^{he}_i \mp \mathbf{b}^{hh}_i \mp \mathbf{b}^{he}_i
\tag{4.45}
$$

where the mass matrices, stiffness matrices, and flux matrices are the same as those in Section 4.1.1. The nonzero entries of the remaining matrices and vectors are given by

$$
\begin{aligned}
[\mathbf{b}^{pp}_i]_k &= \sum_m A^p_m\int_{\partial\Omega_{i,m}\in\partial\Omega^\top}\Psi^p_{i,k}\cdot\hat{\mathbf{n}}_{i,m}\times(\hat{\mathbf{n}}_{i,m}\times\mathbf{b}^p_{i,m})d\mathbf{r}\\
[\mathbf{b}^{p\tilde{p}}_i]_k &= \sum_m B^p_m\int_{\partial\Omega_{i,m}\in\partial\Omega^\top}\Psi^p_{i,k}\cdot\hat{\mathbf{n}}_{i,m}\times\mathbf{b}^{\tilde{p}}_{i,m}d\mathbf{r}\\
[\mathbf{t}^{pp}_i]_k &= \sum_m A^p_m\int_{\partial\Omega_{i,m}\in\partial\Omega}\Psi^p_{i,k}\cdot\hat{\mathbf{n}}_{i,m}\times(\hat{\mathbf{n}}_{i,m}\times\mathbf{i}^p_{i,m})d\mathbf{r}\\
[\mathbf{t}^{p\tilde{p}}_i]_k &= \sum_m B^p_m\int_{\partial\Omega_{i,m}\in\partial\Omega}\Psi^p_{i,k}\cdot\hat{\mathbf{n}}_{i,m}\times\mathbf{i}^{\tilde{p}}_{i,m}d\mathbf{r}
\end{aligned}
\tag{4.46}
$$

In the above expressions, $p \in \{e, h\}$, $\tilde{p} = h$ for $p = e$ and $\tilde{p} = e$ for $p = h$, $A_m^e = 1/(Z_i + \tilde{Z})$, $A_m^h = 1/(Y_i + \tilde{Y})$, $B_m^e = \tilde{Z}/(Z_i + \tilde{Z})$, $B_m^h = \tilde{Y}/(Y_i + \tilde{Y})$, $\mathbf{b}_{i,m}^e = \mathbf{E}_{i,m}^{inc}$, $\mathbf{b}_{i,m}^h = \mathbf{H}_{i,m}^{inc}$, and

$$\mathbf{i}_{i,m}^e = \sum_{i'} \sum_{m'} \left\{ \sum_{l'=1}^K \mu_0 \mathcal{L}_{i',m'} \left(f_{i',l'}^h \hat{\mathbf{n}}_{i',m'} \times \Psi_{i',l'}^h \right) + \sum_{l'=1}^K \mathcal{K}_{i',m'} \left(f_{i',l'}^e \hat{\mathbf{n}}_{i',m'} \times \Psi_{i',k'}^e \right) \right\}$$

$$\mathbf{i}_{i,m}^h = \sum_{i'} \sum_{m'} \left\{ \sum_{l'=1}^K \mathcal{K}_{i',m'} \left(f_{i',l'}^h \hat{\mathbf{n}}_{i',m'} \times \Psi_{i',l'}^h \right) - \sum_{l'=1}^K \varepsilon_0 \mathcal{L}_{i',m'} \left(f_{i',l'}^e \hat{\mathbf{n}}_{i',m'} \times \Psi_{i',l'}^e \right) \right\}$$

4.1.4 Multi-time Stepping Scheme of the DGTDBI

The semi-discrete system in Eqs. (4.44) and (4.45) is numerically integrated in time to obtain the samples of $\mathbf{e}_i$ and $\mathbf{h}_i$. It should be emphasized that the time step size used by the DGTD and TDBI methods does not need to be the same. Let Δt_{DG} represent the time step size of the integration scheme used. To ensure stability, Δt_{DG} has to satisfy a Courant–Friedrichs–Lewy (CFL)-like condition that depends on the order and type of spatial discretization and time integration schemes as described in detail in [6, 8, 9, 20, 34, 46]. On the other hand, the time step size used by the TDBI method, denoted by Δt_{BI}, is only required to resolve the maximum frequency of the excitation and is independent from the spatial discretization [49, 50]. In a typical scenario $\Delta t_{BI} \gg \Delta t_{DG}$; therefore, if one uses a single time step size for the hybrid DGTDBI scheme and sets it to Δt_{DG}, the computational requirements of the TDBI method will increase significantly. Consequently, spatially discretized retarded-time surface integrals $\mathbf{i}_{i,m}^p$ are sampled at $k\Delta t_{BI}$ requiring the computation of $f_{i',l'}^p(k\Delta t_{BI} - R/c_0)$. But during marching their samples at $k\Delta t_{DG}$, i.e. $f_{i',l'}^p(l\Delta t_{DG})$, are available. Therefore, an interpolation scheme is needed to compute $f_{i',l'}^p(k\Delta t_{BI} - R/c_0)$ from $f_{i',l'}^p(l\Delta t_{DG})$. From the opposite side, only samples $\mathbf{i}_{i,m}^p(k\Delta t_{BI})$ are available during time marching. But the time integration scheme requires samples $\mathbf{t}_i^{pp}(l\Delta t_{DG})$ and $\mathbf{t}_i^{p\tilde{p}}(l\Delta t_{DG})$. Therefore, an interpolation scheme is needed to compute them from $\mathbf{i}_{i,m}^p(k\Delta t_{BI})$. Both interpolation operations are carried out using shifted higher-order Lagrange polynomials [49, 50]. Also, the time derivative in operator $\mathcal{L}_{i',m'}$ is moved onto this interpolation function. We should also emphasize here that, to maintain the explicitness of the time marching, which is one of the main advantages of the DGTD scheme over classical finite-element methods, the minimum distance between any two points on $\partial\Gamma$ and $\partial\Omega$ has to be larger than $c_0 \Delta t_{BI}$. A simple illustration of the linear temporal interpolation process is shown in Figure 4.2.

4.1.5 Numerical Examples for the DGTDBI

In this section, we provide two numerical examples to demonstrate the accuracy and the applicability of the DGTDBI. In the first example, the scatterer is a

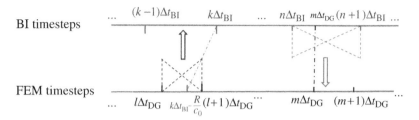

Figure 4.2 The illustration of temporal interpolation process.

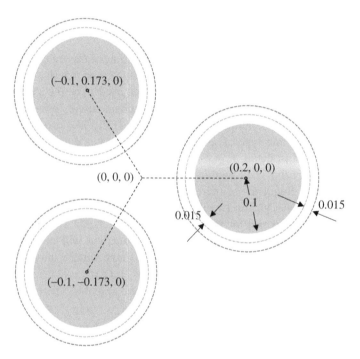

Figure 4.3 Cluster of three PEC spheres. All dimensions are in meters. Reproduced with permission from [24].

cluster of three PEC spheres residing in free space (Figure 4.3). Use of the TDBI approach allows the computation domain to be defined in terms of three equally sized domains, each of which encloses one of the spheres. Disconnected spherical boundaries of these domains and the Huygens surfaces are located 0.03 and 0.015 m away from the sphere surfaces, respectively (Figure 4.3). For the first simulation, the cluster is excited with a plane wave with parameters $f_m = 500$ MHz, $t_0 = 13.66$ ns, and $\tau_m = 2.6$ ns. During the simulation, the transient scattered electric near field computed at the origin is recorded.

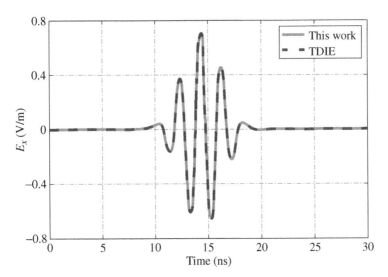

Figure 4.4 Transient scattered electric field computed at the origin of cluster of PEC spheres. Source: Reproduced with permission from Li et al. [24]/IEEE.

Figure 4.4 compares this field to that computed using a time-domain integral equation (TDIE) solver [49, 50], and the results agree well. For the second simulation, $\tau_m = 0.65$ ns while other excitation parameters are kept the same. After the time-domain simulation is completed, radar cross-section (RCS) of the cluster is computed from the Fourier-transformed currents on the Huygens surface. Figures 4.5 and 4.6 plot the RCS on the *xy*- and *xz*-planes computed at 2.53 MHz and 1.003 GHz, respectively. The results agree very well with those obtained from the solution of an in-house method of moments (MoM) solver.

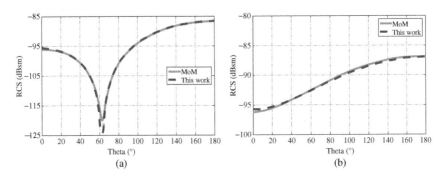

Figure 4.5 RCS on (a) *xy*- and (b) *xz*-planes computed at 2.53 MHz from the solutions of the DGTDBI method and MoM. Source: Reproduced with permission from Li et al. [24]/IEEE.

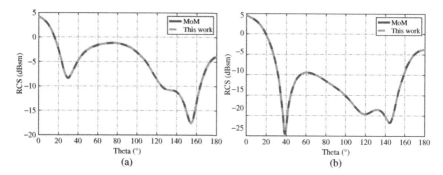

Figure 4.6 RCS on (a) *xy*- and (b) *xz*-planes computed at 1.003 GHz from the solutions of the DGTDBI method and MoM. Source: Reproduced with permission from Li et al. [24]/IEEE.

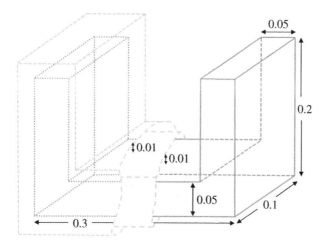

Figure 4.7 U-shaped scatterer. All dimensions are in meters. Source: Reproduced with permission from Li et al. [24]/IEEE.

These two figures clearly demonstrate that the "absorption" enforced by the TDBI approach is accurate even at low frequencies.

The second example is a U-shaped scatterer residing in free space (Figure 4.7). The boundary of the computation domain and the Huygens surface conform to the shape of the scatterer and are located 0.02 and 0.01 m away from its surface, respectively. The scatterer is excited with a plane wave with parameters $f_m = 500$ MHz, $t_0 = 13.66$ ns, and $\tau_m = 2.6$ ns. Two simulations are carried out: (i) The scatterer surface is PEC. (ii) The scatterer is a dielectric body with relative permittivity of 4.0. After the time-domain simulations, bistatic and monostatic RCS of the scatterers are computed. Figures 4.8a,b and 4.9a,b plot the bistatic and monostatic RCS of the

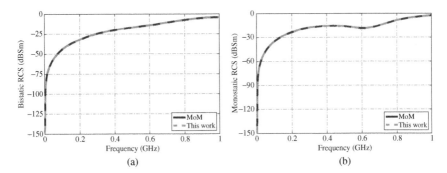

Figure 4.8 The bistatic (a) and monostatic (b) RCS of the PEC U-shaped scatterer computed at a range of frequencies changing from 286.4 kHz to 1.0 GHz. Source: Reproduced with permission from Li et al. [24]/IEEE.

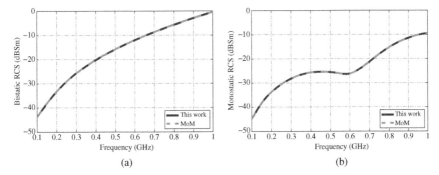

Figure 4.9 The bistatic (a) and monostatic (b) RCS of the dielectric U-shaped scatterer computed at a range of frequencies changing from 100 MHz to 1.0 GHz. Source: Reproduced with permission from Li et al. [24]/IEEE.

PEC and dielectric scatterers computed at a wide range of frequencies, respectively. Plots clearly demonstrate that the results agree very well with those obtained from an MoM solver, which further verifies the accuracy and also low-frequency absorbing capability of the DGTDBI method. Additionally, Figure 4.10a,b compare the RCS of the PEC scatterer on the *xz*- and *yz*-planes computed at 1 GHz to those obtained by the same DGTD method, which uses ABC instead of the TDBI method. It is clearly shown that ABC is not accurate for this structure,

4.1.6 The DGTD Scheme with Nodal Basis Functions

For "traditional" continuous FEM, vector basis functions often have to be used instead of nodal basis functions to avoid spurious solutions [21]. On the other hand, for the DGTD method, it is found that spurious modes dissipate much faster

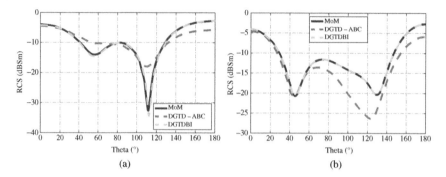

Figure 4.10 The calculated RCS in xz (a) and yz (b) planes with the DGTDBI method and its comparisons from MoM and the DGTD-ABC method at 1.00013 GHz. Source: Reproduced with permission from Li et al. [24]/IEEE.

than physical modes when upwind or penalized fluxes are used, and therefore they can be avoided even without using vector basis functions [8]. In addition, for the DGTD method, even curl-conforming vector basis functions cannot ensure that the solutions are globally divergence-free. From this viewpoint, the main advantages of using vector basis functions in FEM do not hold true for the DGTD method.

From an implementation perspective, generation of high-order nodal basis functions is much easier, and often visualization of simulation results directly obtained at nodes does not require difficult post-processing operations. Furthermore, in many multiphysics problems, all field components are used to realize the coupling between different sets of equations (representing different physical phenomena). In the nodal basis function approach, all field components are expanded separately and the continuity between neighboring elements for each component is recovered through the numerical flux.

All these reasons point out to the fact that nodal basis functions are a good option for the DGTD implementations. In the nodal DG formulation, each Cartesian component of **E** and **H** is expanded using a set of nodal basis functions. A popular choice is the Lagrange polynomials Ψ_l with nodes located at Legendre–Gauss–Lobatto (LGL) quadrature points [8]:

$$E_i^v(\mathbf{r}, t) \simeq \sum_{l=1}^{N_p} E_i^v(\mathbf{r}_l, t)\Psi_l(\mathbf{r}) = \sum_{l=1}^{N_p} e_{i,l}^v(t)\Psi_l(\mathbf{r}) \tag{4.47}$$

$$H_i^v(\mathbf{r}, t) \simeq \sum_{l=1}^{N_p} H_i^v(\mathbf{r}_l, t)\Psi_l(\mathbf{r}) = \sum_{l=1}^{N_p} h_{i,l}^v(t)\Psi_l(\mathbf{r}) \tag{4.48}$$

where i is the element index, $l = 1, \ldots, N_p$, $N_p = (p + 1)(p + 2)(p + 3)/6$ is the number of interpolation nodes, p is the order of the Lagrange polynomials, $\mathbf{r}_l$

denote the locations of the interpolation nodes, $v \in \{x, y, z\}$ denotes the components of vectors in the Cartesian coordinate system, and $e_{i,l}^v(t)$ and $h_{i,l}^v(t)$ are the time-dependent expanding coefficients to be solved for.

Applying Galerkin testing on element i as is done in the vector DG formulation, one can obtain the component-wise semi-discrete form

$$\varepsilon_i \frac{\partial \mathbf{e}_i^x}{\partial t} = \mathbf{D}_i^y \mathbf{h}_i^z - \mathbf{D}_i^z \mathbf{h}_i^y - \mathbf{F}_i \mathbf{f}_i^{e,x} - \mathbf{j}_i^x \tag{4.49}$$

$$\varepsilon_i \frac{\partial \mathbf{e}_i^y}{\partial t} = \mathbf{D}_i^z \mathbf{h}_i^x - \mathbf{D}_i^x \mathbf{h}_i^z - \mathbf{F}_i \mathbf{f}_i^{e,y} - \mathbf{j}_i^y \tag{4.50}$$

$$\varepsilon_i \frac{\partial \mathbf{e}_i^z}{\partial t} = \mathbf{D}_i^x \mathbf{h}_i^y - \mathbf{D}_i^y \mathbf{h}_i^x - \mathbf{F}_i \mathbf{f}_i^{e,z} - \mathbf{j}_i^z \tag{4.51}$$

$$\mu_i \frac{\partial \mathbf{h}_i^x}{\partial t} = -\mathbf{D}_i^y \mathbf{e}_i^z + \mathbf{D}_i^z \mathbf{e}_i^y + \mathbf{F}_i \mathbf{f}_i^{h,x} \tag{4.52}$$

$$\mu_i \frac{\partial \mathbf{h}_i^y}{\partial t} = -\mathbf{D}_i^z \mathbf{e}_i^x + \mathbf{D}_i^x \mathbf{e}_i^z + \mathbf{F}_i \mathbf{f}_i^{h,y} \tag{4.53}$$

$$\mu_i \frac{\partial \mathbf{h}_i^z}{\partial t} = -\mathbf{D}_i^x \mathbf{e}_i^y + \mathbf{D}_i^y \mathbf{e}_i^x + \mathbf{F}_i \mathbf{f}_i^{h,z} \tag{4.54}$$

where $\mathbf{e}_i^v = [e_{i,1}^v, \ldots, e_{i,N_p}^v]^T$ and $\mathbf{h}_i^v = [h_{i,1}^v, \ldots, h_{i,N_p}^v]^T$ are unknown vectors, $\mathbf{j}_i^v = [j_{i,1}^v, \ldots, j_{i,N_p}^v]^T$ is the current density vector, $j_{i,l}^v = j_i^v(\mathbf{r}_l, t)$, and $\mathbf{f}_i^{e,v}$ and $\mathbf{f}_i^{h,v}$ are coefficients corresponding to the v components of $\hat{n} \times [\mathbf{H}_i - \mathbf{H}_i^*]$ and $\hat{n} \times [\mathbf{E}_i - \mathbf{E}_i^*]$, respectively. Here, the numerical flux $\mathbf{H}_i^*$ and $\mathbf{E}_i^*$ are defined the same as in Eqs. (4.19) and (4.20).

The operators $\mathbf{D}_i^v = \mathbf{M}_i^{-1} \mathbf{S}_i^v$ and $\mathbf{F}_i = \mathbf{M}_i^{-1} \mathbf{L}_i$, where $\mathbf{M}_i$ is the mass matrix defined as

$$\mathbf{M}_i(l, k) = \int_{\Omega_i} \Psi_l(\mathbf{r}) \Psi_k(\mathbf{r}) dr \tag{4.55}$$

$\mathbf{S}_i^v$ is the stiffness matrix defined as

$$\mathbf{S}_i^v(l, k) = \int_{\Omega_i} \Psi_l(\mathbf{r}) \partial_v \Psi_k(\mathbf{r}) dr \tag{4.56}$$

$\mathbf{L}_i$ is the surface mass matrix defined as

$$\mathbf{L}_i(l, k) = \oint_{\partial\Omega_i} \Psi_l(\mathbf{r}) \Psi_k(\mathbf{r}) dr \tag{4.57}$$

The semi-discrete system in Eqs. (4.49)–(4.54) can be integrated in time using explicit schemes just like it is done for the semi-discrete system in Eqs. (4.30) and (4.31), which is obtained using vector basis functions.

4.2 Application of the DGTD Method to Real Problems

4.2.1 Graphene-Based Devices

The two-dimensional (2D) materials have received significant attention in the last decade due to their remarkable electrical, thermal, and mechanical properties. Graphene, which belongs to the family of 2D materials, is an atomically thick layer of carbon atoms that are arranged in a honeycomb lattice. In the last few years, graphene has found various applications in different academic research fields as well as industry [51–57]. Simulation of graphene-based electromagnetic devices is one of these research fields. Indeed, several methods, which rely on MoM [58, 59], the finite difference time-domain [60–62], DGTD method [31, 63, 64], and circuit representations [56], have been developed for numerical characterization of transmission, reflection, and absorption of electromagnetic fields on graphene sheets as well as generation of graphene surface plasmon polaritons within the gigahertz-terahertz frequency band. All these methods model the atomically thick graphene using an equivalent surface impedance boundary condition (SIBC) to avoid very fine volumetric discretization of the graphene layer, which would dramatically increase the number of unknowns and decrease the time step size (through CFL condition) in the case of explicit time-domain methods [31, 63, 64].

Surface conductivity of SIBC is either a scalar [56] under electrostatic biasing or a tensor [56, 58] due to quantum hall effect under magnetostatic biasing. In either case, surface conductivity is frequency dependent, which makes its efficient modeling within time-domain methods challenging. In this section, we introduce two different DGTD-based methods, which use finite integration and auxiliary differential equation (ADE) methods, respectively, to take into account the temporal dispersion effects and anisotropic surface conductivity in the time domain [31, 63].

4.2.1.1 A Resistive Boundary Condition to Represent Graphene Within the DGTD Method

SIBC associated with graphene is incorporated within the DGTD method via the re-formulation of the numerical flux. First, the surface conductivity of graphene is approximated by a series of rational basis functions generated using the fast-relaxation vector-fitting (FRVF) method in the Laplace-domain [65] [66]. Then, using inverse Laplace transform, corresponding time-domain matrix equations in integral form (in time t) are obtained. Finally, these matrix equations are solved by time-domain finite integration technique (FIT). For elements not touching the graphene sheet, the well-known Runge–Kutta method is employed to integrate Maxwell's equations as described in Section 4.1.1. We emphasize here again that the use of SIBC significantly reduces the computational requirements of the DGTD method since the volumetric mesh that would be defined inside the thin graphene sheet is now avoided. In the rest of the section, we first present the formulation underlying the DGTD scheme extended to account for graphene's SIBC model and provide various numerical examples to show the extended scheme's accuracy, efficiency, and applicability.

In the formulation described here, the 2D graphene layer is considered as an infinitely thin surface. In the absence of an external magnetostatic biasing field, the surface conductivity of this graphene layer is a scalar denoted by $\sigma_g(\omega, \mu_c, \Gamma, T)$. Using Kubo's formula, σ_g is expressed as [58]:

$$
\sigma_g(\omega, \mu_c, \Gamma, T) = \frac{jq^2}{\pi \hbar (\omega - j2\Gamma)} \int_0^\infty \varepsilon \left[\frac{\partial f_d(\varepsilon)}{\partial \varepsilon} - \frac{\partial f_d(-\varepsilon)}{\partial \varepsilon} \right]
$$
$$
- \frac{jq^2(\omega - j2\Gamma)}{\pi \hbar^2} \int_0^\infty \frac{f_d(-\varepsilon) - f_d(\varepsilon)}{(\omega - j2\Gamma)^2 - 4(\varepsilon/\hbar)^2}
\tag{4.58}
$$

where ε indicates the energy state, $\hbar$ denotes the reduced Planck constant, $-q$ is the charge of electron, and $f_d(\varepsilon) = \left[e^{(\varepsilon - \mu_c)/k_B T} + 1 \right]^{-1}$ is the Fermi-Dirac distribution with Boltzmann constant k_B. In (4.58), the first term corresponds to the intraband contribution that can be expressed as [58]

$$
\sigma_{\text{intra}} = -\frac{jq^2 k_B T}{\pi \hbar^2 (\omega - j2\Gamma)} \left[\frac{\mu_c}{k_B T} + 2 \ln \left(e^{-\mu_c/k_B T} + 1 \right) \right]
\tag{4.59}
$$

while the second term corresponds to the interband contribution, which can be approximated by

$$
\sigma_{\text{inter}} = -\frac{jq^2}{4\pi \hbar} \ln \left[\frac{2|\mu_c| - (\omega - j2\Gamma)\hbar}{2|\mu_c| + (\omega - j2\Gamma)\hbar} \right]
\tag{4.60}
$$

when $k_B T \ll \hbar \omega$ and $|\mu_c|$. On the resistive thin sheet with the finite conductivity σ_g, the electromagnetic fields satisfy the following boundary conditions [63]:

$$
\hat{n} \times (\mathbf{E}^2 - \mathbf{E}^1) = 0
$$
$$
\hat{n} \times (\mathbf{H}^2 - \mathbf{H}^1) = \sigma_g \mathbf{E}_t
\tag{4.61}
$$

where the superscripts 1 and 2 represent the two sides of the graphene sheet, $\hat{n}$ is a unit normal vector pointing from side 1 to side 2, and $\mathbf{E}_t = -\hat{n} \times \left[\hat{n} \times (\mathbf{E}^1 + \mathbf{E}^2)/2 \right]$ is the tangential component of the electric field along the graphene sheet. Note that (4.61) is expressed in frequency domain (or Laplace domain).

Similar to the fundamental boundary conditions, as discussed in Section 4.1.2, the resistive boundary conditions in (4.61) can be incorporated within the DGTD method by reformulating the jump conditions used in the expression of the numerical flux (see (4.13)–(4.20)) in Section 4.1.1 for the jump conditions). More specifically, the jump condition for the magnetic field for characteristics $x_n = 0$ is updated as

$$
\hat{n}_{i f_g} \times (\mathbf{H}_{f_g}^{**} - \mathbf{H}_{f_g}^{*}) = \sigma_g \mathbf{E}_t
\tag{4.62}
$$

for element facets f_g that touch the graphene sheet. Here, i represents the index of these elements. Note that only one facet can touch the graphene sheet for a given element. Using (4.62) together with the other jump conditions (4.13)–(4.20) from Section 4.1.1 (or more specifically their frequency-domain counterparts), the upwind numerical fluxes for facets that touch the graphene sheet are obtained as

$$
\hat{\mathbf{n}}_{i,f_g} \times \mathbf{E}^*_{f_g} = \hat{\mathbf{n}}_{i,f_g} \times \left[\frac{\left(Y_i \mathbf{E}_i + Y_{j,f_g} \mathbf{E}_{j,f_g}\right) + \hat{\mathbf{n}}_{i,f_g} \times \left(\mathbf{H}_{j,f_g} - \mathbf{H}_i\right)}{Y_i + Y_{j,f_g}} \right.
$$
$$
\left. - \frac{\sigma_g \left(\mathbf{E}_i + \mathbf{E}_{j,f_g}\right)}{2\left(Y_i + Y_{j,f_g}\right)} \right] \tag{4.63}
$$

$$
\hat{\mathbf{n}}_{i,f_g} \times \mathbf{H}^*_{f_g} = \hat{\mathbf{n}}_{i,f_g} \times \left[\frac{\left(Z_i \mathbf{H}_i + Z_{j,f_g} \mathbf{H}_{j,f_g}\right) + \hat{\mathbf{n}}_{i,f_g} \times \left(\mathbf{E}_i - \mathbf{E}_{j,f_g}\right)}{Z_i + Z_{j,f_g}} \right.
$$
$$
\left. + \frac{Z_{j,f_g} \sigma_g \hat{\mathbf{n}}_{i,f_g} \times \left(\mathbf{E}_i + \mathbf{E}_{j,f_g}\right)}{2\left(Z_i + Z_{j,f_g}\right)} \right] \tag{4.64}
$$

The numerical flux in (4.63) and (4.64) cannot be directly converted to the time domain since σ_g is frequency dependent (with a rather complicated function). To overcome this issue, we first represent σ_g in terms of a set of rational basis functions in the Laplace domain (using FRVF technique [65, 66]). Then, we use (4.58) in Laplace-domain DG equations. Then, via inverse Laplace transform of the resulting equations, a time marching scheme is obtained for the DGTD method.

The FRVF technique is applied to a set of frequency samples of σ_g to yield the following representation in terms of rational functions [65, 66]:

$$
\sigma_g(s) = \frac{u_1 s^{-1} + u_2 s^{-2} + \cdots + u_P s^{-P}}{d_0 + d_1 s^{-1} + d_2 s^{-2} + \cdots + d_Q s^{-Q}} = \frac{\displaystyle\sum_{p=1}^{P} u_p s^{-p}}{\displaystyle\sum_{q=0}^{Q} d_q s^{-q}} \tag{4.65}
$$

Here, s is the Laplace state variable, u_p and d_q are the pth and the qth coefficients, and P and Q are the orders of numerator and denominator, respectively.

The tested DG equations for Maxwell's curl equations in the Laplace domain are given as

$$
\int_{\Omega_i} \Psi^e_{i,k} \cdot \left[\varepsilon_i \mathbf{E}_i - s^{-1} \nabla \times \mathbf{H}_i\right] d\mathbf{r} = s^{-1} \sum_{f=1}^{4} \int_{\partial\Omega_{i,f}} \Psi^e_{i,k} \cdot \left[\hat{\mathbf{n}}_{i,f} \times (\mathbf{H}^*_f - \mathbf{H}_i)\right] d\mathbf{r} \tag{4.66}
$$

$$\int_{\Omega_i} \Psi^h_{i,k} \cdot \left[\mu_i \mathbf{H}_i + s^{-1} \nabla \times \mathbf{E}_i\right] d\mathbf{r} = s^{-1} \sum_{f=1}^{4} \int_{\partial\Omega_{i,f}} \Psi^h_{i,k} \cdot \left[\hat{\mathbf{n}}_{i,f} \times (\mathbf{E}_i - \mathbf{E}^*_f)\right] d\mathbf{r} \qquad (4.67)$$

Note that on the right-hand side of (4.66) and (4.67), when $f = f_g$, the numerical flux expression in (4.63) and (4.64) is used. For other facets, numerical flux stays the same as those in (4.19) and (4.20).

Inserting (4.63)–(4.65) in (4.66) and (4.67), and using the Fourier transform pair

$$s^{-n} F(s) \leftrightarrow \underbrace{\int \cdots \int}_{n} f(\tau) \underbrace{d\tau \cdots d\tau}_{n} \qquad (4.68)$$

in the resulting equation yields the following time-domain matrix equations [63]:

$$\mathbf{M}^e_i \left(d_0 \mathbf{e}_{i,0} + d_1 \mathbf{e}_{i,1} + \cdots + d_Q \mathbf{e}_{i,Q}\right) - \mathbf{S}^e_i \left(d_0 \mathbf{h}_{i,1} + d_1 \mathbf{h}_{i,2} + \cdots + d_Q \mathbf{h}_{i,Q+1}\right)$$
$$= \sum_{f=1}^{4} \left(\mathbf{F}^{ee}_{ii,f} \mathbf{e}_{i,f} + \mathbf{F}^{ee}_{ij,f} \mathbf{e}_{j,f} + \mathbf{F}^{eh}_{ii,f} \mathbf{h}_{i,f} + \mathbf{F}^{eh}_{ij,f} \mathbf{h}_{i,f}\right) + \underbrace{\mathbf{F}^{ee,\sigma_g}_{ij} \mathbf{e}^{\sigma_g}_j + \mathbf{F}^{ee,\sigma_g}_{ii} \mathbf{e}^{\sigma_g}_i}_{\mathbf{F}^{e,\sigma_g}} \qquad (4.69)$$

$$\mathbf{M}^h_i \left(d_0 \mathbf{h}_{i,0} + d_1 \mathbf{h}_{i,1} + \cdots + d_Q \mathbf{h}_{i,Q}\right) + \mathbf{S}^h_i \left(d_0 \mathbf{e}_{i,1} + d_1 \mathbf{e}_{i,2} + \cdots + d_Q \mathbf{e}_{i,Q+1}\right)$$
$$= \sum_{f=1}^{4} \left(\mathbf{F}^{hh}_{ii,f} \mathbf{h}_{i,f} + \mathbf{F}^{hh}_{ij,f} \mathbf{h}_{i,f} + \mathbf{F}^{he}_{ii,f} \mathbf{e}_{i,f} + \mathbf{F}^{he}_{ij,f} \mathbf{e}_{i,f}\right) + \underbrace{\mathbf{F}^{he,\sigma_g}_{ij} \mathbf{e}^{\sigma_g}_j + \mathbf{F}^{he,\sigma_g}_{ii} \mathbf{e}^{\sigma_g}_i}_{\mathbf{F}^{h,\sigma_g}} \qquad (4.70)$$

where $\mathbf{M}^{e/h}_i$ and $\mathbf{S}^{e/h}_i$ are mass and stiffness matrices and $\mathbf{F}^{ee}$, $\mathbf{F}^{eh}$, $\mathbf{F}^{hh}$, and $\mathbf{F}^{he}$ are the flux matrices. Matrix entries for those matrices can be found in Section 4.1.1. The entries of the matrices $\mathbf{F}^{ee,\sigma_g}_{ii}$, $\mathbf{F}^{ee,\sigma_g}_{ij}$ $\mathbf{F}^{he,\sigma_g}_{ii}$, and $\mathbf{F}^{he,\sigma_g}_{ij}$ are given by:

$$[\mathbf{F}^{ee,\sigma_g}_{ii}]_{kl} = \frac{Z_{j,f_g}}{2(Z_i + Z_{j,f_g})} \int_{\partial\Omega_{i,j}} \Psi^e_{i,k} \cdot \hat{\mathbf{n}}_{i,f_g} \times (\hat{\mathbf{n}}_{i,f_g} \times \Psi^e_{i,l}) \, d\mathbf{r}$$

$$[\mathbf{F}^{ee,\sigma_g}_{ij}]_{kl} = \frac{Z_{j,f_g}}{2(Z_i + Z_{j,f_g})} \int_{\partial\Omega_{i,j}} \Psi^e_{i,k} \cdot \hat{\mathbf{n}}_{i,f_g} \times (\hat{\mathbf{n}}_{i,f_g} \times \Psi^e_{j,l}) \, d\mathbf{r}$$

$$[\mathbf{F}^{he,\sigma_g}_{ii}]_{kl} = \frac{1}{2(Y_i + Y_{j,f_k})} \int_{\partial\Omega_{i,j}} \Psi^h_{i,k} \cdot (\hat{\mathbf{n}}_{i,f_g} \times \Psi^e_{i,l}) \, d\mathbf{r}$$

$$[\mathbf{F}^{he,\sigma_g}_{ij}]_{kl} = \frac{1}{2(Y_i + Y_{j,f_g})} \int_{\partial\Omega_{i,j}} \Psi^h_{i,k} \cdot (\hat{\mathbf{n}}_{i,f_g} \times \Psi^e_{j,l}) \, d\mathbf{r}.$$

In (4.69), (4.70), flux matrices $\mathbf{F}^{\sigma_g}_e = \mathbf{F}^{ee,\sigma_g}_{ij} + \mathbf{F}^{ee,\sigma_g}_{ii}$, $\mathbf{F}^{\sigma_g}_h = \mathbf{F}^{he,\sigma_g}_{ij} + \mathbf{F}^{he,\sigma_g}_{ii}$ are nonzero only for faces $(f = f_g)$ touching the graphene sheet; otherwise $(f \neq f_g)$ they are zero. Moreover, terms $\mathbf{e}_{i,q}$ and $\mathbf{h}_{i,q}$ are column vectors storing the unknown coefficients

e_i^k and h_i^l and are defined as

$$
\mathbf{e}_{i,q}^k = \underbrace{\int \cdots \int e_i^k(\tau)d\tau \cdots d\tau}_{q}
$$

$$
\mathbf{h}_{i,q}^l = \underbrace{\int \cdots \int h_i^l(\tau)d\tau \cdots d\tau}_{q}
$$

$$(4.71)$$

The expressions for other column vectors are given as

$$
(\mathbf{e}/\mathbf{h})_{if} = \sum_{q=0}^{Q} d_q (\mathbf{e}_{if}/\mathbf{h}_{if})_{q+1}, \quad (\mathbf{e}/\mathbf{h})_{jf} = \sum_{q=0}^{Q} d_q (\mathbf{e}_{jf}/\mathbf{h}_{jf})_{q+1}
$$

$$
\mathbf{e}_i^{\sigma_g} = \sum_{p=1}^{P} u_p \mathbf{e}_{i,p+1}, \quad \mathbf{e}_j^{\sigma_g} = \sum_{p=1}^{P} u_p \mathbf{e}_{jf_g,p+1}
$$

$$(4.72)$$

Note that all terms in (4.72) involve integrals in time. To enable the solution of (4.69) and (4.70) using a time marching scheme, these integrals in time are discretized using an FIT scheme. This scheme uses the rectangular rule [59], which yields the discretized version of (4.71) with order p or q at $t = (n+1)\delta t$ as:

$$
\mathbf{e}_{i,p/q}^{n+1} = (\delta t)_{p/q} \sum_{k_{p/q}=0}^{n} \cdots \sum_{k_2=0}^{k_3} \sum_{k_1=0}^{k_2} \mathbf{e}_{i,k_1+1}
$$

$$
\mathbf{h}_{i,p/q}^{n+1} = (\delta t)_{p/q} \sum_{k_{p/q}=0}^{n} \cdots \sum_{k_2=0}^{k_3} \sum_{k_1=0}^{k_2} \mathbf{h}_{i,k_1+1}
$$

$$(4.73)$$

$$
\mathbf{e}_{i,0}^{n+1} = \mathbf{e}_i^{n+1}, \quad \mathbf{h}_{i,0}^{n+1} = \mathbf{h}_i^{n+1}.
$$

To maintain the explicit nature of the time marching scheme, the terms involving contributions from neighboring elements are discretized using a forward rectangular rule. This yields

$$
\mathbf{e}_{j,p/q}^{n+1} = (\delta t)_{p/q} \sum_{k_{p/q}=0}^{n} \cdots \sum_{k_2=0}^{k_3} \sum_{k_1=0}^{k_2} \mathbf{e}_{j,k_1}
$$

$$
\mathbf{h}_{j,p/q}^{n+1} = (\delta t)_{p/q} \sum_{k_{p/q}=0}^{n} \cdots \sum_{k_2=0}^{k_3} \sum_{k_1=0}^{k_2} \mathbf{h}_{j,k_1}
$$

$$(4.74)$$

After a lengthy mathematical manipulation, (4.69) and (4.70) are expressed in a more compact form as a matrix system at $t = (n+1)\delta t$:

$$
\begin{pmatrix} \tilde{\mathbf{M}}_i^e & \tilde{\mathbf{S}}_i^e \\ \tilde{\mathbf{S}}_i^h & \tilde{\mathbf{M}}_i^h \end{pmatrix} \begin{bmatrix} \mathbf{e}_i^{n+1} \\ \mathbf{h}_i^{n+1} \end{bmatrix} = \begin{pmatrix} \tilde{\mathbf{F}}_i^e \\ \tilde{\mathbf{F}}_i^h \end{pmatrix}
$$

$$(4.75)$$

Here, the matrix blocks are given by

$$\tilde{\mathbf{M}}_i^e = \sum_{q=0}^{Q} d_q(\delta t)_q \mathbf{M}_i^e - \sum_{q=0}^{Q} d_q(\delta t)_{q+1} \mathbf{F}_{ii}^{ee} - \sum_{p=1}^{P} u_p(\delta t)_{p+1} \mathbf{F}_{ii}^{ee,\sigma_g}$$

$$\tilde{\mathbf{S}}_i^e = -\sum_{q=0}^{Q} d_q(\delta t)_{q+1} \mathbf{S}_i^e + \sum_{q=0}^{Q} d_q(\delta t)_{q+1} \mathbf{F}_{ii}^{eh}$$

$$\tilde{\mathbf{M}}_i^h = \sum_{q=0}^{Q} d_q(\delta t)_q \mathbf{M}_i^h - \sum_{q=0}^{Q} d_q(\delta t)_{q+1} \mathbf{F}_{ii}^{hh} \tag{4.76}$$

$$\tilde{\mathbf{S}}_i^h = \sum_{q=0}^{Q} d_q(\delta t)_q \mathbf{S}_i^h - \sum_{q=0}^{Q} d_q(\delta t)_{q+1} \mathbf{F}_{ii}^{he} - \sum_{p=1}^{P} u_p(\delta t)_{p+1} \mathbf{F}_{ii}^{he,\sigma_g}$$

and the column vectors are given by

$$\tilde{\mathbf{F}}_i^e = -\left\{ \sum_{q=0}^{Q} d_q(\delta t)_q \tilde{\mathbf{e}}_{i,q}^{n+1} \right\} \mathbf{M}_i^e + \left\{ \sum_{q=0}^{Q} d_q(\delta t)_{q+1} \tilde{\mathbf{h}}_{i,q+1}^{n+1} \right\} \mathbf{S}_i^e$$

$$+ \left\{ \sum_{q=0}^{Q} d_q(\delta t)_{q+1} \tilde{\mathbf{e}}_{i,q+1}^{n+1} \right\} \mathbf{F}_{ii}^{ee} + \left\{ \sum_{q=0}^{Q} d_q(\delta t)_{q+1} \mathbf{e}_{j,q+1}^{n+1} \right\} \mathbf{F}_{ij}^{ee}$$

$$+ \left\{ \sum_{q=0}^{Q} d_q(\delta t)_{q+1} \tilde{\mathbf{h}}_{i,q+1}^{n+1} \right\} \mathbf{F}_{ii}^{eh} + \left\{ \sum_{q=0}^{Q} d_q(\delta t)_{q+1} \mathbf{h}_{j,q+1}^{n+1} \right\} \mathbf{F}_{ij}^{eh}$$

$$+ \left\{ \sum_{p=1}^{P} u_p(\delta t)_{p+1} \tilde{\mathbf{e}}_{i,p+1}^{n+1} \right\} \mathbf{F}_{ii}^{ee,\sigma_g} + \left\{ \sum_{p=1}^{P} u_p(\delta t)_{p+1} \mathbf{e}_{j,p+1}^{n+1} \right\} \mathbf{F}_{ij}^{ee,\sigma_g}$$

$$\tilde{\mathbf{F}}_i^h = -\left\{ \sum_{q=0}^{Q} d_q(\delta t)_q \tilde{\mathbf{h}}_{i,q}^{n+1} \right\} \mathbf{M}_i^h - \left\{ \sum_{q=0}^{Q} d_q(\delta t)_{q+1} \tilde{\mathbf{e}}_{i,q+1}^{n+1} \right\} \mathbf{S}_i^h$$

$$+ \left\{ \sum_{q=0}^{Q} d_q(\delta t)_{q+1} \tilde{\mathbf{h}}_{i,q+1}^{n+1} \right\} \mathbf{F}_{ii}^{hh} + \left\{ \sum_{q=0}^{Q} d_q(\delta t)_{q+1} \mathbf{h}_{j,q+1}^{n+1} \right\} \mathbf{F}_{ij}^{hh}$$

$$+ \left\{ \sum_{q=0}^{Q} d_q(\delta t)_{q+1} \tilde{\mathbf{e}}_{i,q+1}^{n+1} \right\} \mathbf{F}_{ii}^{he} + \left\{ \sum_{q=0}^{Q} d_q(\delta t)_{q+1} \mathbf{e}_{j,q+1}^{n+1} \right\} \mathbf{F}_{ij}^{he}$$

$$+ \left\{ \sum_{p=1}^{P} u_p(\delta t)_{p+1} \tilde{\mathbf{e}}_{i,n+1}^{n+1} \right\} \mathbf{F}_{ii}^{he,\sigma_g} + \left\{ \sum_{p=1}^{P} u_p(\delta t)_{p+1} \mathbf{e}_{i,n+1}^{n+1} \right\} \mathbf{F}_{ii}^{he,\sigma_g}$$

where

$$\tilde{\mathbf{e}}_{i,p/q}^{n+1} = \mathbf{e}_{i,p/q}^{n+1} - (\delta t)_{p/q} \mathbf{e}_i^{n+1}$$

$$\tilde{\mathbf{h}}_{i,p/q}^{n+1} = \mathbf{h}_{i,p/q}^{n+1} - (\delta t)_{p/q} \mathbf{h}_i^{n+1}$$

$$\tilde{\mathbf{e}}_{i,0}^{n+1} = 0 , \quad \tilde{\mathbf{h}}_{i,0}^{n+1} = 0$$

To efficiently evaluate $(e/h)_{i,p/q}^{n+1}$, $(e/h)_{j,p/q}^{n+1}$, and $(\tilde{e}/\tilde{h})_{i,p/q}^{n+1}$, the following recursive schemes can be used [59]:

$$(e/h)_{i/j,p/q}^{n+1} = (e/h)_{i/j,p/q}^{n} + \delta t \cdot (e/h)_{i/j,p-1/q-1}^{n+1}$$

$$(\tilde{e}/\tilde{h})_{i,p/q}^{n+1} = (\tilde{e}/\tilde{h})_{i,p/q}^{n} + \delta t \cdot (\tilde{e}/\tilde{h})_{i,p-1/q-1}^{n+1}$$

(4.77)

In the remainder of this section, two numerical examples are provided to demonstrate the accuracy and the applicability of the DGTD scheme described above. In both examples, the excitation is a plane wave with electric field $\mathbf{E}^{inc}(\mathbf{r}, t) = \hat{\mathbf{y}}g(t - \hat{\mathbf{k}} \cdot \mathbf{r}/c_0)$, and $g(t) = \exp(-(t - t_0)^2/\tau_m^2)\cos(2\pi f_m(t - t_0))$ is a Gaussian pulse with modulation frequency f_m, duration τ_m, and delay t_0.

In the first example, the scatterer is an infinitely large graphene sheet that resides in free space on the xy-plane. The temperature used in the expression of σ_g is set to $T = 300$ K while μ_c and Γ are changed for two different simulations. Similarly, $\hat{\mathbf{k}} = \hat{\mathbf{z}}$ for both of the simulations while τ_m, f_m, and t_0 are changed for the two simulations.

For the first simulation, $\mu_c = 0.3$ eV, the scattering rate $\Gamma = 0.41$ meV$/\hbar$. The FRVF scheme with four poles is applied to the samples of σ_g in a frequency range between microwave and terahertz. In Figure 4.11, the comparison between the fitted values and the original data is shown for frequencies between 500 MHz and 10 THz. Excellent agreement is observed, which shows that FRVF can produce an accurate representation of σ_g in terms of rational functions in a wide range of frequencies. The reflection, transmission, and absorption coefficients (denoted by Γ_R, Γ_T, and Γ_A, respectively) are computed using the DGTD scheme. The results are shown in Figure 4.12. For comparison, the analytical solutions calculated by the method described in [58] are also presented in the same figure. Good agreement between the two sets of results is observed.

For the second simulation, the transmission of a plane wave through the graphene sheet with $\mu_c = 0.12$ eV, $\Gamma = 2 \times 10^{12}$ Hz is studied. The FRVF with seven poles is used to represent σ_g in the frequency range between 500 GHz and 100 THz. Note that we use seven poles here since σ_g has a jump around 60 THz. The transmission coefficient computed using the DGTD scheme and the analytical expression are compared in Figure 4.13. Again, very good agreement is observed. We should mention here that, for these two simulations, the computation domain is truncated using the Silver-Muller ABC [33] (see (4.32) in Section 4.1.2.1), which yields very accurate results since the wave is normally incident on the truncation surface.

In the second example, the DGTD scheme is validated for a three-dimensional (3D) problem. Two simulations are carried out. In the first simulation, the scatterer is a $5 \times 10 \, \mu m^2$ graphene patch that resides in free space on the xy-plane [67]. Parameters of σ_g are $T = 300$ K, $\mu_c = 0$ eV, and $\Gamma = \frac{1}{2\tau}$ with $\tau = 10^{-13}$ s, and the parameters of excitation are $f_m = 2.5$ THz, $\hat{\mathbf{k}} = \hat{\mathbf{z}}$, $\tau_m = 1.274 \times 10^{-13}$ s, and

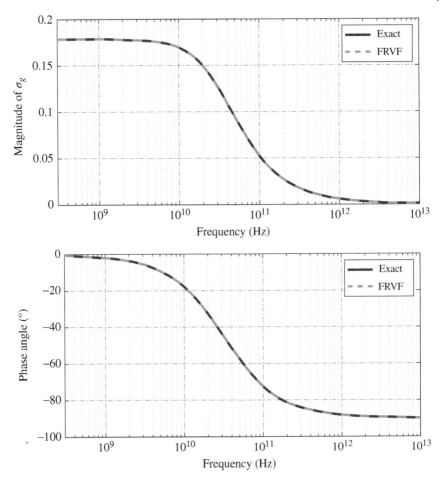

Figure 4.11 Fitted magnitude and phase of the surface conductivity σ_g from 500 MHz to 10 THz with four poles. Source: Reproduced with permission from Li et al. [63]/IEEE.

$t_0 = 3\tau_m$. The FRVF scheme with three poles and residues is used to represent σ_g in the frequency range between 500 MHz and 5 THz. For comparison, the normalized extinction cross-section (ECS) between 0.1 and 4 THz is computed using the DGTD scheme as shown in Figure 4.14. The reference result in the same figure is obtained by using an integral equation solver [67]. The good agreement between the results demonstrates the accuracy and applicability of the DGTD scheme for 3D examples.

In the second simulation, a $50 \times 50\ \mu m^2$ graphene patch that resides in free space on the xy-plane is characterized. Parameters of σ_g are $T = 300\ K$,

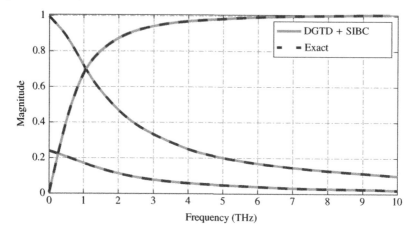

Figure 4.12 The calculated magnitudes of coefficients Γ_T, Γ_R, and Γ_A by the proposed algorithm as well as the theoretical data for $\mu_c = 0.3$ eV. Source: Reproduced with permission from Li et al. [63]/IEEE.

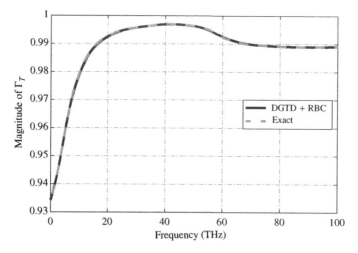

Figure 4.13 Magnitude of calculated transmission coefficient Γ_T by the proposed algorithm as well as theoretical data from terahertz to near-infrared region. Source: Reproduced with permission from Li et al. [63]/IEEE.

$\Gamma = 2.5$ meV$/\hbar$, $f_m = 2.5$ THz, and $\tau_m = 6.37 \times 10^{-14}$ seconds. Scattering and absorption cross-sections of the graphene patch are computed for different values of μ_c. In Figure 4.15, the normalized cross-sections between 1 and 10 THz are shown for $\mu_c = 0.5$, 1.0, and 1.5 eV. As expected, plasmon resonances are observed at various frequencies. The increase in μ_c results in the

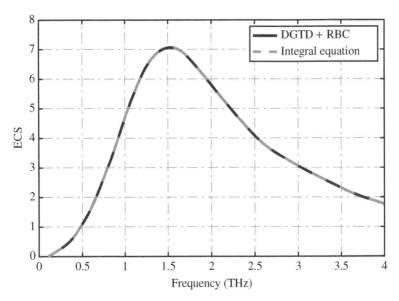

Figure 4.14 Normalized ECS versus frequency for a freestanding graphene patch in [67] and its counterpart calculated by the integral equation method. Source: Reproduced with permission from Li et al. [63]/IEEE.

upshift of resonant frequencies. In addition, the RCS between 1 and 10 THz is presented in Figure 4.16. It is found that the peaks in the RCS happen at the plasmon resonant frequencies, which result from the near-field enhancement. In Figure 4.17, the normalized near-field patterns for the y-component of the electric field obtained in the simulation with $\mu_c = 1.5$ eV at three plasmon frequencies $f_1 = 1.76$ THz, $f_2 = 4.98$ THz, and $f_3 = 6.97$ THz are shown. Noticeable near-field enhancement is observed at the resonant frequencies. In Figure 4.18, the normalized far-field scattered patterns in E- and H-planes at the fundamental plasmon frequency $f_1 = 1.76$ THz are also presented. It is interestingly found that the far-field patterns resemble those of conventional short dipoles, which is in line with the assertion in [54]. The nonsymmetrical pattern in H-plane is attributed to the boundary condition in (4.61). We should mention here that, for these two simulations for 3D problems, the computation domain is truncated using the TDBI method as explained in Section 4.1.3 to maintain the accuracy of the solution.

4.2.1.2 A Resistive Boundary Condition and an Auxiliary Equation Method to Represent Magnetized Graphene Within the DGTD Method

The Section 4.2.1.1 introduces a scheme to analyze the graphene with a scalar surface conductivity, which is observed in the absence of an external magnetostatic

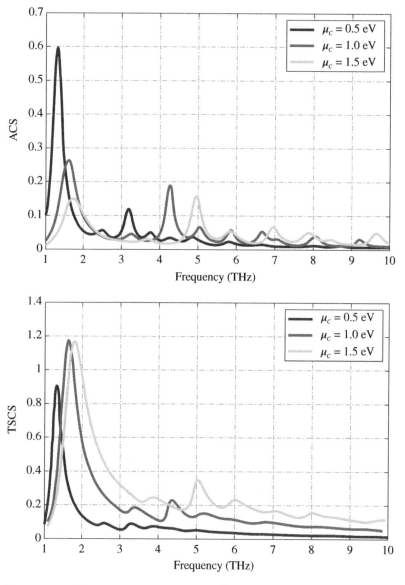

Figure 4.15 TSCS and ACS of the 50×50 μm^2 freestanding graphene patch corresponding to different chemical potentials $\mu_c = 0.5, 1.0,$ and 1.5 eV. Source: Reproduced with permission from Li et al. [63]/IEEE.

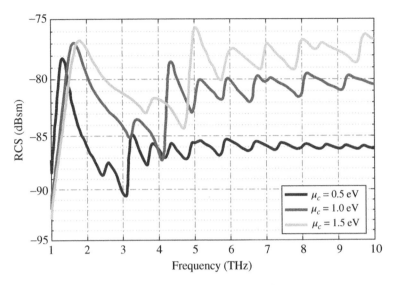

Figure 4.16 Calculated forward RCS of the $50 \times 50\mu m^2$ freestanding graphene patch corresponding to different chemical potentials $\mu_c = 0.5, 1.0,$ and $1.5 eV$. Reproduced with permission from [63].

biasing. However, when there is a biasing magnetic field, the surface conductivity becomes a tensor due to the quantum Hall effect [56, 58]. Direct incorporation of this anisotropic resistive boundary condition (RBC) (representing the graphene sheet) within the DGTD method would result in a numerical flux in a tensor form. In addition, still, the dispersion of the surface conductivity has to be accounted for. To address these problems, in this section, we introduce another DGTD scheme that can account for both the anisotropy and the dispersion of the surface conductivity. In this scheme, a surface polarization current is introduced over the graphene sheet [31, 68]. This current satisfies an ADE that involves the electric field. With the ADE and the polarization current, the anisotropic and the dispersive RBC is converted into an isotropic and nondispersive one. Therefore, the time-domain convolution is avoided, and the corresponding numerical flux in the presence of the new RBC is made isotropic.

Suppose a graphene sheet is placed at the $z = 0$ plane and biased by a static magnetic field perpendicular to the graphene plane, i.e. $\mathbf{B}_0 = \hat{z}B_0$. Because of the Lorentz force, the surface conductivity exhibits anisotropy [69]. To have a fundamental insight into this phenomenon, the motion of an electron in the presence of an electric field $\mathbf{E}$ is studied. First, an x- polarized electric field $\mathbf{E} = \hat{x}E_x$ is considered. In this case, the electron is accelerated along the x-direction due to the electric force $\mathbf{F}_e = -e\mathbf{E}$ with $-e$ denoting the electron charge. Simultaneously, the moving electron with velocity v will be exposed to a Lorentz force $\mathbf{F}_m = -ev \times \mathbf{B}_0$ along

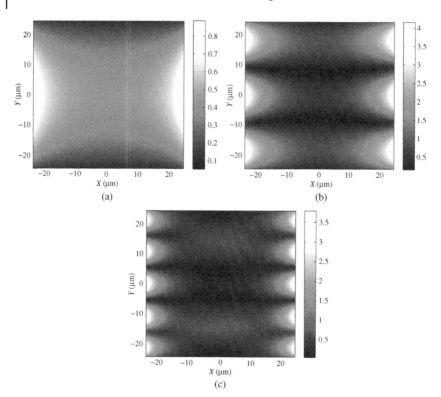

Figure 4.17 Normalized near-field distribution of E_y over the graphene sheet at
(a) $f_1 = 1.76$ THz, (b) $f_2 = 4.98$ THz, and (c) $f_3 = 6.97$ THz for $\mu_c = 1.5$ eV. Source:
Reproduced with permission from Li et al. [63]/IEEE.

the y-direction. As a result, two current components along the x- and y-directions
will be generated [69]:

$$\mathbf{J} = \sigma_{xx}E_x\hat{\mathbf{x}} + \sigma_{yx}E_x\hat{\mathbf{y}} \tag{4.78}$$

where σ_{xx} and σ_{yx} are conductivities parallel and perpendicular to the electric field
E, respectively. Similarly, for the case with an electric field $\mathbf{E} = \hat{\mathbf{y}}E_y$, two current
components will be generated:

$$\mathbf{J} = -\sigma_{xy}E_y\hat{\mathbf{x}} + \sigma_{yy}E_y\hat{\mathbf{y}} \tag{4.79}$$

For graphene with same properties in all directions, the conductivities must sat-
isfy $\sigma_{xx} = \sigma_{yy}$ and $\sigma_{yx} = \sigma_{xy}$. Based on this, the generated currents in (4.78) and
(4.79) can be combined together and rewritten in a more compact form as matrix
equation [69]:

$$\mathbf{J} = \overline{\overline{\sigma}}_g \cdot \mathbf{E} \tag{4.80}$$

Figure 4.18 Normalized far-field pattern in *E*- and *H*-plane at $f_1 = 1.76$ THz for $\mu_c = 1.5$ eV. Source: Reproduced with permission from Li et al. [63]/IEEE.

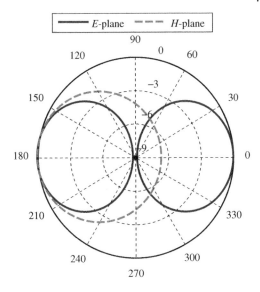

with $\mathbf{E} = [E_x, E_y, 0]^T$ and

$$\overline{\overline{\sigma}}_g = \begin{bmatrix} \sigma_{xx} & -\sigma_{yx} & 0 \\ \sigma_{yx} & \sigma_{xx} & 0 \\ 0 & 0 & 0 \end{bmatrix} \tag{4.81}$$

The analytical expressions for the surface conductivities σ_{xx} and σ_{yx} consisting of intraband and interband contributions can be obtained based on Kubo's formula [70] (see also (4.59) and (4.60) in Section 4.2.1.1). For frequencies in the terahertz range, the intraband term is dominant over the interband term [69]. Thus, only the intraband term is considered here. Based on the fact that the probability of electron transitions between Landau levels around the Fermi level μ_c is the strongest over others, the expressions for σ_{xx} and σ_{yx} can be approximated by a Drude-like model [68, 69]:

$$\sigma_{xx}(\omega, \mu_c, \tau, T, B_0) = \sigma_0 \frac{1 + j\omega\tau}{(\omega_c\tau)^2 + (1 + j\omega\tau)^2}$$

$$\sigma_{yx}(\omega, \mu_c, \tau, T, B_0) = \sigma_0 \frac{\omega_c\tau}{(\omega_c\tau)^2 + (1 + j\omega\tau)^2} \tag{4.82}$$

with

$$\sigma_0 = \frac{e^2 \tau k_B T}{\pi \hbar^2} \left[\frac{\mu_c}{k_B T} + 2\ln(e^{-\mu_c/k_B T} + 1) \right] \tag{4.83}$$

where we choose $T = 300$ K in this chapter, τ is the scattering time, and $\omega_c \approx eB_0 v_F^2 / \mu_c$ is the cyclotron frequency with $v_F \approx 10^6$ m/s denoting the Fermi velocity.

As done in the Section 4.2.1.1, the graphene layer is assumed to be an infinitesimally thin conductive sheet. Therefore, the following boundary conditions over the graphene sheet have to be satisfied:

$$\hat{n} \times (\mathbf{E}^2 - \mathbf{E}^1) = 0$$
$$\hat{n} \times (\mathbf{H}^2 - \mathbf{H}^1) = \bar{\bar{\sigma}}_g \cdot \mathbf{E} \tag{4.84}$$

Note that (4.81) is expressed in the frequency domain and the only difference with respect to (4.58) in Section 4.2.1.1 is that the surface conductivity is a tensor.

Since the incorporation of boundary conditions within the DGTD method is enforced by redefining the numerical flux, the anisotropic boundary condition in (4.84) would result in a tensor form flux expression that is simultaneously dispersive, which no doubt complicates the problem significantly (the convolution in the time domain, is required and the different components of electromagnetic fields are entangled with each other). To address these challenges, an auxiliary polarization surface current $\mathbf{J}(\omega)$ is introduced by rewriting (4.84) as $\hat{n} \times (\mathbf{H}^2 - \mathbf{H}^1) = \mathbf{J}(\omega)$ with $\mathbf{J}$ governed by the ADE

$$J_x(\omega) = \sigma_{xx}E_x - \sigma_{yx}E_y$$
$$J_y(\omega) = \sigma_{yx}E_x + \sigma_{xx}E_y \tag{4.85}$$

After some mathematical manipulations, (4.85) can be rewritten as

$$j\omega J_x(\omega) = -2\Gamma J_x(\omega) - \omega_c J_y(\omega) + 2\Gamma\sigma_0 E_x(\omega)$$
$$j\omega J_y(\omega) = -2\Gamma J_y(\omega) + \omega_c J_x(\omega) + 2\Gamma\sigma_0 E_y(\omega) \tag{4.86}$$

where $\Gamma = 1/2\tau$ denotes the phenomenological scattering rate. Through the inverse Fourier transform, the time-domain counterparts of (4.86) are obtained as

$$\frac{\partial \mathbf{J}}{\partial t} = \bar{\bar{C}} \cdot \mathbf{J} + 2\Gamma\sigma_0 \mathbf{E} \tag{4.87}$$

with

$$\bar{\bar{C}} = \begin{bmatrix} -2\Gamma & -\omega_c \\ \omega_c & -2\Gamma \end{bmatrix} \tag{4.88}$$

With this auxiliary polarization surface current $\mathbf{J}$ and the ADE, the anisotropic and dispersive RBC in (4.81) is transformed to an isotropic and nondispersive one in (4.88). Then, the incorporation of this scalar RBC within the DGTD method can be facilitated in the conventional way by reformulating the numerical flux, which will not be repeated again in this section.

For the DGTD formulation, the electric field and magnetic field are approximated in the same way as described in the Section 4.2.1.1. Other than that, the polarization current $\mathbf{J}$ is approximated by 2D vector basis functions $\varphi_i(\mathbf{r})$:

$$\mathbf{J}_i = \sum_{l=1}^{K} c_{i,l}(t)\varphi_{i,l}(\mathbf{r}) \tag{4.89}$$

By applying DG testing to Maxwell's curl equations and the ADE, we obtain

$$\int_{\Omega_i} \Psi_{i,k}^e \cdot \left[\varepsilon_i \frac{\partial \mathbf{E}_i}{\partial t} - \nabla \times \mathbf{H}_i \right] d\mathbf{r} = \sum_{f=1}^{4} \int_{\partial \Omega_{i,f}} \Psi_{i,k}^e \cdot \left[\hat{\mathbf{n}}_{i,f} \times (\mathbf{H}_f^* - \mathbf{H}_i) \right] d\mathbf{r}$$

$$\int_{\Omega_i} \Psi_{i,k}^h \cdot \left[\mu_i \frac{\partial \mathbf{H}_i}{\partial t} + \nabla \times \mathbf{E}_i \right] d\mathbf{r} = \sum_{f=1}^{4} \int_{\partial \Omega_{i,f}} \Psi_{i,k}^h \cdot \left[\hat{\mathbf{n}}_{i,f} \times (\mathbf{E}_i - \mathbf{E}_f^*) \right] d\mathbf{r} \qquad (4.90)$$

$$\int_{\partial \Omega_{i,f_g}} \varphi_{i,k} \cdot \frac{\partial \mathbf{J}_i}{\partial t} d\mathbf{r} = \int_{\partial \Omega_{i,f_g}} \varphi_{i,k} \cdot \overline{\overline{C}} \cdot \mathbf{J}_i d\mathbf{r} + 2\Gamma \sigma_0 \int_{\partial \Omega_{i,f_g}} \varphi_{i,k} \cdot \mathbf{E}_i d\mathbf{r}.$$

The semi-discrete matrix equations for the ADE can then be obtained as

$$\overline{\mathbf{J}}_i \frac{\partial \mathbf{c}_i}{\partial t} = \overline{\mathbf{M}}_c \mathbf{c}_i + 2\Gamma \sigma_0 \overline{\mathbf{M}}_v \mathbf{e}_i \qquad (4.91)$$

where

$$\begin{aligned}
\left[\mathbf{M}_c \right]_{kl} &= \int_{\partial \Omega_{i,f_g}} \varphi_{i,k} \cdot \overline{\overline{C}} \cdot \varphi_{i,l} \, d\mathbf{r} \\
\left[\mathbf{M}_v \right]_{kl} &= \int_{\partial \Omega_{i,f_g}} \varphi_{i,k} \cdot \Psi_{i,l}^e \, d\mathbf{r} \\
\left[\overline{\mathbf{J}}^i \right]_{kl} &= \int_{\partial \Omega_{i,f_g}} \varphi_{i,k} \cdot \varphi_{i,l} \, d\mathbf{r}
\end{aligned} \qquad (4.92)$$

We only give the matrix related to the ADE since other matrices are same as those introduced in the Section 4.2.1.1 [31]. In the remainder of this section, two numerical examples are provided to demonstrate the accuracy and the applicability of the DGTD scheme described above.

As the first example, we use the DGTD scheme to study the Faraday rotation and surface plasmon polarization by varying the chemical potential and the magnetostatic biasing. An infinitely large graphene sheet placed on the xy-plane is biased by a z-directed magnetostatic field $\mathbf{B}_0 = \hat{z} B_0$. The graphene sheet is excited in a plane wave with electric field $\mathbf{E}^{inc}(\mathbf{r}, t) = \hat{x} g(t - \hat{\mathbf{k}} \cdot \mathbf{r}/c_0)$, where $\hat{\mathbf{k}} = \hat{z}$ is the propagation direction and $g(t) = \exp(-(t - t_0)^2/\tau_m^2) \cos(2\pi f_m (t - t_0))$ is a Gaussian pulse, where the pulse's modulation frequency, duration, and delay are set to $f_m = 5$ THz, $\tau_m = 6.37 \times 10^{-13}$ seconds, and $t_0 = 5t_m$, respectively. The first-order SM-ABC [71] (see also (4.32)) is used to truncate the computation domain along the z-direction, which is sufficient for this example since all waves are normally incident to the truncation boundary. In Figure 4.19, the total transmission coefficient, Faraday rotation angle, and cross-polarized transmission coefficient obtained using the DGTD scheme are presented. For comparison, the analytical results for the total transmission coefficient and Faraday rotation angle are also provided in Figure 4.19. Very good agreement is observed between the two sets of results. The results show that the magnetostatic biasing has significant impacts on the electromagnetic waves propagation through a graphene sheet.

The next example validates the accuracy of the DGTD scheme for 3D examples. First, a 5×10 μm unmagnetized graphene patch in [67] is revisited. The same

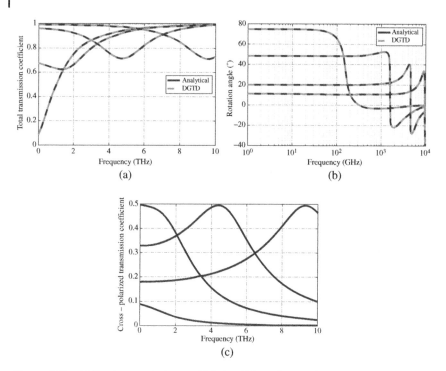

Figure 4.19 (a) Total transmission coefficients, (b) Faraday rotation angle, and (c) cross-polarized transmission coefficients versus frequency for different magnetostatic biasing. Source: Reproduced with permission from Li and Jiang [31]/IEEE.

excitation is employed as in the previous example. The computation domain is truncated using the TDBI method as explained in Section 4.1.3 to maintain the accuracy of the solution. In Figure 4.20, the ECS computed using the DGTD scheme and an integral equation solver [67] are compared. The figure shows that the results agree well. Then, a $20 \times 100 \ \mu m^2$ graphene patch under biased by a static magnetic field $\mathbf{B}_0 = 0.25\hat{z}$ is analyzed. To show the impacts of chemical potentials on the graphene resonance, the scattering and the ECS obtained under different μ_c by letting the scattering rate $\Gamma = 2.5$ meV$/\hbar$ are presented in Figure 4.21. It is noted that the resonant frequencies shift upward with higher chemical potentials, and the resonance becomes much stronger. Next, to study the effect of the substrate material on the plasmon resonance, the graphene patch is located on a substrate with thickness $h = 2\mu m$. The materials tested are silicon (Si), silicon-nitrate (Si_3N_4), and silicon-dioxide (SiO_2) (all of which are commonly used by experimental researchers). The ECS computed under $\mu_c = 0.5$ eV for these three different materials is shown in Figure 4.22. It is noted that the first two resonant peaks of the ECS shift to low frequencies with the increasing

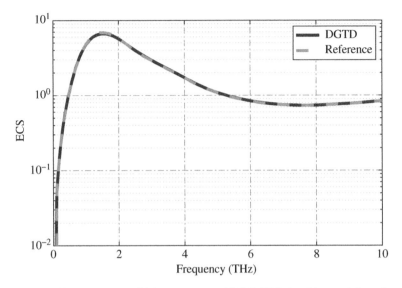

Figure 4.20 Calculated ECS by the proposed DGTD-RBC algorithm and the reference result by the integral equation (IE) method in [67]. Source: Reproduced with permission from Li and Jiang [31]/IEEE.

permittivity due to the physical dimension of the graphene becoming larger compared with wavelength $\lambda = \lambda_0 / \sqrt{\varepsilon_r}$.

4.2.2 Multiphysics Simulation of Optoelectronic Devices

The several advantages the DG methods come with (as explained in Section 4.1.6) make them very suitable for multiphysics simulations. Indeed, in the last decade, they have been applied to many multiphysics problems in various fields, such as coupled simulations of electromagnetic/thermal interactions, electromagnetic/plasma interactions, and finally electromagnetic/semiconductor carrier interactions [72, 73]. In general, multiphysics simulations involve solving different partial differential equations simultaneously, and those partial differential equations are usually characterized by different scales in space and time. The DG method permits various types of mesh elements for discretization and leads to easy implementations of high-order spatial basis functions and high-order time integration schemes. Furthermore, it allows non-conformal meshes, adaptive h/p-refinement, and local time stepping. These properties can be utilized to greatly reduce computational requirements when dealing with multiphysics problems [8, 13, 15, 19, 74, 75].

This section presents the application of the DG method in multiphysics simulations of photoconductive devices. The operation of these devices relies on the

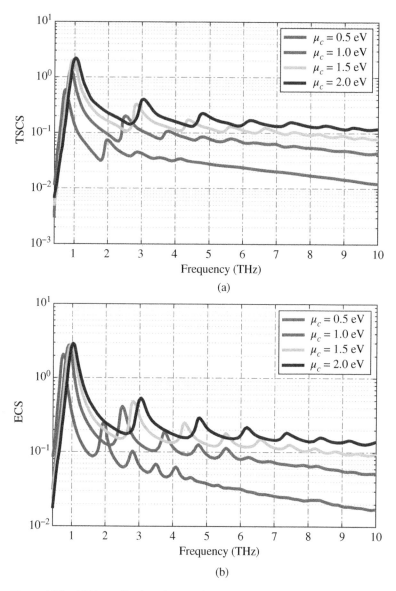

Figure 4.21 (a) Normalized total scattering cross-section (TSCS) and (b) ECS versus different chemical potentials. Source: Reproduced with permission from Li and Jiang [31]/IEEE.

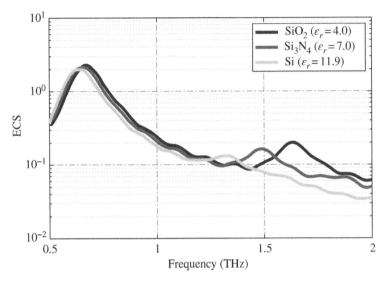

Figure 4.22 Normalized ECS for substrates Si, SiO$_2$, and Si$_3$N$_4$. Source: Reproduced with permission from Li and Jiang [31]/IEEE.

interactions between electromagnetic fields and semiconductor carriers. A numerical tool has to account for full-wave electromagnetic interactions, carrier dynamics, and their nonlinear coupling. We note here although the DG framework presented in this section is developed for simulation of photoconductive devices, it can be readily applied to similar optoelectronic devices (such as phototransistors and photovoltaic devices) with only minor modifications.

The operation of photoconductive devices can be analyzed into two stages. Initially, a bias voltage is applied to the electrodes. The resulting static electric field changes the carrier distribution. The redistributed carriers, in turn, affect the electric potential distribution. The device reaches a nonequilibrium steady-state mathematically described by a coupled system of Poisson and stationary drift-diffusion equations [75–77]

$$\nabla \cdot [\varepsilon(\mathbf{r})\mathbf{E}^s(\mathbf{r})] = q[C(\mathbf{r}) + n_h^s(\mathbf{r}) - n_e^s(\mathbf{r})] \tag{4.93}$$

$$\nabla \cdot \mathbf{J}_c^s(\mathbf{r}) = \pm qR^s(n_e^s, n_h^s) \tag{4.94}$$

$$\mathbf{J}_c^s(\mathbf{r}) = q\mu_c(\mathbf{E}^s)\mathbf{E}^s(\mathbf{r})n_c^s(\mathbf{r}) \pm qd_c(\mathbf{E}^s)\nabla n_c^s(\mathbf{r}) \tag{4.95}$$

where $\mathbf{E}^s(\mathbf{r})$ is the static electric field, $\varepsilon(\mathbf{r})$ is the dielectric permittivity, q is the electron charge, $C(\mathbf{r})$ is the doping concentration, the superscript "s" stands for static, the subscript $c \in \{e, h\}$ represents the carrier type and hereinafter the upper and lower signs should be selected for electron ($c = e$) and hole ($c = h$), respectively, $n_c^s(\mathbf{r})$ is the carrier density, $R(n_e, n_h)$ is the recombination rate, and $\mu_c(\mathbf{E}^s)$ and $d_c(\mathbf{E}^s)$ are the mobility and diffusion coefficient, respectively. From the Einstein relation, $d_c(\mathbf{E}^s) = V_T \mu_c(\mathbf{E}^s)$, where $V_T = k_B T/q$ is the thermal voltage, k_B is the Boltzmann constant, and T is the absolute temperature.

Two most common recombination processes are the trap-assisted recombination described by the Shockley–Read–Hall (SRH) model [75–77]

$$R_{\text{SRH}}(n_e^s, n_h^s) = \frac{n_e^s(\mathbf{r})n_h^s(\mathbf{r}) - n_i^2}{\tau_e(n_{h1} + n_h^s(\mathbf{r})) + \tau_h(n_{e1} + n_e^s(\mathbf{r}))}$$

and the three-particle band-to-band transition described by the Auger model

$$R_{\text{Auger}}(n_e^s, n_h^s) = [n_e^s(\mathbf{r})n_h^s(\mathbf{r}) - n_i^2][C_e^A n_e^s(\mathbf{r}) + C_h^A n_h^s(\mathbf{r})]$$

Here, n_i is the intrinsic carrier concentration, τ_e and τ_h are the carrier lifetimes, n_{e1} and n_{h1} are SRH model parameters related to the trap energy level, and C_e^A and C_h^A are the Auger coefficients. Depending on the semiconductor material, more recombination models can be included [76]. The net recombination rate $R^s(n_e^s, n_h^s)$ is given by the summation of all recombination terms.

To accurately model semiconductor devices, it is important to include proper mobility models with respect to the semiconductor materials and device operating conditions [76, 78–81]. For a photoconductive device working under a high-bias voltage, it is important to consider the high-field mobility model that accounts for the carrier velocity saturation effect [76, 79–81]. Here, the Caughey–Thomas model [76, 82] is used

$$\mu_c(\mathbf{E}^s) = \mu_c^0 \left[1 + \left(\frac{\mu_c^0 E_\|(\mathbf{r})}{V_c^{\text{sat}}} \right)^{\beta_c} \right]^{\beta_c^{-1}}$$

where $E_\|(\mathbf{r})$ is amplitude of the electric field intensity parallel to the current flow, μ_c^0 is the low-field mobility, and V_c^{sat} and β_c are fitting parameters obtained from experimental data.

A transient stage starts when an optical electromagnetic excitation is incident on the device. The photoconductive material absorbs the optical electromagnetic wave energy induced on the device and generates carriers. The carriers are driven by both the bias electric field and the optical electromagnetic fields. The carrier dynamics and electromagnetic wave interactions are mathematically described by a coupled system of the time-dependent Maxwell and drift-diffusion equations [19]:

$$\mu(\mathbf{r})\frac{\partial \mathbf{H}(\mathbf{r}, t)}{\partial t} = -\nabla \times \mathbf{E}(\mathbf{r}, t) \tag{4.96}$$

$$\varepsilon(\mathbf{r})\frac{\partial \mathbf{E}(\mathbf{r}, t)}{\partial t} = \nabla \times \mathbf{H}(\mathbf{r}, t) - [\mathbf{J}_e(\mathbf{r}, t) + \mathbf{J}_h(\mathbf{r}, t)] \tag{4.97}$$

$$q\frac{\partial n_c(\mathbf{r}, t)}{\partial t} = \pm\nabla \cdot \mathbf{J}_c(\mathbf{r}, t) - q[R(n_e, n_h) - G(\mathbf{E}, \mathbf{H})] \tag{4.98}$$

$$\mathbf{J}_c(\mathbf{r}, t) = q\mu_c(\mathbf{E})\mathbf{E}(\mathbf{r}, t)n_c(\mathbf{r}, t) \pm qd_c(\mathbf{E})\nabla n_c(\mathbf{r}, t) \tag{4.99}$$

Here, $\mathbf{E}(\mathbf{r}, t)$ and $\mathbf{H}(\mathbf{r}, t)$ are the total electric and magnetic field intensities, $n_e(\mathbf{r}, t)$ and $n_h(\mathbf{r}, t)$ are the total electron and hole densities, $\mathbf{J}_e(\mathbf{r}, t)$ and $\mathbf{J}_h(\mathbf{r}, t)$ are the total current densities due to electron and hole movement, $\varepsilon(\mathbf{r})$ and $\mu(\mathbf{r})$ are the dielectric permittivity and permeability, $R(n_e, n_h)$ and $G(\mathbf{E}, \mathbf{H})$ are the recombination and generation rates, $\mu_c(\mathbf{E})$ and $d_c(\mathbf{E})$ are mobility and diffusion coefficients, respectively.

Since the bias voltage persists during the transient stage, and the boundary conditions for Poisson and drift-diffusion equations, e.g. the Dirichlet boundary conditions on the electrodes, do not change, it is assumed that $\mathbf{E}^s(\mathbf{r})$ and $n_c^s(\mathbf{r})$ solved from the Poisson-drift-diffusion system are valid throughout the transient stage. Therefore, the total field intensities and carrier and current densities can be separated into stationary and transient components as $\mathbf{E}(\mathbf{r}, t) = \mathbf{E}^s(\mathbf{r}) + \mathbf{E}^t(\mathbf{r}, t)$, $\mathbf{H}(\mathbf{r}, t) = \mathbf{H}^s(\mathbf{r}) + \mathbf{H}^t(\mathbf{r}, t)$, $n_c(\mathbf{r}, t) = n_c^s(\mathbf{r}) + n_c^t(\mathbf{r}, t)$, and $\mathbf{J}_c(\mathbf{r}, t) = \mathbf{J}_c^s(\mathbf{r}) + \mathbf{J}_c^t(\mathbf{r}, t)$, respectively. Here, the superscript "t" stands for transient components. Similarly, the recombination rate $R(n_e, n_h)$ is decomposed into stationary and transient components as $R(n_e, n_h) = R^s(n_e^s, n_h^s) + R^t(n_e^t, n_h^t)$. The mobility is assumed to be a function of $\mathbf{E}^s(\mathbf{r})$ only, i.e. $\mu_c(\mathbf{E}) \approx \mu_c(\mathbf{E}^s)$, $c \in \{e, h\}$. With these definitions, it is straightforward to eliminate the static components from the left hand of (4.96)–(4.99), yielding a reduced coupled system of time-dependent Maxwell and drift-diffusion equations [19, 83]:

$$\mu(\mathbf{r})\frac{\partial \mathbf{H}^t(\mathbf{r}, t)}{\partial t} = -\nabla \times \mathbf{E}^t(\mathbf{r}, t) \tag{4.100}$$

$$\varepsilon(\mathbf{r})\frac{\partial \mathbf{E}^t(\mathbf{r}, t)}{\partial t} = \nabla \times \mathbf{H}^t(\mathbf{r}, t) - [\mathbf{J}_e^t(\mathbf{r}, t) + \mathbf{J}_h^t(\mathbf{r}, t)] \tag{4.101}$$

$$q\frac{\partial n_c^t(\mathbf{r}, t)}{\partial t} = \pm\nabla \cdot \mathbf{J}_c^t(\mathbf{r}, t) - q[R^t(n_e^t, n_h^t) - G(\mathbf{E}^t, \mathbf{H}^t)] \tag{4.102}$$

$$\mathbf{J}_c^t(\mathbf{r}, t) = q\mu_c(\mathbf{E}^s)([\mathbf{E}^s(\mathbf{r}) + \mathbf{E}^t(\mathbf{r}, t)]n_c^t(\mathbf{r}, t) \tag{4.103}$$
$$+ \mathbf{E}^t(\mathbf{r}, t)n_c^s(\mathbf{r}, t)) \pm qd_c(\mathbf{E}^s)\nabla n_c^t(\mathbf{r}, t)$$

Here, the generation rate describes the generation of carriers upon absorption of optical electromagnetic wave energy [77, 84, 85]

$$G(\mathbf{E}^t, \mathbf{H}^t) = \eta\frac{P^{abs}(\mathbf{r}, t)}{E^{ph}}$$

where η is the intrinsic quantum efficiency (number of electron–hole pairs generated by each absorbed photon), $P^{abs}(\mathbf{r}, t)$ is the absorbed power density of optical waves, $E^{ph} = hv$ is the photon energy, h is the Planck constant, and v is the frequency of the optical wave. Here, v must be high enough such that E^{ph} is large enough to excite electrons, e.g. E^{ph} should be larger than the bandgap energy E^g in direct bandgap semiconductors [77, 84, 85]. Apparently, $P^{abs}(\mathbf{r}, t)$ only depends on the optical electromagnetic fields, and hence $G(\mathbf{E}^t, \mathbf{H}^t)$ is only a function of $\mathbf{E}^t$ and $\mathbf{H}^t$. A simple approximation for $P^{abs}(\mathbf{r}, t)$ is $P^{abs}(\mathbf{r}, t) \approx \alpha \left| \mathbf{E}^t(\mathbf{r}, t) \times \mathbf{H}^t(\mathbf{r}, t) \right|$, where α is the photon absorption coefficient.

A complete multiphysics framework for modeling photoconductive devices consists of a steady-state solver that solves the Poisson-drift-diffusion system (4.93)–(4.95) and a transient solver that solves the Maxwell-drift-diffusion system (4.96)–(4.99). The steady-state solutions are used as inputs to the transient solver. The steady-state solver uses the Gummel iteration method to take the nonlinearity into account [75]. At every iteration of this method, one has to solve two partial differential equations, i.e. the nonlinear Poisson equation and the linearized drift-diffusion equation. The transient solver directly solves the time-dependent Maxwell's equations and drift-diffusion equations using explicit time integration methods and takes into account their coupling by alternately feeding the updated solutions into each other during the time marching. Since the carrier response is much slower than the time variation of the electromagnetic waves, the drift-diffusion equations are updated using a larger time step size (typically 10 times larger). This mixed-step explicit time integration scheme can greatly reduce the computational cost. The time marching scheme is illustrated in Figure 4.23 [19]. For illustration, the time step size for the DD equations is assumed to be twice the step size for Maxwell's equations. The time steps of the two systems are synchronized at time $T = T'$. The generation rate G^T is first calculated from $\mathbf{E}^{t,T}$ and $\mathbf{H}^{t,T}$ and then used to update the carrier densities $n_c^{t,T'}$ (from step $T' - \Delta T'$ to T'). Then, $n_c^{t,T'}$ are used to compute the current densities $\mathbf{J}_c^{t,T}$ and

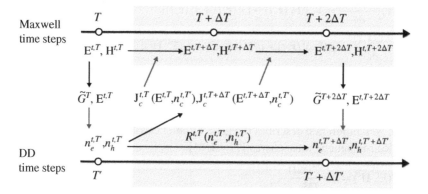

Figure 4.23 Time marching scheme. Source: Chen and Bagci [19]/IEEE/Public Domian CC BY.

$\mathbf{J}_c^{t,T+\Delta T}$, $\mathbf{J}_c^{t,T}$ and $\mathbf{J}_c^{t,T+\Delta T}$ are then used to update Maxwell's equations at steps T and $T + \Delta T$ to produce $\{\mathbf{E}^{t,T+\Delta T}, \mathbf{H}^{t,T+\Delta T}\}$ and $\{\mathbf{E}^{t,T+2\Delta T}, \mathbf{H}^{t,T+2\Delta T}\}$, respectively. The time steps of the two systems match again at time $T + 2\Delta T = T' + \Delta T'$ and the process described here is repeated. Note that $\tilde{G}^{T+2\Delta T}$ is calculated as the average value $(G^{T+\Delta T} + G^{T+2\Delta T})/2$.

The above multiphysics framework calls for solving four partial differential equations, namely, the nonlinear Poisson equation, the stationary drift-diffusion equation [75], the time-dependent Maxwell's equations, and the time-dependent drift-diffusion equations [19]. As discussed before, the DG method is preferred for multiphysics problems because of its flexibility in discretizing geometries and equations, which helps to capture the multiscale features involved in the coupled systems of equations. A special challenge in modeling semiconductor carrier dynamics (by solving drift-diffusion equation) is that they belong to the group of so-called convection-dominated convection-diffusion problems. An efficient solution to these problems has been a challenging research topic in computational mathematics for more than 40 years [7]. The difficulty stems from the existence of exponential boundary layers in the solution, which makes numerical methods unstable unless extremely fine meshes are used. To this end, FEM requires additional techniques to ensure stability, and FVM (generalized from the Scharfetter–Gummel scheme) poses restrictions on the meshes [75]. The DG method seems to be favorable for this type of problem since it is free from these issues while providing high-order accuracy. As discussed previously, since all electric field components described by Poisson and Maxwell's equations are used in the drift-diffusion equations, nodal basis functions are preferred here. With these discussions, all the four partial differential equations are discretized using a unified nodal DG framework. DG discretization of Maxwell's equations is the same as that introduced at the beginning of this chapter. In this section, DG formulations for the remaining three equations are briefly introduced.

The time-dependent drift-diffusion equation is solved using the local DG (LDG) method. Since the electron and hole drift-diffusion equations only differ by the sign in front of the drift term, we only discuss the electron drift-diffusion equation. Also note that the subscript "e" (meaning electron) and the superscript "t" (meaning transient) are dropped from the variables to simplify the notation. Since $\mathbf{E}^s(\mathbf{r})$ is provided as an input to the time-domain simulation, variables that are functions of it are assumed to change with $\mathbf{r}$. Furthermore, as the three drift terms in (4.103) are treated in the same way, for brevity, they are combined into one term and denoted by $\mathbf{v}(\mathbf{r}, t)n(\mathbf{r}, t)$. Under those notation simplifications, the electron drift-diffusion equations (4.102) and (4.103) are expressed as the following initial-boundary value problem:

$$\frac{\partial n(\mathbf{r}, t)}{\partial t} = \nabla \cdot [d(\mathbf{r})\mathbf{q}(\mathbf{r}, t)] + \nabla \cdot [\mathbf{v}(\mathbf{r}, t)n(\mathbf{r}, t)] - R(\mathbf{r}, t), \mathbf{r} \in \Omega \quad (4.104)$$

$$\mathbf{q}(\mathbf{r}, t) = \nabla n(\mathbf{r}, t), \mathbf{r} \in \Omega \tag{4.105}$$

$$n(\mathbf{r}, t) = f_D(\mathbf{r}), \mathbf{r} \in \partial\Omega_D \tag{4.106}$$

$$\hat{\mathbf{n}}(\mathbf{r}) \cdot [d(\mathbf{r})\mathbf{q}(\mathbf{r}, t) + \mathbf{v}(\mathbf{r}, t)n(\mathbf{r}, t)] = f_R(\mathbf{r}), \mathbf{r} \in \partial\Omega_R \tag{4.107}$$

Here, $\mathbf{q}(\mathbf{r}, t)$ is an auxiliary variable introduced to reduce the order of the spatial derivative in the diffusion term, $R(\mathbf{r}, t) \equiv R^t(n_e^t, n_h^t) - G(\mathbf{E}^t, \mathbf{H}^t)$, $\partial\Omega_D$ and $\partial\Omega_R$ represent the surfaces where Dirichlet and Robin boundary conditions are enforced and f_D and f_R are the coefficients associated with these boundary conditions, respectively, and $\hat{\mathbf{n}}(\mathbf{r})$ denotes the outward normal vector $\partial\Omega_R$. For the problems considered in this work, $\partial\Omega_D$ represents the electrode/semiconductor interfaces and $f_D = 0$; and $\partial\Omega_R$ represents the semiconductor/insulator interfaces and $f_R = 0$ indicating that there are no carrier spills out of those interfaces [86].

DG discretization of the above initial-boundary value problem is similar to the previous process described for Maxwell's equations. Here, the nodal basis functions (see Section 4.1.6) are used. Testing (4.104) and (4.105) with $\Psi_l(\mathbf{r})$ on element i and applying the divergence theorem yield the weak form

$$\int_{\Omega_i} \frac{n_i(\mathbf{r}, t)}{\partial t} \Psi_l(\mathbf{r}) dV = -\int_{\Omega_i} d(\mathbf{r})\mathbf{q}_i(\mathbf{r}, t) \cdot \nabla\Psi_l(\mathbf{r}) dV + \oint_{\partial\Omega_i} \hat{\mathbf{n}}(\mathbf{r}) \cdot (d\mathbf{q})^* \Psi_l(\mathbf{r}) dS$$

$$-\int_{\Omega_i} \mathbf{v}(\mathbf{r}, t)n_i(\mathbf{r}, t) \cdot \nabla\Psi_l(\mathbf{r}) dV + \oint_{\partial\Omega_i} \hat{\mathbf{n}}(\mathbf{r}) \cdot (\mathbf{v}n)^* \Psi_l(\mathbf{r}) dS - \int_{\Omega_i} R(\mathbf{r}, t)\Psi_l(\mathbf{r}) dV$$
$$\tag{4.108}$$

$$\int_{\Omega_i} q_i^\nu(\mathbf{r}, t)\Psi_l(\mathbf{r}) dV = -\int_{\Omega_i} n_i(\mathbf{r}, t)\partial_\nu \Psi_l(\mathbf{r}) dV + \oint_{\partial\Omega_i} \hat{n}_\nu(\mathbf{r})n^* \Psi_l(\mathbf{r}) dS \tag{4.109}$$

Here, n^*, $(d\mathbf{q})^*$, and $(\mathbf{v}n)^*$ are numerical fluxes defined as

$$n^* = \{n\} + 0.5\hat{\beta} \cdot \hat{\mathbf{n}} [[n]] \tag{4.110}$$

$$(d\mathbf{q})^* = \{d\mathbf{q}\} - 0.5\hat{\beta}(\hat{\mathbf{n}} \cdot [[d\mathbf{q}]]) \tag{4.111}$$

$$(\mathbf{v}n)^* = \{\mathbf{v}n\} + \alpha\hat{\mathbf{n}} [[n]] \tag{4.112}$$

where $\{\odot\} = 0.5(\odot^- + \odot^+)$ and $[[\odot]] = \odot^- - \odot^+$, are "average" and "jump" operators, $\odot$ is a scalar or a vector variable. Superscripts "-" and "+" refer to variables defined in element i and in its neighboring element, respectively. For the diffusion term, (4.110) and (4.111) are the so-called LDG alternate flux [2]. The vector $\hat{\beta}$ determines the upwind direction of n and $(d\mathbf{q})$. In LDG, opposite directions are chosen for n and $(d\mathbf{q})$, while the precise direction of each variable is not important [2, 7, 8]. Here, we choose $\hat{\beta} = \hat{\mathbf{n}}$ on each element surface. 4.112 is the local Lax–Friedrichs flux [8], with $\alpha = \max(|\hat{\mathbf{n}} \cdot \mathbf{v}^-|, |\hat{\mathbf{n}} \cdot \mathbf{v}^+|)/2$ [8], which mimics the path of the information propagation. On $\partial\Omega_D$, $n^* = f_D$, $(d\mathbf{q})^* = (d\mathbf{q})^-$ and $(\mathbf{v}n)^* = \mathbf{v}^- f_D$, and on $\partial\Omega_R$, $n^* = n^-$ and $(d\mathbf{q})^* + (\mathbf{v}n)^* = f_R$.

Expanding $n_i(\mathbf{r}, t)$ and $q_k^\nu(\mathbf{r}, t)$, $\nu \in \{x, y, z\}$, using the set of Lagrange polynomials

$$n_i(\mathbf{r}, t) \simeq \sum_{l=1}^{N_p} n_i(\mathbf{r}_l, t)\Psi_l(\mathbf{r}) = \sum_{l=1}^{N_p} n_{i,l}(t)\Psi_l(\mathbf{r}) \tag{4.113}$$

$$q_i^\nu(\mathbf{r}, t) \simeq \sum_{l=1}^{N_p} q_i^\nu(\mathbf{r}_l, t)\Psi_l(\mathbf{r}) = \sum_{l=1}^{N_p} q_{i,l}^\nu(t)\Psi_l(\mathbf{r}) \tag{4.114}$$

yields the semi-discretized form

$$\mathbf{M}_i \frac{\mathbf{n}_i}{\partial t} = \mathbf{C}_i \mathbf{n}_i + \mathbf{C}_{ii'} \mathbf{n}_{k'} + \mathbf{D}_i \overline{d}_i \mathbf{q}_i + \mathbf{D}_{ii'} \overline{d}_{k'} \mathbf{q}_{i'} - \mathbf{B}_i^n \tag{4.115}$$

$$\mathbf{M}_i^{\mathbf{q}} \mathbf{q}_i = \mathbf{G}_i \mathbf{n}_k + \mathbf{G}_{ii'} \mathbf{n}_{i'} + \mathbf{B}_i^{\mathbf{q}} \tag{4.116}$$

where the global unknown vectors are defined as $\mathbf{n}_i = [n_{i,1}, \dots, n_{i,N_p}]^T$ and $\mathbf{q}_i = [\mathbf{q}_i^x, \mathbf{q}_i^y, \mathbf{q}_i^z]^T$, with $\mathbf{q}_i^\nu = [q_{i,1}^\nu, \dots, q_{i,N_p}^\nu]$, $\nu \in \{x, y, z\}$.

In (4.115)–(4.116), $\mathbf{M}_i$ and $\mathbf{M}_i^{\mathbf{q}}$ are mass matrices. $\mathbf{M}_i^{\mathbf{q}}$ is a 3×3 block diagonal matrix with blocks $\mathbf{M}_i$ defined in (4.55). $\overline{\overline{d}}_k$ is a diagonal matrix with entries $d_1, \dots, d_N$, where $d_i = (d_i^x, d_i^y, d_i^z)$ and $d_i^\nu(i) = d_i(\mathbf{r}_l)$, $l = 1, \dots, N_p$, $\nu \in \{x, y, z\}$. We note that $d(\mathbf{r})$ is assumed isotropic and constant in each element.

Matrices $\mathbf{G}_i$ and $\mathbf{G}_{ii'}$ and $\mathbf{D}_i$ and $\mathbf{D}_{ii'}$ correspond to the gradient and the divergence operators, respectively. For the LDG flux, $\mathbf{D}_i = -\mathbf{G}_i^T$ and $\mathbf{D}_{ii'} = -\mathbf{G}_{ii'}^T$. $\mathbf{G}_i$ is a $3N_p \times N_p$ matrix, and it has contributions from the volume and surface integral terms on the right-hand side of 4.109: $\mathbf{G}_i = \mathbf{G}_i^V + \mathbf{G}_i^S$, $\mathbf{G}_i^V = [\tilde{\mathbf{S}}_i^x \ \tilde{\mathbf{S}}_i^y \ \tilde{\mathbf{S}}_i^z]^T$, and $\mathbf{G}_i^S = [\mathbf{L}_i^x \ \mathbf{L}_i^y \ \mathbf{L}_i^z]^T$, where $\tilde{\mathbf{S}}_i^\nu = -[\mathbf{S}_i^\nu]^T$ and

$$\mathbf{L}_i^\nu(l, k) = \frac{1 + \text{sign}(\hat{\beta} \cdot \hat{\mathbf{n}})}{2} \theta_i(k) \oint_{\partial\Omega_{ii'}} \hat{n}_\nu(\mathbf{r})\Psi_l(\mathbf{r})\Psi_k(\mathbf{r}) dS$$

Here, $\partial\Omega_{ii'}$ denotes the interface connecting element i and i' and $\theta_i(k)$ selects the interpolation nodes on the interface

$$\theta_i(k) = \begin{cases} 1, & \mathbf{r}_k \in \Omega_i, \mathbf{r}_k \in \partial\Omega_{ii'} \\ 0, & \text{otherwise} \end{cases}$$

Matrix $\mathbf{G}_{ii'}$ corresponds to the surface integral term in (4.109), which involves the unknowns from neighboring elements of element i: $\mathbf{G}_{ii'} = [\mathbf{L}_{i'}^x \ \mathbf{L}_{i'}^y \ \mathbf{L}_{i'}^z]^T$, where

$$\mathbf{L}_{i'}^\nu(l, k) = \frac{1 - \text{sign}(\hat{\beta} \cdot \hat{\mathbf{n}})}{2} \theta_{i'}(j) \oint_{\partial\Omega_{ii'}} \hat{n}_\nu(\mathbf{r})\Psi_l(\mathbf{r})\Psi_k(\mathbf{r}) dS$$

Similarly, matrix $\mathbf{C}_i$ has contributions from the volume integral and the surface integral related to the drift term on the right-hand side of (4.108), $\mathbf{C}_i = \mathbf{C}_i^V + \mathbf{C}_i^S$, where

$$\mathbf{C}_i^V(l, k) = -\sum_v \int_{\Omega_i} v_v(\mathbf{r}, t) \partial_v \Psi_l(\mathbf{r}) \Psi_k(\mathbf{r}) dV$$

and

$$\mathbf{C}_i^S(l, k) = \theta_i(k) \oint_{\partial \Omega_{ii'}} \left[\frac{1}{2} \sum_v \hat{n}_v(\mathbf{r}) v_v(\mathbf{r}, t) + \alpha(\mathbf{r}, t) \right] \Psi_l(\mathbf{r}) \Psi_k(\mathbf{r}) dS$$

Matrix $\mathbf{C}_{ii'}$ is from the last surface integral term in (4.108), which involves the unknowns from neighboring elements of the element i:

$$\mathbf{C}_{ii'}(l, k) = \theta_{i'}(k) \oint_{\partial \Omega_{ii'}} \left[\frac{1}{2} \sum_v \hat{n}_v(\mathbf{r}) v_v(\mathbf{r}, t) - \alpha(\mathbf{r}, t) \right] \Psi_l(\mathbf{r}) \Psi_k(\mathbf{r}) dS$$

Matrices $\mathbf{B}_i^n$ and $\mathbf{B}_i^q$ are contributed from the force term and boundary conditions (where element k' does not exist) and are expressed as

$$\mathbf{B}_i^n(k) = \int_{\Omega_i} R(\mathbf{r}, t) \Psi_l(\mathbf{r}) dV + \oint_{\partial \Omega_i \cap \partial \Omega_R} f_R(\mathbf{r}) \Psi_l(\mathbf{r}) dS$$

$$+ \oint_{\partial \Omega_i \cap \partial \Omega_D} \hat{n}(\mathbf{r}) \cdot \mathbf{v}(\mathbf{r}, t) f_D(\mathbf{r}) \Psi_l(\mathbf{r}) dS$$

$$\mathbf{B}_i^{q,v}(k) = \oint_{\partial \Omega_i \cap \partial \Omega_D} \hat{n}_v(\mathbf{r}) f_D(\mathbf{r}) \Psi_l(\mathbf{r}) dS$$

An explicit third-order total-variation-diminishing Runge–Kutta method [87] is used for integrating (4.115)–(4.116) in time. With initial value $n(\mathbf{r}, t = 0) = 0$, time samples of the unknown vector $\mathbf{n}_i$ are obtained step by step in time.

The nonlinear Poisson equation and the stationary drift-diffusion equation are discretized using similar LDG schemes. Details of their corresponding DG formulations can be found in [75]. One should note that these two equations are time independent and belong to elliptic-type partial differential equations. Solving them calls for solving global linear systems. Fortunately, for photoconductive device simulation, one only needs to solve the steady-state once (for each bias voltage), and the solution can be reused in different transient studies.

To demonstrate the applicability of the above DG-based multiphysics framework, two practical device simulation examples are presented below. We first consider a conventional face-to-face dipole photoconductive device [88]. This device has a relatively simple structure, and it can be simulated using the FDTD [89] or FEM-based [90, 91] approach, where the carrier generation due to the optical electromagnetic wave excitation is modeled using closed-form analytical expressions [89–91].

Since the focus of the device study is the photocurrent response, we only consider the central gap region of the device. The cross-section of the device is shown

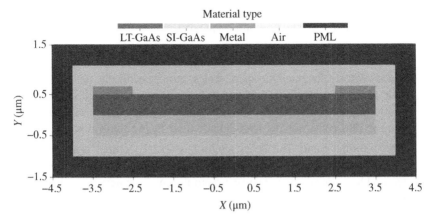

Figure 4.24 Cross-section of the conventional PCD considered in the first example. Source: Chen and Bagci [19]/ IEEE / Public Domian CC BY.

in Figure 4.24. The width and height of the semiconductor and substrate layers are 7 μm and 0.5 μm, respectively. Both electrodes are made of gold and have a width of 1 μm and a height of 0.19 μm. In the transient simulation, the permittivity of gold is represented using the Drude model [41]. The Drude model parameters of gold at the optical frequencies [92] as well as other physical parameters are shown in Table 4.3.

In multiphysics simulations, proper boundary conditions must be set for each of the above DG solvers, and the device should be meshed with respect to the characteristic space and time scales corresponding to each equation. The stationary and time-dependent drift-diffusion equations are solved only within the photoconductive semiconductor (LT-GaAs) layer, while the Poisson equation and the time-dependent Maxwell's equations are solved in the whole domain. The boundary conditions are as follows:

(i) **Stationary and time-dependent drift-diffusion equations**: On electrode/semiconductor interfaces, metal contacts are assumed ideal Ohmic contacts and Dirichlet boundary condition $n_e = (C + \sqrt{C^2 + 4n_i^2})/2$, $n_h = n_i^2/n_e$, which is derived using the local charge neutrality [76, 86], is used. On semiconductor/insulator interfaces, no carrier spills happen, which corresponds to the homogeneous Robin boundary condition $\hat{n} \cdot J_{e(h)} - 0$ [83, 86].

(ii) **Poisson equation**: On electrode surfaces, assuming ideal Ohmic contact, Dirichlet boundary condition $\varphi = V_{external} + V_T \ln(n_e^s/n_i)$ is enforced [76]. On truncation boundary of the computation domain, homogeneous Neumann boundary condition $\hat{n} \cdot \nabla\varphi = 0$ is enforced under the assumption that the static fields reaching the boundary are small and do not change the solution near the device.

(iii) **Maxwell's equations**: The computation domain is "wrapped around" by PMLs [93] truncated by first-order ABCs [8].

The characteristic space and time characteristic scales are as follows:

(i) **Space scales**: Characteristic space scales, more specifically, the electromagnetic wavelength at the optical and THz frequencies, the skin depth of the electromagnetic wave in the metal l_d [92], Debye length l_D [76], and the Peclet number [94] constrain the edge length used in the mesh discretizing the computation domain. The Debye length $l_D = \sqrt{2\varepsilon V_T / (qn_c)}$ is a measure of the carrier density variation in space [76]. Here, n_c is an estimate of maximum achievable carrier density. The Peclet number $\Delta_d C_P$, where Δ_d is the largest edge length of the local mesh element and $C_P = |\mu_c \mathbf{E}|/(2D_c)$, has to be less than 1 to ensure the stability of the convection-diffusion component in the drift-diffusion equations [94]. We should note here that the potential distribution is smooth; therefore, the mesh using the edge length determined by the space scales described above discretizes it accurately. The values of these space-scale parameters for the materials used in this example (see Table 4.3) and the corresponding required edge lengths are provided in Table 4.1. We should note that the last number in Table 4.1 is the smallest geometry dimension and the mesh used for this example represents the geometry very accurately.

(ii) **Time scales**: The Courant-Friedrichs–Lewy (CFL) stability conditions of Maxwell's equations [8] and diffusion and drift components of the drift-diffusion equations [95, 96] constrain the time step sizes used in the transient solution. Table 4.2 lists the values of the maximum time step sizes allowed by these three CFL conditions with the edge length obtained from the characteristic space scale discussion above. We should note that the time step sizes required to resolve the periods of the optical and THz electromagnetic

Table 4.1 Length scales.

Quantity	Value (nm)	Required mesh size (nm)
Optical wavelength	800	~ 200
THz wavelength	3×10^5	$\sim 10^5$
l_d	25	~ 10
l_D	30	~ 10
C_P^{-1}	10	~ 10
Geometrical details	100	~ 100

Source: Chen and Bagci [19]/ IEEE / Public Domian CC BY.

Table 4.2 Time scales.

Quantity	Required time step size (ps)
CFL (Maxwell's equations)	$\sim 10^{-7}$
CFL (DD drift term)	$\sim 10^{-6}$
CFL (DD diffusion term)	$\sim 10^{-6}$

Source: Chen and Bagci [19]/ IEEE / Public Domian CC BY.

waves are larger than those required by their CFL conditions. Table 4.2 clearly shows that the drift-diffusion equations can be integrated using a larger time step size than the one required for the integration of Maxwell's equations.

Using these scales as a guide, the device is discretized with mesh sizes in the range [10,200] nm, where the finest meshes are used near the boundaries of the semiconductor layer and near the metallic electrodes, the maximum mesh size in the semiconductor layer and the substrate is 70 nm, and the maximum mesh size in the rest of the computation domain is 200 nm. The time step sizes for the Maxwell and drift-diffusion equations are selected to be 10^{-7} ps and 5×10^{-7} ps, respectively.

The steady-state solutions are first solved from the Poisson-drift-diffusion solver and used as inputs in the transient simulation. Here, we focus on the transient simulation. An aperture source located at $y = 0.8$ μm generates optical electromagnetic waves propagating from top to bottom. The pulse shape parameters are given in Table 4.3. The intensity of the electromagnetic field on the aperture has a Gaussian distribution with beam width 3 μm.

Figure 4.25a,b shows $|H_z^t(\mathbf{r}, t)|$ and $n_e^t(\mathbf{r}, t)$ at several time instants, respectively, and Figure 4.26a,b shows the time signatures of the optical electromagnetic wave excitation and the generated THz current and their spectrum obtained using Fourier transform. Figures 4.25a,b and 4.26a also serve as a reference for the following discussion.

At $t = 0.05$ ps, the optical electromagnetic wave generated by the aperture arrives at the semiconductor layer. Because the permittivity of LT-GaAs is high, a large part of the incident field's energy is reflected back (see the reflected wave above the air-semiconductor interface and behind the aperture). At the same time, in the LT-GaAs layer, the incident field's energy entering the device is partially absorbed and carriers are generated near the air-semiconductor interface. At $t = 0.25$ ps, the incident field reaches its pulse peak and the electron density increases to $\sim 10^{11}$ cm^{-3}. The short incident field pulse passes quickly and, after $t = 0.5$ ps, only some scattered fields reside in the high permittivity region due to internal reflections. During that time, the electron density keeps increasing until

Table 4.3 Physical parameters used for the PCD examples.

Laser[a]	$f_c = 375$ THz, $f_w = 25$ THz, Power $= 0.63$ mW
LT-GaAs	$\epsilon_r = 13.26$, $\mu_r = 1.0$, $\alpha = 1$ μm^{-1}, $\eta = 1.0$
SI-GaAs	$\epsilon_r = 13.26$, $\mu_r = 1.0$
Metal[b]	$\epsilon_\infty = 1$, $\omega_p = 9.03/\hbar$, $\gamma = 0.053q/\hbar$
Temperature	300 K
V_{bias}	10 V
C	1.3×10^{16} cm^{-3}
n_i	9×10^6 cm^{-3}
Mobility	$\mu_e^0 = 8000$ cm^2/V/s, $\mu_h^0 = 400$ cm^2/V/s
	$V_e^{sat} = 1.725 \times 10^7$ cm/s, $V_h^{sat} = 0.9 \times 10^7$ cm/s
	$\beta_e = 1.82$, $\beta_h = 1.75$
	$\tau_e = 0.3$ ps, $\tau_h = 0.4$ ps
Recombination	$n_{e1} = n_{h1} = 4.5 \times 10^6$ cm^{-3}
	$C_e^A = C_h^A = 7 \times 10^{-30}$ cm^6/s

a) f_c is the center frequency and f_w is the bandwidth.
b) Drude model parameters [92].
Source: Chen and Bagci [19]/ IEEE / Public Domian CC BY.

the excitation pulse decays to 20% of its peak value at $t \approx 0.4$ ps, after $t \approx 0.4$ ps, $n_e^t(\mathbf{r}, t)$ decays slowly due to the recombination.

Comparing the electron density distributions at different time instants, it can be clearly observed that electrons move toward the anode on the left side. The picture of holes (not shown) is similar but with holes moving toward the cathode on the right side with a lower speed. The resulting current shown in Figure 4.26a,b match well with the result presented in [89], where the latter has been shown to agree with experimental results very well [89]. Because of the simplicity of the single interface scattering, the recorded electromagnetic field intensity in the semiconductor layer has almost the same pulse shape as the optical excitation signal. This explains why the analytical generation rate [89] works very well.

Next, a plasmonic photoconductive device is simulated for comparison. The cross-section of the device is shown in Figure 4.27. For this example, gold nanostructures are added between the two electrodes. The periodicity, the duty cycle, and the thickness of the nanograting are 180 nm, 5/9, and 190 nm, respectively. The surfaces of the nanostructures are modeled as floating potential surfaces for the stationary state simulation (i.e. for solution of the Poisson equation) [75, 97]. Here, finer meshes are used near and inside the nanostructures to correctly resolve the geometry and the exponential decay of the plasmonic fields. In this region, the mesh has a minimum edge length of 3 nm and the maximum edge length of the

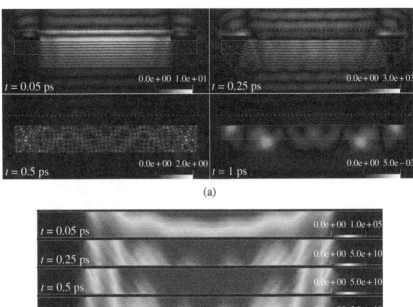

(a)

(b)

Figure 4.25 (a) $|H_z(\mathbf{r}, t)|$ computed using the DG scheme at different time instants. The dotted line shows the aperture of optical EM wave excitation. The gray box indicates the LT-GaAs region. (b) $n_e(\mathbf{r}, t)$ computed using the DG scheme at different time instants. Source: Chen and Bagci [19]/ IEEE / Public Domian CC BY.

mesh is allowed to reach 70 nm in the GaAs layers and 200 nm in the rest of the computation domain. The corresponding time step size used for the time integration of Maxwell's equations is 0.3×10^{-7} ps. Other simulation parameters and boundary conditions are same as those in the previous example.

Figure 4.28a,b shows $|H_z^t(\mathbf{r}, t)|$ and $n_e^t(\mathbf{r}, t)$ at several time instants, respectively. For comparison, Figure 4.28 uses the same color scale as Figure 4.25 for the previous example.

Figure 4.28 shows that the transient electromagnetic fields on the plasmonic photoconductive device are much stronger compared with the conventional

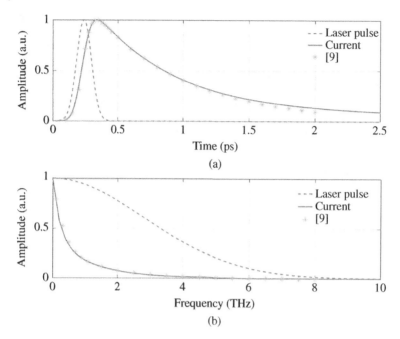

Figure 4.26 (a) The time signature and (b) the spectrum of the optical EM wave excitation and the generated THz current. Source: Chen and Bagci [19]/ IEEE / Public Domian CC BY.

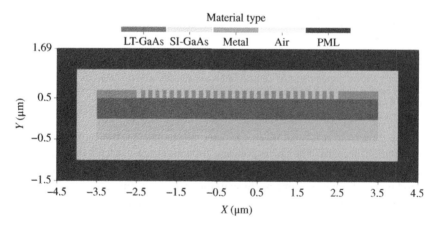

Figure 4.27 Cross-section of the plasmonic PCD. Source: Chen and Bagci [19]/ IEEE / Public Domian CC BY.

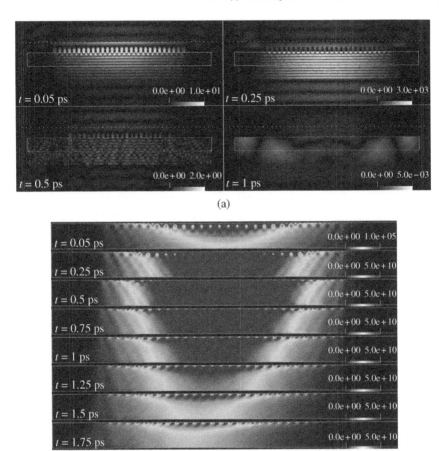

Figure 4.28 (a) $|H_z(\mathbf{r}, t)|$ computed using the DG scheme at different time instants. The dotted line shows the aperture of optical EM wave excitation. The gray box indicates the LT-GaAs region. (b) $n_e(\mathbf{r}, t)$ computed using the DG scheme at different time instants. Reproduced with permission from [19].

photoconductive device. In addition, plasmonic mode patterns are observed at $t = 0.05$ ps and $t = 0.25$ ps. Accordingly, the level of electron density in Figure 4.28 is much higher and shows an inhomogeneous pattern. As time marching goes on, electrons drift toward the anode. Figure 4.29a,b compare the time signatures of the current induced on the plasmonic photoconductive device and the conventional one (previous example) and their spectrum obtained using Fourier transform. The figures clearly show that the inclusion of the nanostructures in the photoconductive device design enhances the current by almost seven times.

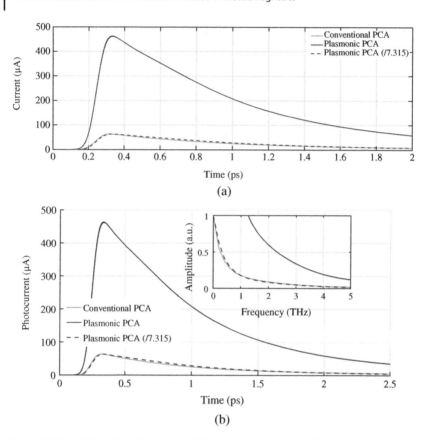

Figure 4.29 (a) The time signature and (b) the spectrum of the THz current generated on the plasmonic PCD and on the conventional PCD. Source: Chen and Bagci [19]/ IEEE / Public Domian CC BY.

The above examples show the applicability of the multiphysics DG framework. However, due to the high computational cost, only the central gap region of the device is considered. Two extensions toward more efficient simulation of photoconductive devices have been reported. In [83], a unit-cell scheme is presented for fast characterization of nanostructure-enhanced photoconductive devices. By using a "potential-drop" boundary condition, this scheme models a biased photoconductive device within a unit cell of the nanostructure and thus greatly reduces the computational cost. Another challenge in modeling full-structure photoconductive devices is the inclusion of both optical and terahertz (THz) frequency electromagnetic waves. The wavelength of optical and THz waves differs by almost 400 times. Correspondingly, the THz antenna is typically 400 times larger than the size of the nanostructure. This scale difference makes the direct simulation of full-structure photoconductive devices almost impossible. This challenge has been addressed by a DG-based dual-mesh scheme [98], where the optical and the THz waves are treated separately in two DG solvers and their two-way couplings are taken into account using high-order interpolations. With this dual-mesh scheme, the space-time-dependent conductivity in photoconductive materials can be strictly modeled in the THz antenna simulation. Thus, the impedance matching between different photoconductive structures and THz antennas can be strictly analyzed.

References

1 Reed, W.H. and Hill, T.R. (1973). Triangular mesh methods for the neutron transport equation (No. LA-UR-73-479; CONF-730414-2). In: *Proceedings of the American Nuclear Society*. Los Alamos Scientific Lab., N. Mex.(USA), 1–23.

2 Cockburn, B., Karniadakis, G.E., and Shu, C.-W. (2000). The development of discontinuous Galerkin methods. In: *The Development of Discontinuous Galerkin Methodsof Discontinuous Galerkin Methods*, 3–50. Berlin, Heidelberg: Springer.

3 Bourdel, F., Mazet, P.-A., and Helluy, P. (1991). Resolution of the non-stationary or harmonic Maxwell equations by a discontinuous finite element method: application to an E.M.I. (electromagnetic impulse) case. *Proceedings of the 10th International Conference on Computer Methods Application and Science Engineering*, pp. 405–422, USA, Mar 1991.

4 Hesthaven, J.S. and Warburton, T. (2002). Nodal high-order methods on unstructured grids: I. Time-domain solution of Maxwell's equations. *Journal of Computational Physics* 181 (1): 186–221.

5 Kabakian, A.V., Shankar, V., and Hall, W.F. (2004). Unstructured grid-based discontinuous Galerkin method for broadband electromagnetic simulations. *Journal of Scientific Computing* 20 (3): 405–431.

6 Fezoui, L., Lanteri, S., Lohrengel, S., and Piperno, S. (2005). Convergence and stability of a discontinuous Galerkin time-domain method for the 3D heterogeneous Maxwell equations on unstructured meshes. *ESAIM: Mathematical Modelling and Numerical Analysis* 39 (6): 1149–1176.

7 Shu, C.W. (2016). Discontinuous Galerkin methods for time-dependent convection dominated problems: basics, recent developments and comparison with other methods. In: *Building Bridges: Connections and Challenges in Modern Approaches to Numerical Partial Differential Equations*, 371–399. Cham: Springer International Publishing.

8 Hesthaven, J.S. and Warburton, T. (2008). *Nodal Discontinuous Galerkin Methods, Texts in Applied Mathematics*, vol. 54. New York: Springer.

9 Dosopoulos, S. and Lee, J.F. (2010). Interior penalty discontinuous Galerkin finite element method for the time-dependent first order Maxwell's equations. *IEEE Transactions on Antennas and Propagation* 58 (12): 4085–4090.

10 García de Abajo, F.J. (2013). Graphene nanophotonics. *Science* 339 (6122): 917–918.

11 Li, P., Jiang, L.J., and Bagci, H. (2014). Cosimulation of electromagnetics-circuit systems exploiting DGTD and MNA. *IEEE Transactions on Components, Packaging and Manufacturing Technology* 4 (6): 1052–1061.

12 Lin, C.P., Yan, S., Arslanbekov, R.R. et al. (2016). A DGTD algorithm with dynamic h-adaptation and local time-stepping for solving Maxwell's equations. *2016 IEEE Antennas and Propagation Society International Symposium APSURSI 2016 - Proceedings*, pp. 2079–2080, October 2016.

13 Harmon, M., Gamba, I.M., and Ren, K. (2016). Numerical algorithms based on Galerkin methods for the modeling of reactive interfaces in photoelectrochemical (PEC) solar cells. *Journal of Computational Physics* 327: 140–167.

14 Angulo, L., Alvarez, J., Pantoja, M. et al. (2015). Discontinuous Galerkin time domain methods in computational electrodynamics: state of the art. *Forum for Electromagnetic Research Methods and Application Technologies (FERMAT)* 10 (4): 1–24.

15 Yan, S., Greenwood, A.D., and Jin, J.-M. (2016). Modeling of plasma formation during high-power microwave breakdown in air using the discontinuous Galerkin time-domain method. *IEEE Journal on Multiscale and Multiphysics Computational Techniques* 1: 2–13.

16 Ren, Q., Bian, Y., Kang, L. et al. (2017). Leap-frog continuous-discontinuous Galerkin time domain method for nanoarchitectures with the Drude model. *Journal of Lightwave Technology* 35 (22): 4888–4896. https://doi.org/10.1109/JLT.2017.2760913.

17 Chen, L., Sirenko, K., and Bagci, H. (2018). Multiphysics analysis of plasmonic photomixers under periodic boundary conditions using discontinuous Galerkin time domain method. *Progress in Electromagnetics Research Symposium* (1–4 August 2018), in Toyama, Japan.

18 Sirenko, K., Sirenko, Y., and Bagci, H. (2018). Exact absorbing boundary conditions for periodic three-dimensional structures: derivation and implementation in discontinuous Galerkin time-domain method. *IEEE Journal on Multiscale and Multiphysics Computational Techniques* 3: 108–120.

19 Chen, L. and Bagci, H. (2020). Multiphysics simulation of plasmonic photoconductive devices using discontinuous Galerkin methods. *IEEE Journal on Multiscale and Multiphysics Computational Techniques* 5: 188–200.

20 Gedney, S.D., Luo, C., Guernsey, B. et al. (2007). The discontinuous Galerkin finite element time domain method (DGFETD): a high order, globally-explicit method for parallel computation. *IEEE International Symposium on Electromagnetic Compatibility.*

21 Jin, J.M. (2002). *The Finite Element Method in Electromagnetics*, 2e. New York: Wiley.

22 Li, P. and Jiang, L.J. (2013). A hybrid electromagnetics-circuit simulation method exploiting discontinuous Galerkin finite element time domain method. *IEEE Microwave and Wireless Components Letters* 23 (3): 113–115.

23 Li, P. and Jiang, L.J. (2013). Integration of arbitrary lumped multiport circuit networks into the discontinuous Galerkin time-domain analysis. *IEEE Transactions on Microwave Theory and Techniques* 61 (7): 2525–2534.

24 Li, P., Shi, Y., Jiang, L.J., and Bagci, H. (2014). A hybrid time-domain discontinuous Galerkin-boundary integral method for electromagnetic scattering analysis. *IEEE Transactions on Antennas and Propagation* 62 (5): 2841–2846.

25 Li, P. and Jiang, L.J. (2013). Simulation of electromagnetic waves in the magnetized cold plasma by a DGFETD method. *IEEE Antennas and Wireless Propagation Letters* 12: 1244–1247.

26 Klöckner, A., Warburton, T., Bridge, J., and Hesthaven, J.S. (2009). Nodal discontinuous Galerkin methods on graphics processors. *Journal of Computational Physics* 228 (21): 7863–7882.

27 Dosopoulos, S., Gardiner, J.D., and Lee, J.-F. (2011). An MPI/GPU parallelization of an interior penalty discontinuous Galerkin time domain method for Maxwell's equations. *Radio Science* 46 (3): 1–10.

28 Sun, Q., Zhan, Q., Ren, Q., and Liu, Q.H. (2017). Wave equation-based implicit subdomain DGTD method for modeling of electrically small problems. *IEEE Transactions on Microwave Theory and Techniques* 65 (4): 1111–1119.

29 Sankaran, K. (2007). Accurate domain truncation techniques for time-domain conformal methods. PhD thesis. ETH ZUTICH.

30 Jeffrey, I. (2011). Finite-volume simulations of Maxwell's equations on unstructured grids. PhD thesis. University of Manitoba.

31 Li, P. and Jiang, L.J. (2015). Modeling of magnetized graphene from microwave to THz range by DGTD with a scalar RBC and an ADE. *IEEE Transactions on Antennas and Propagation* 63 (10): 4458–4467.

32 Liu, M., Sirenko, K., and Bagci, H. (2012). An efficient discontinuous Galerkin finite element method for highly accurate solution of Maxwell equations. *IEEE Transactions on Antennas and Propagation* 60 (8): 3992–3998.

33 Mur, G. (1981). Absorbing boundary conditions for the finite-difference approximation of the time-domain electromagnetic-field equations. *IEEE Transactions on Electromagnetic Compatibility* EMC-23 (4): 377–382.

34 Dosopoulos, S. and Lee, J.F. (2010). Interconnect and lumped elements modeling in interior penalty discontinuous Galerkin time-domain methods. *Journal of Computational Physics* 229 (22): 8521–8536.

35 Hagstrom, T., Warburton, T., and Givoli, D. (2010). Radiation boundary conditions for time-dependent waves based on complete plane wave expansions. *Journal of Computational and Applied Mathematics* 234 (6): 1988–1995.

36 Sirenko, K., Liu, M., and Bagci, H. (2013). Incorporation of exact boundary conditions into a discontinuous Galerkin finite element method for accurately solving 2D time-dependent Maxwell equations. *IEEE Transactions on Antennas and Propagation* 61 (1): 472–477.

37 Higdon, R.L. (1986). Absorbing boundary conditions for difference approximations to the multi-dimensional wave equation. *Mathematics of Computation* 47 (176): 437.

38 Berenger, J.P. (1994). A perfectly matched layer for the absorption of electromagnetic waves. *Journal of Computational Physics* 114 (2): 185–200.

39 Gedney, S.D. (1996). An anisotropic perfectly matched layer-absorbing medium for the truncation of FDTD lattices. *IEEE Transactions on Antennas and Propagation* 44 (12): 1630–1639.

40 Chen, L., Ozakin, M.B., Ahmed, S., and Bagci, H. (2020). A memory-efficient implementation of perfectly matched layer with smoothly-varying coefficients in discontinuous Galerkin time-domain method. *IEEE Transactions on Antennas and Propagation* 69 (6): 3605–3610.

41 Gedney, S.D., Young, J.C., Kramer, T.C., and Roden, J.A. (2012). A discontinuous Galerkin finite element time-domain method modeling of dispersive media. *IEEE Transactions on Antennas and Propagation* 60 (4): 1969–1977.

42 Gedney, S.D. and Zhao, B. (2010). An auxiliary differential equation formulation for the complex-frequency shifted PML. *IEEE Transactions on Antennas and Propagation* 58 (3): 838–847.

43 Abarbanel, S., Gottlieb, D., and Hesthaven, J.S. (2002). Long time behavior of the perfectly matched layer equations in computational electromagnetics. *Journal of Scientific Computing* 17 (1–4): 405–422.

44 Dong, M., Li, P., and Bagci, H. (2020). An explicit time domain finite element boundary integral method with element level domain decomposition for electromagnetic scattering analysis. *14th European Conference on Antennas Propagation, EuCAP 2020*, March 2020.

45 Dong, M., Chen, L., Jiang, L. et al. (2022). An explicit time domain finite element boundary integral method for analysis of electromagnetic scattering. *IEEE Transactions on Antennas and Propagation* 1. https://doi.org/10.1109/TAP.2022.3142319.

46 Alvarez, J., Angulo, L.D., Pantoja, M.F. et al. (2010). Source and boundary implementation in vector and scalar DGTD. *IEEE Transactions on Antennas and Propagation* 58 (6): 1997–2003.

47 Jiao, D., Ergin, A.A., Shanker, B. et al. (2001). A fast higher-order time-domain finite element-boundary integral method for 3-D electromagnetic scattering analysis. *IEEE Antennas and Propagation Society AP-S International Symposium*, Volume 4, pp. 334–337.

48 Shanker, B., Lu, M., Ergin, A.A., and Michielssen, E. (2005). Plane-wave time-domain accelerated radiation boundary kernels for FDTD analysis of 3-D electromagnetic phenomena. *IEEE Transactions on Antennas and Propagation* 53 (11): 3704–3716.

49 Yilmaz, A.E., Jin, J.M., and Michielssen, E. (2004). Time domain adaptive integral method for surface integral equations. *IEEE Transactions on Antennas and Propagation* 52 (10): 2692–2708.

50 Bagci, H., Yilmaz, A.E., Jin, J.M., and Michielssen, E. (2007). Fast and rigorous analysis of EMC/EMI phenomena on electrically large and complex cable-loaded structures. *IEEE Transactions on Electromagnetic Compatibility* 49 (2): 361–381.

51 Sharma, P., Perruisseau-Carrier, J., Moldovan, C., and Ionescu, A.M. (2014). Electromagnetic performance of RF NEMS graphene capacitive switches. *IEEE Transactions on Nanotechnology* 13 (1): 70–79.

52 Geim, A.K. and Novoselov, K.S. (2007). The rise of graphene. *Nature Materials* 6 (3): 183–191.

53 Schwierz, F. (2010). Graphene transistors. *Nature Nanotechnology* 5 (7): 487–496.

54 Tamagnone, M., Gómez-Díaz, J.S., Mosig, J.R., and Perruisseau-Carrier, J. (2012). Analysis and design of terahertz antennas based on plasmonic resonant graphene sheets. *Journal of Applied Physics* 112 (11): 114915.

55 Koppens, F.H.L., Chang, D.E., and García de Abajo, F.J. (2011). Graphene plasmonics: a platform for strong light-matter interactions. *Nano Letters* 11 (8): 3370–3377.

56 Lovat, G. (2012). Equivalent circuit for electromagnetic interaction and transmission through graphene sheets. *IEEE Transactions on Electromagnetic Compatibility* 54 (1): 101–109.

57 Chen, P.Y., Argyropoulos, C., and Alu, A. (2013). Terahertz antenna phase shifters using integrally-gated graphene transmission-lines. *IEEE Transactions on Antennas and Propagation* 61 (4): 1528–1537.

58 Hanson, G.W. (2008). Dyadic Green's functions and guided surface waves for a surface conductivity model of graphene. *Journal of Applied Physics* 103 (6): 064302.

59 Shapoval, O.V., Gomez-Diaz, J.S., Perruisseau-Carrier, J. et al. (2013). Integral equation analysis of plane wave scattering by coplanar graphene-strip gratings in the THz range. *IEEE Transactions on Terahertz Science and Technology* 3 (5): 666–674.

60 Lin, H., Pantoja, M.F., Angulo, L.D. et al. (2012). FDTD modeling of graphene devices using complex conjugate dispersion material model. *IEEE Microwave and Wireless Components Letters* 22 (12): 612–614.

61 Nayyeri, V., Soleimani, M., and Ramahi, O.M. (2013). Wideband modeling of graphene using the finite-difference time-domain method. *IEEE Transactions on Antennas and Propagation* 61 (12): 6107–6114.

62 Feizi, M., Nayyeri, V., and Ramahi, O.M. (2018). Modeling magnetized graphene in the finite-difference time-domain method using an anisotropic surface boundary condition. *IEEE Transactions on Antennas and Propagation* 66 (1): 233–241.

63 Li, P., Jiang, L.J., and Bagci, H. (2015). A resistive boundary condition enhanced DGTD scheme for the transient analysis of graphene. *IEEE Transactions on Antennas and Propagation* 63 (7): 3065–3076.

64 Li, P., Jiang, L.J., and Bagci, H. (2018). Discontinuous Galerkin time-domain modeling of graphene nanoribbon incorporating the spatial dispersion effects. *IEEE Transactions on Antennas and Propagation* 66 (7): 3590–3598.

65 Gustavsen, B. and Semlyen, A. (1999). Rational approximation of frequency domain responses by vector fitting. *IEEE Transactions on Power Delivery* 14 (3): 1052–1059.

66 Gustavsen, B. (2006). Improving the pole relocating properties of vector fitting. *IEEE Transactions on Power Delivery* 21 (3): 1587–1592.

67 Llatser, I., Kremers, C., Cabellos-Aparicio, A. et al. (2012). Graphene-based nano-patch antenna for terahertz radiation. *Photonics and Nanostructures - Fundamentals and Applications* 10 (4): 353–358.

68 Wang, X.H., Yin, W.Y., and Chen, Z. (2013). Matrix exponential FDTD modeling of magnetized graphene sheet. *IEEE Antennas and Wireless Propagation Letters* 12: 1129–1132.

69 Sounas, D.L. and Caloz, C. (2012). Gyrotropy and nonreciprocity of graphene for microwave applications. *IEEE Transactions on Microwave Theory and Techniques* 60 (4): 901–914.

70 Kubo, R. (1957). Statistical-mechanical theory of irreversible processes. I. General theory and simple applications to magnetic and conduction problems. *Journal of the Physical Society of Japan* 12 (6): 570–586.

71 Dosopoulos, S., Zhao, B., and Lee, J.F. (2013). Non-conformal and parallel discontinuous Galerkin time domain method for Maxwell's equations: EM analysis of IC packages. *Journal of Computational Physics* 238: 48–70.

72 Homsi, L., Geuzaine, C., and Noels, L. (2017). A coupled electro-thermal discontinuous Galerkin method. *Journal of Computational Physics* 348: 231–258. https://doi.org/10.1016/j.jcp.2017.07.028.

73 Dong, Y., Tang, M., Li, P., and Mao, J. (2019). Transient electromagnetic-thermal simulation of dispersive media using DGTD method. *IEEE Transactions on Electromagnetic Compatibility* 61 (4): 1305–1313.

74 Jacobs, G.B. and Hesthaven, J.S. (2006). High-order nodal discontinuous Galerkin particle-in-cell method on unstructured grids. *Journal of Computational Physics* 214 (1): 96–121.

75 Chen, L. and Bagci, H. (2020). Steady-state simulation of semiconductor devices using discontinuous Galerkin methods. *IEEE Access* 8: 16203–16215.

76 Vasileska, D., Goodnick, S.M., and Klimeck, G. (2010). *Computational Electronics: Semiclassical and Quantum Device Modeling and Simulation*. Boca Raton, FL: CRC Press.

77 Chuang, S.L. (2012). *Physics of Photonic Devices*, vol. 80. Wiley.

78 Selberherr, S. (1984). *Analysis and Simulation of Semiconductor Devices*. New York, Wien: Springer-Verlag.

79 ATLAS User's Manual: device simulation software (2016). *Silvaco, Int. St. Clara, CA*.

80 Minimos-NT User Manual (2017). *Inst. Microelectron. (TU Vienna) Glob. TCAD Solut. GmbH, St. Clara, CA*.

81 Semiconductor Module User's Guide, version 5.3a (2017). *COMSOL Multiphysics*.

82 Moreno, E., Pantoja, M.F., Ruiz, F.G. et al. (2014). On the numerical modeling of terahertz photoconductive antennas. *Journal of Infrared, Millimeter, and Terahertz Waves* 35 (5): 432–444.

83 Chen, L., Sirenko, K., Li, P., and Bagci, H. (2021). Efficient discontinuous Galerkin scheme for analyzing nanostructured photoconductive devices. *Optics Express* 29 (9): 12903–12917.

84 Saleh, B.E.A. and Teich, M.C. (2019). *Fundamentals of Photonics*. Hoboken, NJ: Wiley.

85 Piprek, J. (ed.) (2018). *Handbook of Optoelectronic Device Modeling and Simulation*. Boca Raton, FL: CRC Press.

86 Schroeder, D. (1994). *Modelling of Interface Carrier Transport for Device Simulation*. Wien: Springer-Verlag.

87 Shu, C.W. and Osher, S. (1988). Efficient implementation of essentially non-oscillatory shock-capturing schemes. *Journal of Computational Physics* 77 (2): 439–471.

88 Tani, M., Matsuura, S., Sakai, K., and Nakashima, S.-i. (1997). Emission characterstics of photoconductive antennas based on low-temperature-grown GaAs and semi-insulating GaAs. *Applied Optics* 36 (30): 7853–7859.

89 Moreno, E., Pantoja, M.F., Garcia, S.G. et al. (2014). Time-domain numerical modeling of {TH}z photoconductive antennas. *IEEE Transactions on Terahertz Science and Technology* 4 (4): 490–500.

90 Burford, N. and El-Shenawee, M. (2016). Computational modeling of plasmonic thin-film terahertz photoconductive antennas. *Journal of the Optical Society of America B* 33 (4): 748–759.

91 Bashirpour, M., Ghorbani, S., Kolahdouz, M. et al. (2017). Significant performance improvement of a terahertz photoconductive antenna using a hybrid structure. *RSC Advances* 7 (83): 53010–53017.

92 Olmon, R.L., Slovick, B., Johnson, T.W. et al. (2012). Optical dielectric function of gold. *Physical Review B* 86 (23): 235147.

93 Gedney, S.D., Luo, C., Roden, J.A. et al. (2009). The discontinuous Galerkin finite-element time-domain method solution of Maxwell's equations. *Applied Computational Electromagnetics Society Journal* 24 (2): 129.

94 Trangenstein, J.A. (2013). *Numerical Solution of Elliptic and Parabolic Partial Differential Equations with CD-ROM*. Cambridge: Cambridge University Press.

95 Liu, Y.X. and Shu, C.-W. (2016). Analysis of the local discontinuous Galerkin method for the drift-diffusion model of semiconductor devices. *Science China Mathematics* 59 (1): 115–140.

96 Wang, H., Shu, C.-W., and Zhang, Q. (2015). Stability and error estimates of local discontinuous Galerkin methods with implicit-explicit time-marching for advection-diffusion problems. *SIAM Journal on Numerical Analysis* 53 (1): 206–227.

97 Chen, L., Dong, M., and Bagci, H. (2020). Modeling floating potential conductors using discontinuous Galerkin method. *IEEE Access* 8: 7531–7538.

98 Chen, L. and Bagci, H. (2020). A dual-mesh framework for multiphysics simulation of photoconductive terahertz devices. *2020 33rd General Assembly and Scientific Symposium of the International Union of Radio Science URSI GASS 2020*, August 2020.

5

Adaptive Discontinuous Galerkin Time-Domain Method for the Modeling and Simulation of Electromagnetic and Multiphysics Problems

Su Yan

Department of Electrical Engineering and Computer Science, Howard University, Washington, DC, USA

5.1 Introduction

The time-domain solutions of Maxwell's equations can be obtained with various numerical methods developed by the computational electromagnetics (CEM) community. Of them, the finite-difference time-domain (FDTD) method [1] has been most widely used due to its simplicity and high efficiency. Also because of its scalar representation of each field component on structured mesh grids, the continuity of the electromagnetic (EM) fields can be preserved, which is especially important when the EM fields are coupled to particle motion. However, the stair-case approximation of the solution domain, the finite-difference approximation of the fields and their derivatives, and the explicit leapfrog time-marching scheme used in the FDTD method result in low-order accuracy, which requires an extremely dense mesh grid and an extremely tiny time step size in a simulation to achieve the desired accuracy. On the contrary, another widely used method, the time-domain finite-element method (TDFEM) [2], which employs an unstructured mesh for the geometric discretization, high-order vector basis functions for the field expansion, and an implicit time-marching scheme, is able to achieve high-order accuracy and unconditional stability. Unfortunately, the implicit scheme results in a globally coupled system that has to be solved at every time step, which leads to a reduction in efficiency compared to the FDTD method. More importantly, because of the application of the vector basis functions, only the field components that are tangential to the elemental interfaces are continuous, and those that are normal to the interfaces are much less accurate. When coupled with a particle solver, discontinuous normal components of the fields will impose an unpredictable amount of force to the particles and result in a spurious solution or even a numerical breakdown of the simulation. To overcome

Advances in Time-Domain Computational Electromagnetic Methods, First Edition.
Edited by Qiang Ren, Su Yan, and Atef Z. Elsherbeni.

the aforementioned issues of both methods, the nodal discontinuous Galerkin time-domain (DGTD) method [3–12] is used to solve for the EM fields. Different from the vector DGTD [13, 14] traditionally used in the CEM community, the nodal DGTD method employs nodal interpolatory basis functions to represent the physical quantities, which guarantees the field continuity in all three spatial directions. Using an unstructured mesh, high-order nodal basis functions, and an explicit high-order time-marching scheme, the nodal DGTD method achieves high-order accuracy in the geometric discretization of the solution domain and in the spatial and the temporal discretization of the field solution. In addition, the nodal DGTD method also enjoys the following advantages: (i) high parallel efficiency on a massively parallel platform and (ii) high flexibility and efficiency by using dynamic *hp*-adaptation and a multirate time integration (MTI) scheme.

We demonstrate the numerical simulation of one of the most challenging multiphysics problems, EM–plasma interactions, using adaptive DGTD methods. The DGTD method is used to simulate both EM fields and plasma fluids and to account for their mutual couplings in a self-consistent manner. Based on the DGTD method, dynamic adaptation algorithms [15, 16], which adjust either the local mesh size or the local polynomial order in a real-time simulation, are employed to provide a sufficient numerical resolution of the physics while keeping the total computational cost low. To alleviate constraints of the time step size of an explicit time integrator, a MTI method is employed to advance the physics in time, which permits the application of different time step sizes in elements with different sizes or polynomial orders. The incorporation of the dynamic adaptation and MTI techniques makes the DGTD method a very powerful simulation tool that can be used to investigate very complicated EM–plasma interaction problems. Numerical examples are presented in this chapter to demonstrate the accuracy, efficiency, and flexibility of the proposed algorithm in the simulation of high-power microwave (HPM) air breakdown and the resulting physical phenomena. While the EM-plasma interaction presented in this chapter is one of the most challenging multiphysics problems, examples of other EM-related multiphysics problems can be found in the review paper [17].

5.2 Nodal Discontinuous Galerkin Time-Domain Method

In the nodal DGTD method, physical quantities are defined on each node with vector quantities expanded into *x*-, *y*-, and *z*-components. Consider Maxwell's equations

$$\frac{\partial \mathbf{H}}{\partial t} = -\frac{1}{\mu} \nabla \times \mathbf{E}$$

$$\frac{\partial \mathbf{E}}{\partial t} = \frac{1}{\varepsilon} (\nabla \times \mathbf{H} - \mathbf{J}_c) \tag{5.1}$$

subject to the following boundary conditions:

$$\hat{\mathbf{n}} \times \mathbf{E} = 0, \quad \hat{\mathbf{n}} \cdot \mathbf{H} = 0, \qquad\qquad r \in \Gamma_{\text{PEC}} \qquad\qquad (5.2)$$

$$\hat{\mathbf{n}} \times \mathbf{H} = 0, \quad \hat{\mathbf{n}} \cdot \mathbf{E} = 0, \qquad\qquad r \in \Gamma_{\text{PMC}} \qquad\qquad (5.3)$$

$$\hat{\mathbf{n}} \times \mathbf{E} + Z\ \hat{\mathbf{n}} \times \hat{\mathbf{n}} \times \mathbf{H} = 0, \quad \hat{\mathbf{n}} \times \mathbf{H} - Y\ \hat{\mathbf{n}} \times \hat{\mathbf{n}} \times \mathbf{E} = 0, \quad r \in \Gamma_{\text{ABC}} \quad (5.4)$$

where Γ_{PEC}, Γ_{PMC}, and Γ_{ABC} denote the boundaries for the perfect electric conductor (PEC), the perfect magnetic conductor (PMC), and the absorbing boundary condition (ABC), respectively.

Testing these equations with Lagrange polynomials $\mathbf{l}_i = \hat{\mathbf{w}} l_i$ ($\hat{\mathbf{w}} = \hat{\mathbf{x}}, \hat{\mathbf{y}},$ or $\hat{\mathbf{z}}$) on each tetrahedron element V_e and applying the divergence theorem twice yield the strong form

$$\int_{V_e} \mathbf{l}_i \cdot \frac{\partial \mathbf{H}}{\partial t} dV = -\frac{1}{\mu} \left[\int_{V_e} \mathbf{l}_i \cdot \nabla \times \mathbf{E} dV + \oint_{\partial V_e} \mathbf{l}_i \cdot \hat{\mathbf{n}} \times (\mathbf{E}^* - \mathbf{E}) dS \right]$$

$$\int_{V_e} \mathbf{l}_i \cdot \frac{\partial \mathbf{E}}{\partial t} dV = \frac{1}{\varepsilon} \left[\int_{V_e} \mathbf{l}_i \cdot \nabla \times \mathbf{H} dV + \oint_{\partial V_e} \mathbf{l}_i \cdot \hat{\mathbf{n}} \times (\mathbf{H}^* - \mathbf{H}) dS - \int_{V_e} \mathbf{l}_i \cdot \mathbf{J}_c dV \right]$$

$$(5.5)$$

where $\mathbf{E}^*$ and $\mathbf{H}^*$ are the numerical fluxes defined on the boundary faces of each tetrahedron element. Two choices of these numerical fluxes are commonly applied: the central flux and the upwind flux.

To facilitate the description, we define the addition $\langle \cdot \rangle$ and jump $[\![\cdot]\!]$ functions as follows:

$$\langle a \rangle = a^+ + a^-, \quad [\![a]\!] = a^+ - a^- \qquad\qquad (5.6)$$

where the superscripts $-$ and $+$ indicate the inside and the outside of a surface, respectively. Note that the above definitions apply to both cases where a is a scalar or a vector variable.

- Central flux

$$\mathbf{E}^* = \frac{1}{2}\langle \mathbf{E} \rangle, \quad \mathbf{H}^* = \frac{1}{2}\langle \mathbf{H} \rangle \qquad\qquad (5.7)$$

- Upwind flux

$$\mathbf{E}^* = \frac{1}{\langle Y \rangle}(\langle Y\,\mathbf{E} \rangle + \hat{\mathbf{n}} \times [\![\mathbf{H}]\!])$$

$$\mathbf{H}^* = \frac{1}{\langle Z \rangle}(\langle Z\,\mathbf{H} \rangle - \hat{\mathbf{n}} \times [\![\mathbf{E}]\!]) \qquad\qquad (5.8)$$

Using either of the these numerical fluxes, and expanding $\mathbf{E}$ and $\mathbf{H}$ in terms of pth-order Lagrange polynomial basis functions

$$\mathbf{E} = \sum_{j=1}^{N_p} l_j (\hat{\mathbf{x}} E_{xj} + \hat{\mathbf{y}} E_{yj} + \hat{\mathbf{z}} E_{zj}), \qquad \mathbf{H} = \sum_{j=1}^{N_p} l_j (\hat{\mathbf{x}} H_{xj} + \hat{\mathbf{y}} H_{yj} + \hat{\mathbf{z}} H_{zj}) \qquad (5.9)$$

where $N_p = \prod_{i=1}^{d}(p+i)/d!$ stands for the number of degrees of freedom (DoFs) in a d-dimensional element, the strong form of Maxwell's equations can be converted into the matrix form representation

$$M^e \frac{\partial}{\partial t} H_x = -\frac{1}{\mu}\left(S_y^e E_z - S_z^e E_y + M_f^e F_x^E\right)$$

$$M^e \frac{\partial}{\partial t} H_y = -\frac{1}{\mu}\left(S_z^e E_x - S_x^e E_z + M_f^e F_y^E\right)$$

$$M^e \frac{\partial}{\partial t} H_z = -\frac{1}{\mu}\left(S_x^e E_y - S_y^e E_x + M_f^e F_z^E\right)$$

$$M^e \frac{\partial}{\partial t} E_x = \frac{1}{\varepsilon}\left(S_y^e H_z - S_z^e H_y + M_f^e F_x^H - M^e J_{cx}\right)$$

$$M^e \frac{\partial}{\partial t} E_y = \frac{1}{\varepsilon}\left(S_z^e H_x - S_x^e H_z + M_f^e F_y^H - M^e J_{cy}\right)$$

$$M^e \frac{\partial}{\partial t} E_z = \frac{1}{\varepsilon}\left(S_x^e H_y - S_y^e H_x + M_f^e F_z^H - M^e J_{cz}\right) \tag{5.10}$$

where the flux terms are given by

$$\mathbf{F}^E = \frac{1}{2}\hat{\mathbf{n}} \times [\![\mathbf{E}]\!]$$

$$\mathbf{F}^H = \frac{1}{2}\hat{\mathbf{n}} \times [\![\mathbf{H}]\!] \tag{5.11}$$

for the central flux, and

$$\mathbf{F}^E = \frac{1}{\langle Y \rangle}(Y^+\hat{\mathbf{n}} \times [\![\mathbf{E}]\!] + \hat{\mathbf{n}} \times \hat{\mathbf{n}} \times [\![\mathbf{H}]\!])$$

$$\mathbf{F}^H = \frac{1}{\langle Z \rangle}(Z^+\hat{\mathbf{n}} \times [\![\mathbf{H}]\!] - \hat{\mathbf{n}} \times \hat{\mathbf{n}} \times [\![\mathbf{E}]\!]) \tag{5.12}$$

for the upwind flux, while

$$[M^e]_{ij} = \int_{V_e} l_i l_j dV$$

$$[S_w^e]_{ij} = \int_{V_e} l_i \frac{\partial l_j}{\partial w} dV$$

$$[M_f^e]_{ij} = \int_{\partial V_e} l_i l_j dS \tag{5.13}$$

are mass, stiffness, and facial mass matrices defined in each tetrahedron element V_e and its surface ∂V_e, respectively. Here, w represents x, y, or z.

5.2.1 High-Order Spatial Discretization

In this section, high-order hierarchical basis functions and high-order interpolatory basis functions are introduced, both of which are used in the implementation of the DGTD method. Since tetrahedron elements are used in the geometric discretization of the solution domain, basis functions that are defined on tetrahedrons will be introduced. Those defined on different types of elements can also be defined in a similar manner.

5.2.1.1 Definition of Basis Functions: Modal Basis and Nodal Basis

Unlike the definition of high-order vector basis functions widely used by the CEMs community, the definition of high-order scalar basis functions is slightly different. High-order hierarchical basis functions, which are more commonly known as the *modal basis*, are a set of orthonormal polynomials defined on a simplex, which is a tetrahedron in 3-D. Specifically, the following orthonormal polynomial of order $p \geq i + j + k \geq 0$ is defined using the well-known Jacobi polynomials

$$\psi_{ijk}(r, s, t) = \sqrt{8} P_i^{0,0}(a) P_j^{2i+1,0}(b)(1-b)^j P_k^{2i+2j+2,0}(c)(1-c)^{i+j} \tag{5.14}$$

where $a = -2\frac{1+r}{s+t} - 1$, $b = 2\frac{1+s}{1-t} - 1$, and $c = t$, with the 3-D simplex defined by r, s, $t \geq -1, r + s + t \leq -1$. The Jacobi polynomials are given by

$$P_n^{\alpha,\beta}(z) = \frac{(-1)^n}{2^n n!}(1-z)^{-\alpha}(1+z)^{-\beta}\frac{d^n}{dz^n}\{(1-z)^\alpha(1+z)^\beta(1-z^2)^n\} \tag{5.15}$$

When $\alpha = \beta = 0$, the Jacobi polynomial reduces to the Legendre polynomials

$$P_n(z) = \frac{1}{2^n n!}\frac{d^n}{dz^n}\{(z^2 - 1)^n\} \tag{5.16}$$

Note that the modal basis functions are hierarchical in nature, meaning that a set of higher order modal basis contains all of the lower order modal basis (similar to the Fourier basis).

The physical quantity can therefore be expanded in terms of the modal basis as (here the subscripts i, j, k are lumped into n for simplicity)

$$u(\mathbf{x}, t) = \sum_{n=1}^{N_p} \hat{u}_n(t)\psi_n(\mathbf{x}) \tag{5.17}$$

The definition of high-order interpolatory basis functions has multiple choices. For example, one can use monomials

$$\zeta^i \eta^j \xi^k, \quad i, j, k \geq 0, i + j + k \leq p \tag{5.18}$$

or polynomials as the basis functions. Here, the widely used Lagrange polynomials are used as the high-order interpolatory basis functions. One unique property of Lagrange polynomials is that their value is 1 only at one interpolating node (which is known as its definition node), but zeroes on all other $N_p - 1$ interpolating nodes. Hence, Lagrange basis functions are better known as *nodal basis*, as opposed to the modal basis introduced earlier. This property has two advantageous consequences. One is the expansion coefficient of each Lagrange basis function stands for the value of the physical quantity at the definition node of the basis function. The other one is when using the Lobatto nodes to define the Lagrange basis functions (introduced in the next section), the value of a Lagrange basis function defined in a d-dimensional simplex is purely zero on any of its lower d'-dimensional simplex subset ($0 \le d' < d$), if the definition node of the basis function is not in that d'-dimensional simplex. For example, for the Lagrange basis defined inside the volume of a tetrahedron, its value is purely zero on all the four faces, six edges, and four vertices of the tetrahedron. This property is especially useful when implementing the DG method because when trying to collect the values for the numerical flux, all that matters is the DoFs that are defined on the common triangle of two neighboring tetrahedrons, and all other DoFs have zero contribution.

The difficulty of applying the Lagrange basis functions in a high-order polynomial expansion is that there are no general explicit expressions for high-order Lagrange polynomials. In a calculation that involves Lagrange polynomials, one can relate the Lagrange basis function (nodal basis) to the orthonormal basis function (modal basis) as

$$u(\mathbf{x}, t) = \sum_{i=1}^{N_p} u(\mathbf{x}_i, t) l_i(\mathbf{x}) = \sum_{n=1}^{N_p} \hat{u}_n(t) \psi_n(\mathbf{x}) \tag{5.19}$$

where l_i denotes the 3-D Lagrange polynomials defined on the interpolating node $\mathbf{x}_i$. Matching the values of u at all the interpolating nodes $\mathbf{x}_i$ ($i = 1, ..., N_p$) yields

$$\mathcal{V}\,\hat{\mathbf{u}} = \mathbf{u} \tag{5.20}$$

where $\mathcal{V}_{nm} = \psi_m(\mathbf{x}_n)$ is the generalized Vandermonde matrix, $\hat{\mathbf{u}} = \{\hat{u}_1, ..., \hat{u}_{N_p}\}^{\mathrm{T}}$, and $\mathbf{u} = \{u_1, ..., u_{N_p}\}^{\mathrm{T}}$. The Lagrange polynomials can thus be expressed as

$$l_n(x) = \sum_{m=1}^{N_p} (\mathcal{V}^{-1})_{mn} \psi_m(\mathbf{x}), \text{ or } l = \mathcal{V}^{-\mathrm{T}} \psi \tag{5.21}$$

and can be used to calculate the elemental matrices, which will be discussed in the following section.

5.2.1.2 Choice of Interpolating Nodes

A significant aspect of applying the Lagrange basis functions is how to choose the interpolating nodes. If a set of equi-spaced nodes is used to define the Lagrange basis functions, the famous Runge phenomenon will lead to a highly oscillating function value when very high-order polynomials are used. The oscillation becomes even worse near the boundary of the definition domain of the polynomial, which would completely ruin the DG method where the numerical flux plays a central role of the algorithm. To remove the Runge phenomenon, a set of nonuniformly spaced nodes is generally used to define the Lagrange polynomial, which are often the root set of certain polynomials defined in the same simplex. Examples of such polynomials include Legendre polynomials, Chebyshev polynomials of the first or the second kind, and Lobatto polynomials. For the application of the DG method, the Lobatto nodes are especially attractive, since the node set includes those on the boundary of a simplex, which becomes handy when calculating the numerical flux. Unfortunately, the Lobatto nodes in 3-D simplex are not well defined. There are several approaches to obtain the Lobatto nodes. For example, to minimize the Lebesgue constant, a quantity that measures the interpolation quality of a given polynomial, or to maximize the magnitude of the determinant of the Vandermonde matrix using either constraint optimization method or the Monte Carlo method. In this section, a set of optimized Lobatto nodes in 3-D simplex is used, as shown in Figure 5.1 [9]. The optimized set of nodes produces a small Lebesgue constant compared to other node sets available in the literature and reduces to the 1-D Gauss–Lobatto–Legendre (GLL) node set on the edges of the tetrahedron.

Multiple advantages can be achieved by choosing an optimal set of interpolating nodes. The first one is that it results in a smaller Lebesgue constant and, therefore, a better interpolation quality. Second, it also reduces the spectral radius of the facial matrix $M^{-1}T_f$, where T_f is the same as M_f, with its number of rows expanded to N_p to allow the matrix–matrix product with M^{-1}. Since the spectral radius $\rho(M^{-1}T_f)$ determines the time step size required to maintain stability, a smaller spectral radius results in a larger time step size and, consequently, a more efficient algorithm. Figure 5.2 shows the comparison made between the widely used hierarchical vector basis functions and the orthonormal basis functions used in this work. From Figure 5.2a, the spectral radius of the orthonormal basis functions is smaller than that of the vector basis and, hence, results in a larger step size as seen in Figure 5.2b.

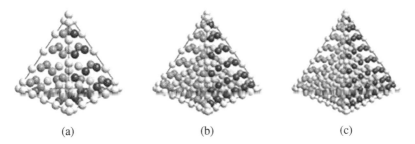

(a) (b) (c)

Figure 5.1 Optimized interpolating nodes in tetrahedron [9]. (a) $p = 6$, (b) $p = 8$, and (c) $p = 10$.

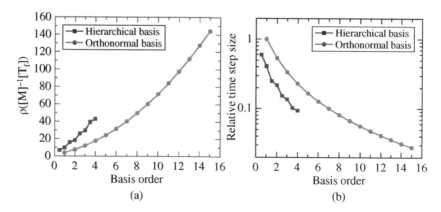

Figure 5.2 (a) Spectral radius of the facial matrix versus polynomial basis order.
(b) Time step size required to maintain stability versus polynomial basis order. The time
step sizes are normalized against that of the first-order orthonormal basis function.

5.2.1.3 Elemental Matrices in the DG Method

After the high-order basis functions and the associated definition nodes are
obtained, the elemental matrices in the DG method can be explicitly expressed.
In this section, the mass, stiffness, and facial mass matrices are calculated.
We start from the calculation of the mass matrix. Using the relation $1 = \mathcal{V}^{-T}\psi$,
one can express the mass matrix as

$$M^e = J^e \int_I 11^T dV = J^e \mathcal{V}^{-T} \int_I \psi\,\psi^T dV\,\mathcal{V}^{-1} \tag{5.22}$$

where I is the reference simplex in 3-D and J^e is the Jacobian matrix mapping from
the reference simplex to the real tetrahedron, and is reduced to a single scalar if the
geometric discretization is linear (1st-order tetrahedron). Since the modal basis is
orthonormal, the mass matrix can be written in an explicit form as

$$M^e = J^e(\mathcal{V}\mathcal{V}^T)^{-1} = J^e M \tag{5.23}$$

Note that in the explicit form of the mass matrix, no numerical quadrature
is required. The facial mass matrix can be obtained in a similar manner, the
only difference is that everything in the explicit expression reduces from 3-D to
2-D as

$$M_f^e = J_f^e\left(\mathcal{V}_f\mathcal{V}_f^T\right)^{-1} = J_f^e M_f \tag{5.24}$$

The stiffness matrix involves the partial derivative with respect to x, y, or z. Here, we take S_x^e, for example, to illustrate its calculation.

$$
\begin{aligned}
\left[S_x^e\right]_{ij} &= \int_{V_e} l_i \frac{\partial l_j}{\partial x} dV = \int_{V_e} l_i \left(\frac{\partial l_j}{\partial r} \frac{\partial r}{\partial x} + \frac{\partial l_j}{\partial s} \frac{\partial s}{\partial x} + \frac{\partial l_j}{\partial t} \frac{\partial t}{\partial x} \right) dV \\
&= r_x \int_I l_i \frac{\partial l_j}{\partial r} dV + s_x \int_I l_i \frac{\partial l_j}{\partial s} dV + t_x \int_I l_i \frac{\partial l_j}{\partial t} dV
\end{aligned}
\tag{5.25}
$$

In derivation (5.25), the derivatives r_x, s_x, and t_x are constants if linear tetrahedron elements are used, and the integration terms can be further written as

$$
\begin{aligned}
\int_I l_i \frac{\partial l_j}{\partial r} dV &= \int_I l_i \left(\sum_{n=1}^{N_p} \frac{\partial l_j}{\partial r} \bigg|_{r_n} l_n \right) dV = \sum_{n=1}^{N_p} \left(\int_I l_i l_n dV \frac{\partial l_j}{\partial r} \bigg|_{r_n} \right) \\
&= \sum_{n=1}^{N_p} M_{in} D_{r,nj}
\end{aligned}
\tag{5.26}
$$

where the interpolation of $\frac{\partial l_j}{\partial r}$ using Lagrange polynomials is applied, and $D_{r,nj}$ is the differential matrix with respect to the variable r

$$
D_{r,nj} = \frac{\partial l_j}{\partial r} \bigg|_{r_n}
\tag{5.27}
$$

which can also be expressed analytically using the relation between the Lagrange and the orthonormal basis functions. Details will not be given here. The stiffness matrix can finally be expressed as

$$
S_x^e = M(r_x D_r + s_x D_s + t_x D_t)
\tag{5.28}
$$

and S_y^e and S_z^e can be expressed similarly.

5.2.2 High-Order Temporal Discretization

After the system equations are discretized with high-order basis functions in space, a set of ODEs can be obtained and formally written as

$$
\frac{\partial u}{\partial t} = F(t, u), \qquad u(t_0) = u_0
\tag{5.29}
$$

The system of ODEs can be solved using the well-known explicit high-order Runge–Kutta methods [18], for instance, the classic fourth order Runge–Kutta method, to achieve high-order accuracy in time. However, the traditional Runge–Kutta methods are, in general, not strong-stability preserving (SSP), and the SSP coefficient $C \approx 1$, which is the ratio between the maximum time step size for a given time integrator and that for the forward Euler's method. Here, we introduce the SSP Runge–Kutta (SSPRK) method [19], which was also known as the total variation diminishing (TVD) Runge–Kutta method [20, 21] in the past. The SSP property means that if the embedded first-order Euler's method is stable, the high-order SSPRK method is also stable, which diminishes the total variation in the solution. In the meantime, the SSP coefficient can be made larger to allow a larger time step size. For example, the low-storage four-stage third-order SSPRK is given by

$$q_1 = u^n$$
$$q_2 = q_1 + \frac{h}{2}F(q_1)$$
$$q_2 = q_2 + \frac{h}{2}F(q_2)$$
$$q_2 = \frac{2}{3}q_1 + \frac{1}{3}\left[q_2 + \frac{h}{2}F(q_2)\right]$$
$$u^{n+1} = q_2 + \frac{h}{2}F(q_2) \tag{5.30}$$

and the SSP coefficient $C \approx 2$.

5.3 Modeling and Simulation of Electromagnetic–Plasma Interaction

In this section, the physical models for EM–plasma interactions are first presented, with emphasis on different plasma fluid models and the nonlinear coupling between EM waves and plasma fluids. This is followed by the discussion of the numerical modeling scheme, the DGTD method.

5.3.1 Physical Models of EM–Plasma Interactions

The behavior of EM fields and waves is governed by Maxwell's equations in the time domain

$$\nabla \times \mathbf{E} = -\frac{\partial \mathbf{B}}{\partial t}$$
$$\nabla \times \mathbf{H} = \frac{\partial \mathbf{D}}{\partial t} + \mathbf{J}$$
$$\nabla \cdot \mathbf{D} = \rho$$
$$\nabla \cdot \mathbf{B} = 0 \tag{5.31}$$

where $\mathbf{E}$ and $\mathbf{D}$ stand for the electric field and flux, respectively, and $\mathbf{H}$ and $\mathbf{B}$ stand for the magnetic field and flux, respectively, and $\mathbf{J}$ and ρ stand for the electric current and charge densities, respectively.

Under a high pressure and high collision frequency condition, the plasma behavior can be described using the macroscopically averaged quantities, n, the plasma density, and $\mathbf{U}$, the plasma velocity, which are governed by the plasma diffusion and the collisional momentum transfer equations [22–25]:

$$\frac{\partial n}{\partial t} - \nabla \cdot (D_{\text{eff}} \nabla n) = (v_i - v_a)n$$

$$\frac{\partial \mathbf{U}}{\partial t} + v_m \mathbf{U} = \frac{1}{m_e} \mathbf{F} \tag{5.32}$$

In this model, all these plasma transport coefficients, including the ionization frequency v_i, the attachment frequency v_a, and the effective diffusion coefficient D_{eff}, are nonlinear functions of the reduced effective electric field

$$\frac{E_{\text{eff}}}{p_{\text{amb}}} = \frac{E_{\text{rms}}}{p_{\text{amb}} \sqrt{1 + \omega^2/v_m^2}} \tag{5.33}$$

where p_{amb} is the pressure of ambient air and E_{rms} is the root-mean-square value of the local electric field averaged over one period of the EM wave with an angular frequency ω. Such nonlinear relations between the reduced effective field and the plasma parameters are obtained by empirical models based on the curve fitting of measured data [23]. In (5.32), $v_m = 5.3 \times 10^9 p_{\text{amb}}$ denotes the electron-neutral collision frequency, m_e denotes the electron mass at rest, and $\mathbf{F} = q_e(\mathbf{E} + \mathbf{U} \times \mathbf{B})$ denotes the Lorentz force exerted by the EM fields to the electrons with charge q_e. Once the plasma density and velocity are solved, the coupling from the plasma fluids to the EM fields can be calculated as the plasma current $\mathbf{J} = -q_e n \mathbf{U}$, which serves as the source of the secondary EM fields.

In addition to the plasma diffusion model, a five-moment plasma fluid model can be used to describe the generation and evolution of the plasma. In this model, the electron density n, velocity $\mathbf{U}$, and mean energy $\mathcal{E}$ are governed by the following three equations [26–28]:

$$\frac{\partial n}{\partial t} + \nabla \cdot (nU) = (v_i - v_a)n$$

$$\frac{\partial nU}{\partial t} + \nabla \cdot \left(nUU + \frac{1}{m_e} P\mathbb{I}\right) = \frac{n}{m_e} F - v_m nU$$

$$\frac{\partial n\mathcal{E}}{\partial t} + \nabla \cdot [U(n\mathcal{E} + P)] = q_e \mathbf{E} \cdot n\mathbf{U} - Q_e n \tag{5.34}$$

which are also known as Euler's equations. In (5.34), $\mathbb{I}$ stands for the identity tensor and P the electron partial pressure [26]. In this model, the plasma transport coefficients, v_i, v_a, v_m, and Q_e (the energy loss frequency) are all nonlinear functions of local electron energy $\mathcal{E}$ and are obtained by integrating the non-Maxwellian electron energy distribution function (EEDF) with collision cross sections, which take into account various collision reactions during the highly nonequilibrium ionization process [29]. These parameters are usually pre-calculated and tabulated as functions of the mean electron energy $\mathcal{E}$. According to the third equation in (5.34), the electrons gain energy from the electric field through Joule heating $q_e \mathbf{E} \cdot n\mathbf{U} = -\mathbf{E} \cdot \mathbf{J}$.

5.3.2 Numerical Modeling of EM–Plasma Interactions

The complete description of EM–plasma interactions is obtained by coupling Maxwell's Eqs. (5.31) with either the plasma diffusion model (5.32) or the plasma five-moment model (5.34), which are then solved in a self-consistent manner to account for their interaction. In numerical simulations of pure EM problems, usually only the two curl equations in all four Maxwell's equations are solved, the other two divergence equations are ignored. This works fine if the current continuity condition $\partial \rho/\partial t = -\nabla \cdot \mathbf{J}$ is satisfied but can result in erroneous solutions in an EM–plasma simulation, where the plasma density and velocity are solved separately. To correct the numerical error introduced by the violation of the current continuity condition, the purely hyperbolic Maxwells equations and the divergence-cleaning technique [30–39] presented in the preceding section can be used whenever necessary.

The DGTD method can be employed to solve Maxwell's, diffusion, and Euler's equations, which can be rewritten uniformly into the following conservation form:

$$\mathfrak{D}\frac{\partial \mathbf{G}}{\partial t} + \nabla \cdot \mathfrak{F}(\mathbf{G}) = \mathbf{S}(\mathbf{G}) \tag{5.35}$$

with $\mathfrak{D}$ being the material parameter tensor, $\mathbf{G}$ the unknown quantity vector, $\mathfrak{F}$ the physical flux tensor, and $\mathbf{S}$ the source term. The conservation equation can be discretized and converted into a strong form by integrating with the testing function l_i in each mesh element V_e and applying the divergence theorem twice, which yields

$$\int_{V_e} l_i\, \mathfrak{D}\frac{\partial \mathbf{G}}{\partial t}\, dV = -\int_{V_e} l_i \nabla \cdot \mathfrak{F} dV - \oint_{\partial V_e} l_i\, (\mathfrak{F}^* - \mathfrak{F}) \cdot \hat{\mathbf{n}}\, dS + \int_{V_e} l_i\, \mathbf{S}\, dV \tag{5.36}$$

where $\mathfrak{F}^*$ is the numerical flux, whose choice depends on the mathematical property of the equation. After expanding each unknown quantity with Lagrange polynomials, the partial differential equation in its strong form (5.36) can be expressed into an ordinary differential equation (ODE) [22, 39], which can be integrated in time using an explicit high-order time integration method [18, 19].

- Maxwell's equations
  For the first equation in (5.31), $\mathfrak{D} = \mu$, $\mathbf{G} = (H_x, H_y, H_z)^{\mathrm{T}}$ (T stands for transpose), the flux tensor is

$$\mathfrak{F} = \begin{pmatrix} 0 & E_z & -E_y \\ -E_z & 0 & E_x \\ E_y & -E_x & 0 \end{pmatrix} \tag{5.37}$$

and the source term is $\mathbf{S} = (0, 0, 0)^{\mathrm{T}}$. Similarly, for the second equation in (5.31), $\mathfrak{D} = \varepsilon$, $\mathbf{G} = (E_x, E_y, E_z)^{\mathrm{T}}$, the flux tensor is

$$\mathfrak{F} = \begin{pmatrix} 0 & -H_z & H_y \\ H_z & 0 & -H_x \\ -H_y & H_x & 0 \end{pmatrix} \tag{5.38}$$

and the source term is $\mathbf{S} = (J_x, J_y, J_z)^{\mathrm{T}} = (-q_e n U_x, -q_e n U_y, -q_e n U_z)^{\mathrm{T}}$. Due to the hyperbolic property of Maxwell's equations, the mixed flux

$$(\mathfrak{F}^* - \mathfrak{F}) \cdot \hat{\mathbf{n}} = \frac{1}{\langle Z \rangle}(Z^+ \hat{\mathbf{n}} \times [\![\mathbf{H}]\!] - \tau \hat{\mathbf{n}} \times \hat{\mathbf{n}} \times [\![\mathbf{E}]\!]) \tag{5.39}$$

and

$$(\mathfrak{F}^* - \mathfrak{F}) \cdot \hat{\mathbf{n}} = \frac{1}{\langle Y \rangle}(Y^+ \hat{\mathbf{n}} \times [\![\mathbf{E}]\!] + \tau \hat{\mathbf{n}} \times \hat{\mathbf{n}} \times [\![\mathbf{H}]\!]) \tag{5.40}$$

can be used [10, 22, 40] for the first two equations in (5.31), respectively, to propagate the fields from one element to its neighbors. The combination factor, τ, can be adjusted such that $\tau = 0$ corresponds to the central flux and $\tau = 1$ corresponds to the upwind flux. A good choice is $\tau = 0.025$, which can remove spurious solutions effectively without introducing too much undesired numerical dissipation [41]. In the above expression, $Z = \sqrt{\mu/\varepsilon}$ and $Y = 1/Z$ denote the intrinsic impedance and admittance of the background medium, respectively. The definitions of the addition $\langle \cdot \rangle$ and jump $[\![\cdot]\!]$ functions are defined in the preceding section.

- Diffusion equation

To solve the diffusion Eq. (5.32), an intermediate variable is first introduced as $\mathbf{q} = \nabla n$ such that the diffusion equation can be rewritten as [7, 22, 42]

$$\frac{\partial n}{\partial t} - \nabla \cdot (D_{\mathrm{eff}} \mathbf{q}) = (v_i - v_a)n \tag{5.41}$$

In terms of the conservation form (5.35), $\mathfrak{D} = 1$, $\mathbf{G} = n$, the flux tensor is $\mathfrak{F} = D_{\mathrm{eff}} \mathbf{q}$, and the source term is $\mathbf{S} = (v_i - v_a)n$. Using the central flux formulation, the total flux is given as

$$(\mathfrak{F}^* - \mathfrak{F}) \cdot \hat{\mathbf{n}} = \frac{1}{2}[\![D_{\mathrm{eff}} \mathbf{q}]\!] \tag{5.42}$$

which connects the neighboring elements and permits the solution of (5.41) once the value of the intermediate variable $\mathbf{q}$ is known. This is achieved through a similar discretization process by applying the discontinuous Galerkin method

$$\int_{V_e} l_i q_u \, dV = \int_{V_e} l_i \{\nabla n\}_u \, dV + \oint_{\partial V_e} l_i \, \hat{\mathbf{n}} \cdot \hat{\mathbf{u}}(n^* - n) dS \tag{5.43}$$

where the subscript u stands for x, y, or z and $\hat{\mathbf{u}}$ stands for the corresponding unit vector $\hat{\mathbf{x}}$, $\hat{\mathbf{y}}$, or $\hat{\mathbf{z}}$. The central flux formulation leads to $n^* - n = \frac{1}{2}[\![n]\!]$, which can be used to solve (5.43) for $\mathbf{q}$.

The momentum transfer equation in (5.32) describes a simple initial value problem (IVP) governed by a first-order ODE, which can be solved in a point-wise manner on all the definition points of the Lagrange polynomials.

- Euler's equations

Euler's equations (5.34) are already in the conservation form, where $\mathfrak{D}$ is an identity tensor, $G = (n, nU_x, nU_y, nU_z, n\mathcal{E})^T$. The primitive variables can be recovered as $U = nU/n$, $\mathcal{E} = n\mathcal{E}/n$, and $P = (\gamma - 1)\left(n\mathcal{E} - \frac{1}{2}m_e nU^2\right)$, which are used in the calculation of the flux tensor

$$
\mathfrak{F} = \begin{pmatrix}
nU_x & nU_y & nU_z \\
nU_x^2 + \dfrac{1}{m_e}P & nU_xU_y & nU_xU_z \\
nU_xU_y & nU_y^2 + \dfrac{1}{m_e}P & nU_yU_z \\
nU_xU_z & nU_yU_z & nU_z^2 + \dfrac{1}{m_e}P \\
U_x(n\mathcal{E} + P) & U_y(n\mathcal{E} + P) & U_z(n\mathcal{E} + P)
\end{pmatrix}
\tag{5.44}
$$

The source term is simply the right-hand side of Euler's equations. To solve Euler's equations in the strong form, the local Lax–Friedrichs flux [10, 26] can be used:

$$
(\mathfrak{F}^* - \mathfrak{F}) \cdot \hat{\mathbf{n}} = \frac{1}{2}\{ [\![\mathfrak{F}]\!] \cdot \hat{\mathbf{n}} - \lambda [\![G]\!] \}
\tag{5.45}
$$

where

$$
\lambda = \max_{g \in \{G^+, G^-\}} \left\{ U(g) + \sqrt{\gamma \frac{P(g)}{m_e n(g)}} \right\}
\tag{5.46}
$$

is the characteristic velocity of Euler's system, which consists of the electron fluid velocity U and the electron acoustic velocity $\sqrt{\gamma P/m_e n}$.

- Multiphysics coupling

The coupling mechanism between EM fields and plasma fluids was discussed in the previous section. In numerical simulations, however, the frequency of numerical coupling operations depends on the characteristic temporal scale of each physics. While the temporal scale of the EM fields is determined by the EM period, which is on the order of picoseconds at a frequency of 100 GHz, the temporal scale of the plasma fluids is different for different models. When the plasma diffusion model is used, the characteristic temporal scale is determined by the plasma diffusion velocity, which is on the order of nanoseconds. In this case, there is no need to couple Maxwell's equations with the diffusion equation at every time step of the simulation. Instead, the EM simulation is first performed for one period of time, followed by the calculation of the effective field intensity using (5.33). The plasma transport coefficients can then be updated and used to advance the plasma density for one time step by solving the plasma diffusion equation. With the updated plasma density, the plasma current density can be evaluated and used as the EM source to integrate Maxwell's equations for the next period. When the plasma five-moment model is used, the plasma density, velocity, and energy oscillate and evolve on the same temporal scale as the EM fields. As a result, the mutual coupling between Maxwell's and Euler's equations takes place at every time step, which are solved simultaneously using the explicit Runge–Kutta method [18, 19] for their time-domain responses. In the implementation of a multistage Runge–Kutta method, the mutual coupling is performed at each stage of the time integration.

5.4 Dynamic Adaptation Algorithm

With the finite element method, h-, p-, and hp-refinements can be performed to achieve the desired numerical accuracy without significantly increasing the computational cost. Due to the use of numerical fluxes, the DGTD algorithm can be made even more flexible in a sense that it permits the change of element sizes and/or polynomial orders during a simulation to resolve the local variation of physics. This feature has been investigated in terms of the dynamic h-adaptation algorithm for pure EM problems [15] and the dynamic p-adaptation algorithm for both EM and multiphysics problems [16]. In this section, these two algorithms are reviewed, with emphasis on the adaptation criteria and the information mapping schemes.

The most important step in developing a dynamic adaptation algorithm is to find a good and cheap adaptation criterion to determine whether the local element size and/or polynomial order need to be adjusted (either increase or decrease) or not. Such an adaptation criterion needs two distinct features for it to be practically applicable. First, it should rely only on the information that is locally available at a given time step in the simulation so that the adaptation decision can be made locally and efficiently. Second, it needs to be related either directly or indirectly to the local numerical accuracy to provide a basis for determining the appropriate mesh size and/or the polynomial order. These two features are explained and explored in the context of dynamic h- and p-adaptation in this section.

5.4.1 Dynamic h-Adaptation

Dynamic h-adaptation refers to the adaptive adjustment of the local element size during simulation. In a pure EM simulation, it is usually required that the element size is smaller than $\lambda/20$ to provide a good spatial resolution to the EM field variation if the first-order polynomial is used to represent the electric or magnetic field. Here, λ denotes the EM wavelength, which is the characteristic length of the EM physics. In a self-consistent simulation of EM–plasma interactions, the local element size must be small enough to resolve the smallest physical feature, which, in this case, is the characteristic length of the plasma, since the plasma varies on a smaller spatial scale than the EM field. According to plasma physics [23, 27], the local characteristic length can be estimated as $L = \|\nabla n/n\|^{-1}$. It is, therefore, appropriate to enforce the following requirement to the element size h

$$h \leq \frac{L}{\xi} \tag{5.47}$$

where $\xi > 0$ is an integer that controls the level of accuracy of the spatial representation and specifies the number of segments used to resolve the characteristic length L. Typically, ξ takes a value between 2 and 4 to achieve a good balance between the numerical accuracy and the simulation efficiency. During the numerical simulation, the characteristic length in each element can be evaluated at every time step on which the plasma equations are solved. The local element size is then checked against the local characteristic length using (5.47) to determine whether an element refinement or coarsening is needed.

An alternative choice of the adaptation criterion is based on the local variation of the plasma density $\|\nabla n\|$. Since smaller elements are needed to resolve physics with faster variation, a larger density gradient requires a denser mesh grid to capture the physical variation. In other words, it is required that an element is small enough such that the density difference across the element does not exceed a preset threshold ζ. Apparently, the threshold ζ is problem dependent and can be specified based on the understanding of a given physical process. Using the finite difference approximation in 1-D, for example, where only the x direction is involved, the requirement reads $|n(x+h) - n(x)| \leq \zeta$. When divided by the element size h,

$$\|\nabla n\| \approx \frac{|n(x+h) - n(x)|}{h} \leq \frac{\zeta}{h} \tag{5.48}$$

As a result,

$$h \leq \frac{\zeta}{\|\nabla n\|} \tag{5.49}$$

is the adaptation criterion that must be enforced in each element.

Once an element is refined or coarsened at a certain step during the simulation, inter-elemental mapping can be established through the calculation of the numerical flux along the common edge/surface that interfaces two neighboring elements. As illustrated in Figure 5.3, the shaded level 1 element has one level 2 and two level 3 neighbors across its upper boundary. In the calculation of the numerical flux along this boundary, the surface integral is broken into three pieces with smaller supports. Using the calculation of the surface integral in (5.43) as

Figure 5.3 Illustration of an adaptively refined nonconformal grid.

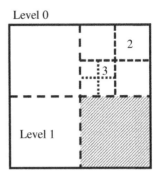

an example, when the central flux is used, the surface integral along the upper boundary S_1 becomes ($\hat{\mathbf{n}} = \hat{\mathbf{y}}$ and $\hat{\mathbf{u}} = \hat{\mathbf{y}}$)

$$\frac{1}{2}\int_{S_1} l_i\,(n^+ - n^-)dS = \frac{1}{2}\left[\int_{S_1} l_i\,n^+dS - \int_{S_1} l_i\,n^-dS\right]$$

$$= \frac{1}{2}\left[\int_{S_{12}} l_i\,n^+dS + \int_{S^1_{13}} l_i\,n^+dS + \int_{S^2_{13}} l_i\,n^+dS - \int_{S_1} l_i\,n^-dS\right]$$

$$(5.50)$$

In this expression, S_{12} stands for the interface between the level 1 and level 2 elements, S^i_{13} ($i = 1, 2$) stands for the interface between the level 1 and the ith level 3 elements such that $S_1 = S_{12} \bigcup S^1_{13} \bigcup S^2_{13}$, n^- stands for the plasma density evaluated along S_1 inside the level 1 element, and n^+ stands for that evaluated along the corresponding surface of integration inside the neighboring element.

5.4.2 Dynamic *p*-Adaptation

To provide a better spatial resolution, increasing the polynomial order is more effective than decreasing the element size, if the physical quantity varies smoothly. A dynamic *p*-adaptation algorithm was developed by using the information of the expansion coefficients of the high-order Lagrange interpolatory polynomial representation of the physics [16] since a physical quantity with a certain order of varia tion can be accurately represented by polynomials of the same order. To determine the appropriate order of expansion, the interpolatory polynomial representation is first converted into a hierarchical polynomial representation

$$g(\mathbf{r}, t) = \sum_{m=1}^{N_p} g_m(\mathbf{r}_m, t)l_m(\mathbf{r}) = \sum_{n=0}^{N_p-1} \hat{g}_n(t)\psi_n(\mathbf{r}) \qquad (5.51)$$

where l_m denotes the mth Lagrange polynomial defined on the interpolating node $\mathbf{r}_m$, ψ_n denotes the nth hierarchical Jacobi polynomials [10] defined in triangles in 2-D and tetrahedra in 3-D, and g_m and $\hat{g}_n$ are the corresponding expansion coefficients. Matching the values of g at all the interpolating nodes $\mathbf{r}_m$ ($m = 1, \ldots, N_p$) yields

$$\mathcal{V}\hat{\mathbf{g}} = \mathbf{g} \tag{5.52}$$

where $\mathcal{V}_{mn} = \psi_n(\mathbf{r}_m)$ denotes the generalized Vandermonde matrix, $\hat{\mathbf{g}} = [\hat{g}_1, \hat{g}_2, \ldots, \hat{g}_{N_p}]^T$ and $\mathbf{g} = [g_1, g_2, \ldots, g_{N_p}]^T$. Therefore, the nodal basis functions can be expressed in terms of the modal basis functions as

$$l_m(\mathbf{r}) = \sum_{n=1}^{N_p} [\mathcal{V}^{-1}]_{nm} \psi_n(\mathbf{r}), \quad \text{or} \quad l = \mathcal{V}^{-T}\psi \tag{5.53}$$

where $l = [l_1, l_2, \ldots, l_{N_p}]^T$ and $\psi = [\psi_1, \psi_2, \ldots, \psi_{N_p}]^T$. At a given time step, the hierarchical polynomial coefficients $\hat{g}_n$ can be converted from the readily available interpolatory polynomial coefficients g_m using (5.52). Since the hierarchical coefficients represent the amplitude of variation of different orders, by examining their amplitudes, the appropriate order p can be determined if

$$|\hat{g}_{p+1}| < \kappa \max_{n=0}^{N_p-1}\{|\hat{g}_n|\} \tag{5.54}$$

where κ is a preset parameter that controls the level of accuracy of the high-order representation.

After the local polynomial orders of all the elements are determined, the information transfer in (5.43) to a local element with order p from its neighbor with order p' is calculated as

$$\int_S l_i^p (n^+ - n^-) dS = \int_S l_i^p\, n^+ dS - \int_S l_i^p\, n^- dS \tag{5.55}$$

where l_i^p is the ith Lagrange polynomial of order p, S is the common surface that interfaces these two elements, and

$$n^+ = \sum_{j=1}^{N_p'} n_j^+ l_j^{p'}, \quad n^- = \sum_{j=1}^{N_p} n_j^- l_j^p. \tag{5.56}$$

In (5.56), $l_j^{p'}$ and l_j^p are the jth Lagrange polynomial of order p' and p, respectively, and n_j^+ and n_j^- are the corresponding expansion coefficients. While the n^- term can be calculated easily, the integration of n^+ is little different. This integration describes the incoming flux from the neighboring element and can be calculated as

$$\int_S l_i^p\, n^+ dS = \sum_j \left[\int_S l_i^p l_j^{p'}\, dS\, n_j^+ \right] = \left\{ \tilde{\mathbb{M}}_f \left\{ n_{p'}^+ \right\} \right\}_i \tag{5.57}$$

where the incoming flux vector $\{n_{p'}^+\}$ consists of the values of n^+ on the definition nodes of l_j, which are of order p', and the facial mass matrix is defined as

$$
\tilde{\mathbb{M}}_f = \int_S l_p l_{p'}^T dS = J_f \int_\Gamma l_p l_{p'}^T dS = J_f \mathcal{V}_{p,2D}^{-T} \left[\int_\Gamma \psi_p \psi_{p'}^T dS \right] \mathcal{V}_{p',2D}^{-1} = J_f \mathcal{V}_{p,2D}^{-T} \tilde{\mathbb{I}}_{2D} \mathcal{V}_{p',2D}^{-1}
$$

(5.58)

where

$$
\tilde{\mathbb{I}}_{2D} = \begin{cases}
\mathbb{I}_{N_{2D}}, & p = p' \\
\left[\mathbb{I}_{N_{2D}}, \mathbb{O}_{N_{2D},(N'_{2D}-N_{2D})} \right], & p < p' \\
\begin{bmatrix} \mathbb{I}_{N'_{2D}} \\ \mathbb{O}_{(N_{2D}-N'_{2D}),N'_{2D}} \end{bmatrix}, & p > p'
\end{cases}
$$

(5.59)

In (5.59), $\mathbb{I}_{N_{2D}}$ stands for an $N_{2D} \times N_{2D}$ identity matrix with N_{2D} being the number of DoFs on the elemental surface with the polynomial order p, $\mathbb{O}_{N_{2D},(N'_{2D}-N_{2D})}$ stands for an $N_{2D} \times (N'_{2D} - N_{2D})$ zero matrix. The facial mass matrix can be further expressed as

$$
\tilde{\mathbb{M}}_f = \left(J_f \mathcal{V}_{p,2D}^{-T} \mathcal{V}_{p,2D}^{-1} \right) \left(\mathcal{V}_{p,2D} \tilde{\mathbb{I}}_{2D} \mathcal{V}_{p',2D}^{-1} \right) = \mathbb{M}_f \mathbb{E}_f
$$

(5.60)

where $\mathbb{E}_f = \mathcal{V}_{p,2D} \tilde{\mathbb{I}}_{2D} \mathcal{V}_{p',2D}^{-1}$ is defined as the facial elevation matrix. The incoming flux can, therefore, be written as

$$
\tilde{\mathbb{M}}_f \{ n_{p'}^+ \} = \mathbb{M}_f \mathbb{E}_f \{ n_{p'}^+ \} = \mathbb{M}_f \{ \tilde{n}_{p'}^+ \}
$$

(5.61)

In the DGTD implementation, the incoming flux $\{n_{p'}^+\}$ is first elevated to the correct polynomial order $\{\tilde{n}_{p'}^+\} = \mathbb{E}_f\{n_{p'}^+\}$, which has the same dimension as n^-. Then, they can be used in the calculation of the total numerical flux.

5.5 Multirate Time Integration Technique

With the development and implementation of dynamic adaptation algorithms, the element size or polynomial order can be adjusted adaptively during the numerical simulation. This is advantageous in a sense that good spatial resolution is achieved while the computational cost at each time step is minimized. The use of time-varying element sizes or polynomial orders necessitates the employment of an explicit time integration method to avoid the repeated global matrix assembly and factorization required by an implicit time integrator when the element size or polynomial order changes. However, explicit integration methods are subject to the stability condition known as the Courant–Friedrichs–Lewy (CFL) condition [43, 44], which limits the time step size by the smallest element size and the highest polynomial order presented in a spatial representation. In an EM–plasma simulation, the adaptively adjusted element size can become very small and the adaptively adjusted polynomial order can become very high, leading to a global time step size that is extremely small and a computational cost that is unnecessarily high. To alleviate the global time step constraint, a MTI technique is adopted from [45], which is based on an explicit multistage Runge–Kutta method and permits different time step sizes in different elements.

The solution of an IVP

$$\frac{d\mathbf{y}}{dt} = \mathbf{f}(t_0, \mathbf{y}_0), \quad \mathbf{y}(t_0) = \mathbf{y}_0 \tag{5.62}$$

can be obtained by an s-stage explicit Runge–Kutta method as [18]

$$\mathbf{y}_{n+1} = \mathbf{y}_n + \Delta t \sum_{i=1}^{s} b_i \mathbf{k}_i \tag{5.63}$$

where

$$\mathbf{k}_i = \mathbf{f}\left(t_n + c_i \Delta t, \ \mathbf{y}_n + \Delta t \sum_{j=1}^{i-1} a_{ij} \mathbf{k}_j \right) \tag{5.64}$$

with $i = 1, 2, \ldots, s$ and the Runge–Kutta matrix $[a]_{ij}$, weights b_i and nodes c_i are designed to obtain different Runge–Kutta schemes [18]. For an explicit Runge–Kutta method, $c_1 = 0$ and $\sum_{j=1}^{i-1} a_{ij} = c_i$ for $i = 2, 3, \ldots, s$. From (5.63) and (5.64), the stage vectors $\mathbf{k}_i$ related to the time step size Δt can be calculated from the first ith-order time derivatives of the unknown vector $\mathbf{y}^{(i)}$ as

$$\{\mathbf{k}\}_{\Delta t} = [\gamma][P_{\Delta t}]\{\mathbf{y}^{(i)}\} \tag{5.65}$$

where $\{\mathbf{k}\}_{\Delta t} = [\mathbf{k}_1, \mathbf{k}_2, \ldots, \mathbf{k}_s]_{\Delta t}^{\mathrm{T}}$, $\{\mathbf{y}^{(i)}\} = [\mathbf{y}^{(1)}, \mathbf{y}^{(2)}, \ldots, \mathbf{y}^{(s)}]^{\mathrm{T}}$, diagonal matrix $[P_{\Delta t}] = \mathrm{diag}[1, \Delta t, \ldots, \Delta t^{s-1}]$,

$$[\gamma] = \begin{bmatrix} 1 & & & \\ 1 & \gamma_{22} & & \\ \vdots & \vdots & \ddots & \\ 1 & \gamma_{s2} & \cdots & \gamma_{ss} \end{bmatrix} \tag{5.66}$$

and γ_{ij} can be calculated recursively as

$$\gamma_{i1} = 1$$

$$\gamma_{ij} = \sum_{k=j-1}^{i-1} a_{ik} \gamma_{k,j-1} \tag{5.67}$$

for $1 \leq j \leq i \leq s$.

Consider two adjacent elements with the size of element 1 twice as large as that of element 2. If both elements employ polynomials of the same order, the time step size $\Delta t_1 = 2\Delta t_2$. At a synchronized step, where $t = t_n$, to advance the solutions in element 1 from t_n to $t_{n+1} = t_n + \Delta t_1$, the stage vectors needed by element 1 that transfer information from element 2 can be calculated as

$$\{\tilde{\mathbf{k}}\}_{\Delta t_1} = [\gamma][P_{\Delta t_1}][P_{\Delta t_2}]^{-1}[\gamma]^{-1}\{\mathbf{k}\}_{\Delta t_2} \tag{5.68}$$

The physical quantities in element 1 are then integrated with time step size Δt_1 using the stage vectors $\{\mathbf{k}\}_{\Delta t_1}$ calculated with information contained in element 1 and $\{\tilde{\mathbf{k}}\}_{\Delta t_1}$ calculated with information contained in element 2. Similarly, to advance the solutions in element 2 from t_n to $t_{n+1/2} = t_n + \Delta t_2$, the stage vectors needed by element 2 that transfer information from element 1 can be calculated as

$$\{\tilde{\mathbf{k}}\}_{\Delta t_2} = [\gamma][P_{\Delta t_2}][P_{\Delta t_1}]^{-1}[\gamma]^{-1}\{\mathbf{k}\}_{\Delta t_1} \tag{5.69}$$

Similarly, the physical quantities in element 2 are then integrated.

At an intermediate step, where $t = t_{n+1/2}$, only the solutions in element 2 need to be integrated. This requires the stage vectors in element 1 at $t_{n+1/2}$ and can be calculated by first extrapolating the solutions and their time derivatives in element 1 using the Taylor series

$$\mathbf{y}^{(i)}(t_{n+1/2}) = \sum_{j=i}^{s} \frac{\Delta t_2^{j-i}}{(j-i)!}\,\mathbf{y}^{(j)}(t_n), \qquad (i = 0, 1, \ldots, s) \tag{5.70}$$

The stage vectors $\{\tilde{\mathbf{k}}\}_{\Delta t_2}$ at $t_{n+1/2}$ can then be evaluated by applying (5.65) to $\mathbf{y}^{(i)}(t_{n+1/2})$.

5.6 Numerical Examples

In this section, several numerical examples are presented to demonstrate the performance of the dynamic h- and p-adaptation algorithms in simulating pure EM as well as multiphysics problems. The MTI technique is also applied to accelerate the simulations.

5.6.1 Scattering from a Cone Sphere with a Slot

To show the performance of the dynamic h-adaptation, the scattering from a benchmark object, a PEC cone sphere with a slot, is considered to further demonstrate the DGTD with the dynamic grid. The PEC scatterer is 1.378 m in length and has a 1.27-cm-wide and 1.27-cm-deep slot around the bottom of the cone. It is illuminated by an incident plane wave with a modulated Gaussian profile. The central frequency of the incident wave is 1.5 GHz and the pulse width is 0.53 ns. To resolve its sharp tip and narrow slot, extremely tiny elements are required. If a uniform time step size were applied, the total computational cost

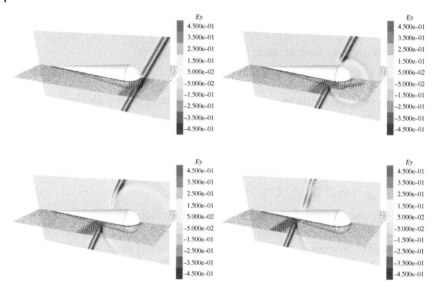

Figure 5.4 Snapshots of the y-component of the electric field and the corresponding dynamically adaptive mesh at (a) 6.75, (b) 7.50, (c) 8.50, and (d) 9.50 ns.

would increase dramatically. In this simulation, a dynamic grid is employed by using elements with three different sizes from $h_0 = \lambda_{\min}/3.66$ to $h_2 = h_0/4$. On top of the dynamic grid, the MTI technique is used to permit different time step sizes for different elements. As a result, the simulation can be performed very efficiently, with the total number of mesh elements changing dynamically from 482,482 to 577,633 during the entire simulation, and the local time-step sizes ranging from 3.75 to 15.00 ps. Shown in Figure 5.4 are the electric field distributions at 6.75, 7.50, 8.50, and 9.50 ns, together with the corresponding dynamically refined grids. Apparently, both the incident and the scattered wave fronts, where the fastest oscillations occur, can be tracked in real time, which demonstrates the capability of the proposed method.

5.6.2 Wave Scattering from an Aircraft

To demonstrate the performance of the proposed method in a pure EM simulation, the transient wave scattering from an aircraft is simulated in 2-D. Sitting in the center of a 150×150 m^2 simulation domain in the xy plane, the aircraft has a total length and width of 32.32 m and 28.74 m, respectively, and is illuminated by a z-polarized modulated Gaussian pulse coming from 45° in the xy plane. The incident plane wave has a maximum intensity of 1 V/m, a center frequency of 300 MHz, and a 100% bandwidth. At the highest frequency of interest where

$f_{max} = 450$ MHz, the simulation domain is $225\lambda \times 225\lambda$ in size and is discretized into triangular elements with an average size of 0.33 m. This problem has been simulated using the DGTD method with (a) fixed third-order polynomials and a uniform time step size, (b) adaptive first- to third-order polynomials and a uniform time step size, and (c) adaptive-order (AO) polynomials and the MTI technique. The uniaxial perfectly matched layer (UPML) is used to truncate the simulation domain in all directions.

The electric field and the polynomial order distributions at time $t = 360$, 460, and 560 ns are shown in Figure 5.5. In Figure 5.5a–c, only the fields with a magnitude greater than 0.01 V/m are shown, and the strong multiple reflections between the engine and the fuselage can be observed very clearly. From Figure 5.5d–f, it is clear that the polynomial orders are elevated in the areas with strong field oscillations and lowered in the rest of the simulation domain. As a result, the patterns shown in Figure 5.5a,d resemble each other closely, so do those in Figure 5.5b,e and Figure 5.5c,f. The electric field recorded at 460 ns along the dashed line in Figure 5.5b is shown in Figure 5.6, with a comparison made between the fixed order and the adaptive order with MTI technique cases. The root-mean-square (RMS) error between these two cases is 4.65%, which

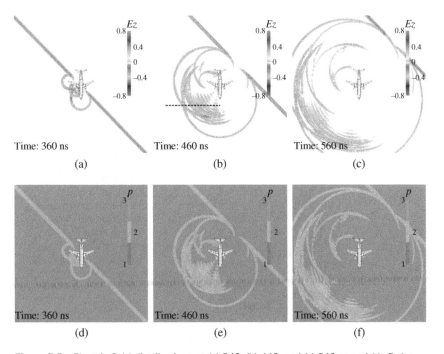

Figure 5.5 Electric field distributions at (a) 360, (b) 460, and (c) 560 ns and (d–f) the corresponding polynomial order distributions.

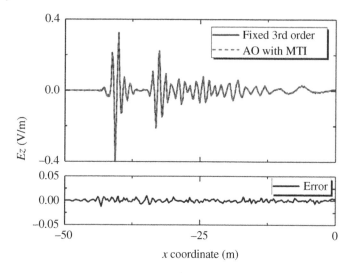

Figure 5.6 Comparison of the electric field distribution at 460 ns along the dashed line in Figure 5.5b. Results are obtained using the fixed third-order polynomials and adaptive-order (AO) polynomials with the multirate time integration (MTI) technique. The RMS error is 4 : 65%.

demonstrates the accuracy of the proposed method. The computational time needed to complete the simulation for 200 cycles (measured at 300 MHz, where 1 cycle =3.33 ns) is presented in Figure 5.7. The simulation with the fixed-order polynomials order takes a total of 351.1 minutes to complete. By using the dynamic p-adaptation and the uniform time step size, the total computational time is reduced to 263.2 minutes. When the dynamic p-adaptation is used together

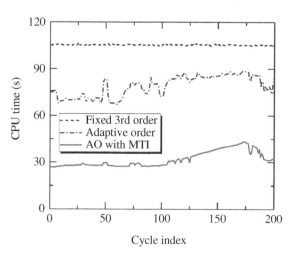

Figure 5.7 Comparison of CPU time consumptions among the DGTD simulations using fixed-order polynomials, adaptive-order (AO) polynomials, and AO polynomials with the MTI technique.

with the MTI technique, the total central processing unit (CPU) time is further reduced to 106.2 minutes, which corresponds to the speedup of 3.3 and 2.5 times compared to the other two cases, respectively. Note that the speedup in fully 3-D simulations can easily achieve over two orders of magnitude due to a much larger ratio between the numbers of elements with low- and high-order polynomials.

5.6.3 Plasma Formation and EM Shielding

To demonstrate the advantages of the dynamic h-adaptation and MTI techniques in multiphysics modeling, a numerical example of HPM air breakdown phenomenon in a waveguide is simulated with the plasma diffusion model and presented in this section. The computational domain is a parallel-plate waveguide made of perfectly electric conductors, which is truncated from the left and right sides with the ABC. In the center of the waveguide, two metallic walls are placed between the parallel plates, which are used to enhance the incident EM waves and trigger the HPM air breakdown. The geometry and parameters of the model are shown in Figure 5.8.

In the physical setting of this example, a 25 GHz, 2 MV/m, and vertically polarized plane wave with a tapered sinusoidal temporal profile is incident from the left boundary and propagates toward the right boundary. The computational domain is filled with 100-Torr low pressure air and initial electrons distributing uniformly with the initial density $n = 10^{15}/\text{m}^3$. Due to the field enhancement between the two metallic insertions, the strong air breakdown will take place around the aperture between the metallic walls. During the numerical simulation, the smallest element size should be smaller than $\lambda/500$ ($\lambda = 12$ mm stands for the EM wavelength in free space) to resolve the very large plasma density gradient and capture

Figure 5.8 Geometry model of a parallel-plate waveguide with two metallic insertions.

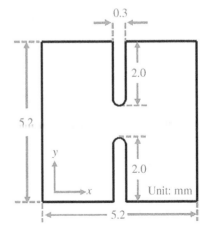

accurately the breakdown process. However, for most of the simulation domain, where air breakdown does not take place, the electron distribution is relatively uniform throughout the entire simulation. The use of a uniformly dense grid will waste a lot of computational resources.

In this simulation, a dynamic grid is employed to permit adaptive h-refinement where the adaptation criteria (5.47) with $\xi = 2$ and (5.49) with $\zeta = 10^{18}$ are employed to demonstrate and compare their performance. Square elements with three different sizes $h_0 = \lambda/150$, $h_1 = h_0/2$, and $h_2 = h_0/4$ are permitted for the adaptation. Initially, the computational domain is discretized by the h_0 elements, which are refined if the adaptation criterion is met. On top of the dynamic grids, the MTI technique is adopted to permit different time step sizes for elements of different sizes. With the use of criterion (5.47), the total number of elements changes from 8436 to 25,544 during the entire simulation, while criterion (5.49) is used it changes from 8436 to 25,032. With the application of the MTI, the local time step sizes range from 10 to 40 ps.

Figure 5.9 shows the effective electric field distribution at 0.284, 0.404, and 0.644 ns, from which the EM wave shielding effect during the HPM air breakdown process can be observed. This is due to the oscillation of the dense electron bulk formed from the air ionization, which acts like a piece of good conductor that shields the incident EM waves. The electron density distributions at the afore-mentioned time instants along with the corresponding mesh grids are shown in Figure 5.10 with criteria (5.47) and (5.49), from which it is clear that the mesh changes dynamically to capture the variation of the electron density distribution. However, some distinct differences are observed when comparing these figures. When the characteristic length in (5.47) is used as the adaptation criterion as shown in Figure 5.10a–c, more grids are refined as air breaks down. After the plasma bulk is fully developed, some small grids inside the plasma bulk are coars-ened back to larger elements. This is because the characteristic length is large both outside and inside the plasma bulk where the plasma density varies slowly but

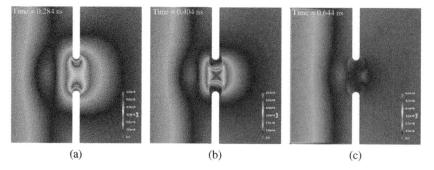

(a)   (b)   (c)

Figure 5.9 Effective field distribution at (a) 0.284, (b) 0.404, and (c) 0.644 ns.

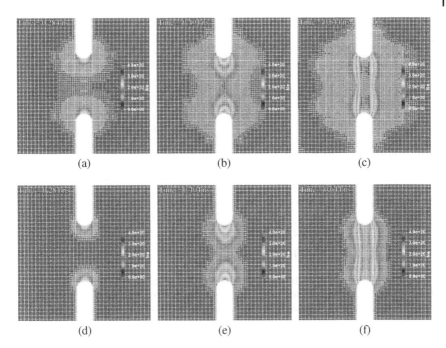

Figure 5.10 Zoomed-in view of the electron density distribution around the insertions and the corresponding dynamically adaptive mesh grids at (a and d) 0.284, (b and e) 0.404, and (c and f) 0.644. The results in the first row are obtained using criterion (5.47) and those in the second row are obtained using criterion (5.49).

becomes small around the plasma edge where the density changes drastically. When the absolute value of the plasma density variation is used as the adaptation criterion as in (5.49), only the grids with a high plasma density get refined. In this example, since the local plasma density increases monotonically during the breakdown, those refined grids are not coarsened back throughout the simulation, as can be seen in Figure 5.10d–f. But the total number of refined grids is smaller than that in Figure 5.10a–c.

To demonstrate the efficiency and accuracy of the dynamic h-adaptation and MTI techniques, this example is simulated with three other settings. In the first setting, a static uniform mesh is employed with the element size fixed as h_2 and the smallest allowed time step size is used in all elements throughout the entire simulation. In the second and the third, only the dynamic h-adaptation with criteria (5.47) and (5.49) are used, respectively, without the application of the MTI. The computational time for all five cases are first presented in Figure 5.11, from which it is clear that the dynamic h-adaptation technique with both criteria can reduce the computational time significantly compared to the uniform case. The total speedups of the computation at the end of the simulation are 3.8 and

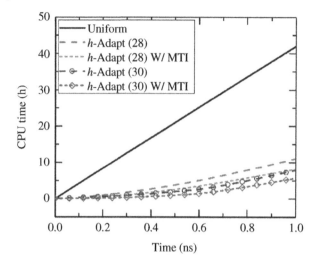

Figure 5.11 Comparison of the cumulative computational time when the uniformly dense grid without the multirate time integration, the dynamic *h*-adaptation without the multirate time integration, and the dynamic *h*-adaptation with the multirate time integration methods are used.

5.3 times when criteria (5.47) and (5.49) are used, respectively. When the dynamic *h*-adaptation is applied together with the MTI, the total computational time can be further reduced, and the total speedups reach 5.1 and 7.4 times compared to the uniform case. Due to the denser grids around the aperture, the simulation with criterion (5.47) is little more expensive when the simulation is terminated at 1 ns. However, if the simulation proceeds at a later time, the electron density to the left of the aperture will continue to increase due to the air ionization by the high-intensity standing wave formed in this region. If criterion (5.49) is used for the mesh adaptation, the grids in the entire left half of the simulation domain will be refined, which will increase the computational time significantly. As a summary, when the size of the plasma bulk is small, criterion (5.49) is more efficient than (5.47). When the plasma bulk is large in its size and uniform in its distribution, criterion (5.47) becomes more efficient. The choice of a "better" criterion is problem dependent.

The electric field and electron density are recorded in the center between the two metallic insertions and compared among all study cases. It is found that the results from both adaptation criteria agree very well, so only those from criterion (5.49) are presented in Figure 5.12. Here, the comparisons are made among the numerical results obtained from the uniformly dense grid without the MTI, the dynamic *h*-adaptation without the MTI, and the dynamic *h*-adaptation with the MTI methods. Clearly, excellent agreement is observed, indicating that the

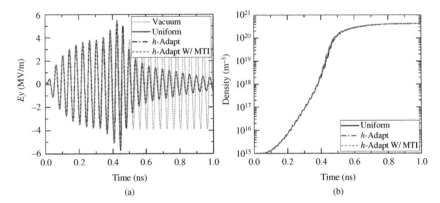

Figure 5.12 Comparison of the numerical solutions recorded at the center point between the two metallic insertions as a function of time. Simulation results are obtained from the uniformly dense grid without the multirate time integration, the dynamic *h*-adaptation without the multirate time integration, and the dynamic *h*-adaptation with the multirate time integration methods. (a) The electric field, where an extra result calculated in the vacuum case without air breakdown is given as a reference (dotted line). (b) The time evolution of the electron density.

dynamic *h*-adaptation and MTI methods achieve the same numerical accuracy as the uniformly dense mesh grid and the uniformly small time step size. In Figure 5.12a, the electric field recorded in a vacuum case, where no air breakdown can take place, is also presented as a reference. Physically, without air breakdown, the electric field keeps oscillating with the same amplitude throughout the simulation. With air breakdown, due to the exponential increase in the electron density as shown in Figure 5.12b, the electrons start to react more and more strongly and eventually become a piece of EM shield when the plasma frequency becomes comparable to the frequency of the incident EM fields.

5.6.4 HPM Air Discharge and Formation of Plasma Filamentary Array

In this section, air discharge and plasma filamentary array formation around two perfectly conducting cylinders are investigated numerically using the five-moment model and the dynamic *p*-adaptation with MTI. The simulation domain is illustrated in Figure 5.13, where two conducting cylinders with the same diameter of 0.5 mm are placed in free space with a gap of 0.1 mm in between. A 5.5 MV/m, 200 GHz, and *y*-polarized plane wave is incident along the *z*-direction. The entire space is filled with 100 Torr, 300 K ambient air. The uniformly distributed electrons have an initial density of 10^6 m^{-3} and an initial temperature same as the ambient air. The boundary of the solution domain is truncated by a perfectly matched layer [2]. The simulation domain shown in

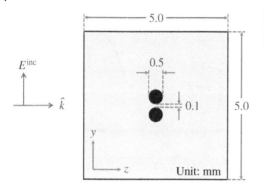

Figure 5.13 Illustration of the problem setting with two conducting cylinders.

Figure 5.13 is discretized into triangular elements with an average size of 5.04 µm, resulting in a total of 495,350 elements. To resolve the large density gradient at the plasma front, the Lagrange polynomial orders are allowed to vary from 2 to 8 with the *p*-adaptation threshold set as $\kappa = 10^{-3}$, meaning that the higher order variations with a relative strength smaller than 1/1000 has been neglected. Since the conversion between the nodal and modal representations is very cheap, this criterion can be applied in a real-time simulation with a low overhead.

The dynamic adaptation of the polynomial orders is investigated and justified by comparing the polynomial order distribution with plasma density distribution and plasma characteristic length distribution, respectively. The comparisons are presented in Figure 5.14 at two time instants 0.5 and 1.2 ns. It is observed from Figure 5.14a,b, in which the polynomial order distributions are overlaid with the equi-density contours of the plasma, that in the regions where the equi-density contours cluster the polynomial orders are elevated. This is because in these regions the plasma density gradients are large and higher order polynomial representations are needed to resolve the fast-changing physics. From Figure 5.14c,d, where the polynomial order distributions are presented together with the characteristic length distribution of the plasma (shown as its reciprocal $1/L = \|\nabla n/n\|$), a very good correspondence between the two are observed, meaning that in regions with a smaller characteristic length a higher polynomial order is needed. These investigations successfully demonstrate the use of the dynamic *p*-adaptation in resolving physical quantities with faster variations and smaller spatial scales.

The HPM air discharge and plasma filamentary array formation are investigated numerically. When the incident HPM illuminates the metallic cylinders, the EM field gets intensified in the gap between the two cylinders as shown in Figure 5.15a. Through the Lorentz force and the Joule heating, a large amount of EM energy is transformed into the electron energy shown in Figure 5.15e, which initiates the ionization process in this region. As presented in Figure 5.16b, the strong ionization first appears in the gap between the cylinders. This high-density plasma

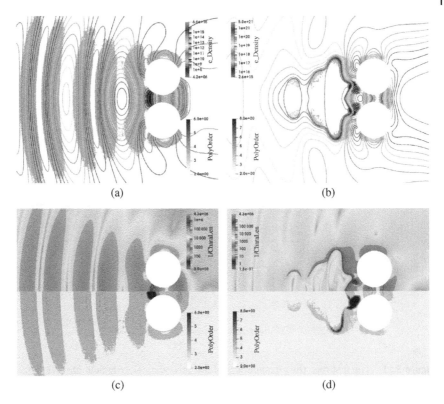

Figure 5.14 Investigation of the dynamically adjusted polynomial order distribution at 0.5 ns [(a) and (c)] and 1.2 ns [(b) and (d)]. (a and b) Polynomial order distribution overlaid with the equi-density contours of the plasma. (c and d) Polynomial order distribution compared with the reciprocal of the characteristic length of the plasma.

region, along with the two cylinders, reflects the incident microwave and results in an area with a high field intensity and high electron energy in the lit region, which leads to fast ionization in this area. Due to the standing wave pattern developed by the wave reflection, the plasma bulk elongates mainly in the direction of the incident wave polarization (the y-direction). At 1.00 ns, the plasma density is high enough to shield the incident wave, which results in a decreased electron energy in the plasma bulk (Figure 5.15b,f). When the electron density of the newly developed plasma bulk exceeds the critical value of $5 \times 10^{20} \mathrm{m}^{-3}$, which corresponds to the plasma frequency of $f_p = 200$ GHz, the plasma bulk starts to shield and reflect the incident microwave again. When this process continues, an array of plasma filaments will gradually develop and the plasma front will propagate against the incident direction of the HPM, as shown in Figure 5.16c–h. The corresponding secondary field and electron energy distributions are presented in Figure 5.15b–d,f–h,

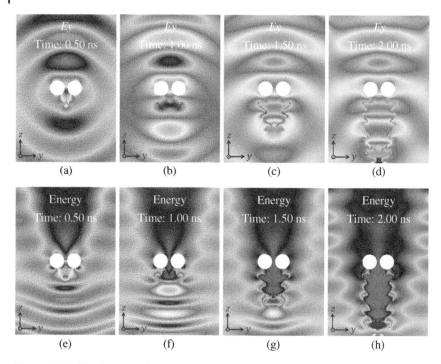

Figure 5.15 The first row: the y-component of the secondary electric field observed at (a) 0.50; (b) 1.00; (c) 1.50; and (d) 2.00 ns. The second row: electron energy distribution observed at (e) 0.50; (f) 1.00; (g) 1.50; and (h) 2.00 ns.

respectively. Qualitatively, the filamentary array generated using the five-moment model agrees with the experimental observation reported in [46]. According to [46], the plasma filamentary pattern is very sharp at a higher air pressure (near atmospheric pressure) and becomes more blurry when the air pressure is relatively low (less than 200 Torr). Note in these figures, the electric field, electron energy, and electron density change abruptly near the edge of the plasma bulk and very smoothly inside and outside the contour. This is an illustration of the multiscale feature caused by the self-consistent EM–plasma interaction, which can only be captured by a numerical method with a high spatial resolution.

To investigate the physical process quantitatively, the electron density and mean energy are plotted along the z-axis in Figure 5.17. It is very clear that the electron density increases fastest at the location with highest energy. When the electron density reaches the critical value, the plasma bulk shields the incident HPM and the electron energy drops. This figure is a quantitative presentation of the dynamics of the formation and evolution of the plasma filamentary array

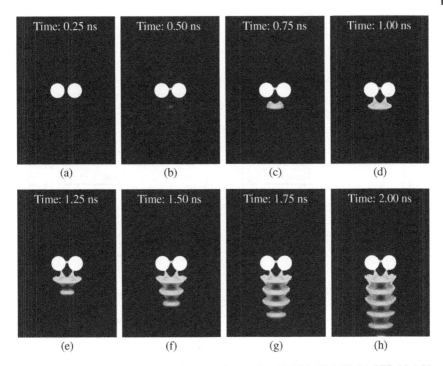

Figure 5.16 Electron density distribution observed at (a) 0.25; (b) 0.50; (c) 0.75; (d) 1.00; (e) 1.25; (f) 1.50; (g) 1.75; and (h) 2.00 ns. White color denotes the larger value.

along the z-axis. More quantitative observations can be made from the numerical simulation results. Shown in Figure 5.18 is the spacing of the fully developed plasma filaments at 2.00 ns. The average spacing between two adjacent plasma filaments is 0.388 mm, which is about a quarter wavelength $\lambda/4 = 0.375$ mm. This is because the strong field reflection from the plasma results in a standing wave pattern in the lit region, which generates plasma filaments with a similar standing wave pattern. Similar observations were made in experiments [47, 48]. The propagation of plasma front is recorded at different plasma densities of 10^{17}m^{-3}, 10^{19}m^{-3}, and 10^{21}m^{-3}. The results are presented in Figure 5.19. From this figure, the mean propagation velocity of the plasma filaments can be estimated at about $V_z = -2000$ km/s as indicated by the gray straight line, which is on the same order ($\sim 10^6$ m/s) of the theoretical estimation [23] of $2\sqrt{D_e v_i}$, where D_e and v_i are the free electron diffusion and ionization coefficients, respectively. Note that the theoretical estimation is valid in the static case, and the numerical simulation is conducted at 200 GHz. Therefore, the agreement is only on the same order of magnitude.

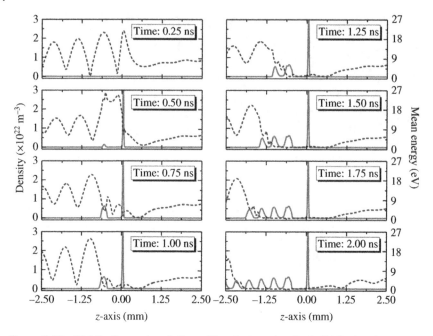

Figure 5.17 Distribution and evolution of the electron density (solid line) and mean energy (dashed line) along the z-axis recorded at different time instants.

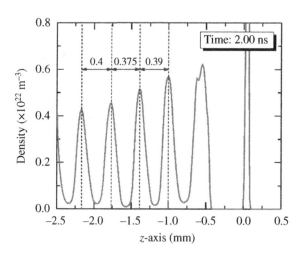

Figure 5.18 Spacing of the fully developed plasma filaments at 2.00 ns.

Figure 5.19
Propagation of the plasma front recorded at different plasma densities. The estimated mean propagation velocity is about 2000 km/s.

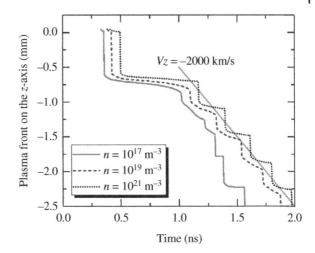

5.7 Conclusion

In this chapter, the DGTD method with dynamic adaptation and MTI techniques for simulating pure EM and multiphysics problems was presented. In describing the multiphysics interactions, the EM fields are governed by Maxwell's equations and the plasma fluids are depicted by either the plasma diffusion or the five-moment fluid model. The DGTD method is employed to solve the multiphysics system and to consider their mutual coupling in a self-consistent manner. Due to the inherent domain decomposition nature, the physical quantities of interest in each mesh element can be solved and integrated in time using the information within the element and its direct neighbors. A dynamic adaptation technique is developed based on this unique feature, which permits the dynamic change of the local element size or polynomial order in a real-time simulation based on the local variation of the physics. It has been shown that the dynamic adaptation technique is able to provide a very good numerical accuracy of the solution while minimizing the total computational cost in resolving the spatial variation of the physics. On top of the dynamic adaptation technique, a MTI method is further used to permit different time step sizes in elements with different sizes or polynomial orders. This alleviates the global constraint on the time step size when an explicit time integrator is used and reduces the total computational time significantly. Numerical examples are provided to demonstrate the performance of the proposed methods in the simulation of EM scattering and EM–plasma interactions. It has been shown that complicated physical phenomena, including the HPM air breakdown, plasma shielding effect, HPM tail erosion, and plasma filamentary pattern formation, can be simulated successfully with the numerical method presented in this chapter, through which the key physical characteristics can be investigated in detail.

References

1 Taflove, A. and Hagness, S.C. (2005). *Computational Electrodynamics: The Finite-Difference Time-Domain Method*, 3e. Norwood, MA: Artech House.

2 Jin, J.M. (2014). *The Finite Element Method in Electromagnetics*, 3e. New York: Wiley–IEEE Press.

3 Baumann, C.E. and Oden, J.T. (1999). A discontinuous *hp* finite element method for convection-diffusion problems. *Comput. Meth. Appl. Mech. Eng.* 175: 311–341.

4 Cockburn, B., Karniadakis, G.E., and Shu, C.-W. (2000). The development of discontinuous Galerkin methods. In: *Discontinuous Galerkin Methods. Theory, Computation and Applications*, Lecture Notes in Computational Science and Engineering, vol. 11 (ed. B. Cockburn, G.E. Karniadakis and C.-W. Shu), 3–50. New York: Springer-Verlag.

5 Cockburn, B., Lin, S.Y., and Shu, C.-W. (1989). TVB Runge–Kutta local projection discontinuous Galerkin finite element method for conservation laws III: one dimensional systems. *J. Comput. Phys.* 84: 90–113.

6 Cockburn, B. and Shu, C.-W. (1989). TVB Runge–Kutta local projection discontinuous Galerkin finite element method for scalar conservation laws II: general framework. *Math. Comput.* 52: 411–435.

7 Cockburn, B. and Shu, C.-W. (1998). The local discontinuous Galerkin method for time-dependent convection-diffusion systems. *SIAM J. Numer. Anal.* 35: 2440–2463.

8 Cockburn, B. and Shu, C.-W. (1998). The Runge–Kutta discontinuous Galerkin finite element method for conservation laws V: multidimensional systems. *J. Comput. Phys.* 141: 199–224.

9 Warburton, T. (2006). An explicit construction of interpolation nodes on the simplex. *J. Eng. Math.* 56 (3): 247–262.

10 Hesthaven, J.S. and Warburton, T. (2008). *Nodal Discontinuous Galerkin Methods Algorithms: Analysis, and Applications*. New York: Springer.

11 Chang, C.-P., Chen, G., Yan, S., and Jin, J.-M. (2017). Waveport modeling for the DGTD simulation of electromagnetic devices. *Int. J. Numer. Modell. Electron. Networks Devices Fields* 31 (4): 1–9.

12 Chen, G., Zhao, L., Yu, W. et al. (2018). A general scheme for the DGTD modeling and S-parameter extraction of inhomogeneous waveports. *IEEE Trans. Microwave Theory Tech.* 66 (4): 1701–1712.

13 Lu, T., Zhang, P.W., and Cai, W. (2004). Discontinuous Galerkin methods for dispersive and lossy Maxwell's equations and PML boundary conditions. *J. Comput. Phys.* 200 (2): 549–580.

14 Gedney, S.D., Luo, C., Roden, J.A. et al. (2009). The discontinuous Galerkin finite-element time-domain method solution of Maxwell's equations. *Appl. Comput. Electromagn. Soc. J.* 24 (2): 129–142.

15 Yan, S., Lin, C.-P., Arslanbekov, R.R. et al. (2017). A discontinuous Galerkin time-domain method with dynamically adaptive Cartesian meshes for computational electromagnetics. *IEEE Trans. Antennas Propag.* 65 (6): 3122–3133.

16 Yan, S. and Jin, J.-M. (2017). A dynamic p-adaptive DGTD algorithm for electromagnetic and multiphysics simulations. *IEEE Trans. Antennas Propag.* 65 (5): 2446–2459.

17 Jin, J.M. and Yan, S. (2019). Multiphysics modeling in computational electromagnetics: technical challenges and potential solutions. *IEEE Antennas Propag. Mag.* 61 (2): 14–26.

18 Butcher, J.C. (2003). *Numerical Methods for Ordinary Differential Equations.* Wiley.

19 Gottlieb, S., Ketcheson, D., and Shu, C.W. (2011). *Strong Stability Preserving Runge–Kutta and Multistep Time Discretizations.* Singapore, NJ: World Scientific.

20 Shu, C.W. (1988). Total-variation diminishing time discretizations. *SIAM J. Sci. Stat. Comput.* 9: 1073–1084.

21 Shu, C.W. and Osher, S. (1988). Efficient implementation of essentially non-oscillatory shock-capturing schemes. *J. Comput. Phys.* 77: 439–471.

22 Yan, S., Greenwood, A.D., and Jin, J.-M. (2016). Modeling of plasma formation during high-power microwave breakdown in air using the discontinuous Galerkin time-domain method (invited paper). *IEEE J. Multiscale Multiphys. Comput. Techn.* 1: 2–13.

23 Boeuf, J.P., Chaudhury, B., and Zhu, G.Q. (2010). Theory and modelling of self-organization and propagation of filamentary plasma arrays in microwave breakdown at atmospheric pressure. *Phys. Rev. Lett.* 104: 015002.

24 Chaudhury, B. and Boeuf, J.P. (2010). Computational studies of filamentary pattern formation in a high power microwave breakdown generated air plasma. *IEEE Trans. Plasma Sci.* 38: 2281–2288.

25 Chaudhury, B., Boeuf, J.P., and Zhu, G.Q. (2010). Pattern formation and propagation during microwave breakdown. *Phys. Plasmas* 17: 123505.

26 Yan, S., Greenwood, A.D., and Jin, J.-M. (2018). Simulation of high-power microwave air breakdown modeled by a coupled Maxwell–Euler system with a non-Maxwellian EEDF. *IEEE Trans. Antennas Propag.* 66 (4): 1882–1893.

27 Bittencourt, J.A. (2004). *Fundamentals of Plasma Physics.* New York: Springer-Verlag.

28 Birdsall, C.K. and Langdon, A.B. (2005). *Plasma Physics Via Computer Simulation.* New York: Taylor & Francis.

29 Hagelaar, G.J.M. and Pitchford, L.C. (2005). Solving the Boltzmann equation to obtain electron transport coefficients and rate coefficients for fluid models. *Plasma Sources Sci. Technol.* 14 (4): 722–733.

30 Boris, J.P. (1970). Relativistic plasma simulations – optimization of a hybrid code. In: *Proceedings of the Conference on the Numerical Simulation of Plasmas (4th),* 3–67. Washington, DC: NRL Washington.

31 Marder, B. (1987). A method incorporating Gauss' law into electromagnetic PIC codes. *J. Comput. Phys.* 68: 48–55.

32 Munz, C.-D., Omnes, P., and Schneider, R. (2000). A three-dimensional finite-volume solver for the Maxwell equations with divergence cleaning on unstructured meshes. *Comput. Phys. Commun.* 130: 83–117.

33 Munz, C.-D., Omnes, P., Schneider, R. et al. (2000). Divergence correction techniques for Maxwell solvers based on a hyperbolic model. *J. Comput. Phys.* 161: 484–511.

34 Jacobs, G.B. and Hesthaven, J.S. (2006). High-order nodal discontinuous Galerkin particle-in-cell method on unstructured grids. *J. Comput. Phys.* 214 (1): 96–121.

35 Jacobs, G.B. and Hesthaven, J.S. (2009). Implicit-explicit time integration of a high-order particle-in-cell method with hyperbolic divergence cleaning. *Comput. Phys. Commun.* 180: 1760–1767.

36 Pfeiffer, M., Munz, C.-D., and Fasoulas, S. (2015). Hyperbolic divergence cleaning, the electrostatic limit, and potential boundary conditions for particle-in-cell codes. *J. Comput. Phys.* 294: 547–561.

37 Dedner, A., Kemm, F., Kröner, D. et al. (2002). Hyperbolic divergence cleaning for the MHD equations. *J. Comput. Phys.* 175 (2): 645–673.

38 Tricco, T.S. and Price, D.J. (2012). Constrained hyperbolic divergence cleaning for smoothed particle magnetohydrodynamics. *J. Comput. Phys.* 231 (21): 7214–7236.

39 Yan, S. and Jin, J.-M. (2017). A continuity-preserving and divergence-cleaning algorithm based on purely and damped hyperbolic Maxwell equations in inhomogeneous media. *J. Comput. Phys.* 334: 392–418.

40 Angulo, L.D., Alvarez, J., Pantoja, M.F. et al. (2015). Discontinuous Galerkin time domain methods in computational electrodynamics: state of the art. *Forum Electromagn. Res. Methods Appl. Technol.* 10: 1–24.

41 Alvarez, J., Angulo, L.D., Bretones, A.R., and Garcia, S.G. (2012). A spurious-free discontinuous Galerkin time-domain method for the accurate modeling of microwave filters. *IEEE Trans. Microwave Theory Tech.* 60 (8): 2359–2369.

42 Zhang, M. and Shu, C.-W. (2003). An analysis of three different formulations of the discontinuous Galerkin method for diffusion equations. *Math. Models Methods Appl. Sci.* 13 (3): 395–413.

43 LeVeque, R.J. (2002). *Finite Volume Methods for Hyperbolic Problems.* Cambridge: Cambridge University Press.

44 Cockburn, B. and Shu, C.W. (2001). Runge–Kutta discontinuous Galerkin methods for convection dominated problems. *J. Sci. Comput.* 16: 173–261.

45 Liu, L., Li, X., and Hu, F.Q. (2010). Nonuniform time-step Runge–Kutta discontinuous Galerkin method for computational aeroacoustics. *J. Comput. Phys.* 229 (19): 6874–6897.

46 Cook, A., Shapiro, M., and Temkin, R. (2010). Pressure dependence of plasma structure in microwave gas breakdown at 110 GHz. *Appl. Phys. Lett.* 97: 011504.

47 Hidaka, Y., Choi, E.M., Mastovsky, I. et al. (2008). Observation of large arrays of plasma filaments in air breakdown by 1.5-MW 110-GHz gyrotron pulses. *Phys. Rev. Lett.* 100: 035003.

48 Hidaka, Y., Choi, E.M., Mastovsky, I. et al. (2009). Plasma structures observed in gas breakdown using a 1.5 MW, 110 GHz pulsed gyrotron. *Phys. Plasmas* 16: 055702.

6

DGTD Method for Periodic and Quasi-Periodic Structures

Pengfei Wen[1], Chao Li[1], Qiang Ren[1], and Jiefu Chen[2]

[1]School of Electronics and Information Engineering, Beihang University, Beijing China
[2]Department of Electrical and Computer Engineering, University of Houston, Houston, United States

6.1 Introduction

6.1.1 Background

Periodic/quasi-periodic structures have been widely used in many fields, such as multi-input and multi-output (MIMO) antennas [1–5], phased array antennas [6], reflect array and transmit array, metamaterials and metasurfaces [7–9], as shown in Figure 6.1. For fast and accurate analysis of large periodic/quasi-periodic arrays, an advanced numerical simulation strategy is required, which has always been one of the challenging problems in computational electromagnetics. The difficulties in modeling large periodic/quasi-periodic array primarily stem from the following features:

(1) A large number of cells results in a huge number of unknowns.
(2) There may be geometrically a small structure inside the array cells, and the size of its finest part is much smaller than the size of the entire array, which reveals typical multiscale characteristics.
(3) The geometry of the cells on the lattice of the array may change gradually along certain directions, such as the reflect arrays and phase gradient metasurfaces. Or a few types of cells with different geometries are "randomly" scattered on the lattice, such as the coding metasurfaces. These two types of arrays are termed as quasi-periodic arrays.
(4) Some antenna arrays are required to be conformal to the platform, such as smart skins and random, which make the formation not a traditional planar structure.

Advances in Time-Domain Computational Electromagnetic Methods, First Edition.
Edited by Qiang Ren, Su Yan, and Atef Z. Elsherbeni.

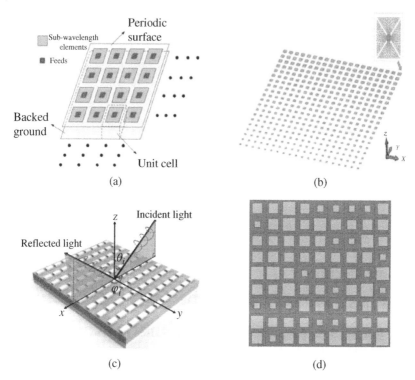

Figure 6.1 Application examples of periodic/quasi-periodic structures. (a) A broadband tightly coupled with patch array antenna. Source: Yang et al. [10]. Reproduced with permission of IEEE. (b) Quasi-periodic reflect array with Malta cross. Source: Dang et al. [11]. Reproduced with permission of IEEE. (c) Broadband plasmonic metasurface-enabled waveplates. Source: Jiang et al. [12]. Reproduced with permission of Nature Research. (d) 2-bit coding metasurface composed of 9 ×9 unit cells. Source: Zhang et al. [13]. Reproduced with permission of IEEE.

In view of the above-listed difficulties, the numerical simulation of large arrays usually employs approximate methods. When the lattice size is large and all the cells are identical in geometry (aka *"periodic array"*), only one cell needs to be simulated using the periodic boundary condition. This strategy converts the problem of modeling the entire array into modeling one cell of the array, which can significantly reduce the cost of computing resources. The infinity approximation of a finite array ignores the boundary truncation effect and assumes that the current (field) distribution on each cell is the same. However, this is obviously not true for a finite-sized array as shown in Figure 6.2, so the numerical errors will be introduced. In addition, this approximation requires that the structure of each cell to be exactly the same, which limits its scope of application. For quasi-periodic arrays, such as the phase gradient metasurfaces, reflect arrays, and transmit

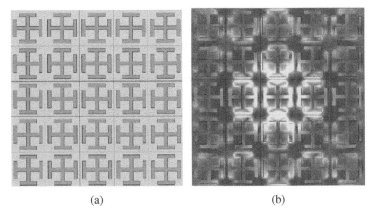

(a)   (b)

Figure 6.2 Truncation effect of a finite array. (a) The frequency selected surface with Jerusalem cells. (b) The different current distributions on the array due to the boundary effect.

arrays, the phase of each type of cell is extracted and used in the subsequent analysis, design, and synthesis, instead of modeling the actual geometry of the array. However, the periodic boundary condition is used again here to extract the phases. This will introduce numerical errors not only from the boundary truncation but also from the improper assumption that each cell has the same neighboring cells.

To overcome the accuracy defects of the abovementioned approximate methods based on the periodic boundary condition, full-wave methods (such as finite-difference time domain [FDTD], method of moment [MoM], finite-element method [FEM]) and hybrid methods (such as MoM–PO) have been improved and applied to the analysis of finite periodic/quasi-periodic array. Many research groups have gained significant achievements in the forward modeling techniques and inverse design strategies combined with optimization algorithms.

The most popular approach applied to model period/quasi-periodic arrays is the MoMs, including both the frequency-domain MoM [14–16] and spatial-domain MoM [17, 18]. To decease requirements of $O(N^2)$ storage and $O(N^3)$ operations in the basic MoM, the acceleration schemes for periodic/quasi-periodic array can be classified into three categories: (i) fast algorithms using the acceleration of matrix–vector production techniques including fast Fourier transformation (FFT) [19, 20], adaptive integral method [21], pre-corrected fast Fourier transformation (pFFT) [22], multilevel fast multipole algorithm (MLFMA) [23], single-level dual-rank IE-QR [24]; (ii) combining MoM with asymptotic methods [25, 26], such as physical optics (PO); and (iii) reduction of DoFs via modified basis functions, such as entire-domain basis functions [27], synthetic basis functions [28], character basis functions [29–32]. Besides, Li and Chew used equivalence principal algorithm (EPA) to model multiscale structures with antenna arrays [33]. Cui and colleagues proposed MoM based on sub-entire domain basis functions [34], which was further developed with acceleration techniques, such as simplified sub-entire domain (SSED) [35] method and the SSED–CG–FFT [36] method by Cui et al. and Wang et al. Ansari-Oghol-Beig et al. and Salary et al. employed the model order reduction (MOR) method to analyze the quasi-periodic metasurfaces successfully [7, 37]. However, when the materials of cells in the array are not uniform (such as patch antennas with substrates), Green's functions for layered medium or volume-surface integral equations are required, which make the algorithm a complex one and not generic.

Compared to MoM, differential equation–based methods for finite array simulation are less popular mainly due to two reasons. First, volume discretization results in a large number of unknowns, which hinder simulation of large finite arrays. Second, numerical dispersion deteriorates accuracy. However, they enjoy superior advantages in handling heterogeneous media. The conventional FDTD method can be applied to model the antenna array, but it does not make full use of the repetition or the similarity between the array cells. In addition, the intrinsic characteristic of the staircase error of the FDTD method makes it unsuitable for the antenna array with curved surfaces and arcs. FEMs are also used for the simulation of antenna arrays [38], including domain decomposition-based FEMs [39] and finite-element boundary integral methods [40]. This type of method is suitable for inhomogeneous media and complex media. But FEM-based methods for finite large arrays suffer from low efficiency.

Aiming at efficient simulation of the response of large finite periodic/quasi-periodic arrays (such as pattern analysis, mutual coupling analysis of array cells), we propose a memory-efficient discontinuous Galerkin time-domain (DGTD) method [41–51], which is suitable for finite large periodic/quasi-periodic arrays [52, 53]. Note the memory-efficient DGTD method is an advanced version of the subdomain-level DGTD method. This method treats each cell in the array (not each element in the mesh) as a subdomain, unlike the element-level DGTD method [54–56]. To achieve a linear relationship between the memory cost and the number of the cell types (instead of the number of the cells) of the array, it uses the property that the cells with identical geometries lead to the same volumetric system matrices. This method has the following advantages:

(1) The memory cost has a linear relationship with the *types* of cells instead of the *number* of cells.
(2) The array cells can contain inhomogeneous media and fine structures.
(3) Direct solver can be used, and the solution time and accuracy are controllable.
(4) It is suitable for conducting broadband analysis.

6.1.2 Overview of the Sections

The second section introduces the conventional subdomain-level DGTD method. In the third section, the detailed theory, discretized formulations, and time integrations of the memory-efficient DGTD method are described. Its accuracy and efficiency are validated by modeling a proof-of-idea cavity and a periodic patch antenna array. In the fourth section, the memory-efficient DGTD method is further extended to allow quasi-periodic arrays, which is validated by two quasi-periodic patch antenna arrays, and this method is summarized in the last section.

6.2 The Subdomain-Level DGTD Method

The subdomain-level DGTD method obtains the weak forms of Maxwell's equations through Galerkin's weighting method, which relaxes the continuity of field quantities between cells, while exchanging information between adjacent discretized elements through the numerical flux. For example, Riemann solver (upwind flux) is widely used to ensure accurate and stable data communication. In this way, the original matrix equation containing all the unknowns in the computational domain (the way in the finite-element time-domain method) is transformed into multiple matrix equations with much smaller system matrices. In other words, a large-scale or multiscale structure can be segmented into multiple subdomains according to its structural characteristics, as shown in Figure 6.3.

6.2.1 Discretized System

The governing equation of the subdomain-level DGTD method are first-order Maxwell's equations in which the electric field intensity **E** and the magnetic field intensity **H** are variables.

$$\varepsilon \frac{\partial \mathbf{E}}{\partial t} + \sigma_e \mathbf{E} - \nabla \times \mathbf{H} = -\mathbf{J}_S \tag{6.1}$$

$$\mu \frac{\partial \mathbf{H}}{\partial t} + \sigma_m \mathbf{H} + \nabla \times \mathbf{E} = -\mathbf{M}_S \tag{6.2}$$

Domain decomposition

Figure 6.3 Schematic of domain decomposition in the subdomain-level DGTD method.

where $\mathbf{J}_S$ and $\mathbf{M}_S$ are electric and magnetic current densities. ε, μ, σ_e, σ_m denote the material's permittivity, permeability, electric conductivity, and magnetic conductivity, respectively.

The Galerkin's weak forms of Maxwell's equations are

$$\int_V \mathbf{\Phi} \cdot \left(\varepsilon \frac{\partial \mathbf{E}}{\partial t} + \sigma_e \mathbf{E} - \nabla \times \mathbf{H} + \mathbf{J}_s \right) dV = 0 \tag{6.3}$$

$$\int_V \mathbf{\Psi} \cdot \left(\mu \frac{\partial \mathbf{H}}{\partial t} + \sigma_m \mathbf{H} + \nabla \times \mathbf{E} + \mathbf{M}_s \right) dV = 0 \tag{6.4}$$

where V denotes the volume of a subdomain, $\mathbf{\Phi}$ and $\mathbf{\Psi}$ denote the testing basis functions. Via integration by parts, we have

$$\int_V \mathbf{\Phi} \cdot \left(\varepsilon \frac{\partial \mathbf{E}}{\partial t} + \sigma_e \mathbf{E} + \mathbf{J}_s \right) dV = \int_V \nabla \times \mathbf{\Phi} \cdot \mathbf{H} dV + \int_S \mathbf{\Phi} \cdot (\mathbf{n} \times \mathbf{H}) dS \tag{6.5}$$

$$\int_V \mathbf{\Psi} \cdot \left(\mu \frac{\partial \mathbf{H}}{\partial t} + \sigma_m \mathbf{H} + \mathbf{M}_s \right) dV = -\int_V \nabla \times \mathbf{\Psi} \cdot \mathbf{E} dV - \int_S \mathbf{\Psi} \cdot (\mathbf{n} \times \mathbf{E}) dS \tag{6.6}$$

If the i-th subdomain is adjacent to the j-th subdomain, the Riemann solver between them reads

$$\left(Z_e^{(i)} + Z_e^{(j)} \right) (\mathbf{n} \times \mathbf{E}) = \mathbf{n} \times \left(Z_e^{(i)} \mathbf{E}^{(i)} + Z_e^{(j)} \mathbf{E}^{(j)} \right) - \mathbf{n} \times \mathbf{n} \times (\mathbf{H}^{(i)} - \mathbf{H}^{(j)}) \tag{6.7}$$

$$\left(Z_h^{(i)} + Z_h^{(j)} \right) (\mathbf{n} \times \mathbf{H}) = \mathbf{n} \times \left(Z_h^{(i)} \mathbf{H}^{(i)} + Z_h^{(j)} \mathbf{H}^{(j)} \right) + \mathbf{n} \times \mathbf{n} \times (\mathbf{E}^{(i)} - \mathbf{E}^{(j)}) \tag{6.8}$$

where

$$Z_h^{(i)} = 1/Z_e^{(i)} = \sqrt{\mu^{(i)}/\varepsilon^{(i)}}, \quad Z_h^{(j)} = 1/Z_e^{(j)} = \sqrt{\mu^{(j)}/\varepsilon^{(j)}} \tag{6.9}$$

The large or multi-scale structures can be divided into N subdomains to enhance the calculation efficiency. The discretized system of the subdomain-level DGTD method is

$$\mathbf{M}_{ee}^{(i)} \frac{d\mathbf{e}^{(i)}}{dt} = \mathbf{K}_{eh}^{(i)} \mathbf{h}^{(i)} + \mathbf{C}_{ee}^{(i)} \mathbf{e}^{(i)} + \mathbf{j}^{(i)} + \sum_{j=1}^N \left(\mathbf{L}_{ee}^{(ij)} \mathbf{e}^{(j)} + \mathbf{L}_{eh}^{(ij)} \mathbf{h}^{(j)} \right), \quad i = 1, \cdots, N \tag{6.10}$$

$$\mathbf{M}_{hh}^{(i)} \frac{d\mathbf{h}^{(i)}}{dt} = \mathbf{K}_{he}^{(i)} \mathbf{e}^{(i)} + \mathbf{C}_{hh}^{(i)} \mathbf{h}^{(i)} + \mathbf{m}^{(i)} + \sum_{i=1}^N \left(\mathbf{L}_{he}^{(ij)} \mathbf{e}^{(j)} + \mathbf{L}_{hh}^{(ij)} \mathbf{h}^{(j)} \right), \quad i = 1, \cdots, N \tag{6.11}$$

where $\mathbf{e}^{(i)}$ and $\mathbf{h}^{(i)}$ are the coefficients of the expanding basis functions for the electric and magnetic field intensities. $\mathbf{M}_{ee}^{(i)}$ and $\mathbf{M}_{hh}^{(i)}$ are the mass matrices, while $\mathbf{K}_{eh}^{(i)}$ and $\mathbf{K}_{he}^{(i)}$ are the stiffness matrices, and $\mathbf{C}_{ee}^{(i)}$ and $\mathbf{C}_{hh}^{(i)}$ are the damping matrices. $\mathbf{j}^{(i)}$ and $\mathbf{m}^{(i)}$ denote the excitation source. The coupling matrices $\mathbf{L}_{ee}^{(ij)}$, $\mathbf{L}_{eh}^{(ij)}$, $\mathbf{L}_{he}^{(ij)}$, and $\mathbf{L}_{hh}^{(ij)}$ depend on the properties of both sides across the shared interface.

6.2.2 Time Stepping Schemes

In the subdomain-level DGTD method, the computational burden in inversion or factorization of system matrices is alleviated by decomposing a big system of matrix equations into a bunch of smaller systems of matrix equations. The fourth-order explicit Runge–Kutta (Ex-RK) method is employed for the subdomain-level DGTD method for the time integration of each subdomain. For convenience, the discrete systems of i-th subdomain can be described as

$$\mathbf{M}^{(i)}\frac{d\mathbf{v}^{(i)}}{dt} = \mathbf{G}^{(i)}\mathbf{v}^{(i)} + \mathbf{f}^{(i)} + \sum_{j=1}^{N}\mathbf{L}^{(ij)}\mathbf{v}^{(j)}, \qquad i = 1, \cdots, N \tag{6.12}$$

where the mass, stiffness, damping matrices, coupling matrices, source, and unknown vectors can be further split into the electrical and magnetic components:

$$\mathbf{M}^{(i)} = \begin{bmatrix} \mathbf{M}_{ee}^{(i)} & 0 \\ 0 & \mathbf{M}_{hh}^{(i)} \end{bmatrix}, \mathbf{G}^{(p)} = \begin{bmatrix} \mathbf{C}_{ee}^{(i)} & \mathbf{K}_{eh}^{(i)} \\ \mathbf{K}_{he}^{(i)} & \mathbf{C}_{hh}^{(i)} \end{bmatrix} \tag{6.13}$$

$$\mathbf{v}^{(i)} = \begin{bmatrix} \mathbf{e}^{(i)} \\ \mathbf{h}^{(i)} \end{bmatrix}, \mathbf{f}^{(p,q)} = \begin{bmatrix} \mathbf{j}^{(i)} \\ \mathbf{m}^{(i)} \end{bmatrix} \tag{6.14}$$

$$\mathbf{L}^{(ii)} = \begin{bmatrix} \mathbf{C}_{ee}^{(i)} + \mathbf{L}_{ee}^{(ii)} & \mathbf{K}_{eh}^{(i)} + \mathbf{L}_{eh}^{(ii)} \\ \mathbf{K}_{he}^{(i)} + \mathbf{L}_{he}^{(ii)} & \mathbf{C}_{hh}^{(i)} + \mathbf{L}_{hh}^{(ii)} \end{bmatrix}, \quad \mathbf{L}^{(ij)} = \begin{bmatrix} \mathbf{L}_{ee}^{(ij)} & \mathbf{L}_{eh}^{(ij)} \\ \mathbf{L}_{he}^{(ij)} & \mathbf{L}_{hh}^{(ij)} \end{bmatrix}, \text{for } i \neq j \tag{6.15}$$

If we have all the field values of the n-th time step, the solution of the $(n+1)$-th time step of the i-th subdomain can be obtained by the s-order Runge–Kutta method as

$$\mathbf{v}_{n+1}^{(i)} = \mathbf{v}_n^{(i)} + \Delta t\sum_{k=1}^{s}b_k\mathbf{u}_k^{(i)} \tag{6.16}$$

$$\mathbf{M}^{(i)}\mathbf{u}_k^{(i)} = \sum_{j=1}^{N}\mathbf{L}^{(ij)}\left(\mathbf{v}_n^{(j)} + \Delta t\sum_{l=1}^{k-1}a_{k,l}\mathbf{u}_l^{(j)}\right) + \mathbf{f}^{(i)}(t_n + c_k\Delta t) \tag{6.17}$$

where coefficients $a_{k,l}$, b_k, and c_k are from the Butcher tableau of the Runge–Kutta scheme, Δt is the time interval of time stepping. In the time integration process, only the mass matrix $\mathbf{M}^{(i)}$ needs to be inverted or decomposed. This step can be effectively performed by LU decomposition.

6.3 Memory-Efficient DGTD Method for Periodic Structures

Chen proposed a memory-efficient DGTD method for the first time and successfully employed it for photonic crystal waveguide modeling. Compared with the conventional subdomain-level DGTD method, it significantly reduces memory consumption [52]. In the memory-efficient DGTD method, the simulation structures are divided into multiple subdomains (also referred to as "cells") with identical geometry, as shown in Figure 6.4. Only one cell requires discretization, and its mesh can be duplicated for all other cells. Meanwhile, the nonconformal mesh guarantees the flexibility of the method. In this way, only the volumetric system matrices of one cell have to be stored in memory instead of the matrices of all the cells as in the conventional subdomain-level DGTD method. This will reduce memory consumption significantly, i.e. the memory consumption linearly increases with the number of the cell types instead of the number of the cells.

However, the memory-efficient DGTD method in [52] uses the Ex-RK scheme in time stepping, which is constrained by the Courant–Friederichs–Lewy (CFL) condition. Especially for multiscale and/or electrically small problems where the electrically fine structures exist, an extremely small time interval will result in an unaffordable number of steps in time integration. Therefore, the implicit Runge–Kutta (Im-RK) time stepping scheme is introduced into the memory-efficient DGTD method to break the limitation of the CFL condition. Only the electrically small subdomains with fine mesh will adopt the Im-RK time stepping method, while the electrically large subdomains with coarse mesh will be solved by the Ex-RK time stepping method to obtain a leverage between time and memory costs. The block Gauss–Seidel algorithm is used for iterations among adjacent implicit subdomains during time stepping. The proposed method can model multiscale finite periodic arrays with high accuracy and efficiency.

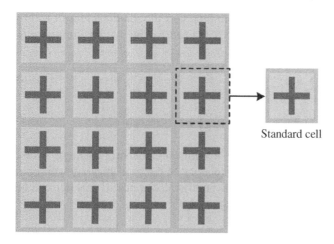

Standard cell

Figure 6.4 Typical finite period structures.

6.3.1 Discretized System

6.3.1.1 Discretized System of Periodic Structures

The memory-efficient DGTD method is applied to simulate finite periodic structures. The volume and surface integrations for each cell need to be separated in this method. The mass matrix ($\mathbf{M}$), stiffness matrix ($\mathbf{K}$), and damping matrix ($\mathbf{C}$) are the same for every cell in the structure; thus they need to be stored only once regardless of the array size. But the boundary condition of a certain cell can be classified into three types: (i) an interface between two cells; (ii) an interface between a cell and a conventional subdomain (such as perfect matched layer (PML)); and (iii) PEC or perfect magnetic conductor (PMC). Thus, the coupling matrices $\mathbf{L}_{ee}^{(ij)}$, $\mathbf{L}_{eh}^{(ij)}$, $\mathbf{L}_{he}^{(ij)}$, and $\mathbf{L}_{hh}^{(ij)}$ of each cell depend on both sides of the interface.

According to the location, the boundary surfaces of a cell can be classified into six scenarios, namely, $x-$, $x+$, $y-$, $y+$, $z-$, and $z+$, as illustrated in Figure 6.5. The matrices of the discretized system in the memory-efficient DGTD method are as follows:

$$\mathbf{M}_{ee}^{(i)}\frac{d\mathbf{e}^{(i)}}{dt} = \mathbf{K}_{eh}^{(i)}\mathbf{h}^{(i)} + \mathbf{C}_{ee}^{(i)}\mathbf{e}^{(i)} + \mathbf{j}^{(i)} + \sum_{j=1}^{N}\left[\sum_{s=1}^{6}\left(\mathbf{L}_{ee}^{(ij)(s)}\mathbf{e}^{(j)(s)} + \mathbf{L}_{eh}^{(ij)(s)}\mathbf{h}^{(j)(s)}\right)\right]$$

(6.18)

$$\mathbf{M}_{hh}^{(i)}\frac{d\mathbf{h}^{(i)}}{dt} = \mathbf{K}_{he}^{(i)}\mathbf{e}^{(i)} + \mathbf{C}_{hh}^{(i)}\mathbf{h}^{(i)} + \mathbf{m}^{(i)} + \sum_{j=1}^{N}\left[\sum_{s=1}^{6}\left(\mathbf{L}_{he}^{(ij)(s)}\mathbf{e}^{(j)(s)} + \mathbf{L}_{hh}^{(ij)(s)}\mathbf{h}^{(j)(s)}\right)\right]$$

(6.19)

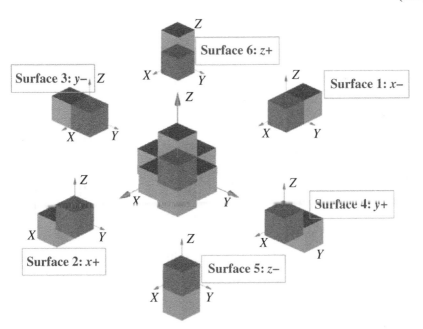

Figure 6.5 One volume and six surface integrations of a cell.

where $i = 1, 2, \cdots, N$, and the matrices $\mathbf{L}_{ee}^{(ij)(s)}$ denote the coupling of the electric field intensity of two adjacent subdomains on the s surface, and so forth for all other coupling matrices. The definitions of system matrices are the same as those in the previous section.

6.3.1.2 Discretized System of Embedded Periodic Structures

To improve the computing efficiency of a finite array with multiscale components, the fine structures within the unit cell can be isolated as a new unit cell (referred to as the internal unit cell) and the remaining parts of the unit cell are referred to as the external unit cell as shown in Figure 6.6. Assume there are N subdomains in the embedded periodic structures, including N_B for the buffer zone, N_P for the PML, and C for the periodic array including C_{ext} external unit cells and C_{int} internal unit cells. All the subdomains are numbered as $1, 2, ..., C_{ext}, C_{ext} + 1,$ $..., C, C + 1, ..., N$. The discretized system equation of the embedded structure is expressed as follows:

$$
\begin{bmatrix} \mathbf{M}^{(1)} & & & & & \\ & \ddots & & & & \\ & & \mathbf{M}^{(1)} & & & \\ & & & \ddots & & \\ & & & & \mathbf{M}^{(C_{ext})} & \\ & & & & & \ddots \\ & & & & & & \mathbf{M}^{(C_{ext})} \end{bmatrix} \frac{d}{dt} \begin{bmatrix} \mathbf{v}^{(1)} \\ \vdots \\ \mathbf{v}^{(C_{ext})} \\ \mathbf{v}^{(C_{ext}+1)} \\ \vdots \\ \mathbf{v}^{(C)} \end{bmatrix} = \begin{bmatrix} \mathbf{f}^{(1)} \\ \vdots \\ \mathbf{f}^{(C_{ext})} \\ \mathbf{f}^{(C_{ext}+1)} \\ \vdots \\ \mathbf{f}^{(C)} \end{bmatrix}
$$

$$
+ \left(\begin{bmatrix} \mathbf{G}^{(1)} & & & & & \\ & \ddots & & & & \\ & & \mathbf{G}^{(1)} & & & \\ & & & \ddots & & \\ & & & & \mathbf{G}^{(C_{ext})} & \\ & & & & & \ddots \\ & & & & & & \mathbf{G}^{(C_{ext})} \end{bmatrix} \right.
$$

$$
\left. + \begin{bmatrix} \mathbf{L}^{(1,1)} & \cdots & \mathbf{L}^{(1,C_{ext})} & \mathbf{L}^{(1,C_{ext}+1)} & \cdots & 0 \\ \vdots & \ddots & \vdots & \vdots & \ddots & \vdots \\ \mathbf{L}^{(C_{ext},1)} & \cdots & \mathbf{L}^{(C_{ext},C_{ext})} & 0 & \cdots & \mathbf{L}^{(C,C_{ext})} \\ \mathbf{L}^{(C_{ext}+1,1)} & \cdots & 0 & \mathbf{L}^{(C_{ext}+1,C_{ext}+1)} & \cdots & 0 \\ \vdots & \ddots & \vdots & \vdots & \ddots & \vdots \\ 0 & \cdots & \mathbf{L}^{(C,C_{ext})} & 0 & \cdots & \mathbf{L}^{(C,C)} \end{bmatrix} \right) \begin{bmatrix} \mathbf{v}^{(1)} \\ \vdots \\ \mathbf{v}^{(C_{ext})} \\ \mathbf{v}^{(C_{ext}+1)} \\ \vdots \\ \mathbf{v}^{(C)} \end{bmatrix}
$$

$$\tag{6.20}$$

where $N = C + N_B + N_P$, $C = C_{ext} + C_{int}$, the specific definitions of other matrices are the same as above equation.

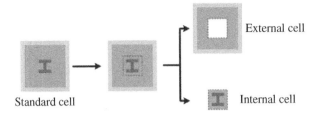

Figure 6.6 Embedded model of a cell in the periodic structure.

6.3.2 Time Stepping Schemes

The process of solving explicit subdomains by factorizing mass matrix $\mathbf{M}^{(i)}$ is exactly the same as that in the Ex-RK time stepping scheme for the conventional subdomain-level DGTD method. However, for the small size part of the multiscale structure, the implicit time stepping method will be adopted, which is introduced in detail in what follows.

The fourth-order ImEx-RK scheme is employed for the memory-efficient DGTD method. Assuming that all the field values of the i-th cell for the n-th time step are known, the field values of the i-th cell for the $(n+1)$-th time step can be calculated using the ImEx-RK scheme with s ($s = 4$ in this chapter) stages.

$$\mathbf{v}_{n+1}^{(i)} = \mathbf{v}_n^{(i)} + \Delta t \sum_{k=1}^{s} b_k \mathbf{u}_k^{(i)}, i = 1, 2, \cdots, N \tag{6.21}$$

where

$$N = C + N_B + N_P, \quad C = C_{\text{ext}} + C_{\text{int}} \tag{6.22}$$

$$\mathbf{M}^{(i)} \mathbf{u}_k^{(i)} = \sum_{j=1}^{C} \mathbf{L}^{(ij)} \left(\mathbf{v}_n^{(j)} + \Delta t \sum_{l=1}^{k} a_{k,l}^{\text{im}} \mathbf{u}_l^{(j)} \right)$$

$$+ \sum_{j=C+1}^{N} \mathbf{L}^{(ij)} \left(\mathbf{v}_n^{(j)} + \Delta t \sum_{l=1}^{k-1} a_{k,l}^{\text{ex}} \mathbf{u}_l^{(j)} \right) + \mathbf{f}^{(i)}(t_n + c_k \Delta t) \tag{6.23}$$

and the coefficients $a_{k,l}^{\text{ex}}$, $a_{k,l}^{\text{im}}$, b_k, and c_k are from the Butcher tableau of the Runge–Kutta scheme. The above formula can be shortened to the following form:

$$\mathbf{R}\mathbf{u}_k = \mathbf{q}_{\text{im}} \tag{6.24}$$

For the external cells, the coupling matrix needs to be classified into three types: the buffer zone, adjacent external cells, and the internal cell inside it. However, the internal cells are only adjacent to the surrounding external cells. The detailed expressions of the matrices are:

$$\mathbf{R} = \mathbf{P} - \Delta t a_{k,k}^{\text{im}} \mathbf{D} \tag{6.25}$$

$$\mathbf{q}_{\text{im}} = \mathbf{D} \begin{bmatrix} \mathbf{r}^{(1)} \\ \vdots \\ \mathbf{r}^{(C_{\text{ext}})} \\ \mathbf{r}^{(C_{\text{ext}}+1)} \\ \vdots \\ \mathbf{r}^{(C)} \end{bmatrix} + \begin{bmatrix} \mathbf{q}^{(1)} \\ \vdots \\ \mathbf{q}^{(C_{\text{ext}})} \\ \mathbf{q}^{(C_{\text{ext}}+1)} \\ \vdots \\ \mathbf{q}^{(C)} \end{bmatrix} \tag{6.26}$$

$$\mathbf{u}_k = \begin{bmatrix} \mathbf{u}_k^{(1)} & \cdots & \mathbf{u}_k^{(C_{\text{ext}})} & \mathbf{u}_k^{(C_{\text{ext}}+1)} & \cdots & \mathbf{u}_k^{(C)} \end{bmatrix}^T \tag{6.27}$$

where $\mathbf{P}$ and $\mathbf{D}$ have $C \times C$ blocks.

$$\mathbf{P} = \begin{bmatrix} \mathbf{M}^{(1)} & & & & & \\ & \ddots & & & & \\ & & \mathbf{M}^{(1)} & & & \\ & & & \ddots & & \\ & & & & \mathbf{M}^{(C_{\text{ext}})} & \\ & & & & & \ddots \\ & & & & & & \mathbf{M}^{(C_{\text{ext}})} \end{bmatrix} \tag{6.28}$$

$$\mathbf{D} = \begin{bmatrix} \mathbf{L}^{(1,1)} & \cdots & \mathbf{L}^{(1,C_{\text{ext}})} & \mathbf{L}^{(1,C_{\text{ext}}+1)} & \cdots & 0 \\ \vdots & \ddots & \vdots & \vdots & \ddots & \vdots \\ \mathbf{L}^{(C_{\text{ext}},1)} & \cdots & \mathbf{L}^{(C_{\text{ext}},C_{\text{ext}})} & 0 & \cdots & \mathbf{L}^{(C,C_{\text{ext}})} \\ \mathbf{L}^{(C_{\text{ext}}+1,1)} & \cdots & 0 & \mathbf{L}^{(C_{\text{ext}}+1,C_{\text{ext}}+1)} & \cdots & 0 \\ \vdots & \ddots & \vdots & \vdots & \ddots & \vdots \\ 0 & \cdots & \mathbf{L}^{(C,C_{\text{ext}})} & 0 & \cdots & \mathbf{L}^{(C,C)} \end{bmatrix} \tag{6.29}$$

$$\mathbf{r}^{(i)} = \mathbf{v}_n^{(i)} + \Delta t \sum_{l=1}^{k-1} a_{k,l}^{\text{im}} \mathbf{u}_l^{(i)} \tag{6.30}$$

$$\mathbf{q}^{(i)} = \sum_{j=C+1}^{N} \mathbf{L}^{(ij)} \left(\mathbf{v}_n^{(j)} + \Delta t \sum_{l=1}^{k-1} a_{k,l}^{\text{ex}} \mathbf{u}_l^{(j)} \right) + \mathbf{f}^{(i)}(t_n + c_k \Delta t) \qquad i = 1, 2, \cdots, C \tag{6.31}$$

If the entire system is directly solved, the simulation efficiency will be greatly reduced and the region decomposition, and also, memory advantages of the memory-efficient DGTD method will be lost. The block Gauss–Seidel method, which is a subdomain-level iterative solver, is employed for time integration of the memory-efficient DGTD method with the ImEx-RK scheme. The pseudo code of the block Gauss–Seidel method can be expressed as:

$$
\begin{cases}
\text{while (result does not converge)} \\
\quad \text{for } i = 1 : N \\
\qquad \text{for } i = 1 : C \\
\qquad\quad \tilde{\mathbf{q}}_{\text{im}}^{(i)} = \mathbf{q}_{\text{im}}^{(i)} \\
\qquad\quad \text{for } j = 1 : i - 1 \\
\qquad\qquad \tilde{\mathbf{q}}_{\text{im}}^{(i)} = \tilde{\mathbf{q}}_{\text{im}}^{(i)} + \Delta t a_{k,k}^{\text{im}} \mathbf{L}^{(ij)} \mathbf{u}_k^{(j)} \\
\qquad\quad \text{end} \\
\qquad\quad \text{for } j = i + 1 : C \\
\qquad\qquad \tilde{\mathbf{q}}_{\text{im}}^{(i)} = \tilde{\mathbf{q}}_{\text{im}}^{(i)} + \Delta t a_{k,k}^{\text{im}} \mathbf{L}^{(ij)} \mathbf{u}_k^{(j)} \\
\qquad\quad \text{end} \\
\qquad\quad \text{solve } \left(\mathbf{P}^{(i,i)} - \Delta t a_{k,k}^{\text{im}} \mathbf{L}^{(ii)} \right) \mathbf{u}_k^{(i)} = \tilde{\mathbf{q}}_{\text{im}}^{(i)} \\
\qquad \text{end} \\
\qquad \text{for } i = C + 1 : N \\
\qquad\quad \tilde{\mathbf{q}}^{(i)} = \mathbf{q}^{(i)} \\
\qquad\quad \text{solve } \mathbf{M}^{(i)} \mathbf{u}_k^{(i)} = \tilde{\mathbf{q}}^{(i)} \\
\qquad \text{end} \\
\quad \text{end} \\
\quad \text{Check if the result reaches convergence} \\
\text{end}
\end{cases}
\tag{6.32}
$$

where vector $\mathbf{q}_{\text{im}}^{(i)}$ stands for the i-th block in $\mathbf{q}_{\text{im}}$, and the submatrix $\mathbf{P}^{(i,\,i)}$ represents a block in the i-th (block) row and j-th (block) column of matrix $\mathbf{P}$.

6.3.3 Numerical Results

6.3.3.1 PEC Cavity with Periodic Structures

As shown in Figure 6.7, six PEC cavities with different sizes are cut into cubic cells, which can be regarded as periodic structures. The edge length of each cell is 0.6 m. The number of cells in the six PEC cavities are 16, 32, 48, 64, 80, and 96.

In this case, we chose the Blackman–Harris window (BHW) pulse with a central frequency of 0.3 GHz as the excitation signal. To improve the simulation efficiency, electric and magnetic fields adopted the 4th-order and 3rd-order basis functions, respectively. The x component of the electric field intensity was recorded in the point receiver fixed at (2.1, 0.8, 0.8) m. The results of the memory-efficient DGTD method compared with the FDTD method and the traditional subdomain-level DGTD method are shown in Figure 6.8. In Figures 6.8, 6.9, and 6.15, Tables 6.1–6.3,

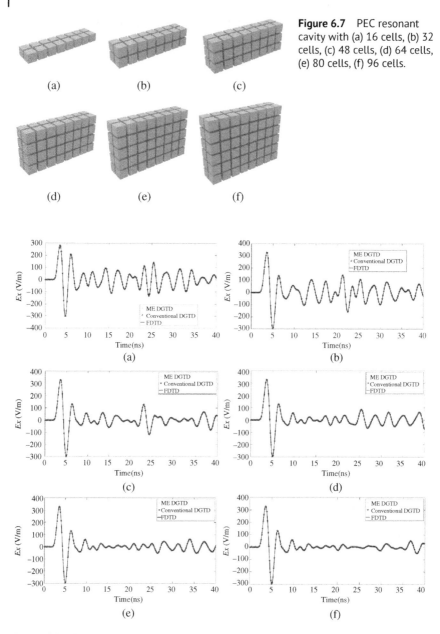

Figure 6.7 PEC resonant cavity with (a) 16 cells, (b) 32 cells, (c) 48 cells, (d) 64 cells, (e) 80 cells, (f) 96 cells.

Figure 6.8 The *x* component of the electric field recorded at the receiver for the cavities with (a) 16 cells, (b) 32 cells, (c) 48 cells, (d) 64 cells, (e) 80 cells, and (f) 96 cells.

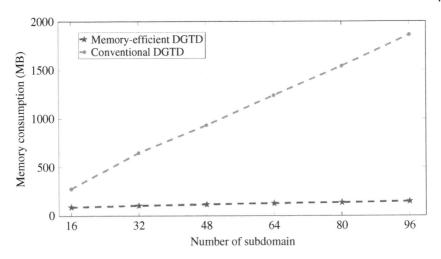

Figure 6.9 Memory consumption for the cavities with different number of cells.

Table 6.1 Simulation time and memory consumption by the memory-efficient DGTD and conventional subdomain-level DGTD methods for six cavities with different number of cells.

Scheme	Number of cells	Simulation time (ns)	Memory consumption (MB)
Conventional DGTD	16	157.71	278.0
	32	333.24	64S.0
	48	534.46	930.6
	64	808.43	1239.8
	80	996.92	1538.8
	96	1003.24	1858.3
Memory-efficient DGTD	16	190.25	90.3
	32	405.70	106.5
	48	610.56	118.4
	64	852.51	129.8
	00	1092.6/	13/.2
	96	1352.00	148.9

Table 6.2 Simulation time and memory used by conventional subdomain-level DGTD, memory-efficient subdomain-level DGTD, FDTD methods, and CST for the periodic patch antenna array.

Scheme	Simulation time (h)	Memory consumption (GB)
Conventional DGTD	24.25	12.94
Memory-efficient DGTD	23.53	2.74
FDTD	10.09	33.25
CST	6.77	35.06

Source: Wen et al. [53]. Reproduced with permission of IEEE.

Table 6.3 Resource consumption of the conventional subdomain-level DGTD method with Ex-RK and ImEx-RK schemes, and the memory-efficient DGTD method with Ex-RK and ImEx-RK schemes.

Method	Time stepping	Δt (ps)	Memory (GB)	CPU time (h)
Conventional DGTD	Ex-RK	0.1	10.01	21.09
	ImEx-RK	2	25.31	6.34
Memory-efficient DGTD	Ex-RK	0.1	1.23	22.26
	ImEx-RK	2	3.63	5.21

the results from the conventional subdomain-level DGTD method will be labeled as *conventional DGTD* for short.

The simulation time and memory consumption are listed in Table 6.1. As the cavity size becomes larger, the simulation time increases. Because the time integration scheme for both the conventional subdomain-level DGTD and memory-efficient DGTD methods is Ex-RK, the overall simulation time should be close. The only advantage of the memory-efficient DGTD method in terms of simulation time (compared to the conventional subdomain-level DGTD method) is that it is a little faster in the preprocessing part. For the conventional subdomain-level DGTD method, the memory increases linearly with the cavity size (i.e. 16 cells, 32 cells, 48 cells, 64 cells, 80 cells, and 96 cells) as exhibited in Figure 6.9. However, the memory-efficient DGTD method is insensitive to the problem size since it only stores one set of volumetric system matrices. This advantage is tremendous when the cell number is very large, such as 148.9 MB versus 1858.3 MB for 96 cells.

6.3.3.2 Periodic Patch Antenna Arrays
The electromagnetic coupling of a patch antenna array can be studied by the proposed method. There are 16 same patch antenna cells in this array. As shown in Figure 6.10a, the top and bottom are the metal and the middle layer is the substrate

with a permittivity of 3.4, respectively. In Figure 6.10b,c, the specific dimensions of the example are described in detail.

The central frequency of the BHW pulse is 3.875 GHz for the lumped voltage source. This array has 16 lumped ports, one of which is the active source in the lower-left corner. To obtain higher precision, the lumped ports with drastic field changes use fine mesh. Consequently, this leads to huge memory consumption during the simulation process if the conformal mesh is used. To handle this problem, one unit cell is split into an internal cell, which is meshed with dense elements for the lumped port part, and an external cell, which is meshed relatively sparser using high-order elements. The internal and the external cells are linked with nonconformal interfaces; thus the total memory consumption will be reduced.

The scattering parameters are calculated by the memory-efficient DGTD method from 20,000 times records with a time step interval of 0.1 ps. By comparing the results of the FDTD method and Computer Simulation Technology (CST) in Figure 6.11, it can be concluded that the memory-efficient DGTD method has good accuracy. Meanwhile, we can see that the proposed method has obvious advantages in memory consumption compared with the conventional subdomain-level DGTD method, FDTD method and CST. The overall memory consumption of the memory-efficient DGTD method is less than that of the FDTD method and CST due to the use of high-order basis functions. For simulation

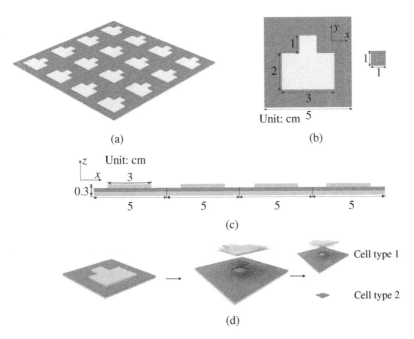

Figure 6.10 Patch antenna array. (a) Overview. (b) Unit cell. (c) Side view of the antenna array. (d) One cell is split into an external cell and an internal cell. Source: Wen et al. [53]. Reproduced with permission of IEEE.

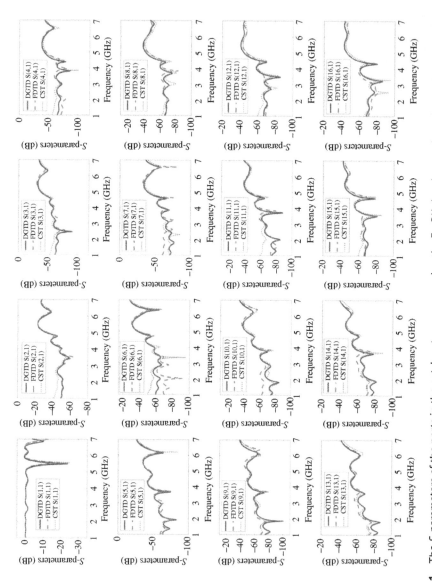

Figure 6.11 The S-parameters of the ports in the patch antenna array using the Ex-RK time integration scheme. Source: Wen et al. [53]. Reproduced with permission of IEEE.

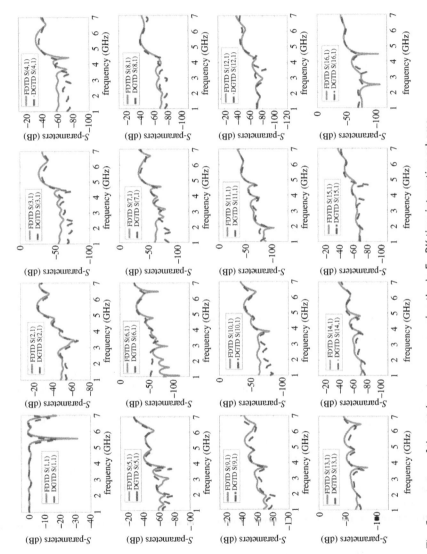

Figure 6.12 The S-parameters of the patch antenna array using the ImEx-RK time integration scheme.

time, both CST and the FDTD method adopt parallel computing techniques, so they are faster than the memory-efficient DGTD method, which is only sequential in the present version (Table 6.2).

The above patch antenna case is also studied using the ImEx-RK time stepping method, in which the Im-RK and Ex-RK time stepping schemes are employed to the patch antenna array and other areas (PML and buffer zone), respectively. Note that in this study, the mesh is slightly different from the abovementioned mesh.

The comparison between the conventional DGTD method with the Ex-RK scheme and ImEx-RK schemes, the memory-efficient DGTD method with the Ex-RK scheme and the ImEx-RK scheme for this case is performed. The effectiveness of the first three configurations has already been validated in previous research and will not be analyzed here. The S-parameters of this array are calculated and compared to the reference results from the FDTD method as illustrated in Figure 6.12, in which good agreement is achieved. As can be observed from Table 6.3, the time step of the ImEx-RK scheme is much larger than that used in the Ex-RK scheme, so the ImEx-RK scheme can shorten the simulation time significantly. Meanwhile, the proposed DGTD method consumes much less memory compared to the conventional DGTD method. In consequence, the memory-efficient DGTD method based on the ImEx-RK time stepping scheme can effectively reduce the amount of computational resources (in terms of both time and memory) required during the simulation.

6.4 Memory-Efficient DGTD Method for Quasi-Periodic Structures

Compared to the conventional DGTD method, the memory-efficient DGTD method can reduce the memory consumption by using the nature of repetition of periodic/quasi-periodic structures. However, this method can analyze only the arrays with a single type of periodic cells. In this section, the memory-efficient DGTD method has been extended to a quasi-periodic scenario, in which multiple types of cells can be arranged almost arbitrarily in the array. This extended version of the memory-efficient subdomain-level DGTD method for quasi-periodic arrays also incorporate implicit time integration schemes and the ability of handling multiscale cases. Through these improvements, the application scope of the algorithm is wider.

6.4.1 Discretized System

6.4.1.1 Discretized System of Quasi-Periodic Structures
It can be observed from Figure 6.13 that the cells in a quasi-periodic array may be different in shapes, sizes, materials, or even meshes of the scatterers

Figure 6.13 Example of the quasi-periodic structure.

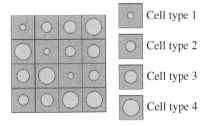

o Cell type 1

Cell type 2

Cell type 3

Cell type 4

(or the drive elements). These cells can be arranged randomly in the lattice of the array. Assume the quasi-periodic array is composed of C types of cells, and the number of each cell is $n_1, n_2, \ldots, n_C$, respectively. Therefore, the periodic array has $N = n_1 + n_2 + n_3 + \ldots + n_c$ cells in total. The discretized system of the quasi-periodic array can be expressed:

$$
\begin{bmatrix} \mathbf{M}^{(1)} \\ & \ddots \\ & & \mathbf{M}^{(1)} \\ & & & \ddots \\ & & & & \mathbf{M}^{(C)} \\ & & & & & \ddots \\ & & & & & & \mathbf{M}^{(C)} \end{bmatrix} \frac{d}{dt} \begin{bmatrix} \mathbf{v}^{(1,1)} \\ \vdots \\ \mathbf{v}^{(1,n_1)} \\ \vdots \\ \mathbf{v}^{(C,1)} \\ \vdots \\ \mathbf{v}^{(C,n_c)} \end{bmatrix} = \begin{bmatrix} \mathbf{f}^{(1,1)} \\ \vdots \\ \mathbf{f}^{(1,n_1)} \\ \vdots \\ \mathbf{f}^{(C,1)} \\ \vdots \\ \mathbf{f}^{(C,n_c)} \end{bmatrix}
$$

$$
+ \begin{bmatrix} \mathbf{L}^{(1,1)}(P^{(1,1)}\mathbf{v}) \\ \vdots \\ \mathbf{L}^{(1,n_1)}\left(P^{(1,n_1)}\mathbf{v}\right) \\ \vdots \\ \mathbf{L}^{(C,1)}(P^{(1,1)}\mathbf{v}) \\ \vdots \\ \mathbf{L}^{(C,n_c)}\left(P^{(1,n_c)}\mathbf{v}\right) \end{bmatrix} + \begin{bmatrix} \mathbf{G}^{(1)} \\ & \ddots \\ & & \mathbf{G}^{(1)} \\ & & & \ddots \\ & & & & \mathbf{G}^{(C)} \\ & & & & & \ddots \\ & & & & & & \mathbf{G}^{(C)} \end{bmatrix} \begin{bmatrix} \mathbf{v}^{(1,1)} \\ \vdots \\ \mathbf{v}^{(1,n_1)} \\ \vdots \\ \mathbf{v}^{(C,1)} \\ \vdots \\ \mathbf{v}^{(C,n_c)} \end{bmatrix}
$$

$$(6.33)$$

where

$$
\mathbf{v} = \begin{bmatrix} \mathbf{v}(1,1)^T & \cdots & \mathbf{v}(1,n_1)^T & \cdots & \mathbf{v}(p,q)^T & \cdots & \mathbf{v}(c,1)^T \cdots & \mathbf{v}(c,n_c)^T \end{bmatrix}^T
$$

$$(6.34)$$

The variables (mass, stiffness, damping matrices, source unknown vectors, and coupling) in the above formula can be expressed as:

$$
\mathbf{M}^{(p)} = \begin{bmatrix} \mathbf{M}_{ee}^p & 0 \\ 0 & \mathbf{M}_{hh}^p \end{bmatrix}, \quad \mathbf{G}^{(p)} = \begin{bmatrix} \mathbf{C}_{ee}^p & \mathbf{K}_{eh}^p \\ \mathbf{K}_{he}^p & \mathbf{C}_{hh}^p \end{bmatrix} \tag{6.35}
$$

$$
\mathbf{v}^{(p,q)} = \begin{bmatrix} \mathbf{e}^{(p,q)} \\ \mathbf{h}^{(p,q)} \end{bmatrix}, \quad \mathbf{f}^{(p,q)} = \begin{bmatrix} \mathbf{j}^{(p,q)} \\ \mathbf{m}^{(p,q)} \end{bmatrix} \tag{6.36}
$$

The coupling matrix $\mathbf{L}^{(p,q)}$ consists of seven blocks. The self-coupling effect is in the first block, where the remaining six blocks are from the effect of neighboring coupling. The details are as follows:

$$\mathbf{L}^{(p,q)} = \begin{bmatrix} \mathbf{L}_s^{(p,q)} & \mathbf{L}_n^{(p,q),1} & \cdots & \mathbf{L}_n^{(p,q),s} & \cdots & \mathbf{L}_n^{(p,q),6} \end{bmatrix} \tag{6.37}$$

$$\mathbf{L}_s^{(p,q)} = \sum_{s=1}^{6} \begin{bmatrix} \mathbf{L}_{ee}^{(p,q),s} & \mathbf{L}_{eh}^{(p,q),s} \\ \mathbf{L}_{he}^{(p,q),s} & \mathbf{L}_{hh}^{(p,q),s} \end{bmatrix}, \quad \mathbf{L}_n^{(p,q),s} = \begin{bmatrix} \mathbf{L}_{ee}^{(p,q),s} & \mathbf{L}_{eh}^{(p,q),s} \\ \mathbf{L}_{he}^{(p,q),s} & \mathbf{L}_{hh}^{(p,q),s} \end{bmatrix} \tag{6.38}$$

where $s = 1, 2, \cdots, 6, \quad p = 1, 2, \cdots, C, \quad q = 1, 2, \cdots, n_p.$

6.4.1.2 Discretized System of Embedded Structures

Sometimes, periodic/quasi-periodic structures have the multiscale property, which hinders efficient electromagnetic simulation. To resolve this problem, we divided each multiscale cell into two subcells (the external cell and the internal cell). Usually, fine geometries of the cell can be all included in the internal cell, while the external cell contains the normal and/or coarse geometries. This divided cell can be referred to as an embedded structure, and its discretized system equation is shown in (6.39).

$$\begin{bmatrix} \mathbf{M}_{ex} & \\ & \mathbf{M}_{in} \end{bmatrix} \frac{d}{dt} \begin{bmatrix} \mathbf{v}_{ex} \\ \mathbf{v}_{in} \end{bmatrix} = \begin{bmatrix} \mathbf{f}_{ex} \\ \mathbf{f}_{in} \end{bmatrix} + \begin{bmatrix} \mathbf{G}_{ex} & \\ & \mathbf{G}_{in} \end{bmatrix} \begin{bmatrix} \mathbf{v}_{ex} \\ \mathbf{v}_{in} \end{bmatrix}$$

$$+ \begin{bmatrix} \mathbf{L}_{ex}^{(1,1)}\mathbf{P}_{ex}^{(1,1)}\mathbf{v}_{ex} + \mathbf{L}_{ex,\,in}^{(1,1)}\mathbf{P}_{ex,\,in}^{(1,1)}\mathbf{v}_{in} \\ \vdots \\ \mathbf{L}_{ex}^{(c,n_c)}\mathbf{P}_{ex}^{(c,n_c)}\mathbf{v}_{ex} + \mathbf{L}_{ex,in}^{(c,n_c)}\mathbf{P}_{ex,in}^{(c,n_c)}\mathbf{v}_{in} \\ \mathbf{L}_{in,\,ex}^{(1,1)}\mathbf{P}_{in,\,ex}^{(1,1)}\mathbf{v}_{ex} + \mathbf{L}_{in}^{(1,1)}\mathbf{P}_{in}^{(1,1)}\mathbf{v}_{in} \\ \vdots \\ \mathbf{L}_{in,\,ex}^{(d,n_d)}\mathbf{P}_{in,\,ex}^{(d,n_d)}\mathbf{v}_{ex} + \mathbf{L}_{in}^{(d,n_d)}\mathbf{P}_{in}^{(d,n_d)}\mathbf{v}_{in} \end{bmatrix} \tag{6.39}$$

where the numbers of the external and internal cell types are c and d, respectively. The matrices are the same as those defined in the preceding subsection, except for the matrices $\mathbf{P}$ and $\mathbf{L}$, which are split into two parts (external and internal) and represented by the subscripts ex and in, respectively. $\mathbf{P}_{ex}$ and $\mathbf{L}_{ex}$ are $7 \times N_{ex}$ and 1×7 block matrices, respectively, while $\mathbf{L}_{in}$, $\mathbf{P}_{in}$, $\mathbf{L}_{in,\,ex}$, and $\mathbf{P}_{in,\,ex}$ are $1 \times 1, 1 \times N_{in}$, 1×2, and $2 \times N_{in}$ block vectors, respectively.

6.4.2 Time Stepping Schemes

The basic principle of the ImEx-RK time stepping method is the scheme of the embedded periodic structure, but it is further extended to the quasi-periodic structure, in which the system matrices are different from those in the periodic structure. The matrices are represented as follows:

$$\mathbf{P} = \begin{bmatrix} \mathbf{A} & \\ & \mathbf{B} \end{bmatrix} \tag{6.40}$$

$$\mathbf{A} = \begin{bmatrix} \mathbf{M}^{(1)} & & & & & \\ & \ddots & & & & \\ & & \mathbf{M}^{(1)} & & & \\ & & & \ddots & & \\ & & & & \mathbf{M}^{(C_{\text{ext}})} & \\ & & & & & \ddots & \\ & & & & & & \mathbf{M}^{(C_{\text{ext}})} \end{bmatrix},$$

$$\mathbf{B} = \begin{bmatrix} \mathbf{M}^{(C_{\text{ext}}+1)} & & & & & \\ & \ddots & & & & \\ & & \mathbf{M}^{(C_{\text{ext}}+1)} & & & \\ & & & \ddots & & \\ & & & & \mathbf{M}^{(C)} & \\ & & & & & \ddots & \\ & & & & & & \mathbf{M}^{(C)} \end{bmatrix}, \tag{6.41}$$

$$\mathbf{D} = \begin{bmatrix} \mathbf{L}_{\text{ex}}^{(1,1)}\mathbf{P}_{\text{ex}}^{(1,1)} + \mathbf{L}_{\text{ex, in}}^{(1,1)}\mathbf{P}_{\text{ex, in}}^{(1,1)} \\ \vdots \\ \mathbf{L}_{\text{ex}}^{(c,n_c)}\mathbf{P}_{\text{ex}}^{(c,n_c)}\mathbf{v}_{\text{ex}} + \mathbf{L}_{\text{ex,in}}^{(c,n_c)}\mathbf{P}_{\text{ex,in}}^{(c,n_c)} \\ \mathbf{L}_{\text{in, ex}}^{(1,1)}\mathbf{P}_{\text{in, ex}}^{(1,1)} + \mathbf{L}_{\text{in}}^{(1,1)}\mathbf{P}_{\text{in}}^{(1,1)} \\ \vdots \\ \mathbf{L}_{\text{in, ex}}^{(d,n_d)}\mathbf{P}_{\text{in, ex}}^{(d,n_d)} + \mathbf{L}_{\text{in}}^{(d,n_d)}\mathbf{P}_{\text{in}}^{(d,n_d)} \end{bmatrix} \tag{6.42}$$

where $N = C + N_B + N_P$, $C = C_{\text{ext}} + C_{\text{int}}$, $C_{\text{ext}} = n_1 + n_2 + \cdots n_c$, $C_{\text{int}} = n_1 + n_2 + \cdots n_d$. N_B denotes the number of subdomains for the buffer zone, while N_P denotes the number of subdomains for PML.

In the time integration of the memory-efficient DGTD method with the ImEx-RK scheme, the block Gauss–Seidel method is employed as a subdomain-level iterative solver.

6.4.3 Numerical Results

6.4.3.1 PEC Cavity Filled with Quasi-Periodic Structures

A PEC cubic box of the size $(3.6 \times 3.6 \times 1.8)$ m^3, which is filled with quasi-periodic structures, is simulated by the memory-efficient DGTD method to reveal its advantages in memory consumption compared to the conventional DGTD

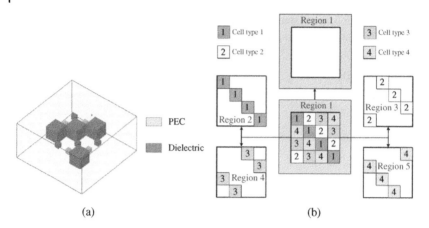

Figure 6.14 PEC cavity filled with quasi-periodic structures. (a) Geometric structure of the model. (b) Detailed arrangements of the cells on the lattice. Source: Wen et al. [53]. Reproduced with permission of IEEE.

method under the same accuracy requirement. A dipole, which is located at (0.8, 0.8, 0.3) m, is excited by a BHW pulse with a central frequency of 0.3 GHz, while the point receiver is fixed at (1.3, 2.3, 1.3) m.

There are 17 subdomains (16 cells and one buffer zone) as illustrated in Figure 6.14a. The 16 cells can be classified into four different cell types according to the structures and materials. Cell type 1 and cell type 4 are air cube and dielectric cube, respectively, while cell type 2 and cell type 3 are air cube filled with PEC and dielectric cube scatterers of the size $(0.2 \times 0.2 \times 0.2)$ m^3, respectively. The specific locations of each cell on the lattice can be obtained from Figure 6.14b. In the memory-efficient DGTD method, we only need to store one volume matrices of each cell type. The buffer zone and cells are simulated by the conventional subdomain-level DGTD and the memory-efficient DGTD methods, respectively, employing 2500 time steps with a time interval of 0.0113 ns.

The result of the memory-efficient DGTD method is in good agreement with that of the conventional subdomain-level DGTD method, as shown in Figure 6.15. The memory requirements for simulations by different methods are illustrated in Table 6.4, where it can be seen that the memory-efficient DGTD method only consumes about 37% of the memory in the conventional subdomain-level DGTD method.

6.4.3.2 Patch Antenna Array with Quasi-Periodic Structures
Quasi-Periodic Patch Antenna Array with the Ex-RK Time Stepping Scheme
Next, a relatively more complex and practical model is investigated to prove the correctness and superiority of the memory-efficient DGTD method for

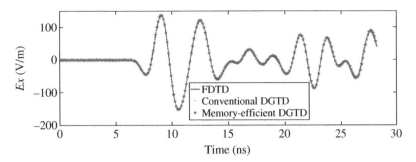

Figure 6.15 The x components of the electric fields. Source: Wen et al. [53]. Reproduced with permission of IEEE.

Table 6.4 Simulation time and memory consumption of quasi-periodic scatters.

Scheme	Simulation time (h)	Memory consumption (MB)
Conventional DGTD	8.33	21421
Memory-efficient DGTD	7.15	7820

Source: Wen et al. [53]. Reproduced with permission of IEEE.

quasi-periodic structures. There are 25 cells in total, including four types of cells as exhibited in Figure 6.16a. Each cell type is a patch antenna with the unique geometry as shown in Figure 6.16b. From Figure 6.16c, the thickness of the ground, the substrate, and the patch can be obtained. In Figure 6.16d, the entire structure is divided into six regions including a 3-mm-thick PML, buffer zone (with a thickness of 3 mm), and patch antenna array (region 1 to region 4). The lumped port parts need dense mesh, so they are isolated separately as new cells (cell type 5). In other words, cell type 5 is separated from cell type 1 to cell type 4 as shown in Figure 6.16e.

The quasi-periodic patch antenna array is simulated by the memory-efficient DGTD method, while the buffer zone and PML are simulated by the conventional subdomain-level DGTD method. The excitation is a voltage BHW pulse with a central frequency of 46.5 GHz ().

This patch antenna array is simulated by the memory-efficient DGTD method and the FDTD method, and the normalized 3D and 2D radiation patterns are compared between these two approaches with good agreement, as shown in Figures 6.17 and 6.18, which validates the effectiveness of the proposed method. This array has also been simulated using the conventional subdomain-level DGTD method and the commercial software package CST; the computational resource overhead is listed in Table 6.5. The advantage of memory consumption of the memory-efficient DGTD method is obvious.

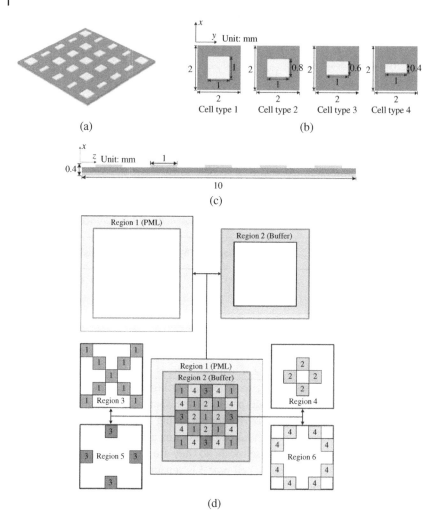

Figure 6.16 Quasi-periodic patch antenna array. (a) Overview. (b) Specific dimensions of different cell types. (c) The side view of the patch antenna array. (d) Divided regions and the locations of the cells. (e) Five types of cells. Source: Wen et al. [53]. Reproduced with permission of IEEE.

Quasi-Periodic Patch Antenna Array with ImEx-RK Time Stepping

To confirm the accuracy and advantages of the memory-efficient DGTD method, the scattering parameters of the quasi-periodic patch antenna array are simulated. The BHW pulse with a central frequency of 3.875 GHz is employed as the excitation of the lumped ports. The model and the detailed information of the patch antenna array are illustrated in Figure 6.19.

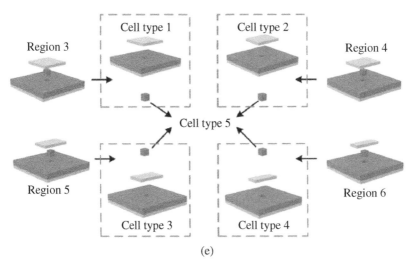

(e)

Figure 6.16 (*Continued*)

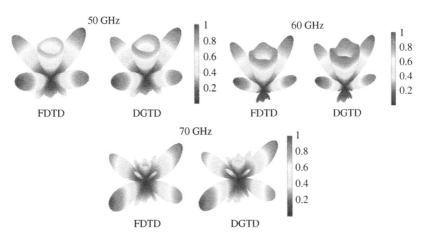

Figure 6.17 Three-dimensional radiation patterns of the quasi-periodic patch antenna array. Source: Wen et al. [53]. Reproduced with permission of IEEE.

The relative permittivity of the substrate is 3 and the thickness is 2 mm, while the thickness of the patches is 1 mm. According to the geometries of the patches, the array is divided into four types of cells as illustrated in Figure 6.19c. The space around the lumped ports, which needs fine mesh, is separated from cell type 1 to cell type 4 to form the embedded cell type 5. The PML and the buffer zone are simulated by the conventional subdomain-level DGTD method with Ex-RK time

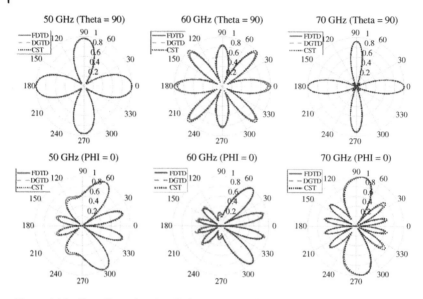

Figure 6.18 Two-dimensional radiation patterns of the quasi-periodic patch antenna array. Source: Wen et al. [53]. Reproduced with permission of IEEE.

Table 6.5 Simulation time and memory consumption of the quasi-periodic patch antenna array.

Scheme	Simulation time (h)	Memory consumption (GB)
Conventional DGTD	36.02	28.56
Memory-efficient DGTD	34.25	7.84
EDTD	11.17	39.67
CST	5.27	42.45

Source: Wen et al. [53]. Reproduced with permission of IEEE.

stepping, while cell type 1 to cell type 5 are simulated using the memory-efficient DGTD method with the ImEx-RK time stepping scheme.

The comparison of the scattering parameters between the memory-efficient DGTD and FDTD methods is shown in Figure 6.20, which exhibits a good agreement. The computational time and memory consumption of the conventional subdomain-level DGTD method and the memory-efficient DGTD method with Ex-RK and ImEx-RK time stepping schemes, respectively, are shown in Table 6.6.

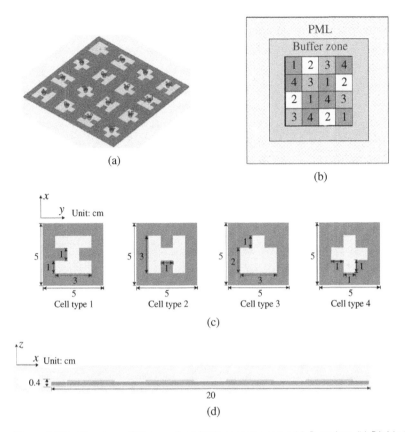

Figure 6.19 Structure of the quasi-periodic antenna array. (a) Overview. (b) Divided regions and the locations of the cells. (c) Specific dimensions of different cell types. (d) The side view of the patch antenna array.

The ImEx-RK scheme can enlarge the time interval by 20 times compared to the Ex-RK scheme, which significantly reduces the central process unit (CPU) time. With the same time stepping scheme, the memory-efficient DGTD method and the conventional subdomain-level DGTD method requires a similar length of simulation time; the tiny advantage of the memory-efficient DGTD method is attributed to a smaller number of matrices to be assembled in the preprocessing procedure. Since the memory-efficient DGTD method only stores one sets of the volumetric system matrices for each cell type, its advantage over the conventional subdomain-level DGTD method in memory consumption is obvious, for both Ex-RK and ImEx-RK time stepping schemes.

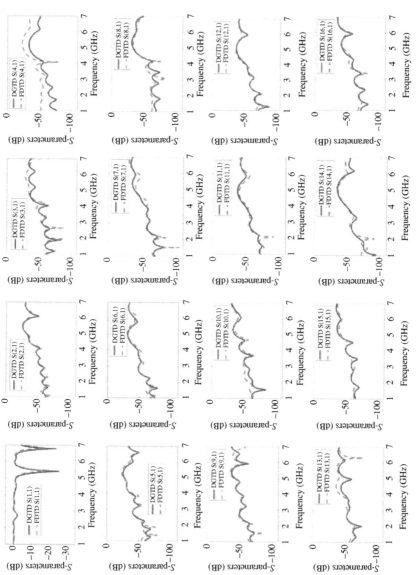

Figure 6.20 The scattering parameters of the quasi-periodic patch antenna array with the ImEx-RK time stepping scheme.

Table 6.6 Resource consumption of the conventional subdomain-level DGTD method and the memory-efficient DGTD method with Ex-RK and ImEx-RK time stepping schemes.

Method	Time stepping	dt (ps)	Menton (GB)	CPU time (h)
Conventional DGTD	Ex-RK	0.1	15.63	23.45
	ImEx-RK	2	32.12	6.88
Memory-efficient DGTD	Ex-RK	0.1	4.48	22.69
	ImEx-RK	2	8.75	6.25

6.5 Conclusions

The memory-efficient DGTD method utilizes the geometrical repeatability of the cells in a periodic/quasi-periodic structure, thus reusing the volumetric system matrices of the same type of cells to reduce the memory consumption significantly compared to the conventional subdomain-level DGTD method. In addition, it inherits multiple advantages from previous research of the DGTD method, such as nonconformal mesh, high-order basis functions, to further minimize the memory cost. To handle the period/quasi-periodic structures with multiscale characteristics (e.g. the thin substrate in a patch antenna array), the technique of embedded cells is developed, in which the geometrically fine and coarse parts are split into internal and external cells, respectively. Fine mesh can be employed for the internal cells to catch fine geometries, while coarse high-order elements for the external cells can maximize the computational efficiency. To surpass the limitation of the CFL condition, the ImEx-RK method is modified to be suitable for periodic/quasi-periodic structures, which leads to a significant reduction in simulation time. Therefore, the memory-efficient DGTD method can achieve low memory consumption and short simulation time simultaneously. This property guarantees it to be an advantageous solver for finite large periodic/quasi-periodic arrays, especially those containing fine structures.

References

1 Yan, S., Xu, L., Zhang, Y. et al. (2019). Order evaluation to new elementary operation approach for MIMO multidimensional systems. *International Journal of Control* 92 (10): 2349–2359.

2 Jensen, M.A. and Wallace, J.W. (2004). A review of antennas and propagation for MIMO wireless communications. *IEEE Transactions on Antennas and Propagation* 52 (11): 2810–2824.

3 Kaiser, T., Zheng, F., and Dimitrov, E. (2009). An overview of ultra-wide-band systems with MIMO. *Proceedings of the IEEE* 97 (2): 285–312.

4 Mabrouk, I.B., Talbi, L., Nedil, M., and Hettak, K. (2012). MIMO–UWB channel characterization within an underground mine gallery. *IEEE Transactions on Antennas and Propagation* 60 (10): 4866–4874.

5 Pan, Y., Cui, Y., and Li, R. (2016). Investigation of a triple-band multibeam MIMO antenna for wireless access points. *IEEE Transactions on Antennas and Propagation* 64 (4): 1234–1241.

6 D'Urso, M., Labate, M.G., and Buonanno, A. (2010). Reducing the number of amplitude controls in radar phased arrays. *IEEE Transactions on Antennas and Propagation* 58 (9): 3060–3064.

7 Ansari-Oghol-Beig, D. and Mosallaei, H. (2015). Array integral equation-fast Fourier transform solver for simulation of supercells and aperiodic penetrable metamaterials. *Journal of Computational and Theoretical Nanoscience* 12 (10): 3864–3878.

8 Kandasamy, K., Majumder, B., Mukherjee, J., and Ray, K. (2015). Low-RCS and polarization-reconfigurable antenna using cross-slot-based metasurface. *IEEE Antennas and Wireless Propagation Letters* 14: 1638–1641.

9 Wong, J.P., Epstein, A., and Eleftheriades, G.V. (2015). Reflectionless wide-angle refracting metasurfaces. *IEEE Antennas and Wireless Propagation Letters* 15: 1293–1296.

10 Yang, X., Qin, P.-Y., Liu, Y. et al. (2017). Analysis and design of a broadband multifeed tightly coupled patch array antenna. *IEEE Antennas and Wireless Propagation Letters* 17 (2): 217–220.

11 Dang, X., Li, M., Yang, F., and Xu, S. (2017). Quasi-periodic array modeling using reduced basis from elemental array. *IEEE Journal on Multiscale and Multiphysics Computational Techniques* 2: 202–208.

12 Jiang, Z.H., Lin, L., Ma, D. et al. (2014). Broadband and wide field-of-view plasmonic metasurface-enabled waveplates. *Scientific Reports* 4 (1): 1–8.

13 Zhang, L., Wan, X., Liu, S. et al. (2017). Realization of low scattering for a high-gain Fabry–Perot antenna using coding metasurface. *IEEE Transactions on Antennas and Propagation* 65 (7): 3374–3383.

14 Newman, E. and Forrai, D. (1987). Scattering from a microstrip patch. *IEEE Transactions on Antennas and Propagation* 35 (3): 245–251.

15 Pozar, D. (1986). Finite phased arrays of rectangular microstrip patches. *IEEE Transactions on Antennas and Propagation* 34 (5): 658–665.

16 King, A.S. and Bow, W.J. (1992). Scattering from a finite array of microstrip patches. *IEEE Transactions on Antennas and Propagation* 40 (7): 770–774.

17 Michalski, K.A. and Hsu, C.-I.G. (1994). RCS computation of coax-loaded microstrip patch antennas of arbitrary shape. *Electromagnetics* 14 (1): 33–62.

18 Ling, F. and Jin, J.M. (1997). Scattering and radiation analysis of microstrip antennas using discrete complex image method and reciprocity theorem. *Microwave and Optical Technology Letters* 16 (4): 212–216.

19 Zhuang, Y., Wu, K.-L., Wu, C., and Litva, J. (1996). A combined full-wave CG-FFT method for rigorous analysis of large microstrip antenna arrays. *IEEE Transactions on Antennas and Propagation* 44 (1): 102–109.

20 Wang, C.-F., Ling, F., and Jin, J.-M. (1998). A fast full-wave analysis of scattering and radiation from large finite arrays of microstrip antennas. *IEEE Transactions on Antennas and Propagation* 46 (10): 1467–1474.

21 Ling, F., Wang, C.-F., and Jin, J.-M. (2000). An efficient algorithm for analyzing large-scale microstrip structures using adaptive integral method combined with discrete complex-image method. *IEEE Transactions on Microwave Theory and Techniques* 48 (5): 832–839.

22 Yuan, N., Yeo, T.S., and Nie, X.-C. (2003). A fast analysis of scattering and radiation of large microstrip antenna arrays. *IEEE Transactions on Antennas and Propagation* 51 (9): 2218–2226.

23 Ling, F., Song, J., and Jin, J.-M. (1999). Multilevel fast multipole algorithm for analysis of large-scale microstrip structures. *IEEE Microwave and Guided Wave Letters* 9 (12): 508–510.

24 Zhao, K. and Lee, J.-F. (2004). A single-level dual rank IE-QR algorithm to model large microstrip antenna arrays. *IEEE Transactions on Antennas and Propagation* 52 (10): 2580–2585.

25 Suter, E. and Mosig, J.R. (2000). A subdomain multilevel approach for the efficient MoM analysis of large planar antennas. *Microwave and Optical Technology Letters* 26 (4): 270–277.

26 Matekovits, L., Laza, V.A., and Vecchi, G. (2007). Analysis of large complex structures with the synthetic-functions approach. *IEEE Transactions on Antennas and Propagation* 55 (9): 2509–2521.

27 Janpugdee, P. and Pathak, P.H. (2006). A DFT-based UTD ray analysis of large finite phased arrays on a grounded substrate. *IEEE Transactions on Antennas and Propagation* 54 (4): 1152–1161.

28 Mahachoklertwattana, P., Pathak, P.H., and Burkholder, R.J. (2008). A fast MoM approach for analyzing large arrays in a grounded multilayered medium. *Radio Science* 43 (06): 1–11.

29 Prakash, V. and Mittra, R. (2003). Characteristic basis function method: a new technique for efficient solution of method of moments matrix equations. *Microwave and Optical Technology Letters* 36 (2): 95–100.

30 Yeo, J., Prakash, V., and Mittra, R. (2003). Efficient analysis of a class of microstrip antennas using the characteristic basis function method (CBFM). *Microwave and Optical Technology Letters* 39 (6): 456–464.

31 Wan, J., Lei, J., and Liang, C.-H. (2005). An efficient analysis of large-scale periodic microstrip antenna arrays using the characteristic basis function method. *Progress In Electromagnetics Research* 50: 61–81.

32 Wang, X., Werner, D.H., and Turpin, J.P. (2013). Investigation of scattering properties of large-scale aperiodic tilings using a combination of the characteristic basis function and adaptive integral methods. *IEEE Transactions on Antennas and Propagation* 61 (6): 3149–3160.

33 Li, M.-K. and Chew, W.C. (2008). Multiscale simulation of complex structures using equivalence principle algorithm with high-order field point sampling scheme. *IEEE Transactions on Antennas and Propagation* 56 (8): 2389–2397.

34 Lu, W.B., Cui, T.J., Qian, Z.G. et al. (2004). Accurate analysis of large-scale periodic structures using an efficient sub-entire-domain basis function method. *IEEE Transactions on Antennas and Propagation* 52 (11): 3078–3085.

35 Lu, W.B., Cui, T.J., Yin, X.X. et al. (2005). Fast algorithms for large-scale periodic structures using subentire domain basis functions. *IEEE Transactions on Antennas and Propagation* 53 (3): 1154–1162.

36 Du, P., Wang, B.-Z., and Li, H. (2008). An extended sub-entire domain basis function method for finite periodic structures. *IEEE Antennas and Wireless Propagation Letters* 7: 404–407.

37 Salary, M.M., Forouzmand, A., and Mosallaei, H. (2017). Model order reduction of large-scale metasurfaces using a hierarchical dipole approximation. *ACS Photonics* 4 (1): 63–75.

38 Manges, J.B., Silvestro, J.W., and Zhao, K. (2012). Finite-element analysis of infinite and finite arrays. *International Journal of Microwave and Wireless Technologies* 4 (3): 357.

39 Zhao, K., Rawat, V., Lee, S.-C., and Lee, J.-F. (2007). A domain decomposition method with nonconformal meshes for finite periodic and semi-periodic structures. *IEEE Transactions on Antennas and Propagation* 55 (9): 2559–2570.

40 Gao, H.-W., Peng, Z., and Sheng, X.-Q. (2017). A geometry-aware domain decomposition preconditioning for hybrid finite element-boundary integral method. *IEEE Transactions on Antennas and Propagation* 65 (4): 1875–1885.

41 Gedney, S.D., Kramer, T., Luo, C. et al. (ed.) (2008). The discontinuous Galerkin finite element time domain method (DGFETD). In: *2008 IEEE International Symposium on Electromagnetic Compatibility*, 1–4. IEEE.

42 Hu, F.-G. and Wang, C.-F. (2012). Modeling of waveguide structures using DG-FETD method with higher order tetrahedral elements. *IEEE Transactions on Microwave Theory and Techniques* 60 (7): 2046–2054.

43 Ren, Q., Zhan, Q., and Liu, Q.H. (2017). An improved subdomain level nonconformal discontinuous Galerkin time domain (DGTD) method for materials with full-tensor constitutive parameters. *IEEE Photonics Journal* 9 (2): 1–13.

44 Ren, Q., Tobón, L.E., and Liu, Q.H. (2013). A new 2D non-spurious discontinuous Galerkin finite element time domain (DG-FETD) method for Maxwell's equations. *Progress In Electromagnetics Research* 143: 385–404.

45 Hu, F.-G. and Wang, C.-F. (2013). Higher-order DG-FETD modeling of wideband antennas with resistive loading. *IEEE Antennas and Wireless Propagation Letters* 12: 1025–1028.

46 Chen, J., Liu, Q.H., Chai, M., and Mix, J.A. (2009). A nonspurious 3-D vector discontinuous Galerkin finite-element time-domain method. *IEEE Microwave and Wireless Components Letters* 20 (1): 1–3.

47 Ren, Q., Sun, Q., Tobón, L. et al. (2016). EB scheme-based hybrid SE-FE DGTD method for multiscale EM simulations. *IEEE Transactions on Antennas and Propagation* 64 (9): 4088–4091.

48 Chen, J., Tobon, L.E., Chai, M. et al. (2011). Efficient implicit–explicit time stepping scheme with domain decomposition for multiscale modeling of layered structures. *IEEE Transactions on Components, Packaging and Manufacturing Technology* 1 (9): 1438–1446.

49 Ren, Q., Tobón, L.E., Sun, Q., and Liu, Q.H. (2015). A new 3-D nonspurious discontinuous Galerkin spectral element time-domain (DG-SETD) method for Maxwell's equations. *IEEE Transactions on Antennas and Propagation* 63 (6): 2585–2594.

50 Chen, J. and Liu, Q.H. (2012). Discontinuous Galerkin time-domain methods for multiscale electromagnetic simulations: a review. *Proceedings of the IEEE* 101 (2): 242–254.

51 Ren, Q., Bian, Y., Kang, L. et al. (2017). Leap-frog continuous–discontinuous Galerkin time domain method for nanoarchitectures with the Drude model. *Journal of Lightwave Technology* 35 (22): 4888–4896.

52 Chen, J. (2014). A memory efficient discontinuous Galerkin finite-element time-domain scheme for simulations of finite periodic structures. *Microwave and Optical Technology Letters* 56 (8): 1929–1933.

53 Wen, P., Ren, Q., Chen, J. et al. (2020). Improved memory-efficient subdomain level discontinuous Galerkin time domain method for periodic/quasi-periodic structures. *IEEE Transactions on Antennas and Propagation* 68 (11): 7471–7479.

54 Alvarez, J., Angulo, L.D., Bretones, A.R., and Garcia, S.G. (2012). 3-D discontinuous Galerkin time-domain method for anisotropic materials. *IEEE Antennas and Wireless Propagation Letters* 11: 1182–1185.

55 Li, P., Jiang, L.J., and Bağci, H. (2016). Transient analysis of dispersive power-ground plate pairs with arbitrarily shaped antipads by the DGTD method with wave port excitation. *IEEE Transactions on Electromagnetic Compatibility* 59 (1): 172–183.

56 Yan, S. and Jin, J.-M. (2017). A dynamic p-adaptive DGTD algorithm for electromagnetic and multiphysics simulations. *IEEE Transactions on Antennas and Propagation* 65 (5): 2446–2459.

Part III

Time-Domain Integral Equation Methods for Scattering Analysis

7

Explicit Marching-on-in-time Solvers for Second-kind Time Domain Integral Equations

Rui Chen[1], Sadeed B. Sayed[2*], Huseyin A. Ulku[3*], and Hakan Bagci[1*]*

[1]*Division of Computer, Electrical, and Mathematical Science and Engineering (CEMSE), King Abdullah University of Science and Technology (KAUST), Thuwal, Saudi Arabia*
[2]*School of Electrical and Electronic Engineering, Nanyang Technological University, Singapore*
[3]*Department of Electronics Engineering, Gebze Technical University, Kocaeli, Turkey*

7.1 Introduction

Transient electromagnetic scattering can be analyzed using time domain (surface or volume) integral equations (TDIEs) [1–43] as an alternative to finite difference time domain [44–47], time domain finite element [48–51], and time domain discontinuous Galerkin methods [52–62]. TDIE solvers have several advantages over these differential equation-based methods [63]: (i) they only discretize the surface or the volume of the scatterer, (ii) they implicitly enforce the radiation condition without using any (approximate) absorbing boundary conditions or perfectly matched layers (PMLs), and (iii) the time step size they use has to resolve only the maximum frequency of the excitation and is not limited by a Courant–Friedrichs–Lewy (CFL)-like condition.

Just like their frequency-domain counterparts, TDIEs are formulated using surface equivalence theorem for perfect electrically conducting (PEC) and homogeneous dielectric scatterers and using volume equivalence theorem for inhomogeneous scatterers [63]. The variables of interest (unknowns) in these formulations are the equivalent surface current and volume field/flux, respectively. The scattered field is expressed as spatio-temporal convolution of the variable of interest with the background or the corresponding unbounded medium Green function. For surface formulations, TDIE is obtained by enforcing the boundary conditions, which relate the incident and scattered fields, on the surface of

*Corresponding Authors: Rui Chen, rui.chen@kaust.edu.sa; Sadeed Bin Sayed, sadeed.sayed @ntu.edu.sg; Huseyin Arda Ulku, haulku@gtu.edu.tr; Hakan Bagci, hakan.bagci@kaust.edu.sa

Advances in Time-Domain Computational Electromagnetic Methods, First Edition.
Edited by Qiang Ren, Su Yan, and Atef Z. Elsherbeni.
© 2023 The Institute of Electrical and Electronics Engineers, Inc. Published 2023 by John Wiley & Sons, Inc.

the scatterer. Similarly, for volume formulations, TDIE is obtained using the fundamental field relation enforced in the volume of scatterers. Just like their frequency-domain counterparts, depending on the formulation (in space), TDIEs can be of the first-kind or the second-kind.

TDIEs are often numerically solved using the marching-on-in-time (MOT) scheme [1–38, 64–70]. The variable of interest is expanded using local basis functions in space and time. Inserting this expansion into TDIE and testing the resulting equation in space and time yield a lower triangular system of equations that can be solved using backward substitution. This yields a time-marching scheme, where at every time step a smaller system of equation, henceforth termed the MOT system, is solved for the unknown space-time expansion coefficients describing the instantaneous discretized variable of interest. The right-hand side (RHS) of this MOT system consists of the tested incident and scattered fields produced by the "past" values of the variable of interest. After the coefficients associated with the "current" time step are obtained, they are used in the computation of the RHS for the next time steps. This process, which is repeated until all expansion coefficients are obtained, is known as the MOT scheme.

There are different choices for basis functions in space and time: Rao-Wilton-Glisson (RWG) [71] and Schaubert–Wilton-Glisson (SWG) [72] functions are often used in space for surface and volume formulations, respectively. The piecewise Lagrange polynomial interpolation functions [6, 11, 20] or approximate prolate spheroidal wave functions (APSWFs) [28, 73–75] are often preferred as the basis function in time. For testing, Galerkin method is almost always used in space [76]. In time, point testing is often used [77] but there are few MOT schemes where Galerkin testing is used [76].

Depending on the spatial and temporal discretization techniques and the time step size, MOT schemes can be implicit [1–28] or explicit [29–35, 38]. Implicit schemes call for the inversion of the MOT system at every time step. The time step size they use has to resolve only the maximum frequency of the excitation and is not limited by a CFL-like condition. The sparseness of the MOT matrix system (to be inverted) depends on the time step size. When the time step size is small (under high-frequency excitation), the MOT system is sparse and an iterative solver can be used to solve it efficiently. In this case, the dominant computational cost comes from the computation of the tested scattered field on the RHS of the MOT system. Note that the cost can be reduced by using (multilevel) plane wave time domain (PWTD) methods [15, 16, 18–20, 24–27, 33, 78–84] or fast Fourier transformation (FFT)-based techniques [7–9, 11–13, 85, 86]. On the other hand, when the time step size is large (under low-frequency excitation), the MOT system becomes dense (even full). In this case, the computational cost of the MOT system solution becomes more dominant over the cost of computing the tested scattered field on the RHS of the system.

The traditional explicit MOT schemes [29–33] often use one-point quadrature rules for computing test and basis integrals in space and finite differences for computing derivatives in time. Even though these schemes do not require inversion of the MOT matrix system during time marching, maintaining the stability of the solution is almost always more challenging than that of the implicit schemes. Reducing the time step size helps with stability but this comes at the cost of increased computational requirements.

It should be noted here that implicit MOT schemes also suffer from stability issues but these are as severe as those associated with the explicit schemes. We believe these issues and the methods developed to address them are not within the scope of this chapter and therefore we refer the reader to [1–4, 74–77, 87–102] and the references therein.

This chapter describes a group of explicit MOT schemes recently developed by the authors to solve the second-kind TDIEs of electromagnetics [34, 35, 38]. This new class of explicit MOT solvers casts the second-kind TDIE in the form of ordinary differential equations (ODEs) that relate the variable of interest (surface current, volume field) to its temporal derivative. The variable of interest is expanded using the basis function in space. Inserting this expansion into TDIE and testing the resulting equations (in space) yield a system of ODEs in unknown time-dependent expansion coefficients. A predictor-corrector algorithm, $PE(CE)^m$, is used to integrate this system in time for these coefficients. To facilitate the computation of the retarded-time integrals, which express the scattered field in terms of the variable of interest, at discrete time steps as required by the $PE(CE)^m$, the piecewise Lagrange polynomial interpolation functions [6, 11, 20] are used.

Unlike the implicit MOT schemes, the resulting explicit MOT scheme calls for the solution of a system with a (spatial) Gram matrix at the evaluation (E) step. Three different explicit MOT solvers are described here: (i) EXP-MOT1 solves the time domain magnetic field integral equation (TD-MFIE) on PEC scatterers and uses the RWG basis functions for discretization in space [34], (ii) EXP-MOT2 also solves TD-MFIE but uses a higher order Nyström method [103–107] in space [38], and (iii) EXP-MOT3 solves the time domain magnetic field volume integral equation (TD-MFVIE) on dielectric scatterers and uses fully linear curl-conforming (FLC) basis functions [108, 109] and Galerkin or point testing for discretization in space [35].

Depending on the type of spatial testing functions and schemes, the Gram matrix in the MOT system has different sparsity structures. For EXP-MOT1, Galerkin testing yields a Gram matrix that is sparse and well conditioned but it does not have a fixed sparsity structure. Therefore, an iterative solver should be used to invert the MOT system at every time step. For EXP-MOT2, the Nyström method yields a block-diagonal Gram matrix that is inverted and stored efficiently before the time marching starts. For EXP-MOT3, Galerkin testing yields a sparse and well conditioned but does not have fixed sparsity pattern (has to be inverted using an iterative solver) but point testing yields a Gram matrix with four diagonal blocks, which is inverted and stored efficiently before the time marching starts.

Unlike the conventional explicit MOT schemes described previously, these explicit MOT solvers use the same time step size as their implicit counterparts without sacrificing from the accuracy and the stability of the solution. In addition, they are significantly faster under low-frequency excitations (i.e. for large time step sizes) since the matrices inverted by the implicit MOT schemes are dense for large time step sizes (unlike the Gram matrices described above). Numerical experiments support these conclusions.

This remainder of this chapter is organized as follows. Sections 7.2.1, 7.2.2, and 7.3 present the formulations for the three solvers. Sections 7.4 and 7.5 describe in detail how the explicit and implicit MOT schemes are executed. Sections 7.6 and 7.7 compare the accuracy and the computational cost of these schemes to their implicit counterparts. Section 7.8 lists several remarks. Section 7.9 presents several sets of numerical results that demonstrate the accuracy, stability, and efficiency of the explicit MOT schemes for PEC and dielectric scatterers. Finally, Section 7.10 provides the summary of the work and draws several future research directions.

7.2 TD-MFIE and Its Discretization

Let S represent the surface of a PEC scatterer located in an unbounded homogeneous background medium with permittivity ε and permeability μ. The scatterer is excited by an incident magnetic field $\mathbf{H}^{\mathrm{inc}}(\mathbf{r}, t)$, which is assumed band-limited to $f_{\max}$ and vanishingly small for $\forall \mathbf{r} \in S$ and $t \leq 0$. An electric current $\mathbf{J}(\mathbf{r}, t)$ is induced on S, which generates a scattered magnetic field $\mathbf{H}^{\mathrm{sca}}(\mathbf{r}, t)$ as

$$\mathbf{H}^{\mathrm{sca}}(\mathbf{r}, t) = \int_S \nabla \times \frac{\mathbf{J}(\mathbf{r}', t - R/c)}{4\pi R} ds' \tag{7.1}$$

Here, $R = |\mathbf{r} - \mathbf{r}'|$ and $c = 1/\sqrt{\mu\varepsilon}$ is the speed of light in the background medium. Substituting (7.1) into the boundary condition on S, i.e. $\partial_t \mathbf{J}(\mathbf{r}, t) = \partial_t \hat{\mathbf{n}}(\mathbf{r}) \times [\mathbf{H}^{\mathrm{inc}}(\mathbf{r}, t) + \mathbf{H}^{\mathrm{sca}}(\mathbf{r}, t)]$, yields TD-MFIE as follows [17, 34]:

$$\frac{1}{2}\partial_t \mathbf{J}(\mathbf{r}, t) = \hat{\mathbf{n}}(\mathbf{r}) \times \partial_t \mathbf{H}^{\mathrm{inc}}(\mathbf{r}, t) + \hat{\mathbf{n}}(\mathbf{r}) \times \int_S \nabla \times \frac{\partial_t \mathbf{J}(\mathbf{r}', t - R/c)}{4\pi R} ds' \tag{7.2}$$

Here, $\hat{\mathbf{n}}(\mathbf{r})$ is the unit normal vector pointing outward and ∂_t denotes the temporal derivative.

7.2.1 Discretization Using RWG Basis Functions

To numerically solve (7.2), $\mathbf{J}(\mathbf{r}, t)$ is discretized in space using the RWG basis functions [71] $\mathbf{f}_n(\mathbf{r})$, $n = 1, \ldots, N_s$, as follows:

$$\mathbf{J}(\mathbf{r}, t) = \sum_{n=1}^{N_s} \{\mathbf{I}(t)\}_n \mathbf{f}_n(\mathbf{r}) \tag{7.3}$$

where $\{\mathbf{I}(t)\}_n = I_n(t)$ are time-dependent expansion coefficients. Substituting (7.3) into (7.2) and testing the resulting equation with $\mathbf{f}_m(\mathbf{r})$, $m = 1, \ldots, N_s$, yield a linear system:

$$\mathbf{G}\partial_t \mathbf{I}(t) = \mathbf{V}^{\text{inc}}(t) - \mathbf{V}^{\text{sca}}(t) \tag{7.4}$$

where $\mathbf{G}$ is the Gram matrix of size $N_s \times N_s$, and $\mathbf{V}^{\text{inc}}(t)$ and $\mathbf{V}^{\text{sca}}(t)$ are time-dependent incident and scattered magnetic field vectors of length N_s, respectively. The entries of $\mathbf{G}$, $\mathbf{V}^{\text{inc}}(t)$, and $\mathbf{V}^{\text{sca}}(t)$ are as follows:

$$\{\mathbf{G}\}_{m,n} = \frac{1}{2} \int_{S_m} \mathbf{f}_m(\mathbf{r}) \cdot \mathbf{f}_n(\mathbf{r}) ds \tag{7.5}$$

$$\{\mathbf{V}^{\text{inc}}(t)\}_m = \int_{S_m} \mathbf{f}_m(\mathbf{r}) \cdot \hat{\mathbf{n}}(\mathbf{r}) \times \partial_t \mathbf{H}^{\text{inc}}(\mathbf{r}, t) ds \tag{7.6}$$

$$\{\mathbf{V}^{\text{sca}}(t)\}_m = -\sum_{n=1}^{N_s} \int_{S_m} \mathbf{f}_m(\mathbf{r}) \cdot \hat{\mathbf{n}}(\mathbf{r}) \times \partial_t [\{\mathbf{I}(t)\}_n * \mathbf{H}_n(\mathbf{r}, t)] ds \tag{7.7}$$

In (7.7), $\mathbf{H}_n(\mathbf{r}, t)$ is the magnetic field because of impulsively excited $\mathbf{f}_n(\mathbf{r})$ and can be calculated analytically [99, 100].

To approximate $\mathbf{I}(t)$, (7.4) is sampled as

$$\mathbf{G}\dot{\mathbf{I}}_h = \mathbf{V}_h^{\text{inc}} - \mathbf{V}_h^{\text{sca}}, \quad h = 1, \ldots, N_t \tag{7.8}$$

Here, $\dot{\mathbf{I}}_h = \partial_t \mathbf{I}(t)|_{t=h\Delta t}$, $\mathbf{I}_h = \mathbf{I}(h\Delta t)$, $\mathbf{V}_h^{\text{inc}} = \mathbf{V}^{\text{inc}}(h\Delta t)$. $\mathbf{V}_h^{\text{inc}}$ can be evaluated as

$$\{\mathbf{V}_h^{\text{inc}}\}_m = \int_{S_m} \mathbf{f}_m(\mathbf{r}) \cdot \hat{\mathbf{n}}(\mathbf{r}) \times \partial_t \mathbf{H}^{\text{inc}}(\mathbf{r}, t)\Big|_{t=h\Delta t} ds \tag{7.9}$$

To facilitate the calculation of $\mathbf{V}_h^{\text{sca}}$ using (7.7), $\mathbf{I}(t)$ in (7.4) is interpolated as

$$\mathbf{I}(t) = \sum_{l=1}^{N_t} \mathbf{I}_l T(t - l\Delta t) \tag{7.10}$$

Here, Δt is the time step size, N_t is the number of time steps, and $T(t)$ is constructed using piecewise Lagrange polynomials [6, 11, 20], where $T(t) = 0$ for $t \leq \Delta t$ and $t \geq T_{\text{max}} \Delta t$, and T_{max} is the order of Lagrange polynomials.

Substituting (7.10) in (7.7) and sampling the resulting expression at $t = h\Delta t$ yield

$$\mathbf{V}_h^{\text{sca}} = \sum_{l=1}^{h} \mathbf{Z}_{h-l}^{\text{exp}} \mathbf{I}_l \tag{7.11}$$

where $\mathbf{Z}_{h-l}^{\mathrm{exp}}$ are explicit MOT matrices of size $N_s \times N_s$ with entries

$$\{\mathbf{Z}_{h-l}^{\mathrm{exp}}\}_{m,n} = -\int_{S_m} \mathbf{f}_m(\mathbf{r}) \cdot \hat{\mathbf{n}}(\mathbf{r}) \times [\partial_t T(t - l\Delta t) * \mathbf{H}_n(\mathbf{r}, t)]_{t=h\Delta t} \, ds \tag{7.12}$$

Substituting (7.12) into (7.8) yields the explicit MOT system for TD-MFIE discretized using the RWG basis functions as [34]

$$\mathbf{G}\dot{\mathbf{I}}_h = \mathbf{V}_h^{\mathrm{inc}} - \sum_{l=1}^{h} \mathbf{Z}_{h-l}^{\mathrm{exp}}\mathbf{I}_l, \quad h = 1, \dots, N_t \tag{7.13}$$

7.2.2 Discretization Using the Nyström Method

In (7.2), $\mathbf{J}(\mathbf{r}, t)$ can be alternatively expanded using [103]

$$\mathbf{J}(\mathbf{r}, t) = \sum_{p=1}^{N_p} \sum_{i=1}^{N_n} [\{\mathbf{I}(t)\}_{ip}^u \mathbf{u}(\mathbf{r}) + \{\mathbf{I}(t)\}_{ip}^v \mathbf{v}(\mathbf{r})] \vartheta^{-1}(\mathbf{r}) \ell_{ip}(\mathbf{r}) \tag{7.14}$$

Here, N_p and N_n are the numbers of the curvilinear triangles discretizing S and the interpolation nodes on each triangle, respectively, $\ell_{ip}(\mathbf{r})$ is the Lagrange interpolator defined at node $\mathbf{r}_{ip}$ (node i on patch p), $\vartheta(\mathbf{r})$ is the Jacobian between the Cartesian coordinate system and the (u, v) reference plane, and $\{\mathbf{I}(t)\}_{ip}^u$ and $\{\mathbf{I}(t)\}_{ip}^v$ are the time-dependent expansion coefficients in the tangential directions of $\mathbf{u}(\mathbf{r}_{ip})$ and $\mathbf{v}(\mathbf{r}_{ip})$, respectively. To account for the time retardation in (7.2), $\{\mathbf{I}(t)\}_{ip}^b$, $b \in \{u, v\}$ are approximated as

$$\{\mathbf{I}(t)\}_{ip}^b = \sum_{l=1}^{N_t} I_l\big|_{ip}^b T(t - l\Delta t) \tag{7.15}$$

Here, $I_l\big|_{ip}^b = \{\mathbf{I}(l\Delta t)\}_{ip}^b, b \in \{u, v\}$.

Inserting (7.14) into (7.2) and testing the resulting equation in space with $\mathbf{u}(\mathbf{r}_{jq})$ and $\mathbf{v}(\mathbf{r}_{jq}), j = 1, \dots, N_n, q = 1, \dots, N_p$ yield a time-dependent ODE system, which has to be sampled at $t = h\Delta t$ to carry out the time integration. Consequently, (7.15) is used on $\{\mathbf{I}(t - R/c)\}_{ip}^b, b \in \{u, v\}$, which results in a fully discretized explicit MOT system as [38]

$$\mathbf{G}\dot{\mathbf{I}}_h = \mathbf{V}_h^{\mathrm{inc}} - \sum_{l=1}^{h} \mathbf{Z}_{h-l}^{\mathrm{exp}}\mathbf{I}_l, \quad h = 1, \dots, N_t \tag{7.16}$$

Here, the Gram matrix $\mathbf{G}$ is block-diagonal and is given by

$$\mathbf{G} = \begin{bmatrix} G|_{11,11}^{uu} & G|_{11,11}^{uv} \\ G|_{11,11}^{vu} & G|_{11,11}^{vv} \\ & & \ddots \\ & & & G|_{N_nN_p,N_nN_p}^{uu} & G|_{N_nN_p,N_nN_p}^{uv} \\ & & & G|_{N_nN_p,N_nN_p}^{vu} & G|_{N_nN_p,N_nN_p}^{vv} \end{bmatrix} \tag{7.17}$$

with its entries expressed as

$$G|_{jq,ip}^{ab} = \frac{1}{2}\mathbf{a}(\mathbf{r}_{jq}) \cdot \mathbf{b}(\mathbf{r}_{ip})\vartheta^{-1}(\mathbf{r}_{ip}) \tag{7.18}$$

$a, b \in \{u, v\}$, $\mathbf{a}, \mathbf{b} \in \{\mathbf{u}, \mathbf{v}\}$, and the explicit MOT matrices $\mathbf{Z}_{h-l}^{\exp}$ are given by

$$\mathbf{Z}_{h-l}^{\exp} = \begin{bmatrix} Z_{h-l}^{\exp}\Big|_{11,11}^{uu} & Z_{h-l}^{\exp}\Big|_{11,11}^{uv} & \cdots & Z_{h-l}^{\exp}\Big|_{11,N_nN_p}^{uu} & Z_{h-l}^{\exp}\Big|_{11,N_nN_p}^{uv} \\ Z_{h-l}^{\exp}\Big|_{11,11}^{vu} & Z_{h-l}^{\exp}\Big|_{11,11}^{vv} & \cdots & Z_{h-l}^{\exp}\Big|_{11,N_nN_p}^{vu} & Z_{h-l}^{\exp}\Big|_{11,N_nN_p}^{vv} \\ \vdots & \vdots & \ddots & \vdots & \vdots \\ Z_{h-l}^{\exp}\Big|_{N_nN_p,11}^{uu} & Z_{h-l}^{\exp}\Big|_{N_nN_p,11}^{uv} & \cdots & Z_{h-l}^{\exp}\Big|_{N_nN_p,N_nN_p}^{uu} & Z_{h-l}^{\exp}\Big|_{N_nN_p,N_nN_p}^{uv} \\ Z_{h-l}^{\exp}\Big|_{N_nN_p,11}^{vu} & Z_{h-l}^{\exp}\Big|_{N_nN_p,11}^{vv} & \cdots & Z_{h-l}^{\exp}\Big|_{N_nN_p,N_nN_p}^{vu} & Z_{h-l}^{\exp}\Big|_{N_nN_p,N_nN_p}^{vv} \end{bmatrix} \tag{7.19}$$

with their entries expressed as

$$Z_{h-l}^{\exp}\Big|_{jq,ip}^{ab} = \mathbf{a}(\mathbf{r}_{jq}) \cdot \hat{\mathbf{n}}(\mathbf{r}_{jq}) \times \int_{S_p} \vartheta^{-1}(\mathbf{r}')\ell_{ip}\left(\mathbf{r}'\right)$$

$$\times \frac{\mathbf{R}}{4\pi R^3}\left[\partial_t T(t) + \frac{R}{c}\partial_t^2 T(t)\right]_{t=(h-l)\Delta t - R/c} \times \mathbf{b}(\mathbf{r}')ds' \tag{7.20}$$

where S_p is the support area of patch p, $\mathbf{R} = \mathbf{r}_{jq} - \mathbf{r}'$, $R = |\mathbf{r}_{jq} - \mathbf{r}'|$, and the tested excitation vectors $\mathbf{V}_h^{\text{inc}}$ are given by

$$\mathbf{V}_h^{\text{inc}} = \left[V_h^{\text{inc}}\Big|_{11}^{u}, V_h^{\text{inc}}\Big|_{11}^{v}, V_h^{\text{inc}}\Big|_{21}^{u}, V_h^{\text{inc}}\Big|_{21}^{v}, \ldots, V_h^{\text{inc}}\Big|_{N_iN_p}^{u}, V_h^{\text{inc}}\Big|_{N_iN_p}^{v}\right]^T$$

with their entries expressed as $V_h^{\text{inc}}\Big|_{jq}^{a} = \mathbf{a}(\mathbf{r}_{jq}) \cdot \hat{\mathbf{n}}(\mathbf{r}_{jq}) \times \partial_t \mathbf{H}^{\text{inc}}(\mathbf{r}_{jq}, h\Delta t)$, and finally the unknown coefficient vectors $\mathbf{I}_l$ are given by

$$\mathbf{I}_l = \left[I_l\Big|_{11}^{u}, I_l\Big|_{11}^{v}, I_l\Big|_{21}^{u}, I_l\Big|_{21}^{v}, \ldots, I_l\Big|_{N_iN_p}^{u}, I_l\Big|_{N_iN_p}^{v}\right]^T$$

and $\dot{\mathbf{I}}_h$ store samples of the time derivative of the coefficients as

$$\dot{\mathbf{I}}_h = \left[\dot{I}_h\Big|_{11}^{u}, \dot{I}_h\Big|_{11}^{v}, \dot{I}_h\Big|_{21}^{u}, \dot{I}_h\Big|_{21}^{v}, \ldots, \dot{I}_h\Big|_{N_iN_p}^{u}, \dot{I}_h\Big|_{N_iN_p}^{v}\right]^T$$

with $\dot{I}_h\Big|_{ip}^{b} = \{\partial_t \mathbf{I}(h\Delta t)\}_{ip}^{b}$.

7.3 TD-MFVIE and Its Discretization Using FLC Basis Functions

Let V denote the volume of a dielectric scatterer with permittivity $\tilde{\varepsilon}(\mathbf{r})$ and permeability μ residing in an unbounded homogeneous background medium with permittivity ε and permeability μ. Upon excitation of $\mathbf{H}^{\text{inc}}(\mathbf{r}, t)$ (vanishingly small for $\forall \mathbf{r} \in V$ and $t \leq 0$), $\mathbf{J}(\mathbf{r}, t)$ is induced inside V, which generates scattered magnetic field as

$$\mathbf{H}^{\text{sca}}(\mathbf{r}, t) = \int_V \nabla \times \frac{\mathbf{J}(\mathbf{r}', t - R/c)}{4\pi R} dv' \tag{7.21}$$

Here, $\mathbf{J}(\mathbf{r}, t)$ is expressed as the curl of total magnetic field $\mathbf{H}(\mathbf{r}, t)$ by

$$\mathbf{J}(\mathbf{r}, t) = \kappa(\mathbf{r}) \nabla \times \mathbf{H}(\mathbf{r}, t) \tag{7.22}$$

where $\kappa(\mathbf{r}) = 1 - \varepsilon/\tilde{\varepsilon}(\mathbf{r})$. Inserting (7.21) and (7.22) into the time derivative form of $\mathbf{H}(\mathbf{r}, t) = \mathbf{H}^{\text{inc}}(\mathbf{r}, t) + \mathbf{H}^{\text{sca}}(\mathbf{r}, t)$ yields TD-MFVIE [35] as

$$\partial_t \mathbf{H}^{\text{inc}}(\mathbf{r}, t) = \partial_t \mathbf{H}(\mathbf{r}, t) + \frac{1}{4\pi} \int_V \kappa\left(\mathbf{r}'\right) \hat{\mathbf{R}}$$
$$\times \left[\frac{\partial_{t'}^2 \nabla' \times \mathbf{H}(\mathbf{r}', t')}{cR} + \frac{\partial_{t'} \nabla' \times \mathbf{H}(\mathbf{r}', t')}{R^2} \right]_{t'=t-R/c} dv' \tag{7.23}$$

where $\hat{\mathbf{R}} = (\mathbf{r} - \mathbf{r}')/R$.

To numerically solve (7.23), V is discretized into tetrahedrons with N_e edges. Then, $\mathbf{H}(\mathbf{r}, t)$ is expanded using the FLC basis functions [108, 109] as

$$\mathbf{H}(\mathbf{r}, t) = \sum_{n=1}^{N_e} \left[\{\mathbf{I}^1(t)\}_n \mathbf{f}_n^1(\mathbf{r}) + \{\mathbf{I}^2(t)\}_n \mathbf{f}_n^2(\mathbf{r}) \right] \tag{7.24}$$

Here, $\mathbf{f}_n^1(\mathbf{r})$ and $\mathbf{f}_n^2(\mathbf{r})$ are the first-order irrotational edge functions [109] and the lowest mixed-order solenoidal edge functions [108], and $\{\mathbf{I}^1(t)\}_n$ and $\{\mathbf{I}^2(t)\}_n$ are the corresponding coefficients, respectively. $\mathbf{f}_n^s(\mathbf{r})$, $s \in \{1, 2\}$ can be expressed as

$$\mathbf{f}_n^s(\mathbf{r}) = \begin{cases} \lambda_n^{d_n^1}(\mathbf{r}) \nabla \lambda_n^{d_n^2}(\mathbf{r}) \pm \lambda_n^{d_n^2}(\mathbf{r}) \nabla \lambda_n^{d_n^1}(\mathbf{r}), \mathbf{r} \in S_n \\ 0, \mathbf{r} \notin S_n \end{cases} \tag{7.25}$$

where sign "+" for $s = 1$ and sign "−" for $s = 2$, $S_n = \cup_{q=1}^{Q_n} S_n^q$ is the support of $\mathbf{f}_n^s(\mathbf{r})$, Q_n is the number of tetrahedrons sharing the nth edge with two nodes d_n^1 and d_n^2, and $\lambda_n^d(\mathbf{r})$, $d \in \{d_n^1, d_n^2\}$ are the barycentric coordinate functions. It can be shown that $\nabla \times \mathbf{f}_n^1(\mathbf{r}) = \mathbf{0}$ and $\nabla \times \mathbf{f}_n^2(\mathbf{r}) \neq \mathbf{0}$.

To facilitate the computation of the retarded-time integral in (7.23), $\{\mathbf{I}^s(t)\}_n$, $s \in \{1, 2\}$ are approximated as

$$\{\mathbf{I}^s(t)\}_n = \sum_{l=1}^{N_t} \{\mathbf{I}_l^s\}_n T(t - l\Delta t) \tag{7.26}$$

Here, $\mathbf{I}_l^s$ is the sample of $\mathbf{I}^s(t)$ at $t = l\Delta t$, i.e. $\mathbf{I}_l^s = \mathbf{I}^s(l\Delta t)$.

Substituting (7.26) in (7.23) and testing the resulting equation with $\mathbf{t}_m^1(\mathbf{r})$ and $\mathbf{t}_m^2(\mathbf{r})$, $m = 1, \ldots, N_e$, yield an ODE matrix system of size $2N_e \times 2N_e$ as

$$\begin{bmatrix} \mathbf{G}_{11} & \mathbf{G}_{12} \\ \mathbf{G}_{21} & \mathbf{G}_{22} \end{bmatrix} \begin{bmatrix} \ddot{\mathbf{I}}^1(t) \\ \ddot{\mathbf{I}}^2(t) \end{bmatrix} = \begin{bmatrix} \mathbf{V}^{\text{inc},1}(t) \\ \mathbf{V}^{\text{inc},2}(t) \end{bmatrix} - \begin{bmatrix} \mathbf{V}^{\text{sca},1}(t) \\ \mathbf{V}^{\text{sca},2}(t) \end{bmatrix} \tag{7.27}$$

Here, $\mathbf{G}_{ps}$, $p, s \in \{1, 2\}$ are $N_e \times N_e$ block matrices of the Gram matrix $\mathbf{G}$ with the entries

$$\{\mathbf{G}_{ps}\}_{m,n} = \int_{P_m^p} \mathbf{t}_m^p(\mathbf{r}) \cdot \mathbf{f}_n^s(\mathbf{r}) dv \qquad (7.28)$$

where P_m^p is the support of $\mathbf{t}_m^p(\mathbf{r})$, $p \in \{1, 2\}$.

Two sets of choices are used for $\mathbf{t}_m^1(\mathbf{r})$ and $\mathbf{t}_m^2(\mathbf{r})$ resulting in Galerkin or point testing. Different choices of $\mathbf{t}_m^p(\mathbf{r})$, $p \in \{1, 2\}$ yield different sparseness structures in $\mathbf{G}$ [as computed using (7.28)] as explained below.

(1) **Point testing**: For point testing, $\mathbf{t}_m^p(\mathbf{r}) = \hat{\mathbf{q}}_m \delta(\mathbf{r} - \mathbf{r}_m^p)$, $p \in \{1, 2\}$, where $\hat{\mathbf{q}}_m$ is a unit vector pointing from d_m^1 to d_m^2 and $\mathbf{r}_m^p$ are Gaussian quadrature points on edge m. Substituting these expressions for $\mathbf{t}_m^p(\mathbf{r})$ into (7.28) and it is noted that, the tangential component of $\mathbf{f}_n^1(\mathbf{r})$ linearly changes from -1 to 1 while that of $\mathbf{f}_n^2(\mathbf{r})$ is 1 on the whole edge n, yield

$$\mathbf{G}_{11} = \mathbf{G}_{21} = \mathbf{I}$$
$$\mathbf{G}_{12} = -\mathbf{G}_{22} = \frac{1}{\sqrt{3}}\mathbf{I} \qquad (7.29)$$

and the inverse of $\mathbf{G}$ is

$$\begin{bmatrix} \mathbf{G}_{11} & \mathbf{G}_{12} \\ \mathbf{G}_{21} & \mathbf{G}_{22} \end{bmatrix}^{-1} = \frac{1}{2}\begin{bmatrix} \mathbf{I} & \mathbf{I} \\ \sqrt{3}\mathbf{I} & -\sqrt{3}\mathbf{I} \end{bmatrix} \qquad (7.30)$$

(2) **Galerkin testing**: For Galerkin testing, $\mathbf{t}_m^1(\mathbf{r}) = \mathbf{f}_m^1(\mathbf{r})$ and $\mathbf{t}_m^2(\mathbf{r}) = \mathbf{f}_m^2(\mathbf{r})$. Using these expressions in (7.28), a summation of second-order polynomial integrals over tetrahedrons, can be obtained, which are computed using a Gaussian quadrature rule [110, 111].

In (7.27), $\mathbf{V}^{\text{inc},p}(t)$ and $\mathbf{V}^{\text{sca},p}(t)$, $p \in \{1, 2\}$ are vectors of spatially tested incident and scattered magnetic fields, respectively. Their entries are given by

$$\{\mathbf{V}^{\text{inc},p}(t)\}_m = \int_{P_m^p} \mathbf{t}_m^p(\mathbf{r}) \cdot \partial_t \mathbf{H}^{\text{inc}}(\mathbf{r}, t) dv \qquad (7.31)$$

$$\{\mathbf{V}^{\text{sca},p}(t)\}_m = -\frac{1}{4\pi} \sum_{n=1}^{N} \int_{P_m^p} \mathbf{t}_m^p(\mathbf{r}) \cdot \sum_{q=1}^{Q_n} \kappa_n^q \int_{S_n^q} \hat{\mathbf{R}} \times \nabla'$$
$$\times \mathbf{f}_n^2(\mathbf{r}') \left[\frac{\partial_{t'}^2 \{\mathbf{I}^2(t')\}_n}{cR} + \frac{\partial_{t'}\{\mathbf{I}^2(t')\}_n}{R^2} \right]_{t'=t-R/c} dv' \, dv \qquad (7.32)$$

In (7.32), $\kappa(\mathbf{r})$ is assumed to be a constant in the tetrahedron S_n^q, i.e. $\kappa_n^q = \kappa(\mathbf{r}_n^q)$, where $\mathbf{r}_n^q$ is the center of S_n^q.

The method used for computing $\mathbf{I}_h^s = \mathbf{I}^s(h\Delta t)$, $s \in \{1, 2\}$ calls for sampling (7.27) in time as

$$\begin{bmatrix} \mathbf{G}_{11} & \mathbf{G}_{12} \\ \mathbf{G}_{21} & \mathbf{G}_{22} \end{bmatrix}\begin{bmatrix} \mathbf{i}_h^1 \\ \mathbf{i}_h^2 \end{bmatrix} = \begin{bmatrix} \mathbf{V}_h^{\text{inc},1} \\ \mathbf{V}_h^{\text{inc},2} \end{bmatrix} - \begin{bmatrix} \mathbf{V}_h^{\text{sca},1} \\ \mathbf{V}_h^{\text{sca},2} \end{bmatrix} \qquad (7.33)$$

where $h = 1, \dots, N_t$, $\mathbf{I}_h^s = \mathbf{I}^s(h\Delta t)$, $s \in \{1, 2\}$, $\mathbf{V}_h^{\text{inc},p} = \mathbf{V}^{\text{inc},p}(h\Delta t)$ using (7.31), and $\mathbf{V}_h^{\text{sca},p} = \mathbf{V}^{\text{sca},p}(h\Delta t)$, $p \in \{1, 2\}$. To compute $\mathbf{V}_h^{\text{sca},p}$, inserting (7.26) with $s = 2$ into (7.32) and calculating the resulting expression at $t = h\Delta t$ yield

$$\mathbf{V}_h^{\text{sca},p} = \sum_{l=1}^{h} \mathbf{M}_{h-l}^p \mathbf{I}_l^2, \quad p \in \{1, 2\} \tag{7.34}$$

where the entries of explicit MOT matrices $\mathbf{M}_{h-l}^p$ are

$$\{\mathbf{M}_{h-l}^p\}_{m,n} = -\frac{1}{4\pi} \int_{P_m^p} \mathbf{t}_m^p(\mathbf{r}) \cdot \sum_{q=1}^{Q_n} \kappa_n^q \int_{S_n^q} \hat{\mathbf{R}} \times \nabla'$$

$$\times \mathbf{f}_n^2(\mathbf{r}') \left[\frac{\partial_{t'}^2 T(t')}{cR} + \frac{\partial_{t'} T(t')}{R^2} \right]_{t'=(h-l)\Delta t - R/c} dv' \, dv \tag{7.35}$$

Inserting (7.34) into (7.33) yields

$$\underbrace{\begin{bmatrix} \mathbf{G}_{11} & \mathbf{G}_{12} \\ \mathbf{G}_{21} & \mathbf{G}_{22} \end{bmatrix}}_{\mathbf{G}} \underbrace{\begin{bmatrix} \mathbf{I}_h^1 \\ \mathbf{I}_h^2 \end{bmatrix}}_{\mathbf{I}_h} = \underbrace{\begin{bmatrix} \mathbf{V}_h^{\text{inc},1} \\ \mathbf{V}_h^{\text{inc},2} \end{bmatrix}}_{\mathbf{V}_h^{\text{inc}}} - \sum_{l=1}^{h} \underbrace{\begin{bmatrix} 0 & \mathbf{M}_{h-l}^1 \\ 0 & \mathbf{M}_{h-l}^2 \end{bmatrix}}_{\mathbf{Z}_{h-l}^{\text{exp}}} \underbrace{\begin{bmatrix} \mathbf{I}_l^1 \\ \mathbf{I}_l^2 \end{bmatrix}}_{\mathbf{I}_l} \tag{7.36}$$

The explicit MOT system (7.36) can be written in a compact form as [35]

$$\mathbf{G}\mathbf{I}_h = \mathbf{V}_h^{\text{inc}} - \sum_{l=1}^{h} \mathbf{Z}_{h-l}^{\text{exp}} \mathbf{I}_l, \quad h = 1, \dots, N_t \tag{7.37}$$

7.4 Predictor–Corrector Scheme

The explicit MOT systems (7.13), (7.16), and (7.37) for TD-MFIE discretized using the RWG basis functions, TD-MFIE discretized using the Nyström method, and TD-MFVIE discretized using the FLC basis functions with Galerkin or point testing, respectively, can be rewritten in a uniform expression as

$$\mathbf{G}\mathbf{I}_h = \mathbf{V}_h^{\text{inc}} - \sum_{l=1}^{h} \mathbf{Z}_{h-l}^{\text{exp}} \mathbf{I}_l, \quad h = 1, \dots, N_t \tag{7.38}$$

The explicit MOT system (7.38) is integrated in time using a PE(CE)^{m}-type K-step scheme [34, 35, 38] as described below.

Loop for time marching $h = 1, \dots, N_t$.

Step 1: Calculate the fixed components on the RHS of (7.38) that do not vary within one time step as

$$\mathbf{V}_h^{\text{fix}} = \mathbf{V}_h^{\text{inc}} - \sum_{l=1}^{h-1} \mathbf{Z}_{h-l}^{\text{exp}} \mathbf{I}_l \tag{7.39}$$

Step 2: Prediction (P) step. $\mathbf{I}_h$ is predicted from $\mathbf{I}_l$ and $\dot{\mathbf{I}}_l$, $l = h - K, \ldots, h - 1$ as

$$\mathbf{I}_h = \sum_{k=1}^{K} [\{\mathbf{p}\}_k \mathbf{I}_{h-1+k-K} + \{\mathbf{p}\}_{K+k} \dot{\mathbf{I}}_{h-1+k-K}] \tag{7.40}$$

Step 3: Evaluation (E) step.
   Step 3.1: Calculate the RHS vector of (7.38) using $\mathbf{I}_h$ in (7.40) as

$$\mathbf{R}_h = \mathbf{V}_h^{\text{fix}} - \mathbf{Z}_0^{\text{exp}} \mathbf{I}_h \tag{7.41}$$

   Step 3.2: Evaluate $\dot{\mathbf{I}}_h$ by solving

$$\mathbf{G}\dot{\mathbf{I}}_h = \mathbf{R}_h \tag{7.42}$$

Step 4: Set $\dot{\mathbf{I}}_h^{(0)} = \dot{\mathbf{I}}_h$. Repeat Steps 4.1–4.3 until convergence ($n = 1, \ldots, m$).
   Step 4.1: Correction (C) step. Calculate $\mathbf{I}_h^{(n)}$ using $\mathbf{I}_l$ and $\dot{\mathbf{I}}_l$, $l = h - K, \ldots, h - 1$, and $\dot{\mathbf{I}}_h^{(n-1)}$ as

$$\mathbf{I}_h^{(n)} = \sum_{k=1}^{K} [\{\mathbf{c}\}_k \mathbf{I}_{h-1+k-K} + \{\mathbf{c}\}_{K+k} \dot{\mathbf{I}}_{h-1+k-K}] + \{\mathbf{c}\}_{2K+1} \dot{\mathbf{I}}_h^{(n-1)} \tag{7.43}$$

   Step 4.2: Evaluation (E) step.
      Step 4.2.1: Calculate $\mathbf{R}_h^{(n)}$ using $\mathbf{I}_h^{(n)}$ in (7.43) as

$$\mathbf{R}_h^{(n)} = \mathbf{V}_h^{\text{fix}} - \mathbf{Z}_0^{\text{exp}} \mathbf{I}_h^{(n)} \tag{7.44}$$

      Step 4.2.2: Evaluate $\dot{\mathbf{I}}_h^{(n)}$ by solving

$$\mathbf{G}\dot{\mathbf{I}}_h^{(n)} = \mathbf{R}_h^{(n)} \tag{7.45}$$

   Step 4.3: Check convergence between $\mathbf{I}_h^{(n)}$ and $\mathbf{I}_h^{(n-1)}$.
Step 5: Upon convergence in Step 4.3 at iteration m, set $\mathbf{I}_h = \mathbf{I}_h^{(m)}$ and $\dot{\mathbf{I}}_h = \dot{\mathbf{I}}_h^{(m)}$.

End loop for time marching.

In (7.40) and (7.43), $\mathbf{p}$ and $\mathbf{c}$ are predictor and corrector coefficients of dimension $2K$ and $2K + 1$, respectively. They can be computed using different methods. If polynomial interpolation/extrapolation on the temporal samples of the solution are used, $\mathbf{p}$ and $\mathbf{c}$ can be obtained analytically and the resulting PE(CE)^m scheme would for example be named Adam–Moulton, Adam–Bashforth, or backward difference methods [112] depending on the order of the interpolation/extrapolation scheme used. $\mathbf{p}$ and $\mathbf{c}$ can also be obtained using a numerical scheme as described in [113]. This scheme expresses the ODE solution in terms of decaying and oscillating exponentials. This expansion's coefficients are never computed explicitly but the expansion is used in an error controllable scheme to obtain $\mathbf{p}$ and $\mathbf{c}$. The resulting PE(CE)^m scheme is still in the form of its traditional (polynomial-based) counterparts but does not require construction of higher order polynomials to achieve high accuracy.

At the start of time marching, it is assumed that $\mathbf{I}_l = \mathbf{0}$ and $\dot{\mathbf{I}}_l = \mathbf{0}$, for $l = 1 - K, \ldots, 0$. This assumption does not introduce any significant errors in the solution since $\mathbf{H}^{\text{inc}}(\mathbf{r}, t)$ is band-limited and vanishingly small for $\forall \mathbf{r} \in \{S, V\}$ and $t \le 0$. For other types of excitations, one can use the Euler method or spectral-deferred correction-type methods for the coefficient initialization [112, 114].

The method used for solving (7.42) and (7.45) is chosen depending on the sparsity structure of $\mathbf{G}$. Note that $\mathbf{G}$ is always sparse regardless of the value of Δt. For TD-MFIE discretized using the Nyström method and TD-MFVIE discretized using the FLC basis functions with point testing, $\mathbf{G}$ is block-diagonal and has four diagonal blocks, respectively. Therefore, for these two cases $\mathbf{G}^{-1}$ is computed and stored very efficiently before the beginning of time marching. During the time marching, $\dot{\mathbf{I}}_h$ and $\dot{\mathbf{I}}_h^{(n)}$ are computed by simply multiplying $\mathbf{G}^{-1}$ with the RHS vector of (7.42) and (7.45), respectively. On the other hand, for TD-MFIE discretized using the RWG basis functions and TD-MFVIE discretized using the FLC basis functions with Galerkin testing, $\mathbf{G}$ is sparse and well conditioned without a specific structure. Therefore, an iterative scheme is used to solve (7.42) and (7.45).

7.5 Implicit MOT Scheme

Inserting (7.3) and (7.10) into (7.2), (7.14) and (7.15) into (7.2), and (7.24) and (7.26) into (7.23) yield the discretized TD-MFIE using the RWG basis functions, the discretized TD-MFIE using the Nyström method, and the discretized TD-MFVIE using the FLC basis functions in space and time, respectively.

Testing the resulting equations with $\mathbf{f}_m(\mathbf{r})$, $m = 1, \ldots, N_s$, $\mathbf{u}(\mathbf{r}_{jq})$ and $\mathbf{v}(\mathbf{r}_{jq})$, $j = 1, \ldots, N_n$, $q = 1, \ldots, N_p$, and $\mathbf{t}_m^1(\mathbf{r})$ and $\mathbf{t}_m^2(\mathbf{r})$, $m = 1, \ldots, N_e$ at times $t = h\Delta t$, $h = 1, \ldots, N_t$ yields the implicit counterparts of (7.13), (7.16), and (7.37), respectively. All these implicit MOT systems can be written in the same form as

$$\mathbf{Z}_0^{\text{imp}} \mathbf{I}_h = \mathbf{V}_h^{\text{inc}} - \sum_{l=1}^{h-1} \mathbf{Z}_{h-l}^{\text{imp}} \mathbf{I}_l, \quad h = 1, \ldots, N_t \tag{7.46}$$

Here, the implicit MOT matrices $\mathbf{Z}_{h-l}^{\text{imp}}$ are given by

$$\mathbf{Z}_{h-l}^{\text{imp}} = \mathbf{G} \, \partial_t T(t)\big|_{t=(h-l)\Delta t} + \mathbf{Z}_{h-l}^{\text{exp}} \tag{7.47}$$

In (7.46), $\mathbf{I}_h$, $h = 1, \ldots, N_t$ are obtained recursively using a time marching scheme. First, at $t = \Delta t$, $\mathbf{I}_1$ is calculated by solving $\mathbf{Z}_0^{\text{imp}} \mathbf{I}_1 = \mathbf{V}_1^{\text{inc}}$ [$h = 1$ in (7.46)]. Then, at $t = 2\Delta t$, $\mathbf{I}_2$ is found by solving $\mathbf{Z}_0^{\text{imp}} \mathbf{I}_2 = \mathbf{V}_2^{\text{inc}} - \mathbf{Z}_1^{\text{imp}} \mathbf{I}_1$ [$h = 2$ in (7.46)]. Next, at $t = 3\Delta t$, $\mathbf{I}_3$ is computed by solving $\mathbf{Z}_0^{\text{imp}} \mathbf{I}_3 = \mathbf{V}_3^{\text{inc}} - \mathbf{Z}_2^{\text{imp}} \mathbf{I}_1 - \mathbf{Z}_1^{\text{imp}} \mathbf{I}_2$ [$h = 3$ in (7.46)] and so on. During this time-marching scheme, the solution of (7.46) required at each time step is always obtained using an iterative solver.

7.6 Comparison of Implicit and Explicit Solutions

In this section, it is demonstrated that, for the same Δt, the corrector updates of the explicit MOT scheme yield the same result as the iterative solution of the implicit MOT system upon convergence if $T(t)$ (constructed using piecewise Lagrange polynomials) has the same order as the Lagrange polynomial interpolation used for deriving $\mathbf{c}$.

For example, it is assumed that $\mathbf{c}$ is obtained using a third-order backward difference method [112], which yields

$$\mathbf{I}_h - \frac{18}{11}\mathbf{I}_{h-1} + \frac{9}{11}\mathbf{I}_{h-2} - \frac{2}{11}\mathbf{I}_{h-3} = \frac{6}{11}\Delta t \dot{\mathbf{I}}_h \tag{7.48}$$

Assume that the corrector updates of the PE(CE)^m scheme have converged, substituting (7.48) into (7.38) yields

$$\frac{11}{6\Delta t}\mathbf{G}\mathbf{I}_h - \frac{3}{\Delta t}\mathbf{G}\mathbf{I}_{h-1} + \frac{3}{2\Delta t}\mathbf{G}\mathbf{I}_{h-2} - \frac{1}{3\Delta t}\mathbf{G}\mathbf{I}_{h-3}$$
$$= \mathbf{V}_h^{\text{inc}} - \sum_{l=1}^{h}\mathbf{Z}_{h-l}^{\text{exp}}\mathbf{I}_l, \quad h = 1, \dots, N_t \tag{7.49}$$

Merging the terms with $\mathbf{I}_{h-3}$, $\mathbf{I}_{h-2}$, and $\mathbf{I}_{h-1}$ to the RHS and those with $\mathbf{I}_h$ to the left-hand side yields

$$\tilde{\mathbf{Z}}_0^{\text{exp}}\mathbf{I}_h = \mathbf{V}_h^{\text{inc}} - \sum_{l=1}^{h-1}\tilde{\mathbf{Z}}_{h-l}^{\text{exp}}\mathbf{I}_l, \quad h = 1, \dots, N_t \tag{7.50}$$

where $\tilde{\mathbf{Z}}_0^{\text{exp}} = 11/(6\Delta t)\mathbf{G} + \mathbf{Z}_0^{\text{exp}}$, $\tilde{\mathbf{Z}}_1^{\text{exp}} = -3/(\Delta t)\mathbf{G} + \mathbf{Z}_1^{\text{exp}}$, $\tilde{\mathbf{Z}}_2^{\text{exp}} = 3/(2\Delta t)\mathbf{G} + \mathbf{Z}_2^{\text{exp}}$, $\tilde{\mathbf{Z}}_3^{\text{exp}} = -1/(3\Delta t)\mathbf{G} + \mathbf{Z}_3^{\text{exp}}$, and $\tilde{\mathbf{Z}}_{h-l}^{\text{exp}} = \mathbf{Z}_{h-l}^{\text{exp}}, l = 1, \dots, h-4$.

If $T(t)$ is constructed using third-order Lagrange polynomials [20], i.e. the same order as the one used for deriving (7.48), then $\partial_t T(t)|_{t=0} = 11/(6\Delta t)$, $\partial_t T(t)|_{t=\Delta t} = -3/(\Delta t)$, $\partial_t T(t)|_{t=2\Delta t} = 3/(2\Delta t)$, and $\partial_t T(t)|_{t=3\Delta t} = -1/(3\Delta t)$. Substituting these expressions into (7.47) yields $\mathbf{Z}_0^{\text{imp}} = 11/(6\Delta t)\mathbf{G} + \mathbf{Z}_0^{\text{exp}}$, $\mathbf{Z}_1^{\text{imp}} = -3/(\Delta t)\mathbf{G} + \mathbf{Z}_1^{\text{exp}}$, $\mathbf{Z}_2^{\text{imp}} = 3/(2\Delta t)\mathbf{G} + \mathbf{Z}_2^{\text{exp}}$, $\mathbf{Z}_3^{\text{imp}} = -1/(3\Delta t)\mathbf{G} + \mathbf{Z}_3^{\text{exp}}$, and $\mathbf{Z}_{h-l}^{\text{imp}} = \mathbf{Z}_{h-l}^{\text{exp}}, l = 1, \dots, h-4$, which demonstrates that the system in (7.50) is the same as that in (7.46).

To this end, it is shown that, upon convergence, the corrector updates generate the same result as the iterative solution of (7.46). This conclusion is demonstrated by numerical results in Section 7.9. Additionally, the numerical results also indicate that the explicit MOT scheme can use the same Δt as its implicit counterpart without sacrificing the stability of the solution.

7.7 Computational Complexity Analysis

The computational complexity of the explicit MOT scheme is compared to that of its implicit counterpart in this section. Since the explicit MOT solver can use the same time step size as its implicit counterpart with the same level of solution accuracy (as shown in Section 7.9), only the computational costs per time step are analyzed here.

Let $C_{\text{fix}}^{\text{imp}}$ and $C_{\text{fix}}^{\text{exp}}$ denote the cost of computing $\mathbf{V}_h^{\text{fix}} = \mathbf{V}_h^{\text{inc}} - \sum_{l=1}^{h-1} \mathbf{Z}_{h-l}^{\text{exp}} \mathbf{I}_l$ [see (7.39)] by the explicit MOT scheme and $\mathbf{V}_h^{\text{inc}} - \sum_{l=1}^{h-1} \mathbf{Z}_{h-l}^{\text{imp}} \mathbf{I}_l$ [see (7.46)] by its implicit counterpart, respectively, at the time step h. Since the cost of computing $\sum_{l=1}^{h-1} \mathbf{Z}_{h-l}^{\text{exp}} \mathbf{I}_l$ is same as that of computing $\sum_{l=1}^{h-1} \mathbf{Z}_{h-l}^{\text{imp}} \mathbf{I}_l$, then $C_{\text{fix}}^{\text{imp}} = C_{\text{fix}}^{\text{exp}}$. It should be noted that, regardless of Δt, the computation of $\sum_{l=1}^{h-1} \mathbf{Z}_{h-l}^{\text{exp}} \mathbf{I}_l$ or $\sum_{l=1}^{h-1} \mathbf{Z}_{h-l}^{\text{imp}} \mathbf{I}_l$ at any time step $h > \lfloor D_{\max}/(c\Delta t)\rfloor + T_{\max}$ needs $O(N^2)$ operations, where $N \in \{2N_p N_n, N_s, 2N_e\}$ denotes the number of unknowns and $D_{\max}$ represents the maximum distance between any two points on S for PEC objects or in V for dielectric scatterers.

Let $C_{\text{tot}}^{\text{imp}} = C_{\text{sol}}^{\text{imp}} + C_{\text{fix}}^{\text{imp}}$ and $C_{\text{tot}}^{\text{exp}} = C_{\text{sol}}^{\text{exp}} + C_{\text{fix}}^{\text{exp}}$ represent the total computational cost at the time step h by the implicit and explicit MOT solvers, respectively. Here, $C_{\text{sol}}^{\text{exp}}$ is the cost for implementing Steps 2–4 in Section 7.4, which is given by

$$C_{\text{sol}}^{\text{exp}} \sim O(\{m[2K+1] + 2K\}N) + O([m+1]\alpha N) + O([m+1]\beta N) \qquad (7.51)$$

where the first term is the cost of summations in (7.40) and (7.43), the second term is the cost of matrix-vector multiplications in (7.41) and (7.44), and the last term is the cost of matrix-vector multiplications in (7.42) and (7.45). Here, α is the sparseness factor of $\mathbf{Z}_0^{\text{exp}}$ and β depends on the method used for solving (7.42) and (7.45). For TD-MFIE discretized using the Nyström method and TD-MFVIE discretized using the FLC basis functions with point testing, (7.42) and (7.45) are solved by multiplying $\mathbf{G}^{-1}$ with the RHS vectors, resulting in $\beta = 2$. For TD-MFIE discretized using the RWG basis functions and TD-MFVIE discretized using the FLC basis functions with Galerkin testing, (7.42) and (7.45) are solved using an iterative solver. This results in $\beta = N_{\text{iter}}^{\text{G}} F_{\text{iter}}^{\text{G}} \delta$, where $N_{\text{iter}}^{\text{G}}$ is the number of iterations, $F_{\text{iter}}^{\text{G}}$ is the number of matrix-vector multiplications at each iteration, and δ is the sparseness factor of $\mathbf{G}$.

On the other hand, for the implicit MOT scheme, (7.46) is always solved by an iterative scheme resulting in the computational cost as

$$C_{\text{sol}}^{\text{imp}} \sim O(N_{\text{iter}}^{\text{imp}} F_{\text{iter}}^{\text{imp}} \gamma N) \qquad (7.52)$$

where γ is the sparseness factor of $\mathbf{Z}_0^{\text{imp}}$, $N_{\text{iter}}^{\text{imp}}$ is the number of iterations, and $F_{\text{iter}}^{\text{imp}}$ is the number of matrix-vector multiplications $\mathbf{Z}_0^{\text{imp}} \mathbf{I}_h^{(n)}$ at every iteration ($\mathbf{I}_h^{(n)}$ is the solution vector at iteration n).

Under high-frequency excitations (i.e. $\Delta t \ll D_{\max}/c$), $\mathbf{Z}_0^{\text{exp}}$ and $\mathbf{Z}_0^{\text{imp}}$ are both very sparse, i.e. $\alpha \ll N$ and $\gamma \ll N$. Therefore, $C_{\text{sol}}^{\text{imp}}$ and $C_{\text{sol}}^{\text{exp}}$ scale as $C_{\text{sol}}^{\text{imp}} \sim O(N_{\text{iter}}^{\text{imp}} F_{\text{iter}}^{\text{imp}} \gamma N)$ and $C_{\text{sol}}^{\text{exp}} \sim O(mKN) + O(m\alpha N) + O(m\beta N)$, respectively. The comparison of $C_{\text{sol}}^{\text{imp}}$ and $C_{\text{sol}}^{\text{exp}}$ depends on actual values of $m, \gamma, N_{\text{iter}}^{\text{imp}}, F_{\text{iter}}^{\text{imp}}, K, \alpha$, and β. Note that $C_{\text{sol}}^{\text{imp}} \ll C_{\text{fix}}^{\text{imp}}$ and $C_{\text{sol}}^{\text{exp}} \ll C_{\text{fix}}^{\text{exp}}$, which results in $C_{\text{tot}}^{\text{imp}} \approx C_{\text{fix}}^{\text{imp}}$ and $C_{\text{tot}}^{\text{exp}} \approx C_{\text{fix}}^{\text{exp}}$. This means that the implicit and explicit solvers require similar total execution times under high-frequency excitations (as shown in Section 7.9).

Under low-frequency excitations (i.e. $\Delta t \approx D_{\text{max}}/c$), $\mathbf{Z}_0^{\text{exp}}$ and $\mathbf{Z}_0^{\text{imp}}$ become denser (even full), which results in $\alpha \approx N$ and $\gamma \approx N$. Since $K \ll N$ and $N_{\text{iter}}^{\text{G}} F_{\text{iter}}^{\text{G}} \delta \ll N$, the scales $C_{\text{sol}}^{\text{imp}} \sim O(N_{\text{iter}}^{\text{imp}} F_{\text{iter}}^{\text{imp}} N^2)$ and $C_{\text{sol}}^{\text{exp}} \sim O(mN^2)$ result in $C_{\text{sol}}^{\text{imp}} \gg C_{\text{fix}}^{\text{imp}}$ and $C_{\text{sol}}^{\text{exp}} \gg C_{\text{fix}}^{\text{exp}}$, respectively. Consequently, under low-frequency excitations, $C_{\text{tot}}^{\text{imp}} \approx C_{\text{sol}}^{\text{imp}}$ and $C_{\text{tot}}^{\text{exp}} \approx C_{\text{sol}}^{\text{exp}}$. The explicit MOT scheme is faster than its implicit counterpart as long as $m < N_{\text{iter}}^{\text{imp}} F_{\text{iter}}^{\text{imp}}$ (as shown in Section 7.9).

It should be noted that both $C_{\text{fix}}^{\text{imp}}$ and $C_{\text{fix}}^{\text{exp}}$ can be reduced using PWTD [15, 16, 18–20, 24–27, 33, 78–84] or FFT-based algorithms [7–9, 11–13, 85, 86], which still ensures $C_{\text{fix}}^{\text{imp}} = C_{\text{fix}}^{\text{exp}}$. Similarly, FFT-based algorithms have been developed to reduce the cost of matrix-vector multiplications $\mathbf{Z}_0^{\text{imp}} \mathbf{I}_h^{(n)}$ [10]. The same methods can account for computing the matrix-vector multiplications in (7.41) and (7.44) as well. Therefore, the above conclusions on the computational complexity are still applicable even when acceleration algorithms are used.

7.8 Remarks

Several comments on the explicit and implicit MOT schemes are provided below.

(1) In (7.2) and (7.23), temporal derivative forms of TD-MFIE and TD-MFVIE are used, respectively, which are required to cast TD-MFIE and TD-MFVIE into the forms of an ODE. Note that, when an MOT scheme is used for the solution, a DC component often appears in the solution even if the zero initial condition is enforced at the start of time marching. This DC component is caused by numerical errors, especially from the solution of the MOT system [115]. The numerical results presented in Section 7.9 show that the DC component is stable with a small amplitude, which can be reduced by increasing the accuracy of the solution.

(2) The temporal interpolation function used above is discretely causal, i.e. $T(t) = 0$ for $t \leq -\Delta t$. Therefore, during time marching, $\mathbf{I}_l$, $l > h$ ["future" samples of $\mathbf{I}(t)$] do not contribute to the computation of $\mathbf{I}_h$. Actually, $T(t)$ can also be noncausal, e.g. APSWF [73], which results in $\mathbf{Z}_{h-l}^{\text{exp}} \neq \mathbf{0}$, $l > h$ and $\mathbf{I}_l$, $l > h$ are needed to compute $\mathbf{I}_h$. The causality of the time marching can be restored using extrapolation schemes [28, 74, 75].

(3) The explicit MOT matrices $\mathbf{Z}_{h-l}^{\text{exp}} = \mathbf{0}$ for $h - l > \lfloor D_{\text{max}}/(c\Delta t) \rfloor + T_{\text{max}}$. With the increase of Δt, the number of nonzero matrices $\mathbf{Z}_{h-l}^{\text{exp}}$ decreases while they become denser, e.g. when $D_{\text{max}}/(c\Delta t) < 1$, every nonzero $\mathbf{Z}_{h-l}^{\text{exp}}$ is a full matrix. With the decrease of Δt, the number of nonzero matrices $\mathbf{Z}_{h-l}^{\text{exp}}$ increases while they become sparser.

(4) In (7.47), only nonzero entries of $\mathbf{G}$ contribute to $\mathbf{Z}_{h-l}^{\text{imp}}$ [see (7.5), (7.18), and (7.28)] and when $\partial_t T(t)|_{t=(h-l)\Delta t} \neq 0$ [only for $h - l \in \{0, 1, \ldots, T_{\max}\}$]. For TD-MFIE, $\mathbf{Z}_{h-l}^{\text{imp}}$ and $\mathbf{Z}_{h-l}^{\text{exp}}$ have the same sparsity structure since $\mathbf{G}$ contributes to $\mathbf{Z}_{h-l}^{\text{imp}}$ in which the entries of $\mathbf{Z}_{h-l}^{\text{exp}}$ are already nonzero. For TD-MFVIE, adding $\mathbf{G}_{12}$ and $\mathbf{G}_{22}$ to $\mathbf{M}_{h-l}^1$ and $\mathbf{M}_{h-l}^2$, respectively, does not change the sparsity structure of $\mathbf{Z}_{h-l}^{\text{imp}}$. On the other hand, the sparsity level of $\mathbf{Z}_{h-l}^{\text{imp}}$ decreases with the inclusive of $\mathbf{G}_{11}$ and $\mathbf{G}_{21}$, which results in the cost of computing $\sum_{l=1}^{h-1} \mathbf{Z}_{h-l}^{\text{imp}} \mathbf{I}_l$ [see (7.46)] higher than that of computing $\sum_{l=1}^{h-1} \mathbf{Z}_{h-l}^{\text{exp}} \mathbf{I}_l$ [see (7.39)]. However, this difference can be ignored in the computational complexity comparison since $T_{\max} \ll D_{\max}/(c\Delta t)$ for small Δt and $\mathbf{G}_{11}$ and $\mathbf{G}_{21}$ are sparser than $\mathbf{M}_{h-l}^1$ and $\mathbf{M}_{h-l}^2$ for large Δt.

7.9 Numerical Results

This section presents numerical results that demonstrate the accuracy, stability, and efficiency of the proposed explicit MOT schemes. In all examples, the scatterer residing in free space ($\mu = \mu_0$, $\varepsilon = \varepsilon_0$, and $c = c_0$) is excited by a plane wave with magnetic field

$$\mathbf{H}^{\text{inc}}(\mathbf{r}, t) = \hat{\mathbf{p}}\sqrt{\varepsilon/\mu}G(t - \hat{\mathbf{k}} \cdot \mathbf{r}/c) \tag{7.53}$$

where $\hat{\mathbf{p}}$ is the direction of polarization, $\hat{\mathbf{k}}$ is the direction of propagation, and $G(t) = \cos[2\pi f_0(t - t_0)] \exp[-(t - t_0)^2/(2\sigma^2)]$ is a Gaussian pulse. Here, f_0 is the center frequency, t_0 is the delay, σ is the pulse duration, $f_{\max} = f_0 + f_{\text{bw}}$ is the effective maximum frequency, and f_{bw} is the bandwidth.

A transpose-free quasi-minimal residual (TFQMR) method [116] is used to iteratively solve (7.42) and (7.45) for TD-MFIE discretized using the RWG basis functions and TD-MFVIE discretized using the FLC basis functions with Galerkin testing as well as the implicit MOT system (7.46). The convergence criteria for TFQMR iterations and correction updates Step 4.3 in the PE(CE)^{m} scheme are defined as $\left\| \mathbf{I}_h^{(n)} - \mathbf{I}_h^{(n-1)} \right\| < \chi^{\text{type}}$ for TD-MFIE and as $\left\| \mathbf{I}_h^{(n)} - \mathbf{I}_h^{(n-1)} \right\| < \chi^{\text{type}} \left\| \mathbf{I}_{h-1} \right\|$ for TD-MFVIE, where $\chi^{\text{type}} \in \{\chi^{\text{TFQMR}}, \chi^{\text{PECE}}\}$ is a threshold parameter.

Once the time domain simulations are completed, the solution coefficients (surface currents for TD-MFIE and magnetic field for TD-MFVIE) are Fourier transformed at frequency f. Then, Fourier transformed solutions are normalized by the Fourier transform of $G(t)$ at f. This yields the time-harmonic solution coefficients, which are then used to compute radar cross section (RCS) of the scatterer (see, for example, [117] for details of these operations). To quantify the accuracy of simulations, the L_2-norm relative error in RCS is defined as

$$\sigma_{\text{err}}^{\text{type}} = \sqrt{\frac{\sum_{n=0}^{N_\theta} \left| \sigma^{\text{type}}(n\Delta\theta, \varphi, f) - \sigma^{\text{ref}}(n\Delta\theta, \varphi, f) \right|^2}{\sum_{n=0}^{N_\theta} \left| \sigma^{\text{ref}}(n\Delta\theta, \varphi, f) \right|^2}} \tag{7.54}$$

where $\Delta\theta$ is the sampling step along $\hat{\theta}$ direction, $\varphi = 0°$ in all examples, and $\sigma^{\text{type}}(\theta, \varphi, f)$ and $\sigma^{\text{ref}}(\theta, \varphi, f)$ represent the RCS computed using the MOT schemes and a reference solver. Reference solver is either the Mie series [118], a frequency-domain surface integral equation solver [71], or a frequency-domain volume integral equation solver [72].

7.9.1 TD-MFIE Discretized Using RWG Basis Functions

For all the examples in this section, $\hat{\mathbf{k}} = -\hat{\mathbf{z}}$, $\hat{\mathbf{p}} = -\hat{\mathbf{y}}$, $\sigma = 7/(2\pi f_{\text{bw}})$, $t_0 = 3.5\sigma + 10/(f_0 + f_{\text{bw}})$, and $T_{\text{max}} = 3$. The coefficients $\mathbf{p}$ and $\mathbf{c}$ are obtained using a numerical scheme [113], which requires the solution of two minimum norm least square problems with the precision τ [34]. $\sigma^{\text{Mie}}(\theta, \varphi, f)$, $\sigma^{\text{MOM}}(\theta, \varphi, f)$, $\sigma^{\text{imp}}(\theta, \varphi, f)$, and $\sigma^{\text{exp}}(\theta, \varphi, f)$ represent the RCS computed using the Mie series solution and the solutions obtained by a frequency-domain surface integral equation solver [71], the implicit MOT scheme, and the explicit MOT scheme, respectively.

For the first example, scattering from a unit PEC sphere is analyzed for high-frequency excitations. The sphere is meshed into $N_s = 18,750$ RWG basis functions. The excitation has $f_0 = 300$ MHz and $f_{\text{bw}} = 200$ MHz with the time step $\Delta t = 0.1$ ns. $K = 22$ and $\tau = 10^{-4}$ for obtaining $\mathbf{p}$ and $\mathbf{c}$, $\chi^{\text{PECE}} = 10^{-14}$, and $\chi^{\text{TFQMR}} = 10^{-16}$. Figure 7.1 shows that the solution coefficient obtained using both MOT solvers agree very well. Figure 7.2 compares $\sigma^{\text{Mie}}(\theta, \varphi, f)$, $\sigma^{\text{imp}}(\theta, \varphi, f)$, and $\sigma^{\text{exp}}(\theta, \varphi, f)$ for $\theta = [0°, 180°]$, $\varphi = 0°$, and $f = 300$ MHz. This figure demonstrates the accuracy of both MOT solvers.

To investigate the convergence of the solution, Δt is swept between 0.05 and 0.4 ns for the above excitation. For each set of Δt, both MOT schemes are executed to calculate the RCS values. Figure 7.3 shows $\sigma_{\text{err}}^{\text{imp}}$ and $\sigma_{\text{err}}^{\text{exp}}$ versus Δt. Several observations are listed in order: (i) Both $\sigma_{\text{err}}^{\text{imp}}$ and $\sigma_{\text{err}}^{\text{exp}}$ saturate when $\Delta t < 0.1$ ns. This is because of that errors from spatial discretization dominate those from temporal discretization for smaller Δt. (ii) Both $\sigma_{\text{err}}^{\text{imp}}$ and $\sigma_{\text{err}}^{\text{exp}}$ follow $O(\Delta t^3)$ curve since $T(t)$ is constructed using third-order Lagrange polynomials in this section. (iii) For a given Δt, $\sigma_{\text{err}}^{\text{exp}}$ is slightly smaller than $\sigma_{\text{err}}^{\text{imp}}$. It means that the temporal derivative on $\mathbf{J}(\mathbf{r}, t)$ in (7.2) is approximated more accurately by the explicit MOT scheme.

For the second example, scattering from a flower-shaped PEC scatterer [75] is analyzed. The parameters for this example are $N_s = 23,550$, $f_0 = 200$ MHz, $f_{\text{bw}} = 150$ MHz, and $\Delta t = 0.1428$ ns. $K = 22$ and $\tau = 10^{-4}$ for obtaining $\mathbf{p}$ and $\mathbf{c}$, $\chi^{\text{PECE}} = 10^{-14}$, and $\chi^{\text{TFQMR}} = 10^{-16}$. Figure 7.4 compares the coefficient of an RWG basis function using the explicit and implicit MOT solvers; both coefficients match very well. Figure 7.5a–c shows snapshots of the surface current at times $300\Delta t$, $400\Delta t$, and $450\Delta t$. Figure 7.6 compares $\sigma^{\text{imp}}(\theta, \varphi, f)$ and $\sigma^{\text{exp}}(\theta, \varphi, f)$ to $\sigma^{\text{MOM}}(\theta, \varphi, f)$ at $f = 240$ MHz for $\theta = [0°, 180°]$ and $\varphi = 0°$. Again, these results

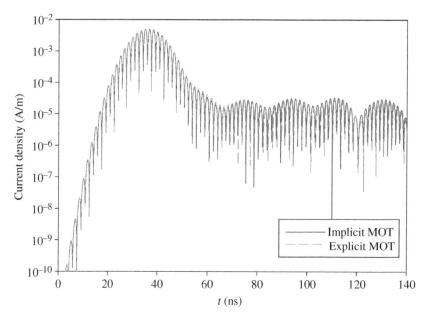

Figure 7.1 Solution coefficient obtained by the explicit and the implicit MOT solvers. Source: Reproduced with permission from [34].

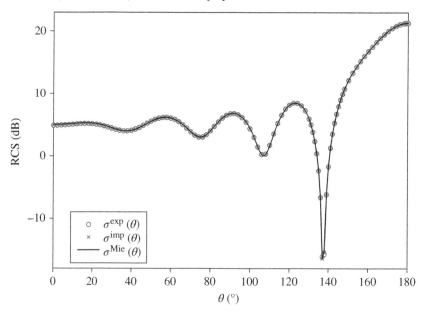

Figure 7.2 Comparison of $\sigma^{\text{Mie}}(\theta, \varphi, f)$, $\sigma^{\text{imp}}(\theta, \varphi, f)$, and $\sigma^{\text{exp}}(\theta, \varphi, f)$ for $\theta = [0°, 180°]$, $\varphi = 0°$, and $f = 300$ MHz. Source: Reproduced with permission from [34].

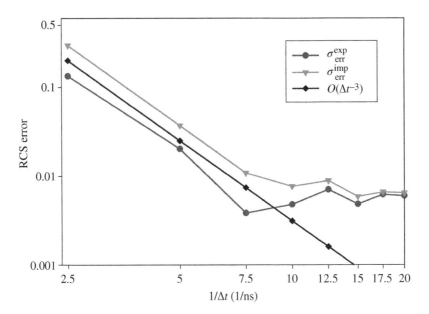

Figure 7.3 σ_{err}^{imp} and σ_{err}^{exp} versus Δt. Reproduced with permission from [34].

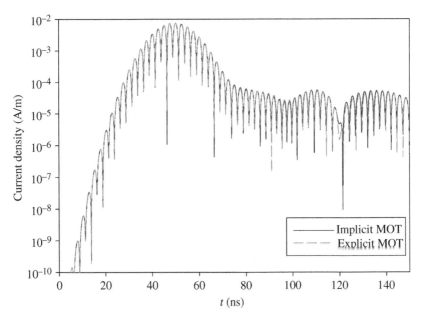

Figure 7.4 Comparison of the coefficient of an RWG basis function using the explicit and implicit MOT solvers. Source: Reproduced with permission from [34].

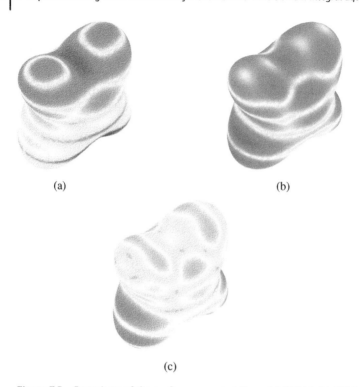

(a)   (b)

(c)

Figure 7.5 Snapshots of the surface current at times (a) 300Δt, (b) 400Δt, and (c) 450Δt. Source: Reproduced with permission from [34].

demonstrate the accuracy of both MOT schemes and show that solution of the explicit MOT solver has the same level of accuracy as that of its implicit counterpart with the same Δt.

For the next set of simulations, the computational efficiency of the explicit MOT scheme is demonstrated by comparing its time cost to that of its implicit counterpart. Two different sets of meshes for a unit PEC sphere and four different sets of excitations are considered: $N_s = \{2430, 7680\}$ and $f_0 = \{10, 25, 50, 100\}$ MHz with $f_{bw} = 0.75 f_0$ and $\Delta t = 1/(17.5 f_0)$. The explicit MOT solver uses two different sets of coefficients **p** and **c** resulting in $K = 22$ and $\tau = 10^{-4}$ (Explicit-I) and $K = 7$ and $\tau = 10^{-13}$ (Explicit-II), respectively. For the explicit MOT solver, the tolerance value of Step 4.3 in the PE(CE)^m scheme is set to $\chi^{PECE} = 10^{-13}$ and that of iteratively solving (7.42) and (7.45) is $\chi^{TFQMR} = 10^{-16}$. For the implicit MOT solver, the tolerance of TFQMR iterations is $\chi^{TFQMR} = 10^{-13}$. All simulations are run for $N_t = 500$ time steps. Table 7.1 compares the time costs of different stages of both

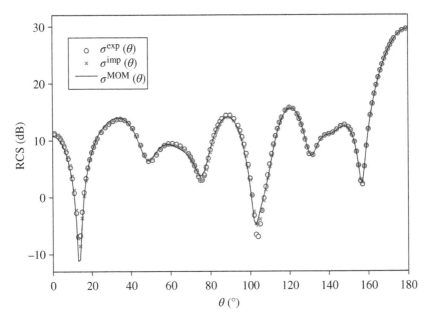

Figure 7.6 Comparison of $\sigma^{imp}(\theta, \varphi, f)$ and $\sigma^{exp}(\theta, \varphi, f)$ to $\sigma^{MOM}(\theta, \varphi, f)$ at $f = 240$ MHz for $\theta = [0°, 180°]$ and $\varphi = 0°$. Source: Reproduced with permission from [34].

MOT schemes. In this table, T^{imp}_{TFQMR} is the total time required to iteratively solve (7.46) by the TFQMR solver for $h = 1, \ldots, N_t$. T^{exp}_{PECE} is the total time for implementing Steps 2–4 in Section 7.4 for $h = 1, \ldots, N_t$. T^{imp}_{tot} and T^{exp}_{tot} are the total execution times of the implicit and explicit MOT schemes, respectively. $\{N^{imp}_{iter} F^{imp}_{iter}\}_{avg}$ and m_{avg} are the average values of $N^{imp}_{iter} F^{imp}_{iter}$ and m over N_t time steps, respectively.

As seen from Table 7.1, the explicit MOT solver is faster than its implicit counterpart for large Δt since $m_{avg} < \{N^{imp}_{iter} F^{imp}_{iter}\}_{avg}$. In this regime of Δt, T^{exp}_{tot} is almost half of T^{imp}_{tot}. Table 7.1 also shows that for large Δt, T^{exp}_{PECE} is almost one third of T^{imp}_{TFQMR}. However, for small Δt, both MOT solvers require roughly equal execution time. Table 7.1 also indicates that increasing the accuracy of **p** and **c** decreases the value of m_{avg}, therefore, Explicit-II is more efficient than Explicit-I. In Table 7.2, σ^{imp}_{err}, σ^{expI}_{err}, and σ^{expII}_{err} at $f = f_0$ for $\theta = [0°, 180°]$ and $\varphi = 0°$ are computed by the implicit, "Explicit-I," and "Explicit-II" MOT solvers, respectively. As expected, the explicit MOT solver is slightly more accurate than the implicit one for a given value of Δt. It means that for a given level of accuracy, the explicit MOT scheme is more efficient since it can obtain this accuracy using a slightly larger Δt.

Table 7.1 Comparison of the time costs[a] of different stages of both MOT schemes.

N_s	f_0	Implicit			Explicit-I					Explicit-II				
		T^{imp}_{TFQMR}	T^{imp}_{tot}	$\{N^{imp}_{iter} F^{imp}_{iter}\}_{avg}$	T^{exp}_{PECE}	T^{exp}_{tot}	m_{avg}	$\dfrac{T^{imp}_{TFQMR}}{T^{exp}_{PECE}}$	$\dfrac{T^{imp}_{tot}}{T^{exp}_{tot}}$	T^{exp}_{PECE}	T^{exp}_{tot}	m_{avg}	$\dfrac{T^{imp}_{TFQMR}}{T^{exp}_{PECE}}$	$\dfrac{T^{imp}_{tot}}{T^{exp}_{tot}}$
2430	10	173	297	18.81	115	210	10.7	1.50	1.41	76	173	6.43	2.26	1.71
2430	25	38	158	18.50	19	134	4.93	1.96	1.17	16	132	4.25	2.31	1.19
2430	50	12	162	18.38	9	158	3.89	1.34	1.02	8	157	3.44	1.49	1.03
2430	100	7	263	21.8	7	262	4.83	0.92	1.00	6	257	3.14	1.17	1.02
7680	10	1674	2813	19.52	1025	1908	10.97	1.63	1.47	656	1536	6.66	2.55	1.83
7680	25	312	1377	18.71	112	1121	4.96	2.77	1.22	102	1118	4.31	2.94	1.20
7680	50	101	1454	18.82	40	1369	3.80	2.53	1.06	35	1345	3.37	2.98	1.09
7680	100	43	2094	22.54	28	2035	4.64	1.51	1.02	21	2069	3.09	2.06	0.97

a) The unit of time is second.
Source: Reproduced with permission from [34].

Table 7.2 Comparison of σ_{err}^{imp}, σ_{err}^{expl}, and σ_{err}^{expll} at $f = f_0$ for $\theta = [0°, 180°]$ and $\varphi = 0°$.

N_s	f_0 (MHz)	Implicit σ_{err}^{imp}	Explicit-I σ_{err}^{expl}	Explicit-II σ_{err}^{expll}
2430	10	0.0365	0.0211	0.0246
2430	25	0.0316	0.0097	0.0156
2430	50	0.0374	0.0063	0.0129
2430	100	0.0216	0.0188	0.0160
7680	10	0.0242	0.0085	0.0120
7680	25	0.0192	0.0040	0.0034
7680	50	0.0313	0.0108	0.0069
7680	100	0.0104	0.0078	0.0047

Source: Reproduced with permission from [34].

7.9.2 TD-MFIE Discretized Using the Nyström Method

In this section, the scatterer is a unit PEC sphere centered at the origin, $\hat{\mathbf{k}} = \hat{\mathbf{z}}$, $\hat{\mathbf{p}} = \hat{\mathbf{y}}$, $\sigma = 7/(2\pi f_{bw})$, and $t_0 = 3.5\sigma + 10/(f_0 + f_{bw})$. The order of $\ell_{ip}(\mathbf{r})$ is two ($N_n = 6$) [103], and the order of Lagrange polynomials for constructing $T(t)$ is three ($T_{max} = 3$) [20]. The coefficients $\mathbf{p}$ and $\mathbf{c}$ are obtained using the Adams–Bashforth and third-order backward difference methods [112], respectively, resulting in $K = 4$. Convergence thresholds are $\chi^{PECE} = \chi^{TFQMR} = 10^{-13}$. $\sigma^{Mie}(\theta, \varphi, f)$, $\sigma^{imp}(\theta, \varphi, f)$, and $\sigma^{exp}(\theta, \varphi, f)$ represent the RCS computed using the Mie series solution and the solutions obtained by the implicit MOT (I-MOT) scheme and explicit MOT (E-MOT) scheme, respectively.

For the first example, $N_p = 1126$ (resulting in $N = 13,512$), $f_0 = 300$ MHz, $f_{bw} = 200$ MHz, $N_t = 1600$, and $\Delta t = 0.1$ ns $[1/(20 f_{max})]$. Figure 7.7a,b compare the $\left|\{\mathbf{I}(t)\}_{ip}^u\right|$ and $\left|\{\mathbf{I}(t)\}_{ip}^v\right|$ computed by the implicit and explicit MOT schemes at $\mathbf{r}_{ip} = (0.97, -0.245, 0.005)$ m ($p = 1014$, $i = 6$), respectively. The results match each other. Figure 7.8 plots $\sigma^{Mie}(\theta, \varphi, f)$, $\sigma^{imp}(\theta, \varphi, f)$, and $\sigma^{exp}(\theta, \varphi, f)$ versus θ for $\varphi = 0°$ at $f \in \{230, 300, 350\}$ MHz. It can be seen that the results obtained by both MOT schemes are accurate within the pulse band.

For the second set of simulations, two meshes of the scatterer are used: $N_p = 202$ ($N = 2424$), and $N_p = 622$ ($N = 7464$). For each mesh, seven sets of f_0 are used: $f_0 \in \{5, 10, 25, 50, 100, 200, 400\}$ MHz and $f_{bw} = 0.75 f_0$. Both MOT solvers are executed for $N_t = 500$ time steps with $\Delta t = 1/(17.5 f_0) = 1/(10 f_{max})$.

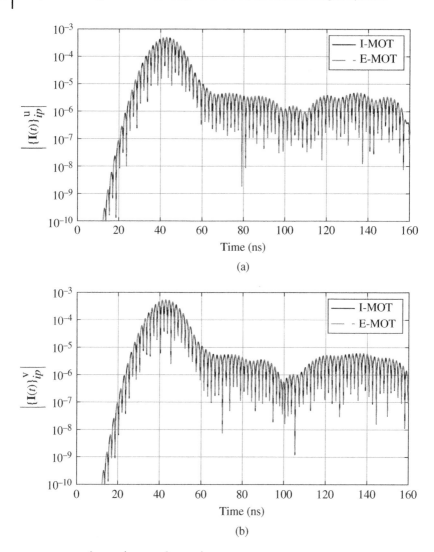

Figure 7.7 (a) $\left|\{I(t)\}_{ip}^{u}\right|$ and (b) $\left|\{I(t)\}_{ip}^{v}\right|$ computed by the implicit and explicit MOT schemes at $\mathbf{r}_{ip} = (0.97, -0.245, 0.005)$ m ($p = 1014, i = 6$). Source: Reproduced with permission from [38].

The sparseness factors of the matrices $\mathbf{Z}_0^{\exp}$ and $\mathbf{Z}_0^{\mathrm{imp}}$ for these sets of f_0 and meshes are $\alpha = \{2424, 1777, 303, 143, 61, 38, 13\}$ for $N_p = 202$ and $\gamma = \{7464, 5471, 881, 340, 116, 54, 32\}$ for $N_p = 622$, respectively. For each set of excitation and mesh, $\sigma^{\mathrm{Mie}}(\theta, \varphi, f)$, $\sigma^{\mathrm{imp}}(\theta, \varphi, f)$, and $\sigma^{\exp}(\theta, \varphi, f)$ are calculated for $\theta = [0°, 180°]$, $\varphi = 0°$, and $f = f_0$ with $N_\theta = 360$ and $\Delta\theta = 0.5°$. Table 7.3 shows

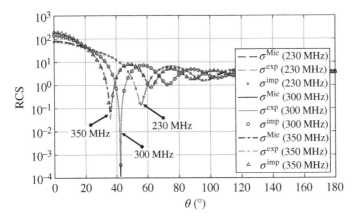

Figure 7.8 $\sigma^{\text{Mie}}(\theta, \varphi, f)$, $\sigma^{\text{imp}}(\theta, \varphi, f)$, and $\sigma^{\text{exp}}(\theta, \varphi, f)$ versus θ for $\varphi = 0°$ at $f \in \{230, 300, 350\}$ MHz. Source: Reproduced with permission from [38].

that both MOT schemes produce the same accuracy of solutions. Table 7.3 also compares the time costs of different stages of both MOT schemes. In this table, $T_{\text{fix}}^{\text{imp}}$ and $T_{\text{fix}}^{\text{exp}}$ are the total times required for computing $\mathbf{V}_h^{\text{inc}} - \sum_{l=1}^{h-1} \mathbf{Z}_{h-l}^{\text{imp}} \mathbf{I}_l$ and $\mathbf{V}_h^{\text{inc}} - \sum_{l=1}^{h-1} \mathbf{Z}_{h-l}^{\text{exp}} \mathbf{I}_l$ for $h = 1, \dots, N_t$, by the implicit and explicit MOT schemes, respectively. It is clearly shown in Table 7.3 that $T_{\text{fix}}^{\text{imp}} \approx T_{\text{fix}}^{\text{exp}}$. Additionally, for all sets of Δt, $T_{\text{PECE}}^{\text{exp}} < T_{\text{TFQMR}}^{\text{imp}}$ since $m_{\text{avg}} \ll \{N_{\text{iter}}^{\text{imp}} F_{\text{iter}}^{\text{imp}}\}_{\text{avg}}$, which results in different consequences for simulations with large and small Δt. For large Δt (i.e. low-frequency excitation), $T_{\text{TFQMR}}^{\text{imp}} \gg T_{\text{fix}}^{\text{imp}}$, $T_{\text{PECE}}^{\text{exp}} \gg T_{\text{fix}}^{\text{exp}}$, and $T_{\text{fix}}^{\text{imp}} \approx T_{\text{fix}}^{\text{exp}}$ yield $T_{\text{tot}}^{\text{imp}} > T_{\text{tot}}^{\text{exp}}$. Indeed, Table 7.3 shows that the explicit MOT solver is almost four times faster than its implicit counterpart. As Δt decreases (i.e. the frequency increases), $T_{\text{fix}}^{\text{imp}}$ and $T_{\text{fix}}^{\text{exp}}$ becomes larger than $T_{\text{TFQMR}}^{\text{imp}}$ and $T_{\text{PECE}}^{\text{exp}}$, and $T_{\text{PECE}}^{\text{exp}} < T_{\text{TFQMR}}^{\text{imp}}$ does not reflect on $T_{\text{tot}}^{\text{exp}}$ and $T_{\text{tot}}^{\text{imp}}$ anymore. Indeed, Table 7.3 shows that $T_{\text{tot}}^{\text{exp}} \approx T_{\text{tot}}^{\text{imp}}$ as Δt gets smaller. All the results presented in Table 7.3 demonstrate the computational complexity analysis in Section 7.7.

For the next set of simulations, $N_p = 622$ ($N = 7464$), $f_0 = 50$ MHz, $f_{\text{bw}} = 37.5$ MHz, and both MOT solvers are executed for Δt changed between 2.29 ns [$1/(5f_{\text{max}})$] and 0.29 ns [$1/(40f_{\text{max}})$]. $\sigma_{\text{err}}^{\text{imp}}$ and $\sigma_{\text{err}}^{\text{exp}}$ are calculated for $f \in \{35, 50, 65\}$ MHz, $N_\theta = 360$, $\Delta\theta = 0.5°$, and $\varphi = 0°$ using (7.54), Figure 7.9 plots $\sigma_{\text{err}}^{\text{imp}}$ and $\sigma_{\text{err}}^{\text{exp}}$ versus $1/\Delta t$ for $f \in \{35, 50, 65\}$ MHz. Again, solutions of both MOT schemes have the same accuracy. In addition, Figure 7.9 shows that the accuracy of both MOT schemes is in the order of $O(\Delta t^3)$ for larger values of Δt (note that $T_{\text{max}} = 3$ used in this section), however, for smaller values of Δt, the accuracy is limited by the second-order Nyström discretization, therefore, the convergence becomes slightly slower than $O(\Delta t^3)$.

Table 7.3 Comparison of the time costs of both MOT solvers.

		Implicit MOT					Explicit MOT						
N	f_0	σ_{err}^{imp}	T_{fix}^{imp} (s)	T_{TFQMR}^{imp} (s)	T_{tot}^{imp} (s)	$\{N_{iter}^{imp}F_{iter}^{imp}\}_{avg}$	σ_{err}^{exp}	T_{fix}^{exp} (s)	T_{PECE}^{exp} (s)	T_{tot}^{exp} (s)	m_{avg}	$\dfrac{T_{TFQMR}^{imp}}{T_{PECE}^{exp}}$	$\dfrac{T_{tot}^{imp}}{T_{tot}^{exp}}$
2424	5	0.011	61.81	763.61	825.53	66.66	0.011	65.43	129.02	194.56	9.78	5.92	4.24
2424	10	0.007	67.67	567.24	635.00	68.02	0.007	68.05	67.68	135.83	7.19	8.38	4.68
2424	25	0.004	83.87	100.29	184.21	70.15	0.004	83.65	7.40	91.14	4.19	13.55	2.02
2424	50	0.025	89.84	55.09	144.99	80.26	0.025	90.02	3.46	93.57	3.92	15.92	1.55
2424	100	0.047	91.70	28.56	120.31	92.53	0.047	91.71	2.07	93.86	5.20	13.80	1.28
2424	200	0.046	93.85	19.52	113.41	96.40	0.046	93.99	1.33	95.42	4.77	14.68	1.19
2424	400	0.068	99.47	8.61	108.13	99.80	0.068	99.62	0.63	100.33	4.32	13.67	1.08
7464	5	0.010	583.40	6266.59	6850.92	57.62	0.010	585.11	1155.56	1741.13	9.79	5.42	3.93
7464	10	0.006	638.02	4650.56	5289.34	58.56	0.006	637.91	640.10	1278.45	7.20	7.27	4.14
7464	25	0.004	779.17	784.12	1563.65	60.78	0.004	780.93	64.87	846.22	4.06	12.09	1.85
7464	50	0.025	828.96	349.04	1178.32	70.56	0.025	829.76	22.90	853.07	3.56	15.24	1.38
7464	100	0.047	850.05	136.80	987.14	79.08	0.047	851.62	10.80	862.83	4.85	12.67	1.14
7464	200	0.046	857.49	69.68	927.46	81.46	0.046	851.57	5.59	857.57	4.68	12.47	1.08
7464	400	0.066	882.09	44.40	926.77	82.91	0.066	884.64	3.70	888.75	4.46	12.00	1.04

Source: Reproduced with permission from [38].

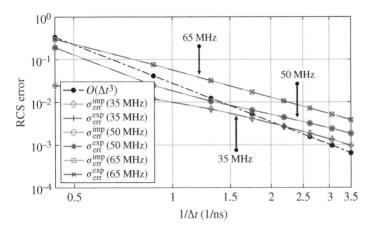

Figure 7.9 σ_{err}^{imp} and σ_{err}^{exp} versus $1/\Delta t$ for $f \in \{35, 50, 65\}$ MHz. Source: Reproduced with permission from [38].

7.9.3 TD-MFVIE Discretized Using FLC Basis Functions

For all examples in this section, excitation parameters are $\hat{\mathbf{k}} = \hat{\mathbf{z}}$, $\hat{\mathbf{p}} = \hat{\mathbf{y}}$, and $\sigma = 3/(2\pi f_{bw})$. The order of Lagrange polynomials for constructing $T(t)$ is four ($T_{max} = 4$) [11]. The coefficients $\mathbf{p}$ and $\mathbf{c}$ are obtained using the Adams–Bashforth and backward difference methods [112], respectively. The volume integrals in (7.28), (7.31), and (7.35) are computed using a third-order Gauss–Legendre quadrature rule [110, 111]. All the TFQMR iterations in this section are diagonally preconditioned. The TFQMR solver used for the implicit MOT scheme starts with an initial guess $\mathbf{I}_h = 2\mathbf{I}_{h-1} - \mathbf{I}_{h-2}$ at time step h. In the legends of figures, the names/types of the MOT schemes are abbreviated as follows: implicit MOT scheme with Galerkin testing ([MOT]$_{GT}^{imp}$), implicit MOT scheme with point testing ([MOT]$_{PT}^{imp}$), explicit MOT scheme with Galerkin testing ([MOT]$_{GT}^{exp}$), and explicit MOT scheme with point testing ([MOT]$_{PT}^{exp}$). Similarly, the RCS computed from the solutions obtained by the schemes are denoted by $\sigma_{GT}^{imp}(\theta, \varphi, f)$, $\sigma_{PT}^{imp}(\theta, \varphi, f)$, $\sigma_{GT}^{exp}(\theta, \varphi, f)$, and $\sigma_{PT}^{exp}(\theta, \varphi, f)$, respectively. $\sigma^{FD}(\theta, \varphi, f)$ refers to the RCS computed from the solution obtained by a frequency-domain electric field volume integral equation solver [72] that uses the same mesh of the scatterer as the MOT schemes but uses the SWG basis functions for discretization.

For the first set of simulations, scattering from a unit dielectric sphere with permittivity $10\varepsilon_0$ is analyzed. The sphere is meshed with 5350 tetrahedrons resulting in $N_e = 13,494$. The excitation parameters are $f_0 = 10$ MHz and $f_{bw} = 5$ MHz. The average, minimum, and maximum edge length of the mesh are $l_{avg} = \lambda_{min}/33.28$, $l_{min} = \lambda_{min}/62.0$, and $l_{max} = \lambda_{min}/19.76$, respectively, where

$\lambda_{\min} = c/(\sqrt{10}f_{\max})$ is the wavelength at $f_{\max}$ inside the scatterer. Convergence thresholds are $\chi^{\text{PECE}} = \chi^{\text{TFQMR}} = 10^{-6}$. The implicit and explicit MOT schemes for the TD-MFVIE with Galerkin and point testing are executed for $N_t = 210$ with $\Delta t = 6.667$ ns $[1/(10f_{\max})]$, $N_t = 140$ with $\Delta t = 10$ ns $[0.15/f_{\max}]$, and $N_t = 105$ with $\Delta t = 13.333$ ns $[1/(5f_{\max})]$, respectively. Figure 7.10a compares $\mathbf{H}(\mathbf{r}, t)$ computed using all four MOT schemes at $\mathbf{r} = (0.51, -0.64, 0.12)$ m for the simulation with $\Delta t = 6.667$ ns. All the results agree well. For the same simulation with $\Delta t = 6.667$ ns, Figure 7.10b plots the number of $m_{\max}$ ($m_{\max} = m$) and $N_{\text{iter}}^{\text{imp}}$ required by the explicit and implicit MOT schemes at every time step, respectively. As seen from this figure, the average number of m is less than that of $N_{\text{iter}}^{\text{imp}}$ over all the time steps. For the same simulation with $\Delta t = 6.667$ ns, Figure 7.10c compares $\sigma_{\text{PT}}^{\text{imp}}(\theta, \varphi, f)$, $\sigma_{\text{GT}}^{\text{imp}}(\theta, \varphi, f)$, $\sigma_{\text{PT}}^{\text{exp}}(\theta, \varphi, f)$, and $\sigma_{\text{GT}}^{\text{exp}}(\theta, \varphi, f)$ to $\sigma^{\text{Mie}}(\theta, \varphi, f)$ for $\theta = [0°, 180°]$, $\varphi = 0°$, and $f = f_0$. It shows that solutions of all four MOT schemes have almost the same accuracy. Table 7.4 presents computation times required by all four MOT schemes for three sets of Δt. In Table 7.4, T_{fill} represents

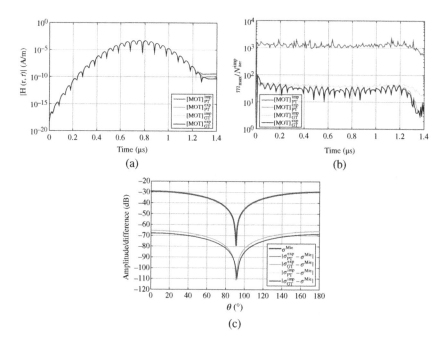

Figure 7.10 (a) Comparison of $\mathbf{H}(\mathbf{r}, t)$ computed using all four MOT schemes at $\mathbf{r} = (0.51, -0.64, 0.12)$ m for the simulation with $\Delta t = 6.667$ ns. (b) The number of m and $N_{\text{iter}}^{\text{imp}}$ required by the explicit and implicit MOT schemes at every time step. (c) Comparison of $\sigma_{\text{PT}}^{\text{imp}}(\theta, \varphi, f)$, $\sigma_{\text{GT}}^{\text{imp}}(\theta, \varphi, f)$, $\sigma_{\text{PT}}^{\text{exp}}(\theta, \varphi, f)$, and $\sigma_{\text{GT}}^{\text{exp}}(\theta, \varphi, f)$ to $\sigma^{\text{Mie}}(\theta, \varphi, f)$ for $\theta = [0°, 180°]$, $\varphi = 0°$, and $f = f_0$. Source: Reproduced with permission from [35].

Table 7.4 Computation times required by all four MOT schemes for three sets of Δt.

Δt (ns)	N_t	MOT	T_{fill} (s)	T_{tot} (s)	T_{all} (s)	$\sum N_{iter}^{imp} / \sum m$	RCS error
6.667	210	$[MOT]_{PT}^{imp}$	55	12,847	12,902	258,121	0.0152
6.667	210	$[MOT]_{PT}^{exp}$	57	297	354	6,152	0.0152
6.667	210	$[MOT]_{GT}^{imp}$	965	499	1464	8,277	0.0114
6.667	210	$[MOT]_{GT}^{exp}$	966	253	1,219	6,089	0.0114
10	140	$[MOT]_{PT}^{imp}$	55	9,274	9,328	185,833	0.0357
10	140	$[MOT]_{PT}^{exp}$	57	206	263	6,476	0.0357
10	140	$[MOT]_{GT}^{imp}$	963	295	1,258	6,888	0.0364
10	140	$[MOT]_{GT}^{exp}$	964	248	1,211	6,383	0.0364
13.333	105	$[MOT]_{PT}^{imp}$	55	6,636	6,691	145,932	0.109
13.333	105	$[MOT]_{PT}^{exp}$	57	200	257	6,637	0.109
13.333	105	$[MOT]_{GT}^{imp}$	962	256	1,218	4,926	0.110
13.333	105	$[MOT]_{GT}^{exp}$	963	238	1,201	4,517	0.110

Source: Reproduced with permission from [35].

the time required for computing all the relevant matrices and $T_{all} = T_{fill} + T_{tot}$. It can be seen that, the explicit MOT scheme for the TD-MFVIE with point testing is significantly faster than the other three. Additionally, Table 7.4 provides the RCS error values for all the simulations. It shows that all four MOT schemes have the same level of accuracy for a given set of simulations and the accuracy of solutions increases with the decrease of Δt.

For the second set of simulations, the same scatterer, discretization, and excitation are used as the first set of simulations. All four MOT schemes are executed for $N_t = 1200$ with $\Delta t = 6.667$ ns $[1/(10 f_{max})]$ with three different sets of threshold parameters $\chi^{PECE} = \chi^{TFQMR} \in \{10^{-6}, 10^{-8}, 10^{-10}\}$. Figure 7.11a–c plots $\mathbf{H}(\mathbf{r}, t)$ computed using all four MOT schemes at $\mathbf{r} = (0.51, -0.64, 0.12)$ m for $\chi^{PECE} = \chi^{TFQMR} = 10^{-6}$, $\chi^{PECE} = \chi^{TFQMR} = 10^{-8}$, and $\chi^{PECE} = \chi^{TFQMR} = 10^{-10}$, respectively. As seen from these figures, all four MOT schemes generate stable results with a very small DC component. Note that, the amplitude of this DC component can be further suppressed with smaller χ^{PECE} and χ^{TFQMR}.

For the next set of simulations, the permittivity of the unit dielectric sphere is increased to $50\varepsilon_0$. The sphere is discretized using 11,697 tetrahedrons resulting in $N_e = 28,970$. The excitation parameters are $f_0 = 10$ MHz and $f_{bw} = 5$ MHz. The edge sizes of the mesh are $l_{avg} = \lambda_{min}/19.51$, $l_{min} = \lambda_{min}/39.34$, and $l_{max} = \lambda_{min}/11.55$, respectively, and $\lambda_{min} = c/(\sqrt{50} f_{max})$. Since only the explicit MOT scheme for the TD-MFVIE with Galerkin testing generates stable solution

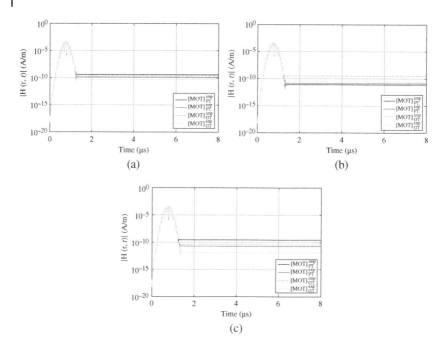

Figure 7.11 $H(\mathbf{r}, t)$ computed at $\mathbf{r} = (0.51, -0.64, 0.12)$ m using all four MOT schemes for (a) $\chi^{\text{PECE}} = \chi^{\text{TFQMR}} = 10^{-6}$, (b) $\chi^{\text{PECE}} = \chi^{\text{TFQMR}} = 10^{-8}$, and (c) $\chi^{\text{PECE}} = \chi^{\text{TFQMR}} = 10^{-10}$. Reproduced with permission from [35].

for this example, it is executed for $N_t = 600$ with $\Delta t = 6.667$ ns $[1/(10 f_{\max})]$, $N_t = 400$ with $\Delta t = 10$ ns $[0.15/f_{\max}]$, and $N_t = 300$ with $\Delta t = 13.333$ ns $[1/(5 f_{\max})]$, respectively. Convergence thresholds are $\chi^{\text{PECE}} = \chi^{\text{TFQMR}} = 10^{-6}$. Figure 7.12a plots $H(\mathbf{r}, t)$ calculated using the explicit MOT scheme for TD-MFVIE with Galerkin testing at $\mathbf{r} = (0.51, -0.64, 0.12)$ m for the simulation with $\Delta t = 6.667$ ns. As seen from this figure, stable results are obtained. For the same simulation with $\Delta t = 6.667$ ns, Figure 7.12b compares $\sigma_{\text{GT}}^{\exp}(\theta, \varphi, f)$ with $\sigma^{\text{Mie}}(\theta, \varphi, f)$ for $\theta = [0°, 180°]$, $\varphi = 0°$, and $f = f_0$. The accuracy of the solution obtained using the explicit MOT solver for TD-MFVIE with Galerkin testing is verified. Table 7.5 provides the execution times and RCS errors of the explicit MOT scheme for the TD-MFVIE with Galerkin testing for all three simulations. It shows that the accuracy increases with smaller Δt.

For the last set of simulations, scattering from a piecewise dielectric slab is analyzed. The permittivity values of left and right half volumes of the slab are $3\varepsilon_0$ and $9\varepsilon_0$, respectively. The slab is meshed using 7905 tetrahedrons resulting in $N_e = 20,570$. The excitation parameters are $f_0 = 10$ MHz and $f_{\text{bw}} = 5$ MHz.

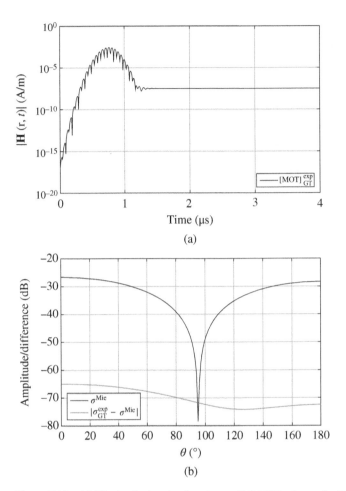

Figure 7.12 (a) $\mathbf{H}(\mathbf{r}, t)$ calculated using the explicit MOT scheme for TD-MFVIE with Galerkin testing at $\mathbf{r} = (0.51, -0.64, 0.12)$ m for the simulation with $\Delta t = 6.667$ ns. (b) Comparison of $\sigma_{GT}^{exp}(\theta, \varphi, f)$ with $\sigma^{Mie}(\theta, \varphi, f)$ for $\theta = [0°, 180°]$, $\varphi = 0°$, and $f = f_0$. Source: Reproduced with permission from [35].

Table 7.5 Execution times and RCS errors of $[\mathrm{MOT}]_{GT}^{exp}$ for all three simulations.

Δt (ns)	N_t	T_{fill} (s)	T_{tot} (s)	T_{all} [s]	$\sum m$	RCS error
6.667	600	3623	7387	11,009	42,322	0.012
10	400	3623	7120	10,743	42,378	0.051
13.33	300	3625	6343	9,968	38,671	0.129

Source: Reproduced with permission from [35].

The edge sizes of the discretization are $l_{avg} = \lambda_{min}/31.7$, $l_{min} = \lambda_{min}/66.6$, and $l_{max} = \lambda_{min}/19$, respectively, and $\lambda_{min} = c/(\sqrt{9}f_{max})$ is the wavelength inside the right half volume of the slab. All four MOT schemes are executed for $N_t = 210$ with $\Delta t = 6.667$ ns $[1/(10f_{max})]$, $N_t = 140$ with $\Delta t = 10$ ns $[0.15/f_{max}]$, and $N_t = 105$ with $\Delta t = 13.333$ ns $[1/(5f_{max})]$, respectively. Convergence thresholds are $\chi^{PECE} = \chi^{TFQMR} = 10^{-6}$. Figure 7.13a plots $\mathbf{H}(\mathbf{r}, t)$ obtained from all four MOT solvers at $\mathbf{r} = (0.23, 0.14, 0.57)$ m for the simulation with $\Delta t = 6.667$ ns. The results agree very well. For the same simulation with $\Delta t = 6.667$ ns, Figure 7.13b compares the numbers of m required by the explicit MOT scheme and N_{iter}^{imp} required by the implicit MOT scheme at every time step. It shows that the average number of m is less than that of N_{iter}^{imp} over all the time steps. For the same simulation with $\Delta t = 6.667$ ns, Figure 7.13c compares $\sigma_{PT}^{imp}(\theta, \varphi, f)$, $\sigma_{GT}^{imp}(\theta, \varphi, f)$, $\sigma_{PT}^{exp}(\theta, \varphi, f)$, and $\sigma_{GT}^{exp}(\theta, \varphi, f)$ to $\sigma^{FD}(\theta, \varphi, f)$ for $\theta = [0°, 180°]$, $\varphi = 0°$, and $f = f_0$. This figure shows that all four MOT schemes produce the solutions of the same level of accuracy. In Table 7.6, execution times required by all four MOT schemes

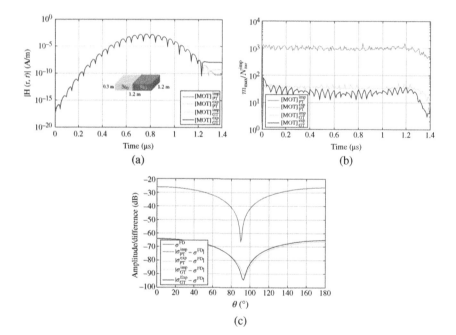

(a)

(b)

(c)

Figure 7.13 (a) H(r, t) obtained at r = (0.23, 0.14, 0.57) m from all four MOT solvers for the simulation with $\Delta t = 6.667$ ns. (b) Comparison of the numbers of m required by the explicit MOT scheme and N_{iter}^{imp} required by the implicit MOT scheme at every time step. (c) Comparison of $\sigma_{PT}^{imp}(\theta, \varphi, f)$, $\sigma_{GT}^{imp}(\theta, \varphi, f)$, $\sigma_{PT}^{exp}(\theta, \varphi, f)$, and $\sigma_{GT}^{exp}(\theta, \varphi, f)$ to $\sigma^{FD}(\theta, \varphi, f)$ for $\theta = [0°, 180°]$, $\varphi = 0°$, and $f = f_0$. Source: Reproduced with permission from [35].

Table 7.6 Execution times required by all four MOT solvers for three sets of Δt.

Δt (ns)	N_t	MOT	T_{fill} (s)	T_{tot} (s)	T_{all} (s)	$\sum N_{iter}^{imp} / \sum m$	RCS error
6.667	210	$[\mathrm{MOT}]_{PT}^{imp}$	108	23,003	23,111	204,307	0.0113
6.667	210	$[\mathrm{MOT}]_{PT}^{exp}$	111	346	457	5,179	0.0113
6.667	210	$[\mathrm{MOT}]_{GT}^{imp}$	1730	917	2,647	7,390	0.0123
6.667	210	$[\mathrm{MOT}]_{GT}^{exp}$	1734	493	2,227	5,170	0.0123
10	140	$[\mathrm{MOT}]_{PT}^{imp}$	107	18,380	18,487	162,433	0.0458
10	140	$[\mathrm{MOT}]_{PT}^{exp}$	111	340	451	5,292	0.0458
10	140	$[\mathrm{MOT}]_{GT}^{imp}$	1730	642	2,372	5,171	0.0465
10	140	$[\mathrm{MOT}]_{GT}^{exp}$	1735	441	2,176	5,223	0.0465
13.333	105	$[\mathrm{MOT}]_{PT}^{imp}$	107	14,314	14,421	127,842	0.117
13.333	105	$[\mathrm{MOT}]_{PT}^{exp}$	111	345	456	5,559	0.117
13.333	105	$[\mathrm{MOT}]_{GT}^{imp}$	1728	522	2,250	4,377	0.118
13.333	105	$[\mathrm{MOT}]_{GT}^{exp}$	1729	431	2,160	5,128	0.118

Source: Reproduced with permission from [35].

for three sets of Δt are presented. As expected, the explicit MOT scheme for TD-MFVIE with point testing is much faster than the others. In addition, the RCS error values presented in Table 7.6 also show that all four MOT schemes have the same level of accuracy for the same simulation and the accuracy can be improved by decreasing Δt.

7.10 Conclusion

Explicit MOT schemes, which are developed for solving second-kind TDIEs (more specifically TD-MFIE for PEC scatterers and TD-MFVIE for dielectric scatterers), are described in detail. These schemes cast TDIE in the form of ODEs, which relate the variable of interest (surface current for TD-MFIE and magnetic field for TD-MFVIE) to its temporal derivative. Inserting the spatial expansions of the variable of interest in TDIE and spatially testing the resulting equation yield a system of ODEs in time-dependent unknown expansion coefficients. The ODE systems are integrated in time using a PE(CE)^m-type K-step scheme to yield these expansion coefficients. Unlike the implicit MOT schemes, the resulting explicit MOT scheme calls for the solution of a system with a (spatial) Gram matrix at the evaluation (E) step. This Gram matrix is sparse and well conditioned regardless of the time step size, and therefore can always be inverted very efficiently.

Numerical results demonstrate that the explicit MOT solvers described in this chapter use the same time step size as their implicit counterparts without sacrificing from the accuracy and the stability of the solution. In addition, they are significantly faster under low-frequency excitations (i.e. for large time step sizes) since the matrices inverted by the implicit MOT schemes are dense for large time step sizes (unlike the Gram matrices).

Several extensions of these explicit MOT schemes are possible. The first extension is to use the explicit MOT scheme for solving the time domain combined field integral equation (TD-CFIE) for PEC scatterers to avoid internal resonance issues. The second extension involves using mixed discretization techniques to improve the accuracy of the TD-MFIE solution at low frequencies [76, 119–123]. The third extension is to formulate and develop an explicit MOT scheme for analyzing electromagnetic scattering from nonlinear dielectric objects.

References

1 Bluck, M.J. and Walker, S.P. (1997). Time-domain BIE analysis of large three-dimensional electromagnetic scattering problems. *IEEE Transactions on Antennas and Propagation* 45 (5): 894–901.

2 Dodson, S.J., Walker, S.P., and Bluck, M.J. (1998). Implicitness and stability of time domain integral equation scattering analyses. *Applied Computational Electromagnetics Society Journal* 13: 291–301.

3 Rynne, B.P. (1985). Stability and convergence of time marching methods in scattering problems. *IMA Journal of Applied Mathematics* 35 (3): 297–310.

4 Rynne, B.P. and Smith, P.D. (1990). Stability of time marching algorithms for the electric field integral equation. *Journal of Electromagnetic Waves and Applications* 4 (12): 1181–1205.

5 Walker, S.P., Bluck, M.J., and Chatzis, I. (2002). The stability of integral equation time-domain scattering computations for three-dimensional scattering; similarities and differences between electrodynamic and elastodynamic computations. *International Journal of Numerical Modelling: Electronic Networks, Devices and Fields* 15 (5–6): 459–474.

6 Manara, G., Monorchio, A., and Reggiannini, R. (1997). A space-time discretization criterion for a stable time-marching solution of the electric field integral equation. *IEEE Transactions on Antennas and Propagation* 45 (3): 527–532.

7 Yilmaz, A.E., Weile, D.S., Shanker, B. et al. (2002). Fast analysis of transient scattering in lossy media. *IEEE Antennas and Wireless Propagation Letters* 1: 14–17.

8 Yilmaz, A.E., Jin, J.-M., and Michielssen, E. (2004). Time domain adaptive integral method for surface integral equations. *IEEE Transactions on Antennas and Propagation* 52 (10): 2692–2708.

9 Yilmaz, A.E., Jin, J.-M., and Michielssen, E. (2005). A parallel FFT accelerated transient field-circuit simulator. *IEEE Transactions on Microwave Theory and Techniques* 53 (9): 2851–2865.

10 Yilmaz, A.E., Jin, J.-M., and Michielssen, E. (2007). Analysis of low-frequency electromagnetic transients by an extended time-domain adaptive integral method. *IEEE Transactions on Advanced Packaging* 30 (2): 301–312.

11 Bagci, H., Yilmaz, A.E., Lomakin, V., and Michielssen, E. (2005). Fast solution of mixed-potential time-domain integral equations for half-space environments. *IEEE Transactions on Geoscience and Remote Sensing* 43 (2): 269–279.

12 Bagci, H., Yilmaz, A.E., Jin, J.-M., and Michielssen, E. (2007). Fast and rigorous analysis of EMC/EMI phenomena on electrically large and complex cable-loaded structures. *IEEE Transactions on Electromagnetic Compatibility* 49 (2): 361–381.

13 Bagci, H., Yilmaz, A.E., and Michielssen, E. (2010). An FFT-accelerated time-domain multiconductor transmission line simulator. *IEEE Transactions on Electromagnetic Compatibility* 52 (1): 199–214.

14 Bagci, H., Andriulli, F.P., Vipiana, F. et al. (2010). A well-conditioned integral-equation formulation for efficient transient analysis of electrically small microelectronic devices. *IEEE Transactions on Advanced Packaging* 33 (2): 468–480.

15 Ergin, A.A., Shanker, B., and Michielssen, E. (1998). Fast evaluation of three-dimensional transient wave fields using diagonal translation operators. *Journal of Computational Physics* 146 (1): 157–180.

16 Ergin, A.A., Shanker, B., and Michielssen, E. (1999). The plane-wave time-domain algorithm for the fast analysis of transient wave phenomena. *IEEE Antennas and Propagation Magazine* 41 (4): 39–52.

17 Shanker, B., Ergin, A.A., Aygun, K., and Michielssen, E. (2000). Analysis of transient electromagnetic scattering from closed surfaces using a combined field integral equation. *IEEE Transactions on Antennas and Propagation* 48 (7): 1064–1074.

18 Shanker, B., Ergin, A.A., Aygun, K., and Michielssen, E. (2000). Analysis of transient electromagnetic scattering phenomena using a two-level plane wave time-domain algorithm. *IEEE Transactions on Antennas and Propagation* 48 (4): 510–523.

19 Shanker, B., Ergin, A.A., Lu, M., and Michielssen, E. (2003). Fast analysis of transient electromagnetic scattering phenomena using the multilevel plane wave time domain algorithm. *IEEE Transactions on Antennas and Propagation* 51 (3): 628–641.

20 Aygun, K., Shanker, B., Ergin, A.A., and Michielssen, E. (2002). A two-level plane wave time-domain algorithm for fast analysis of EMC/EMI problems. *IEEE Transactions on Electromagnetic Compatibility* 44 (1): 152–164.

21 Chen, N.-W., Aygün, K., and Michielssen, E. (2001). Integral-equation-based analysis of transient scattering and radiation from conducting bodies at very low frequencies. *IEE Proceedings-Microwaves, Antennas and Propagation* 148 (6): 381–387.

22 Andriulli, F.P., Bagci, H., Vipiana, F. et al. (2007). A marching-on-in-time hierarchical scheme for the solution of the time domain electric field integral equation. *IEEE Transactions on Antennas and Propagation* 55 (12): 3734–3738.

23 Andriulli, F.P., Bagci, H., Vipiana, F. et al. (2009). Analysis and regularization of the TD-EFIE low-frequency breakdown. *IEEE Transactions on Antennas and Propagation* 57 (7): 2034–2046.

24 Liu, Y., Yücel, A.C., Bağcı, H., and Michielssen, E. (2015). A scalable parallel PWTD-accelerated SIE solver for analyzing transient scattering from electrically large objects. *IEEE Transactions on Antennas and Propagation* 64 (2): 663–674.

25 Gres, N.T., Ergin, A.A., Michielssen, E., and Shanker, B. (2001). Volume-integral-equation-based analysis of transient electromagnetic scattering from three-dimensional inhomogeneous dielectric objects. *Radio Science* 36 (3): 379–386.

26 Kobidze, G., Gao, J., Shanker, B., and Michielssen, E. (2005). A fast time domain integral equation based scheme for analyzing scattering from dispersive objects. *IEEE Transactions on Antennas and Propagation* 53 (3): 1215–1226.

27 Shanker, B., Aygün, K., and Michielssen, E. (2004). Fast analysis of transient scattering from lossy inhomogeneous dielectric bodies. *Radio Science* 39 (2): 1–14.

28 Sayed, S.B., Ülkü, H.A., and Bağcı, H. (2015). A stable marching on-in-time scheme for solving the time-domain electric field volume integral equation on high-contrast scatterers. *IEEE Transactions on Antennas and Propagation* 63 (7): 3098–3110.

29 Rao, S.M. and Wilton, D.R. (1991). Transient scattering by conducting surfaces of arbitrary shape. *IEEE Transactions on Antennas and Propagation* 39 (1): 56–61.

30 Rao, S.M. (1999). *Time Domain Electromagnetics*. Elsevier.

31 Jung, B.H. and Sarkar, T.K. (2002). Time-domain CFIE for the analysis of transient scattering from arbitrarily shaped 3D conducting objects. *Microwave and Optical Technology Letters* 34 (4): 289–296.

32 Al-Jarro, A., Salem, M.A., Bagci, H. et al. (2012). Explicit solution of the time domain volume integral equation using a stable predictor-corrector scheme. *IEEE Transactions on Antennas and Propagation* 60 (11): 5203–5214.

33 Liu, Y., Al-Jarro, A., Bağcı, H., and Michielssen, E. (2016). Parallel PWTD-accelerated explicit solution of the time-domain electric field volume integral equation. *IEEE Transactions on Antennas and Propagation* 64 (6): 2378–2388.

34 Ülkü, H.A., Bağcı, H., and Michielssen, E. (2013). Marching on-in-time solution of the time domain magnetic field integral equation using a predictor-corrector scheme. *IEEE Transactions on Antennas and Propagation* 61 (8): 4120–4131.

35 Sayed, S.B., Ulku, H.A., and Bagci, H. (2019). Explicit time marching schemes for solving the magnetic field volume integral equation. *IEEE Transactions on Antennas and Propagation* 68 (3): 2224–2237.

36 Chen, R. and Bagci, H. (2018). An explicit MOT scheme for solving the Nyström-discretized TD-MFIE. *IEEE International Symposium on Antennas and Propagation and USNC/URSI National Radio Science Meeting*, pp. 2443–2444.

37 Chen, R. and Bagci, H. (2020). Explicit solution of time domain scalar potential surface integral equations for penetrable scatterers. *IEEE International Symposium on Antennas and Propagation and USNC-URSI National Radio Science Meeting*, pp. 1001–1002.

38 Chen, R., Sayed, S.B., Ulku, H.A., and Bagci, H. (2021). An explicit time marching scheme for efficient solution of the magnetic field integral equation at low frequencies. *IEEE Transactions on Antennas and Propagation* 69 (2): 1213–1218.

39 Jung, B.H., Chung, Y.-S., and Sarkar, T.K. (2003). Time-domain EFIE, MFIE, and CFIE formulations using Laguerre polynomials as temporal basis functions for the analysis of transient scattering from arbitrary shaped conducting structures. *Progress in Electromagnetics Research* 39: 1–45.

40 Jung, B.H., Sarkar, T.K., Chung, Y.-S. et al. (2004). Transient electromagnetic scattering from dielectric objects using the electric field integral equation with Laguerre polynomials as temporal basis functions. *IEEE Transactions on Antennas and Propagation* 52 (9): 2329–2340.

41 Ji, Z., Sarkar, T.K., Jung, B.H. et al. (2004). A stable solution of time domain electric field integral equation for thin-wire antennas using the Laguerre polynomials. *IEEE Transactions on Antennas and Propagation* 52 (10): 2641–2649.

42 Ji, Z., Sarkar, T.K., Jung, B.H. et al. (2006). Solving time domain electric field integral equation without the time variable. *IEEE Transactions on Antennas and Propagation* 54 (1): 258–262.

43 Chung, Y.-S., Sarkar, T.K., Jung, B.H. et al. (2004). Solution of time domain electric field integral equation using the Laguerre polynomials. *IEEE Transactions on Antennas and Propagation* 52 (9): 2319–2328.

44 Taflove, A. and Hagness, S.C. (2005). *Computational Electrodynamics: The Finite-difference Time-domain Method*. Artech House.

45 Sullivan, D.M. (2013). *Electromagnetic Simulation using the FDTD Method*. Wiley.

46 Berenger, J.-P. (1996). Three-dimensional perfectly matched layer for the absorption of electromagnetic waves. *Journal of Computational Physics* 127 (2): 363–379.

47 Sirenko, K., Pazynin, V., Sirenko, Y.K., and Bagci, H. (2011). An FFT-accelerated FDTD scheme with exact absorbing conditions for characterizing axially symmetric resonant structures. *Progress in Electromagnetics Research* 111: 331–364.

48 Jin, J.-M. (2015). *The Finite Element Method in Electromagnetics*. Wiley.

49 Jiao, D., Lu, M., Michielssen, E., and Jin, J.-M. (2001). A fast time-domain finite element-boundary integral method for electromagnetic analysis. *IEEE Transactions on Antennas and Propagation* 49 (10): 1453–1461.

50 Jiao, D., Jin, J.-M., Michielssen, E., and Riley, D.J. (2003). Time-domain finite-element simulation of three-dimensional scattering and radiation problems using perfectly matched layers. *IEEE Transactions on Antennas and Propagation* 51 (2): 296–305.

51 Yilmaz, A.E., Lou, Z., Michielssen, E., and Jin, J.-M. (2007). A single-boundary implicit and FFT-accelerated time-domain finite element-boundary integral solver. *IEEE Transactions on Antennas and Propagation* 55 (5): 1382–1397.

52 Hesthaven, J.S. and Warburton, T. (2007). *Nodal Discontinuous Galerkin Methods: Algorithms, Analysis, and Applications*. Springer Science & Business Media.

53 Sirenko, K., Liu, M., and Bagci, H. (2012). Incorporation of exact boundary conditions into a discontinuous Galerkin finite element method for accurately solving 2D time-dependent Maxwell equations. *IEEE Transactions on Antennas and Propagation* 61 (1): 472–477.

54 Sirenko, K., Sirenko, Y., and Bağcı, H. (2018). Exact absorbing boundary conditions for periodic three-dimensional structures: derivation and implementation in discontinuous Galerkin time-domain method. *IEEE Journal on Multiscale and Multiphysics Computational Techniques* 3: 108–120.

55 Chen, L. and Bagci, H. (2020). Steady-state simulation of semiconductor devices using discontinuous Galerkin methods. *IEEE Access* 8: 16203–16215.

56 Chen, L., Dong, M., and Bagci, H. (2020). Modeling floating potential conductors using discontinuous Galerkin method. *IEEE Access* 8: 7531–7538.

57 Li, P., Shi, Y., Jiang, L.J., and Bağcı, H. (2014). A hybrid time-domain discontinuous Galerkin-boundary integral method for electromagnetic scattering analysis. *IEEE Transactions on Antennas and Propagation* 62 (5): 2841–2846.

58 Li, P., Jiang, L.J., and Bağci, H. (2016). Transient analysis of dispersive power-ground plate pairs with arbitrarily shaped antipads by the DGTD method with wave port excitation. *IEEE Transactions on Electromagnetic Compatibility* 59 (1): 172–183.

59 Liu, M., Sirenko, K., and Bagci, H. (2012). An efficient discontinuous Galerkin finite element method for highly accurate solution of Maxwell equations. *IEEE Transactions on Antennas and Propagation* 60 (8): 3992–3998.

60 Chen, J. and Liu, Q.H. (2012). Discontinuous Galerkin time-domain methods for multiscale electromagnetic simulations: a review. *Proceedings of the IEEE* 101 (2): 242–254.

61 Gedney, S.D., Young, J.C., Kramer, T.C., and Roden, J.A. (2012). A discontinuous Galerkin finite element time-domain method modeling of dispersive media. *IEEE Transactions on Antennas and Propagation* 60 (4): 1969–1977.

62 Miller, N.C., Baczewski, A.D., Albrecht, J.D., and Shanker, B. (2014). A discontinuous Galerkin time domain framework for periodic structures subject to oblique excitation. *IEEE Transactions on Antennas and Propagation* 62 (8): 4386–4391.

63 Jin, J.-M. (2011). *Theory and Computation of Electromagnetic Fields*. Wiley.

64 Chen, R., Alharthi, N., Sayed, S.B. et al. (2017). An explicit time marching scheme for solving surface integral equations of acoustics. *URSI General Assembly and Scientific Symposium*.

65 Chen, R., Sayed, S.B., and Bagci, H. (2018). An explicit marching-on-in-time scheme for solving the Kirchhoff integral equation. *IEEE International Symposium on Antennas and Propagation and USNC/URSI National Radio Science Meeting*, pp. 2409–2410.

66 Chen, R. and Bagci, H. (2019). A higher-order explicit marching-on-in-time for analysis of transient acoustic scattering from rigid objects. *IEEE International Symposium on Antennas and Propagation and USNC-URSI National Radio Science Meeting*, pp. 179–180.

67 Chen, R., Sayed, S.B., Alharthi, N. et al. (2019). An explicit marching-on-in-time scheme for solving the time domain Kirchhoff integral equation. *The Journal of the Acoustical Society of America* 146 (3): 2068–2079.

68 Chen, R. and Bagci, H. (2020). An explicit time marching scheme to solve surface integral equations for acoustically penetrable scatterers. *URSI General Assembly and Scientific Symposium*.

69 Chen, R. and Bagci, H. (2021). On the internal resonance modes of time domain surface integral equations for acoustic transmission problems. *URSI General Assembly and Scientific Symposium*.

70 Chen, R. (2021). Transient analysis of electromagnetic and acoustic scattering using second-kind surface integral equations. PhD dissertation. Thuwal, Saudi Arabia: King Abdullah University of Science and Technology.

71 Rao, S., Wilton, D., and Glisson, A. (1982). Electromagnetic scattering by surfaces of arbitrary shape. *IEEE Transactions on Antennas and Propagation* 30 (3): 409–418.

72 Schaubert, D., Wilton, D., and Glisson, A. (1984). A tetrahedral modeling method for electromagnetic scattering by arbitrarily shaped inhomogeneous dielectric bodies. *IEEE Transactions on Antennas and Propagation* 32 (1): 77–85.

73 Knab, J. (1979). Interpolation of band-limited functions using the approximate prolate series (Corresp.). *IEEE Transactions on Information Theory* 25 (6): 717–720.

74 Weile, D.S., Pisharody, G., Chen, N.-W. et al. (2004). A novel scheme for the solution of the time-domain integral equations of electromagnetics. *IEEE Transactions on Antennas and Propagation* 52 (1): 283–295.

75 Wildman, R.A., Pisharody, G., Weile, D.S. et al. (2004). An accurate scheme for the solution of the time-domain integral equations of electromagnetics using higher order vector bases and bandlimited extrapolation. *IEEE Transactions on Antennas and Propagation* 52 (11): 2973–2984.

76 Beghein, Y., Cools, K., Bagci, H., and De Zutter, D. (2012). A space-time mixed Galerkin marching-on-in-time scheme for the time-domain combined field integral equation. *IEEE Transactions on Antennas and Propagation* 61 (3): 1228–1238.

77 Davies, P.J. and Duncan, D.B. (2004). Stability and convergence of collocation schemes for retarded potential integral equations. *SIAM Journal on Numerical Analysis* 42 (3): 1167–1188.

78 Shanker, B., Ergin, A.A., and Michielssen, E. (2002). Plane-wave–time-domain-enhanced marching-on-in-time scheme for analyzing scattering from homogeneous dielectric structures. *JOSA A* 19 (4): 716–726.

79 Liu, Y., Yücel, A.C., Lomakin, V., and Michielssen, E. (2014). Graphics processing unit implementation of multilevel plane-wave time-domain algorithm. *IEEE Antennas and Wireless Propagation Letters* 13: 1671–1675.

80 Liu, Y., Yücel, A.C., Bağcı, H. et al. (2018). A wavelet-enhanced PWTD-accelerated time-domain integral equation solver for analysis of transient scattering from electrically large conducting objects. *IEEE Transactions on Antennas and Propagation* 66 (5): 2458–2470.

81 Jiang, P.-L. and Michielssen, E. (2005). Temporal acceleration of time-domain integral-equation solvers for electromagnetic scattering from objects residing in lossy media. *Microwave and Optical Technology Letters* 44 (3): 223–230.

82 Lu, M., Wang, J., Ergin, A.A., and Michielssen, E. (2000). Fast evaluation of two-dimensional transient wave fields. *Journal of Computational Physics* 158 (2): 161–185.

83 Lu, M., Yegin, K., Michielssen, E., and Shanker, B. (2004). Fast time domain integral equation solvers for analyzing two-dimensional scattering phenomena; Part I: temporal acceleration. *Electromagnetics* 24 (6): 425–449.

84 Lu, M., Michielssen, E., and Shanker, B. (2004). Fast time domain integral equation solvers for analyzing two-dimensional scattering phenomena; Part II: full PWTD acceleration. *Electromagnetics* 24 (6): 451–470.

85 Yilmaz, A.E., Weile, D.S., Jin, H.-M., and Michielssen, E. (2002). A hierarchical FFT algorithm (HIL-FFT) for the fast analysis of transient electromagnetic scattering phenomena. *IEEE Transactions on Antennas and Propagation* 50 (7): 971–982.

86 Kaur, G. and Yilmaz, A.E. (2015). Envelope-tracking adaptive integral method for band-pass transient scattering analysis. *IEEE Transactions on Antennas and Propagation* 63 (5): 2215–2227.

87 Smith, P.D. (1990). Instabilities in time marching methods for scattering: cause and rectification. *Electromagnetics* 10 (4): 439–451.

88 Vechinski, D.A. and Rao, S.M. (1992). A stable procedure to calculate the transient scattering by conducting surfaces of arbitrary shape. *IEEE Transactions on Antennas and Propagation* 40 (6): 661–665.

89 Sadigh, A. and Arvas, E. (1993). Treating the instabilities in marching-on-in-time method from a different perspective (electromagnetic scattering). *IEEE Transactions on Antennas and Propagation* 41 (12): 1695–1702.

90 Davies, P.J. (1998). A stability analysis of a time marching scheme for the general surface electric field integral equation. *Applied Numerical Mathematics* 27 (1): 33–57.

91 Wang, X., Wildman, R.A., Weile, D.S., and Monk, P. (2008). A finite difference delay modeling approach to the discretization of the time domain integral equations of electromagnetics. *IEEE Transactions on Antennas and Propagation* 56 (8): 2442–2452.

92 Wang, X. and Weile, D.S. (2010). Electromagnetic scattering from dispersive dielectric scatterers using the finite difference delay modeling method. *IEEE Transactions on Antennas and Propagation* 58 (5): 1720–1730.

93 Wang, X. and Weile, D.S. (2011). Implicit Runge-Kutta methods for the discretization of time domain integral equations. *IEEE Transactions on Antennas and Propagation* 59 (12): 4651–4663.

94 Pisharody, G. and Weile, D.S. (2005). Robust solution of time-domain integral equations using loop-tree decomposition and bandlimited extrapolation. *IEEE Transactions on Antennas and Propagation* 53 (6): 2089–2098.

95 Pisharody, G. and Weile, D.S. (2005). Electromagnetic scattering from perfect electric conductors using an augmented time-domain integral-equation technique. *Microwave and Optical Technology Letters* 45 (1): 26–31.

96 Andriulli, F.P., Cools, K., Olyslager, F., and Michielssen, E. (2009). Time domain Calderón identities and their application to the integral equation analysis of scattering by PEC objects part II: stability. *IEEE Transactions on Antennas and Propagation* 57 (8): 2365–2375.

97 Shi, Y., Xia, M.-Y., Chen, R.-S. et al. (2010). Stable electric field TDIE solvers via quasi-exact evaluation of MOT matrix elements. *IEEE Transactions on Antennas and Propagation* 59 (2): 574–585.

98 Wang, X., Shi, Y., Lu, M. et al. (2021). Stable and accurate marching-on-in-time solvers of time domain EFIE, MFIE, and CFIE based on quasi-exact integration technique. *IEEE Transactions on Antennas and Propagation* 69 (4): 2218–2229.

99 Ülkü, H.A. and Ergin, A.A. (2007). Analytical evaluation of transient magnetic fields due to RWG current bases. *IEEE Transactions on Antennas and Propagation* 55 (12): 3565–3575.

100 Ulku, H.A. and Ergin, A.A. (2011). Application of analytical retarded-time potential expressions to the solution of time domain integral equations. *IEEE Transactions on Antennas and Propagation* 59 (11): 4123–4131.

101 Ülkü, H.A. and Ergin, A.A. (2010). On the singularity of the closed-form expression of the magnetic field in time domain. *IEEE Transactions on Antennas and Propagation* 59 (2): 691–694.

102 Ulku, H.A., Ergin, A.A., and Dikmen, F. (2011). On the evaluation of retarded-time potentials for SWG bases. *IEEE Antennas and Wireless Propagation Letters* 10: 187–190.

103 Kang, G., Song, J., Chew, W.C. et al. (2001). A novel grid-robust higher order vector basis function for the method of moments. *IEEE Transactions on Antennas and Propagation* 49 (6): 908–915.

104 Chen, R. and Bagci, H. (2019). On higher-order Nyström discretization of scalar potential integral equation for penetrable scatterers. *International Applied Computational Electromagnetics Society Symposium (ACES)*, pp. 1–2.

105 Alharthi, N., Chen, R., Bagci, H., and Keyes, D. (2017). A comparative study of singularity treatment schemes in higher-order Nyström method for acoustic scattering. *URSI General Assembly and Scientific Symposium*.

106 Al-Harthi, N., Alomairy, R., Akbudak, K. et al. (2020). Solving acoustic boundary integral equations using high performance tile low-rank LU factorization. *International Conference on High Performance Computing*, pp. 209–229.

107 Abduljabbar, M., Farhan, M.A., Al-Harthi, N. et al. (2019). Extreme scale FMM-accelerated boundary integral equation solver for wave scattering. *SIAM Journal on Scientific Computing* 41 (3): C245–C268.

108 Webb, J.P. (1999). Hierarchal vector basis functions of arbitrary order for triangular and tetrahedral finite elements. *IEEE Transactions on Antennas and Propagation* 47 (8): 1244–1253.

109 Peterson, A.F. (2013). Efficient solenoidal discretization of the volume EFIE for electromagnetic scattering from dielectric objects. *IEEE Transactions on Antennas and Propagation* 62 (3): 1475–1478.

110 Davis, P.J. and Rabinowitz, P. (2007). *Methods of Numerical Integration*. Courier Corporation.

111 Holdych, D.J., Noble, D.R., and Secor, R.B. (2008). Quadrature rules for triangular and tetrahedral elements with generalized functions. *International Journal for Numerical Methods in Engineering* 73 (9): 1310–1327.

112 Hairer, E., Nørsett, S.P., and Wanner, G. (1993). *Solving Ordinary Differential Equations I: Nonstiff Problems*. Springer-Verlag.

113 Glaser, A. and Rokhlin, V. (2009). A new class of highly accurate solvers for ordinary differential equations. *Journal of Scientific Computing* 38 (3): 368–399.

114 Dutt, A., Greengard, L., and Rokhlin, V. (2000). Spectral deferred correction methods for ordinary differential equations. *BIT Numerical Mathematics* 40 (2): 241–266.

115 Shi, Y., Bağcı, H., and Lu, M. (2014). On the static loop modes in the marching-on-in-time solution of the time-domain electric field integral equation. *IEEE Antennas and Wireless Propagation Letters* 13: 317–320.

116 Freund, R.W. (1993). A transpose-free quasi-minimal residual algorithm for non-hermitian linear systems. *SIAM Journal on Scientific Computing* 14 (2): 470–482.

117 Uysal, I.E., Ülkü, H.A., and Bağci, H. (2016). Transient analysis of electromagnetic wave interactions on plasmonic nanostructures using a surface integral equation solver. *JOSA A* 33 (9): 1747–1759.

118 Harrington, R.F. (1961). *Time-harmonic Electromagnetic Fields*. McGraw-Hill.

119 Ylä-Oijala, P., Kiminki, S.P., Cools, K. et al. (2012). Mixed discretization schemes for electromagnetic surface integral equations. *International Journal of Numerical Modelling: Electronic Networks, Devices and Fields* 25 (5–6): 525–540.

120 Cools, K., Andriulli, F.P., De Zutter, D., and Michielssen, E. (2011). Accurate and conforming mixed discretization of the MFIE. *IEEE Antennas and Wireless Propagation Letters* 10: 528–531.

121 Ülkü, H.A., Bogaert, I., Cools, K. et al. (2017). Mixed discretization of the time-domain MFIE at low frequencies. *IEEE Antennas and Wireless Propagation Letters* 16: 1565–1568.

122 Bogaert, I., Cools, K., Andriulli, F.P., and Bağcı, H. (2013). Low-frequency scaling of the standard and mixed magnetic field and Müller integral equations. *IEEE Transactions on Antennas and Propagation* 62 (2): 822–831.

123 Chen, R. and Bagci, H. (2020). On the low-frequency behavior of decoupled vector potential integral equations for perfect electrically conducting scatterers. *URSI General Assembly and Scientific Symposium.*

8

Convolution Quadrature Time Domain Integral Equation Methods for Electromagnetic Scattering

Alexandre Dély[1], Adrien Merlini[2], Kristof Cools[3], and Francesco P. Andriulli[4]

[1] *Department of Electronics, Politecnico di Torino, Torino, Italy*
[2] *Microwave department, IMT Atlantique, Brest, France*
[3] *Department of Information Tech., Ghent University, Belgium*
[4] *Department of Electronics, Politecnico di Torino, Torino, Italy*

8.1 Introduction

Time domain boundary element methods are the tools of choice when faced with the task of modeling systems that are strongly radiating and include nonlinear components. In the time domain, boundary integral equations take the form of retarded potential equations. An operator valued distribution is convolved with the unknown trace and should equate the trace of the incident field.

These convolutional equations need to be discretized so that approximate solutions can be computed. A number of avenues for this discretization are available, all with their advantages and disadvantages.

In order to be fit for purpose, the discretization procedure should adhere to a number of restrictions. First, the discretization should preserve the time translation symmetry of the continuous system. If this is not the case, discrete interactions do not solely depend on the time delay between emitting and observing a signal, but also on the precise time the signal was emitted. For each possible emission time, a separate set of interaction matrices has to be stored in memory, increasing the storage requirements of the method beyond what is available, even in the most advanced HPC facilities. Second, the discretization should be causal: field values at a given time should only depend on source values at earlier times. More generally and more abstractly: the linear system resulting upon discretization should be lower triangular with respect to the temporal indices. Note that this allows generalizations such as the march-on-in-order method where temporal degrees of freedom span the entire time interval of interest but where notwithstanding a lower triangular – and thus easy to solve – linear system is produced.

Advances in Time-Domain Computational Electromagnetic Methods, First Edition.
Edited by Qiang Ren, Su Yan, and Atef Z. Elsherbeni.

We briefly summarize the most common discretization schemes that are compatible with the above constraints. Perhaps the most natural thing to do, especially for practitioners coming from frequency domain finite element or boundary element methods, is to build space-time variational formulations. The unknown trace is expanded as a linear combination of products of spatial and temporal basis functions. The temporal basis functions are shifted copies of a reference function centered near $t = 0$. These techniques date back as far as 1968 [1]. Modern baseline implementation techniques can be found in [2]. The tail of the temporal reference solution does not extend left of $t = -\Delta t$. These two properties guarantee the scheme is convolutional and causal. The advantages of space-time Galerkin techniques are that they introduce no numerical dispersion at all, that they rely on real floating point operations solely and thus are executed rapidly by common use processors that are designed for such applications, and that tailored fast methods and compression techniques have been developed for this class of solvers. The disadvantage is that in order to arrive at stable results, fairly complicated, kernel dependent, semi-analytical quadrature rules have to be used in addition to Helmholtz splitting and rescaling techniques that will be discussed in detail in this chapter.

It is possible to use nonlocal temporal basis functions and still arrive at a scheme that allows stepping to incrementally build the solution. This is done in the class of marching-in-order solvers. They rely on the special properties of exponentially modulated Laguerre polynomials [3]. The advantage of this scheme is that the solution does not suffer from late-time instabilities, as it is constructed as a linear combination of exponentially decaying basis functions. The disadvantage is that one needs a priori estimated for the temporal support of the solution and the polynomial content of the solution, which are in general not easy to come by. Marching-in-order solvers cannot be used in systems that contain nonlinear components. These solvers require stepping in time to allow the coupling to work.

The third class of discretizations is the central focus of this chapter: convolution quadrature based discretizations. These discretizations are based on the intuitive idea that solutions of ODEs can be computed both as the convolution of the right-hand side with a fundamental solution and as the result of applying some time stepping scheme. This idea has been provided with a solid mathematical basis in [4, 5]. Alternatively and equivalently, convolution quadrature methods are the result of approximating the Laplace variable s by a polynomial in the shift operator $e^{s\Delta t}$ and computing the inverse Laplace transform. These approximations can be linked to corresponding techniques for the numerical solution of ODEs. Whereas initially these ideas have been explored for single stage ODE solvers, more recent and accurate convolution quadrature are based on multi-stage Runge–Kutta methods [6]. In the control engineering and signal processing communities, this technique is sometimes referred to as the Tustin transform [7].

When applied to the computation of retarded potentials, this methodology results in the recursive computation of field values based on past source samples. The use of convolutional quadrature for the solution of time domain boundary integral equation has first been explored in elastodynamics [8]. Convolution quadrature methods for boundary integral equations based on Runge–Kutta ODE solvers have been subsequently described for acoustics [9] and electromagnetics [10]. The advantage of this method are that their stability properties are much better understood than those of space-time Galerkin schemes, that all interaction integrals can be done with field tested methods well documented in literature, and that the solver can with relative ease be built on top of existing frequency domain boundary element methods. The main disadvantage is that convolution quadrature schemes, just like time stepping schemes for the solution of ODEs, invariably introduce dispersion errors. These errors increase for broad band inputs.

This chapter aims to provide an overview of the underpinnings of the convolution quadrature method and its application in computational electromagnetics. It starts with the relationship between the Laplace transforms and Z-transform and where approximations are introduced to reduce a continuous convolution to a discrete convolution. The text explains how this approximation can be built based on simple time stepping schemes for the solution of ODEs but also for certain subclasses of implicit Runge-Kutta (IRK) rules. A key ingredient in these convolution quadrature methods is evaluation of frequency domain integral operators at matrix valued frequencies. We will sketch how this can be integrated on top of existing frequency domain boundary element method implementations.

Like space-time Galerkin methods, convolution quadrature methods can be susceptible to instabilities. We follow the classification of instabilities in three categories: low frequency or DC instabilities, resonant instabilities, and high frequency instabilities. In this chapter, we focus on how to eliminate the former two. DC instabilities originate from the existence of constant-in-time regime solutions to the homogeneous equation. Using a Helmholtz splitting and rescaling approach, the scattering problem can be rewritten in an equivalent form that no longer supports these constant-in-time solutions. As a result, the method is immune from DC instabilities, even in the (unavoidable) presence of quadrature errors. Resonant instabilities are related to the existence of interior solutions to the scattering problem. It will be detailed in this chapter how to – starting from a convolution quadrature boundary element method that has been treated for DC instabilities – build a formulation that in addition is no longer susceptible to resonant instabilities. This can be achieved by the careful construction of a combined field formulation in which both the electric and magnetic term respect the Helmholtz structure of the unknown traces.

The resulting time domain solver has a combination of desirable properties: (i) it can be easily implemented on top of existing boundary element methods,

(ii) it is not susceptible to DC instabilities, and (iii) it is not susceptible to resonant instabilities. The accuracy and robustness of the solver will be demonstrated on benchmark and real-life examples.

8.2 Background and Notations

8.2.1 Time Domain Integral Equations

We consider a perfect electric conductor (PEC) in an homogeneous medium. The boundary of the PEC is denoted Γ and its outwards pointing normal $\hat{n}$. The characteristic impedance of the medium is η, the speed of light is c and the permeability is $\mu = \eta/c$. Our objective is to compute the electromagnetic field $(\boldsymbol{E}^{\text{sca}}(\boldsymbol{r}, t), \boldsymbol{H}^{\text{sca}}(\boldsymbol{r}, t))$ scattered by the PEC scatterer when it is illuminated by an incident electromagnetic field $(\boldsymbol{E}^{\text{inc}}(\boldsymbol{r}, t), \boldsymbol{H}^{\text{inc}}(\boldsymbol{r}, t))$. We denote by $\boldsymbol{J}$ the surface current density (in A/m) that is the rotated trace of the total magnetic field

$$\boldsymbol{J}(\boldsymbol{r}, t) = \hat{\boldsymbol{n}}(\boldsymbol{r}) \times (\boldsymbol{H}^{\text{inc}}(\boldsymbol{r}, t) + \boldsymbol{H}^{\text{sca}}(\boldsymbol{r}, t)), \quad \boldsymbol{r} \in \Gamma \tag{8.1}$$

and because the scatterer is a PEC, the tangential trace of the total electric field is zero on Γ, i.e.

$$\hat{\boldsymbol{n}}(\boldsymbol{r}) \times (\boldsymbol{E}^{\text{inc}}(\boldsymbol{r}, t) + \boldsymbol{E}^{\text{sca}}(\boldsymbol{r}, t)) = \boldsymbol{0}, \quad \forall \boldsymbol{r} \in \Gamma, \forall t \tag{8.2}$$

The current is the solution of the electric field integral equation (EFIE) and magnetic field integral equation (MFIE), respectively

$$\eta \mathcal{T} \boldsymbol{J}(\boldsymbol{r}, t) = -\hat{\boldsymbol{n}}(\boldsymbol{r}) \times \boldsymbol{E}^{\text{inc}}(\boldsymbol{r}, t) \tag{8.3}$$

$$\left(\frac{\mathcal{I}}{2} - \mathcal{K} \right) \boldsymbol{J}(\boldsymbol{r}, t) = \hat{\boldsymbol{n}}(\boldsymbol{r}) \times \boldsymbol{H}^{\text{inc}}(\boldsymbol{r}, t) \tag{8.4}$$

and of their linear combination, the combined field integral equation (CFIE)

$$\left(-\alpha \mathcal{T} + (1 - \alpha)\hat{\boldsymbol{n}}(\boldsymbol{r}) \times \left(\frac{\mathcal{I}}{2} - \mathcal{K} \right) \right) \boldsymbol{J}(\boldsymbol{r}, t) = \frac{\alpha}{\eta} \hat{\boldsymbol{n}}(\boldsymbol{r}) \times \boldsymbol{E}^{\text{inc}}(\boldsymbol{r}, t)$$
$$+ (1 - \alpha)\hat{\boldsymbol{n}}(\boldsymbol{r}) \times \hat{\boldsymbol{n}}(\boldsymbol{r}) \times \boldsymbol{H}^{\text{inc}}(\boldsymbol{r}, t) \tag{8.5}$$

where α is a constant parameter and the surface integral operators are defined by

$$\mathcal{T} \boldsymbol{f}(\boldsymbol{r}, t) = -\frac{1}{c} \frac{\partial}{\partial t} \mathcal{T}_s \boldsymbol{f}(\boldsymbol{r}, t) + c \int_{t'=-\infty}^{t} \mathcal{T}_h \boldsymbol{f}(\boldsymbol{r}, t') dt' \tag{8.6}$$

$$\mathcal{T}_s \boldsymbol{f}(\boldsymbol{r}, t) = \hat{\boldsymbol{n}}(\boldsymbol{r}) \times \iint_{\boldsymbol{r}' \in \Gamma} \frac{1}{4\pi |\boldsymbol{r} - \boldsymbol{r}'|} \boldsymbol{f} \left(\boldsymbol{r}', t - \frac{|\boldsymbol{r} - \boldsymbol{r}'|}{c} \right) dS' \tag{8.7}$$

$$\mathcal{T}_h \boldsymbol{f}(\boldsymbol{r}, t) = \hat{\boldsymbol{n}}(\boldsymbol{r}) \times \nabla \iint_{\boldsymbol{r}' \in \Gamma} \frac{1}{4\pi |\boldsymbol{r} - \boldsymbol{r}'|} \nabla' \cdot \boldsymbol{f} \left(\boldsymbol{r}', t - \frac{|\boldsymbol{r} - \boldsymbol{r}'|}{c} \right) dS' \tag{8.8}$$

$$\mathcal{K}f(r,t) = \hat{n}(r) \times \nabla \times \iint_{r' \in \Gamma} \frac{1}{4\pi|r-r'|} f\left(r', t - \frac{|r-r'|}{c}\right) dS' \qquad (8.9)$$

$$= \hat{n}(r) \times p.v. \iint_{r' \in \Gamma} \frac{r'-r}{4\pi|r-r'|^2} \times \left[\frac{1}{|r-r'|} f\left(r', t - \frac{|r-r'|}{c}\right)\right.$$

$$\left. + \frac{1}{c}\frac{\partial f}{\partial t}\left(r', t - \frac{|r-r'|}{c}\right)\right] dS' \qquad (8.10)$$

$$\mathcal{I}f(r,t) = f(r,t) \qquad (8.11)$$

We assume that there is a time t_0 such that $\forall t < t_0, \forall r \in \Gamma, J(r,t) = 0$, and the incident fields $E^{inc}(r,t)$ and $H^{inc}(r,t)$ also vanish for all r in the neighborhood of Γ.

Once the current has be obtained, the scattered fields can be computed using the vector potential A as

$$E^{sca}(r,t) = -\frac{1}{c}\frac{\partial A}{\partial t}(r,t) + c \int_{t'=-\infty}^{t} \nabla\nabla \cdot A(r,t') dt' \qquad (8.12)$$

$$H^{sca}(r,t) = \frac{1}{\mu}\nabla \times A(r,t) \qquad (8.13)$$

$$A(r,t) = \mu \iint_{r' \in \Gamma} \frac{1}{4\pi|r-r'|} J\left(r', t - \frac{|r-r'|}{c}\right) dS' \qquad (8.14)$$

Note that the explicit expression of the surface integral operators in the time domain are not required and are provided for completeness for the scheme we will present, only their Laplace domain expressions are, as Section 8.3 will show.

8.3 Solution Using Convolution Quadrature

One could think of discretizing the time variable in the above equations directly, as it is done, for example, in standard time domain space-time marching-on-in-time discretization approaches [11]. In convolution Quadrature schemes, per contra, the equations are not directly discretized in time they will instead undergo a sequence of transformations. First, the time domain integral equation is transformed into Laplace domain. The equation in Laplace variable is then approximated as a rational function that results in an equation in the Z domain. Finally, the inverse Z-transform produces a discrete convolution. This is summarized in the diagram in Figure 8.1.

8.3.1 Laplace Transform

The Laplace transform [12] maps a function x in the time domain to the function $\mathcal{L}(x)$ in the Laplace domain. It is defined for all $s \in \mathbb{C}$ in its region of convergence as

$$\mathcal{L}(x)(s) = \int_{t=0}^{+\infty} x(t)e^{-st} dt = X(s) \qquad (8.15)$$

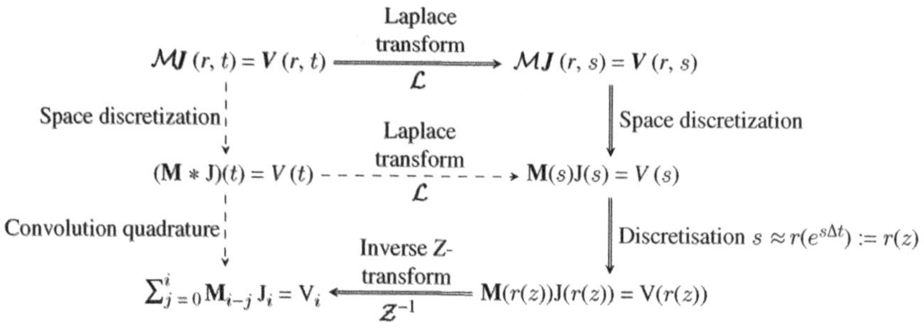

$$
\begin{array}{ccc}
\mathcal{M}\mathcal{J}(r,t) = V(r,t) & \xrightarrow[\mathcal{L}]{\text{Laplace transform}} & \mathcal{M}\mathcal{J}(r,s) = V(r,s) \\
\downarrow\text{Space discretization} & \xrightarrow[\mathcal{L}]{\text{Laplace transform}} & \downarrow\text{Space discretization} \\
(M * J)(t) = V(t) & \dashrightarrow & M(s)J(s) = V(s) \\
\downarrow\text{Convolution quadrature} & \xleftarrow[\mathcal{Z}^{-1}]{\text{Inverse Z-transform}} & \downarrow\text{Discretisation } s \approx r(e^{s\Delta t}) := r(z) \\
\sum_{j=0}^{i} M_{i-j}\, J_i = V_i & & M(r(z))J(r(z)) = V(r(z))
\end{array}
$$

Figure 8.1 Sequence of transform for the convolution quadrature solution of time domain integral equations.

The inverse Laplace transform of X is

$$
\mathcal{L}^{-1}(X)(t) = \frac{1}{2\pi i} \int_{s=\sigma-i\infty}^{\sigma+i\infty} X(s)e^{st}\, ds = x(t) \tag{8.16}
$$

where the integration is performed on the line $\Re(s) = \sigma$ in the region of convergence and with all the singularities of $X(s)$ on its left.

The passage from time domain to Laplace domain is straightforward using the following properties of the Laplace transform for the time differentiation, time integration, and time delay [7].

$$
\mathcal{L}\left(t \mapsto \frac{d}{dt}x(t)\right)(s) = s\mathcal{L}(x)(s) \tag{8.17}
$$

$$
\mathcal{L}\left(t \mapsto \int_{t'=-\infty}^{t} x(t')dt'\right)(s) = \frac{1}{s}\mathcal{L}(x)(s) \tag{8.18}
$$

$$
\mathcal{L}\left(t \mapsto x(t-\tau)\right)(s) = e^{-s\tau}\mathcal{L}(x)(s) \tag{8.19}
$$

assuming $x(t) = 0$ when $t \leq 0$.

8.3.2 Laplace Domain Integral Equations

The time domain integral equations and operators can be transformed to Laplace domain. The expressions of the Laplace-transformed equations remain similar to their time-domain counterpart, albeit with different definitions for the Laplace domain operators. The Laplace domain EFIE, MFIE and CFIE respectively are

$$
\eta\mathcal{T}J(r,s) = -\hat{n}(r) \times E^{\text{inc}}(r,s) \tag{8.20}
$$

$$
\left(\frac{I}{2} - \mathcal{K}\right)J(r,s) = \hat{n}(r) \times H^{\text{inc}}(r,s) \tag{8.21}
$$

and

$$\left(-\alpha\mathcal{T} + (1-\alpha)\hat{n}(r) \times \left(\frac{I}{2} - \mathcal{K}\right)\right) J(r,s)$$

$$= \frac{\alpha}{\eta}\hat{n}(r) \times E^{\text{inc}}(r,s) + (1-\alpha)\hat{n}(r) \times \hat{n}(r) \times H^{\text{inc}}(r,s) \tag{8.22}$$

where s is the Laplace variable and the Laplace domain surface integral operators are

$$\mathcal{T}f(r,s) = -\frac{s}{c}\mathcal{T}_s f(r,s) + \frac{c}{s}\mathcal{T}_h f(r,s) \tag{8.23}$$

$$\mathcal{T}_s f(r,s) = \hat{n}(r) \times \iint_{r' \in \Gamma} \frac{e^{-s|r-r'|/c}}{4\pi|r-r'|} f(r',s)\, dS' \tag{8.24}$$

$$\mathcal{T}_h f(r,s) = \hat{n}(r) \times \nabla \iint_{r' \in \Gamma} \frac{e^{-s|r-r'|/c}}{4\pi|r-r'|} \nabla' \cdot f(r',s)\, dS' \tag{8.25}$$

$$\mathcal{K}f(r,s) = \hat{n}(r) \times \iint_{r' \in \Gamma} \nabla\left(\frac{e^{-s|r-r'|/c}}{4\pi|r-r'|}\right) \times f(r',s)\, dS' \tag{8.26}$$

Note that the frequency domain (time harmonics) operators have exactly the same form. They can be retrieved by setting $s = i\omega$ (respectively $s = -i\omega$) to obtain the $e^{i\omega t}$ (respectively $e^{-i\omega t}$) harmonic convention. Note that this expression is built from frequency domain boundary integral operators evaluated in complex valued frequencies. This is quite relevant since code to build these Laplace discrete representations of these operators can be easily derived from existing frequency domain boundary element method implementations.

8.3.3 Z-Transform

The Z-transform $\mathcal{Z}$ maps a discrete sequence $(X_n)_{n\in\mathbb{N}}$ into a complex function $\mathcal{Z}(X)$ defined for all $z \in \mathbb{C}$ in its region of convergence as

$$\mathcal{Z}(X)(z) = \sum_{n=0}^{+\infty} X_n z^{-n} \tag{8.27}$$

The inverse transform of X is the sequence $(\mathcal{Z}^{-1}(X)_n)_{n\in\mathbb{N}}$ that is defined as

$$\mathcal{Z}^{-1}(X)_n = \frac{1}{2\pi i} \oint_{z\in C} X(z) z^{n-1}\, dz \tag{8.28}$$

where the counterclockwise contour C is in the region of convergence of X and goes around 0. In particular, when C is a circle of radius ρ, the inverse Z-transform can be approximated by the trapezoidal rule on Q sub-intervals of $[-\pi, \pi]$

$$\mathcal{Z}^{-1}(X)_n = \frac{\rho^n}{2\pi} \int_{\theta=-\pi}^{\pi} X(\rho e^{i\theta}) e^{i\theta n}\, d\theta \tag{8.29}$$

$$\approx \frac{\rho^n}{Q} \sum_{q=0}^{Q-1} X(\rho e^{2\pi i \frac{q}{Q}}) e^{2\pi i \frac{q}{Q} n} \tag{8.30}$$

There is a close relation between the Z-transform and the Laplace transform, as the former can be viewed as the discrete counterpart of the latter. Consider a function of time x, a sampling step Δt, and define X and Y as

$$X(t) = x(t) \sum_{n=0}^{+\infty} \delta(t - n\Delta t) \tag{8.31}$$

$$Y_n = x(n\Delta t) \tag{8.32}$$

X is a function of time that samples the function x by multiplying it with Dirac deltas, while $Y = (Y_n)_{n=0}^{\infty}$ is a sequence of samples of x. The relation between X and Y is simply

$$\mathcal{L}(X)(s) = \int_0^{\infty} x(t) \sum_{n=0}^{\infty} \delta(t - n\Delta t) e^{st} \, dt \tag{8.33}$$

$$= \sum_{n=0}^{\infty} x(n\Delta t) e^{ns\Delta t} \tag{8.34}$$

$$= \sum_{n=0}^{\infty} Y_n e^{n(s\Delta t)} \tag{8.35}$$

$$= \mathcal{Z}(Y)(e^{s\Delta t}) \tag{8.36}$$

In the other direction this implies that a function of s that can be written as a series in $e^{s\Delta t}$ is the Laplace transform of a sampled function of time.

8.3.4 Runge–Kutta Methods

The Runge–Kutta methods are numerical approaches used to solve ordinary differential equations [13]. Specifically consider

$$\begin{cases} \dfrac{dy(t)}{dt} = F(t, y) \\ y(t_0) = y_0 \end{cases} \tag{8.37}$$

where y is the unknown function of t whose derivative is a function of t and y, and y_0 is the initial condition. The approximate solution $y_i \approx y(t_i)$ is computed iteratively at the time $t_i = t_0 + i\Delta t$ where Δt is the time step. Let's denote p the number of stages of the Runge–Kutta method we use. At each iteration, the solution at the next time step y_{i+1} is computed by summing the current solution with p interpolants of the slope noted $[F_i]_k$ ($k \in [1, p]$), each having a weight noted $[b]_k$ (b is a column vector).

The slopes $[F_i]_k$ can be interpreted as approximations of values of $F(t, y(t))$ at time steps between t_i and t_{i+1}. In general, for implicit methods, they are the solutions to the system of nonlinear equations

$$[F_i]_k = F(t_i + [c]_k \Delta t, [Y_i]_k) \tag{8.38}$$

$$[Y_i]_k = y_i + \Delta t \sum_{l=1}^{p} [A]_{kl} [F_i]_l \tag{8.39}$$

Once these slope values are known, the approximate solution y_{i+1} at time t_{i+1} is computed as

$$y_{i+1} = y_i + \Delta t \sum_{k=1}^{p} [b]_k [F_i]_k \tag{8.40}$$

Rewriting (8.39) and (8.40) in vector form yields

$$Y_i = y_i 1_p + \Delta t A F_i, \quad \text{with} \quad [F_i]_k = F(t_i + [c]_k \Delta t, [Y_i]_k) \tag{8.41}$$

$$y_{i+1} = y_i + \Delta t b^T F_i \tag{8.42}$$

In the above, the matrix $A \in \mathbb{R}^{p \times p}$, and the vectors $b, c \in \mathbb{R}^p$ fully determine the Runge–Kutta method. They are often displayed as a Butcher tableau [13] (see also the examples in the appendix):

$$\frac{c \mid A}{\mid b^T} = \begin{array}{c|ccc} [c]_1 & [A]_{11} & \cdots & [A]_{1p} \\ \vdots & \vdots & \ddots & \vdots \\ [c]_p & [A]_{p1} & \cdots & [A]_{pp} \\ \hline & [b]_1 & \cdots & [b]_p \end{array} \tag{8.43}$$

When A is strictly lower triangular ($[A]_{kl} = 0$ for all $l \geq k$), then the method is said to be explicit because each $[F_i]_k$ can be computed from the all the previous $[F_i]_l$ ($l < k$) sequentially (for example, this is the case for the most popular "RK" method [14, 15]). Otherwise, the method is implicit if there exist a $[A]_{kl} \neq 0$ with $l \geq k$. In this chapter we consider only IRK methods.

The order r of a method determines the local error as $\Delta t \to 0$

$$y_1 - y(t_0 + \Delta t) = O(\Delta t^{r+1}) \tag{8.44}$$

Higher orders impose more algebraic conditions on A, b, and c to be satisfied [16]. This of course comes at the price of larger computational efforts, but results in methods with faster convergence rates; the use of implicit over explicit schemes, moreover, results in more favorable stability properties [13]. For example, the order 1 condition imposes that the weights sum to 1:

$$\sum_{i=1}^{p} [b]_i = 1 \tag{8.45}$$

the order 2 add the constraint

$$\sum_{i=1}^{p} [b]_i [c]_i = \frac{1}{2} \tag{8.46}$$

and so on.

Also, in the following we make use of the inverse of A, which cannot exist if the method is explicit (since A is a strict lower triangular matrix in that case). In addition, we will only use methods that verify $b^T A^{-1} = [0, \ldots, 0, 1]$ (i.e. b^T is equal to the last row of A). These methods are called stiffly accurate [17]. In particular

$$b^T A^{-1} 1_p = 1 \tag{8.47}$$

where we denote 1_p the $\mathbb{R}^p$ vector of all 1. This property will be very useful to simplify some expressions in the following. On the one hand, the terms $b^T A^{-1} 1_p - 1$ will cancel and $b^T A^{-1} X$ is simply the last element of any vector X of $\mathbb{C}^p$.

When the RK method is implicit, from (8.43) we obtain $\Delta t F_i = A^{-1}(Y_i - y_i 1_p)$. Substituting it in the first equation and using (8.47) we obtain

$$y_{i+1} = b^T A^{-1} Y_i + y_i (b^T A^{-1} 1_p - 1) \tag{8.48}$$
$$= b^T A^{-1} Y_i$$

Finally, we will also assume that $[c]_p = 1$. The Radau IIA and Lobatto IIIC methods that verify all these assumptions are presented in the appendix.

8.3.5 Solution of a Differential Equation Using Runge–Kutta Methods

In this section, we use RK methods to solve a particular differential equation that will be fundamental for what it will follow.

Consider the system of equations

$$\begin{cases} \dfrac{dy(t)}{dt} = sy(t) + g(t) \\ y(0) = 0 \end{cases} \tag{8.49}$$

where g is a known function and s is a complex scalar. With the notations of Section 3.4 we have $F(y, t) = sy(t) + g(t)$ and $F_i = sY_i + g_i$. Z-transforming (8.42) and (8.41) results in

$$z\mathcal{Z}(y)(z) - zy_0 = \mathcal{Z}(y)(z) + \Delta t b^T \mathcal{Z}(F)(z) \tag{8.50}$$

$$\mathcal{Z}(Y)(z) = \mathcal{Z}(y)(z) 1_p + \Delta t A \mathcal{Z}(F)(z) \tag{8.51}$$

after eliminating $\mathcal{Z}(y)(z)$ and enforcing $y_0 = y(0) = 0$ we obtain

$$\mathcal{Z}(Y)(z) = \Delta t \left(A + \frac{1_p b^T}{z - 1} \right) \mathcal{Z}(F)(z) \tag{8.52}$$

Then, using $\mathcal{Z}(F) = s\mathcal{Z}(Y) + \mathcal{Z}(g)$ and solving for $\mathcal{Z}(Y)$ we have

$$\mathcal{Z}(Y)(z) = \left(s(z) - sI_p \right)^{-1} \mathcal{Z}(g)(z) \tag{8.53}$$

where I_p is the $\mathbb{R}^{p \times p}$ identity matrix and by exploiting the Sherman–Morrison formula

$$s(z) = \frac{1}{\Delta t} \left(A + \frac{1_p b^T}{z - 1} \right)^{-1} \tag{8.54}$$

$$= \frac{1}{\Delta t} \left(A^{-1} - \frac{A^{-1} 1_p b^T A^{-1}}{z - 1 + b^T A^{-1} 1_p} \right) \tag{8.55}$$

Using the property that $b^T A^{-1} 1_p = 1$, it simplifies further to

$$s(z) = \frac{1}{\Delta t} \left(A^{-1} - A^{-1} 1_p b^T A^{-1} z^{-1} \right) \tag{8.56}$$

8.3.6 Convolution Quadrature Using Runge–Kutta Methods

As seen in the following, the discretization of time domain integral equations in Section 8.2.1 will yield matrix-valued temporal convolutions in the form $(M * J)(t) = V(t)$. For this reason it will be important to find an approximation scheme for the convolution product

$$v(t) = (f * g)(t) = \int_{u=0}^{t} f(t-u)g(u)du \tag{8.57}$$

that will have the form

$$v((j+1)\Delta t) \approx v_{j+1} = \sum_{i=0}^{j} w_{i-j} g_j \tag{8.58}$$

where the samples g depend on g and the weight w depend on the Laplace transform of f. In fact, f does not need to be known explicitly and only $\mathcal{L}(f)$ will be required. We first substitute f by the inverse Laplace transform of $\mathcal{L}(f)$ obtaining

$$v(t) = \frac{1}{2\pi i} \int_{s=\sigma-i\infty}^{\sigma+i\infty} \mathcal{L}(f)(s) \int_{u=0}^{t} e^{s(t-u)} g(u) du\, ds \tag{8.59}$$

It should be noted that the inner integral in (8.59) is the solution of the differential equation (8.49) that can be solved numerically via a RK method as

$$y(t) = \int_{u=0}^{t} e^{s(t-u)} g(u) du \tag{8.60}$$

$$y((j+1)\Delta t) \approx y_{j+1} = b^T A^{-1} Y_j(s) \tag{8.61}$$

So by using the approximation (8.61) in (8.59) we get

$$v_{j+1} = \frac{1}{2\pi i} \int_{s=\sigma-i\infty}^{\sigma+i\infty} \mathcal{L}(f)(s) b^T A^{-1} Y_j(s) ds \tag{8.62}$$

Z-transforming this equation and using (8.53) results in

$$\mathcal{Z}(v_{j+1})(z) = \frac{1}{2\pi i} \int_{s=\sigma-i\infty}^{\sigma+i\infty} \mathcal{L}(f)(s) b^T A^{-1} \mathcal{Z}(Y(s))(z) ds \tag{8.63}$$

$$- \frac{1}{2\pi i} \int_{s=\sigma-i\infty}^{\sigma+i\infty} \mathcal{L}(f)(s) b^T A^{-1} \left(s(z) - s I_p \right)^{-1} \mathcal{Z}(y)(z) ds$$

The integral over s is exactly $\mathcal{L}(f)(s)$, evaluated at $s(z)$, as computed by using the Cauchy formula [18]:

$$\mathcal{L}(f)(s(z)) = \frac{1}{2\pi i} \int_{s=\sigma-i\infty}^{\sigma+i\infty} \mathcal{L}(f)(s) \left(s(z) - s I_p \right)^{-1} ds \tag{8.64}$$

In practice, functions at matrix valued arguments, when the matrix is diagonalizable, are computed by performing an eigenvalue decomposition, and evaluating the function at each of the eigenvalues [18]. Indeed, consider a function $X(s)$ that possesses a power series expansion

$$X(s) = \sum_{n=0}^{+\infty} \frac{s^n}{n!} \frac{d^n X}{ds^n}(0) \tag{8.65}$$

We can compute $X(\mathbf{s})$ with a matrix valued argument $\mathbf{s} \in \mathbb{C}^{p \times p}$ whose eigenvalue decomposition is $\mathbf{s} = \mathbf{UDU}^{-1}$ as the matrix

$$X(\mathbf{s}) = \sum_{n=0}^{+\infty} \frac{\mathbf{s}^n}{n!} \frac{d^n X}{ds^n}(0) = \mathbf{U} \begin{pmatrix} X([\mathbf{D}]_{11}) & & (0) \\ & \ddots & \\ (0) & & X([\mathbf{D}]_{pp}) \end{pmatrix} \mathbf{U}^{-1} \tag{8.66}$$

Now, after substituting (8.64), (8.63) becomes

$$\mathcal{Z}(v_{i+1})(z) = \mathbf{b}^{\mathsf{T}} \mathbf{A}^{-1} \mathcal{L}(f)(\mathbf{s}(z)) \mathcal{Z}(g)(z) \tag{8.67}$$

to which the inverse Z-transform can be applied, transforming the product of two z valued functions into the discrete convolution of sequences, yielding

$$v_{i+1} = \sum_{j=0}^{i} \mathbf{b}^{\mathsf{T}} \mathbf{A}^{-1} \mathcal{Z}^{-1}(\mathcal{L}(f)(\mathbf{s}(z)))_{i-j} g_j \tag{8.68}$$

$$= \sum_{j=0}^{i} w_{i-j} g_j \tag{8.69}$$

where

$$w_j = \mathbf{b}^{\mathsf{T}} \mathbf{A}^{-1} \mathbf{W}_j \tag{8.70}$$

$$\mathbf{W}_j = \mathcal{Z}^{-1}(z \mapsto \mathcal{L}(f)(\mathbf{s}(z)))_j \tag{8.71}$$

$$[g_j]_k = g(t_j + [c]_k \Delta t) \tag{8.72}$$

Note that the (inverse) Z-transform of a vector or a matrix is defined component-wise. Since we use RK methods such that $\mathbf{b}^{\mathsf{T}} \mathbf{A}^{-1} = (0, \dots, 0, 1)$, w_j is the last row of $\mathcal{Z}^{-1}(\mathcal{L}(f)(\mathbf{s}(z)))_j$. When we use the convolution quadrature to solve the discretized boundary integral equations, we already know the result of the convolution that is given by the excitation. Instead, we want to compute the source. In the above example it corresponds to computing the g_j from the knowledge of the v_i and w_i, that is pN_t unknowns for N_t equations, thus more constraints should be found. To this purpose, we introduce the sequence of $\mathbb{R}^p$ vectors v_i such that

$$[v_i]_k = v(t_i + [c]_k \Delta t), \quad k \in [1, p] \tag{8.73}$$

The vector $[v_i]_k$ contains samples of the right-hand side at stage times between consecutive time steps. In [6], it is shown that IRK solution methods for convolutional equations converge with this choice for the right-hand side. Moreover, the vector $[v_i]_k$ verifies $b^T A^{-1} v_i = v_{i+1}$, assuming that $[c]_p = 1$ (which is the case for several common RK methods such as the Radau IIA [19] and Lobatto IIIC that are given in the appendix). Thus, we now have

$$v_i = \sum_{j=0}^{i} W_{i-j} g_j \tag{8.74}$$

that is pN_t equations for pN_t unknowns that can be solved by marching on in time. Conversely, once all the g_j have been obtained using (8.74), they clearly verify the convolution quadrature (8.68) by applying $b^T A^{-1}$ on the left of each side of (8.74).

8.3.7 Discretization of Boundary Integral Equations

8.3.7.1 Space Discretization

The RWG [20] basis functions (also known as order 0 Raviart-Thomas basis functions [21]) can be used to discretize the current in space. They will be noted (f_n) where $n \in [1, N_s]$, and N_s, the number of spatial unknowns, is equal to the number of edges. The coefficients of the expansion depend on time

$$J(r, t) \approx \sum_{n=1}^{N} [J(t)]_n f_n(r) \tag{8.75}$$

Assume that $J(r, t)$ is solution of the following integral equation

$$\mathcal{M} J(r, t) = V(r, t) \tag{8.76}$$

where $\mathcal{M}$ is a boundary integral operator (in particular, that can be the operator of the differentiated EFIE, MFIE, or CFIE) and V is the excitation.

Differentiated TD PEC-EFIE

We discretize the time domain PEC-EFIE that has been differentiated. The reason why the EFIE is differentiated is to remove the time integration of the gradient of the scalar potential in $\mathcal{T}$

$$\eta \frac{\partial}{\partial t} \mathcal{T} J(r, t) = -\hat{n}(r) \times \frac{\partial}{\partial t} E^{inc}(r, t) \tag{8.77}$$

The integral equation is tested with the set of rotated RWG basis functions $(\hat{n} \times f_m)$. The partially discretized system can be formally written as

$$(T * J)(t) = V(t) \tag{8.78}$$

where

$$[T(t)]_{mn} = \left\langle \hat{n} \times f_m, \eta \frac{\partial}{\partial t} \mathcal{T}(f_n \delta)(t) \right\rangle \tag{8.79}$$

$$= \eta \iint_{r \in \Gamma} \hat{n}(r) \times f_m(r) \cdot \frac{\partial}{\partial t} \mathcal{T}(f_n \delta)(r, t) dS \tag{8.80}$$

$$[V(t)]_m = \left\langle \hat{n} \times f_m, -\hat{n} \times \frac{\partial}{\partial t} E^{\mathrm{inc}}(t) \right\rangle \tag{8.81}$$

$$= -\iint_{r \in \Gamma} f_m(r) \cdot \frac{\partial}{\partial t} E^{\mathrm{inc}}(r, t) dS \tag{8.82}$$

Here $\delta(t)$ is the Dirac delta distribution.

Note that in practice these partially discretized matrices are not computed. Instead, their Laplace transform will be computed. We note their Laplace transform $\mathbf{T}(s) = \mathcal{L}(t \mapsto \mathbf{T}(t))(s)$, whose elements are

$$[\mathbf{T}(s)]_{mn} = \left\langle \hat{n} \times f_m, \eta s T f_n(s) \right\rangle \tag{8.83}$$

$$= \eta s \iint_{r \in \Gamma} \hat{n}(r) \times f_m(r) \cdot T f_n(r, s) dS \tag{8.84}$$

The computation of these matrix elements is very standard in frequency domain [11].

TD PEC-MFIE

The MFIE in time domain is

$$\left(\frac{I}{2} - \mathcal{K} \right) J(r, t) = \hat{n}(r) \times H^{\mathrm{inc}}(r, t) \tag{8.85}$$

We test the equation with the set of rotated BC [22] basis functions ($\hat{n} \times g_m$).

We want to point out that the MFIE is traditionally discretized by testing it with RWG basis functions [10]. Even though it is easier to implement and computationally it costs less than the discretization with BC, that discretization of the MFIE with RWG is not conforming and yields results that are less accurate than the EFIE, contrary to the discretization with BC functions that has the same accuracy as the EFIE [23]. Therefore, we think that the conforming discretization with BC is far more reliable.

The MFIE in the Laplace domain is

$$\left(\frac{I}{2} - \mathcal{K} \right) J(r, s) = \hat{n}(r) \times H^{\mathrm{inc}}(r, s) \tag{8.86}$$

and it is discretized as

$$(\mathbf{G}/2 - \mathbf{K}(s)) J(s) = \mathbf{H}(s) \tag{8.87}$$

where

$$[\mathbf{G}]_{mn} = \left\langle \hat{n} \times g_m, f_n \right\rangle \tag{8.88}$$

$$= \iint_{r \in \Gamma} \hat{n}(r) \times g_m(r) \cdot f_n(r) dS \tag{8.89}$$

$$[\mathbf{K}(s)]_{mn} = \left\langle \hat{n} \times g_m, \mathcal{K} f_n(s) \right\rangle \tag{8.90}$$

$$= \iint_{r \in \Gamma} \hat{n}(r) \times g_m(r) \cdot \mathcal{K} f_n(r, s) dS \tag{8.91}$$

TD PEC-CFIE

The CFIE is the sum of the EFIE scaled by $-\alpha/\eta$ with the MFIE "scaled" by $(1 - \alpha)\hat{n}\times$, where $0 < \alpha < 1$ is a chosen constant. It reads

$$\left(-\alpha\mathcal{T} + (1 - \alpha)\hat{n} \times \left(\frac{\mathcal{I}}{2} - \mathcal{K}\right)\right)J = \frac{\alpha}{\eta}\hat{n} \times E^{\text{inc}} + (1 - \alpha)\hat{n} \times \hat{n} \times H^{\text{inc}} \quad (8.92)$$

The $\hat{n} \times \mathcal{I}$ operator that is applied to the MFIE is discretized with RWG as source and rotated RWG as testing to be consistent with the testing of the EFIE. It simplifies to the RWG gram matrix.

$$[\mathbf{N}]_{mn} = \left\langle \hat{n} \times f_m, (\hat{n} \times \mathcal{I})f_n \right\rangle \quad (8.93)$$

$$= \iint_{r\in\Gamma} f_m(r) \cdot f_n(r)dS \quad (8.94)$$

The operator $\hat{n} \times \mathcal{I}$ has f as a source, and the MFIE is tested with $\hat{n} \times g$, therefore in the passage between the two discretizations we need to use the inverse of a matrix whose elements (m, n) are given by $\left\langle \hat{n} \times g_m, f_n \right\rangle$ that is $\mathbf{G}$.

The discretized CFIE is

$$\left(-\alpha\mathbf{T} + (1 - \alpha)\mathbf{N}\mathbf{G}^{-1}\left(\mathbf{G}/2 - \mathbf{K}(s)\right)\right)J(s) = \alpha/\eta\mathbf{E}(s) + (1 - \alpha)\mathbf{N}\mathbf{G}^{-1}\mathbf{H}(s) \quad (8.95)$$

In time domain, one will use the differentiated version that corresponds to multiplying both sides by s in the Laplace domain, again to remove the time integral in the operator $\mathcal{T}$.

8.3.7.2 Time Discretization

The fully discretized system involves a sequence of solution vectors (J_i), each i corresponding to a different time step. The vectors in the sequence contain N elements that correspond to the N RWG coefficients of the current. Since we use a p stages RK method, there are p vectors per time step. These p $\mathbb{R}^N$ vectors can also be seen as one $\mathbb{R}^{pN}$ vector containing N blocks of p coefficients.

Similarly, we have a sequence of excitation vectors in the RHS (V_i). As for the current, the p $\mathbb{R}^N$ vectors of each time step are interleaved in the same $\mathbb{R}^{pN}$ vector containing N blocks of p coefficients. Finally, there is a sequence of interaction matrices $(\mathbf{M}_i)$. Each matrix contains $N \times N$ blocks, each block corresponding to the interaction of one source with one testing. To take into account the p stages of the RK method, each block is a small $p \times p$ matrix.

Applying the convolution quadrature to (8.78), with $\mathbf{M} = \mathbf{T}$ in this case, would result in

$$V_i = \sum_{j=0}^{i} \mathbf{M}_{i-j}J_j \quad (8.96)$$

where

$$[J_j]_{k+(n-1)p} = [J(t_j + [c]_k \Delta t)]_n \tag{8.97}$$

$$J(r, t_j + [c]_k \Delta t) \approx \sum_{n=1}^{N} [J_j]_{k+(n-1)p} f_n(r) \tag{8.98}$$

$$[V_i]_{k+(n-1)p} = [V(t_i + [c]_k \Delta t)]_n \tag{8.99}$$

$$[M_i]_{k+(m-1)p,l+(n-1)p} = [\mathcal{Z}^{-1}(z \mapsto \mathcal{L}(t \mapsto M(t))(s(z)))_i]_{k+(m-1)p,l+(n-1)p} \tag{8.100}$$

$$= [\mathcal{Z}^{-1}(z \mapsto \mathcal{L}(t \mapsto [M(t)]_{mn})(s(z)))_i]_{kl} \tag{8.101}$$

In the following, to simplify the notations we will write it as $M_i = \mathcal{Z}^{-1}(M(s(z)))_i$.

8.3.8 Computation of the Interaction Matrices

This section describes how to compute the Laplace domain interaction matrix $M(s)$ evaluated at a matrix valued Laplace parameter $s \in \mathbb{C}^{p \times p}$ from the interaction matrices $M(s)$ evaluated at scalar Laplace parameters s. Assume that the matrix parameter s has the eigenvalue decomposition $s = UDU^{-1}$ where D is a diagonal matrix of the eigenvalues and U columns are the respective eigenvectors of s. Then $M(s)$ can be constructed from the p matrices $M([D]_{kk})$ $(1 \leq k \leq p)$ where $[D]_{kk}$ is an eigenvalue of s. The matrix $M(s)$ is constructed block-wise where each block is $p \times p$

$$[M(s)]_{k+(m-1)p,l+(n-1)p} = [U\tilde{M}_{mn}U^{-1}]_{kl} \tag{8.102}$$

$$\tilde{M}_{mn} = \begin{pmatrix} [M([D]_{11})]_{mn} & & & (0) \\ & [M([D]_{22})]_{mn} & & \\ & & \ddots & \\ (0) & & & [M([D]_{pp})]_{mn} \end{pmatrix} \tag{8.103}$$

With the interaction matrices as a function of the Z-transform variable in hand, the time domain interaction matrices M_j can be computed by a simple discretization of the complex contour integral for the inverse Z-transform given by (8.29):

$$M_j = \frac{1}{Q} \sum_{q=0}^{Q-1} M(s(z_q)) z_q^j \quad \text{with } z_q = \rho e^{2\pi i \frac{q}{Q}} \tag{8.104}$$

8.3.9 Marching-on-in-Time (MOT)

The fully discretized convolution (8.96) is rewritten such that is can be solved for J_i

$$M_0 J_i = V_i - \sum_{j=1}^{i} M_j J_{i-j} \tag{8.105}$$

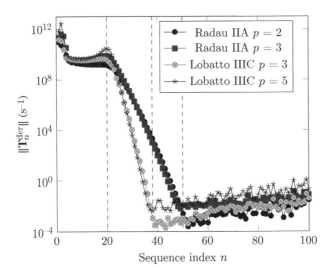

Figure 8.2 Norm of the interaction matrices in the differentiated EFIE for different RK rules (on a sphere of diameter $D = 2$ m and $\Delta t = 3.3 \times 10^{-10}$ s). Starting from $n = D/(c\Delta t) \approx 20$, the norm of the interaction matrices decreases exponentially with n. As a result N_{conv} can be chosen such that for all $n > N_{conv}$, $||\mathbf{T}_n^{der}||$ is below the machine precision, and in fact the computed matrices are numerical noise. Note that the N_{conv} for the Radau IIA rules is larger than the one for the Lobatto IIIC rules.

The number of terms in the sum in the RHS is equal to the number of time steps already computed. Fortunately, the norm of the interaction matrices $||\mathbf{M}_i||$ decreases exponentially as $i \to +\infty$ for the differentiated EFIE, MFIE, and CFIE [10]. So there is a number of terms N_{conv} in the convolution after which the remaining terms can be truncated since they go below the machine precision. This N_{conv} is in the same order of magnitude as $D/(c\Delta t)$ where $D = \max\limits_{r,r' \in \Gamma} |r - r'|$ is the diameter of the scatterer. A valid choice for N_{conv} is discussed below (see Figure 8.2). The resulting MOT scheme reads

$$\mathbf{M}_0 \mathsf{J}_i = \mathsf{V}_i - \sum_{j=1}^{N_{conv}} \mathbf{M}_j \mathsf{J}_{i-j} \tag{8.106}$$

where $\mathsf{J}_i = 0$ whenever $i < 0$.

8.3.10 Examples

8.3.10.1 Differentiated EFIE
The differentiated EFIE in the Laplace domain is

$$\eta s \mathcal{T} J(r,s) = -\hat{n}(r) \times s E^{inc}(r,s) \tag{8.107}$$

After applying the convolution quadrature, it results in the MoT (8.105) with $V_i = -E_i^{der}$ and $M_j = \eta T_j^{der}$ where we have defined

$$[E_i^{der}]_{k+(m-1)p} = \iint_{r\in\Gamma} f_m(r) \cdot \frac{\partial E^{inc}}{\partial t}(r, t_i + [c]_k \Delta t) dS \tag{8.108}$$

and the sequence of interaction matrices $T_j^{der} = \mathcal{Z}^{-1}(\tilde{s}(z)T(s(z)))_j$. $\tilde{s}$ is defined as

$$\tilde{s} = I_N \otimes s = \begin{pmatrix} s & & (0) \\ & \ddots & \\ (0) & & s \end{pmatrix} \tag{8.109}$$

with N the number of spatial degrees of freedom, and $T(s(z))$ is computed as described in Section 8.3.8. Therefore, the MoT can be written explicitly as

$$\eta T_0^{der} J_i = -E_i^{der} - \eta \sum_{j=1}^{i} T_j^{der} J_{i-j} \tag{8.110}$$

8.3.10.2 MFIE

Applying the convolution quadrature to the MFIE also yields a MoT that has the form of (8.105). The management of the gram matrix is particular since it needs to be augmented to take into account the RK stages. Therefore, we define

$$\tilde{G} = G \otimes I_p = \begin{pmatrix} [G]_{1,1}I_p & \cdots & [G]_{1,N_s}I_p \\ \vdots & \ddots & \vdots \\ [G]_{N_s,1}I_p & \cdots & [G]_{N_s,N_s}I_p \end{pmatrix} \tag{8.111}$$

Also, since $\tilde{G}$ is a constant of time, its inverse Z-transform is $\mathcal{Z}^{-1}(\tilde{G})_i = \delta_{i0}\tilde{G}$, where δ_{ij} is a Kronecker delta. Then, noting $K_j = \mathcal{Z}^{-1}(K(s(z)))_j$, we have the MoT (8.105) with $M_j = \delta_{0j}\tilde{G}/2 - K_j$ and $V_i = H_i$

$$[H_i]_{k+(m-1)p} = \iint_{r\in\Gamma} g_m(r) \cdot H^{inc}(r, t_i + [c]_k \Delta t) dS \tag{8.112}$$

or more explicitly

$$\left(\frac{1}{2}\tilde{G} - K_0\right) J_i = H_i + \sum_{j=1}^{i} K_j J_{i-j} \tag{8.113}$$

8.3.10.3 Differentiated MFIE

We introduce the differentiated MFIE as an intermediate step for building the differentiated CFIE. Here we focus especially on the consequences in the MoT of the differentiation. We define $\tilde{s}_i = \mathcal{Z}^{-1}(\tilde{s}(z))_i$. From (8.56), we obtain that the nonzero elements of this sequence are

$$\tilde{s}_0 = \frac{1}{\Delta t}\tilde{A}^{-1} \tag{8.114}$$

$$\tilde{s}_1 = -\frac{1}{\Delta t}\tilde{A}^{-1}\tilde{1}_p\tilde{b}^T\tilde{A}^{-1} \tag{8.115}$$

where

$$\tilde{\mathbf{A}} = \mathbf{I}_N \otimes \mathbf{A} = \begin{pmatrix} \mathbf{A} & & (0) \\ & \ddots & \\ (0) & & \mathbf{A} \end{pmatrix} \tag{8.116}$$

$$\tilde{\mathbf{1}}_p \tilde{\mathbf{b}}^{\mathsf{T}} = \mathbf{I}_N \otimes (\mathbf{1}_p \mathbf{b}^{\mathsf{T}}) \tag{8.117}$$

Then defining $\mathbf{K}_j^{\mathrm{der}} = \mathcal{Z}^{-1}(\tilde{\mathbf{s}}(z)\mathbf{K}(\mathbf{s}(z)))_j$ and

$$[\mathsf{H}_i^{\mathrm{der}}]_{k+(m-1)p} = \iint_{r\in\Gamma} g_m(\boldsymbol{r}) \cdot \frac{\partial \boldsymbol{H}^{\mathrm{inc}}}{\partial t}(\boldsymbol{r}, t_i + [\mathsf{c}]_k \Delta t) dS \tag{8.118}$$

the MoT takes the form (8.105) with $\mathbf{M}_j = \tilde{\mathbf{G}}\tilde{\mathbf{s}}_j/2 - \mathbf{K}_j^{\mathrm{der}}$ and $\mathsf{V}_i = \mathsf{H}_i^{\mathrm{der}}$, or more explicitly

$$\left(\frac{1}{2}\tilde{\mathbf{G}}\tilde{\mathbf{s}}_0 - \mathbf{K}_0^{\mathrm{der}} \right) \mathsf{J}_i = \mathsf{H}_i^{\mathrm{der}} - \frac{1}{2}\tilde{\mathbf{G}}\tilde{\mathbf{s}}_1 \mathsf{J}_{i-1} + \sum_{j=1}^{i} \mathbf{K}_j^{\mathrm{der}} \mathsf{J}_{i-j} \tag{8.119}$$

8.3.10.4 Differentiated CFIE

Here the differentiated MFIE is left multiplied by $\tilde{\mathbf{N}}\tilde{\mathbf{G}}^{-1}$ and combined with the differentiated EFIE. Similarly to (8.111), we have used the notation $\tilde{\mathbf{N}} = \mathbf{N} \otimes \mathbf{I}_p$ (we recall the reader that $\mathbf{N}$ is defined in (8.93)).

$$\tilde{\mathbf{N}}\tilde{\mathbf{G}}^{-1}\left(\frac{1}{2}\tilde{\mathbf{G}}\tilde{\mathbf{s}}_0 - \mathbf{K}_0^{\mathrm{der}} \right) \mathsf{J}_i = \tilde{\mathbf{N}}\tilde{\mathbf{G}}^{-1}\left(\mathsf{H}_i^{\mathrm{der}} - \frac{1}{2}\tilde{\mathbf{G}}\tilde{\mathbf{s}}_1 \mathsf{J}_{i-1} + \sum_{j=1}^{i} \mathbf{K}_j^{\mathrm{der}} \mathsf{J}_{i-j} \right) \tag{8.120}$$

$$-\mathbf{T}_0^{\mathrm{der}} \mathsf{J}_i = \frac{1}{\eta}\mathsf{E}_i^{\mathrm{der}} + \sum_{j=1}^{i} \mathbf{T}_j^{\mathrm{der}} \mathsf{J}_{i-j} \tag{8.121}$$

Again the MoT takes the form (8.105) with

$$\mathbf{M}_j = -\alpha\mathbf{T}_j^{\mathrm{der}} + (1-\alpha)\tilde{\mathbf{N}}\tilde{\mathbf{G}}^{-1}\left(\frac{1}{2}\tilde{\mathbf{G}}\tilde{\mathbf{s}}_j - \mathbf{K}_j^{\mathrm{der}} \right) \tag{8.122}$$

$$\mathsf{V}_j = \frac{\alpha}{\eta}\mathsf{E}_i^{\mathrm{der}} + (1-\alpha)\tilde{\mathbf{N}}\tilde{\mathbf{G}}^{-1}\mathsf{H}_i^{\mathrm{der}} \tag{8.123}$$

i.e. in expanded form

$$\left(-\alpha\mathbf{T}_0^{\mathrm{der}} + (1-\alpha)\tilde{\mathbf{N}}\tilde{\mathbf{G}}^{-1}\left(\frac{1}{2}\tilde{\mathbf{G}}\tilde{\mathbf{s}}_0 - \mathbf{K}_0^{\mathrm{der}} \right) \right) \mathsf{J}_i = \alpha\left(\frac{1}{\eta}\mathsf{E}_i^{\mathrm{der}} + \sum_{j=1}^{i} \mathbf{T}_j^{\mathrm{der}} \mathsf{J}_{i-j} \right)$$

$$+ (1-\alpha)\tilde{\mathbf{N}}\tilde{\mathbf{G}}^{-1}\left(\mathsf{H}_i^{\mathrm{der}} - \frac{1}{2}\tilde{\mathbf{G}}\tilde{\mathbf{s}}_1 \mathsf{J}_{i-1} + \sum_{j=1}^{i} \mathbf{K}_j^{\mathrm{der}} \mathsf{J}_{i-j} \right) \tag{8.124}$$

8.4 Implementation Details

In this section, we describe how to build the above time domain solvers starting from software components commonly encountered in implementations of frequency domain boundary element methods. We first present a baseline implementation, and then we will explain how to properly choose the different simulation parameters.

8.4.1 Building a Time Domain Solver from a Frequency Domain Code: Baseline Implementation of the MOT

We describe here a step-by-step procedure to solve a time domain problem by taking advantage of a frequency domain code that can handle complex wave numbers. We first describe a baseline procedure whose implementation should be straightforward and then we will discuss the possibilities for improvements for speed and memory:

1. Compute the Z-domain parameter $z_q = \rho e^{2\pi i \frac{q}{Q}}$ for each $q \in [0, Q-1]$.
2. Compute the matrix valued Laplace parameter $\mathbf{s}_q = \frac{1}{\Delta t}\left(\mathbf{A} + \frac{1 b^{\mathsf{T}}}{z_q - 1}\right)^{-1}$ for each z_q.
3. Compute the eigenvalue decomposition $\mathbf{s}_q = \mathbf{U}_q \mathbf{D}_q \mathbf{U}_q^{-1}$ for each $\mathbf{s}_q$.
4. For each eigenvalue $s_{q,k} = [\mathbf{D}_q]_{kk}$ ($q \in [0, Q-1], k \in [1, p]$) compute the boundary element matrices $\mathbf{M}(s_{q,k})$ that discretizes the main operator with Laplace parameter (or equivalently complex frequency) $s_{q,k}$. This is the step where an implementation of a frequency domain boundary element method (handling complex frequency values) can be used.
5. Build the matrices $\mathbf{M}(\mathbf{s}_q)$ as described in Section 8.3.8.
6. Compute the sequence of interaction matrices $\mathbf{M}_n$ ($n \in [0, N_{\mathrm{conv}}]$) by inverse Z-transforming the boundary element matrices.

$$\mathbf{M}_n = \frac{1}{Q}\sum_{q=0}^{Q-1}\mathbf{M}(\mathbf{s}_q)z_q^n \tag{8.125}$$

7. Solve the MoT for each time step $i \in [1, N_t]$, i.e.:
   - Compute the excitation field vector $\mathbf{V}_i$ using (8.99).
   - Compute the RHS $\mathbf{V}_i - \sum_{j=1}^{N_{\mathrm{conv}}} \mathbf{M}_j \mathbf{J}_{i-j}$.
   - Solve the system (8.105) for the current $\mathbf{J}_i$.

Note that the sequence of interaction matrices $\mathbf{M}_n$ is real even though it is the sum of complex matrices. By taking advantage of the fact that z_{Q-q} is the complex conjugate of z_q and in turn $\mathbf{s}_{Q-q}$ and $\mathbf{M}(\mathbf{s}_{Q-q})$ are the conjugates of $\mathbf{s}_q$ and $\mathbf{M}(\mathbf{s}_q)$, respectively, only half of the Laplace domain matrices need to be computed. In other words, to halve the cost of computing $\mathbf{M}_n$, the following expression can be used in (8.104)

$$\mathbf{M}(\mathbf{s}_q)z_q^n + \mathbf{M}(\mathbf{s}_{Q-q})z_{Q-q}^n = 2\Re(\mathbf{M}(\mathbf{s}_q)z_q^n) \tag{8.126}$$

Further improvements using fast solvers are discussed in Section 8.5.1.2, and preconditioning techniques are discussed in Section 8.5.2.4.

Also, note that the excitation field V_i is computed at each time step. In principle it can depend on the solution of the previous time steps, especially to model nonlinear boundary conditions or other physical effects.

8.4.2 Choice of the Simulation Parameters

8.4.2.1 Choice of the RK Method

The accuracy of a RK method can be characterized by its order r. Each step introduces a local truncation error in the solution proportional to $O(\Delta t^{r+1})$. After many steps, the local errors accumulate and result in a global truncation error proportional to $O(\Delta t^r)$ and higher order RK methods require higher number of stages p [13]. In determining the order of the Runge–Kutta method, a balance is required between the targeted accuracy and the computational costs. Low-order RK methods result in smaller matrices and fewer samples in the complex Z-plane. Higher order RK methods deliver more accurate results and can be used with electrically large targets and wide-band excitations, but require significantly more computational resources and storage.

8.4.2.2 Choice of the Time Step and the Discretization Density

Since we use a multistage method, we simplify the analysis by assuming a sampling step $\delta t = \Delta t/p$, where Δt is the time step and p is the number of stages. According to the Nyquist–Shannon sampling theorem, a signal that is bandlimited with no frequency higher than f_{max} can be perfectly reconstructed with a sampling step $\delta t \leq 1/(2f_{max})$. Unfortunately, it is also well known from Fourier analysis that a nonzero bandlimited signal necessarily has an infinite support in time, i.e. N_t would have to be infinite. The opposite is also true, i.e. a nonzero signal with a finite support in time cannot be bandlimited. In practice, most signals are neither timelimited nor bandlimited, but they are localized enough in time and frequency such that their frequency content higher than $1/(2\delta t)$ is negligible, and their amplitude is small enough (e.g. below the machine precision) away from an interval $[t_0, t_0 + N_t \Delta t]$. The typical example of such a signal that is localized both in frequency and time is a Gaussian whose amplitude in time (respectively frequency) decays superexponentially away from the peak amplitude (respectively central frequency). Therefore a valid choice for the time step is $\Delta t = p/(2f_{max}\psi_t)$ where $\psi_t > 1$ is the oversampling factor.

Assuming a uniform discretization and noting h the average edge length of the mesh approximating scatterer, the sampling theorem requires $h \leq \lambda_{sol,min}/2 = 1/(2f_{sol,max})$, where $f_{sol,max}$ denotes the maximal spatial frequency (again, up to the machine precision) at which the solution changes and $\lambda_{sol,min}$ is the corresponding wavelength. If the excitation contains only low frequencies, i.e. for large time steps, the geometry of the scatterer is the one that determines the frequency at which the solution changes, and therefore the discretization should be chosen to accurately represent the geometry of the scatterer. If the excitation contains high frequencies, i.e. for small time steps, the maximum spatial frequency $f_{\text{sol,max}}$ at which the solution varies is proportional to the maximum frequency f_{max} of the excitation, as $f_{\text{sol,max}} = f_{\text{max}}/c$. Thus, in this case, a valid choice for the edge lengths is $h = c/(2f_{\text{max}}\psi_s)$ where $\psi_s > 1$ is the spatial oversampling factor.

8.4.2.3 Choice of the Inverse Z-Transform Parameters

The number of complex frequencies (Laplace parameters) to evaluate is equal to Qp where Q is the number of integration points used for the inverse $\mathcal{Z}$-transform and p is the number of stages in the RK method. ρ is the radius of the circle that is used as an integration contour for the inverse Z-transform. We need $\rho > 1$ from the derivations.

Assuming that the Laplace domain elements of $\mathbf{M}(s)$ can be computed with a precision ϵ, then the choice of Q and ρ will affect the precision of the elements of the time domain interaction matrices $\mathbf{M}_i = \mathcal{Z}^{-1}(\mathbf{M}(\mathbf{s}))_i$ $(0 \leq i \leq N_{\text{conv}})$ as follow (see [5, 6]):

- to obtain an error $O(\sqrt{\epsilon})$ on the $\mathbf{M}_i$ elements, it is sufficient to choose $Q = N_{\text{conv}}$ and $\rho = \epsilon^{-1/(2Q)}$.
- to obtain an error $O(\epsilon)$ on the $\mathbf{M}_i$ elements, choose $Q \geq N_{\text{conv}} \log(1/\epsilon)$ and $\rho = e^{\gamma \Delta t}$ where γ is a constant that depends on the integral operator being discretized as it is explained in [5, 6].

8.5 Acceleration, Preconditioning, and Stabilizations

8.5.1 Computational Complexity and Fast Solver Acceleration

In the following, the computational complexity is estimated straightforwardly using the parameters N_s (the number of spatial unknowns), N_t (the number of time steps), p (the number of stages in the RK method), Q (the number of integration points for the inverse $\mathcal{Z}$-transform), N_{conv} (the number of terms in the convolution in the RHS) and N_{iter} (the number of iterations in the iterative solver).

8.5.1.1 Complexity Analysis of a Naive Implementation

In this section, the storage and computational requirements of a naive implementations are deduced. This means no gains are taken into account that could be achieved by relying on the matrix sparsity pattern, their (semi-)diagonalized representations, or the possibility to compress along the spatial indices using fast multipole methods or H-matrix methods. In following sections, this state of affairs is improved upon.

In term of storage, there are $N_{\text{conv}} + 1$ real matrices $\mathbf{Z}_n$ ($n \in [0, N_{\text{conv}}]$) to store and each has dimensions $pN_s \times pN_s$, i.e. $O(N_{\text{conv}}p^2N_s^2)$ in memory. To be computed, they require Qp complex matrices $\mathbf{Z}(s_{q,k})$ ($q \in [0, Q-1], k \in [1, p]$), each of size $N_s \times N_s$, i.e. $O(QpN_s^2)$ in memory. The solution surface current corresponds to N_t vectors of dimension N_s, i.e. $O(N_tN_s)$ in memory.

In term of computation, the matrices $\mathbf{Z}(s_{q,k})$ require $O(QpN_s^2)$ operations, assuming a constant cost for the computation of a single matrix element. The matrices $\mathbf{Z}_n$ require $O(QN_{\text{conv}}p^2N_s^2)$ operations. At each time step the RHS requires $O(N_{\text{conv}}p^2N_s^2)$ operations to be computed, i.e. $O(N_tN_{\text{conv}}p^2N_s^2)$ operations for all the time steps. Finally, the system $\mathbf{Z}_0J_i = \text{RHS}$ has to be solved for each time step. Assuming that an iterative solver is used and that N_{iter} iterations are required to solve the system, it requires $O(N_{\text{iter}}N_t)$ matrix vector products in total.

8.5.1.2 Acceleration with Fast Solvers

The computational cost increases quadratically with the number of spatial unknowns N_s. Fortunately, there are well known frequency domain algorithms to compress the MoM matrices and multiply them with a vector in $O(N_s \log(N_s))$ memory and time complexity such as the ACA [24, 25] in low frequency, or the FMM [11, 26] in high frequency. Fortunately, these algorithms also work in case complex frequencies are used (typically used for simulating lossy media in frequency domain).

The computational cost of the Qp matrices $\mathbf{Z}(s_{q,k})$ ($q \in [0, Q-1], k \in [1, p]$) is therefore $O(QpN_s \log(N_s))$ in memory. The matrices $\mathbf{Z}(\mathbf{s}_q)$ and $\mathbf{Z}_n$ are not stored but their matrix vector products are done on the fly using (8.102) and (8.104). Then the cost of one matrix vector product with $\mathbf{Z}_n$ is $O(QpN_s \log(N_s))$. Since there are $O(N_{\text{iter}}N_t)$ MVP to perform to solve the linear system for all the time step and there are $O(N_{\text{conv}}N_t)$ MVP to compute for all the RHS, the overall computation complexity is $O(Qp(N_{\text{iter}} + N_{\text{conv}})N_tN_s \log(N_s))$.

Note however that this approach, although it is easy to implement from existing code, would lose efficiency when Δt is very small. Indeed, assuming that all the other parameters are fixed, both Q and N_{conv} grow proportionally to $1/\Delta t$, as $\Delta t \to 0$. Instead, other techniques must be used to further accelerate the solver [27, 28].

8.5.2 Ill-Conditioning and Instabilities

8.5.2.1 Interior Resonances and CFIE

The problem of resonances is a well-known issue of frequency domain integral equations. On closed scatterers, each of the EFIE operator and the MFIE operator have infinitely (countable) many frequencies for which they are not invertible [29]. As a result, at the resonant frequencies, there are nonphysical currents in the nullspaces of these operators. In other words, the solution of the EFIE (respectively MFIE) is not unique at these frequencies since the spurious currents can be added to the true physical solution and still verify the EFIE (respectively MFIE). In addition to the nonuniqueness of the solution, the boundary element matrices that discretize the operators have unbounded condition numbers around these frequencies. Fortunately, it is possible to combine the EFIE and the MFIE in a CFIE such that the CFIE operator is always invertible [29]. This is a high frequency problem, in the sense that there are no resonances when the wavelength is larger than some characteristic length of the scatterer (for example, on a sphere of radius R with wavenumber k, the resonances cannot happen if $kR < 2.74$). In other words, the CFIE is really needed only when the frequency is large enough.

The above mentioned problem also has consequences in the time domain. There exist time dependent currents that are in the nullspace of the TD-EFIE or TD-MFIE, so they can corrupt the true physical solution. It is sufficient to take any linear combination of the time harmonic spurious currents to obtain a current in time domain that is in the nullspace of the EFIE or MFIE. Thus, if the time step is small enough to allow the discretization of a spurious current, then the CFIE should be used.

As an illustration of the problem, we have computed in Figure 8.3 the surface current density on the surface of a sphere illuminated by a Gaussian plane wave with a time step small enough to clearly see that only the CFIE provides an acceptable solution while the EFIE and MFIE are quickly corrupted.

8.5.2.2 DC Instability

The DC (direct current) instability is a kind of problem that is present only in time domain, without any frequency domain counterpart away from statics. We consider a surface current density J_{cs} that is constant in time and solenoidal, i.e. it verifies $\frac{\partial}{\partial t} J_{cs} = 0$ and $\nabla \cdot J_{cs} = 0$. The constancy in time implies that $\frac{\partial}{\partial t} \mathcal{T}_s J_{cs} = \mathbf{0}$ while the solenoidal property implies $\mathcal{T}_h J_{cs} = \mathbf{0}$. Therefore, $\mathcal{T} J_{cs} = \mathbf{0}$. This means that if a current J is solution or the EFIE (8.3), then $J + J_{cs}$ is also a solution. In order to enforce $J_{cs} = \mathbf{0}$, we impose that the solution of the EFIE is equal to $\mathbf{0}$ for all $t < t_0$. Note that in the case of the time differentiated EFIE (8.77), also solenoidal currents that are linear in time are in the nullspace of the main operator, i.e. if $\frac{\partial^2}{\partial t^2} J_{ls} = \mathbf{0}$ and $\nabla \cdot J_{ls} = 0$ then $\frac{\partial}{\partial t} \mathcal{T} J_{ls} = \mathbf{0}$.

Unfortunately, even though the continuous equation has a unique solution, this nullspace will corrupt the numerical solution due to errors in approximations and floating point finite precision. For example, if the numerical solution contains at some point even a tiny constant (or linear in time) solenoidal part, this non-physical current will affect the late time solution because it cannot be resolved by the discrete convolutional operator and thus does not affect the right-hand sides that are used to compute the solution at subsequent time steps.

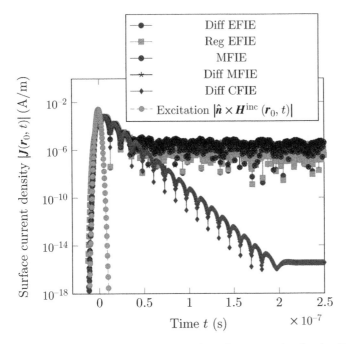

Figure 8.3 Current density on the surface of a sphere of radius 1 m illuminated by a Gaussian pulse plane wave, obtained for different formulations. The time step Δt is small enough, such that only the CFIE yields an acceptable solution. The EFIE and MFIE are corrupted by the spurious currents.

Typically, this problem manifests as a saturated solution in the late time step when the excitation fields have already vanished, as illustrated on the Figure 8.4.

Instabilities can be better understood by studying the MOT system resulting from the differentiated EFIE without any excitation

$$\mathbf{Z}_0 J_i = -\sum_{j=1}^{N_{\text{conv}}} \mathbf{Z}_j J_{i-j} \tag{8.127}$$

This MOT can be rewritten as the following recurrence relation

$$\tilde{J}_i = \tilde{\mathbf{Z}}_C \tilde{J}_{i-1} \tag{8.128}$$

where $\tilde{J}_i$ denote a block vector containing consecutive solutions and $\tilde{\mathbf{Z}}_C$ is a block matrix called the companion matrix defined as

$$\tilde{J}_i = \begin{pmatrix} J_i \\ J_{i-1} \\ \vdots \\ J_{i-(N_{\text{conv}}-1)} \end{pmatrix} \tag{8.129}$$

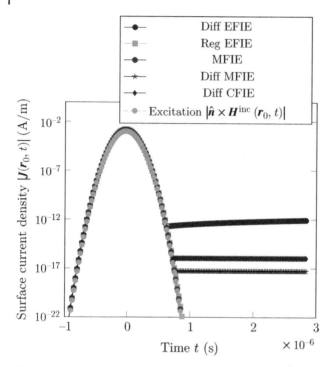

Figure 8.4 Current density on the surface of a sphere of radius 1 m illuminated by a Gaussian plane wave, obtained for different formulations. The differentiated formulations all have a DC instability because constant currents are in the nullspaces of their operators. It takes the form of a wrong solution that does not vanish in the late time steps.

$$
\tilde{\mathbf{Z}}_C = \begin{pmatrix} -\mathbf{Z}_0^{-1}\mathbf{Z}_1 & -\mathbf{Z}_0^{-1}\mathbf{Z}_2 & \cdots & -\mathbf{Z}_0^{-1}\mathbf{Z}_{N_{\text{conv}}-1} & -\mathbf{Z}_0^{-1}\mathbf{Z}_{N_{\text{conv}}} \\ \mathbf{I} & & & & \\ & \mathbf{I} & & (\mathbf{0}) & \\ & & \ddots & & \\ & (\mathbf{0}) & & \mathbf{I} & \end{pmatrix} \tag{8.130}
$$

The study of the eigenvalues of $\tilde{\mathbf{Z}}_C$ noted λ_n indicate the asymptotic behavior of the solution when $i \to \infty$:

- If all the eigenvalues are inside the unit circle then the system is stable ($|\lambda_n| < 1$ for all n). Any numerical perturbation in the solution is guaranteed to decay exponentially.
- If an eigenvalue is outside the unit circle then the system is unstable ($|\lambda_n| > 1$ for some n). A numerical perturbation in the solution is guaranteed to grow exponentially (if it has some components along the eigenvector associated with λ_n).

In practice, for the differentiated PEC-EFIE, there is a cluster of eigenvalues around 1 (compare Figure 8.5 with Figure 8.6, with the maximum eigenvalue in

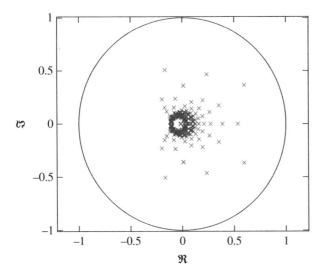

Figure 8.5 Eigenvalues of $\tilde{Z}_C$ for the stabilized EFIE on a sphere. All the eigenvalues are inside the unit circle, thus the stabilized EFIE does not suffer from DC instability.

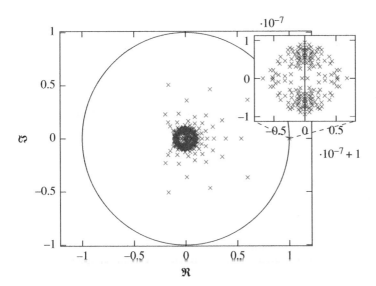

Figure 8.6 Eigenvalues of $\tilde{Z}_C$ for the differentiated EFIE on a sphere. The cluster of eigenvalues at 1 is responsible of the DC instability.

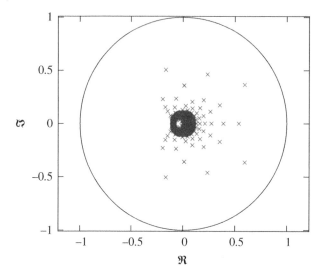

Figure 8.7 Eigenvalues of $\tilde{\mathbf{Z}}_C$ for the MFIE on a sphere. All the eigenvalues are inside the unit circle, thus the MFIE does not suffer from DC instability.

absolute value being larger than 1, although it is usually so close to 1 that the error has almost a constant amplitude.

A solution to the DC instability of the EFIE is presented in Section 8.5.2.4. Among the formulations presented here all the differentiated one (EFIE, MFIE, and CFIE) exhibit DC instabilities. Note that the MFIE does not suffer from DC instabilities on simply connected geometries (refer to Figure 8.7), while the solution can be corrupted by a static null-space for nonsimply connected structures (structures containing handles line) [30]. This can be seen in the case of a Torus in Figure 8.8.

8.5.2.3 Large Time Step Breakdown

The large time step breakdown is the time domain counterpart of the low frequency breakdown. In order to analyze this breakdown, we need to decompose the operators. We introduce the connectivity matrices $\mathbf{\Sigma} \in \mathbb{R}^{N_s \times N_f}$ and $\mathbf{\Lambda} \in \mathbb{R}^{N_s \times N_v}$ (N_s is the number of edges (equal to the number of spatial unknowns), N_f is the number of faces and N_v is the number of vertices) whose element are

$$[\mathbf{\Sigma}]_{mn} = \begin{cases} +1, & \text{if the edge } m \text{ is on the boundary of the face } n \text{ clockwise} \\ -1, & \text{if the edge } m \text{ is on the boundary of the face } n \text{ counterclockwise} \\ 0, & \text{otherwise} \end{cases}$$

$$(8.131)$$

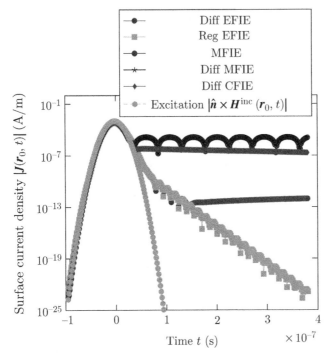

Figure 8.8 Current density on the surface of a torus of outer radius 1.5 m and inner radius 0.5 m illuminated by a Gaussian plane wave, obtained for different formulations.

$$[\Lambda]_{mn} = \begin{cases} +1, & \text{if the edge } m \text{ leaves the vertex } n \\ -1, & \text{if the edge } m \text{ arrive at the vertex } n \\ 0, & \text{otherwise} \end{cases} \qquad (8.132)$$

The connectivity matrices can be seen as basis transformation matrices: Σ transforms a current in the local star basis to the RWG basis while Λ transforms a current in the local loop basis to the RWG basis. The local star basis functions are based on faces and are nonsolenoidal while the local loop basis functions are based on vertices and are solenoidal. The coefficients of loops and stars are a complete basis for simply connected geometries, the spaces spanned by their coefficients will be denoted by Λ and Σ, respectively. In the case of multiply connected geometry, instead, a further set of basis functions coefficients needs to be added representing the discrete counterpart of the harmonic subspace of standard Helmholtz decompositions [30]. This subspace, denoted by H in the following, is solenoidal, quasi-irrotational [30], and it is orthogonal to the Σ and Λ. Contrary to the Σ and Λ, it is not as straightforward to construct functions in H, but fortunately we won't have to. Indeed, it is sufficient for our analysis (and later for preconditioning) to build projectors on Σ and its orthogonal complement. We note $\mathbf{P}^\Sigma$ the projector on nonsolenoidal functions, and its complementary $\mathbf{P}^{\Lambda H}$ is the projector on solenoidal functions [31]

$$\mathbf{P}^\Sigma = \Sigma(\Sigma^T \Sigma)^+ \Sigma^T \tag{8.133}$$

$$\mathbf{P}^{\Lambda H} = \mathbf{I} - \mathbf{P}^\Sigma \tag{8.134}$$

where $+$ is the Moore–Penrose pseudoinverse.

The projectors are then used to decompose the MoM matrix of the EFIE operator

$$\mathbf{T}(s) = \left(\mathbf{P}^{\Lambda H} + \mathbf{P}^\Sigma \right) \left(-\frac{s}{c}\mathbf{T}_s(s) + \frac{c}{s}\mathbf{T}_h(s) \right) \left(\mathbf{P}^{\Lambda H} + \mathbf{P}^\Sigma \right) \tag{8.135}$$

$$= \left(\mathbf{P}^{\Lambda H} \quad \mathbf{P}^\Sigma \right) \begin{pmatrix} -\frac{s}{c}\mathbf{T}_s(s) & -\frac{s}{c}\mathbf{T}_s(s) \\ -\frac{s}{c}\mathbf{T}_s(s) & -\frac{s}{c}\mathbf{T}_s(s) + \frac{c}{s}\mathbf{T}_h(s) \end{pmatrix} \begin{pmatrix} \mathbf{P}^{\Lambda H} \\ \mathbf{P}^\Sigma \end{pmatrix} \tag{8.136}$$

$$= \left(\mathbf{P}^{\Lambda H} \quad \mathbf{P}^\Sigma \right) \begin{pmatrix} \propto (s) & \propto (s) \\ \propto (s) & \propto (s^{-1}) \end{pmatrix} \begin{pmatrix} \mathbf{P}^{\Lambda H} \\ \mathbf{P}^\Sigma \end{pmatrix} \tag{8.137}$$

where $\propto (f(s))$ denotes a matrix whose norm is proportional to $f(s)$ as $s \to 0$. By applying the convolution quadrature, we have $s \propto 1/\Delta t$, so

$$\mathbf{T}_0 = \left(\tilde{\mathbf{P}}^{\Lambda H} \quad \tilde{\mathbf{P}}^\Sigma \right) \begin{pmatrix} \propto (1/\Delta t) & \propto (1/\Delta t) \\ \propto (1/\Delta t) & \propto (\Delta t) \end{pmatrix} \begin{pmatrix} \tilde{\mathbf{P}}^{\Lambda H} \\ \tilde{\mathbf{P}}^\Sigma \end{pmatrix} \tag{8.138}$$

where $\propto (f(\Delta t))$ denotes a matrix whose norm is proportional to $f(\Delta t)$ as $\Delta t \to +\infty$. A block wise inversion yields

$$\mathbf{T}_0^{-1} = \left(\tilde{\mathbf{P}}^{\Lambda H} \quad \tilde{\mathbf{P}}^\Sigma \right) \begin{pmatrix} \propto (\Delta t) & \propto (1/\Delta t) \\ \propto (1/\Delta t) & \propto (1/\Delta t) \end{pmatrix} \begin{pmatrix} \tilde{\mathbf{P}}^{\Lambda H} \\ \tilde{\mathbf{P}}^\Sigma \end{pmatrix} \tag{8.139}$$

Therefore, we obtain that $\mathrm{cond}(\mathbf{T}_0) = ||\mathbf{T}_0|| \, ||\mathbf{T}_0^{-1}|| \propto \Delta t^2$.

8.5.2.4 Treatment of the LF Breakdown and DC Instability

As it is explained in Section 8.5.2.3, the large time step breakdown is the time domain analogue of the low frequency breakdown. Therefore, its treatment is inspired from the techniques developed to solve the low frequency breakdown in frequency domain.

Preconditioning is a very useful technique to solve ill-conditioned linear systems. Although general purpose preconditioners exist, researchers have developed more efficient preconditioning strategies that take into account the properties of the matrices that arise from the BEM discretization. A class of preconditioning techniques consists in decomposing the system and the solution into its solenoidal component and its nonsolenoidal component. Here we use the quasi-Helmholtz projectors [31] $\mathbf{P}^{\Lambda H}$ and $\mathbf{P}^\Sigma$ to obtain such a decomposition.

Then, a carefully chosen frequency rescaling of the different components of the equation using left and right preconditioners results in a well-conditioned system. A division or a multiplication by the frequency in frequency domain corresponds to an integration or a differentiation in time domain. This strategy has been applied to time Galerkin [32] as well as convolution quadrature [33]. We present the latter in the following.

By differentiating the nonsolenoidal component on the right-hand side, we get rid of the time integration of the scalar potential part. By integrating the solenoidal part on the left-hand side, we get rid of the time derivative of the vector potential that is responsible of the DC instability by allowing some constant currents in the nullspace of the EFIE. Overall, this results in the following matrix in the left-hand side of the equation in the Laplace domain:

$$\mathbf{L}(s)\mathbf{T}(s)\mathbf{R}(s) = \left(\frac{c}{sa}\mathbf{P}^{\Lambda H} + \mathbf{P}^{\Sigma}\right)\left(-\frac{s}{c}\mathbf{T}_s(s) + \frac{c}{s}\mathbf{T}_h(s)\right)\left(\mathbf{P}^{\Lambda H} + \frac{sa}{c}\mathbf{P}^{\Sigma}\right) \quad (8.140)$$

$$= \left(\begin{array}{cc} \mathbf{P}^{\Lambda H} & \mathbf{P}^{\Sigma} \end{array}\right)\left(\begin{array}{cc} -\frac{1}{a}\mathbf{T}_s(s) & -\frac{s}{c}\mathbf{T}_s(s) \\ -\frac{s}{c}\mathbf{T}_s(s) & -\frac{as^2}{c^2}\mathbf{T}_s(s) + a\mathbf{T}_h(s) \end{array}\right)\left(\begin{array}{c} \mathbf{P}^{\Lambda H} \\ \mathbf{P}^{\Sigma} \end{array}\right) \quad (8.141)$$

$$= \left(\begin{array}{cc} \mathbf{P}^{\Lambda H} & \mathbf{P}^{\Sigma} \end{array}\right)\left(\begin{array}{cc} -\frac{1}{a}\mathbf{T}_s(s) & 0 \\ 0 & a\mathbf{T}_h(s) \end{array}\right)\left(\begin{array}{c} \mathbf{P}^{\Lambda H} \\ \mathbf{P}^{\Sigma} \end{array}\right) + O(s) \quad (8.142)$$

where a is a length proportional to the diameter of the object. The preconditioned system in the Laplace domain reads

$$\mathbf{Z}^{\text{reg}}(s)\mathbf{Y}(s) = \mathbf{V}(s) \quad (8.143)$$

where

$$\mathbf{Z}^{\text{reg}}(s) = \eta\mathbf{L}(s)\mathbf{T}(s)\mathbf{R}(s) \quad (8.144)$$

$$\mathbf{V}(s) = \mathbf{L}(s)\mathbf{E}(s) \quad (8.145)$$

$$\mathbf{Y}(s) = \mathbf{L}(s)\mathbf{E}(s) \quad (8.146)$$

and the original solution can be reconstructed as

$$\mathbf{J}(s) = \mathbf{R}(s)\mathbf{Y}(s) \quad (8.147)$$

The preconditioned RHS requires the computation of a primitive of the original RHS which can be done analytically or numerically

$$\mathbf{V}(s) = \left(\frac{c}{sa}\mathbf{P}^{\Lambda H} + \mathbf{P}^{\Sigma}\right)\mathbf{E}(s) \quad (8.148)$$

$$\mathbf{V}_i = \frac{c}{a}\tilde{\mathbf{P}}^{\Lambda H}\mathbf{E}_i^{\text{prim}} + \tilde{\mathbf{P}}^{\Sigma}\mathbf{E}_i \quad (8.149)$$

where $\frac{\partial E^{\mathrm{prim}}}{\partial t} = E$ and

$$[E_i]_{(m-1)p+k} = -\langle f_m, E(t_i + [c]_k \Delta t) \rangle \tag{8.150}$$

$$[E_i^{\mathrm{prim}}]_{(m-1)p+k} = -\langle f_m, E^{\mathrm{prim}}(t_i + [c]_k \Delta t) \rangle \tag{8.151}$$

Alternatively, if E^{prim} is not available, then one can compute E_i^{prim} from E_i using the IRK scheme. This is achieved by using the following recurrence relation that simply results from inverse Z-transforming $\tilde{s}(z)\mathcal{Z}(E^{\mathrm{prim}})(z) = \mathcal{Z}(E)(z)$ which is the discrete equivalent of $s\mathcal{L}(E^{\mathrm{prim}})(s) = \mathcal{L}(E)(s)$

$$E_{i+1}^{\mathrm{prim}} = E_i^{\mathrm{prim}} + \Delta t(\tilde{A}(E_{i+1} - E_i) + \tilde{1}_p \tilde{b}^{\mathsf{T}} E_i) \tag{8.152}$$

Note that the choice of the constant of integration for E^{prim} should not affect V_i, because the constant part cancels when $\tilde{P}^{\Lambda H}$ is applied on E_i^{prim}. However, it should be chosen such that E^{prim} goes to 0 when $t \to \pm\infty$ to avoid loss of significant digits in the early or late time steps. The stabilized MoT then reads

$$Z_0^{\mathrm{reg}} Y_i = V_i - \sum_{j=1}^{N_{\mathrm{conv}}} Z_j^{\mathrm{reg}} Y_{i-j} \tag{8.153}$$

When the system has been solved, we obtain the auxiliary solution Y_i from which the original solution can be retrieved as

$$J(s) = \left(P^{\Lambda H} + \frac{sa}{c} P^{\Sigma} \right) Y(s) \tag{8.154}$$

$$J_i = \tilde{P}^{\Lambda H} Y_i + \frac{a}{c\Delta t} \tilde{P}^{\Sigma} \tilde{A}^{-1}(Y_i - \tilde{1}_p \tilde{b}^{\mathsf{T}} \tilde{A}^{-1} Y_{i-1}) \tag{8.155}$$

where

$$\tilde{P}^{\Sigma} = P^{\Sigma} \otimes I_p = \begin{pmatrix} [P^{\Sigma}]_{1,1} I_p & \cdots & [P^{\Sigma}]_{1,N_s} I_p \\ \vdots & \ddots & \vdots \\ [P^{\Sigma}]_{N_s,1} I_p & \cdots & [P^{\Sigma}]_{N_s,N_s} I_p \end{pmatrix} \tag{8.156}$$

$$\tilde{P}^{\Lambda H} = P^{\Lambda H} \otimes I_p \tag{8.157}$$

The fact that the approach solves the DC instability problem is further confirmed in Figure 8.5 where the absence of instability eigenvalues near one is evident.

8.6 Details of the Numerical Examples Used in the Chapter

In the examples of this chapter we have used plane waves as excitation. Specifically we used Gaussian pulses and modulated Gaussian PW. They are parametrized by the polarization $\hat{p}$, the direction of propagation $\hat{k}$ (with $\hat{k} \cdot \hat{p} = 0$), the peak

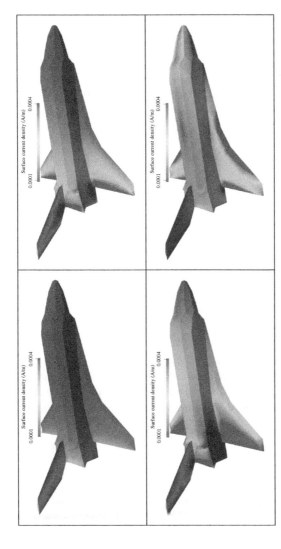

Figure 8.9 Surface current density induced on a shuttle model at different time steps.

amplitude of the electric field E_0, a characteristic time σ (that is proportional to the inverse of the bandwidth) and a central frequency f_0 (for the modulated PW). Also, for the large time step stable formulation, we give their primitive $E_\pm^{\text{prim}}$ that goes to 0 as $t \to \pm\infty$. We note the retarded time

$$\tau(r, t) = t - \frac{\hat{k} \cdot r}{c} \tag{8.158}$$

Then the PW are:

- Modulated Gaussian plane wave

$$E^{\text{inc}}(r, t) = E_0 \cos(2\pi f_0 \tau(r, t)) \exp\left(-\frac{\tau(r, t)^2}{2\sigma^2}\right) \hat{p} \tag{8.159}$$

$$E_\pm^{\text{prim}}(r, t) = \mp\Re\left(\text{erfc}\left(\pm\frac{\tau(r, t) + 2\pi i f_0 \sigma^2}{\sqrt{2}\sigma}\right)\right) E_0 \sqrt{\frac{\pi}{2}}\sigma \, \exp\left(-2\pi^2 f_0^2 \sigma^2\right) \hat{p} \tag{8.160}$$

- Gaussian pulse plane wave

$$E^{\text{inc}}(r, t) = E_0 \exp\left(-\frac{\tau(r, t)^2}{2\sigma^2}\right) \hat{p} \tag{8.161}$$

$$E_\pm^{\text{prim}}(r, t) = \mp\text{erfc}\left(\pm\frac{\tau(r, t)}{\sqrt{2}\sigma}\right) E_0 \sqrt{\frac{\pi}{2}}\sigma \, \hat{p} \tag{8.162}$$

In both cases, the magnetic field is given by

$$H^{\text{inc}}(r, t) = \frac{1}{\eta}\hat{k} \times E^{\text{inc}}(r, t) \tag{8.163}$$

where η is the impedance of the medium.

Regarding the geometries we have used spheres of radius 1 m as a benchmark, a torus of outer radius 1.5 m and inner radius 0.5 m as nonsimply connected structure, and a shuttle model, whose simulated currents at different time steps are shown in Figure 8.9.

8.7 Conclusions

This chapter dealt with Time-Domain Integral Equations discretized with convolution quadrature (CQ) strategy based on Implicit Runge–Kutta methods. A general description of the approach delineating the discretization spaces, the strategies for the computation of the linear system matrices, and the associated Marching-on-in-time solutions has been presented together with implementation-related details. After presenting the main regularization and preconditioning strategies relevant to the topic, a set of numerical results has shown the practical relevance of the presented techniques.

8.A List of RK Methods

8.A.1 Radau IIA Methods

These methods have an accuracy of order $2p - 1$ [13]. The elements of the array c are the roots of the order p polynomial $\frac{d^{p-1}}{dx^{p-1}}(x^{p-1}(x-1)^p)$ or equivalently, the roots of $Q(x) = P_p(2x-1) - P_{p-1}(2x-1)$ where P_n is the order n Legendre polynomial. The array $b^\top$ is simply the last row of $\mathbf{A}$. The elements of the matrix $\mathbf{A}$ are given by

$$[\mathbf{A}]_{lk} = \frac{1}{Q'([c]_k)} \int_{x=0}^{[c]_l} \frac{Q(x)}{x - [c]_k} dx \tag{8.A.1}$$

We have listed only the methods that have a simple algebraic expression, but higher order methods can be easily obtained numerically using the formulas above.

$$
\begin{array}{c|cc}
\frac{1}{3} & \frac{5}{12} & -\frac{1}{12} \\
1 & \frac{3}{4} & \frac{1}{4} \\
\hline
 & \frac{3}{4} & \frac{1}{4}
\end{array}
$$

2 stages Radau IIA

$$
\begin{array}{c|ccc}
\frac{4-\sqrt{6}}{10} & \frac{88-7\sqrt{6}}{360} & \frac{296-169\sqrt{6}}{1800} & \frac{-2+3\sqrt{6}}{225} \\
\frac{4+\sqrt{6}}{10} & \frac{296+169\sqrt{6}}{1800} & \frac{88+7\sqrt{6}}{360} & \frac{-2-3\sqrt{6}}{225} \\
1 & \frac{16-\sqrt{6}}{36} & \frac{16+\sqrt{6}}{36} & \frac{1}{9} \\
\hline
 & \frac{16-\sqrt{6}}{36} & \frac{16+\sqrt{6}}{36} & \frac{1}{9}
\end{array}
$$

3 stages Radau IIA

8.A.2 Lobatto IIIC

These methods have order $2p - 2$. The elements of the array c are the roots of the order p polynomial $\frac{d^{p-2}}{dx^{p-2}}(x^{p-1}(x-1)^{p-1})$ or equivalently, the roots of $Q(x) = P_p(2x-1) - P_{p-2}(2x-1)$ where P_n is the order n Legendre polynomial. The array $b^\top$ is the last row of $\mathbf{A}$ that is given by

$$[b]_k = \frac{1}{Q'([c]_k)} \int_{x=0}^{1} \frac{Q(x)}{x - [c]_k} dx = \frac{1}{p(p-1)P_{p-1}(2[c]_k - 1)^2} \tag{8.A.2}$$

The matrix $\mathbf{A}$ is defined as the unique matrix that verifies $[\mathbf{A}]_{i1} = [b]_1$ for all $i = 1, \ldots, p$ and

$$\sum_{j=1}^{p} [\mathbf{A}]_{ij}[c]_j^{k-1} = \frac{[c]_i^k}{k} \tag{8.A.3}$$

for all $i = 1, \ldots, p$ and $k = 1, \ldots, p - 1$ [34, 35], i.e.

$$
A = \begin{pmatrix} [b]_1 & [c]_1 & [c]_1^2/2 & \cdots & [c]_1^{p-1}/(p-1) \\ [b]_1 & [c]_2 & [c]_2^2/2 & \cdots & [c]_2^{p-1}/(p-1) \\ \vdots & \vdots & \vdots & \ddots & \vdots \\ [b]_1 & [c]_p & [c]_p^2/2 & \cdots & [c]_p^{p-1}/(p-1) \end{pmatrix} \begin{pmatrix} 1 & [c]_1^0 & [c]_1^1 & \cdots & [c]_1^{p-2} \\ 0 & [c]_2^0 & [c]_2^1 & \cdots & [c]_2^{p-2} \\ \vdots & \vdots & \vdots & \ddots & \vdots \\ 0 & [c]_p^0 & [c]_p^1 & \cdots & [c]_p^{p-2} \end{pmatrix}^{-1}
$$

(8.A.4)

Again, we have listed the methods that have a simple algebraic expression.

$$
\begin{array}{c|cc}
0 & \frac{1}{2} & -\frac{1}{2} \\
1 & \frac{1}{2} & \frac{1}{2} \\
\hline
 & \frac{1}{2} & \frac{1}{2}
\end{array}
$$

2 stages Lobatto IIIC

$$
\begin{array}{c|ccc}
0 & \frac{1}{6} & -\frac{1}{3} & \frac{1}{6} \\
\frac{1}{2} & \frac{1}{6} & \frac{5}{12} & -\frac{1}{12} \\
1 & \frac{1}{6} & \frac{2}{3} & \frac{1}{6} \\
\hline
 & \frac{1}{6} & \frac{2}{3} & \frac{1}{6}
\end{array}
$$

3 stages Lobatto IIIC

$$
\begin{array}{c|cccc}
0 & \frac{1}{12} & -\frac{\sqrt{5}}{12} & \frac{\sqrt{5}}{12} & -\frac{1}{12} \\
\frac{1}{2} - \frac{\sqrt{5}}{10} & \frac{1}{12} & \frac{1}{4} & \frac{10-7\sqrt{5}}{60} & \frac{\sqrt{5}}{60} \\
\frac{1}{2} + \frac{\sqrt{5}}{10} & \frac{1}{12} & \frac{10+7\sqrt{5}}{60} & \frac{1}{4} & -\frac{\sqrt{5}}{60} \\
1 & \frac{1}{12} & \frac{5}{12} & \frac{5}{12} & \frac{1}{12} \\
\hline
 & \frac{1}{12} & \frac{5}{12} & \frac{5}{12} & \frac{1}{12}
\end{array}
$$

4 stages Lobatto IIIC

$$
\begin{array}{c|ccccc}
0 & \frac{1}{20} & -\frac{7}{60} & \frac{2}{15} & -\frac{7}{60} & \frac{1}{20} \\
\frac{1}{2} - \frac{\sqrt{21}}{14} & \frac{1}{20} & \frac{29}{180} & \frac{47-15\sqrt{21}}{315} & \frac{203-30\sqrt{21}}{1260} & -\frac{3}{140} \\
\frac{1}{2} & \frac{1}{20} & \frac{329+105\sqrt{21}}{2880} & \frac{73}{360} & \frac{329-105\sqrt{21}}{2880} & \frac{3}{160} \\
\frac{1}{2} + \frac{\sqrt{21}}{14} & \frac{1}{20} & \frac{203-30\sqrt{21}}{1260} & \frac{47+15\sqrt{21}}{315} & \frac{29}{180} & -\frac{3}{140} \\
1 & \frac{1}{20} & \frac{49}{180} & \frac{16}{45} & \frac{49}{180} & \frac{1}{20} \\
\hline
 & \frac{1}{20} & \frac{49}{180} & \frac{16}{45} & \frac{49}{180} & \frac{1}{20}
\end{array}
$$

5 stages Lobatto IIIC

References

1 Bennett, C.L. Jr. (1968). *A Technique for Computing Approximate Electromagnetic Impulse Response of Conducting Bodies*. Purdue University.

2 Rao, S.M. (1999). *Time Domain Electromagnetics*. Elsevier.

3 Chung, Y.-S., Sarkar, T., Jung, B.H. et al. (2004). Solution of time domain electric field integral equation using the Laguerre polynomials. *IEEE Transactions on Antennas and Propagation* 52 (9): 2319–2328. https://doi.org/10.1109/TAP.2004.835248.

4 Lubich, C. (1988). Convolution quadrature and discretized operational calculus. I. *Numerische Mathematik* 52 (2): 129–145.

5 Lubich, C. (1988). Convolution quadrature and discretized operational calculus. II. *Numerische Mathematik* 52 (4): 413–425.

6 Lubich, C. and Ostermann, A. (1993). Runge–Kutta methods for parabolic equations and convolution quadrature. *Mathematics of Computation* 60 (201): 105–131.

7 Oppenheim, A.V. (2009). *Discrete-Time Signal Processing*, English, 3e. Upper Saddle River, NJ: Pearson. ISBN: 978-0-13-198842-2.

8 Schanz, M. and Antes, H. (1997). A new visco-and elastodynamic time domain boundary element formulation. *Computational Mechanics* 20 (5): 452–459.

9 Banjai, L., Messner, M., and Schanz, M. (2012). Runge–Kutta convolution quadrature for the boundary element method. *Computer Methods in Applied Mechanics and Engineering* 245: 90–101.

10 Wang, X. and Weile, D.S. (2011). Implicit Runge–Kutta methods for the discretization of time domain integral equations. *IEEE Transactions on Antennas and Propagation* 59 (12): 4651–4663.

11 Chew, W.C., Michielssen, E., Song, J., and Jin, J.-M. (2001). *Fast and Efficient Algorithms in Computational Electromagnetics*. Artech House, Inc.

12 Schwartz, L. (1966). *Mathematics for the Physical Sciences*. Hermann.

13 Butcher, J.C. (2016). *Numerical Methods for Ordinary Differential Equations*. Wiley.

14 Runge, C. (1895). Über die numerische auflösung von differentialgleichungen. *Mathematische Annalen* 46 (2): 167–178.

15 Kutta, W. (1901). Beitrag zur naherungsweisen integration totaler differentialgleichungen. *Z. Math. Phys.* 46: 435–453.

16 Hairer, E., Lubich, C., and Wanner, G. (2006). Structure-preserving algorithms for ordinary differential equations. *Geom. Numer. Integr.* 31.

17 Skvortsov, L.M. (2003). Accuracy of Runge–Kutta methods applied to stiff problems. *Comput. Math. Math. Phys.* 43 (9): 1320–1330.

18 Higham, N.J. (2008). *Functions of Matrices: Theory and Computation.* SIAM.

19 Banjai, L. and Lubich, C. (2019). Runge–Kutta convolution coercivity and its use for time-dependent boundary integral equations. *IMA Journal of Numerical Analysis* 39 (3): 1134–1157.

20 Rao, S., Wilton, D., and Glisson, A. (1982). Electromagnetic scattering by surfaces of arbitrary shape. *IEEE Transactions on Antennas and Propagation* 30 (3): 409–418.

21 Raviart, P.-A. and Thomas, J.-M. (1977). A mixed finite element method for 2-nd order elliptic problems. In: (eds. Ilio Galligani and Enrico Magenes) *Mathematical Aspects of Finite Element Methods*, 292–315. Berlin: Springer.

22 Buffa, A. and Christiansen, S. (2007). A dual finite element complex on the barycentric refinement. *Mathematics of Computation* 76 (260): 1743–1769.

23 Cools, K., Andriulli, F., De Zutter, D., and Michielssen, E. (2011). Accurate and conforming mixed discretization of the mfie. *IEEE Antennas and Wireless Propagation Letters* 10: 528–531.

24 Bebendorf, M. (2008). *Hierarchical Matrices: A Means to Efficiently Solve Elliptic Boundary Value Problems*, vol. 63. Springer Science & Business Media.

25 Zhao, K., Vouvakis, M.N., and Lee, J.-F. (2005). The adaptive cross approximation algorithm for accelerated method of moments computations of emc problems. *IEEE Transactions on Electromagnetic Compatibility* 47 (4): 763–773.

26 Song, J.M. and Chew, W.C. (1995). Multilevel fast-multipole algorithm for solving combined field integral equations of electromagnetic scattering. *Microwave and Optical Technology Letters* 10 (1): 14–19.

27 Maruyama, T., Saitoh, T., Bui, T., and Hirose, S. (2016). Transient elastic wave analysis of 3-D large-scale cavities by fast multipole BEM using implicit Runge–Kutta convolution quadrature. *Computer Methods in Applied Mechanics and Engineering* 303: 231–259.

28 Banjai, L. and Kachanovska, M. (2014). Fast convolution quadrature for the wave equation in three dimensions. *Journal of Computational Physics* 279: 103–126.

29 Colton, D. and Kress, R. (2013). *Integral Equation Methods in Scattering Theory.* SIAM.

30 Cools, K., Andriulli, F.P., Olyslager, F., and Michielssen, E. (2009). Nullspaces of MFIE and Calderon preconditioned EFIE operators applied to toroidal surfaces. *IEEE Transactions on Antennas and Propagation* 57 (10): 3205–3215.

31 Andriulli, F.P., Cools, K., Bogaert, I., and Michielssen, E. (2012). On a well-conditioned electric field integral operator for multiply connected geometries. *IEEE Transactions on Antennas and Propagation* 61 (4): 2077–2087.

32 Beghein, Y., Cools, K., and Andriulli, F.P. (2015). A DC stable and large-time step well-balanced TD-EFIE based on quasi-Helmholtz projectors. *IEEE Transactions on Antennas and Propagation* 63 (7): 3087–3097.

33 Dély, A., Andriulli, F.P., and Cools, K. (2019). Large time step and DC stable TD-EFIE discretized with implicit Runge–Kutta methods. *IEEE Transactions on Antennas and Propagation* 68 (2): 976–985.

34 Chipman, F. (1971). A-stable Runge–Kutta processes. *BIT Numerical Mathematics* 11 (4): 384–388.

35 Wanner, G. and Hairer, E. (1996). *Solving Ordinary Differential Equations II*, vol. 375. Berlin, Heidelberg: Springer.

9

Solving Electromagnetic Scattering Problems Using Impulse Responses

Gaobiao Xiao, Yuyang Hu, Xuezhe Tian, Shifeng Huang, and Rui Liu

Department of Electronic Engineering, Shanghai Jiao Tong University, Shanghai, P.R. China

9.1 Introduction

Time-domain solvers are very important tools for analyzing transient scattering problems [1–3]. Although not as popular as the finite-difference time-domain (FDTD) method, the integral-equation-based time-domain (TD-IE) methods, such as the time-domain electric field integral equation (TD-EFIE), time-domain magnetic field integral equation (TD-MFIE), and time-domain combined field integral equation (TD-CFIE) methods, have also been widely used, especially for wide band applications [4–6]. TD-IEs are often solved using the marching-on-in-time (MOT) scheme, where all fields and currents are expanded with some kinds of spatial basis functions and a set of temporal basis functions. The solutions of the system can be calculated step by step temporally. Since TD-IE solvers require to discretize only the scatterer and implicitly enforce the radiation condition without the necessity for implementing absorbing boundary conditions to terminate the computation domain, they are usually preferred over differential equation solvers for scattering problems in an open environment. The time step size in TD-IEs is usually restricted only by the maximum frequency of the excitation and not subject to a constraint like the Courant–Friedrichs–Lewy (CFL) criterion. Therefore, TD-IEs may adopt a larger time step size and may have better computational efficiency than FDTD methods.

However, the application of MOT TD-IE solvers is still far less widespread than other time domain solvers such as the FDTD method. The major issue is the late time instability problem associated with the MOT TD-IEs, which is very complicated [7–10]. It is almost impossible to find a simple criterion like the CFL constraint in the FDTD method under which MOT TD-IEs are stable.

Factors related to the MOT late time instability may be roughly divided into three types. The first one comes from the discretization errors. Currently, the coupling

Advances in Time-Domain Computational Electromagnetic Methods, First Edition.
Edited by Qiang Ren, Su Yan, and Atef Z. Elsherbeni.
© 2023 The Institute of Electrical and Electronics Engineers, Inc. Published 2023 by John Wiley & Sons, Inc.

coefficients between spatial bases in MOT can be evaluated with high accuracy [11–16]. However, it has not been verified that high accuracy can guarantee stable MOT TD-IE solutions. The second type is rooted in the operator spectrum. Theoretic analysis shows that the TD-EFIE spectrum is always blended with solutions corresponding to spurious static solenoidal currents and interior resonance currents. The TD-MFIE also admits spurious interior resonance solutions but does not admit static solenoidal currents, while the TD-CFIE has completely eliminated spurious interior resonance solutions and static solenoidal solutions. It is well known that spurious static solenoidal currents and interior resonance currents have a significant influence on the stability properties of the MOT TD-IEs [17–22]. For the sake of convenience, the late time instability caused by spurious static solenoidal currents and the interior resonance currents are denoted as DC instability and interior resonance instability, respectively. Since the behaviors of the TD-EFIE and the TD-MFIE near the interior resonance frequencies are quite different, the late time stability properties of the two TD-IEs are consequently quite different. The last type is related to numerical algorithms. Implicit MOT schemes are generally more stable than explicit MOT schemes [23, 24]. Methods using filtering or averaging techniques [25–27] and those using high-order or smooth temporal basis functions [28, 29] can effectively improve the stability properties of the TD-IEs.

The DC instability can be subdued with techniques of preconditioning TD-EFIEs by making use of Calderón identities [19, 20, 30–34] because the square of the EFIE operator does not have eigenvalues accumulating at zero or at infinity. With the time-domain Calderón identities derived in [19] starting from formulas for causal transient electromagnetic fields, a Calderón-preconditioned TD-EFIE can give rise to system matrices that are well conditioned and are independent of the discretization density. A key technique in implementing the Calderón preconditioner is to create a non-singular Gram matrix to bridge the domain and range of the EFIE operator. For time-domain surface integral equation (TD-SIE) formulations, the Buffa–Christiansen (BC) basis functions [35] defined on barycentrically refined triangular patch meshes are div-conforming and quasi-curl-conforming. The Gram matrix linking the BC basis functions and the standard Rao–Wilton–Glisson (RWG) [36] basis functions is well conditioned. The Calderón-preconditioned MOT TD-EFIE formulated in this way is demonstrated to be stable and effective.

In the frequency domain (FD), in order to remedy the low-frequency instability, EFIE is usually augmented with the normal boundary condition or the current continuity equation to restrict the behaviors of the induced currents. The counterparts of these augmented electric field integral equation (AEFIE) have found their application in the time domain to improve the low-frequency property of the traditional TD-EFIE [37], where the current continuity equation is imposed on the traditional TD-EFIE. In [38], a loop-tree decomposition method for TD-EFIE is proposed to obtain a low-frequency stable MOT solver. It employs a spatial discretization based on the loop-tree decomposition of the divergence-conforming vector bases to separately tackle the equations tested with solenoidal spatial testing functions from the other equations. In [39], the Helmholtz decomposition is implemented using the loop flower basis functions. In these methods, the filtering technique is adopted to eliminate the high-frequency oscillations, which is computationally efficient despite a slight loss of accuracy. Basically, these methods can regularize the eigenvalues at zero and are able to get DC-stable MOT systems.

Another technique to remedy the instability issue is to solve the TD-IEs with the marching-on-in-degree (MOD) method [40]. MOD stabilizes the TD-IEs by employing a set of associated Laguerre functions as time-domain basis functions, using an exact Galerkin testing procedure and then solving the system of equations recursively. Since the associated Laguerre functions decay exponentially in time, MOD solvers are immune to the late-time instability problem existing in MOT methods. However, when the MOD method marches to a high polynomial degree, the MOD method may become unstable if the oscillatory integrals on triangular domains are not addressed properly. A stabilized method was proposed in [41], which is not sensitive to algorithm parameters.

In this chapter, a kind of impulse response is defined for TD-SIEs. The stability properties of TD-IEs are investigated from the properties of their impulse responses. The spurious solutions to the TD-IEs can be clearly recognized in the spectra of the impulse responses of the TD-EFIE and the TD-MFIE. No resonance solutions are included in the impulse responses to the TD-CFIE. Especially, static solenoidal currents are blended in the solutions to the TD-EFIE, with their main components flowing through the exciting element.

The TD-IEs are linear dynamic systems. Their stability properties can be predicted rigorously by the spectral radii or the eigenvalues of the corresponding discrete systems. In this chapter, the eigenvalues corresponding to the spurious static solenoidal currents and the interior resonance currents are re-checked according to the power conservation law [42]. In the MOT TD-EFIE, these spurious currents are non-radiating currents [18, 21]. If the discretization errors are ignored, the eigenvalues caused by the interior resonances should be located on the unit circle in the complex eigenvalue plane. Unfortunately, discretization errors always exist and will shift some of the eigenvalues outside the unit circle, making the TD-EFIE unstable, as has been well discussed in [20]. However, in the TD-MFIE, the interior resonance currents will radiate fields into space outside the scatterer. The radiated energy can only come from the incident fields. According to the power conservation law, it is obvious that the interior resonance currents should decay to zero after the incident fields disappear, and the eigenvalues due to the interior resonances should reside inside the unit circle. Therefore, the TD-MFIE is basically stable.

As for the MOT TD-CFIE, analysis shows that it is basically stable although it has eigenvalues on the unit circle.

The domain decomposition method (DDM) is efficient to handle a large multi-scale system. It allows us to divide a complex and large-scale system into subdomains and solve the electromagnetic fields in each subdomain independently. By taking into account the mutual couplings among all subdomains, the electromagnetic characteristics of the whole system can be analyzed. In the frequency domain, DDM methods based on the Huygens equivalence principle, such as the generalized transition matrix (GTM) method [43], equivalence principle algorithm (EPA) [44], non-conformal DDM [45], and linear embedding via Green's operators (LEGOs) [46], have shown superiority in analyzing complex systems, such as phased antenna arrays, band-gap structures, coupled components in integrated packages, as well as anisotropic bodies. This chapter will discuss the DDM in the time domain for TD-IEs, which will be termed as the time-domain generalized transition matrix (TD-GTM) algorithm.

Similar to the GTM method in the frequency domain (FD-GTM), the TD-GTM method is basically developed for domain decomposition, and it consists of two main steps. The first step is to create equivalent models for all subdomains, namely, the TD-GTM models. A TD-GTM model should characterize the time-domain electromagnetic property of the corresponding subdomain. The TD-GTM models can be transplanted to apply for other subdomains that have identical structures and inclusions. The impulse responses can be truncated, which provides a convenient way for creating the TD-GTM models. The second step is to address the mutual couplings between subdomains. Similar to the generalized surface integral equation (GSIE) in the frequency domain [47], a time-domain generalized surface integral equation (TD-GSIE) is derived to link the interactions among the equivalence surface currents of all subdomains. The time-domain solution of the whole system can be obtained by solving the TD-GSIE.

This chapter is organized as follows. The impulse responses to the TD-EFIE, TD-MFIE, and TD-CFIE are defined and discussed in Section 9.2. The behaviors of TD-SIEs at resonances are analyzed in Section 9.3. The stability analysis of TD-SIEs based on eigenvalue distribution is presented in Section 9.4. The analytical expressions for retarded potentials are discussed in Section 9.5. The numerical verification of the stability characteristics of TD-SIEs is provided in Section 9.6. The transient analysis technique based on truncated impulse responses is illustrated in Section 9.7, together with a discussion on the truncation effect. The TD-GTM and TD-GSIE for DDM in the time domain are discussed in Section 9.8, with a brief conclusion in Section 9.9.

9.2 Impulse Responses

Consider in the free space a perfect electrically conducting (PEC) scatterer with boundary S. When illuminated by an incident field of $E^{in}(r, t)$, the scattered electric field is

$$E^s(r, t) = -\frac{\partial}{\partial t}A(r, t) - \nabla V(r, t) \tag{9.1}$$

with the two potentials expressed by

$$A(r, t) = \frac{\mu_0}{4\pi}\int_S \frac{J_s(r', t - R/v)}{R}dS' \tag{9.2}$$

$$V(r, t) = \frac{1}{4\pi\varepsilon_0}\int_S \frac{\rho_s(r', t - R/v)}{R}dS' \tag{9.3}$$

where the sources satisfy the current continuity law

$$\rho_s(r, t) = -\int_{-\infty}^t \nabla \cdot J_s(r, \tau)d\tau \tag{9.4}$$

$J_s(r, t)$ and $\rho_s(r, t)$ are the surface current density and the surface charge density on S at the point r and time t, respectively. v stands for the wave velocity in the free space, and $R = |r - r'|$ is the distance between points r and r'. μ_0 and ε_0 are, respectively, the permeability and the permittivity of the free space. The surface current can be obtained by solving the following electric field integral equation:

$$\frac{\partial}{\partial t}A(r, t)\bigg|_{tan} + \nabla V(r, t)|_{tan} = E^{in}(r, t)\bigg|_{tan} \tag{9.5}$$

Expanding the surface sources with RWG [36] basis functions f_n gives

$$J_s(r, t) = \sum_{n=1}^N I_n(t)f_n(r) \tag{9.6}$$

$$\rho_s(r, t) = -\sum_{n=1}^N \int_{-\infty}^t I_n(\xi)d\xi\nabla \cdot f_n(r) \tag{9.7}$$

where N is the number of the RWG basis functions. Testing Eq. (9.5) with $f_m(r)$ gives

$$\sum_{n=1}^N \left[\frac{\partial A_{mn}(t)}{\partial t} + \int_{-\infty}^t \Phi_{mn}(\xi)d\xi\right] * I_n(t) = e_m^{in}(t) \tag{9.8}$$

The asterisk means temporal convolution. Equation (9.8) can be expressed in its matrix form:

$$\overline{D}(t) * I(t) = e^{in}(t) \tag{9.9}$$

where $I(t)$ is the transpose of $[I_1(t), \cdots, I_N(t)]$, and $e^{in}(t)$ is the transpose of $\left[e_1^{in}(t), \cdots, e_N^{in}(t)\right]$. The entries of $\overline{D}(t)$ consist of the contribution from the two potentials,

$$D_{mn}(t) = \frac{\partial A_{mn}(t)}{\partial t} + \int_{-\infty}^{t} \Phi_{mn}(\xi)d\xi \tag{9.10}$$

and the two potentials are functions with respect to t,

$$A_{mn}(t) = \begin{cases} \dfrac{\mu_0}{4\pi} \displaystyle\int_{S_m} \boldsymbol{f}_m(\mathbf{r}) \cdot \int_{S_n} \dfrac{\delta(t - R/v)}{R} \boldsymbol{f}_n(\mathbf{r}')dS'dS, & \text{for } t_{\min} \le t \le t_{\max} \\ 0, & \text{else} \end{cases} \tag{9.11}$$

$$\Phi_{mn}(t) = \begin{cases} \dfrac{1}{4\pi\varepsilon_0} \displaystyle\int_{S_m} \nabla \cdot \boldsymbol{f}_m(\mathbf{r}) \int_{S_n} \dfrac{\delta(t - R/v)}{R} \nabla' \cdot \boldsymbol{f}_n(\mathbf{r}')dS'dS, & \text{for } t_{\min} \le t \le t_{\max} \\ 0, & \text{else} \end{cases} \tag{9.12}$$

$t_{\min}$ and $t_{\max}$, respectively, denote the traveling time that the wave propagates from the n-th basis to the m-th basis along the shortest and the longest straight path between them. The spatial surface integrations take their Cauchy principal values. Note that for RWG spatial basis functions, the two potentials can be evaluated semi-analytically, as will be demonstrated in the numerical examples. The right-hand side of Eq. (9.9) can be represented as the convolution of the input signal and the Dirac delta function $\delta(t)$:

$$\overline{D}(t) * I(t) = e^{in}(t) = [\overline{U}\delta(t)] * e^{in}(t) \tag{9.13}$$

where $\overline{U}$ is the identity matrix. The response to $\overline{U}\delta(t)$ is defined as the impulse response [21] and denoted by $\overline{I}^{imp}(t)$. The impulse response satisfies

$$\overline{D}(t) * \overline{I}^{imp}(t) = \overline{U}\delta(t) \tag{9.14}$$

The n-th column of $\overline{I}^{imp}(t)$ corresponds to the solution of Eq. (9.14) under the excitation that only the n-th element of $e^{in}(t)$ is $\delta(t)$, while all other elements are zeros. The surface currents due to an arbitrary excitation $e^{in}(t)$ can then be calculated by the temporal convolution

$$I(t) = \overline{I}^{imp}(t) * e^{in}(t) \tag{9.15}$$

Note that $\overline{I}^{imp}(t)$ can be accurately determined for a given mesh structure equipped with RWG bases. However, the most commonly used strategy is to expand all variables with temporal basis functions $T_k(t)$ and sample them with time step Δt. The convolution in Eq. (9.15) can then be performed with summation

$$\mathbf{I}_j = \sum_{k=0}^{j} \overline{\mathbf{I}}_k^{\text{imp}} \cdot \mathbf{e}_{j-k}^{\text{in}} \tag{9.16}$$

For the sake of simplicity, $\mathbf{I}(j\Delta t)$ is denoted by $\mathbf{I}_j$ in Eq. (9.16), $\overline{\mathbf{I}}^{\text{imp}}(k\Delta t)$ by $\overline{\mathbf{I}}_k^{\text{imp}}$, and $\mathbf{e}^{\text{in}}(j\Delta t - k\Delta t)$ by $\mathbf{e}_{j-k}^{\text{in}}$. Obviously, $\overline{\mathbf{I}}^{\text{imp}}(k\Delta t)$ can be calculated column by column using the discretized system of Eq. (9.9), which is the following matrix equation:

$$\sum_{k=0}^{j} \overline{\mathbf{D}}_k \cdot \mathbf{I}_{j-k} = \mathbf{e}_j^{\text{in}} \tag{9.17}$$

The entries of the coefficient matrices are

$$\overline{\mathbf{D}}_k(m,n) = A_{mn}(t) * \frac{\partial T_k(t)}{\partial t} + \Phi_{mn}(t) * \int_{-\infty}^{t} T_k(\xi)d\xi \tag{9.18}$$

The n-th column of $\overline{\mathbf{I}}^{\text{imp}}(k\Delta t)$ contains the current at the k-th time step under excitation of

$$\mathbf{e}_{k'}^{\text{in}}(n') = \begin{cases} 1, & n' = n, k' = 1 \\ 0, & \text{else} \end{cases} \tag{9.19}$$

The impulse responses of TD-MFIE and TD-CFIE can be obtained in a similar way. The TD-MFIE formulation here is expressed as

$$[-\hat{\mathbf{n}} \times \mathbf{H}^s(\mathbf{r},t) + \mathbf{J}_s(\mathbf{r},t)]\big|_{\text{tan}} = \hat{\mathbf{n}} \times \mathbf{H}^{\text{in}}(\mathbf{r},t)\big|_{\text{tan}} \tag{9.20}$$

The TD-CFIE is the combination of TD-EFIE and TD-MFIE:

$$-[\mathbf{E}^s(\mathbf{r},t) + \eta\hat{\mathbf{n}} \times \mathbf{H}^s(\mathbf{r},t)]\big|_{\text{tan}} + \eta\mathbf{J}_s(\mathbf{r},t) = [\mathbf{E}^{\text{in}}(\mathbf{r},t) + \eta\hat{\mathbf{n}} \times \mathbf{H}^{\text{in}}(\mathbf{r},t)]\big|_{\text{tan}} \tag{9.21}$$

where η is the intrinsic impedance of the medium. The scattered magnetic field is calculated from the vector potential using the relation of $\mathbf{H}^s(\mathbf{r},t) = \mu_0^{-1}\nabla \times \mathbf{A}(\mathbf{r},t)$. The matrix equation of TD-MFIE can be written as

$$\sum_{k=0}^{j} \overline{\mathbf{F}}_k \cdot \mathbf{I}_{j-k} = \mathbf{h}_j^{\text{in}} \tag{9.22}$$

The entries of $\overline{\mathbf{F}}_k$ for $k > 0$ are convolutions,

$$\overline{\mathbf{F}}_k(m,n) = F_{mn}(t) * T_k(t) + [tF_{mn}(t)] * \frac{\partial T_k(t)}{\partial t} \tag{9.23}$$

where

$$F_{mn}(t) = \begin{cases} \dfrac{1}{4\pi} \displaystyle\int_{S_m} \boldsymbol{f}_m(\boldsymbol{r}) \cdot \hat{\boldsymbol{n}} \times \int_{S_n} \delta(t - R/v) \dfrac{\boldsymbol{R}}{R^3} \times \boldsymbol{f}_n(\boldsymbol{r}') dS' dS, & \text{for } t_{\min} \leq t \leq t_{\max} \\ 0, & \text{else} \end{cases}$$

(9.24)

If $k = 0$, a residual matrix $(0.5\overline{\mathbf{G}})$ due to the singularity of the Green's function has to be added to Eq. (9.23), where $\overline{\mathbf{G}}$ is the Gram matrix of the spatial basis functions. Note that the gradient operation over the Dirac delta function in Eq. (9.23) is transformed to the differentiation operation on the current. Therefore, evaluation of surface integrals containing derivatives of the Dirac delta function is avoided.

The impulse responses can be calculated using the above methods. Take a PEC cuboid with sides of 0.6, 0.8, and 1 m as an example. The surface mesh has 200 triangles, resulting in 300 RWG basis functions. Assume that only the #10 RWG basis has an excitation with unit amplitude. The impulse responses at the #10 RWG are shown in Figure 9.1a–c. It can be seen that the impulse response associated with the TD-CFIE approaches zero smoothly. The impulse response associated with the TD-MFIE also approaches zero but has small oscillating ripples and converges more slowly than that of the TD-CFIE. In contrast, the impulse response to the TD-EFIE approaches a nonzero constant value, with oscillating ripples like those in the TD-MFIE.

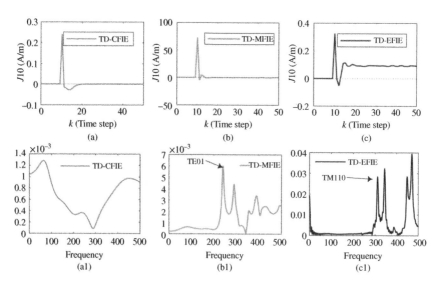

Figure 9.1 Transient impulse responses and their spectra. (a) (a1) TD-CFIE, (b) (b1) TD-MFIE (multiplied by η^{-1}), (c) (c1) TD-EFIE.

Figure 9.2 Static current patterns in the impulse response of the TD-EFIE. The shadowed part is the excitation RWG. (a) Excited at internal RWGs. (b) Excited at RWGs touching the boundary.

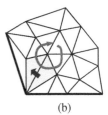

(a)   (b)

The typical impulse responses between two nonadjacent RWG bases are also checked. Their spectra are calculated using fast Fourier transform and are shown in Figure 9.1a1–c1. It can be seen that the spectrum of the impulse response associated with the TD-CFIE is smooth, while those of the TD-MFIE and the TD-EFIE have many spikes at frequencies corresponding to the interior resonances. By comparing with the theoretic results, the modes TE101 (240 MHz) and TM110 (313 MHz) can be recognized, as labeled in the figures. It can be seen that a spike appears at DC in the TD-EFIE spectrum, which is corresponding to the static solenoidal current shown in Figure 9.1c. It is obvious that there is no DC spike in the TD-MFIE spectrum.

The impulse responses clearly demonstrate that static surface currents are mixed in the solution of the TD-EFIE. It can be further examined that the main part of the static current usually forms one or two dominant loops flowing through the exciting RWG, while the static currents on all the meshes elsewhere are very small and are almost negligible, as shown in Figure 9.2. If the exciting RWG does not touch the boundary of an open surface, there will be two dominant loop currents, each one surrounding one endpoint of the common edge of the exciting RWG; otherwise, there will be only one dominant current loop.

9.3 Behavior at the Interior Resonance Frequencies

The interior resonance phenomenon is conveniently investigated in the FD, and the resultant conclusion is certainly helpful for explaining the late time stability properties of TD-SIEs. Consider a PEC scatterer enclosed by surface S. It is excited by an incident field $E^{in}(r)$, $H^{in}(r)$, as shown in Figure 9.3. The exterior region V is hence the problem region. According to the equivalence principle, the interior area V_1 may be assumed to be filled with any kind of medium that is convenient for solving the original problem. For example, the inner side of surface S (hereafter denoted by S^-) can be assumed to be a PEC wall, or a perfect magnetically conducting (PMC) wall, or even an impedance wall. If the boundary condition at S seen from region V remains unchanged, these assumptions will not affect the solutions to the original problem in the exterior domain V.

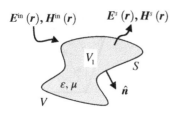

$E^{in}(r), H^{in}(r)$ $E^{s}(r), H^{s}(r)$ **Figure 9.3** Scattering problem of a PEC scatterer.

The induced surface current density J_s^{ind} on surface S must satisfy the following EFIE and MFIE simultaneously:

$$j\omega\mu\hat{n} \times \mathcal{L}\{J_s^{ind}\} = \hat{n} \times E^{in}(r) \tag{9.25}$$

$$-\hat{n} \times \mathcal{K}\{J_s^{ind}\} + \frac{1}{2}J_s^{ind}(r) = \hat{n} \times H^{in}(r) \tag{9.26}$$

The operators are defined in their usual way as follows:

$$\mathcal{K}\{f\} = P.V. \oint_S \overline{G}_m(r, r') \cdot f(r')dS' \tag{9.27}$$

$$\mathcal{L}\{f\} = P.V. \oint_S \overline{G}_e(r, r') \cdot f(r')dS' \tag{9.28}$$

where $\overline{G}_e(r, r')$ and $\overline{G}_m(r, r')$ are the electric and magnetic dyadic Green's functions in region V, respectively. The residual term associated with the singularity of Green's function is separated from the integral in Eq. (9.26). The notation $P.V.$ is the Cauchy principal value.

Equations (9.25) and (9.26) are derived from the fact that the electric field and magnetic field in the interior region of the PEC scatterer are zeroes. Consequently, the tangential components of the electric field and the magnetic field on S^- should all be zeroes. Therefore, the surface S^- acts not only like a PEC wall but also like a PMC wall. The induced current density on S is uniquely determined for a specified incident wave and disappears along with the incident waves. However, when MFIE or EFIE is applied alone, it cannot guarantee that both the tangential component of the electric field and the magnetic field are zeroes. As pointed out in [18, 21], nontrivial solutions to the homogeneous equations associated with EFIE and MFIE will exist at frequencies that equal the resonance frequencies of the interior cavity bounded by a PEC surface S^- or a PMC surface S^-. In these situations, the surface current at S will contain two parts: one part is the induced current J_s^{ind}, depending on the incident wave; the other part is related to nontrivial solutions to the associated homogeneous equation. Denote

$$J_s(r) = J_s^{ind}(r) + J_{sn}^e \tag{9.29}$$

for EFIE, and

$$J_s(r) = J_s^{ind}(r) + J_{sn}^h \tag{9.30}$$

for MFIE. The vectors J_{sn}^e and J_{sn}^h are, respectively, the nontrivial solutions to the homogeneous equations associated with EFIE and MFIE, which are termed as eigencurrents in [17]. The subscript "sn" is used to represent the nontrivial solutions of the n-th interior resonance.

The spatial distribution of the nontrivial solutions J_{sn}^e and J_{sn}^h is independent of incident fields. Hence, these currents can be considered as independent current sources. However, the behavior of J_{sn}^e and J_{sn}^h are quite different from each other. The nontrivial solutions J_{sn}^e associated with EFIE satisfy

$$j\omega\mu\hat{n} \times \mathcal{L}\{J_{sn}^e\} = 0 \tag{9.31}$$

If interpreted explicitly, Eq. (9.31) means that at frequencies corresponding to its eigenvalues, there exist nonzero surface current sources on S, the tangential components of whose radiated electric fields on S are zeroes. Although J_{sn}^e can be seen as independent current sources, they cannot radiate waves to the exterior region V. This can be checked straightforwardly using the Huygens equivalence principle. On the other hand, the surface currents at frequencies equal to the resonance frequencies of the interior cavity bounded by a PEC surface S are also subject to Eq. (9.31). Therefore, nontrivial solutions J_{sn}^e actually also correspond to currents associated with interior resonance modes of the virtual cavity bounded by a PEC surface S^-. Because the tangential component of the electric field by the nonzero J_{sn}^e on S^- is zero, the surface S^- can be regarded as a PEC wall. Intuitively, J_{sn}^e are current sources infinitely approaching a closed PEC surface; hence, it is correct to predict that J_{sn}^e cannot radiate fields to the exterior region V.

On the other hand, the nontrivial solutions J_{sn}^h to the homogeneous equation associated to MFIE satisfy

$$-\hat{n} \times \mathcal{K}\{J_{sn}^h\} + \frac{1}{2}J_{sn}^h(r) = 0, \quad r \in S \tag{9.32}$$

Since J_{sn}^h may exist without excitation, they can also be treated as independent sources just like J_{sn}^e. Similarly, J_{sn}^h can be considered as current sources placed infinitely close to a closed PMC surface, and it can definitely radiate fields to the exterior region V. Meanwhile, it can be verified that at resonance modes of a cavity formed by a PMC surface S^-, the associated resonance surface current on the cavity wall satisfies exactly the same equation of Eq. (9.32). Therefore, nontrivial solutions to Eq. (9.32) are corresponding to the resonance modes of the interior cavity bounded by a PMC wall.

It is important to note that solenoidal static currents satisfy Eq. (9.31), but do not satisfy Eq. (9.32). This means that EFIE admits interior resonance at $\omega = 0$, or DC resonance, while MFIE has no interior resonance at DC.

Theoretically, the interior resonance occurs only at discrete frequencies, so the spectrum should be discrete single lines. However, when solving the SIEs with numerical methods, the spectrum may be blurred due to numerical errors like the discretization errors, and the interior resonance might occur in a narrow band in the vicinity of the discrete resonance frequencies.

The equivalent circuit model can be used to investigate this problem [48]. In the case of EFIE, the tangential components of the total electric fields at interior resonance frequencies should vanish in theory, so there should be no ohmic loss on the PEC surface S. The interior cavity associated with the scatterer in this situation can be modeled as an ideal resonator. However, due to the discretization errors, the calculated tangential components of the total electric fields may have nonzero small values, which will lead to a pseudo-numerical loss P_{loss} on the surface S, which can be approximately denoted as

$$P_{\text{loss}} = \oint_S \left[E^{\text{in}}(r) - j\omega\mu\mathcal{L}\left\{ J_s^{\text{ind}} + J_{sn}^e \right\} \right] \cdot J_{sn}^{e*}(r)dS \tag{9.33}$$

Here, the superscript "*" means conjugate. This pseudo-numerical loss makes the interior cavity a dumping resonator. Assume that P_{loss} is dissipated by an effective resistor R_{num}^e, then the behavior of the interior resonator may be approximately described using the equivalent *RLC* circuit model shown in Figure 9.4a. The Q-factor of the cavity resonator will not be infinitely large because of the effect of R_{num}^e. Therefore, the corresponding resonance current is finite, and the spectrum at the interior resonance may span to occupy a narrow band. If a finer mesh structure is adopted to get higher numerical accuracy, R_{num}^e tends to become smaller, leading to a larger Q-factor, narrower band, and larger resonance current. This feature is in agreement with the observation in [17].

In the case of MFIE, the resonance current may radiate fields to the exterior region. This can be regarded as a radiation loss seen from the interior cavity. Therefore, the interior resonance associated with MFIE can be described using the equivalent *RLC* circuit model in Figure 9.4b, with R_{num}^h accounting for the numerical loss and R_r accounting for the radiation loss. Because R_r is usually much larger than R_{num}^h, the Q-factor associated with the interior cavity in the case of MFIE is much smaller than that in EFIE. Consequently, comparing to the case of EFIE, the spectrum band associated with MFIE at a resonance frequency is much wider and the amplitude of the resonance current is much smaller.

Although the spatial distribution of the resonance currents is independent on the incident fields, their excitation is strongly dependent on the incident fields.

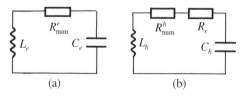

(a) (b)

Figure 9.4 (a) Circuit model for interior resonance associated with EFIE. (b) Circuit model for interior resonance associated with MFIE.

Consider the scattered fields by an incident field at a certain interior resonance frequency. As has been pointed out previously, the total surface current contains an induced current and a resonance current. In the case of EFIE, we can write

$$\hat{n} \times \left[E^{in}(\mathbf{r}) - j\omega\mu\mathcal{L}\left\{ J_s^{ind} + J_{sn}^e \right\} \right] = 0, \quad \mathbf{r} \in S \tag{9.34}$$

When solving this equation with method of moment (MoM), the condition number of the coefficient matrix tends to be very large at interior frequencies, as can also be predicted using the *RLC* circuit mode shown in Figure 9.4a. The solution may be unstable if the iteration solver is used. However, a stable and unique solution can still be found using the pseudo-inversion technique. Theoretically, the scattered fields are not influenced by the resonance current in the case of EFIE.

The induced current can be separated from the total current by

$$\hat{n} \times \left[H^{in}(\mathbf{r}) + H^s \left(J_s^{ind} + J_{sn}^e \right) \right] = J_s^{ind}, \quad \mathbf{r} \in S \tag{9.35}$$

Power must be conserved in electromagnetic scattering problems. Because J_{sn}^e is independent of the incident fields and does not radiate waves into the exterior region V, the scattered fields are totally caused by J_s^{ind}. In the case of a PEC scatterer, the power conservation law requires that

$$\oint_{S_1} \left[E^{in}(\mathbf{r}) + E^s \left(J_s^{ind} \right) \right] \times \left[H^{in}(\mathbf{r}) + H^s \left(J_s^{ind} \right) \right]^* \cdot \hat{n} dS = 0 \tag{9.36}$$

where S_1 is an arbitrary closed surface enclosing the scatterer. Since the numerical loss is usually negligible, the requirement of Eq. (9.36) is always easy to meet. The power conservation relation poses no limitation to the resonance current J_{sn}^e, the amplitudes of which may possibly become very large if the discretization error is very small.

In the case of MFIE, the total surface current $\left(J_s^{ind} + J_{sn}^h \right)$ satisfies

$$-\hat{n} \times \mathcal{K}\left\{ J_s^{ind} + J_{sn}^h \right\} + \frac{1}{2} \left(J_s^{ind} + J_{sn}^h \right) = \hat{n} \times H^{in}(\mathbf{r}), \quad \mathbf{r} \in S \tag{9.37}$$

The spatial distribution of J_{sn}^h is also independent of the incident fields. However, since J_{sn}^h can radiate waves to the exterior region, their amplitudes are limited by the incident fields according to the power conservation law. Because of this limitation, when solving Eq. (9.37) with MoM, the condition number of the coefficient matrix is always a finite value, and the amplitude of the resonance current should be bounded. This feature can also be predicted using the *RLC* circuit model illustrated in Figure 9.4b.

Consider a PEC cylinder with radius a. Assumed that it is illuminated by a TM plane wave with $E = \hat{a}_z \exp(-jkx)$ (V/m). The surface current density of the n-th interior resonance mode associated with EFIE has only the z-component and can be denoted as $J_{sn}^{TM,e} = \hat{z} j_{sn}^{TM,e}$. The discrete interior resonance frequencies $f_{nl}^{TM,e}$ can be determined according to $k_{nl}^{TM,e} a = v_{nl}$, with v_{nl} being the l-th root of the n-th order Bessel function and $k_{nl}^{TM,e} = 2\pi f_{nl}^{TM,e}\sqrt{\mu\varepsilon}$. In the vicinity of $f_{nl}^{TM,e}$, the total surface current then has the form of $\left(J_s^{ind} + C_n^e j_{sn}^{TM,e}\right)$, where C_n^e is a coefficient depending on the discretization error and the frequency offset from the resonance frequency. It is numerically analyzed in the vicinity of $ka = v_{01}$, sweeping from $kaL = 2.404, 825, 557, 435, 7$ to $kaH = 2.404, 825, 557, 475, 7$ with an incremental step of $\Delta ka = 1.0 \times 10^{-12}$. The circumference of the cylinder is divided into 900 equal-length segments, and 900 roof-top basis functions are used to expand the surface current densities and the rotated tangential field components. Standard Galerkin's scheme is used to discretize the corresponding EFIE. Figure 9.5a shows the magnitude of $2C_n^e$. It has a peak at $ka0$, with its amplitude calculated to be $2C_n^e = 2.079(A/m)$, which is about 400 times of that of the induced surface current. The bi-static radar cross sections (RCSs) at three frequency points are plotted in Figure 9.5b, where the current density and the RCSs are all normalized with wavelength. It can be seen that although very large resonance currents are mixed in the total currents, the calculated RCSs are almost not affected by the resonance currents. That said, $J_{sn}^{TM,e}$ do not radiate fields.

Denote the surface current density of the n-th interior resonance mode associated with the MFIE on the surface of the PEC cylinder as $J_{sn}^{TM,h} = \hat{z} j_{sn}^{TM,h}$. The discrete interior resonance frequencies $f_{nl}^{TM,h}$ can be determined based on $k_{nl}^{TM,h} a = \mu_{nl}$, where μ_{nl} is the l-th root of the derivative function of the n-th order Bessel function. The corresponding normalized resonance surface currents have the form of $j_{sn}^{TM,h}(\phi) = \hat{z} e^{jn\phi}$, and the total surface currents can be expressed as $\left(J_s^{ind} + C_n^h j_{sn}^{TM,h}\right)$, where C_n^h is the coefficient of the surface current associated with the n-th resonance mode. As has been discussed previously, C_n^h is bounded by the power conservation relation. The behavior of the interior resonance mode in the vicinity of $ka = \mu_{11} = 1.841$ is numerically analyzed, with ka changing from

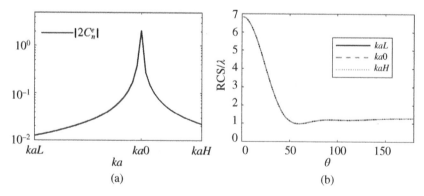

Figure 9.5 The numerical results in the vicinity of $ka = v_{01}$. (a) The amplitude of the resonance surface current. (b) Bi-static RCSs.

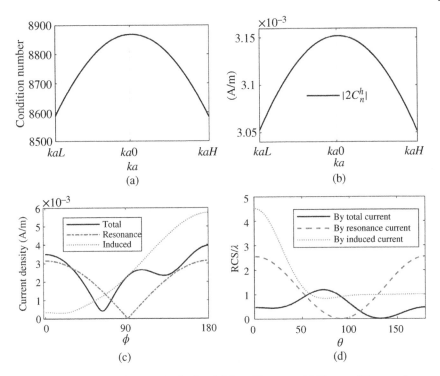

Figure 9.6 The numerical results in the vicinity of $ka = \mu_{11}$. (a) The condition number of the coefficient matrix. (b)The amplitudes of the resonance surface currents. (c) The calculated surface currents at $ka0$. (d) The corresponding bi-static RCSs.

$kaL = 1.841, 202$ to $kaH = 1.841, 282$ at an incremental step of $\Delta ka = 0.000, 002$. The condition number of the coefficient matrix is shown in Figure 9.6a, and the calculated amplitude of the resonance surface current is plotted in Figure 9.6b. Both reach their maximum value at $ka0$. The amplitude of the resonance surface current is smaller than the amplitude of the induced current in this case. The calculated surface currents at $ka0$ are plotted in Figure 9.6c, and the RCSs associated with them are plotted in Figure 9.6d. Apparently, the interior resonance current associated with MFIE can radiate fields.

9.4 Impact on MOT Late Time Instability

Based on the previous discussion, it can be concluded that the TD-EFIE suffers from DC instability together with interior resonance instability. The TD-MFIE is basically immune to DC instability and interior resonance instability.

The TD-CFIE should also be immune to DC instability and interior instability because the TD-CFIE admits no spurious solutions. However, the late time behavior of the TD-CFIE is different from that of the TD-MFIE.

The late time stability properties of the TD-IEs can be predicted more clearly by analyzing the spectra of the TD-IE operators, or the eigenvalues of their corresponding discretized systems. Using the discretization strategy described previously, the TD-EFIE Eq. (9.9) can be transformed into matrix form as

$$\sum_{k=0}^{k_{max}-1} \overline{\mathbf{D}}_k \cdot \mathbf{I}_{j-k} + \overline{\mathbf{D}}_{k_{max}} \cdot \mathbf{Q}_{j-k_{max}} = \mathbf{e}_j^{in} \tag{9.38}$$

where $\mathbf{Q}_j = \sum_{k=0}^{j} \mathbf{I}_k$. For $0 \le k \le k_{max} - 1$, $\overline{\mathbf{D}}_k(m, n)$ can be calculated using Eq. (9.18). If the TD-EFIE has a static solenoidal current $\mathbf{I}_{sol}$, it should satisfy the homogeneous form of Eq. (9.38).

Performing the temporal differentiation on Eq. (9.9) gives the differentiated TD-EFIE system. In this chapter, triangle temporal basis functions are used and the differentiated TD-EFIE is deduced from Eq. (9.38) as follows:

$$\overline{\mathbf{D}}_0 \cdot \mathbf{I}_j + \sum_{k=1}^{k_{max}} (\overline{\mathbf{D}}_k - \overline{\mathbf{D}}_{k-1}) \cdot \mathbf{I}_{j-k} = \mathbf{e}_j^{in} - \mathbf{e}_{j-1}^{in} \tag{9.39}$$

Following the method in [21], we can construct the canonical form of Eq. (9.39) in the absence of incident fields as follows:

$$
\begin{bmatrix} \mathbf{I}_j \\ \mathbf{I}_{j-1} \\ \vdots \\ \mathbf{I}_{j-k_{max}+2} \\ \mathbf{I}_{j-k_{max}+1} \end{bmatrix} =
\begin{bmatrix} \overline{\beta}_1 & \overline{\beta}_2 & \cdots & \overline{\beta}_{k_{max}-1} & \overline{\beta}_{k_{max}} \\ \mathbf{U} & \mathbf{0} & \cdots & \mathbf{0} & \mathbf{0} \\ \vdots & \vdots & \ddots & \vdots & \vdots \\ \mathbf{0} & \cdots & \mathbf{U} & \mathbf{0} & \mathbf{0} \\ \mathbf{0} & \cdots & \mathbf{0} & \mathbf{U} & \mathbf{0} \end{bmatrix}
\begin{bmatrix} \mathbf{I}_{j-1} \\ \mathbf{I}_{j-2} \\ \vdots \\ \mathbf{I}_{j-k_{max}+1} \\ \mathbf{I}_{j-k_{max}} \end{bmatrix} \tag{9.40}
$$

where $\overline{\beta}_k = \overline{\mathbf{D}}_0^{-1} (\overline{\mathbf{D}}_{k-1} - \overline{\mathbf{D}}_k)$. Denote $\mathbf{J}_j = [\mathbf{I}_j, \cdots, \mathbf{I}_{j-k_{max}+1}]^t$, then Eq. (9.40) can be denoted compactly as $\mathbf{I}_j = \overline{\mathbf{Z}}_c \cdot \mathbf{I}_{j-1}$, where $\overline{\mathbf{Z}}_c$ is the companion matrix of the difference equation system. Similarly, we can construct the companion matrix for the non-differentiated system Eq. (9.38) in the following way:

$$
\begin{bmatrix} \mathbf{I}_j \\ \mathbf{I}_{j-1} \\ \vdots \\ \mathbf{I}_{j-k_{max}+2} \\ \mathbf{Q}_{j-k_{max}+1} \end{bmatrix} =
\begin{bmatrix} \beta_1' & \beta_2' & \cdots & \beta_{k_{max}-1}' & \beta_{k_{max}}' \\ \mathbf{U} & \mathbf{0} & \cdots & \mathbf{0} & \mathbf{0} \\ \vdots & \vdots & \ddots & \vdots & \vdots \\ \mathbf{0} & \cdots & \mathbf{U} & \mathbf{0} & \mathbf{0} \\ \mathbf{0} & \cdots & \mathbf{0} & \mathbf{U} & \mathbf{U} \end{bmatrix}
\begin{bmatrix} \mathbf{I}_{j-1} \\ \mathbf{I}_{j-2} \\ \vdots \\ \mathbf{I}_{j-k_{max}+1} \\ \mathbf{Q}_{j-k_{max}} \end{bmatrix} \tag{9.41}
$$

Where $\overline{\beta}_k' = -\overline{\mathbf{D}}_0^{-1} \overline{\mathbf{D}}_k$. Note that the last row in Eq. (9.41) comes from the relationship of $\mathbf{Q}_k = \mathbf{I}_k + \mathbf{Q}_{k-1}$.

The general solutions to the discrete system can be expressed by a polynomial of eigenvalues and eigenvectors. According to the eigenvalues λ of the companion matrix, a discrete system can be divided into three stability types, which was originally proposed in reference [20], with a slight modification here in type 2:

Type 1: There is at least one eigenvalue λ so that $|\lambda| > 1$. The system is unstable, and the solution will grow exponentially.

Type 2: There is at least one λ so that $|\lambda| = 1$; all other eigenvalues are within the unit circle. The system is possibly stable with the amplitude of the solution approaching a constant value or unstable with the solution growing at most at the order of polynomials.

Type 3: $|\lambda| < 1$ holds for all the eigenvalues λ. The system is strictly stable, and the solution will decay to zero after the incident fields disappear.

In practical situations, it is usually required to find a solution with a specified excitation, i.e. the particular solution, which can be obtained by using the method of variation of constants. The TD-IE systems can be considered actually as initial value problems with their initial values determined by the excitations. It is not certain that all eigenvalues contribute to the late time solution. If we can prevent the eigenvectors corresponding to all unstable eigenvalues creeping into the late time solutions, it is possible to get stable results even in an unstable system.

For TD-EFIEs, both DC instability and interior resonance instability are the main concerns. The eigenvalues of TD-EFIE can be calculated using Eq. (9.40) for the differentiated TD-EFIE and Eq. (9.41) for the non-differentiated TD-EFIE. There is no significant difference between the two TD-EFIE systems with regard to the spectrum and the late time instability property. In ideal situations without discretization errors, the eigenvalues corresponding to static solenoidal solutions exactly pile on the point of $\lambda_0 = 1 + i0$, and the TD-EFIE system is of type 2. However, discretization errors are inevitable. The eigenvalues will shift to places around $(1 + i0)$ with part of them moving outside the unit circle on the complex plane of λ. Therefore, the TD-EFIE basically belongs to stability type 1 and is unstable.

If there are no discretization errors, the eigenvalues corresponding to the interior resonances will locate exactly on the unit circle. Assume that ω_i is angular frequency of the i-th interior resonance. The solution at resonance comprises only sinusoidal waves with time convention of $e^{\pm j\omega_i t}$. The eigenvalue λ_i corresponding to this resonance current should satisfy the eigenequation.

$$e^{j\omega_i(k+1)\Delta t} = \lambda_i e^{\pm j\omega_i k\Delta t} \tag{9.42}$$

Hence, there exists a pair of eigenvalues corresponding to one interior resonance state, which is dependent on the resonance frequency and the sampling time step, i.e.

$$\lambda_i = e^{\pm j\omega_i \Delta t} \tag{9.43}$$

If the currents are expanded with temporal basis functions $T_k(t)$ satisfying the translation relationship of $T_k(t) = T(t - k\Delta t)$, it can be proved the eigenvalues will not change with translation because

$$\int e^{j\omega_i t} T_{k+1}(t)dt = \lambda_i \int e^{j\omega_i t} T_k(t)dt \tag{9.44}$$

It is straightforward to check that the eigenvalue satisfying Eq. (9.44) still has the form of Eq. (9.43). This observation indicates that using higher-order temporal basis functions is not a critical measure to construct stable TD-IE solvers since the eigenvalues corresponding to spurious solutions are not very sensitive to the type of temporal basis functions that satisfy the translation relationship.

As discussed previously, the TD-MFIE does not suffer from DC instability. The interior resonances also will not cause late time instability according to their behaviors at the resonances. In the discretized TD-MFIE system in Eq. (9.22), $\overline{\mathbf{F}}_k$ is nonzero usually only in a bounded range of k, $0 \leq k \leq k_{max} - 1$. Unless the discretization errors are too large, the TD-MFIE is essentially of stability type 3.

The TD-CFIE and its differentiated version are stable because they have eliminated spurious DC and interior resonance solutions. However, TD-CFIE belongs to stability type 2 instead of type 3 due to the additional eigenvalue introduced by the numerical algorithm. Write the discrete TD-CFIE system as follows:

$$\sum_{k=0}^{k_{max}-1} \overline{\mathbf{C}}_k \cdot \mathbf{I}_{j-k} + \eta^{-1} \overline{\mathbf{D}}_{k_{max}} \cdot \mathbf{Q}_{j-k_{max}} = \eta^{-1} \mathbf{e}_j^{in} + \mathbf{h}_j^{in} \tag{9.45}$$

where $\overline{\mathbf{C}}_k = \eta^{-1} \overline{\mathbf{D}}_k + \overline{\mathbf{F}}_k$. At first, we can check that both the TD-CFIE and the differentiated TD-CFIE admit static late time solutions. Assume that the incident fields have completely disappeared, and $\mathbf{I}_k = \mathbf{I}_c$ for $j - k_{max} \leq k \leq j - 1$, where $\mathbf{I}_c$ is a constant vector. Making use of Eq. (9.45), $\mathbf{I}_j$ can be calculated iteratively in this case with

$$\mathbf{I}_j = \mathbf{I}_{j-1} + \eta^{-1} \overline{\mathbf{D}}_{k_{max}} \cdot \mathbf{I}_c \tag{9.46}$$

If $\overline{\mathbf{D}}_{k_{max}} \cdot \mathbf{I}_c = 0$, it is obvious that $\mathbf{I}_j = \mathbf{I}_{j-1}$. Following the same deduction, it can be checked that $\mathbf{I}_k = \mathbf{I}_c$ for all $k > j$. Therefore, if there is no excitation, static currents can be solutions to the TD-CFIE. As a result, there is always an eigenvalue at $(1 + i0)$ for TD-CFIE. However, the behavior of the static solutions in the TD-CFIE is largely different from that in the TD-EFIE. In the TD-EFIE and its differentiated version, the static currents are solenoidal and are nontrivial solutions to the system. The solution associated with the eigenvalue λ_{DC} at $(1 + i0)$ has the form of $\mathbf{I}_{k+1} = \lambda_{DC} \mathbf{I}_k$. If at the end of excitation there exists a constant current $\mathbf{I}_c$ satisfying $\overline{\mathbf{D}}_{k_{max}} \cdot \mathbf{I}_c = 0$, then according to Eq. (9.46), this current $\mathbf{I}_c$ will be delivered step by step exactly in the form of $\mathbf{I}_j = \mathbf{I}_{j-1}$ and will not grow. The corresponding eigenvalue will exactly be located at $(1 + i0)$. Hence, the TD-CFIE is a special case of the stability type 2: It is stable with its solution approaching a constant value when $j \to \infty$. Note that the above deduction is also valid for the differentiated TD-CFIE.

9.5 Analytical Expressions for the Retarded-Time Potentials

In order to evaluate the two potentials on RWG bases, there are two basic kinds of retarded-time potential integrals in the TD-IEs, the scalar one and the vector one, which are defined as follows [12]:

$$\Gamma_n(\mathbf{r}, t) = \int_{S_n} \frac{\delta(t - R/c)}{R} dS' \tag{9.47}$$

$$\mathbf{\Gamma}_n(\mathbf{r}, t) = \int_{S_n} \frac{\delta(t - R/c)}{R} (\mathbf{r}' - \mathbf{r}_o) dS' \tag{9.48}$$

where $\mathbf{r}_o$ is the projection of the observation point $\mathbf{r}$ on the source triangle plane. S_n is one triangle patch of the n-th RWG basis that models the object. And c is the speed of light in the investigated medium.

Closed-form expressions for Eqs. (9.47) and (9.48) are provided in [12, 15], which were presented in a more compact and easier-to-program way in [16]. Geometric parameters are depicted in Figure 9.7a,b, where $\mathbf{r}'_1$, $\mathbf{r}'_2$, and $\mathbf{r}'_3$ are the vertexes of a source triangle. The distance between the field point $\mathbf{r}$ and the source triangle plane is denoted by $d = (\mathbf{r} - \mathbf{r}'_1) \cdot \hat{\mathbf{n}}'$, in which $\hat{\mathbf{n}}'$ is the unit normal vector of the source triangle. $\rho = \mathbf{r}' - \mathbf{r}_o$ is the local in-plane coordinates. For the i-th edge of the source triangle, $\hat{\mathbf{u}}'_i$ is its outward unit normal vector, and $\hat{\mathbf{l}}'_i$ is the unit tangential vector, pointing from $\mathbf{r}'^-_i$ to $\mathbf{r}'^+_i$. They are subject to a righthand rule of $\hat{\mathbf{u}}'_i = \hat{\mathbf{l}}'_i \times \hat{\mathbf{n}}'$.

The source triangle is divided into three sub-triangles with a common node $\mathbf{r}_o$, as is shown in Figure 9.7a. The integration is carried out on each sub-triangle respectively. It can be checked that Γ_n and $\mathbf{\Gamma}_n$ are, respectively, related to the length and bisecting vector of the intersecting arcs. The integration of Eqs. (9.47) and (9.48) can be finally expressed by using the vertex coordinates of the RWG patches as

$$\Gamma_n(\mathbf{r}, t) = c \sum_{i=1}^{3} \text{sign}(h_i) \left(\sigma_i^+ \Delta\theta_i^+ - \sigma_i^- \Delta\theta_i^- \right) \tag{9.49}$$

$$\mathbf{\Gamma}_n(\mathbf{r}, t) = 2\rho c \sum_{i=1}^{3} \text{sign}(h_i) \left(\sigma_i^+ \hat{\mathbf{e}}_i^+ \sin\frac{\Delta\theta_i^+}{2} - \sigma_i^- \hat{\mathbf{e}}_i^- \sin\frac{\Delta\theta_i^-}{2} \right) \tag{9.50}$$

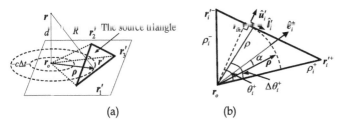

(a)　　　　　　　　(b)

Figure 9.7 Geometric parameter definitions. (a) Observation point $\mathbf{r}$ projecting on the source triangle. (b) Parameters for one sub-triangle.

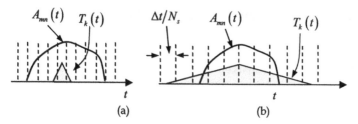

Figure 9.8 Sampling scheme for the temporal convolution. The shadowed triangles are the temporal basis functions used in this paper. (a) Smaller time step. (b) Larger time step.

in which, $\rho = |\boldsymbol{\rho}| = \sqrt{(ct)^2 - d^2}$, $\rho_i^{\pm} = |\boldsymbol{r'}_i^{\pm} - \boldsymbol{r}_0|$, $h_i = (\boldsymbol{r'}_i^{+} - \boldsymbol{r}) \cdot \hat{\boldsymbol{u}}_i'$, $\sigma_i^{\pm} = \text{sign}$ $\left[(\boldsymbol{r'}_i^{\pm} - \boldsymbol{r}) \cdot \hat{\boldsymbol{l}}_i'\right]$, and

$$\Delta \theta_i^{\pm} = \begin{cases} \cos^{-1}\left(\dfrac{|h_i|}{\rho_i^{\pm}}\right), \rho \le |h_i| \\[3mm] \cos^{-1}\left(\dfrac{|h_i|}{\rho_i^{\pm}}\right) - \cos^{-1}\left(\dfrac{|h_i|}{\rho}\right), |h_i| \le \rho \le \rho_i^{\pm} \\[3mm] 0, \rho \ge \rho_i^{\pm} \end{cases} \tag{9.51}$$

$$\hat{\boldsymbol{e}}_i^{\pm} = \text{sign}(h_i)\cos\left(\theta_i^{\pm} - \frac{\Delta \theta_i^{\pm}}{2}\right)\hat{\boldsymbol{u}}_i' + \sigma_i^{\pm}\sin\left(\theta_i^{\pm} - \frac{\Delta \theta_i^{\pm}}{2}\right)\hat{\boldsymbol{l}}_i' \tag{9.52}$$

These formulas can be programmed without the necessity to calculate the intersecting points. The geometric relationships between the concentric time spheres and the source triangle can be judged automatically. With these two formulas, the internal spatial surface integrations involved in $A_{mn}(t)$, $\Phi_{mn}(t)$, and $F_{mn}(t)$ can all be performed analytically while the outer spatial surface integrations can be carried out using Gaussian quadrature.

The convolution operations involved in calculating $A_{mn}(t)$, $\Phi_{mn}(t)$, and $F_{mn}(t)$ can be performed with high accuracy. As shown in Figure 9.8, $A_{mn}(t)$ and $T_k(t)$ are sampled with $\Delta t/N_s$, where N_s can be adaptively selected according to the time step Δt and the time span of $A_{mn}(t)$.

9.6 Numerical Verification of Stability Properties

Two numerical examples are used to demonstrate the stability properties of the TD-IEs. The first example is to analyze a PEC sphere with radius of 1 m. It is placed in free space, with its center located at the origin. The excitation is an x-polarized Gaussian-shaped plane wave pulse with the electric field expressed by

$$\boldsymbol{E}^{\text{in}} = \boldsymbol{E}_0 \frac{1}{\sqrt{\pi}} \exp\left[-\frac{16}{(v\tau)^2}(vt - vt_0 - \hat{\boldsymbol{k}} \cdot \boldsymbol{r})^2\right] \ (\text{V/m}) \tag{9.53}$$

The parameters are chosen as $v\tau = 4$ m, $vt_0 = 6$ m, $\boldsymbol{E}_0 = 1.0\hat{\boldsymbol{a}}_x$, and $\hat{\boldsymbol{k}} = \hat{\boldsymbol{a}}_z$. A coarse mesh structure with 76 triangles is used, which has 114 RWGs. Denote the space–time discretization ratio as $\chi = c\Delta t/l_{\max}$, with $l_{\max}$ being the largest edge length of the triangles in the mesh. The time step is chosen to be 0.5 ns, corresponding to $\chi = 0.182$. The surface currents are calculated using the TD-EFIE, TD-MFIE, and TD-CFIE. Figure 9.9 shows the logarithm of the resultant currents ($\log|J|$) at point $(0.22\,\text{m}, 0.17\,\text{m}, 0.92\,\text{m})$. It can be seen that the TD-EFIE is of type 1, whose solution is stable till 2.4 μs and then grows exponentially. The TD-CFIE is apparently of type 2. Its solution remains strictly constant after 0.2 μs. The TD-MFIE is of type 3, with its solution decaying exponentially. The early time solutions to the three TD-IEs agree very well, as shown in Figure 9.9b. Although not plotted in the figures, it has been checked that the solutions to the differentiated TD-EFIE and the differentiated TD-CFIE almost coincide with those of the TD-EFIE and the TD-CFIE, respectively.

The stability properties of the TD-IEs can be further verified from their eigenvalues. The eigenvalues corresponding to the TD-EFIE are plotted in Figure 9.10a. Among the 1710 eigenvalues, about 80 eigenvalues have modulus larger than 1.

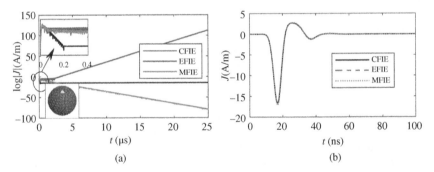

Figure 9.9 PEC sphere. Solutions to TD-EFIE, TD-MFIE, and TD CFIE. (a) Wide-span solutions. (b) Early time solutions.

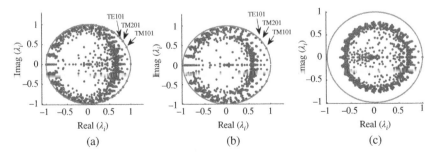

Figure 9.10 Eigenvalues of the TD-IEs [21]. (a) TD-EFIE. (b)TD-MFIE. (c)TD-CFIE.

The eigenvalues of the TD-MFIE system are shown in Figure 9.10b. Their moduli are all smaller than 1. The TD-CFIE system has eigenvalues located at $(1 + i0)$, as shown in Figure 9.10c. Some of the eigenvalues corresponding to interior resonances have been recognized and are denoted in Figure 9.10. For example, the lowest mode is TM101 (133 MHz). Using [43], its eigenvalue is found to be $\lambda_{TM101} = e^{\pm i\omega\Delta t} = 0.91 \pm i0.41$.

The time step Δt has significant influence on the stability properties of the TD-IEs. The TD-IEs have been repeatedly solved with time steps of $\Delta t = 0.01$, 0.5, 3, and 10 ns, corresponding to $\chi = 0.0036, 0.182, 1.09$, and 3.63, respectively. The mesh structure and the integration scheme remain unchanged. In all these situations, similar stability features to the TD-IEs have been observed. The TD-CFIE and the TD-MFIE are always stable. For example, in the case of $\chi = 0.182$, one million steps have been achieved and the solution to the TD-CFIE remains constant. Stable early time solutions are obtained using the TD-EFIE except in the case of $\Delta t = 10$ ns. The currents at the point $(0.22$ m, 0.17 m, 0.92 m) are shown in Figure 9.11. Using a smaller time step will bring higher accuracy but will cost much more CPU time and memory because the number of the nonzero coefficient matrices linearly increase with $(1/\Delta t)$.

The second example is to check a thin almond, where 870 unknowns are analyzed. The time step is 0.02 ns, corresponding to $\chi = 0.378$. The results at a checking point are plotted in Figure 9.12a, at MOT with 0.5 million steps. In this case, the solution to the TD-MFIE converges very fast and almost to zero after 4 μs. The solution to the TD-EFIE is also stable in a very wide time scope but does increase gradually. The early time responses agree very well, as shown in Figure 9.12b.

In the numerical examples, the coupling coefficients are evaluated with relatively high precision. A smaller time step makes the solutions to the TD-IEs more accurate and more stable, but the computational efficiency is worse. As has been pointed previously, if the local-domain temporal basis functions $T_k(t)$ that satisfy the translation relationship of $T_k(t) = T(t - k\Delta t)$ are used, the type of the basis function has no significant influence on the eigenvalues. This conclusion has been

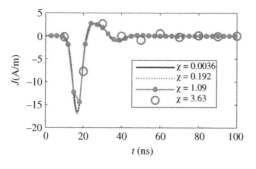

Figure 9.11 Early time solutions to the TD-CFIE with different time steps [21].

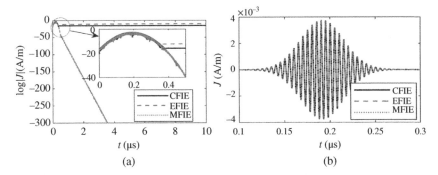

Figure 9.12 PEC almond. (a) Late time responses in logarithm. (b) Early time responses.

justified by the fact that all the numerical results in this chapter are obtained by using low-order triangular temporal basis functions. Some researchers have observed that higher-order local-domain temporal basis functions may help to improve the stability property of the TD-EFIE. However, there is no solid proof to support that higher-order basis functions can definitely guarantee higher stability.

Entire domain temporal basis functions such as Laguerre polynomials [40] are not subject to the translation relationship. If entire domain temporal basis functions are used, the eigenvalues corresponding to the interior resonance currents are changed and can no longer be expressed by Eq. (9.44). In this case, it might be possible to implement absolutely stable TD-EFIE solvers by shifting all eigenvalues within the unit circle.

9.7 Effect of Impulse Response Truncation

The impulse responses approach a constant value in the TD-EFIE, and approach zero in the TD-CFIE and the TD-MFIE. Equation (9.16) provides us a possible way to get stable solutions to the TD-CFIE and the TD-MFIE using their truncated impulse responses [21]. Truncating Eq. (9.16) yields

$$\mathbf{I}_j \approx \sum_{k=0}^{N_{tr}} \overline{\mathbf{I}}_k^{-imp} \cdot \mathbf{e}_{j-k}^{in} \tag{9.54}$$

where N_{tr} is the number of terms of the truncated impulse responses. In the case of the TD-EFIE, the constant term should be kept when performing truncation.

It is a very difficult task to evaluate the influence of truncation on the accuracy. An empirical criterion that is sometimes used in time-domain measurement and FDTD algorithm can be adopted, i.e. the duration of the truncated impulse response can be set to be about three times of the time that wave costs traveling along the longest straight path in the scatterer under consideration.

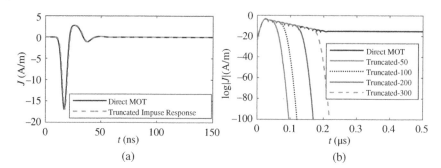

Figure 9.13 The TD-CFIE solutions by truncated impulse responses [21]. (a) Early time solutions. (b) Late time solutions.

The truncation effect on the impulse response of the TD-CFIE is shown using the above example of a PEC sphere with 114 unknowns. The longest path is the diameter, which is 2 m. The truncation criterion is of 20 ns, corresponding to 120 marching steps. We choose $N_{tr} = 150$ in this example. The results are shown in Figure 9.13a. The error due to truncation is small in this example.

The effect of the truncation length in the case of the TD-CFIE is shown in Figure 9.13b. The truncation length is chosen to be 50, 100, 200, and 300, respectively. The solutions are stable and converge exponentially. If longer truncation length is used, the result tends to agree better with that obtained by solving the problem directly with the MOT TD-CFIE solver.

9.8 Domain Decomposition Method Based on Impulse Responses

9.8.1 TD-GTM Model

Consider a subdomain (scatterer) as shown in Figure 9.14. It is assumed to be enclosed by a Huygens equivalent surface S. The incident fields are denoted as

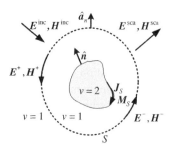

Figure 9.14 A scatterer enclosed by the Huygens surface.

E^{inc} and H^{inc}. Their rotated tangential components on the equivalent surface are denoted as E^+ and H^+, respectively,

$$\begin{bmatrix} E^+ \\ \eta H^+ \end{bmatrix} = \begin{bmatrix} E^{\text{inc}} \times \hat{a}_n \\ \hat{a}_n \times \eta H^{\text{inc}} \end{bmatrix} \tag{9.55}$$

Here $\hat{a}_n$ is the outward unit normal of the equivalent surface.

In order to represent the fields E^{rad} and H^{rad} radiated in the region-v ($v = 1, 2$) by the surface equivalent electric currents J_s and magnetic currents M_s, an operator matrix $[S_v]$ is defined as

$$\begin{bmatrix} E^{\text{rad}} \times \hat{n} \\ \hat{n} \times \eta H^{\text{rad}} \end{bmatrix} = [S_v] \begin{bmatrix} M_s \\ \eta J_s \end{bmatrix} \tag{9.56}$$

in which,

$$[S_v] = \begin{bmatrix} \mathcal{K}_v^M & -\eta^{-1} \mu_v \mathcal{L}_v^J \\ \eta \varepsilon_v \mathcal{L}_v^M & \mathcal{K}_v^J \end{bmatrix} \tag{9.57}$$

$\mathcal{L}_v$ and $\mathcal{K}_v$ are the electric field integral operator and the magnetic field integral operator, respectively, as defined in [13], μ_v is the permeability, and ε_v represents permittivity in region-v. $v = 1$ denotes the uniform background, usually free space.

The rotated tangential fields imposed on the scatterer inside the closed Huygens surface can be considered as being generated by E^+ and H^+ propagating inward to the scatterer, which can be expressed as follows:

$$\begin{bmatrix} E^i \times \hat{n} \\ \hat{n} \times \eta H^i \end{bmatrix} = -[S_1] \begin{bmatrix} E^+ \\ \eta H^+ \end{bmatrix} \tag{9.58}$$

The equivalent electric currents J_s and magnetic currents M_s on the surface of the scatterer can be solved using the MOT procedure based on specific TD-IEs. They will generate the outgoing fields, whose rotated tangential components on S can be expressed by

$$\begin{bmatrix} E^- \\ \eta H^- \end{bmatrix} = [S_1] \begin{bmatrix} M_s \\ \eta J_s \end{bmatrix} \tag{9.59}$$

Based on the Huygens equivalence principle, the scattered fields E^{sca} and H^{sca} outside S can be obtained directly using E^- and H^- as sources. The TD-GTM of the module is defined on the equivalence surface S, which describes the relationship between the outgoing and the ingoing rotated tangential field variables E^+, H^+ and E^-, H^- in the time domain. Eq. (9.59) can be solved with various time-domain solvers according to the property of the scatterer. With the commonly used discretization and testing procedure, we can express the relationship between the discretized outgoing variables $e_{n,j}^-$, $h_{n,j}^-$ and the discretized incident variables $e_{n,j}^{\text{inc}}$, $h_{n,j}^{\text{inc}}$ by convolution,

$$\begin{bmatrix} e^- \\ \eta h^- \end{bmatrix}_j = \sum_{k=0}^{k_{\max}^c - 1} \mathbf{C}_k \begin{bmatrix} e^{\text{inc}} \\ \eta h^{\text{inc}} \end{bmatrix}_{j-k} + \mathbf{Q}_{j-k_{\max}^c}^c \tag{9.60}$$

in which,

$$Q^c_{j-k^c_{max}} = Q^c_{j-k^c_{max}-1} + C_{k^c_{max}} \begin{bmatrix} \mathbf{e}^{inc} \\ \eta \mathbf{h}^{inc} \end{bmatrix}_{j-k^c_{max}} \tag{9.61}$$

Here, $\mathbf{C}_k$ ($k = 0, 1, ..., k^c_{max}$) are defined as the time-domain generalized transition matrices [21], denoted as TD-GTM. The number k^c_{max} is the length of $\mathbf{C}_k$. To obtain the TD-GTM, we set $\mathbf{e}^{inc}_k$ and $\mathbf{h}^{inc}_k$ to be the delta-like impulses [21]:

$$\begin{bmatrix} \mathbf{e}^{inc} \\ \eta \mathbf{h}^{inc} \end{bmatrix}_k = \begin{cases} \boldsymbol{\delta}_m & k = 0 \\ 0 & \text{else} \end{cases}, \quad m = 1, 2, \cdots, N_{es} \tag{9.62}$$

where N_{es} is the number of unknowns on the equivalent surface, $\boldsymbol{\delta}_m$ is a column vector, and its entries are

$$\delta_m(n) = \begin{cases} 1 & n = m \\ 0 & \text{else} \end{cases}, \quad n = 1, 2, \cdots, N_{es} \tag{9.63}$$

The inner solver for the TD-GTM extraction should be resonance free, otherwise, k^c_{max} will be very large. To determine a reasonable k^c_{max}, the sample length ratio (r_{sl}) is introduced, which equals the ratio between $k^c_{max}\Delta t$ and the time that wave travels in the longest path inside S.

From its definition, it can be seen clearly that the TD-GTM is a kind of system response of the subdomain. It is defined on a reference surface containing the subdomain, consisting of the convolution coefficient matrices. The output fields of the system $\mathbf{E}^-, \mathbf{H}^-$ can be naturally obtained by the convolution of the system response and the input fields $\mathbf{E}^+, \mathbf{H}^+$. Obviously, a TD-GTM model comprises three parts: a virtual reference surface, a set of variables (rotated field components), and a TD-GTM defined on the reference surface.

9.8.2 TD-GSIE

After the TD-GTM models for all subdomains are created, TD-GSIE can be established based on TD-GTM models of subdomains to address the mutual couplings among them. Consider a multi-block structure as shown in Figure 9.15. It is separated into M subdomains, and each subdomain is enclosed by a regular, usually simple-shaped, equivalent surface. The interactions between the original subdomains are conveyed through their reference surfaces. The total fields imposed on the reference surface of the m-th module consist of two parts: the original incident fields and the coupling fields from other modules,

$$\begin{bmatrix} \mathbf{E}^+ \\ \eta \mathbf{H}^+ \end{bmatrix}_m = \begin{bmatrix} \mathbf{E}^{inc} \times \hat{a}_n \\ \hat{a}_n \times \eta \mathbf{H}^{inc} \end{bmatrix}_m + \sum_{n=1, n \neq m}^{M} [S_1]_{m,n} \begin{bmatrix} \mathbf{E}^- \\ \eta \mathbf{H}^- \end{bmatrix}_n \tag{9.64}$$

Figure 9.15 Total field imposed on the *m*-th block in a multi-module system.

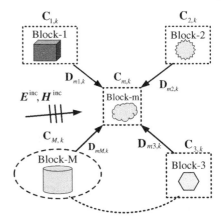

in which $[S_1]_{m,n}$ is the operator matrix that transmits the interaction from the *n*-th to the *m*-th module. After testing the fields on block-*m*, the convolution system can be cast into

$$\begin{bmatrix} \mathbf{e}^- \\ \eta\mathbf{h}^- \end{bmatrix}_{m,j} = \sum_{k=0}^{k_{max}^c - 1} \mathbf{C}_{m,k} \begin{bmatrix} \mathbf{e}^i \\ \eta\mathbf{h}^i \end{bmatrix}_{m,j-k} + \mathbf{Q}^c_{m,j-k_{max}^c} \tag{9.65}$$

in which

$$\begin{bmatrix} \mathbf{e}^i \\ \eta\mathbf{h}^i \end{bmatrix}_{m,j} = \begin{bmatrix} \mathbf{e}^{inc} \\ \eta\mathbf{h}^{inc} \end{bmatrix}_{m,j} + \sum_{m=1,m\neq n}^{M} \left\{ \sum_{k=0}^{k_{max}^{d,mn} - 1} \mathbf{D}_{mn,k} \begin{bmatrix} \mathbf{e}^- \\ \eta\mathbf{h}^- \end{bmatrix}_{n,j-k} + \mathbf{Q}^d_{mn,j-k_{max}^{d,mn}} \right\} \tag{9.66}$$

and,

$$\mathbf{Q}^c_{m,j-k_{max}^c} = \mathbf{Q}^c_{m,j-k_{max}^c} + \mathbf{C}_{m,k_{max}^c} \begin{bmatrix} \mathbf{e}^i \\ \eta\mathbf{h}^i \end{bmatrix}_{m,j-k_{max}^c} \tag{9.67}$$

$$\mathbf{Q}^d_{mn,j-k_{max}^{d,mn}} = \mathbf{Q}^d_{mn,j-k_{max}^{d,mn}-1} + \mathbf{D}_{mn,k_{max}^{d,mn}} \begin{bmatrix} \mathbf{e}^- \\ \eta\mathbf{h}^- \end{bmatrix}_{n,j-k_{max}^{d,mn}} \tag{9.68}$$

Here, $\mathbf{D}_{mn}$ is the discretized matrix of $[S_1]_{m,n}$ and $k_{max}^{d,mn}$ is its length. $\mathbf{Q}^c_{m,k}$ and $\mathbf{Q}^d_{mn,k}$ are the accumulating vectors due to the integration of time in the operators for each module and their interactions. The entries of $\mathbf{D}_{mn}$ have to be evaluated precisely so that the coupling interactions among different modules are transmitted accurately. It's also worthy to note that proper $\mathbf{D}_{mn}$ can be obtained even when $\mathbf{C}_m$ and $\mathbf{C}_n$ are discretized by different time steps, leading to a localized MOT system, which can provide higher flexibility.

9.8.3 Numerical Results

The transient scattering characteristics of a NASA almond array is analyzed using TD-GTM and TD-GSIE. The incident field is set to be the modulated Gaussian plane wave:

$$\boldsymbol{E}^{i}(\boldsymbol{r}, t) = \hat{\boldsymbol{p}} e^{-\gamma^{2}} \cos(2\pi f_{0}\tau) \qquad (9.69)$$

where $\gamma = (\tau - t_{p})/(\sqrt{2}\sigma)$, $\tau = t - (\boldsymbol{r} \cdot \hat{\boldsymbol{k}})/c$, $\sigma = 3/(2\pi f_{bw})$, and $t_{p} = 8\sigma$. f_{0} is the center frequency, f_{bw} is the half bandwidth. $\hat{\boldsymbol{k}}$ is the propagation direction and $\hat{\boldsymbol{p}}$ is the polarization direction. Analytical formulas (9.49) and (9.50) are utilized to evaluate the matrix entries. The time basis function is the triangle basis. Parameters for the incident wave are $f_{0} = 3$ GHz, $f_{bw} = 1$ GHz, $\hat{\boldsymbol{p}} = \hat{\boldsymbol{x}}$ and $\hat{\boldsymbol{k}} = \hat{\boldsymbol{z}}$. The length, width, and height of the almond are 0.1, 0.04, and 0.013 m, respectively. One almond is discretized by 1438 triangles with an average edge length of 0.0027 m, which is composed of 2157 RWG basis functions. For a PEC object, its natural surface can be chosen as the reference surface, and the TD-GTM can be defined to represent the relationship between the surface current and the incident electric field. In this example, the PEC almond is handled like a common scatterer enclosed by a cuboidal reference surface, the size of which is 0.105 m × 0.046 m × 0.016 m and meshed by 450 RWG basis functions with an average edge length of 0.011 m. The details of the structure are shown in Figure 9.16. We choose the time step of 0.015 ns, corresponding to space–time discretization ratio $\chi = 0.16$. The length of TD-GTM is 100, corresponding to sample length ratio $r_{sl} = 3.8$. Moreover, the MOT solver is established based on the TD-CFIE.

The single almond configuration is extended to a 3×3 almond array. The distance between adjacent almonds is 0.11 m in X direction and 0.06 m in Y direction as shown in Figure 9.16. It takes 10.7 hours to calculate the TD-GTM of one of the almond models and can be directly reused by other elements in the periodic array.

Figure 9.17a compares the RCS results of the single model at 3 GHz obtained by both the conventional MoM and the GTM method. The accuracy of the TD-GSIE is verified through the RCS comparison with the MoM method at the

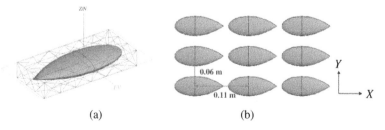

(a)   (b)

Figure 9.16 The configuration for the almond array. (a) Single model. (b) Array model.

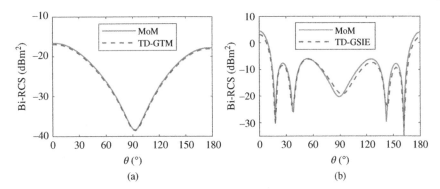

Figure 9.17 RCS results of the 3×3 almond array. (a) Single model. (b) Array model at 3 GHz (azimuth angle $\varphi = 0$).

center frequency 3 GHz in Figure 9.17b. The possible cause to the deviation is the truncation error of the coupling matrix between the equivalent surfaces. The number of unknowns in TD-GSIE is decreased to only 13.4% of the original problem in MoM. This greatly improves efficiency, while ensuring accuracy.

9.9 Conclusions

The late time stability properties of TD-IEs are related to their impulse responses and eigenvalue patterns. Analysis and numerical examples verified that the TD-EFIE is basically unstable because of the effect of the spurious static solenoidal currents and the interior resonance currents. The power conservation law supports that the TD-MFIE is basically stable. Although TD-CFIE has an eigenvalue at $(1 + i0)$, it is stable because the eigenvalue is caused by the numerical algorithm. These observations provide some clues to get stable TD-IE solutions.

The impulse response defined in this chapter basically serves as the system response. The output signal of any input signal can be obtained by the convolution of the input signal and the system response. The introduction of truncated impulse responses provides an effective tool to construct a time-domain equivalence model for a scatterer, the TD-GTM model. TD-GTM, in conjunction with TD-GSIE, formulates a promising time-domain DDM for solving the electromagnetic problems of multi-scale systems.

Some key techniques still need further investigating, such as how to efficiently generate clean impulse responses, how to truncate the impulse responses properly, and how to extract and express the electromagnetic properties of an object efficiently.

Author Biography

Gaobiao Xiao (M'08, SM'20) received the BS degree from Huazhong University of Science and Technology, Wuhan, China, in 1988, the MS degree from the National University of Defense Technology, Changsha, China, in 1991, and the PhD degree from Chiba University, Chiba, Japan, in 2002.

He has been a faculty member since 2004 in the Department of Electronic Engineering, Shanghai Jiao Tong University, Shanghai, China. His research interests are computational electromagnetics, coupled thermo-electromagnetic analysis, microwave filter designs, fibre-optic filter designs, phased arrays, and inverse scattering problems.

Yuyang Hu (S'19) received the BS degree in telecommunications engineering from Xidian University, Xi'an, China, in 2018. He is currently pursuing the PhD degree with the Key Laboratory of Ministry of Education of Design and Electromagnetic Compatibility of High Speed Electronic Systems, Shanghai Jiao Tong University, Shanghai.

His current research interests include computational electromagnetics and its application in scattering and radiation problems.

Xuezhe Tian received his BS degree from Harbin Engineering University in 2011 and the PhD degree in electrical engineering from the Shanghai Jiao Tong University in 2016. From August 2016 to March 2019, he has been working as a Post-Doctoral Researcher in the ElectroScience Laboratory of the Ohio State University. From April 2019 to October 2020, he was working as R&D Engineer at Lorentz Solution, Inc. He started to work as Principal Software Engineer at Cadence Design Systems, Inc. from October 2020.

His research interests are in scientific computing, focusing on the computational electromagnetics and its application in various electromagnetic engineering problems, especially for full-wave scattering/radiation problems, as well as for interconnect modelling on printed circuit board (PCB) and IC package designs.

Shifeng Huang (S'19) received the BS and MS degrees from Wuhan University, Wuhan, China, in 2014 and 2017, respectively. He is currently pursuing the PhD degree in electronic engineering with the Shanghai Jiao Tong University, Shanghai, China.

His current research interests include computational electromagnetics and its application in electromagnetic compatibility and scattering problems.

 **Rui Liu** received the BS degree from the Shanghai Jiao Tong University, Shanghai, China, in 2017. He is currently pursuing the PhD degree in electronic engineering with the Shanghai Jiao Tong University.

His research interests include computational electromagnetics and inverse scattering problems.

References

1 Rao, S.M. and Wilton, D.R. (1991). Transient scattering by conducting surfaces of arbitrary shape. *IEEE Transactions on Antennas and Propagation* 39 (1): 56–61.

2 Shanker, B., Ergin, A.A., Aygun, K., and Michielssen, E. (2000). Analysis of transient electromagnetic scattering from closed surfaces using a combined field integral equation. *IEEE Transactions on Antennas and Propagation* 48 (7): 1064–1074.

3 Yilmaz, A.E., Jin, J.M., and Michielssen, E. (2004). Time domain adaptive integral method for surface integral equations. *IEEE Transactions on Antennas and Propagation* 52 (10): 2692–2708.

4 Bagci, H., Yilmaz, A.E., Jin, J.M., and Michielssen, E. (2007). Fast and rigorous analysis of EMC/EMI phenomena on electrically large and complex cable-loaded structures. *IEEE Transactions on Electromagnetic Compatibility* 49 (2): 361–381.

5 Aygun, K., Fischer, B.C., Meng, J. et al. (2003). A fast hybrid field-circuit simulator for transient analysis of microwave circuits. *IEEE Transactions on Microwave Theory and Techniques* 52 (2): 573–583.

6 Tian, X., Xiao, G.B., and Xiang, S. (2016). Time-domain generalized transition matrix for transient scattering analysis of arrays. *IEEE Antennas and Wireless Propagation Letters* 15: 238–241.

7 Sadigh, A. and Arvas, E. (1993). Treating the instabilities in marching-on-in-time method from a different perspective. *IEEE Transactions on Antennas and Propagation* 41 (12): 1695–1702.

8 Pray, J., Nair, N V., and Shanker, B. (2012). Stability properties of the time domain electric field integral equation using a separable approximation for the convolution with the retarded potential. *IEEE Transactions on Antennas and Propagation* 60 (8): 3772–3781.

9 Walker, S.P., Bluck, M.J., and Chatzis, I. (2000). The stability of integral equation time domain computations for three-dimensional scattering; similarities and differences between electrodynamic and elastodynamic computations. *International Journal Numerical Modelling* 15: 459–474.

10 Adams, R.J. (2004). Physical and analytical properties of a stabilized electric field integral equation. *IEEE Transactions on Antennas and Propagation* 52 (2): 362–372.

11 Lu, M. and Michielssen, E. (2000). Closed form evaluation of time domain fields due to Rao–Wilton–Glisson sources for use in marching-on-in-time based EFIE solvers. In: *IEEE Antennas and Propagation Society International Symposium*, 74–77. IEEE.

12 Yücel, A.C. and Ergin, A.A. (2006). Exact evaluation of retarded-time potential integrals for the RWG bases. *IEEE Transactions on Antennas and Propagation* 54 (5): 1496–1502.

13 Shanker, B., Lu, M., Yuan, J., and Michielssen, E. (2009). Time domain integral equation analysis of scattering from composite bodies via exact evaluation of radiation fields. *IEEE Transactions on Antennas and Propagation* 57 (5): 1506–1520.

14 Shi, Y., Xia, M.-Y., Chen, R.-S. et al. (2011). Stable electric field TD-IE solvers via quasi-exact evaluation of MOT matrix elements. *IEEE Transactions on Antennas and Propagation* 59 (2): 574–585.

15 Ülkü, H.A. and Ergin, A.A. (2011). Application of analytical retarded-time potential expressions to the solution of time domain integral equations. *IEEE Transactions on Antennas and Propagation* 59 (11): 4123–4130.

16 Tian, X., Xiao, G.B., and Xiang, S. (2014). Application of analytical expressions for retarded-time potentials in analyzing the transient scattering by dielectric objects. *IEEE Antennas and Wireless Propagation Letters* 13: 1313–1316.

17 Peterson, F. (1990). The interior resonance problem associated with surface integral equations of electromagnetics: numerical consequences and a survey of remedies. *Electromagnetics* 10 (3): 293–312.

18 Chew, W.C. and Song, J.M. (2007). Gedanken experiments to understand the internal resonance problems of electromagnetic scattering. *Electromagnetics* 27 (8): 457–471.

19 Cools, K., Andriulli, F.P., Olyslager, F., and Michielssen, E. (2009). Time-domain Calderón identities and their application to the integral equations analysis of scattering by PEC objects part I: preconditioning. *IEEE Transactions on Antennas and Propagation* 57 (8): 2352–2364.

20 Andriulli, F.P., Cools, K., Olyslager, F., and Michielssen, E. (2009). Time-domain Calderón identities and their application to the integral equations analysis of scattering by PEC objects part II: stability. *IEEE Transactions on Antennas and Propagation* 57 (8): 2365–2375.

21 Xiao, G.B., Tian, X., Luo, W., and Fang, J. (2015). The impulse responses and the late time stability properties of time domain integral equations. *IET Microwaves, Antennas and Propagation* 9 (7): 603–610.

22 Shi, Y., Bagci, H., and Lu, M. (2014). On the static loop modes in the marchingon-in-time solution of the time-domain electric field integral equation. *IEEE Antennas and Wireless Propagation Letters* 13: 317–320.

23 Rao, S.M. (1999). Chapter 3, Infinite conducting cylinders. In: *Time Domain Electromagnetics*, Academic Press Series in Engineering (ed. J.D. Irwin), 88–90. San Diego, CA: Academic Press.

24 Dély, A., Andriulli, F.P., and Cools, K. (2020). Large time step and DC stable TD-EFIE discretized with implicit Runge–Kutta methods. *IEEE Transactions on Antennas and Propagation* 68 (2): 976–985.

25 Rynne, B.P. (1985). Stability and convergence of time marching methods in scattering problems. *IMA Journal of Applied Mathematics* 35: 297–310.

26 Rynne, B.P. and Smith, P.D. (1990). Stability of time marching algorithms for the electric field integral equation. *Journal of Electromagnetic Waves and Application.* 4 (12): 1181–1205.

27 Davies, P.J. and Duncan, D.B. (1997). Averaging techniques for time-marching schemes for retarded potential integral equations. *Applied Numerical Mathematics* 23: 291–310.

28 Graglia, R.D., Wilton, D.R., and Peterson, A.F. (1997). Higher order interpolatory vector bases for computational electromagnetics. *IEEE Transactions on Antennas and Propagation* 45 (3): 329–342.

29 Wildman, R.A., Pisharody, G., Weile, D.S. et al. (2004). An accurate scheme for the solution of the time-domain integral equations of electromagnetics using higher order vector bases and bandlimited extrapolation. *IEEE Transactions on Antennas and Propagation* 52 (11): 2973–2984.

30 Andriulli, F.P., Cools, K., Bagci, H. et al. (2008). A multiplicative Calderón preconditioner for the electric field integral equation. *IEEE Transactions on Antennas and Propagation* 56 (8): 2398–2412.

31 Stephanson, M.B. and Lee, J.-F. (2009). Preconditioned electric field integral equation using Calderón identities and dual loop/star basis functions. *IEEE Transactions on Antennas and Propagation* 57 (4): 1274–1279.

32 Valdés, F., Andriulli, F.P., Cools, K., and Michielssen, E. (2011). High-order div- and quasi curl-conforming basis functions for multiplicative Calderón preconditioning of the EFIE. *IEEE Transactions on Antennas and Propagation* 59 (4): 1321–1337.

33 Cools, K., Andriulli, F.P., and Michielssen, E. (2011). A Calderón multiplicative preconditioner for the PMCHWT integral equation. *IEEE Transactions on Antennas and Propagation* 59 (12): 4579–4587.

34 Inchauspe, P.E. and Hanckes, C.J. (2019). Fast Calderón preconditioning for the electric field integral equation. *IEEE Transactions on Antennas and Propagation* 67 (4): 2555–2564.

35 Buffa, A. and Christiansen, S. (2007). A dual finite element complex on the barycentric refinement. *Mathematics of Computation* 76 (260): 1743–1769.

36 Rao, S.M., Wilton, D.R., and Glisson, A.W. (1982). Electromagnetic scattering by surfaces of arbitrary shape. *IEEE Transactions on Antennas and Propagation* 30 (5): 409–418.

37 Tian, X. and Xiao, G.B. (2014). Time-domain augmented electric field integral equation for a robust marching on in time solver. *IET Microwaves, Antennas and Propagation* 8 (9): 688–694.

38 Pisharody, G. and Weile, D.S. (2005). Robust solution of time-domain integral equations using loop-tree decomposition and bandlimited extrapolation. *IEEE Transactions on Antennas and Propagation* 53 (6): 2089–2098.

39 Tian, X., Xiao, G.B., and Fang, J. (2015). Application of loop-flower basis functions in the time-domain electric field integral equation. *IEEE Transactions on Antennas and Propagation* 63 (3): 1178–1181.

40 Chung, Y.S., Sarkar, T.K., Jung, B.H. et al. (2004). Solution of time domain electric field integral equation using the Laguerre polynomials. *IEEE Transactions on Antennas and Propagation* 52 (9): 2319–2328.

41 Zhu, M., Sarkar, T.K., and Chen, H. (2019). A stabilized marching-on-in-degree scheme for the transient solution of the electric field integral equation. *IEEE Transactions on Antennas and Propagation* 67 (5): 3232–3240.

42 Collin, R.E. (1981). Rayleigh scattering and power conservation. *IEEE Transactions on Antennas and Propagation* 29 (5): 795–798.

43 Xiao, G.B., Mao, J., and Yuan, B. (2008). A generalized transition matrix for arbitrarily-shaped scatterers or scatterer groups. *IEEE Transactions on Antennas and Propagation* 56 (12): 3723–3732.

44 Li, M.K. and Chew, W.C. (2007). Wave-field interaction with complex structures using equivalence principle algorithm. *IEEE Transactions on Antennas and Propagation* 55 (1): 130–138.

45 Peng, Z., Wang, X.C., and Lee, J.F. (2011). Integral equation based domain decomposition method for solving electromagnetic wave scattering from non-penetrable objects. *IEEE Transactions on Antennas and Propagation* 59 (9): 3328–3338.

46 van de Water, A.M., de Hon, B.P., van Beurden, M.C. et al. (2005). Linear embedding via Green's operators: a modeling technique for finite electromagnetic band-gap structures. *Physical Review E* 72 (5): 056704-1-11.

47 Xiao, G.B., Mao, J., and Yuan, B. (2009). A generalized surface integral equation formulation for analysis of complex electromagnetic systems. *IEEE Transactions on Antennas and Propagation* 57 (3): 701–710.

48 Hong, L., Tian, G., Fang, J., and Xiao, G.B. (2013). The behavior of MFIE and EFIE at interior resonances and its impact in MOT late time stability. *American Journal of Electromagnetics and Applications* 1 (2): 30–37.

Part IV

Applications of Deep Learning in Time-Domain Methods

10

Time-Domain Electromagnetic Forward and Inverse Modeling Using a Differentiable Programming Platform

Yanyan Hu[1], Yuchen Jin[1], Xuqing Wu[2], and Jiefu Chen[1]

[1] *Department of Electrical and Computer Engineering, University of Houston, Houston, USA*
[2] *Department of Information and Logistics Technology, University of Houston, Houston, USA*

10.1 Introduction

High-fidelity simulations of electromagnetic wave propagation in complex environments by conventional methods such as finite-difference (FD) or finite-element (FE) can be computationally expensive on traditional platforms. This also leads to difficulty for applications that iterate around forward modeling, e.g. uncertainty quantification, optimization, and inversion. In recent years, deep learning approaches have been widely applied in forward and inverse modeling constrained by partial differential equations (PDEs)-constrained forward and inverse modeling. The occurrence of differentiable programming platforms equipped with automatic differentiation (AD) and graphics processing units (GPUs) allows the researchers and scientists to focus on theories, algorithms, and high-performance computing. The improvements in deep learning models including convolutional neural networks (CNNs) [1], recurrent neural networks (RNNs) [2], and generative adversarial networks (GANs) [3] also provide more perspectives to handle data with different properties.

For fast forward modeling using surrogates, there are typically four representative approaches to deploy the deep learning scheme: data-driven networks, physics-constrained networks, networks combined with traditional framework, and the augmented scientific models with machine-learnable structures. In general, data-driven methods rely on supervised learning and require sufficient simulation data on the discretized grids as the target. Long et al. [4] studied the feasibility of a data-driven approach and proposed a PDE–NET to learn evolution of PDEs from data. Sparse regression is adopted in [5] to discover the governing PDEs of a given system. A data-driven discretization method in [6] is introduced to systematically derive discretization for continuous physical

Advances in Time-Domain Computational Electromagnetic Methods, First Edition.
Edited by Qiang Ren, Su Yan, and Atef Z. Elsherbeni.

systems. Champion et al. [7] designed a custom deep autoencoder to discover a coordinate transformation into a reduced space so that the governing equations and the associated coordinate system can be learned simultaneously. Qi et al. [8] modified a multichannel end-to-end CNN to form the EM-net to predict the scattered field by the complex geometries under plane-wave illumination. Besides the aforementioned techniques to improve modeling performance for systems with known equations, deep learning approaches have been used to predict dynamics for complex systems with unknown equations [9, 10].

Currently, physics-constrained or theory-enhanced networks are becoming popular for their less dependency on a large amount of training data and more reliability [11]. Different from pure data-driven methods, the incorporated theories or physics make deep learning techniques more reasonable and interpretable. First, prior knowledge can be incorporated to increase the computation efficiency, e.g. learning universal linear embeddings of the nonlinear dynamics based on Koopman representations [12] and CNNs specially designed on unstructured grids for spherical signals [13]. Second, physical laws can be embedded into the network to provide guidance, for example, a novel loss function that guarantees divergence?free velocity fields is employed to optimize the generative model [14]. Third, the physical knowledge can be introduced by constraint learning, i.e. learning the models by minimizing the violation of the physical constraints [11, 15]. Physics-inform neural network (PINN) [16–18] encodes the governing equations, initial conditions, and boundary conditions as the constraints when training the network, turning an electromagnetic simulation problem into an optimization process.

Different from the aforementioned approaches, the third kind of method aims to adopt deep learning to optimize traditional iterative methods. Mishra [19] recast the standard numerical methods for time-dependent ordinary differential equations (ODEs) and PDEs as multilayer artificial neural networks with a set of trainable parameters. Hsieh et al. [20] learned to modify the updates of an existing iterative solver using a deep neural network and to achieve good generalization. The augmented scientific models start from the idea to exploit scientific structures as a modeling basis for machine learning. Chen et al. [21] proposed the neural ODEs and parameterizes the derivative of the hidden state using a neural network. Rackauckas et al. [22] designed the universal differential equations (UDEs), a mechanistic deterministic model with missing portions defined by some universal approximators (UAs) such as neural network, Chebyshev expansion, or a random forest. Through embedding UA into the differential equation models, the computational cost can be effectively reduced. Applications following this trend include control model estimation of global COVID-19 spread [23], guarantee on asymptotic stability of depth flows [24], Bayesian systems identification [25], etc.

Meanwhile, deep learning techniques have also been widely employed to solve inverse problems. Yao et al. [26] added a fully connected intermediate layer between the U-Net and adopted different channels to represent the real and imaginary parts of the data, to increase the network's capacity of extracting and transform features from the input to the output. Xu et al. [27] proposed a learning-based inversion approach to solve the inverse scattering problems (ISPs), where the physical information is processed in advance and taken as the inputs of the network, making the training easier and the prediction more accurate. Li et al. [28] built the connection between deep neural networks (DNNs) and the conventional iterative inverse algorithms and developed a cascaded "deepNIS" network with three DNN modules, where each module is equivalent to one iteration to update the dielectric distribution toward fine resolution. An induced current learning method (ICLM) is introduced in [29] by incorporating physical expertise inspired by traditional iterative algorithms into the architecture of CNNs. The ICLM defines a combined loss function with multiple labels in a cascaded end-to-end CNN (CEE–CNN) architecture to decrease the nonlinearity of the objective function. It also adopts the skip connections to let the CEE–CNN focus on learning the minor part of the induced current to accelerate the convergence speed. Supervised descent method (SDM) iteratively learns a set of descent directions in the offline training process and updates the models with the learned descent directions as well as data residuals in the prediction stage. It has been applied to magnetotellurics (MT) [30], logging-while-drilling (LWD) [31], and microwave inversion [32]. Moreover, the breakthrough of forward modeling also contributes to inversion. For example, PINNs are used in electrical impedance tomography (EIT) [33] and inversion in nano-optics and metamaterials [34].

With the development of differentiable programming platforms, the theory-guided RNN-based formulation of wave motion has been explored to solve scalar wave equation and the corresponding inverse problems on TensorFlow [35–37]. The performances are promising. Especially, only one single training data set is required when fulfilling the full waveform inversion due to the incorporated theory. However, in [35, 36], only scalar wave equation is solved and the efficiency is not mentioned. In this chapter, the RNN approach is adapted and used to solve the electromagnetic forward and inverse modeling problems on the state-of-the-art differentiable programming platform – PyTorch [38]. PyTorch is an ideal platform for implementing deep-learning-based inversion algorithms due to its performance-focused design and equipment of automatic differentiation [39, 40]. Meanwhile, a native formulation of forward modeling on PyTorch could take full advantage of a differentiable programming framework and avoid bottlenecks caused by interfacing with a third-party forward model during training. In this chapter, FDTD is implemented for the vectorial Maxwell's equations on the PyTorch platform and its performance is compared to a Matlab implementation. The advantage of the approach is also demonstrated in solving inverse problems. The automatic differentiation provided by PyTorch is much more accurate and efficient in computing the gradient than the finite-difference approximation. Meanwhile, existing optimizers of PyTorch can be conveniently applied.

The structure of this chapter is organized as follows. Section 10.2 introduces the detailed RNN-based formulation of wave propagation, where the results of forward simulation and accuracy and efficiency comparisons between PyTorch and Matlab will be demonstrated. In Section 10.3, gradient sensitivity analysis is performed to validate the superior performance of AD over the finite-difference method. Section 10.4 illustrates inversion of electromagnetic data fulfilled by network training. Four gradient-based optimization algorithms are adopted to solve the inverse problem; its performance is shown by the data misfit and normalized model misfit. Section 10.5 concludes the chapter.

10.2 RNN-Based Formulation of Wave Propagation

10.2.1 FDTD Method for Solving Maxwell's Equations

The differential form of Maxwell's equations in three dimensions [41] is formulated as

$$
\nabla \times \boldsymbol{H} = \frac{\partial \boldsymbol{D}}{\partial t} + \sigma \boldsymbol{E} + \boldsymbol{J}_{\text{source}}
$$

$$
\nabla \times \boldsymbol{E} = -\frac{\partial \boldsymbol{B}}{\partial t} \tag{10.1}
$$

where $\boldsymbol{E}$ is the electric field, $\boldsymbol{D} = \varepsilon \boldsymbol{E} = \varepsilon_r \varepsilon_0 \boldsymbol{E}$ is the electric flux density, $\boldsymbol{H}$ is the magnetic field, $\boldsymbol{B} = \mu \boldsymbol{H} = \mu_r \mu_0 \boldsymbol{H}$ is the magnetic flux density, σ is the electric conductivity, and $\boldsymbol{J}_{\text{source}}$ is the electric current density independent of the field energy, in which $\varepsilon, \varepsilon_r$, and ε_0 are the electrical, relative, and free-space permittivities, and μ, μ_r, μ_0 are the magnetic, relative, and free-space permeabilities, respectively.

Here, the transverse-magnetic mode with respect to z (TM_z) in two dimensions is considered, and the vector components of the curl operators of Eq. (10.1) in Cartesian coordinates can be written out. Equation (10.1) is simplified as

$$
\frac{\partial H_x}{\partial t} = -\frac{1}{\mu} \frac{\partial E_z}{\partial y}
$$

$$
\frac{\partial H_y}{\partial t} = \frac{1}{\mu} \frac{\partial E_z}{\partial x}
$$

$$
\frac{\partial E_z}{\partial t} = \frac{1}{\varepsilon} \left[\frac{\partial H_y}{\partial x} - \frac{\partial H_x}{\partial y} - (J_{\text{source}_z} + \sigma E_z) \right] \tag{10.2}
$$

The PDEs in Eq. (10.2) can be solved with the finite-difference method under Yee space lattice [42] for TM$_z$ mode. A space point in a uniform, rectangular lattice is denoted as $(i, j) = (i\Delta x, j\Delta y)$, and a superscript n is adopted to represent a discrete time point $n\Delta t$, where Δx, Δy is the lattice space increments in the x and y coordinate directions, Δt is the time increment, and i, j, n are all integers.

$$H_x^{n+1/2}\left(i, j + \frac{1}{2}\right) = CP\,(m) \cdot H_x^{n-1/2}\left(i, j + \frac{1}{2}\right)$$
$$- CQ\,(m) \cdot \frac{E_z^n\,(i, j + 1) - E_z^n(i, j)}{\Delta y}$$

$$H_y^{n+1/2}\left(i + \frac{1}{2}, j\right) = CP\,(m) \cdot H_y^{n-1/2}\left(i + \frac{1}{2}, j\right)$$
$$- CQ\,(m) \cdot \frac{E_z^n\,(i + 1, j) - E_z^n\,(i, j)}{\Delta x}$$

$$E_z^{n+1}\,(i, j) = CA\,(m) \cdot E_z^n\,(i, j)$$
$$+ CB\,(m) \cdot \left[\frac{H_y^{n+1/2}\left(i + \frac{1}{2}, j\right) - H_y^{n+1/2}\left(i - \frac{1}{2}, j\right)}{\Delta x} \right.$$
$$\left. - \frac{H_x^{n+1/2}\left(i, j + \frac{1}{2}\right) - H_x^{n+1/2}\left(i, j - \frac{1}{2}\right)}{\Delta y} - J_{\text{source}_z}^{n+1/2} \right] \quad (10.3)$$

where

$$CP\,(m) = \frac{\frac{\mu(m)}{\Delta t} - \frac{\sigma^*(m)}{2}}{\frac{\mu(m)}{\Delta t} + \frac{\sigma^*(m)}{2}}$$

$$CQ\,(m) = \frac{1}{\frac{\mu(m)}{\Delta t} + \frac{\sigma^*(m)}{2}}$$

$$CA\,(m) = \frac{\frac{\varepsilon(m)}{\Delta t} - \frac{\sigma(m)}{2}}{\frac{\varepsilon(m)}{\Delta t} + \frac{\sigma(m)}{2}}$$

$$CB\,(m) = \frac{1}{\frac{\varepsilon(m)}{\Delta t} + \frac{\sigma(m)}{2}} \quad (10.4)$$

are parameters related to the materials, and m is set as the same space index as the field at the left-hand side in Eq. (10.3).

According to Eq. (10.3), wave propagation is fulfilled by computing electric and magnetic fields at the next time step with the fields at current time step and injecting the source term. Due to its structural similarity to an RNN, the architecture of the RNN is exploited to implement the electromagnetic forward modeling.

10.2.2 RNN-Based Implementation

RNNs are a powerful model to deal with sequential data, especially for inherently dynamic processes such as speech. With a special structure of feedback connections that retain the internal memory (internal state), RNNs are able to make the output dependent on not only the external input but also on the previous computation and, therefore, capture the sequential information. This capability makes RNNs a great success in the application of machine translation, time series prediction, and speech and handwriting recognition. The typical compact and unfolded structure of RNNs with one hidden state is shown in Figure 10.1, where h_{t-1} is the hidden state at the previous time instant, s_t is the input at the current time instant, $\mathcal{A}s$ is a network that repeatedly take h_{t-1} and s_t as inputs and output d_t. RNNs come in many variants. For example, long short-term memory (LSTM) [43] is one of the most successful designs with the gating mechanism to handle values over arbitrary time intervals and the vanishing gradient problem.

To implement the FDTD scheme in Eq. (10.3), the single RNN cell $\mathcal{A}$ is designed as a finite-difference operator shown in Figure 10.2. Supposing the previous time instant is t, the corresponding input fields are $E_z(t)$, $H_x(t - \Delta t/2)$ and $H_y(t - \Delta t/2)$ since the electric field leads the magnetic fields by $\Delta t/2$ in time. The time index of the source term is the same as the magnetic fields. So given $s(t + \Delta t/2)$, the fields will be updated to time instant $t + \Delta t$ and the cell will output $d(t + \Delta t)$. $d(t + \Delta t)$ represents the union of the sampled electric and magnetic data, where δ_x, δ_y, and δ_z related to the receiver locations are introduced to perform as the gate factor as mentioned in LSTM. Note that the subscripts x, y, and z are used to distinguish the gate factors for the three fields. The spatial difference in Eq. (10.3) on the fields at time instant t is represented by differential operators ∇_x and ∇_y in Figure 10.2. Naturally, these differential operators can be calculated by horizontally or vertically convoluted with a differential operator $[-1, 1]$. However, in PyTorch, tensor slicing is more efficient in calculating ∇_x and ∇_y. In Figure 10.2, the FDTD scheme is deconstructed into

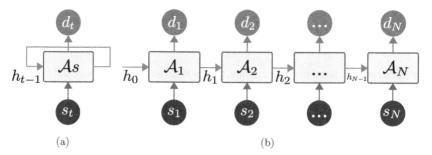

(a) (b)

Figure 10.1 RNN model demonstration. (a) Compact form. (b) Unfolded form.

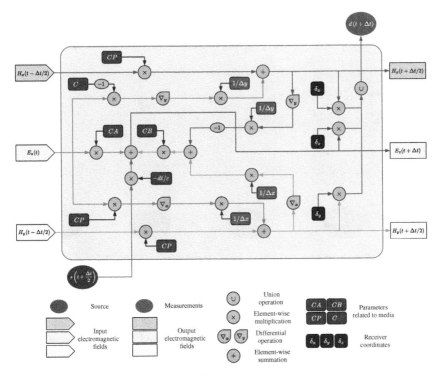

Figure 10.2 Single cell architecture of RNN model for simulating wave propagation using FDTD.

basic PyTorch operators with the material-related parameters (marked with dark gray background) fixed for forward simulation.

Figure 10.3 shows the 2D FDTD implementation using an RNN, where three hidden states are incorporated and updated in each cell using internal hidden states from the previous cell and external inputs. Note that the outputs of each cell only include the sampled $d(t)$. The fields as the hidden states are saved in the memory to compute the next time instant.

Further, the material-related parameters in Figure 10.2 can be set as trainable weights of the RNN model, so that the training of the network is equivalent to solving an inverse problem [36]. The details are illustrated in Section 10.4.

10.2.3 Experiments

10.2.3.1 Accuracy Comparison

For 2D forward simulation, the homogeneous lossless nonmagnetic material is considered and $\varepsilon_r = \mu_r = 1$. The computational domain is set as $x \in [0, 1.6]$, $y \in [0, 3.6]$. A Gaussian pulse with a central frequency $f_c = 600$ MHz is placed

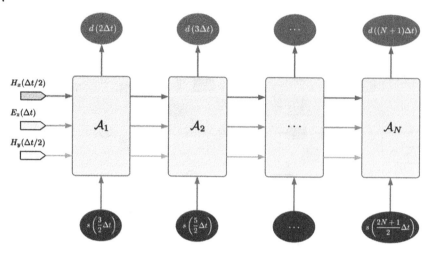

Figure 10.3 RNN model for simulating wave propagation using FDTD.

at $x_s = 0$, $y_s = 1.8$. Two receivers, Receiver 1 and Receiver 2, are placed at $x_{r1} = 0.8, y_{r1} = 0.6$ and $x_{r2} = 0.8, y_{r2} = 3$, respectively. $\Delta x = 0.01$ m, $\Delta y = 0.01$ m, $\Delta t = 10$ ps is chosen to satisfy the accuracy requirement and stability condition. Uniaxial perfectly matched layer (UPML) [44] is used. The number of grids for PML is 20 on the top, bottom, left, and right sides of the computational domain. At each time point, the total computation points are 201×401 for E_z, 201×400 for H_x, and 200×401 for H_y. Compared to the Matlab implementation of FDTD, excellent agreements with simulation results using RNN/PyTorch can be observed in Figure 10.4 (Meep unit system in [45] is adopted to scale the electric and magnetic data to the same level).

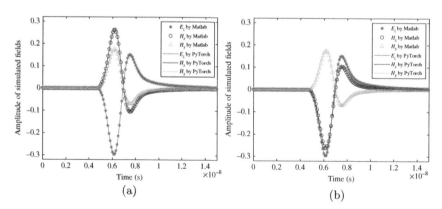

Figure 10.4 Simulation results by PyTorch and Matlab. (a) From Receiver 1. (b) From Receiver 2.

Table 10.1 Computational costs of computing 20,000 time steps on PyTorch and Matlab with different problem scales.

Problem scale	PyTorch	Matlab
200×200	18.2 s	22.4 s
200×400	21.6 s	41.3 s
400×400	25.8 s	79.1 s
800×800	87.5 s	388.7 s
1600×1600	381.1 s	1844.9 s

Note that due to the adoption of UPML, electric flux density $\mathbf{D}$ and magnetic flux density $\mathbf{B}$ are incorporated as intermediate variables when computing the electric and magnetic fields. To explain the structure of the RNN model clearly and concisely, these two intermediate variables are not included in Figures 10.2 and 10.3, but a specific formulation can be found in [41].

10.2.3.2 Efficiency Comparison

In Table 10.1, the computational cost of calculating 20,000 time steps on PyTorch and on Matlab is compared, with different problem scales defined by the number of computation points along the x and y axes. Table 10.1 shows that the computational cost with PyTorch increases much slower than Matlab when the complexity of the problem scales up. Due to the adoption of Yee algorithm for higher accuracy, the fields have to be indexed to perform the difference operation. Compared to Matlab, indexing is much more efficiently handled by PyTorch.

Overall, PyTorch is a computationally competitive platform for the equivalent FDTD implementation. With its AD support, it has a wide range of applications that rely on electromagnetic forward modeling, such as uncertainty quantification, optimization, and inverse problems.

10.3 Gradient Sensitivity Analysis

Accurate and efficient gradient estimation largely determines the performance of gradient-based optimization algorithms [46]. In general, the finite-difference method approximates the gradient by introducing a perturbation and takes the electromagnetic solver as a black box [47]. However, a too large perturbation will lead to truncation error, while a too small perturbation will induce cancellation error. Neither can get an accurate gradient estimation [48]. In addition, it is computationally expensive since the number of calls of the forward simulation is proportional to the number of model parameters. This can be quite costly especially when handling problems with complex configurations and high-dimensional model parameters.

The automatic differentiation technique calculates the analytical derivatives without hand-coding the derivative computation. It performs a nonstandard interpretation of a given computer program to propagate derivatives per the chain rule of differential calculus [49]. Fast implementation of AD has been achieved on differentiable programming platforms such as PyTorch, TensorFlow, and Julia [50]. When solving the inverse problems based on gradient descent algorithms, AD can be exploited to calculate the gradient during backpropagation.

To compare the accuracy and efficiency of AD and the finite-difference method to compute the gradient, we set up an inverse problem and generate a model with certain parameter distribution as shown in Figure 10.5, where part (a) is the true distribution of relative permittivity and part (b) is the initial model with uniform background. The computation domain is set as $x \in [0, 0.6]$, $y \in [0, 0.6]$, the inversion domain is set as $x \in [0.1, 0.5]$, $y \in [0.1, 0.5]$, and two scatters with circular and elliptical shapes are located inside the inversion domain. The corresponding relative permittivity of the background, the non-scatter region inside the inversion domain, and the scatter are 1, 2, and 4, respectively. Four Gaussian pulses (triangles in Figure 10.5) with central frequency $f_c = 600$ MHz and eight receivers (circles in Figure 10.5) are adopted. For each receiver, we sample $N_t = 1500$ time steps. Similarly, UPML is used as the boundary condition.

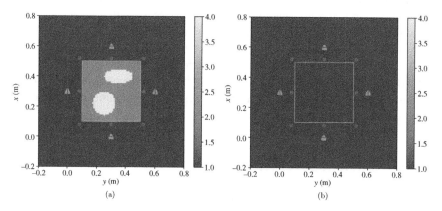

Figure 10.5 The true model. (a) The true distribution of permittivity. (b) The initial model.

Figure 10.6 Gradient computed by AD.

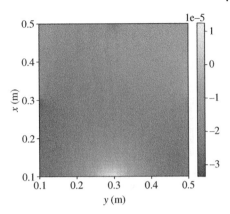

For forward simulation, $\Delta x = 0.01$ m, $\Delta y = 0.01$ m, and $\Delta t = 10$ ps are chosen to satisfy the accuracy requirement and stability condition. The same grids are used in the inversion.

The objective data is synthesized based on Figure 10.5a, and the gradient is computed at the initial model (b). The results are demonstrated in Figure 10.6, Figure 10.7, and Table 10.2. Figure 10.6 shows the gradient computed by AD with respect to the inversion domain. Figure 10.7 shows the corresponding gradients computed by the finite-difference method with different perturbations (Figure 10.7a,c,e), as well as the differences between the gradient computed by AD (Figure 10.7b,d,f). From Figure 10.7, the gradient computed by the finite-difference method is relatively close to that computed by AD when the perturbation $\delta = 0.1$. However, when $\delta = 10$, or $\delta = 0.001$, both of the differences increase due to the aforementioned truncation error or cancellation error, leading to an inaccurate approximation of the gradient. Moreover, according to Table 10.2, the time consumption for computing the gradient for once with AD is only about 2.5 seconds, while with finite-difference method it is about 2000 seconds because of the repetitive callings of forward modeling.

Admittedly, the adjoint state method can be used to compute the gradient with high efficiency [51]. It has been proven effective in plenty of fields, such as geophysical applications [52]. The occurrence of AD is not to replace the adjoint method. Instead, it can probably be combined with the adjoint method to further accelerate the computation. For example, AD is used to substantially ease the development of the adjoint computational fluid dynamics (CFD) code in [53]. However, more exploration is still needed.

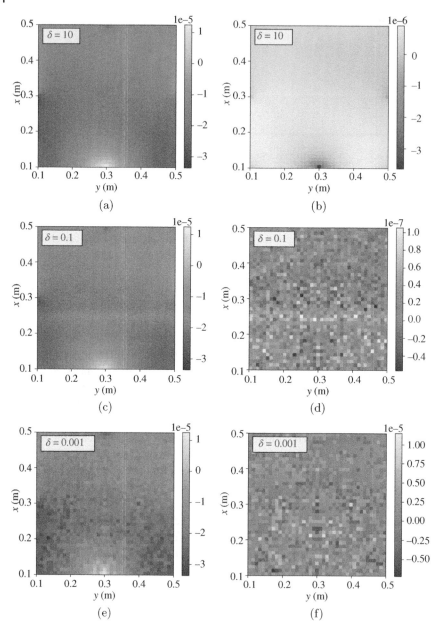

Figure 10.7 Gradients computed by the finite-difference method with different perturbation δ and their differences between the gradient computed by AD. (a), (c), and (e) are the gradients corresponding to $\delta = 10$, $\delta = 0.1$, and $\delta = 0.001$, respectively; (b), (d), and (f) are the differences corresponding to $\delta = 10$, $\delta = 0.1$, and $\delta = 0.001$, respectively.

Table 10.2 Computational costs of gradient computation with AD and the finite-difference method.

	AD	Finite-difference
Time consumption	2.5 s	2000 s

10.4 Electromagnetic Data Inversion Fulfilled by Network Training

10.4.1 RNN Training and Inversion

The training of a specially designed RNN for wave simulation has been proven to be equivalent to gradient-based full waveform inversion in [36]. Similarly, 2D time-domain electromagnetic inversion can be acquired through training the RNN designed in Section 10.2 by choosing the material-related parameters, e.g. the relative permittivity, as the trainable weights and minimizing the loss function $\mathcal{L}$ defined by the mean square error (MSE) between the output d_t of the network and the training data, i.e. the observed data. Specifically,

$$
\mathcal{L} = \frac{\alpha}{N_s N_r (N_t + 1)} \sum_{s=1}^{N_s} \sum_{r=1}^{N_r} \sum_{t=1}^{N_t+1} \| d_{E_z}(s, r, t) - \hat{d}_{E_z}(s, r, t) \|_2^2
$$

$$
+ \frac{\beta}{N_s N_r N_t} \sum_{s=1}^{N_s} \sum_{r=1}^{N_r} \sum_{t=1}^{N_t} \| d_{H_x}(s, r, t) - \hat{d}_{H_x}(s, r, t) \|_2^2
$$

$$
+ \frac{\gamma}{N_s N_r N_t} \sum_{s=1}^{N_s} \sum_{r=1}^{N_r} \sum_{t=1}^{N_t} \| dH_y(s, r, t) - \hat{d}_{H_y}(s, r, t) \|_2^2 \tag{10.5}
$$

where $d_{E_z}(s, r, t)$, $d_{H_x}(s, r, t)$, and $d_{H_y}(s, r, t)$ are the observed data; $\hat{d}_{E_z}(s, r, t)$, $\hat{d}_{H_x}(s, r, t)$, and $\hat{d}_{H_y}(s, r, t)$ are sampled from forward model simulations; α, β, and γ are the weighting coefficients to balance between the electric and magnetic data; N_s and N_r represent the number of sources and receivers, respectively.

With the theory-guided structure equivalent to FDTD, the RNN is well-constrained to sufficiently perform inversion using only a single training data set. Moreover, backpropagation is significantly accelerated by the adoption of automatic differentiation. Admittedly, a large memory is required to store the entire fields, i.e. all the internal states of each RNN cell. Besides, multiple optimizers are provided on PyTorch. We will illustrate four optimizers and demonstrate their effectiveness.

Gradient-based optimization methods are widely used in the applications to leverage machine learning techniques. The basic gradient descent (GD) algorithm sometimes has a slow convergence rate, and a series of stochastic optimization algorithms have been adopted to improve the efficiency [54]. Momentum or gradient descent with momentum [55] computes an exponentially weighted average

[56] of the gradients, smooths out the gradient "oscillation," and, therefore, leads to faster convergence. Adaptive subgradient (AdaGrad) method [57] adapts the learning rate with past gradients in each iteration for every single parameter, leading to good performance for sparse data and improved robustness. Root-mean-square gradient (RMSprop) [58] algorithm can further avoid AdaGrad's diminishing learning rates via dividing the learning rate by an exponentially decaying average of the squared gradients. Adaptive moment (Adam) [59] estimation combines the idea of Momentum and RMSprop by retaining both exponentially decaying average of the past gradients and squared gradients.

To measure the accuracy of inversion, the normalized misfit between the reconstructed parameter model $\hat{M}$ and the true model M is defined as

$$rms_M = \frac{\|\hat{M} - M\|_F^2}{\|M\|_F^2} \tag{10.6}$$

where F is the Frobenius norm.

10.4.2 RNN Training with Different Optimization Algorithms

Hyperparameter tuning is usually the key in machine learning or deep learning methods and is problem-dependent. A large number of trials are needed to determine these hyperparameters. Here, the choice of "learning rate" with different optimization methods to solve the inverse problem is explored. Note that the range of the hyperparameter may be different from the previous experience. For example, the learning rate usually lies between 0 and 1 when every layer of the network is well normalized. However, the learning rate here is physics-dependent and can exceed 1.

To demonstrate the feasibility of inversion under the RNN structure on PyTorch, the same configuration described by Figure 10.5 is adopted. First, the training data set is synthesized based on Figure 10.5a, and then the RNN is trained with four optimization methods, GD, Momentum, RMSprop, and Adam starting from Figure 10.5b, each with four possible learning rates to find the reasonable range. In the experiments, α, β, and γ are set as 1 because the electric and magnetic data are at the same level due to the adoption of the Meep unit system.

10.4.2.1 Gradient Descent
Inversion results using gradient descent method are shown in Figure 10.8, where parts (a)–(d) are the inverted distribution of the relative permittivity, and (e) shows the loss, i.e. the MSE computed according to Eq. (10.5), versus the iteration number. The learning rates are set as 100, 200, 280, and 300. The maximum iteration number is 100. Table 10.3 shows the corresponding normalized model misfit. Figure 10.8e indicates that the loss starts to oscillate when learning rate is between 280 and 300. Among the oscillation-free cases, the loss decreases faster and reaches a lower value when the learning rate is 200, where a relatively closer inversion result to the true model can be seen from Figure 10.8b and a relatively smaller normalized model misfit. For this inverse

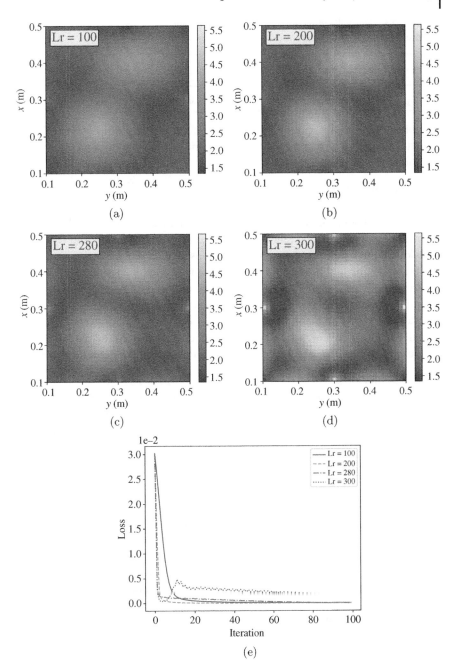

Figure 10.8 Inversion results based on the gradient descent method with different learning rates. (a), (b), (c), and (d) correspond to the learning rate of 100, 200, 280, and 300, respectively. (e) Loss value versus iteration number for each learning rate.

Table 10.3 Normalized model misfit with the gradient descent method using different leaning rates.

Learning rate	100	200	280	300
rms_M	0.173	0.155	0.151	0.248

problem, the performance of optimizing with the gradient descent algorithm is already not bad. The feasibility of other optimization algorithms will be explored in the following parts.

10.4.2.2 Momentum

Inversion results with Momentum method are shown in Figure 10.9 and Table 10.4, where Figure 10.9a–d are the inverted distribution of the relative permittivity with learning rate as 10, 20, 30, and 40, respectively, and (e) shows the loss. From Figure 10.9e, it seems that the oscillation of the inversion loss from Momentum method is severer than that from the gradient method, conflicting with previous analysis. This is because there is only one single training data set in the experiment, not the comment case in machine learning training. Therefore, the advantages of Momentum method cannot be reflected. However, a relatively small loss after convergence and relatively accurate reconstructed model still can be seen in Figure 10.9, which validates the effectiveness of Momentum algorithm.

10.4.2.3 RMSprop

Figure 10.10 and Table 10.5 show the inversion results with RMSprop algorithm with the learning rates set as 0.1, 0.2, 0.3, and 0.4, respectively. From the inverted model, the loss, and the normalized model misfit, RMSprop algorithm can obtain a good inversion when the learning rate is reasonable.

10.4.2.4 Adam

The inversion results using Adam algorithm are demonstrated by Figure 10.11 and Table 10.6 with the learning rates set as 0.25, 0.3, 0.4, and 0.6, respectively. Even though the losses oscillate at the beginning and the normalized model misfits are not the lowest compared with other algorithms, Adam can still achieve a relatively good inversion for the inverse problem.

These trials over different learning rates can provide an intuitive sense of the parameter range for each optimization method. At the same time, these trials validate the effectiveness of the aforementioned algorithms to be conveniently used to solve inverse problems based on the RNN model.

Table 10.4 Normalized model misfit with the Momentum method using different leaning rates.

Learning rate	10	20	30	40
rms_M	0.222	0.177	0.163	0.160

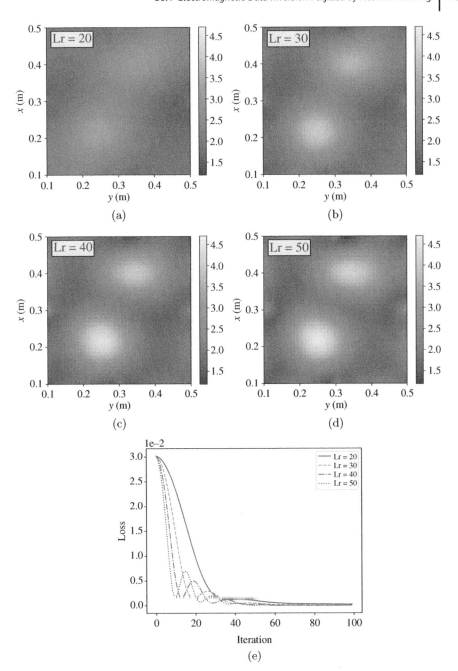

Figure 10.9 Inversion results based on Momentum method with different learning rates. (a), (b), (c), and (d) correspond to learning rates as 20, 30, 40, and 50, respectively. (e) Loss value versus iteration number for each learning rate.

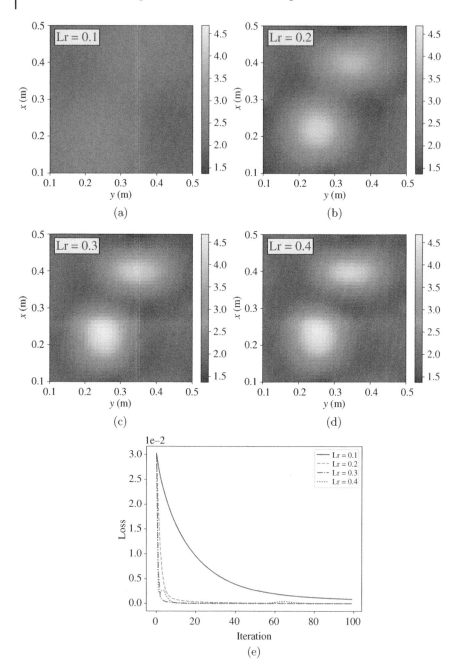

Figure 10.10 Inversion results based on RMSprop method with different learning rates. (a), (b), (c), and (d) correspond to learning rates as 0.1, 0.2, 0.3, and 0.4, respectively. (e) Loss value versus iteration number for each learning rate.

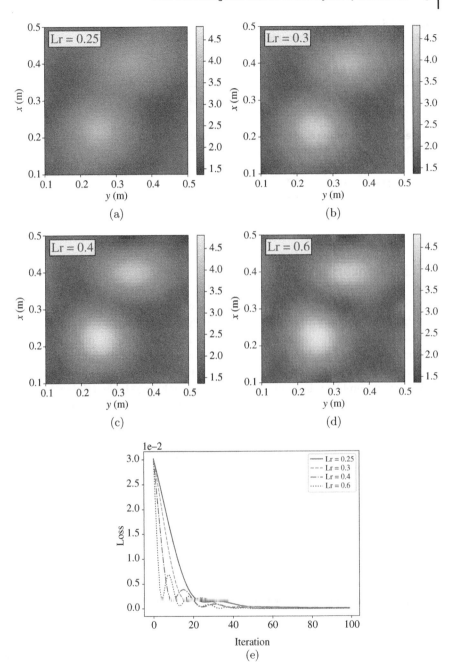

Figure 10.11 Inversion results based on the Adam method with different learning rates. (a), (b), (c), and (d) correspond to learning rates as 0.25, 0.3, 0.4, and 0.6, respectively. (e) Loss value versus iteration number for each learning rate.

Table 10.5 Normalized model misfit with the RMSprop method using different leaning rates.

Learning rate	0.1	0.2	0.3	0.4
rms_M	0.278	0.165	0.150	0.147

Table 10.6 Normalized model misfit with the Adam method using different leaning rates.

Learning rate	0.25	0.3	0.4	0.6
rms_M	0.187	0.167	0.157	0.153

10.5 Conclusion

In this work, an RNN-based formulation of electromagnetic wave propagation on the differentiable programming platform PyTorch is described. It can achieve accurate electromagnetic simulations and outperform Matlab over efficiency. Meanwhile, RNN-based implementation endows this method with the potential to further leverage the advantages of differentiable programming platforms and solve inverse problems by network training with only one single training data set. Multiple optimizers can be directly employed and provide good performance. The incorporated automatic differentiation greatly accelerates accuracy and efficiency of gradient computation, which can be further explored to combine with the adjoint method to achieve even faster computation. In the future, the fast forward modeling provides possibility to explore problems such as uncertainty quantification, optimization, and inversion that iterate around forward modeling. Current RNN-based inversion frameworks can also be extended to other applications such as ground-penetrating radar (GPR).

References

1 Lawrence, S., Giles, C.L., Tsoi, A.C., and Back, A.D. (1997). Face recognition: a convolutional neural-network approach. *IEEE Transactions on Neural Networks* 8 (1): 98–113.

2 Mandic, D.P. (2001). *Recurrent Neural Networks for Prediction: Learning Algorithms, Architectures, and Stability, Wiley Series in Adaptive and Learning Systems for Signal Processing, Communications, and Control*. Chichester: John Wiley. ISBN 0471495174.

3 Radford, A., Metz, L., and Chintala, S. (2015). Unsupervised representation learning with deep convolutional generative adversarial networks. arXiv preprint arXiv:1511.06434.

4 Long, Z., Lu, Y., and Dong, B. (2019). PDE-Net 2.0: Learning PDEs from data with a numeric-symbolic hybrid deep network. *Journal of Computational Physics* 399: 108925.

5 Rudy, S.H., Brunton, S.L., Proctor, J.L., and Kutz, J.N. (2016). Data-driven discovery of partial differential equations. *Science Advances* 3 (4): e1602614.

6 Bar-Sinai, Y., Hoyer, S., Hickey, J., and Brenner, M.P. (2019). Learning data-driven discretizations for partial differential equations. *Proceedings of the National Academy of Sciences of the United States of America* 116 (31): 15344–15349. http://dx.doi.org/10.1073/pnas.1814058116.

7 Champion, K., Lusch, B., Kutz, J.N., and Brunton, S.L. (2019). Data-driven discovery of coordinates and governing equations. *Proceedings of the National Academy of Sciences of the United States of America* 116 (45): 22445–22451. http://dx.doi.org/10.1073/pnas.1906995116.

8 Qi, S., Wang, Y., Li, Y. et al. (2020). Two-dimensional electromagnetic solver based on deep learning technique. *IEEE Journal on Multiscale and Multiphysics Computational Techniques* 5: 83–88.

9 Mardt, A., Pasquali, L., Wu, H., and Noé, F. (2018). VAMPnets for deep learning of molecular kinetics. *Nature Communications* 9 (1). http://dx.doi.org/10.1038/s41467-017-02388-1.

10 Vlachas, P.R., Byeon, W., Wan, Z.Y. et al. (2018). Data-driven forecasting of high-dimensional chaotic systems with long short-term memory networks. *Proceedings of the Royal Society A: Mathematical, Physical and Engineering Sciences* 474 (2213): 20170844. http://dx.doi.org/10.1098/rspa.2017.0844.

11 Zhu, Y., Zabaras, N., Koutsourelakis, P.-S., and Perdikaris, P. (2019). Physics-constrained deep learning for high-dimensional surrogate modeling and uncertainty quantification without labeled data. *Journal of Computational Physics* 394: 56–81. http://dx.doi.org/10.1016/j.jcp.2019.05.024.

12 Lusch, B., Kutz, J.N., and Brunton, S.L. (2018). Deep learning for universal linear embeddings of nonlinear dynamics. *Nature Communications* 9 (1). http://dx.doi.org/10.1038/s41467-018-07210-0.

13 Jiang, C.M., Huang, J., Kashinath, K. et al. (2019). Spherical CNNs on unstructured grids. *International Conference on Learning Representations*. https://openreview.net/forum?id=Bkl-43C9FQ (accessed 1 June 2022).

14 Kim, B., Azevedo, V.C., Thuerey, N. et al. (2019). Deep fluids: a generative network for parameterized fluid simulations. *Computer Graphics Forum* 38 (2): 59–70. http://dx.doi.org/10.1111/cgf.13619.

15 Stewart, R. and Ermon, S. (2017). Label-free supervision of neural networks with physics and domain knowledge. Thirty-First AAAI Conference on Artificial Intelligence.

16 Dissanayake, M.W.M.G. and Phan-Thien, N. (1994). Neural-network-based approximations for solving partial differential equations. *Communications in Numerical Methods in Engineering* 10 (3): 195–201.

17 Raissi, M., Perdikaris, P., and Karniadakis, G.E. (2019). Physics-informed neural networks: a deep learning framework for solving forward and inverse problems involving nonlinear partial differential equations. *Journal of Computational Physics* 378: 686–707.

18 Zhang, P., Hu, Y., Jin, Y. et al. (2021). A Maxwell's equations based deep learning method for time domain electromagnetic simulations. *IEEE Journal on Multiscale and Multiphysics Computational Techniques* 6: 35–40. http://dx.doi.org/10.1109/JMMCT.2021.3057793.

19 Mishra, S. (2018). A machine learning framework for data driven acceleration of computations of differential equations. arXiv preprint arXiv:1807.09519.

20 Hsieh, J.-T., Zhao, S., Eismann, S. et al. (2019). Learning neural PDE solvers with convergence guarantees. arXiv preprint arXiv:1906.01200.

21 Chen, R.T.Q., Rubanova, Y., Bettencourt, J., and Duvenaud, D. (2018). Neural ordinary differential equations. *Advances in Neural Information Processing Systems* 31.

22 Rackauckas, C., Ma, Y., Martensen, J. et al. (2020). Universal differential equations for scientific machine learning. arXiv preprint arXiv:2001.04385.

23 Dandekar, R. and Barbastathis, G. (2020). Neural network aided quarantine control model estimation of global COVID-19 spread. arXiv preprint arXiv:2004.02752.

24 Massaroli, S., Poli, M., Bin, M. et al. (2020). Stable neural flows. arXiv preprint arXiv:2003.08063.

25 Yang, Y., Bhouri, M.A., and Perdikaris, P. (2020). Bayesian differential programming for robust systems identification under uncertainty. *Proceedings of the Royal Society A* 476 (2243): 20200290.

26 Yao, H.M., Jiang, L., and Sha, W.E.I. (2020). Enhanced deep learning approach based on the deep convolutional encoder-decoder architecture for electromagnetic inverse scattering problems. *IEEE Antennas and Wireless Propagation Letters* 19 (7): 1211–1215.

27 Xu, K., Wu, L., Ye, X., and Chen, X. (2020). Deep learning-based inversion methods for solving inverse scattering problems with phaseless data. *IEEE Transactions on Antennas and Propagation*. 68 (11): 7457–7470.

28 Li, L., Wang, L.G., Teixeira, F.L. et al. (2019). DeepNIS: Deep neural network for nonlinear electromagnetic inverse scattering. *IEEE Transactions on Antennas and Propagation* 67 (3): 1819–1825. http://dx.doi.org/10.1109/tap.2018.2885437.

29 Wei, Z. and Chen, X. (2019). Physics-inspired convolutional neural network for solving full-wave inverse scattering problems. *IEEE Transactions on Antennas and Propagation* 67 (9): 6138–6148.

30 Guo, R., Li, M., Yang, F. et al. (2020). Application of supervised descent method for 2D magnetotelluric data inversion. *Geophysics* 85 (4): WA53–WA65. http://dx.doi.org/10.1190/geo2019-0409.1.

31 Hu, Y., Guo, R., Jin, Y. et al. (2020). A supervised descent learning technique for solving directional electromagnetic logging-while-drilling inverse problems. *IEEE Transactions on Geoscience and Remote Sensing.* 58 (11): 8013–8025.

32 Guo, R., Jia, Z., Song, X. et al. (2020). Pixel-and model-based microwave inversion with supervised descent method for dielectric targets. *IEEE Transactions on Antennas and Propagation.* 68 (12): 8114–8126 1.

33 Bar, L. and Sochen, N. (2019). Unsupervised deep learning algorithm for PDE-based forward and inverse problems. arXiv preprint arXiv:1904.05417.

34 Chen, Y., Lu, L., Karniadakis, G.E., and Dal Negro, L. (2020). Physics-informed neural networks for inverse problems in nano-optics and metamaterials. *Optics Express* 28 (8): 11618. http://dx.doi.org/10.1364/oe.384875.

35 Richardson, A. (2018). Seismic full-waveform inversion using deep learning tools and techniques. *arXiv e-prints*, art. arXiv:1801.07232, January 2018.

36 Sun, J., Niu, Z., Innanen, K.A. et al. (2020). A theory-guided deep-learning formulation and optimization of seismic waveform inversion. *Geophysics* 85 (2): R87–R99.

37 Abadi, M., Barham, P., Chen, J. et al. (2016). TensorFlow: A system for large-scale machine learning. *12th USENIX Symposium on Operating Systems Design and Implementation (OSDI 16)*, pp. 265–283. https://www.usenix.org/system/files/conference/osdi16/osdi16-abadi.pdf (accessed 1 June 2022).

38 Ketkar, N. (2017). *Introduction to PyTorch*, 195–208. Berkeley, CA: Apress. ISBN 978-1-4842-2766-4. http://dx.doi.org/10.1007/978-1-4842-2766-4_12.

39 Paszke, A., Gross, S., Massa, F. et al. (2019). PyTorch: An imperative style, high-performance deep learning library. *Advances in Neural Information Processing Systems* 32.

40 Duarte, V., Duarte, D., Fonseca, J., and Montecinos, A. (2020). Benchmarking machine-learning software and hardware for quantitative economics. *Journal of Economic Dynamics and Control* 111: 103796. https://doi.org/10.1016/j.jedc.2019.103796.

41 Taflove, A. and Hagness, S.C. (2005). *Computational Electrodynamics: The Finite-difference Time-domain Method*, Artech House antennas and propagation library. 3e. Boston, MA: Artech House.

42 Yee, K. (1966). Numerical solution of initial boundary value problems involving Maxwell's equations in isotropic media. *IEEE Transactions on Antennas and Propagation* 14 (3): 302–307.

43 Hochreiter, S. and Schmidhuber, J. (1997). Long short-term memory. *Neural Computation* 9 (8): 1735–1780. http://dx.doi.org/10.1162/neco.1997.9.8.1735.

44 Gedney, S.D. (1996). An anisotropic perfectly matched layer-absorbing medium for the truncation of FDTD lattices. *IEEE Transactions on Antennas and Propagation* 44 (12): 1630–1639.

45 Oskooi, A.F., Roundy, D., Ibanescu, M. et al. (2010). MEEP: A flexible free-software package for electromagnetic simulations by the FDTD method. *Computer Physics Communications* 181 (3): 687–702. https://doi.org/10.1016/j.cpc.2009.11.008.

46 Haslinger, J. (2003). *Introduction to shape optimization theory, approximation, and computation*. Advances in design and control; 7. Society for Industrial and Applied Mathematics SIAM, 3600 Market Street, Floor 6, Philadelphia, PA 19104, Philadelphia, PA. ISBN 1-68015-788-4.

47 Toivanen, J.I., Makinen, R.A.E., Jarvenpaa, S. et al. (2009). Electromagnetic sensitivity analysis and shape optimization using method of moments and automatic differentiation. *IEEE Transactions on Antennas and Propagation* 57 (1): 168–175.

48 Jerrell, M. (1997). Automatic differentiation and interval arithmetic for estimation of disequilibrium models. *Computational Economics* 10: 295–316. http://dx.doi.org/10.1023/A:1008633613243.

49 Baydin, A.G., Pearlmutter, B.A., Radul, A.A., and Siskind, J.M. (2015). Automatic differentiation in machine learning: a survey. *Journal of Marchine Learning Research* 18 (2018): 1–43.

50 Bezanson, J., Edelman, A., Karpinski, S., and Shah, V.B. (2017). Julia: A fresh approach to numerical computing. *SIAM Review* 59 (1): 65–98. http://dx.doi.org/10.1137/141000671.

51 Pollini, N., Lavan, O., and Amir, O. (2017). Adjoint sensitivity analysis and optimization of hysteretic dynamic systems with nonlinear viscous dampers. *Structural and Multidisciplinary Optimization* 11. http://dx.doi.org/10.1007/s00158-017-1858-2.

52 Plessix, R.-E. (2006). A review of the adjoint-state method for computing the gradient of a functional with geophysical applications. *Geophysical Journal International* 167 (2): 495–503. http://dx.doi.org/10.1111/j.1365-246X.2006.02978.x.

53 Giles, M., Ghate, D., and Duta, M. (2008). Using automatic differentiation for adjoint CFD code development. *Computational Fluid Dynamics Journal* 16. https://ora.ox.ac.uk/objects/uuid:14ea20eb-d094-48d4-b7bc-4ef65e97566b.

54 Ruder, S. (2016). An overview of gradient descent optimization algorithms. arXiv preprint arXiv:1609.04747.

55 Qian, N. (1999). On the momentum term in gradient descent learning algorithms. *Neural Networks* 12 (1): 145–151. http://dx.doi.org/https://doi.org/10.1016/S0893-6080(98)00116-6.

56 Ross, S.M. (2020). *Introduction to Probability and Statistics for Engineers and Scientists*, 4e. Amsterdam: Academic Press/Elsevier. ISBN 9780123704832.

57 Duchi, J., Hazan, E., and Singer, Y. (2011). Adaptive subgradient methods for online learning and stochastic optimization. *Journal of Machine Learning Research* 12 (61): 2121–2159.

58 Hinton, G., Srivastava, N., and Swersky, K. (2012). Neural networks for machine learning. Lecture 6a: Overview of mini-batch gradient descent. *Cited on* 14 (8): 2.

59 Kingma, D.P. and Ba, J. (2014). Adam: A method for stochastic optimization. arXiv preprint arXiv:1412.6980.

11

Machine Learning Application for Modeling and Design Optimization of High Frequency Structures

Mohamed H. Bakr[1], Shirook Ali[2], and Atef Z. Elsherbeni[3]

[1] Department of Electrical and Computer Engineering, McMaster University, Hamilton, Ontario, Canada
[2] Faculty of Applied Science and Technology, Sheridan College, Brampton, Ontario, Canada
[3] Electrical Engineering Department, Colorado School of Mines, Golden, CO, USA

11.1 Introduction

Computational electromagnetic (CEM) methods are used to get numerical solutions of Maxwell's equations for the given problem parameters (material distributions, boundary conditions, and excitations). These methods can be classified as either time-domain (TD) methods or frequency-domain (FD) methods. The most common TD methods include transmission line modeling (TLM) [1, 2], finite-difference time-domain (FDTD) method [3, 4], and finite-volume time-domain (FVTD) method [5]. These methods provide solutions for the temporal electric and magnetic fields equations in the entire domain. A frequency-rich excitation is usually utilized to compute the response over a wide frequency band. FD methods, such as the finite-element method (FEM) [6, 7] and method of moments (MoMs) [8], are used to solve Maxwell's equations only at one frequency at a time. Frequency domain solvers are repeatedly invoked at different frequencies to cover the desired frequency band. These methods are usually preferred for narrow band applications while TD methods are preferred for wideband applications. All CEM methods discretize the computational domain and/or boundary into cells or elements. Maxwell's equations are to be solved in the computational domain either by solving a complex linear system of equations or through a time-marching scheme.

For small electromagnetic problems, the computational cost associated with solving Maxwell's equations is not severe. However, as the size of the problem grows, the computational cost for solving Maxwell's equations becomes intensive. This is particularly true for electrically long structures with small features that require fine meshing. Applying optimization using these time-intensive solvers

Advances in Time-Domain Computational Electromagnetic Methods, First Edition.
Edited by Qiang Ren, Su Yan, and Atef Z. Elsherbeni.

can be prohibitive as optimizers repeatedly invoke the electromagnetic solvers with perturbed parameter values. Alternative optimization methods such as surrogate model optimization [9, 10], space mapping optimization [11, 12], and adjoint sensitivity analysis (ASA) [13, 14] were proposed to accelerate the optimization phase of time-intensive structures. Surrogate model optimization replaces the time-intensive electromagnetic solver with a much faster model that can be utilized within the optimization loop. These models could be empirical model, circuit-theoretic models, or mathematical models. Space mapping optimization utilizes the existence of a coarse model of the same electromagnetic problem to accelerate the optimization phase. A mapping is established between the parameter spaces of the coarse and fine models. Optimization is applied to the "mapped" coarse model thus significantly cutting down the optimization time. ASA estimates the sensitivities of the objective function with respect to all design parameters using at most one extra simulation. The computational cost of estimating the gradient is reduced from n, where n is the number of parameters, to only 1. In some applications, no extra simulations are needed to evaluate all sensitivities and the problem is labeled as self-adjoint [15]. ASA methods are useful when gradient-based optimization algorithms are involved. These alternative optimization methods were applied in many electromagnetic-based problems including filter design [16], antenna design [17], breast cancer imaging [18], metamaterial design [19], just to mention a few.

Artificial neural networks (ANNs) offer an alternative approach for accelerating electromagnetic modeling. ANNs imitate the behavior of the human brain [20]. Computational units called neurons are connected to each other through synaptic connections. These connections have different weights. By adjusting these weights, an ANN is capable of modeling different relationships between the input and the output. ANNs can be trained to model arbitrary mapping from the input to the output [20]. The initial contributions of the area of ANNs utilized shallow neural networks (see Figure 11.1). These neural networks have one or two hidden layers. The number of neurons in these layers is limited. They are trained with small sets of training data. Shallow ANNs were applied to many electromagnetic-related areas such as the design of microwave filters and devices [21–24]. These neural networks were not easily able to model complex mapping from the input to the output spaces.

Recently, there has been explosion of research in the area of deep neural networks (DNNs). These ANNs have many hidden layers with many hidden neurons. The number of hidden layers could be in the thousands [25]. The large number of weights connecting different layers require large training datasets. Several types of these DNNs were developed that are best suited for recognizing images [26] or for time-domain predictions [27]. Novel training algorithms were developed that achieve the optimal weights using the given training datasets [28, 29]. DNNs were successfully applied to many challenging fields such as image classification and recognition [30], speech recognition [31], data forecasting [32], and language processing [33]. DNNs were also applied to the modeling and design optimization of high frequency structures. They have been applied to many electromagnetic-related fields including the modeling of radio waves and obstructions in tunnels [34, 35], modeling metasurfaces [36], the solution of forward and inverse scattering problems [37, 38], medical imaging [39], antenna array design [40], and direction of arrival estimation [41].

We focus in this chapter on the applications of ANNs to efficient electromagnetic modeling and design optimization. We start by reviewing some of the basic concepts governing ANNs in general and DNNs in specific. Our review is intended for a general reader with no background in machine learning. We discuss the different types of DNN layers and how they were exploited in several recent applications of electromagnetic problems. Finally, we give our expectations of the future advancement of research based on artificial intelligence (AI) for electromagnetics applications.

11.2 Background

11.2.1 Feedforward ANN

The initial configurations of ANN included a single layer shallow neural network (see Figure 11.1). This network has one input layer, one output layer, and a single middle (hidden) layer. The output from every layer is fed as input to the next layer with certain weights thus the name feedforward.

Figure 11.1 A feedforward shallow ANN. The outputs from the hidden layer **V** depend on the inputs to the neural network **X** and the hidden weights **W**^{h}. The output **y** of the shallow ANN is a function of both **V** and the weights of the output layer **W**^{o}.

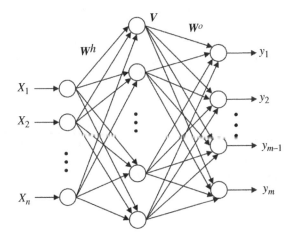

The input to the *j*th neuron in the hidden layer is the weighted sum of all the inputs to the ANN:

$$z_j = \sum_{i=1}^{n} w_{ji}^h x_i \tag{11.1}$$

where w_{ji}^h is the weight of the synaptic link connecting the *i*th input parameter x_i to the *j*th hidden neuron and *n* is the number of input parameters. The vector $\boldsymbol{W}^h$ contains all weights $\{w_{ji}^h, \forall i,j\}$ connected to the neurons of the hidden layer. The excitation z_j of the *j*th hidden neuron triggers a response V_j at the output of that neuron $V_j = \sigma(z_j)$. This response is governed by a nonlinear activation function $\sigma(z)$. Some of the examples of possible activation functions utilized in shallow ANNs include [20]:

$$\text{Sigmoid function: } \sigma(z) = \frac{1}{1 + e^{-z}}$$

$$\text{Arc-tangent function: } \sigma(z) = \left(\frac{2}{\pi}\right) \arctan(z)$$

$$\text{Hyperbolic-tangent function: } \sigma(z) = \frac{e^z - e^{-z}}{e^z + e^{-z}} \tag{11.2}$$

The outputs from the hidden layer are then fed to the output layer. The *j*th output of the ANN, y_j, is a weighted sum of the outputs of the neurons of the hidden layer:

$$y_j = \sum_{i=1}^{N_h} w_{ji}^o V_i, \quad j = 1, 2, \ldots, m \tag{11.3}$$

where N_h is the number of hidden neurons and *m* is the number of ANN's outputs. The vector $\boldsymbol{W}^o$ is the vector of all weights connecting the outputs of the hidden layer to the inputs of the output layer.

Training the neural network involves determining the hidden layer weights $\boldsymbol{W}^h$ and output layer weights $\boldsymbol{W}^o$ that would best model the mapping between the vector of input parameters $\boldsymbol{x}$ and the vectors of output responses $\boldsymbol{y}$. To carry out this training, a dataset that has pairs of inputs and outputs $\{(\boldsymbol{x}_k, \boldsymbol{y}_k), k = 1, 2, \ldots, N_{tr}\}$ is first assembled. The data samples in this dataset represent the desired input–output mapping. This set may be created through actual measurements or through simulations. Every input–output pair (data sample) is presented to the ANN and the corresponding ANN output is determined. A loss function (an objective function) is then evaluated, which represents the difference between the desired output from the ANN and the actual output. The weights of the ANN are iteratively updated to minimize this loss function over all input–output pairs. The gradient of the loss function with respect to all weights are determined through a backpropagation algorithm [20]. Gradient-based algorithms are then used to iteratively adjust the weights [42].

While theoretically it is possible to model any input–output relationship with a shallow ANN, in practice, it was not easy to train them to model complex nonlinear mappings. The training process can get stuck in local minima when handling more complex nonlinear relationships. The area of machine learning did not make the required breakthrough until DNNs were introduced in the past few years. The advancement in computational power and cloud computing made this new class of ANNs feasible for many practical applications.

11.2.2 Deep Neural Networks

DNNs were shown to be more capable of modeling complex nonlinear relationships [43]. Figure 11.2 shows a feedforward DNN which has a large number of hidden layers (hundreds or even thousands). Similar to a shallow feedforward ANN, the outputs from one hidden layer are fed as weighted inputs to the next hidden layer. Because of the large number of layers and the large number of neurons in each layer, DNNs have a large number of weights. They require extensive training datasets to achieve their targets. A class of training algorithms called deep learning algorithms [44] were developed for these massive DNNs.

The activation functions (11.2) used in shallow neural networks may not be suitable for DNNs. The vanishing gradient problem [45] limits the application of these activation functions to ANNs with many layers. The vanishing gradient problem renders the loss function insensitive to the weights as the number of layers increases. Alternative activation functions such as the rectified linear unit (ReLU) can be used. The derivative of this activation function with respect to its argument

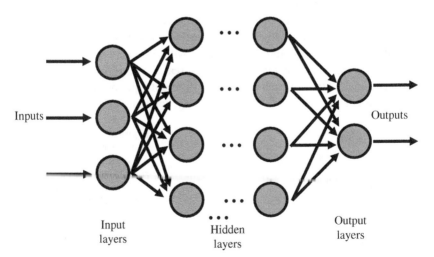

Figure 11.2 A multi-layer feedforward deep neural network.

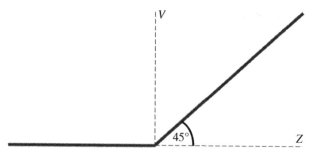

Figure 11.3 The ReLU activation function.

has the value of one for positive values of the input. This eliminates the vanishing gradient problem and makes it more suitable for deep networks. Figure 11.3 shows the ReLU activation function. The ReLU is nondifferentiable at the origin, and this makes the modeled response nonsmooth. Modified ReLU functions were presented as shown, for example, in [46], to smoothen the ANN output.

Several types of layers have been incorporated in DNNs for different applications. Convolutional neural networks (CNNs or Convnets) utilize layers that convolute inputs of large dimensionalities with different filters [47]. These filters allow the extraction of different features from the input. CNNs have been applied successfully in different classification and regression problems involving images. CNNs have several widely used implementations such as Deep CNN-AlexNet, CNN-VGG, CNN-GoogLeNet, CNN-ResNet, and CNN-DenseNet [48]. Figure 11.4 shows a convolutional layer applied to an input image and the resulting two-dimensional output. The size of the input image is 8×8 and it has 3 three different colors. A 3×3 filter is then applied to the image. This filter is a two-dimensional matrix with coefficients that extract certain features. This filter is shifted across the image and is convoluted with all 3×3 regions within the image. Each convolution operation yields one scalar. The result is thus a 6×6 two-dimensional matrix for each color layer. Different types of filters can extract different types of features from a given image (e.g. boundaries).

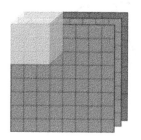
=
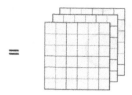

Figure 11.4 An illustration of a convolution layer. A 3×3 filter is applied to all color layers of an 8×8 image yielding three 2-D layers each of size 6×6.

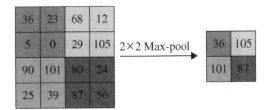

The max pooling (MP) is another layer that is widely used in DNNs. It is used to extract features from two-dimensional images or matrices and reduce their dimensionalities to lower the computational and memory requirements. As the name implies, this layer applies a maximum filter to a set of neighboring values. Figure 11.5 shows how a MP layer operates on a two-dimensional input matrix to reduce its dimensionality and extract its key features. Figure 11.6 shows a cascade of convolutional layers and MP layers. A typical DNN will have cascades of different types of layers.

Another type of DNNs that has been widely used in time series and dynamic applications is recurrent neural networks (RNNs) [49]. RNNs utilize a successfully delayed versions of the controlling variables as inputs to the neural network. The output is also delayed and fed back to the input of the ANN. The long short-term memory (LSTM) was proposed as a variant of RNNs that overcome the observed training issues [50]. They have been utilized in training DNNs for many innovative applications including the data-driven solution of Maxwell's equations [51]. Figure 11.7 shows an example of a RNN. The weights of the RNN are determined by presenting temporal sequences at the input and adjusting the weights to produce the target temporal output.

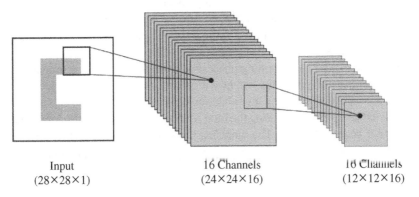

Input
(28×28×1)

16 Channels
(24×24×16)

16 Channels
(12×12×16)

Figure 11.6 An illustration of two layers of a DNN. A convolutional layer with 16 filters (channels) is followed by a 2 × 2 max pooling layer. The 2 × 2 pooling layer reduces the size of each layer in every dimension by one half.

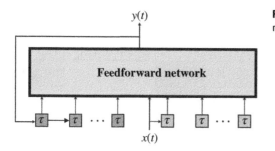

Figure 11.7 A recurrent neural network.

11.2.3 Training of Deep Neural Networks

Similar to shallow neural networks, the training of a DNN is an iterative process through which the optimal weights that describe the relationship between the input and output are determined. There are several learning strategies for neural networks including supervised learning, unsupervised learning, and reinforcement learning [52]. In this chapter, we focus on supervised learning as it is the most widely used one in the area of electromagnetics. The training process starts with a dataset describing the desired mapping from the input to the output. The neural network is trained when its output corresponding to the input vector x_i, $\forall i$, is as close as possible to the desired vector y_i. The main optimizable parameters of a DNN are the weights associated with all layers W. There are other parameters that are selected beforehand and kept fixed during the training phase. These parameters include the number of layers and number of neurons in each layer and are denoted as hyperparameters. If the DNN does not produce a sufficiently accurate model, the hyperparameters are adjusted. The training phase is then repeated with the new hyperparameters until an acceptable model accuracy is achieved.

The dataset is usually divided into three subsets. The training dataset is used in adjusting the weights W until the optimal set of weights W^* is reached. The validation dataset is not used in training but is used to evaluate changes done to hyperparameters after each training phase. For example, the ANN is trained for a certain number of layers. This will give a certain training error. The accuracy improvement is evaluated using the validation dataset when a new layer is added. The third dataset is used for testing the final ANN's ability to generalize for inputs that has not been trained on. The modeling error corresponding to the training dataset is denoted as the training error. The modeling error corresponding to the testing dataset is denoted as the testing error.

An ideally trained neural network has a zero-training error and zero testing error. This means that the training accuracy and generalization accuracy are 100%. This ideal situation is impossible to reach in practical applications. Several other practical scenarios are observed during the training of an ANN. If the training error is large as compared to a reasonable threshold, this means that the ANN is

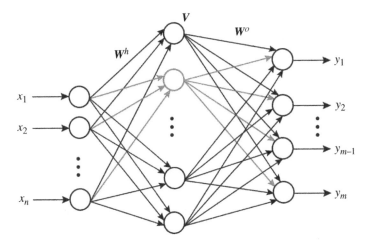

Figure 11.8 An illustration of the drop out step. Neurons (gray color) are randomly selected and removed from the ANN with all their input and output links.

not capable of learning the relationship between the input and output. This situation is called underfitting. The standard approach to handle underfitting is to add more layers or add more neurons in each layer. These changes will hopefully allow the bigger ANN to capture the mapping from the input space to the output space. Another common scenario happens when the training error is relatively low but the testing error is high. This situation is called overfitting. The ANN remembers the training examples but is incapable of generalizing to other inputs it has not been trained on. There are several solutions to address overfitting. The dataset could be expanded if that is a possibility. Another approach is to use a drop out layer [53]. As the name implies, some neurons are randomly selected and completely dropped out from the ANN. All the input links and output links connected to these neurons are removed. This random drop out step effectively creates a different ANN that has a better chance of learning the input–output mapping. The drop out step is used only during the training phase. Once training is over, the ANN topology is fixed during the inference steps. Figure 11.8 shows a possible drop out step. The selected neuron with all its input and output links are removed from the ANN structure.

11.2.4 Supervised Deep Learning Steps

Addressing a deep learning problem follows several steps. These steps are given by the following procedure (see Figure 11.9):

(1) **Preparing the datasets**: A number of input–output pairs are created. This dataset is obtained through measurement or through simulations.

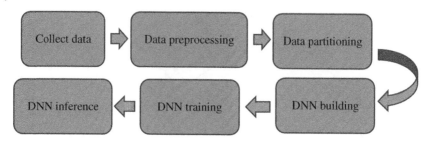

Figure 11.9 The flow sequence of a machine learning problem.

(2) **Data preprocessing**: The input parameter values are usually normalized. For regression problems, the output response values should also be normalized. For classification problems, the output should be labeled.

(3) **Data partioning**: The available data samples are divided into three sets; a training set, a validation set, and a testing set. In some applications only a training set and testing set are used. Usually, 80% or more of the available data samples are used for training.

(4) **ANN construction**: The number of layers, their types, and the number of neurons in every layer are determined.

(5) **ANN training**: The ANN is trained using the training set until a good training error is achieved. If no good training error is achieved, the hyperparameters are changed and training is repeated.

(6) **ANN inference**: The ANN is trained and can be used to predict the model response for inputs not presented to the ANN before.

11.3 Applications of Machine Learning to Electromagnetics

The past few years saw an explosion in the applications of machine learning in many fields. There are several reported publications in CEMs. There are two main trends that can be distinguished in the published research. The first trend focuses on using the ANN as a surrogate model that could be used to replace the time-intensive electromagnetic model. The trained ANN may then be used within an optimization loop to efficiently determine the optimal set of design parameters. In other applications, as will be shown, the trained ANN is used to model the inverse relationship. The inputs to the ANN are the responses while the outputs are the design parameters that generate these responses. This approach eliminates the need for using an optimization algorithm as the optimal responses are directly mapped through the ANN to the optimal design parameters.

Another trend that is starting to emerge over the past few years is to integrate machine learning within the electromagnetic solver itself to accelerate the simulation time. This work is still in its infancy, but it is promising to revolutionize the field of CEMs in the next few years.

11.3.1 ANN-Based Design Optimization

11.3.1.1 Forward ANN Modeling

Figure 11.10a illustrates the classical optimization approach. In this figure, an optimizer (optimization algorithm) is used to drive the electromagnetic solver. The optimizer is initialized with an initial design $x^{(0)}$ and the design specifications. The optimizer then repeatedly invokes the electromagnetic solver for different parameter values. The electromagnetic solver returns the electromagnetic response (S-parameters, electric and magnetic fields, radiation pattern, radar cross section, etc.) and possibly their derivatives to the optimizer. The optimizer uses the responses and their gradients to guide the optimization steps and obtain the optimal design x^*. If the time of one simulation is relatively long or if the number of parameters is large, the optimization phase becomes formidable. In this case, ANNs offer an effective alternative for the optimization of time-intensive structures. First, a dataset is created that sufficiently cover the solution domain. An ANN is then trained using this dataset to achieve good training and testing errors. The trained ANN is then used to replace the electromagnetic solver within the optimization loop as shown in Figure 11.10b. Here the trained ANN is utilized as a surrogate of the time-intensive electromagnetic solver. Figure 11.11 shows how an ANN is used as a surrogate to model the design of a plasmonic power splitter.

There are many reported publications in the literature that adopt this approach for the design of different types of structures. For example, in [54] an ANN is used to accelerate the design of microwave and millimeter-wave filters. The ANN replaces the electromagnetic solver to guide the optimization iterations in situations where the starting point is far from the optimal design. In [55], a trained neural network was used to accelerate the design of a wideband antenna. Key shape parameters were selected as inputs to the ANN. The output of the

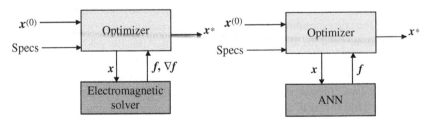

Figure 11.10 Relacing a time-intensive electromagnetic simulation with a trained ANN.

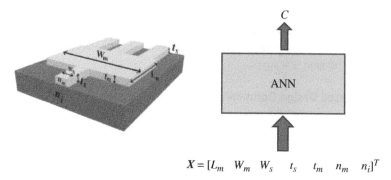

$$X = [L_m \quad W_m \quad W_s \quad t_s \quad t_m \quad n_m \quad n_i]^T$$

Figure 11.11 An illustration of the forward ANN approach. The values of the parameters of a plasmonic power splitter are fed to the input of a trained ANN that models the power coupling coefficient C at one arm of the splitter.

ANN is the reflection coefficient of the antenna. The high frequency solver HFSS was used to simulate the structure. In [56], a modified FDTD, time-domain finite-integration theorem (TDFIT) was developed to increase the modeling accuracy of a millimeter-wave antenna array. As the simulation time of the array integrated with the electrically large radome was prohibitive, an ANN was trained to simulate the broadband transmittance of the whole system. Optimization was then applied to the surrogate ANN model. It was shown in this work that the optimization time was cut to only one half through a hybrid integration between the TDFIT and ANNs.

In [57], an ANN is trained to model the relationship between the input parameters and output responses. Instead of utilizing an optimization algorithm, a backpropagation method is used to determine the vector of optimal parameters x^* corresponding to the vector of desired responses y^*. The authors utilized their approach in optimizing a single-phase transmission line while accounting for magneto-thermal effects.

11.3.1.2 Inverse ANN Modeling

Another approach that was reported in different applications is illustrated in Figure 11.12. The ANN is trained to model the inverse mapping. The outputs of the electromagnetic structure are used as the inputs of the ANN while the output represents the corresponding parameters. The interesting aspect of this method is that there is no need for an optimization algorithm. After the ANN is trained, it maps the desired output y^* to the corresponding optimal set of input parameters x^*. The main problem with the inverse ANN approach is the nonuniqueness of the dataset. Two different sets of parameters x_1 and x_2 can give rise to the same response y. This results in a conflicting dataset and poor training accuracy.

In [58], the authors map different objectives of the antenna (S-parameters, gain, and radiation pattern) into the desired geometrical parameters of the antenna.

Figure 11.12 An inverse ANN is used to map the responses to the corresponding parameters.

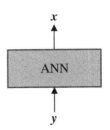

The authors show that the size of the dataset required is relatively small as compared to other forward optimization methods.

In [59], the nonuniqueness problem of the ANN is addressed by utilizing a multi-branch inverse neural network. The nonuniqueness problem of the data is first detected using the derivative of the response with respect to the parameters. Once nonuniqueness is detected, the dataset is classified into different sets based on their derivatives. Separate ANNs are then used to train the dataset of each branch. The authors of [59] applied their approach to the optimization of antenna directivity.

The approaches suggested in [60, 61] utilize a similar approach for the design of a high frequency structure. Instead of utilizing time-domain responses (electric fields, magnetic fields, transmission, reflection) or frequency domain responses (S-parameters, transmissivity, reflectivity) of a structure, the corresponding transfer function is found and its poles and zeros are used as the inputs to the inverse ANN. The output from the ANN are the corresponding values of the geometric and physical parameters. In [60], this approach was adopted for the design of metasurfaces. In [61], the inverse ANN approach was applied to the design of microwave filters.

In [62], an inverse ANN is used to solve the inverse microwave imaging problem. A new type of convolutional layers is introduced that can handle complex responses (FD scattered electric field) over a given observation domain. A new DNN topology made of an encoder followed by a decoder is used to extract the dielectric contrasts at the output. The output from the DNN is a microwave image that is created using the measured scattered fields. Figure 11.13 illustrates this approach.

11.3.1.3 Neuro-Space Mapping

Another modeling approach that exploits neural networks involves integrating ANNs with coarse models to build a surrogate of the time-intensive electromagnetic solver. Figure 11.14a illustrates this approach. The electromagnetic solver may be used to directly simulate the electromagnetic structure. This approach may be time-intensive especially if integrated with an optimizer.

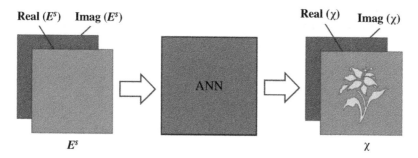

Figure 11.13 An illustration of the approach presented in [62]; the real and imaginary parts of the scattered field are input to the ANN. The ANN is trained to output the real and imaginary parts of the material contrast of the object.

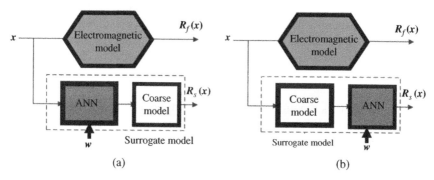

Figure 11.14 The neuro-space mapping approach: (a) input neuro-space mapping approach and (b) output neuro-space mapping.

An alternative approach is to use a coarse model of the same electromagnetic problem. This coarse model may be, for example, a coarse electromagnetic model or a circuit-theoretic model. This model does not have the same accuracy as the electromagnetic solver. An ANN is trained to map the parameters of the fine model x to the corresponding coarse model parameters x_c that has the same output response [63, 64]. This input neuro-space mapping approach is illustrated in Figure 11.14a. In this case, the weights of the ANN maps one set of parameters to another set of parameters. A variation of the neuro-space mapping approach uses the ANN to map the output of the coarse model to the desired electromagnetic output [65]. This approach, which is denoted as output neuro-space mapping, is shown in Figure 11.14b. In this case, the ANN maps the inaccurate coarse model responses to the more accurate fine model responses. The approaches in [63–65] combine ANNs with the coarse model form a surrogate model of the electromagnetic problem. This surrogate model can then be used in applications that require large number of simulations such as design optimizations, yield analysis, statistical analysis, and sensitivity analysis.

11.3.1.4 Image-Based ANN

All the techniques discussed in Sections 11.3.1–11.3.2 deal with an electromagnetic structure as a parameterized topology. The topology is defined by a number of parameters (lengths, widths, depths, material properties). The vector x includes these parameters. Recent advances in computer vision enabled an alternative approach for modeling electromagnetic structures. An image of the EM structure is used to capture its features. This image could be in shades of gray or color to indicate different materials and dimensions. A sequence of convolutional layers, MP layers, drop our layers, and flattened layers are then used to model the corresponding response. In this case, the ANN maps an image of the structure to the desired response. Figure 11.15 shows a possible structure for this image-based ANN.

The approach in [66] models the transmission and reflection from a metasurface using the planar image of the metasurface. The metalized pixels of the metasurface are presented with a different color. The work presented in [67] uses images of all-dielectric metasurfaces to predict its response. This method also uses the wavelength at which the response is modeled as an auxiliary input to the ANN. The recent work [68] uses images of a metasurface with a phase changing material (Vanadium Dioxide). The properties of this material change when it is heated thus allowing for a tunable metasurface. The images used in [68] show different color to denote the hot VO_2 pixels, cold VO_2 pixels, the substrate, and

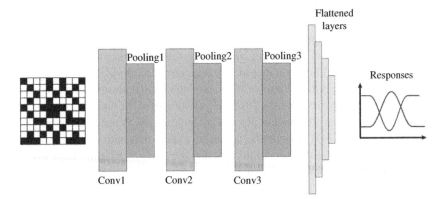

Figure 11.15 An illustration of an Image-based ANN. An image of a metasurface is fed to a DNN to determine the corresponding responses. A cascade of convolutional layers, pooling layers, and flattened layers are used in the DNN-based modeling of the responses. Source: Adapted from Yuze et al. [66].

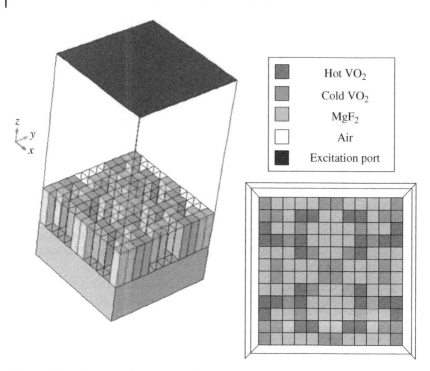

Figure 11.16 A metasurface presented by VO$_2$ cold and hot pixels on a MgF$_2$ substrate [68].

air (see Figure 11.16). Figure 11.17 shows a comparison between the output of the DNN and that of the electromagnetic solver (COMSOL) for a random test metasurface. Good agreement is observed between the response obtained through COMSOL and the response of the image-based ANN surrogate.

A similar approach is adopted in [69] for the solution of scattering problems from nanoscatterers. The DNN receives two images as an input. The first image represents the shape information while the second image represents the material information. The outputs from the DNN are two images of the real and imaginary parts of the field at an observation plane. The dataset used in training this model is created using the FDTD method. The authors show that with relatively small dataset, the DNN was able to accurately predict the fields of arbitrary nanostructures. Figure 11.16 illustrates possible input and output of this DNN.

11.3.2 ANN-Assisted Electromagnetic Modeling

In the previous section, we discussed a number of algorithms that uses ANN as a surrogate of the electromagnetic solver. In all these algorithms the solver

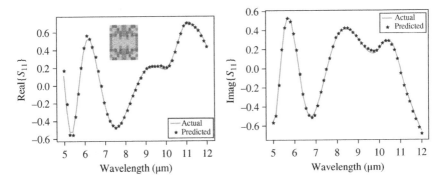

Figure 11.17 The reflection from the VO$_2$ metasurface calculated using COMSOL (–) and using the trained DNN (*) for the shown metasurface [68].

supplies the dataset used to train the ANN. The ANN is not used to accelerate the solver or replace part of its functionality. Another main trend that is gaining interest is to replace some of the repeated time-intensive calculations within the electromagnetic solver by a much faster trained ANN. This approach is still in its infancy, but it has a strong potential. We review in this section some of the key publications in this area.

11.3.2.1 ANN-Based Method of Moments

The method of moments is commonly used to solve for the surface currents of metallic antennas and scatterers. The surface of the metallic object is divided into discrete elements. The induced surface currents over these elements are then expressed in terms of basic functions. A system of equation is used to solve for the unknown surface currents:

$$ZI = V \tag{11.4}$$

where I is the vector of complex current coefficients and V is the excitation vector. The components of the complex matrix Z describe the coupling between different surface elements and are evaluated through time-intensive integrals. The matrix Z includes the submatrices Z_{xx} and Z_{xy} describing the coupling between two current basis functions with parallel or orthogonal orientation. The approach suggested in [70] replaces these calculations with a radial basis function neural network (RBF ANN). Because the coupling between the two elements depends on the distance, orientation, and relative angel between the surface currents, the utilized neural networks in [70], uses the corresponding distance ρ and relative angle φ as inputs. The outputs are the real and imaginary parts of the impedance matrix submatrices Z_{xx}, Z_{xy} (see Figure 11.18). The utilized RBF-ANN are trained first using data collected from actual MoM simulations. The trained ANNs are then used within the MoM solver to accelerate the construction of the impedance matrix Z.

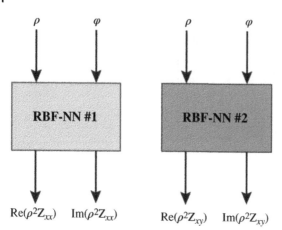

ρ φ ρ φ

RBF-NN #1   RBF-NN #2

$\mathrm{Re}(\rho^2 Z_{xx})$ $\mathrm{Im}(\rho^2 Z_{xx})$ $\mathrm{Re}(\rho^2 Z_{xy})$ $\mathrm{Im}(\rho^2 Z_{xy})$

Figure 11.18 The approach presented in [70]. ANNs are trained to map the distance and relative angle between two current basis functions into the corresponding components of the impedance matrices Z_{xx} and Z_{xy}.

This approach was applied to several planar filters and antennas as discussed in [70]. The results show that good accuracy and speed acceleration were achieved using this approach. The ANN results were compared to those of a commercial MoM solver (momentum) and of an in-house MoM solver. Figure 11.19 shows the S-parameters of a two-element array of rectangular microstrip patches placed along their H-plane [70]. The parameter D is the distance between the two patches. The results obtained using ANNs show good match to the results obtained using only the MoM solver.

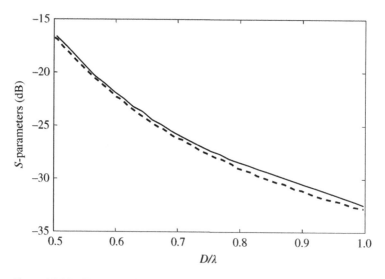

Figure 11.19 The scattering parameter S_{21} results (in dB) for a two-element array of rectangular microstrip patches obtained using an in-house ANN-MoM (--), and the commercial solver Momentum (—) for a sweep of the distance between the patches D. Source: Adapted from Soliman et al. [70].

11.3.2.2 Machine Learning-Assisted FDTD

The FDTD method is widely used in high frequency applications. It is used to simulate wideband response of different electromagnetic structures. Yee's cell is used to model the staggered fields in both time and space [4]. A set of time-marching equations are used to evaluate the electric and magnetic fields as time progresses. Two recent publications attempted to reformulate the FDTD method as a CNN [71] and as a RNN [72]. It is not clear if the computational advantage of these two formulations over the existing FDTD time-marching equations exist for a general FDTD structure.

Another possible application of machine learning to FDTD involves replacing the time-consuming absorbing boundary conditions (ABCs). These ABCs represent numerical boundaries for truncated open structures. Figure 11.20a shows an illustration of a perfectly matched layer (PML) that is used to terminate a 2D FDTD domain from all sides. The PML has usually many FDTD layers with their own time-marching equations. The computational cost in updating the fields within the PML is more involved relative to other cells inside the computational domain. Time-domain diakoptics breaks the computational domain into two or more subdomains that exchange time-domain quantities along the boundaries. Several diakoptics approaches were suggested to reduce the computational cost of the PML. One approach replaces the PML with an analytical model [74]. Another approach utilizes a recursive model to reduce the computational cost of the PML [75].

Recently, an approach was introduced that replaces the PML terminating an FDTD domain with an ANN [73]. At every point on the boundary between the object domain and the PML, the fields are decomposed into two sets. The first set of field components $x^{(k)}$ is used to deduce the second set of field components $y^{(k)}$. An ANN is trained to map the first set to the second set at each point on the

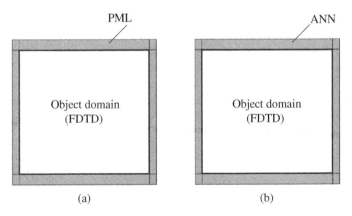

Figure 11.20 An illustration of the approach presented in [73]. The FDTD-based PML is replaced by a trained ANN. Source: Adapted from Yao and Jiang [73].

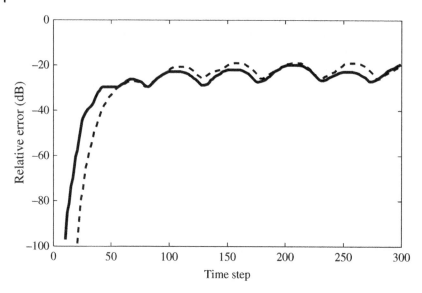

Figure 11.21 The reported relative error in estimating the magnetic field H_z for a TE_z structure; the ANN-based PML error (—) and a 5-cell PML (--). Source: Adapted from Yao and Jiang [73].

interface. The authors in [73] showed that by having both sets of fields at the boundaries, the FDTD update equations for the whole computational domain could be carried out. The hyper tangent basis function given in Eq. (11.2) is used as an activation function for all the neurons. Figure 11.21 illustrates the temporal relative model error in estimating H_z for a TE_z problem. The error utilizing the ANN and the error with a five-layer PML are both shown. The ANN-based model shows good accuracy.

11.3.2.3 ANN-Assisted Variational Methods

Another approach that makes use of the generalization and prediction capabilities of ANNs was presented in [76]. This approach is particularly suited for variational methods such as the FEM and MoM. The basic premise of this method is that a CNN can predict a near exact solution of an electromagnetic problem from an approximate solution that uses a coarser number of N basis functions. The exact solution f of the variational problem is given by:

$$f = \sum_{i=1}^{N} x_i f_i \tag{11.5}$$

where x_i is the unknown coefficient corresponding to the ith basis function f_i. The set of basis function $\{f_i, \forall i\}$ is assumed to be known. Solving for the exact solution

$x = [x_1 \, x_2 \dots x_n]^T$ may be time-intensive. A coarser solution $\check{f}$ is expressed as a sum of a limited number of basis functions:

$$\check{f} = \sum_{i=1}^{\check{N}} \check{x}_i f_i \tag{11.6}$$

where $\check{N} < N$ is the number of utilized basis functions of the coarse model. A CNN is trained to map the coarse solution $\check{x}$ to the exact solution x. The dataset used to train the CNN includes many structures for which both $\check{x}$ and x are evaluated. Figure 11.22 shows the ANN-assisted solution for a dielectric slab problem solved using the FEM method as compared to the exact solution.

11.3.2.4 Physics-Based Unsupervised Learning of Maxwell's Equations

In addition to the previously discussed methods, new class of ANN-based approaches aims at integrating the problem physics with the ANNs. The physics informed neural networks (PINNs) [77–79] integrate physics-based constraints within the training of the ANN to enhance its prediction capabilities. These methods were used to solve several partial differential equations (PDEs) including Burgers' and the Schrödinger equations.

Recently, a new PINNs method is introduced for the solution of time-domain Maxwell's equations [80]. This approach is illustrated in Figure 11.23. The material and source information of a time-domain problem are presented as spatial

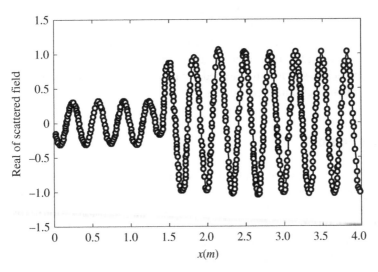

Figure 11.22 Results for the approach reported in [76]; The predicted real part of the scattered field (−) versus the actual field (o) for a dielectric slab sample example. Source: Adapted from Key and Notaroš [76].

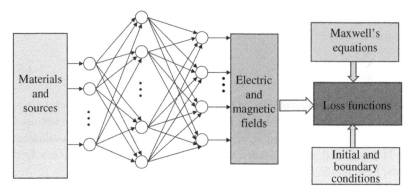

Figure 11.23 The unsupervised learning approach reported in [80]. Source: Adapted from Zhang et al. [80].

and temporal inputs to the ANN. The output from the ANN gives the electric and magnetic field over the domain. These fields are used to evaluate a loss function that estimates how accurately these fields solve Maxwell's equations, satisfy the boundary conditions, and match a given set of reference values. The sensitivities of the loss function with respect to all fields are evaluated through automatic differentiation. Through backpropagation, the sensitivities of the loss function with respect to the weights of ANN are estimated. The weights of the ANN can thus be iteratively adjusted using an optimization algorithm.

The approach used in training the ANN is unsupervised as the fields corresponding to the given materials and sources are not known beforehand. The optimal set of weights corresponds to electric and magnetic fields that satisfy Maxwell's equations and the boundary conditions for the given material and source distributions. This approach was shown to give good results as compared to those obtained using time-domain electromagnetic solvers.

11.4 Discussion

In this chapter, we reviewed some of the key applications of machine learning to the modeling and design optimization of high frequency structures. We showed that there are two main trends in ANN-based high frequency applications. The first trend treats the electromagnetic solver as a black box that supplies the training dataset. The trained ANN using this set is then utilized as a surrogate of the electromagnetic model. The second trend replaces some or all of the time-intensive operations within the electromagnetic solver by a trained ANN. In this case, the ANN becomes part of the solver and not a surrogate. The second approach is invasive and requires access to the internal data structures of the solver.

We expect that more research will be carried out on accelerating electromagnetics using machine learning. We believe that more techniques will be developed to render ANNs an integral part of electromagnetic solvers and not only a possible surrogate. The image-based ANN will find more applications in CEMs. More physics-based models that integrate the physics directly with the ANNs will be developed. We also share the expectations made in [81] that more unsupervised learning approaches will find applications for high frequency structures.

While we are optimistic about the future of ANNs in electromagnetics, we note that there are several obstacles that need to be tackled. The computational cost of the generation of the dataset may discourage the use of ANNs for many practical applications [82]. This computational overhead may, in many cases, overcome the computational advantages gained from using the ANNs. More work is needed on reducing the initial computational overhead so that the ANNs become a more attractive choice in the modeling and design optimization of high frequency structures. Also, the many hyperparameters to be set by the user are also confusing to many. The trial and error in finding the optimal number of layers and the optimal number of neurons in hidden layers is not attractive for many high frequency designers. We expect that more research will be put in automating the selection of the hyperparameters to accelerate the modeling process.

References

1 Christopoulos, C. (1995). *The Transmission-Line Modeling Method: TLM*, 1e. Wiley-IEEE Press.

2 Hoefer, W.J.R. (1985). The transmission-line matrix method theory and applications. *IEEE Transactions on Microwave Theory and Techniques* MTT-33 (2): 882–893.

3 Taflove, A. and Hagness, S.C. (2005). *Computational Electrodynamics: The Finite-Difference Time-Domain Method*. Boston, MA: Artech House Publishers.

4 Elsherbeni, A.Z. and Demir, V. (2016). *The Finite-Difference Time-Domain Method for Electromagnetics with MATLAB Simulations*, ACES Series on Computational Electromagnetics and Engineering, 2e. Edison, NJ: SciTech Publishing, an Imprint of IET.

5 Fumeaux, C., Baumann, D., and Vahldieck, R. (2006). Finite-volume time-domain analysis of a cavity backed Archimedean spiral antenna. *IEEE Transactions on Antennas and Propagation* 54: 844–851.

6 Jin, J.-M. (2014). *The Finite Element Method in Electromagnetics*, 3e. Wiley-IEEE Press.

7 Volakis, J.L., Chatterjee, A., and Kempel, L.C. (1998). *Finite Element Method Electromagnetics: Antennas, Microwave Circuits, and Scattering Applications*, 1e. Wiley-IEEE Press.

8 Harrington, R. (1968). *Field Computation by Moment Methods*. Macmillan.

9 Koziel, S., Ogurtsov, S., Zieniutycz, W., and Bekasiewicz, A. (2015). Design of a planar UWB dipole antenna with an integrated Balun using surrogate-based optimization. *IEEE Antennas and Wireless Propagation Letters* 14: 366–369.

10 Li, Y., Xiao, S., Rotaru, M., and Sykulski, J.K. (2016). A dual Kriging approach with improved points selection algorithm for memory efficient surrogate optimization in electromagnetics. *IEEE Transactions on Magnetics* 52: 1–4.

11 Bakr, M.H., Bandler, J.W., Madsen, K., and Søndergaard, J. (2002). An introduction to the space mapping technique. *Optimization and Engineering* 2: 369–384.

12 Wu, K.-L., Zhao, Y.-J., Wang, J., and Cheng, M.K.K. (2004). An effective dynamic coarse model for optimization design of LTCC RF circuits with aggressive space mapping. *IEEE Transactions on Microwave Theory and Techniques* 52 (1): 393–402.

13 Bakr, M.H., Elsherbeini, A.Z., and Demir, V. (2017). *Adjoint Sensitivity Analysis of High Frequency Structures with MATLAB*. IET.

14 Zhang, Y., Negm, M.H., and Bakr, M.H. (2016). An adjoint variable method for wideband second-order sensitivity analysis through FDTD. *IEEE Transactions on Antennas and Propagation* 64 (2): 675–686.

15 Bakr, M.H., Nikolova, N.K., and Basl, P.A.W. (2005). Self-adjoint S-parameter sensitivities for lossless homogeneous TLM problems. *International Journal of Numerical Modelling: Electronic Networks, Devices and Fields* 18 (6): 441–455.

16 Ismail, M.A., Smith, D., Panariello, A. et al. (2004). EM-based design of large-scale dielectric-resonator filters and multiplexers by space mapping. *IEEE Transactions on Microwave Theory and Techniques* 52 (1): 386–392.

17 Easum, J.A., Nagar, J., Werner, P.L., and Werner, D.H. (2018). Efficient multiobjective antenna optimization with tolerance analysis through the use of surrogate models. *IEEE Transactions on Antennas and Propagation* 66 (12): 6706–6715.

18 Hosseinzadegan, S., Fhager, A., Persson, M. et al. (2021). Discrete dipole approximation-based microwave tomography for fast breast cancer imaging. *IEEE Transactions on Microwave Theory and Techniques* 69 (5): 2741–2752.

19 Selga, J., Rodríguez, A., Boria, V.E., and Martín, F. (2013). Synthesis of split-rings-based artificial transmission lines through a new two-step, fast converging, and robust aggressive space mapping (ASM) algorithm. *IEEE Transactions on Microwave Theory and Techniques* 61 (6): 2295–2308.

20 Haykin, S.O. (1998). *Neural Networks: A Comprehensive Foundation*, 2e. Pearson.

21 Zhang, Q.J., Gupta, K.C., and Devabhaktuni, V.K. (2003). Artificial neural networks for RF and microwave design-from theory to practice. *IEEE Transactions on Microwave Theory and Techniques* 51 (4): 1339–1350.

22 Rayas-Sanchez, J.E. (2004). EM-based optimization of microwave circuits using artificial neural networks: the state-of-the-art. *IEEE Transactions on Microwave Theory and Techniques* 52 (1): 420–435.

23 Watson, P.M. and Gupta, K.C. (1997). Design and optimization of CPW circuits using EM-ANN models for CPW components. *IEEE Transactions on Microwave Theory and Techniques* 45 (12): 2515–2523.

24 Monzo-Cabrera, J., Pedreno-Molina, J.L., Lozano-Gurrrero, A., and Toledo-Moreo, A. (2008). A novel design of a robust ten-port microwave reflectometer with autonomous calibration by using neural networks. *IEEE Transactions on Microwave Theory and Techniques* 56 (12): 2972–2978.

25 LeCun, Y., Bengio, Y., and Hinton, G. (2015). Deep learning. *Nature* 521: 436–444.

26 Venkatesan, R. and Li, B. (2017). *Convolutional Neural Networks in Visual Computing: A Concise Guide*. CRC Press.

27 Hochreiter, S. and Schmidhuber, J. (1997). Long short-term memory. *Neural Computation* 9: 1735–1780.

28 Amari, S.I. (1998). Natural gradient works efficiently in learning. *Neural Computation* 10 (2): 251–276.

29 Kingma, Diederik; Ba, Jimmy (2014). "Adam: A Method for Stochastic Optimization". arXiv:1412.6980 [cs.LG].

30 Bernard, O., Lalande, A., Zotti, C. et al. (2018). Deep learning techniques for automatic MRI cardiac multi-structures segmentation and diagnosis: is the problem solved? *IEEE Transactions on Medical Imaging* 37 (11): 2514–2525. https://doi.org/10.1109/TMI.2018.2837502.

31 Shahamiri, S.R. and Salim, S.S.B. (2014). A multi-views multi-learners approach towards dysarthric speech recognition using multi-nets artificial neural networks. *IEEE Transactions on Neural Systems and Rehabilitation Engineering* 22 (5): 1053–1063.

32 Yu, Y., Cao, J., and Zhu, J. (2019). An LSTM short-term solar irradiance forecasting under complicated weather conditions. *IEEE Access* 7: 145651–145666.

33 Lane, H., Hapke, H., and Howard, C. (2019). *Natural Language Processing In Action: Understanding, Analyzing, and Generating Text with Python*. Manning.

34 Seretis, A., Zhang, X., and Sarris, C.D. (2019). Uncertainty quantification of radio propagation models using artificial neural networks. In: *2019 IEEE International Symposium on Antennas and Propagation and USNC-URSI Radio Science Meeting*, 229–230.

35 Bregar, K., Hrovat, A., Novak, R., and Javornik, T. (2019). Channel impulse response based vehicle analysis in tunnels. In: *2019 13th European Conference on Antennas and Propagation (EuCAP)*, 1–5.

36 Li, J., Li, Y., Cen, Y. et al. (2020). Applications of neural networks for spectrum prediction and inverse design in the terahertz band. *IEEE Photonics Journal* 12 (5): 1–9.

37 Yao, H.M., Sha, W.E.I., and Jiang, L. (2019). Two-step enhanced deep learning approach for electromagnetic inverse scattering problems. *IEEE Antennas and Wireless Propagation Letters* 18 (11): 2254–2258.

38 Ma, Z., Xu, K., Song, R. et al. (2021). Learning-based fast electromagnetic scattering solver through generative adversarial network. *IEEE Transactions on Antennas and Propagation* 69 (4): 2194–2208.

39 Li, M., Guo, R., Zhang, K. et al. (2021). Machine learning in electromagnetics with applications to biomedical imaging: a review. *IEEE Antennas and Propagation Magazine* 63 (3): 39–51.

40 Zardi, F., Nayeri, P., Rocca, P., and Haupt, R. (2021). Artificial intelligence for adaptive and reconfigurable antenna arrays: a review. *IEEE Antennas and Propagation Magazine* 63 (3): 28–38.

41 Liu, Z., Zhang, C., and Yu, P.S. (2018). Direction-of-arrival estimation based on deep neural networks with robustness to array imperfections. *IEEE Transactions on Antennas and Propagation* 66 (12): 7315–7327.

42 Bakr, M.H. (2013). *Nonlinear Optimization in Electrical Engineering with Applications in MATLAB*. IET.

43 M. Telgarsky, "Benefits of depth in neural networks,"arXiv:1602.04485 [cs.LG].

44 Goodfellow, I., Bengio, Y., and Courville, A. (2016). *Deep Learning*, Adaptive Computation and Machine Learning Series. MIT Press.

45 Hochreiter, S. (1998). The vanishing gradient problem during learning recurrent neural nets and problem solutions. *International Journal of Uncertainty, Fuzziness and Knowledge-Based Systems* 6 (2): 107–116.

46 Jin, J., Zhang, C., Feng, F. et al. (2019). Deep neural network technique for high-dimensional microwave modeling and applications to parameter extraction of microwave filters. *IEEE Transactions on Microwave Theory and Techniques* 67 (10): 4140–4155.

47 Jin, K.H., McCann, M.T., Froustey, E., and Unser, M. (2017). Deep convolutional neural network for inverse problems in imaging. *IEEE Transactions on Image Processing* 26 (9): 4509–4522.

48 Khan, A., Sohail, A., Zahoora, U., and Qureshi, A.S. (2020). A survey of the recent architectures of deep convolutional neural networks. *Artificial Intelligence Review* 53: 5455–5516.

49 Mandic, D.P. and Chambers, J.A. (2001). *Recurrent Neural Networks for Prediction: Learning Algorithms, Architectures and Stability*. Wiley.

50 Smagulova, K. and James, A.P. Overview of long short-term memory neural networks. In: *Deep Learning Classifiers with Memristive Networks, Modeling and Optimization in Science and Technologies*, vol. 14 (ed. A. James), 139–153. Cham.: Springer.

51 Fu, H., Cheng, W., and Qin, Y. (2020). Exploration of data-driven methods for multiphysics electromagnetic partial differential equations. In: *2020 IEEE MTT-S International Conference on Numerical Electromagnetic and Multiphysics Modeling and Optimization (NEMO)*, 1–4.

52 Alpaydin, E. (2020). *Introduction to Machine Learning*, 4e. MIT Press.

53 Srivastava, N., Hinton, C.G., Krizhevsky, A. et al. (2014). Dropout: a simple way to prevent neural networks from overfitting. *Journal of Machine Learning Research* 15 (1): 1929–1958.

54 Zhao, P. and Wu, K. (2020). Homotopy optimization of microwave and millimeter-wave filters based on neural network model. *IEEE Transactions on Microwave Theory and Techniques* 68: 1390–1400.

55 Sharma, Y., Zhang, H.H., and Xin, H. (2020). Machine learning techniques for optimizing design of double t-shaped monopole antenna. *IEEE Transactions on Antennas and Propagation* 68: 5658–5663.

56 You, J.W., Tan, S.R., Zhou, X.Y. et al. (2014). A new method to analyze broadband antenna-radome interactions in time-domain. *IEEE Transactions on Antennas and Propagation* 62: 334–344.

57 Belfore, L.A., Arkadan, A.A., and Lenhardt, B.M. (2001). ANN inverse mapping technique applied to electromagnetic design. *IEEE Transactions on Magnetics* 37: 3584–3587.

58 Xiao, L.-Y., Shao, W., Jin, F.-L. et al. (2021). Inverse artificial neural network for multiobjective antenna design. *IEEE Transactions on Antennas and Propagation* 69 (10): 6651–6659.

59 Yuan, L., Yang, X., Wang, C., and Wang, B. (2020). Multibranch artificial neural network modeling for inverse estimation of antenna array directivity. *IEEE Transactions on Antennas and Propagation* 68 (6): 4417–4427.

60 Yuan, L., Wang, L., Yang, X.-S. et al. (2021). An efficient artificial neural network model for inverse design of metasurfaces. *IEEE Antennas and Wireless Propagation Letters* 20 (6): 1013–1017.

61 Zhang, C., Jin, J., Na, W. et al. (2018). Multivalued neural network inverse modeling and applications to microwave filters. *IEEE Transactions on Microwave Theory and Techniques* 66 (8): 3781–3797.

62 Yao, H.M., Jiang, L., and Sha, W.E.I. (2020). Enhanced deep learning approach based on the deep convolutional encoder–decoder architecture for electromagnetic inverse scattering problems. *IEEE Antennas and Wireless Propagation Letters* 19 (7): 1211–1215.

63 Bakr, M.H., Bandler, J.W., Ismail, M.A. et al. (2000). Neural space-mapping optimization for EM-based design. *IEEE Transactions on Microwave Theory and Techniques* 48 (12): 2307–2315.

64 Cao, Y. and Wang, G. (2007). A wideband and scalable model of spiral inductors using space-mapping neural network. *IEEE Transactions on Microwave Theory and Techniques* 55 (12): 2473–2480.

65 Rayas-Sanchez, J.E. and Gutierrez-Ayala, V. (2006). EM-based Monte Carlo analysis and yield prediction of microwave circuits using linear-input neural-output space mapping. *IEEE Transactions on Microwave Theory and Techniques* 54 (12): 4528–4537.

66 Yuze, T., Hai, L., and Qinglin, Z. (2019). On the application of deep learning in modeling metasurface. In: *2019 International Applied Computational Electromagnetics Society Symposium – China (ACES)*, 1–2.

67 Tanriover, I., Hadibrata, W., and Aydin, K. (2020). Physics-based approach for a neural networks enabled design of all-dielectric metasurfaces. *ACS Photonics* 7 (8): 1957–1964.

68 A. Negm, M. Bakr, M. Howlader and S. Ali. (2021) "A Deep-learning Approach for Modeling Phase-change Metasurface in the Mid-infrared," *2021 International Applied Computational Electromagnetics Society Symposium (ACES)*, pp. 1–4.

69 Li, Y., Wang, Y., Qi, S. et al. (2020). Predicting scattering from complex nano-structures via deep learning. *IEEE Access* 8: 139983–139993.

70 Soliman, E.A., Bakr, M.H., and Nikolova, N.K. (2004). Neural networks-method of moments (NN-MoM) for the efficient filling of the coupling matrix. *IEEE Transactions on Antennas and Propagation* 52 (6): 1521–1529.

71 Yao, H.M. and Jiang, L.J. (2018). Machine learning based neural network solving methods for the FDTD method. In: *2018 IEEE International Symposium on Antennas and Propagation & USNC/URSI National Radio Science Meeting*, 2321–2322.

72 Guo, L., Li, M., Xu, S., and Yang, F. (2019). Study on a recurrent convolutional neural network based FDTD method. In: *2019 International Applied Computational Electromagnetics Society Symposium – China (ACES)*, 1–2.

73 Yao, H.M. and Jiang, L. (2019). Machine-learning-based PML for the FDTD method. *IEEE Antennas and Wireless Propagation Letters* 18 (1): 192–196.

74 Chen, Z. (1995). Analytic Johns matrix and its applications in TLM diakoptics. In: *Proceedings of 1995 IEEE MTT-S International Microwave Symposium*, vol. 2, 777–780.

75 Mrozowski, M., Niedzwiecki, M., and Suchomski, P. (1995). A fast recursive highly dispersive absorbing boundary condition using time domain diakoptics and Laguerre polynomials. *IEEE Microwave and Guided Wave Letters* 5 (6): 183–185.

76 Key, C. and Notaroš, B.M. (2020). Data-enabled advancement of computation in engineering: a robust machine learning approach to accelerating variational methods in electromagnetics and other disciplines. *IEEE Antennas and Wireless Propagation Letters* 19 (4): 626–630.

77 Raissi, M., Perdikaris, P., and Karniadakis, G.E. (2019). Physics informed deep learning (part I): data-driven solutions of nonlinear partial differential equations. *Journal of Computational Physics* 378: 686–707.

78 Raissi, M. and Karniadakis, G.E. (2018). Hidden physics models: machine learning of nonlinear partial differential equations. *Journal of Computational Physics* 357: 125–141.

79 Dwivedi, V. and Srinivasan, B. (2020). Physics informed extreme learning machine (PIELM)—a rapid method for the numerical solution of partial. *Neurocomputing* 391: 96–118.

80 Zhang, P., Hu, Y., Jin, Y. et al. (2021). A Maxwell's equations based deep learning method for time domain electromagnetic simulations. *IEEE Journal on Multiscale and Multiphysics Computational Techniques* 6: 35–40.

81 Massa, A., Marcantonio, D., Chen, X. et al. (2019). DNNs as applied to electromagnetics, antennas, and propagation – a review. *IEEE Antennas and Wireless Propagation Letters* 18 (11): 2225–2229.

82 Campbell, S.D., Jenkins, R.P., O'Connor, P.J., and Werner, D. (2021). The explosion of artificial intelligence in antennas and propagation: how deep learning is advancing our state of the art. *IEEE Antennas and Propagation Magazine* 63 (3): 16–27.

Part V

Parallel Computation Schemes for Time-Domain Methods

12

Acceleration of FDTD Code Using MATLAB's Parallel Computing Toolbox

Alec Weiss[1], Atef Z. Elsherbeni[1], Veysel Demir[2], and Mohammed Hadi[1]

[1]*Electrical Engineering Department, Colorado School of Mines, Golden, CO, USA*
[2]*Electrical Engineering Department, Northern Illinois University, Chicago, IL, USA*

12.1   Introduction

Other chapters introduced the core concepts of finite-difference time domain (FDTD) and how different elements, boundaries, and materials can be simulated. Basic simulations with a small number of cells and time steps seem to run in a relatively short period. As simulations become more complicated, using a larger number of cells and time steps, runtimes can rapidly become unfeasible for the problem in hand.

To reduce the computation time of FDTD simulations, parallelization of computation across multiple processors and onto graphical processing units (GPUs) can be performed. Many previous works have successfully shown the speedup of FDTD simulations using parallel processing and GPUs with lower level programming languages such as C and FORTRAN [1–4]. While these works successfully demonstrated the acceleration of the FDTD technique, the languages used require more verbose programming along with a more in-depth knowledge of hardware and the libraries built to run it. Using MATLAB's parallel computing toolbox circumvents this problem by providing an abstraction layer between the user and the complex libraries for parallel computing. Because of this ease of use, research has focused on the specific use of MATLAB for speedups with FDTD [5, 6]. It has been shown that FDTD can be successfully accelerated using only MATLAB's built-in parallelization tools.

Advances in Time-Domain Computational Electromagnetic Methods, First Edition.
Edited by Qiang Ren, Su Yan, and Atef Z. Elsherbeni.

12.2 Parallelization with MATLAB

MATLAB provides multiple built-in options for parallelization [7, 8]. By default, when an operation is performed on a sufficiently large array, MATLAB will automatically perform parallelization on the operation depending on both the size of the array and the function being performed using the central processing unit (CPU) of the computer. This occurs with vectorized operations and elementary operations including *exp()* and trigonometric functions [8], which is referred to as implicit parallelization. While this is by far the easiest to implement as it requires no changes by the user, it comes at the cost of providing almost no control to the user. While implicit parallelization provides a small increase in throughput over an implementation using no parallelization, large gains could be achieved by leveraging explicit parallelization [6].

12.2.1 Explicit Parallelization

Explicit parallelization in MATLAB refers to parallelization that requires the user to explicitly command each parallel worker. To instantiate a worker pool in MATLAB, the *parpool()* command is used. If *parpool()* is called without any arguments, a parallel pool is started with a default number of workers on the system. This value is chosen by MATLAB based on the system specifications. The default value can be changed in the parallel pool settings menu. A number can also be passed into the parallel pool command such as *parpool(2)*, which will start a parallel pool with two workers. This value can be any number between 1 and the maximum number of workers, which again can be changed in the parallel pool settings menu. To access the parallel pool settings, click on the arrow next to the four vertical bars in the bottom corner of the MATLAB interface and select "Parallel Preferences," as shown in Figure 12.1.

Once in the Parallel Preferences menu, the maximum number of parallel pool workers can be set with the "Preferred number of workers in parallel pool" setting.

Figure 12.1 MATLAB's parallel pool settings menu. Source: The MathWorks, Inc.

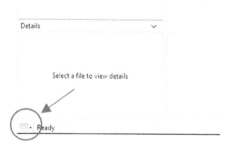

Figure 12.2 Selecting Cluster Profile Manager for additional setting. Source: The MathWorks, Inc.

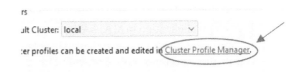

Further settings can be accessed by clicking on the "Cluster Profile Manager" link as shown in Figure 12.2.

From here the "edit" button in the bottom-right corner can be used to edit the default number of workers, the number of threads per worker, and a variety of other options.

Once a parallel pool is started with *parpool()*, a number of different commands are then available in MATLAB. The most basic command is the *parfor* command, which can replace the standard *for* keyword to attempt to parallelize for loops. In FDTD, more fine-grained control is needed over each of the workers, and, therefore, the *spmd* keyword is used. This stands for single program/multiple data (SPMD). This environment will be run independently by each worker in the parallel pool. Each worker can also be instructed to perform specific operations by checking the ID of the worker in the *labindex* variable. The total number of workers in the pool is also available in this environment through the *numlabs* variable. Similar to the message passing interface (MPI) libraries (as in [9]), data is passed explicitly between workers using the *labSend()*, *labRecieve()*, or *labSendRecieve()* commands. The *labSend()* and *labRecieve()* commands require information about the data to send or receive along with the ID (or *labindex*) of the worker for which the data is being sent or received.

With the *spmd* environment, each worker must also contain and handle its own sets of data. This means that before parallel processing can be done, all data that is being processed must be split and distributed across the workers. MATLAB has a few built-in functions to accomplish this with the *distributed()* and *codistributed()* commands. While these functions can be useful for perfect uniform distribution of arrays, in many cases, distribution of arrays must be done manually. Data can be manually distributed by having each worker generate the indices of its required array, and then copying that data into a local array within a *spmd* loop. An example of this can be seen in the code below

```
spmd
        %distribute global_array
        local_array = global_array(index_start:index_end)
end
```

12.2.2 Multi-GPU Processing in MATLAB

While explicit parallelization in MATLAB can provide large speedups compared to implicit parallelization and non-parallelized applications, modern GPUs can provide even further speedups. GPUs provide thousands of parallel computational units that can perform thousands of mathematical operations at once. While the usage of a GPU for acceleration is covered in several other literatures, here we cover the usage of multiple GPUs at once in MATLAB for even faster simulation speeds. MATLAB uses Nvidia's Compute Unified Device Architecture (CUDA) framework [10, 11].

Just like for using multiple processors on the CPU, MATLAB is also capable of performing multi-GPU operation using the *spmd* environment [12]. The distribution of data and communication is performed using the same methods and commands mentioned in the previous section. When a parallel pool is started up, MATLAB automatically maps worker threads to unique GPU devices if they are available. Therefore, if a machine has four unique GPUs, the first four workers in a parallel pool will each be communicating with a different GPU device. The number of GPU devices available can easily be accessed using the *gpuDeviceCount* variable.

Data is transferred onto a single GPU in MATLAB using the *gpuArray()* command. Therefore, to transfer data to multiple GPUs, the *gpuArray()* command can simply be called within a *spmd* block. The code for this operation would look something like the following one:

```
spmd
        %distribute global_array
        local_array = global_array(index_start:index_end);
        %transfer to GPU
        local_array_gpu = gpuArray(local_array);
end
```

12.3 Multi-CPU and Multi-GPU for FDTD Simulation

The rest of this chapter discusses what needs to be performed to change a typical FDTD code, as presented in [13], to perform a parallelized simulation across multiple workers and GPUs. Throughout this chapter, an example problem of a low-pass microstrip filter will be used. The layout of the problem in the FDTD domain can be seen in Figure 12.3.

12.3.1 Distribution of the FDTD Domain

Using the previously discussed tools, an FDTD simulation code can be fully parallelized to take advantage of both multi-CPU and multi-GPU computers. The first

Figure 12.3 The 3D representation of the microstrip low-pass filter problem. Three horizontal planes divide the computational domain into four sub-domains for parallel distribution.

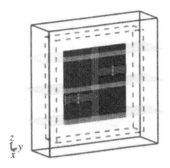

step in the parallelization of a simulation code is to distribute the FDTD domain across each of the workers created by *parpool()*. While the FDTD domain could be split across the workers in all dimensions (x, y, and z), for simplicity, here the splitting of the domain across only a single dimension is described. For speed, the domain is split along the z dimension because the field arrays for the x–y planes are stored linearly in memory. This means that indexing across these x–y planes (across the z dimension) allows memory to be accessed linearly and will drastically decrease the time required to distribute the domain. For simplicity, the domain is split evenly into n subdomains, where n is the number of workers in the parallel pool. The subdomains are distributed in order where the worker with the lowest ID (labindex==1) is given the subdomain at the bottom of the entire simulation domain. The remaining IDs are distributed by counting up on the worker IDs until the last worker is reached (labindex=numlabs), which is given the uppermost subdomain in the simulation.

With this arrangement, the required additions to the FDTD code are as follows:

– Distribute the fields, boundaries, sources, and outputs.
– Add SPMD environments to time marching loop.
– Aggregate the output data after simulation.

A flowchart of these new operations of the FDTD simulation can be seen in Figure 12.4.

12.3.2 Overlapping FDTD Data

At each time step of the FDTD marching loop, every subdomain requires E and H fields information from surrounding domains to perform the field updating equations. For distributed computation (as in the case of multi-CPU and multi-GPU), this means that each core must contain some amount of overlapping information. For the case where the domain is split along the Z direction, E_x, E_y, H_x, and H_y fields must all have an overlap of one cell between the distributed parts of the domain to allow correct updating. During the time marching loop, E fields will be calculated and transferred down by higher subdomains to lower

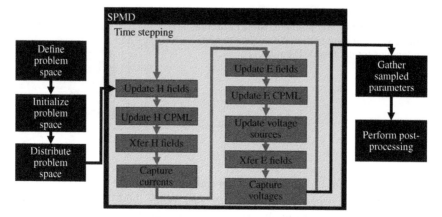

Figure 12.4 Flowchart of parallelized FDTD MATLAB code using MATLAB's parallel computing toolbox.

subdomains, and H fields will be calculated and transferred up by lower sub-domains to higher subdomains. This approach is similar to that used in [6, 14]. Figure 12.5 depicts this division where closer to the boundary, the H_x, and H_y field components are calculated by lower subdomains and transferred up to higher subdomains, while the E_x and E_y field components are calculated by higher subdomains and transferred to lower subdomains after each time step.

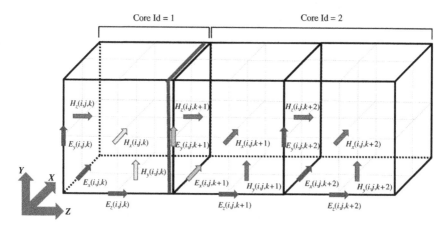

Figure 12.5 Boundaries of distributed sub-domains. The H-field components are calculated by the lower sub-domains and transferred up to the higher sub-domain, while the E-field components are calculated by the higher sub-domains and transferred to the lower sub-domains after each time step.

12.3.2.1 Initializing the Parallel Pool and Distributed Calculations

The first step in distributing the domain is to initialize the parallel pool. This is done by adding a new script to the FDTD code called *initialize_multicore.m*. The listing of this code is as follows:

Initialize_multicore.m

```
%initialize workers for multicore processing (precursor to
%running on multiple GPUs)
fprintf('Initializing Parallel Pool\n');
%check our gpu hardware
if(use_gpu)
    initialize_gpu;
end

if(interactive_mode) %we delete the current pool and make a new one
    delete(gcp('nocreate'));
end
%now get the current pool to see if it exists
mypool = gcp('nocreate');
if(isempty(mypool))
    mypool= parpool(num_threads);
    fprintf('Starting parallel pool with %d threads:',num_threads);
else
    fprintf(' Using Currently Running Pool on
%d workers - ',mypool.NumWorkers);
end
fprintf(' SUCCESS\n');
num_gpu_used = mypool.NumWorkers;
num_threads = mypool.NumWorkers;
%ensure we have the correct number of workers
spmd
fprintf('Hello, I am thread with index %d\n',labindex);
if(use_gpu)
    gpu_data = get_gpu_info(true);
end
end


%% Now find the best axes to split along
%do this by finding the plane size along each axis [x,y,z]
n_vals = [nx,ny,nz];
%now find which axis has the minimum plane size
%[unsplit_axis_length, split_axis_idx] = max(n_vals);
%get the largest axis (smallest planes)
unsplit_axis_length = nz; split_axis_idx = 3; %z-axis


%% Now define our indices to store on each of the gpus.
%We assume we rotate to always split along x axis
```

```
%transfer region size
num_xfer_planes = 1;
%get even split size
split_axis_length = floor(unsplit_axis_length./num_threads);
%get number of extra planes after even splitting
split_extra_length = unsplit_axis_length-(split_axis_length
.*num_threads);
%now actually give each thread their value

cat1_fun = @(x) [x,x(end)+1]; %concatenate +1 to end of array
pre1_fun = @(x) [x(1)-1,x];   %concatenate -1 to beginning of array
cut1_fun = @(x) [x(1:end-1)]; %remove 1 from end of array
spmd
    %first find the number of planes on each core
    num_planes = split_axis_length;
    if(labindex<=split_extra_length)
        num_planes = num_planes + 1;
    end
    num_planes_all = num_planes;
end
num_planes_all = [num_planes{:}];
spmd
    %ensure we're synchronized for the gather coming up
    %now find the start and stop planes for each core
    num_planes_all_local = gather(num_planes_all);
    plane_start = sum(num_planes_all_local(1:labindex-1))+1;
    plane_end = plane_start+num_planes-1;
    %These are the indices stored on each of the different gpus
    %we then overwrite the indices for the axis we are splitting
    %x_idx_ex = 1:nx   ;x_idx_ey = 1:nx+1 ;x_idx_ez = 1:nx+1 ;
    %y_idx_ex = 1:ny+1 ;y_idx_ey = 1:ny   ;y_idx_ez = 1:ny+1 ;
    z_idx_ex = 1:nz+1 ;z_idx_ey = 1:nz+1 ;z_idx_ez = 1:nz   ;
    %x_idx_hx = 1:nx+1 ;x_idx_hy = 1:nx   ;x_idx_hz = 1:nx   ;
    %y_idx_hx = 1:ny   ;y_idx_hy = 1:ny+1 ;y_idx_hz = 1:ny   ;
    z_idx_hx = 1:nz   ;z_idx_hy = 1:nz   ;z_idx_hz = 1:nz+1 ;

    %now generate splits for each dimensional split possibility
    %xfer e fields down and h fields up
    nx_split = nx;
    ny_split = ny;
    nz_split = nz;
    %indices for field indices
    x_idx = 1:nx;
    y_idx = 1:ny;
    z_idx = 1:nz;
    if(num_threads~=1) %if we only are using more than 1 core
        %here we need to xfer Ey,Ez down and Hy,Hz up
        %sizes that don't change
        z_idx_ex = plane_start:plane_end;
        z_idx_ey = plane_start:plane_end;
        z_idx_ez = plane_start:plane_end;
```

```
        z_idx_hx = plane_start:plane_end;
        z_idx_hy = plane_start:plane_end;
        z_idx_hz = plane_start:plane_end;


        if(labindex==num_threads) %the end of the grid
            %H fields lead here again
            %x_idx_hx = pre1_fun(x_idx_hx);
            z_idx_hz = cat1_fun(z_idx_hz); %1:nx+1 (local @ 1:nx+1)
            z_idx_hy = pre1_fun(z_idx_hy); %1:nx+1 (local @ 2:nx)
            z_idx_hx = pre1_fun(z_idx_hx); %1:nx+1 (local @ 2:nx)
            z_idx_ey = cat1_fun(z_idx_ey); %1:nx+1 (local @ 1:nx+1)
            z_idx_ex = cat1_fun(z_idx_ex); %1:nx+1 (local @ 1:nx+1)
        elseif(labindex==1) %extra e planes for xfer down
            % E fields lead here as per usual
            z_idx_ey = cat1_fun(z_idx_ey); %1:nx+1 (local @ 1:nx)
            z_idx_ex = cat1_fun(z_idx_ex); %1:nx+1 (local @ 1:nx)
            z_idx_hz = cat1_fun(z_idx_hz); %1:nx+1 (local @ 1:nx+1)
%this is double calculated on each core
        else %else we have extra e and h planes
            %it is important to note that H-fields lead here
            %(H1,E1,H2,E2,etc in x direction)
            z_idx_ey = cat1_fun(z_idx_ey); %1:nx+1 (local @ 1:nx)
            z_idx_ex = cat1_fun(z_idx_ex); %1:nx+1 (local @ 1:nx)
            z_idx_hz = cat1_fun(z_idx_hz); %1:nx+1 (local @ 1:nx+1)
%this is double calculated on each core
            z_idx_hy = pre1_fun(z_idx_hy); %1:nx+1 (local @ 2:nx+1)
            z_idx_hx = pre1_fun(z_idx_hx); %1:nx+1 (local @ 2:nx+1)
        end
        nz_split = num_planes;
        z_idx = plane_start:plane_end;


    end
    z_idx_local = 1:num_planes;


    field_indices.global.ex = ...
create_linear_index_list(zeros(nx,ny+1,nz+1),x_idx,y_idx,z_idx);
    field_indices.global.ey = ...
create_linear_index_list(zeros(nx+1,ny,nz+1),x_idx,y_idx,z_idx);
    field_indices.global.ez = ...
create_linear_index_list(zeros(nx+1,ny+1,nz),x_idx,y_idx,z_idx);
    field_indices.local.ex  = ...
create_linear_index_list(Ex,x_idx,y_idx,z_idx_local);
    field_indices.local.ey  = ...
create_linear_index_list(Ey,x_idx,y_idx,z_idx_local);
    field_indices.local.ez  = ...
create_linear_index_list(Ez,x_idx,y_idx,z_idx_local);
    field_indices.global.hx = ...
create_linear_index_list(zeros(nx+1,ny,nz),x_idx,y_idx,z_idx);
    field_indices.global.hy = ...
create_linear_index_list(zeros(nx,ny+1,nz),x_idx,y_idx,z_idx);
```

```
    field_indices.global.hz = ...
create_linear_index_list(zeros(nx,ny,nz+1),x_idx,y_idx,z_idx);
    field_indices.local.hx  = ...
create_linear_index_list(Hx,x_idx,y_idx,z_idx_local);
    field_indices.local.hy  = ...
create_linear_index_list(Hy,x_idx,y_idx,z_idx_local);
    field_indices.local.hz  = ...
create_linear_index_list(Hz,x_idx,y_idx,z_idx_local);
    %indices

    field_indices.z_idx.ex = z_idx_ex;
    field_indices.z_idx.ey = z_idx_ey;
    field_indices.z_idx.ez = z_idx_ez;
    field_indices.z_idx.hx = z_idx_hx;
    field_indices.z_idx.hy = z_idx_hy;
    field_indices.z_idx.hz = z_idx_hz;

    %dx,dy,dz
    delta.x = dx; delta.y = dy; delta.z = dz;

    %---------------------------------------------------------%
    % At this point we have the indexes for splitting our grid.  %
    % We now need to split each of the arrays. From there we can %
    % put the whole run_time_marching in a spmd environment,     %
    % which will prevent too much overhead.                      %
    %---------------------------------------------------------%

end

%% Here we use arrayfun for GPU and do not for CPU
if(use_gpu)
    update_e_fields_CPML_ABC = @(Efields,Hfields,CPML_vals,Idx) ...
update_e_fields_CPML_ABC_arrayfun(Efields,Hfields,CPML_vals,Idx);
    update_h_fields_CPML_ABC = @(Efields,Hfields,CPML_vals,Idx) ...
update_h_fields_CPML_ABC_arrayfun(Efields,Hfields,CPML_vals,Idx);
    update_e_fields          = @(Efields,Hfields,ec,Idx) ...
update_e_fields_arrayfun(Efields,Hfields,ec,Idx);
    update_h_fields          = @(Efields,Hfields,hc,Idx) ...
update_h_fields_arrayfun(Efields,Hfields,hc,Idx);
else
    update_e_fields_CPML_ABC = @(Efields,Hfields,CPML_vals,Idx) ...
update_e_fields_CPML_ABC_optimized(Efields,Hfields,CPML_vals,Idx);
    update_h_fields_CPML_ABC = @(Efields,Hfields,CPML_vals,Idx) ...
update_h_fields_CPML_ABC_optimized(Efields,Hfields,CPML_vals,Idx);
    update_e_fields          = @(Efields,Hfields,ec,Idx) ...
update_e_fields_optimized(Efields,Hfields,ec,Idx);
    update_h_fields          = @(Efields,Hfields,hc,Idx) ...
update_h_fields_optimized(Efields,Hfields,hc,Idx);
end
```

This script has a few functionalities. It first initializes the parallel pool, but only if it doesn't already exist. The parallel pools can take a while to start up, so it is typically more efficient to use the current pool than starting a new one. Here it looks for a *use_gpu* flag. If this flag is set to true, the script will also verify that the system has the number of GPUs requested (this is assumed to be the number of workers requested). Before distribution, it is also verified that the z axis is the largest axis as this will reduce the communication required during each time step. If the z axis is not the largest, the domain is rotated such that it is. One final thing to note on this script is the declaration of updating functions toward the end. Using updating functions that rely on *arrayfun()* is highly efficient when using GPUs, but not CPUs. Therefore, two different versions of *update_e_fields.m* are created and the functions to use are selected at the runtime based on the *use_gpu* flag.

Once the overlaps between subdomains are known and the parallel pool initiated, the domain (which is comprised of E and H field components along with their coefficients can be correctly divided. This is also partially done in the *initialize_multicore.m* script. The first step in this division is to calculate the z indices that each worker will contain. This has to be done separately for each E and H field component along with their coefficients because of the previously mentioned overlap. This is performed as a separate step from the actual distribution for two main reasons, simplicity and flexibility. Once the indices for each worker in the z direction are known, the distribution of the grid becomes trivial. Minimal further calculations are required to distribute the FDTD domain across the workers. Having this as a separate step also makes it easier to have more flexible configurations. This could include options like unequal splits between workers without having to change any extra code. The code shown here performs an equal split of the domain but can handle cases where the number of cells in the z dimension are not evenly divisible by the number of workers. If the number of workers does not evenly divide into the number of z cells, an extra adjacent plane is added to the cores with the lowest IDs until all remainder planes are taken care of and the sum of the z cells for each core is equal to the total number of z cells. A secondary script *initialize_indexing.m* then calls 2 further functions *get_xfer_Idx.m* and *get_update_Idx.m*. These functions take the values calculated in *initialize_multicore.m* and calculate the array of indices for both updating the field values and transferring the data along the boundaries. These array coefficients are then packed into structures to organize and easily pass the data into other functions. It should also be noticed that many of these functions are called within *spmd* loops to create indices specific to each worker in the parallel pool. Each worker contains both global and local indices for each of its arrays. This allows calculation in the local domain during updating along with indices for distribution and gathering before and after running the simulation. This index creation step is the same when using either multiple CPUs or multiple GPUs for calculation. The listings for *initialize_indexing.m*, *get_xfer_Idx.m*, and *get_update_Idx.m* are given in the following text.

Initialize_indexing.m

```
spmd
%generate indices used to update and transfer values
Idx      = get_update_Idx(nx_split,ny_split,nz_split);
xfer_Idx = get_xfer_Idx(nx_split,ny_split,nz_split,num_threads);
end
```

get_xfer_Idx.m

```
function [ Idx ] = get_xfer_Idx(nx,ny,nz,num_threads )
%GET_XFER_IDX

    %% H field transfer indices
    %first the labindex to transfer to and receive from
    Idx.h_labindex_rcv = labindex+1;
    Idx.h_labindex_src = labindex-1;
    if(labindex==num_threads) %we don't have rcv
        Idx.h_labindex_rcv = [];
    end
    if(labindex==1) %don't have src
        Idx.h_labindex_src = [];
    end

    %generate indexes for H field transfers
    plane_send_p_start = nz+1; plane_send_p_end = nz+1;
    plane_recv_n_start = 1;    plane_recv_n_end = 1;
    if(labindex==1)
        plane_send_p_start = nz; plane_send_p_end=nz;
        plane_recv_n_start = 0 ; plane_recv_n_end = 0;
        %plane_recv_n_start = 1 ; plane_recv_n_end = 1;
    end
    %now get the axis
    %start with everything being the size of the array
    Idx.x_idx_send_hx_p = 1:nx+1;Idx.x_idx_send_hy_p = 1:nx ...
;Idx.x_idx_send_hz_p = 1:nx  ;
    Idx.y_idx_send_hx_p = 1:ny  ;Idx.y_idx_send_hy_p = ...
1:ny+1;Idx.y_idx_send_hz_p = 1:ny  ;
    Idx.z_idx_send_hx_p = 1:nz  ;Idx.z_idx_send_hy_p = 1:nz ...
;Idx.z_idx_send_hz_p = 1:nz+1;
    Idx.x_idx_recv_hx_n = 1:nx+1;Idx.x_idx_recv_hy_n = 1:nx ...
;Idx.x_idx_recv_hz_n = 1:nx  ;
    Idx.y_idx_recv_hx_n = 1:ny  ;Idx.y_idx_recv_hy_n = ...
1:ny+1;Idx.y_idx_recv_hz_n = 1:ny  ;
```

```
    Idx.z_idx_recv_hx_n = 1:nz   ;Idx.z_idx_recv_hy_n = 1:nz ...
;Idx.z_idx_recv_hz_n = 1:nz+1;
    %for our first core these ids are empty arrays (doesn't receive)
    if(labindex==1)
        Idx.x_idx_recv_hx_n = [];Idx.x_idx_recv_hy_n = ...
[];Idx.x_idx_recv_hz_n = [];
        Idx.y_idx_recv_hx_n = [];Idx.y_idx_recv_hy_n = ...
[];Idx.y_idx_recv_hz_n = [];
    end


    %now transfer along the z axis.
%here we transfer hy=nx+1 and hz=nx+1 up

    Idx.z_idx_send_hx_p = plane_send_p_start:plane_send_p_end;
    Idx.z_idx_send_hy_p = plane_send_p_start:plane_send_p_end;
    Idx.z_idx_recv_hx_n = plane_recv_n_start:plane_recv_n_end;
    Idx.z_idx_recv_hy_n = plane_recv_n_start:plane_recv_n_end;


    %% E field indexes
    %now do the same for our Efields
    %first the labindex to transfer to and receive from
    Idx.e_labindex_rcv = labindex-1;
    Idx.e_labindex_src = labindex+1;
    if(labindex==1) %we don't have rcv
        Idx.e_labindex_rcv = [];
    end
    if(labindex==num_threads) %don't have src
        Idx.e_labindex_src = [];
    end
    %start by gathering planes
    %these are the locations of the planes that are being sent and
    %that are being received
    plane_recv_p_start = nz+1; plane_recv_p_end = nz+1;
    plane_send_n_start = 1   ; plane_send_n_end = 1;

    %now get the axis
    %start with everything being the size of the array
    Idx.x_idx_recv_ex_p = 1:nx   ;Idx.x_idx_recv_ey_p = 1:nx+1;
    Idx.x_idx_recv_ez_p = 1:nx+1;
    Idx.y_idx_recv_ex_p = 1:ny+1;Idx.y_idx_recv_ey_p = 1:ny;
    Idx.y_idx_recv_ez_p = 1:ny+1;
    Idx.z_idx_recv_ex_p = 1:nz+1;Idx.z_idx_recv_ey_p = 1:nz+1;
    Idx.z_idx_recv_ez_p = 1:nz ;
    Idx.x_idx_send_ex_n = 1:nx   ;Idx.x_idx_send_ey_n = 1:nx+1;
    Idx.x_idx_send_ez_n = 1:nx   ;
    Idx.y_idx_send_ex_n = 1:ny+1;Idx.y_idx_send_ey_n = 1:ny  ;
    Idx.y_idx_send_ez_n = 1:ny+1;
    Idx.z_idx_send_ex_n = 1:nz+1;Idx.z_idx_send_ey_n = 1:nz+1;
    Idx.z_idx_send_ez_n = 1:nz+1;
```

```
   if(labindex==num_threads)
      Idx.x_idx_recv_ex_p=[];Idx.x_idx_recv_ey_p= ...
[];Idx.x_idx_recv_ez_p=[];
      Idx.y_idx_recv_ex_p=[];Idx.y_idx_recv_ey_p= ...
[];Idx.y_idx_recv_ez_p=[];
   end


   Idx.z_idx_recv_ex_p = plane_recv_p_start:plane_recv_p_end;
   Idx.z_idx_recv_ey_p = plane_recv_p_start:plane_recv_p_end;
   Idx.z_idx_send_ex_n = plane_send_n_start:plane_send_n_end;
   Idx.z_idx_send_ey_n = plane_send_n_start:plane_send_n_end;

end
```

get_update_Idx.m

```
function [ Idx ] = get_update_Idx( nx_split,ny_split,nz_split )
%GET_UPDATE_IDX Generate Idx structure for time marching update
%equations and some default values

Idx.nx = nx_split; Idx.ny = ny_split; Idx.nz = nz_split;

Idx.ex.z = 1:nz_split;
Idx.ex.hy.zn = 1:nz_split-1;
Idx.ex.hy.zp = 2:nz_split;
Idx.ex.hy.z = 1:nz_split;
Idx.ex.hz.zn = 1:nz_split   ;
Idx.ex.hz.zp = 1:nz_split;
Idx.ex.hy.z = 1:nz_split;

Idx.ey.z = 1:nz_split;
Idx.ey.hx.zn = 1:nz_split-1;
Idx.ey.hx.zp = 2:nz_split;
Idx.ey.hz.zn = 1:nz_split   ;
Idx.ey.hz.zp = 1:nz_split;

Idx.ez.z = 1:nz_split;
Idx.ez.hy.zn = 1:nz_split;
Idx.ez.hy.zp = 1:nz_split;
Idx.ez.hx.zn = 1:nz_split;
Idx.ez.hx.zp = 1:nz_split;

Idx.hx.z = 1:nz_split;
Idx.hy.z = 1:nz_split;
Idx.hz.z = 1:nz_split+1;

%now change our z axis (split_axis)
if(labindex==1) %first value
    Idx.ex.z = 2:nz_split;
    Idx.ex.hz.zn = 2:nz_split   ; Idx.ex.hz.zp = 2:nz_split;
    Idx.ey.z = 2:nz_split;
```

```
    Idx.ey.hz.zn = 2:nz_split   ; Idx.ey.hz.zp = 2:nz_split;

else
    Idx.hy.z = 2:nz_split+1;
    Idx.hx.z = 2:nz_split+1;
    Idx.ex.hy.zn = 1:nz_split; Idx.ex.hy.zp = 2:nz_split+1;
    Idx.ey.hx.zn = 1:nz_split; Idx.ey.hx.zp = 2:nz_split+1;
    Idx.ez.hy.zn = 2:nz_split+1; Idx.ez.hy.zp = 2:nz_split+1;
    Idx.ez.hx.zn = 2:nz_split+1; Idx.ez.hx.zp = 2:nz_split+1;

end

%now get the full coverage values for using diff
Idx.ex.hy.z = unique([Idx.ex.hy.zn,Idx.ex.hy.zp]);
Idx.ex.hz.z = unique([Idx.ex.hz.zn,Idx.ex.hz.zp]);
Idx.ey.hx.z = unique([Idx.ey.hx.zn,Idx.ey.hx.zp]);
Idx.ey.hz.z = unique([Idx.ey.hz.zn,Idx.ey.hz.zp]);
Idx.ez.hy.z = unique([Idx.ez.hy.zn,Idx.ez.hy.zp]);
Idx.ez.hx.z = unique([Idx.ez.hx.zn,Idx.ez.hx.zp]);

end
```

At this point, the edges of each subdomain are known, and these can be plotted on top of the image of the domain using the *display_split_planes.m* script.

12.3.2.2 Distributing the Required Information

With the indices created from the scripts previously mentioned, the domain and all parameters can be distributed. The E and H field components are distributed first. The distribution uses the variables *z_idx_[E,H][x,y,z]*, which contain the indices for the distribution calculated in *initialize_multicore.m*. The script that distributes the field components is given as follows.

Distribute_fdtd_parameters_and_arrays.m

```
fprintf('Distributing E-Fields and H-Fields to workers:');
%to single precision
Hx = single((Hx));
Hy = single((Hy));
Hz = single((Hz));
%E fields
Ex = single((Ex));
Ey = single((Ey));
Ez = single((Ez));
spmd
    %distribute the arrays to the workers
    %xfer e fields up and h fields down
    %this will be performed with composites
    Hx = Hx(:,:,z_idx_hx);
    Hy = Hy(:,:,z_idx_hy);
```

```
Hz = Hz(:,:,z_idx_hz);
Ex = Ex(:,:,z_idx_ex);
Ey = Ey(:,:,z_idx_ey);
Ez = Ez(:,:,z_idx_ez);

%% move to GPU if in use
if(use_gpu)
    Hx = gpuArray((Hx));
    Hy = gpuArray((Hy));
    Hz = gpuArray((Hz));
    Ex = gpuArray((Ex));
    Ey = gpuArray((Ey));
    Ez = gpuArray((Ez));
end
end
fprintf('SUCCESS\n');
```

This script distributes the initialized E and H field arrays across the workers by copying the needed sections of the arrays. This script again looks for the *use_gpu* flag for whether or not a multi-GPU implementation is desired. If *use_gpu==true*, then the distributed E and H field components are copied from their workers to the respective GPUs using the *gpuArray()* command. This same thing is performed for all updating coefficients in *distribute_updating_coefficients.m* script. This file has been left out for brevity but can be updated by the reader in a similar way as described above.

The distribution of sources, lumped elements, and output parameters must be done in a different manner. This is because sources and output parameters are not part of the updating coefficient arrays. To distribute sources (voltage sources for example) and samples (e.g. sampled currents and voltages), each worker first must check whether the location of the sources or samples lies within the subdomain updated by the worker. If the subdomain does not contain the source or sample, a flag is set to indicate that the core does not contain any part of the source or sample. If the source or sample lies partially or fully in the subdomain of the worker, then the following occurs:

- A flag is set to indicate all, or part, of the source or sample is on the workers' subdomain.
- The parts of the source or sample in the subdomain are found and saved to the worker.
- Local coefficient values are set in the case of sources.

With this data separately contained on each worker for each source and sample, the time marching loop only needs to update the locations of workers containing elements. *Distribute_sources_and_lumped_elements.m* currently only supports the distribution of voltage sources. *Distribute_output_parameters.m* supports both sampled currents and voltages, allowing for creation of ports in the simulations. The code for each of these new functions is given below.

Distribute_sources_and_lumped_elements.m

```
fprintf('Distributing sources and lumped elements to workers:');

%---------voltage sources---------------%
%first find out what core it should operate on
voltage_sources_global = voltage_sources;
spmd
    voltage_sources = voltage_sources; %make composite
    for ind=1:length(voltage_sources)
        fi = voltage_sources(ind).field_indices; %get field indices
        dir_str = ['e' sampled_voltages(ind).direction(1)];
% directions
        [ismem,ismem_idx] = ismember(fi,field_indices.global ...
.(dir_str));
        if(any(ismem)) %if the field indices are on this core,
%flag it
            if(any(~ismem)) %do any not exist in this field?
                fprintf("WARNING: Output Split between cores\n");
            end
            ismem_idx = ismem_idx(ismem); %only get ones for this
%core
            voltage_sources(ind).on_this_core = 1;
            voltage_sources(ind).field_indices_on_core = ...
            field_indices.global.(dir_str)(ismem_idx); %get indices
%on core
            voltage_sources(ind).local_field_indices = ...
            create_local_field_indices...
            (voltage_sources(ind).field_indices_on_core, ...
            field_indices,dir_str);

            switch (voltage_sources(ind).direction(1))
                case 'x'
                    voltage_sources(ind).Cexs_local = ...
                    voltage_sources(ind).Cexs(ismem);
                case 'y'
                    voltage_sources(ind).Ceys_local = ...
                    voltage_sources(ind).Ceys(ismem);
                case 'z'
                    voltage_sources(ind).Cezs_local = ...
                    voltage_sources(ind).Cezs(ismem);
            end
        else
            voltage_sources(ind).on_this_core = 0;
        end
    end
end
fprintf('SUCCESS\n');
```

Distribute_output_parameters.m

```
fprintf('Distributing output parameters to workers:');

%% Voltage samples
sampled_voltages_global = sampled_voltages; %save before it becomes a
%composite
spmd
    sampled_voltages = sampled_voltages; %assign to make composite
    for ind=1:length(sampled_voltages)
        fi   = sampled_voltages(ind).field_indices; %get field indices
        dir_str = ['e' sampled_voltages(ind).direction(1)];
        [ismem,ismem_idx] = ismember(fi,field_indices.global.(dir_str));
        if(any(ismem)) %if the field indices are on this core, flag it
            if(any(~ismem)) %do any not exist in this field?
                    fprintf("WARNING: Sampled voltage Split between cores\n");
            end
            ismem_idx = ismem_idx(ismem); %only get ones on this core
            sampled_voltages(ind).on_this_core = 1;
            sampled_voltages(ind).field_indices_on_core = ...
            field_indices.global.(dir_str)(ismem_idx); %get indices on core
            sampled_voltages(ind).local_field_indices = ...
            create_local_field_indices(sampled_voltages(ind) ...
            .field_indices_on_core,field_indices,dir_str);
        else
            sampled_voltages(ind).on_this_core = 0;
        end
    end
end

%% Current sampled
sampled_currents_global = sampled_currents;
spmd
    sampled_currents = sampled_currents;
    for ind=1:length(sampled_currents) %go thru each sampled current
        kscs = sampled_currents(ind).node_indices.ks;
        kecs = sampled_currents(ind).node_indices.ke;
        %z_idx contains planes calculated by this core
        %let's check if our values of ks and ke are on this plane
        local_k   = [find(z_idx>=kscs   & z_idx<=kecs   )];

        [~,local_khx] = ismember(z_idx(local_k),z_idx_hx);
        [~,local_khy] = ismember(z_idx(local_k),z_idx_hy);
        [~,local_khz] = ismember(z_idx(local_k),z_idx_hz);
        if(~isempty(local_k)) %the value lies here
            if(length(local_k)==1) %we only have 1 on this core
                    fprintf("WARNING: Sampled current Split between cores\n");
            end
            sampled_currents(ind).on_this_core = 1;
            sampled_currents(ind).kshx_local = min(local_khx); ...
            sampled_currents(ind).kehx_local = max(local_khx);
            sampled_currents(ind).kshy_local = min(local_khy); ...
            sampled_currents(ind).kehy_local = max(local_khy);
```

```
            sampled_currents(ind).kshz_local = min(local_khz); ...
            sampled_currents(ind).kehz_local = max(local_khz);
        else
            sampled_currents(ind).on_this_core = 0;
        end
    end
end

fprintf('SUCCESS\n');
```

Boundary conditions and components are distributed in a similar manner. Convolutional perfectly matched layers (CPML) boundary are distributed in the script *distribute_boundary_conditions.m*, which in turn calls *distribute_CPML_ABC.m*, which distributes the boundary layers in a similar manner to the FDTD fields and coefficient arrays. Because the CPML layers are only on the edges of the domain, the upper and lower z CPML boundaries will only lie on a maximum of a few subdomains near the edge. This means that a flag is required on each worker to indicate whether it contains part or all of the negative or positive z CPML boundary layers.

Farfield components are again transferred in a similar manner with the *distribute_farfield_arrays.m* script. The previously generated indices for the z dimensions of each subdomain are used to slice the farfield arrays onto their correct workers. Like the boundaries, x–y farfield planes may only lie on a single subdomain, and, therefore, each worker must have a flag for whether it contains all or part of each of the farfield planes.

Finally, to support incident plane waves, the scripts *distribute_incident_fields.m* and *distribute_incident_field_coefficients.m* are run. These scripts use the exact same methods as the distribution of the E and H fields in *distribute_fdtd_parameters_and_arrays.m* and *distribute_updating_coefficients.m*. Again, if the *use_gpu* is set, the fields are transferred to the GPUs with the *gpuArray()* command during this distribution step.

Each of these scripts is then added to the *fdtd_solve.m* script. This ensures that all arrays are distributed across the workers and the parallel pool is up and running before the *run_time_marching_loop.m* script is called. The updated version of *fdtd_solve.m* should then look like the following one:

Fdtd_solve.m

```
%% Paths we use
addpath('fdtd_scripts');
addpath('fdtd_functions');
addpath('fdtd_scripts/multicore_scripts');
```

```
%% Define the problem space from the project directory
project_name = '<project_name>;
addpath(project_name);

%run the scripts
define_problem_space_parameters;
define_geometry;
define_sources_and_lumped_elements;
define_output_parameters;

% initialize the problem space and parameters
initialize_fdtd_material_grid;
display_problem_space;
% display_material_mesh;

if run_simulation

    %% Initialize single core equations
    initialize_fdtd_parameters_and_arrays;
    initialize_sources_and_lumped_elements;
    initialize_updating_coefficients;
    initialize_boundary_conditions;
    initialize_output_parameters;
    initialize_farfield_arrays;
    initialize_display_parameters;

    %% Initialize multicore
    initialize_multicore;
    initialize_indexing;
    display_split_planes; %this must be added after the domain is divided

    %% Distribute data to multicore
    distribute_fdtd_parameters_and_arrays;
    distribute_sources_and_lumped_elements;
    distribute_output_parameters;
    distribute_updating_coefficients;
    distribute_boundary_conditions;
    distribute_farfield_arrays;
    distribute_incident_fields;
    distribute_incident_field_coefficients;

    %% Provide some GPU information
    if(use_gpu)
        gpu_info = get_gpu_info(false);
        fprintf('Running on %s at index %d\n', ...
        gpu_info.device.Name,gpu_info.device.Index);
        fprintf('Using %g GB of %g GB\n',(gpu_info.device.TotalMemory-...
        gpu_info.device.AvailableMemory)/1e9, ...
        (gpu_info.device.TotalMemory)/1e9);
    end

    fprintf('Computing for %g million cells\n',nx*ny*nz/1e6);

%-------TIME EVERYTHING FROM HERE-------------%
tic;
```

```
%% FDTD time marching loop
run_fdtd_time_marching_loop;

%% %%-------END TIMING HERE---------%
setup_and_march_time = toc;
mcps = calculate_million_cells_per_second(...
setup_and_march_time,number_of_time_steps,nx,ny,nz);
fprintf('Throughput in MCPS : %f\n',mcps);

%% Gather results from multicore
time_step = gather_composite_scalar(time_step);
gather_output_parameters;
gather_sources_and_lumped_elements;
gather_farfield_arrays;

%% display simulation results
post_process_and_display_results;

end
```

12.3.3 The Time Marching Loop

Once all arrays required for the simulation are distributed, the FDTD updating equations can then be performed. For this, the updating loop must be wrapped inside the *spmd* environment. This ensures that for each time step each worker correctly updates fields, sources, boundary conditions, and samples.

The original FDTD code uses MATLAB scripts for operations such as updating the E and H fields, updating boundary layers, and taking samples. While this works fine for non-parallelized codes, MATLAB does not allow scripts to be called inside a *spmd* environment. Therefore, all scripts that are called inside the time marching loop such as *update_electric_fields.m* must be converted from scripts to functions.

12.3.3.1 Conversion of Time Marching Steps to Functional Programming for SPMD

To improve the readability of code and make all parameters easier to pass into functions, most domains are packed into structures. E fields, H fields, their coefficients, and incident field values are all packed into the structures *Efields, Hfields, E_coeffs, H_coeffs*, and *incident_field_params*, respectively. This is all performed in the *pack_fields_and_coefficients.m* script with the following code:

```
disp("Packing fields and coefficients into structures")
spmd
%% pack typical E/H and coefficients
%now let's initialize some structures we use for easy passing to functions
Efields.x = Ex; Efields.y = Ey; Efields.z = Ez;
Hfields.x = Hx; Hfields.y = Hy; Hfields.z = Hz;
%now coefficients
```

```
E_coeffs.Cexe = Cexe; E_coeffs.Cexhy = Cexhy; E_coeffs.Cexhz = Cexhz;
E_coeffs.Ceye = Ceye; E_coeffs.Ceyhx = Ceyhx; E_coeffs.Ceyhz = Ceyhz;
E_coeffs.Ceze = Ceze; E_coeffs.Cezhx = Cezhx; E_coeffs.Cezhy = Cezhy;
H_coeffs.Chxh = Chxh; H_coeffs.Chxey = Chxey; H_coeffs.Chxez = Chxez;
H_coeffs.Chyh = Chyh; H_coeffs.Chyex = Chyex; H_coeffs.Chyez = Chyez;
H_coeffs.Chzh = Chzh; H_coeffs.Chzex = Chzex; H_coeffs.Chzey = Chzey;
%now pass a structure with the indexes we are updating (needed for the
%multi-core/GPU)
Hfields.incident_plane_wave = incident_plane_wave; %store this in E and
%H fields
Efields.incident_plane_wave = incident_plane_wave;
%% pack scatter fields if required
if incident_plane_wave.enabled == true
Hfields.xic = Hxic;
Hfields.yic = Hyic;
Hfields.zic = Hzic;
Hfields.xip = Hxip;
Hfields.yip = Hyip;
Hfields.zip = Hzip;
%E fields
Efields.xic = Exic;
Efields.yic = Eyic;
Efields.zic = Ezic;
Efields.xip = Exip;
Efields.yip = Eyip;
Efields.zip = Ezip;
%ic coeffs
E_coeffs.Cexeic  = Cexeic; %Ex
E_coeffs.Ceyeic  = Ceyeic; %Ey
E_coeffs.Cezeic  = Cezeic; %Ez
H_coeffs.Chxhic  = Chxhic; %Hx
H_coeffs.Chyhic  = Chyhic; %Hy
H_coeffs.Chzhic  = Chzhic; %Hz
%ip
E_coeffs.Cexeip  = Cexeip; %Ex
E_coeffs.Ceyeip  = Ceyeip; %Ey
E_coeffs.Cezeip  = Cezeip; %Ez
H_coeffs.Chxhip  = Chxhip; %Hx
H_coeffs.Chyhip  = Chyhip; %Hy
H_coeffs.Chzhip  = Chzhip; %Hz


%% now pack our incident plane wave parameters
incident_field_params.incident_plane_wave = incident_plane_wave;
incident_field_params.waveforms = waveforms;
% [EH][xyz]i0 values
incident_field_params.Exi0 = Exi0;
incident_field_params.Eyi0 = Eyi0;
incident_field_params.Ezi0 = Ezi0;
incident_field_params.Hxi0 = Hxi0;
incident_field_params.Hyi0 = Hyi0;
incident_field_params.Hzi0 = Hzi0;
%k dot r values
incident_field_params.k_dot_r_r0 = k_dot_r0;
incident_field_params.k_dot_r_ex = k_dot_r_ex;
incident_field_params.k_dot_r_ey = k_dot_r_ey;
incident_field_params.k_dot_r_ez = k_dot_r_ez;
```

```
incident_field_params.k_dot_r_hx = k_dot_r_hx;
incident_field_params.k_dot_r_hy = k_dot_r_hy;
incident_field_params.k_dot_r_hz = k_dot_r_hz;
%dt
incident_field_params.dt = dt;

end
end
```

From here, the conversion from scripts to functions can begin. This can be started by converting *update_electric_fields.m* and *update_magnetic_fields.m* to functions. These functions take in the previously mentioned structures containing the E and H field data along with their coefficients as inputs and return updated versions of the E and H field structures. Only minor changes to the updating equations are required. These changes involve using the indices computed before time marching to select the correct indices for computation. The function *update_e_fields.m* is given as follows:

```
function [ Efields ] = update_e_fields (Efields,Hfields,ec,Idx)
%% Update all of the fields in a function
% typically called for multi-cpu
% update with transitions along z axis
nx = Idx.nx;
ny = Idx.ny;
nz = Idx.nz;
nxp1 = nx+1;
nyp1 = ny+1;
nzp1 = nz+1;
Efields.x(1:nx,2:ny,Idx.ex.z) = ...
      ec.Cexe.*Efields.x(1:nx,2:ny,Idx.ex.z) ...
    + ec.Cexhz.*diff(Hfields.z(1:nx,1:ny,Idx.ex.hz.z),1,2) ...
    + ec.Cexhy.*diff(Hfields.y(1:nx,2:ny,Idx.ex.hy.z),1,3);

Efields.y(2:nx,1:ny,Idx.ey.z) = ...
      ec.Ceye.*Efields.y(2:nx,1:ny,Idx.ey.z) ...
    + ec.Ceyhx.*diff(Hfields.x(2:nx,1:ny,Idx.ey.hx.z),1,3) ...
    + ec.Ceyhz.*diff(Hfields.z(1:nx,1:ny,Idx.ey.hz.z),1,1);


Efields.z(2:nx,2:ny,Idx.ez.z) = ...
      ec.Ceze.*Efields.z(2:nx,2:ny,Idx.ez.z) ...
    + ec.Cezhy.*diff(Hfields.y(1:nx,2:ny,Idx.ez.hy.z),1,1) ...
    + ec.Cezhx.*diff(Hfields.x(2:nx,1:ny,Idx.ez.hx.z),1,2);

  if Efields.incident_plane_wave.enabled
      Efields.x = Efields.x + ec.Cexeic .* Efields.xic ...
                  + ec.Cexeip .* Efields.xip;
      Efields.y = Efields.y + ec.Ceyeic .* Efields.yic ...
                  + ec.Ceyeip .* Efields.yip;
      Efields.z = Efields.z + ec.Cezeic .* Efields.zic ...
                  + ec.Cezeip .* Efields.zip;
  end
end
```

These same changes are also implemented for the H fields in the *update_h_ fields.m*.

The same idea continues for all other scripts that are originally contained in the time marching loop. This includes updating sources, taking samples, updating the CPML, and updating incident fields. Again, each script is changed to a function. The functions use the same equations as before and are passed into the structures of data that were packed before the time marching loop. The currently implemented functions in the time marching loop include *update_incident_ fields.m, update_h_fields, update_h_fields_CPML_ABC.m, capture_sampled_ currents.m, update_e_fields.m, update_e_fields_CPML_ABC.m, update_voltage _sources.m, capture_sampled_voltages.m*, and *calculate_j_and_m.m*.

12.3.3.2 Transferring of Overlapping Data

Two new special functions must also be included for the transfer of overlapping regions during time marching loops: *xfer_e_fields.m* and *xfer_h_fields.m*. These functions take the previously calculated field transfer indices from *get_xfer_Idx.m* and transfer the E and H field components specified from these indices. These functions first unpack the correct indices of the x and y components of E and H fields. These functions use MATLAB's *labSendReceive()* command to transfer data across the overlap regions between the overlap data between cores as described in the previous section. The transferred data is then placed into the correct locations on the receiving array. The codes for transferring the E and H fields are nearly identical; thus only the E-field transfer code is shown below:

```
function [ Efields ] = xfer_e_fields( Efields,num_threads,Idx )
%XFER_E_FIELDS
%transfer e fields to other workers (just send down)
%assume we are already in an SPMD environment

cpu_Efields.x_n_plane = ...
(Efields.x(Idx.x_idx_send_ex_n,Idx.y_idx_send_ex_n,Idx.z_idx_send_ex_n));
cpu_Efields.y_n_plane = ...
(Efields.y(Idx.x_idx_send_ey_n,Idx.y_idx_send_ey_n,Idx.z_idx_send_ey_n));

recv_Efields.x = ...
labSendReceive(Idx.e_labindex_rcv,Idx.e_labindex_src,cpu_Efields.x_n_plane);
recv_Efields.y = ...
labSendReceive(Idx.e_labindex_rcv,Idx.e_labindex_src,cpu_Efields.y_n_plane);

  if(labindex~=num_threads)%and finally write them back to our arrays
   Efields.x(Idx.x_idx_recv_ex_p,Idx.y_idx_recv_ex_p,Idx.z_idx_recv_ex_p) ...
   = recv_Efields.x;
   Efields.y(Idx.x_idx_recv_ey_p,Idx.y_idx_recv_ey_p,Idx.z_idx_recv_ey_p)...
   = recv_Efields.y;
  end
end
```

One should remember that both transfer functions for E and H fields must be run during each time step.

12.3.3.3 Final Time Marching Loop Code

With all of these additions to the time marching code, the final *run_fdtd_time_marching_loop.m* is given as follows:

Run_fdtd_time_marching_loop.m

```
disp (['Starting the time marching loop']);
disp(['Total number of time steps : ' ...
    num2str(number_of_time_steps)]);

%pack values into structures
pack_fields_and_coefficients;

start_time = cputime;
current_time = 0;
tic
spmd
%use to profile the parallel code
% mpiprofile('on');
tic
current_time = 0;

for time_step = 1:number_of_time_steps

    current_time = current_time +dt/2;
    %update_incident_fields;
    incident_field_params.current_time = current_time;
    [Efields,Hfields] = ...
    update_incident_fields(Efields,Hfields,incident_field_params);
    [Hfields] = update_h_fields(Efields,Hfields,H_coeffs,Idx);
    [Hfields,CPML_vals] = ...
    update_h_fields_CPML_ABC(Efields,Hfields,CPML_vals,Idx);
    hf_no_xfer = Hfields;
    [Hfields] = xfer_h_fields( Hfields,num_threads,xfer_Idx );
    %capture_sampled_magnetic_fields;
    [sampled_currents] = ...
    capture_sampled_currents(Hfields,sampled_currents,delta,time_step);
    current_time = current_time + dt/2;

    %update and transfer electric fields
    [Efields] = update_e_fields(Efields,Hfields,E_coeffs,Idx);
    [Efields,CPML_vals] = ...
    update_e_fields_CPML_ABC(Efields,Hfields,CPML_vals,Idx);
    [Efields] = update_voltage_sources(Efields,voltage_sources,time_step);
    [Efields] = xfer_e_fields( Efields,num_threads,xfer_Idx );
    [sampled_voltages] = ...
    capture_sampled_voltages(Efields,sampled voltages,time step);
    [Hfields] = xfer_hz_fields(Hfields);
    [nf_ff] = calculate_j_and_m(Efields,Hfields,nf_ff,idx_ff,time_step,dt);

    %display_sampled_parameters;
    if(labindex==1)
        if(mod(time_step,100)==0)
            fprintf('iteration %d complete, approximately %f MCPS\n',...
                time_step,(nx*ny*nz*time_step)./(1e6*(toc)));
```

```
        end
      end
  end
  % use to profile the parallel code
  % mpiprofile('viewer');
  end %spmd

  run_time = toc;
  total_time_in_minutes = (run_time)/60;
  disp(['Total simulation time is ' ...
      num2str(total_time_in_minutes) ' minutes.']);
```

12.3.4 Gathering the Results

Once all time steps of the simulation are completed, all of the final simulation data is contained in the structures that are still distributed. This data is then gathered back into a single non-distributed array. This allows all the post-processing algorithms to remain the same and not have to be rewritten for distributed data. Because the post-processing is not the bottleneck in the FDTD simulations, parallelizing this section is unnecessary. Because this gathering is not in the time marching loop, and can therefore contain its own *spmd* environment, it can be a script as opposed to a function. The gathering of the output parameters and the recorded farfield parameters is done with *gather_output_parameters.m* and *gather_farfield_arrays.m*, respectively. A script *gather_sources_and_lumped_elements.m* is also used to collect the voltage sources back onto a single core. This is used in some post-processing scripts. Within MATLAB, the data on each worker can be accessed in a non-*spmd* environment using cell-array-like indexing. For example, if each worker contains a part of an array named *array_a*, the values available to each worker can be accessed by *array_a{w}*, when not in a *spmd* environment, to access the data of worker with labindex *w*.

The script *gather_output_parameters.m* gathers each of the measured output parameters measured during the simulation. Specifically, this loops through each output sampled voltage and sampled current, checks each worker to see if they contain part of the sample, and reconstructs the sampled values from each of these partial samples on the workers. All of this data is gathered to the *sampled_voltages* and *sampled_current* variables to directly interface with the previously developed post-processing code. The listing for this code is shown next.

Gather_output_parameters.m

```
%% Sampled Voltages
sampled_voltages_comp = sampled_voltages;
%and return the non-composite to our value
%return global saved structure to our value (saved before distribution)
sampled_voltages = sampled_voltages_global;
```

```
%now extract the measured values (assume on same core)
for i=1:number_of_sampled_voltages
    %now check each composite
    voltage_sampled = 0; %double check its only on 1 core
    for core=1:num_threads
        sampled_voltages_core = sampled_voltages_comp{core};
        is_on_core = sampled_voltages_core(i).on_this_core;
        if(is_on_core) %if the sampled voltage is on this core
            if(voltage_sampled)%in case we already sampled it
                fprintf('WARNING: Sampled voltage %d sampled on multiple ...
                        cores\n',i);
                %error('ERROR: Sampled voltage %d sampled on multiple cores',i);
            end
            %if it was on this core then get the sampled voltage and sum
            %with the other cores
            sampled_voltages(i).sampled_value = ...
            sampled_voltages(i).sampled_value + ...
            sampled_voltages_core(i).sampled_value;
            voltage_sampled=voltage_sampled+1;
        end
    end
end

%% Sampled Currents
sampled_currents_comp = sampled_currents;
sampled_currents = sampled_currents_global;
for i=1:number_of_sampled_currents
    %now check each composite
    currents_sampled = 0; %double check its only on 1 core
    for core=1:num_threads
        sampled_currents_core = sampled_currents_comp{core};
        is_on_core = sampled_currents_core(i).on_this_core;
        if(is_on_core) %if the sampled voltage is on this core
            if(currents_sampled)%in case we already sampled it
                fprintf('WARNING: Sampled current %d sampled ...
                        on multiple cores\n',i);
                %error('ERROR: Sampled current %d sampled on multiple cores',i);
            end
            %if it was on this core then get the sampled current and sum
            %with the other cores
            sampled_currents(i).sampled_value = ...
sampled_currents(i).sampled_value+sampled_currents_core(i).sampled_value;
            currents_sampled=currents_sampled+1;
        end
    end
end
```

The data for the farfield arrays is also gathered from the distributed cores. This can be done easily using the same indices that were initially used to transfer the arrays to the workers. The original array can then be filled with the sampled data from the time steps. The naming for this again follows the original non-distributed scheme to allow the usage of post-processing scripts without any change. The script *gather_farfield_arrays.m* is given below.

Gather_farfield_arrays.m

```
fprintf('Gathering Farfield Arrays:');

%we just need to gather the cj and cm values
if(number_of_farfield_frequencies==0)
    return;
end

ui = gather_composite_scalar(ui);
uj = gather_composite_scalar(uj);
uk = gather_composite_scalar(uk);
li = gather_composite_scalar(li);
lj = gather_composite_scalar(lj);
lk = gather_composite_scalar(lk);

%% Global values
cjyxp = zeros(number_of_farfield_frequencies,1,uj-lj,uk-lk);
cjzxp = zeros(number_of_farfield_frequencies,1,uj-lj,uk-lk);
cmyxp = zeros(number_of_farfield_frequencies,1,uj-lj,uk-lk);
cmzxp = zeros(number_of_farfield_frequencies,1,uj-lj,uk-lk);

cjyxn = zeros(number_of_farfield_frequencies,1,uj-lj,uk-lk);
cjzxn = zeros(number_of_farfield_frequencies,1,uj-lj,uk-lk);
cmyxn = zeros(number_of_farfield_frequencies,1,uj-lj,uk-lk);
cmzxn = zeros(number_of_farfield_frequencies,1,uj-lj,uk-lk);

cjxyp = zeros(number_of_farfield_frequencies,ui-li,1,uk-lk);
cjzyp = zeros(number_of_farfield_frequencies,ui-li,1,uk-lk);
cmxyp = zeros(number_of_farfield_frequencies,ui-li,1,uk-lk);
cmzyp = zeros(number_of_farfield_frequencies,ui-li,1,uk-lk);

cjxyn = zeros(number_of_farfield_frequencies,ui-li,1,uk-lk);
cjzyn = zeros(number_of_farfield_frequencies,ui-li,1,uk-lk);
cmxyn = zeros(number_of_farfield_frequencies,ui-li,1,uk-lk);
cmzyn = zeros(number_of_farfield_frequencies,ui-li,1,uk-lk);

cjxzp = zeros(number_of_farfield_frequencies,ui-li,uj-lj,1);
cjyzp = zeros(number_of_farfield_frequencies,ui-li,uj-lj,1);
cmxzp = zeros(number_of_farfield_frequencies,ui-li,uj-lj,1);
cmyzp = zeros(number_of_farfield_frequencies,ui-li,uj-lj,1);

cjxzn = zeros(number_of_farfield_frequencies,ui-li,uj-lj,1);
cjyzn = zeros(number_of_farfield_frequencies,ui-li,uj-lj,1);
cmxzn = zeros(number_of_farfield_frequencies,ui-li,uj-lj,1);
cmyzn = zeros(number_of_farfield_frequencies,ui-li,uj-lj,1);

nf_ff_dist = nf_ff;
idx_ff_dist = idx_ff;
%% loop through each of our frequencies on each thread
for mi=1:number_of_farfield_frequencies
    for t=1:num_threads
        nf_ff = nf_ff_dist{t}; %get cell data
        idx_ff = idx_ff_dist{t}; %get indexing for farfield

        %bring to cpu if needed
        if(use_gpu)
```

```
            nf_ff = move_struct_gpu2cpu(nf_ff);
        end

        if nf_ff.has_xp
        %xp
        cjyxp(mi,1,idx_ff.lj_global-lj+1:idx_ff.uj_global-1-...
lj+1,idx_ff.lk_global-lk+1:idx_ff.uk_global-1-lk+1) = nf_ff.cjyxp(mi,:,:,:);
            cjzxp(mi,1,idx_ff.lj_global-lj+1:idx_ff.uj_global-1-...
lj+1,idx_ff.lk_global-lk+1:idx_ff.uk_global-1-lk+1) = nf_ff.cjzxp(mi,:,:,:);
            cmyxp(mi,1,idx_ff.lj_global-lj+1:idx_ff.uj_global-1-...
lj+1,idx_ff.lk_global-lk+1:idx_ff.uk_global-1-lk+1) = nf_ff.cmyxp(mi,:,:,:);
            cmzxp(mi,1,idx_ff.lj_global-lj+1:idx_ff.uj_global-1-...
lj+1,idx_ff.lk_global-lk+1:idx_ff.uk_global-1-lk+1) = nf_ff.cmzxp(mi,:,:,:);
        end

        if nf_ff.has_xn
        %xn
        cjzxn(mi,1,idx_ff.lj_global-lj+1:idx_ff.uj_global-1-...
lj+1,idx_ff.lk_global-lk+1:idx_ff.uk_global-1-lk+1) = nf_ff.cjzxn(mi,:,:,:);
            cjyxn(mi,1,idx_ff.lj_global-lj+1:idx_ff.uj_global-1-...
lj+1,idx_ff.lk_global-lk+1:idx_ff.uk_global-1-lk+1) = nf_ff.cjyxn(mi,:,:,:);
            cmyxn(mi,1,idx_ff.lj_global-lj+1:idx_ff.uj_global-1-...
lj+1,idx_ff.lk_global-lk+1:idx_ff.uk_global-1-lk+1) = nf_ff.cmyxn(mi,:,:,:);
            cmzxn(mi,1,idx_ff.lj_global-lj+1:idx_ff.uj_global-1-...
lj+1,idx_ff.lk_global-lk+1:idx_ff.uk_global-1-lk+1) = nf_ff.cmzxn(mi,:,:,:);
        end

        if nf_ff.has_yp
        %yp
        cjzyp(mi,idx_ff.li_global-li+1:idx_ff.ui_global-1-...
li+1,1,idx_ff.lk_global-lk+1:idx_ff.uk_global-1-lk+1) = ...
nf_ff.cjzyp(mi,:,:,:);
            cjxyp(mi,idx_ff.li_global-li+1:idx_ff.ui_global-1-...
li+1,1,idx_ff.lk_global-lk+1:idx_ff.uk_global-1-lk+1) = ...
nf_ff.cjxyp(mi,:,:,:);
            cmxyp(mi,idx_ff.li_global-li+1:idx_ff.ui_global-1-...
li+1,1,idx_ff.lk_global-lk+1:idx_ff.uk_global-1-lk+1) = ...
nf_ff.cmxyp(mi,:,:,:);
            cmzyp(mi,idx_ff.li_global-li+1:idx_ff.ui_global-1-...
li+1,1,idx_ff.lk_global-lk+1:idx_ff.uk_global-1-lk+1) = ...
nf_ff.cmzyp(mi,:,:,:);
        end

        if nf_ff.has_yn
        %yn
        cjxyn(mi,idx_ff.li_global-li+1:idx_ff.ui_global-1-...
li+1,1,idx_ff.lk_global-lk+1:idx_ff.uk_global-1-lk+1) = ...
nf_ff.cjxyn(mi,:,:,:);
            cjzyn(mi,idx_ff.li_global-li+1:idx_ff.ui_global-1-...
li+1,1,idx_ff.lk_global-lk+1:idx_ff.uk_global-1-lk+1) = ...
nf_ff.cjzyn(mi,:,:,:);
            cmxyn(mi,idx_ff.li_global-li+1:idx_ff.ui_global-1-...
li+1,1,idx_ff.lk_global-lk+1:idx_ff.uk_global-1-lk+1) = ...
nf_ff.cmxyn(mi,:,:,:);
            cmzyn(mi,idx_ff.li_global-li+1:idx_ff.ui_global-1-...
li+1,1,idx_ff.lk_global-lk+1:idx_ff.uk_global-1-lk+1) = ...
```

```
nf_ff.cmzyn(mi,:,:,:);
        end

        if nf_ff.has_zp
        %zp
        cjxzp(mi,idx_ff.li_global-li+1:idx_ff.ui_global-1-...
li+1,idx_ff.lj_global-lj+1:idx_ff.uj_global-1-lj+1,1) = ...
nf_ff.cjxzp(mi,:,:,:);
            cjyzp(mi,idx_ff.li_global-li+1:idx_ff.ui_global-1-...
li+1,idx_ff.lj_global-lj+1:idx_ff.uj_global-1-lj+1,1) = ...
nf_ff.cjyzp(mi,:,:,:);
            cmxzp(mi,idx_ff.li_global-li+1:idx_ff.ui_global-1-...
li+1,idx_ff.lj_global-lj+1:idx_ff.uj_global-1-lj+1,1) = ...
nf_ff.cmxzp(mi,:,:,:);
            cmyzp(mi,idx_ff.li_global-li+1:idx_ff.ui_global-1-...
li+1,idx_ff.lj_global-lj+1:idx_ff.uj_global-1-lj+1,1) = ...
nf_ff.cmyzp(mi,:,:,:);
        end

        if nf_ff.has_zn
        %zn
        cjxzn(mi,idx_ff.li_global-li+1:idx_ff.ui_global-1-...
li+1,idx_ff.lj_global-lj+1:idx_ff.uj_global-1-lj+1,1) = ...
nf_ff.cjxzn(mi,:,:,:);
            cjyzn(mi,idx_ff.li_global-li+1:idx_ff.ui_global-1-...
li+1,idx_ff.lj_global-lj+1:idx_ff.uj_global-1-lj+1,1) = ...
nf_ff.cjyzn(mi,:,:,:);
            cmxzn(mi,idx_ff.li_global-li+1:idx_ff.ui_global-1-...
li+1,idx_ff.lj_global-lj+1:idx_ff.uj_global-1-lj+1,1) = ...
nf_ff.cmxzn(mi,:,:,:);
            cmyzn(mi,idx_ff.li_global-li+1:idx_ff.ui_global-1-...
li+1,idx_ff.lj_global-lj+1:idx_ff.uj_global-1-lj+1,1) = ...
nf_ff.cmyzn(mi,:,:,:);
        end

    end %mi=1:number_of_farfield_frequencies
end %1:num_threads

fprintf('SUCCESS\n');
```

Because all distributed sampled parameters are named the same as previous implementations, this is the last step required for creating a distributed MATLAB-based FDTD simulation. This implementation allows a variable number of workers as well as the usage of multi-GPU and multi-CPU implementations with the setting of only a single variable.

12.4 Sample Results

The microstrip low-pass filter implementation introduced earlier in this chapter can now be run and timed with this new implementation. The results of a multi-GPU, multi-CPU, and single-CPU can be seen in Figure 12.6. The results

Figure 12.6 Comparison of S21 simulated values for microstrip low-pass filter problem with original single CPU code implementation compared to implementation with 4 CPUs and with 4 GPUs.

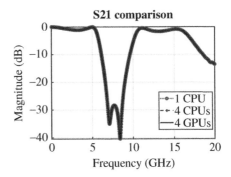

in this figure show that the multi-GPU and multi-CPU implementations produce the same results as the original single CPU implementation.

The overall goal of this parallelization is to decrease the runtime of FDTD simulations with large domain sizes and the number of time steps. For this reason, tests were run to compare speeds with a variety of different hardware configurations. Simulations were run for a variety of FDTD domain sizes. The domain was increased in size by increasing the size of the air gap between the CPML boundary and the device being simulated (low-pass filter configuration). The throughput of the simulation is given in million cells per second (MCPS) as a metric of the code performance.

The multi-CPU implementation was first tested on an AMD Ryzen 2990WX with 128 GB of RAM. The results for these simulation throughputs on a multi-CPU implementation can be seen in Figure 12.7. This plot gives the throughput of the original code as the "Single Core" results, along with the throughputs with a number of cores using the distributed code by means of the *spmd* environment. A "1 CPU core" implementation was tested that used the parallelized code with only a single MATLAB worker. This code was expectedly slower than the "Single Core" results as there was overhead for the *spmd* environment along with the lack of implicit parallelization of operations in the *spmd* environment. It should be noticed that with the increasing number of cores/workers being used, there is a large increase in throughput of the simulation. Each of the curves also has noticeable spikes. These spikes are due to the cache size of the processor allowing a much faster throughput because the simulation does not need to load data from RAM. Also, this spike moves to higher domain sizes as we go from less than 16 to greater than 16 cores. This is because the computer has two physical processors each with 16 physical cores and 32 virtual cores. When the simulation uses more than 16 cores, the cache of the second processor is also utilized, effectively doubling the cache size used in the simulation. As the number of cores is increased beyond the number of physical cores in the system (which in this case is 32), the same increases in throughput are seen as with the physical cores,

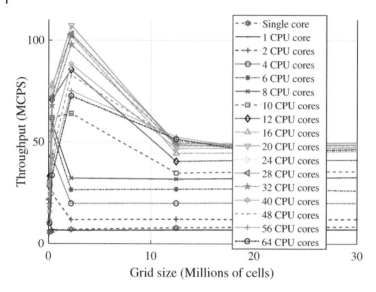

Figure 12.7 Throughput in millions of cells per second for different numbers of cores on two AMD Ryzen 299WX. The original single-core implementation throughput is given by the Single Core results.

and, in some cases, decreases in throughput are observed. This multi-CPU test was also verified with very similar findings on two Intel Xeon E5-2680, each with 8 physical cores and 16 virtual cores. The results for throughput on this computer configuration are given in Figure 12.8.

A similar test was also performed on a multi-GPU configuration. Again, MCPS was used as a metric to test throughput of each of the simulations. Throughput testing was performed on both a system with two Nvidia RTX-2080 GPUs and a system with four Nvidia Titan-Z GPUs. The throughputs for these systems, respectively, can be seen in Figures 12.9 and 12.10. First off, it should be quickly noticed that simply implementing FDTD on the GPU provides a drastic speedup over the CPU for any reasonable domain size. Using the single GPU implementation as a baseline, on both of these two systems for small domain sizes, we see a minimal throughput increase when increasing the number of GPUs due to the communication time required to transfer data from one GPU to another. Unlike direct CUDA libraries, MATLAB provides no control over direct GPU to GPU communication. The implementation here has a side effect that any data communicated between the GPUs must first transfer through the CPU. Because memory must be transferred on each time step, any increase in speed by distributing the grid is offset by the time it takes to gather data from each GPU and send it to another GPU. With the increasing domain size, speed increases become apparent between the different

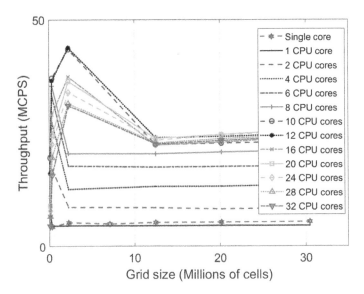

Figure 12.8 Throughput in millions of cells per second for different numbers of cores on two Intel Xeon E5-2680. The original single-core implementation throughput is given by the Single Core results.

Figure 12.9 Throughput in millions of cells per second for different numbers of GPU devices. These tests were performed with up to two Nvidia RTX-2080 GPUs.

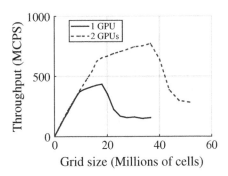

Figure 12.10 Throughput in millions of cells per second for different numbers of GPU devices. These tests were performed with up to four Nvidia Titan-Z GPUs.

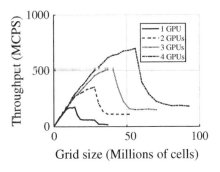

number of GPUs. Eventually, the speed of the GPUs decreases slightly and flattens off for each number of GPUs, which is attributed to the amount of memory on the GPU card. Because in all scenarios (except for those with extremely small domain sizes) increasing the number of GPUs will speed up the simulation, using all available GPUs for the highest possible FDTD throughput is recommended.

12.5 Conclusions

In this chapter, we presented simple modifications to a MATLAB FDTD code to use multiple CPU and GPU devices on a single computing machine. These modifications use MATLAB's explicit parallelization and *spmd* environment to create a parallelized and distributed FDTD simulation. This resulting code can be extended to any number of CPUs or GPUs residing within a single computer system. Sample numerical results for a filter configuration show significant increases in throughput. The parallelized FDTD code can bring runtimes of complex simulations to reasonable time frames for both design and research tasks.

References

1 Inman, M.J., Elsherbeni, A.Z., Maloney, J.G., and Baker, B.N. (2007). GPU based FDTD solver with CPML boundaries. In: *2007 IEEE Antennas and Propagation Society International Symposium*, 5255–5258. https://doi.org/10 .1109/APS.2007.4396732.

2 Hayakawa, K. and Yamano, R. (2015). Multi-core FPGA execution for electromagnetic simulation by FDTD. In: *2015 2nd International Conference on Information Science and Control Engineering*, 829–833. https://doi.org/10.1109/ ICISCE.2015.189.

3 Guo, X.-M., Guo, Q.-X., Zhao, W., and Yu, W. (2012). Parallel FDTD simulation using NUMA acceleration technique. *Progress In Electromagnetics Research Letters* 28: 1–8. https://doi.org/10.2528/PIERL11101706.

4 Nagaoka, T. (2014). Large-scale specific absorption rate computation in various people on GPUs. In: *2014 International Conference on Electromagnetics in Advanced Applications (ICEAA)*, 699–702. https://doi.org/10.1109/ICEAA.2014 .6903947.

5 Diener, J.E. and Elsherbeni, A.Z. (2017). FDTD acceleration using MATLAB parallel computing toolbox and GPU. *Applied Computational Electromagnetics Society (ACES)* 32 (4): 283–288.

6 Weiss, A.J., Elsherbeni, A.Z., Demir, V., and Hadi, M.F. (2019). Using MATLAB's parallel processing toolbox for multi-CPU and multi-GPU accelerated FDTD simulations. *Applied Computational Electromagnetics Society (ACES)* 34 (5): 724–730.

7 "MATLAB Multicore." https://www.mathworks.com/discovery/matlab-multicore.html (accessed 15 June 2022).

8 "Parallel MATLAB: Multiple Processors and Multiple Cores." https://www.mathworks.com/company/newsletters/articles/parallel-matlab-multiple-processors-and-multiple-cores.html (accessed 15 June 2022).

9 "Open MPI: Open Source High Performance Computing." https://www.open-mpi.org (accessed 15 June 2022).

10 "CUDA Zone." *NVIDIA Developer*, 18 Jul 2017. Online at: https://developer.nvidia.com/cuda-zone (accessed 15 June 2022).

11 "MATLAB GPU Computing Support for NVIDIA CUDA Enabled GPUs." https://www.mathworks.com/solutions/gpu-computing.html (accessed 15 June 2022).

12 Mathworks, "Run MATLAB Functions on Multiple GPUs - MATLAB & Simulink Example." Online at: https://www.mathworks.com/help/distcomp/examples/run-matlab-functions-on-multiple-gpus.html (accessed 15 June 2022).

13 Elsherbeni, A. and Demir, V. (2015). *The Finite Difference Time Domain Method for Electromagnetics with MATLAB Simulations*, ACES Series on Computational Electromagnetics and Engineering. Edison, NJ: SciTech Publishing Inc. an Imprint of the IET, Second Edition.

14 Baumeister, P.F., Hater, T., Kraus, J. et al. (2015). A Performance Model for GPU-Accelerated FDTD Applications. In: *2015 IEEE 22nd International Conference on High Performance Computing (HiPC)*, 185–193. https://doi.org/10.1109/HiPC.2015.24.

13

Parallel Subdomain-Level Discontinuous Galerkin Time Domain Method

Jiamei Mi[1], Kaiming Wu[1], Yunfeng Jia[2], Wei Zhang[3], and Qiang Ren[1]

[1]*School of Electronics and Information Engineering, Beihang University, Beijing, China*
[2]*Research Institute of Frontier Science, Beihang University, Beijing, China*
[3]*Sino French Engineering School, Beihang University, Beijing, China*

13.1 Introduction

The discontinuous Galerkin time domain (DGTD) method [1–8] is considered effective and efficient to solve multi-scale problems in computational electromagnetics. Compared with the traditional finite element time domain (FETD) method, the DGTD method introduces the numerical flux to communicate fields between adjacent subdomains/elements; so, it has the advantage of low computational complexity by avoiding solving the sparse matrix equation containing all the degrees of freedom (DoFs) of the entire computational region. The application of the domain decomposition technology means the DGTD method only needs to solve several small matrices in the subdomains, which significantly improves the computational efficiency and has the characteristics of natural parallel computing. In addition, the DGTD method retains the advantages of the complex model with the unstructured mesh of the finite element method and can also calculate the multi-scale problem efficiently. The DGTD method also inherits a variety of strategies in the aspect of time discretization, which can adjust the specific simulation mode according to various specific problems to make it more efficient.

In 1973, W. H. Rood and T. R. Hill of Los Alamos National Key Laboratory first proposed the DGTD method when solving the linear neutron equation [9]. J. S. Hesthaven et al. proposed the discontinuous Galerkin (DG) method based on tetrahedral mesh in 2002 and applied it to the solution of Maxwell's equations [10], which is the first paper of complete theory of DGTD method in the field of computational electromagnetics. Subsequently, different varieties of technologies,

Advances in Time-Domain Computational Electromagnetic Methods, First Edition.
Edited by Qiang Ren, Su Yan, and Atef Z. Elsherbeni.

such as the multi-stage local time stepping scheme, explicit and implicit hybrid time stepping, domain decomposition, and non-conformal mesh, have been proposed to effectively improve the computational efficiency of the DGTD method and reduce the computation cost. However, although the efficiency of the DGTD method has been greatly improved, it is difficult for the existing serial DGTD algorithm to obtain the simulation results of large-scale electromagnetic cases in a timely manner with the trend of increasing complexity and scale of electromagnetic simulation examples, which limits its application scope to a certain extent. Therefore, an efficient parallel technique is indispensable for computational electromagnetics. In this section, a parallel algorithm with an automatic load balancing strategy is proposed for the subdomain-level DGTD method with remarkable reduction of the simulation time.

The DGTD method can be divided into two categories according to the levels of decomposition, namely, the element-level DGTD [11–15] and the subdomain-level DGTD [16–19]. The subdomain-level DGTD are also referred to as DG-FETD or DG-SETD. For the subdomain-level version, the whole computational domain is separated into several pieces, each of which is treated as a subdomain. The electromagnetic fields are continuous within each subdomain, but discontinuous at the interfaces between the adjacent subdomains. The numerical flux serves as the boundary conditions connecting these adjacent subdomains. The element-level DGTD method can be considered as an extreme version of the subdomain-level DGTD method, in which each element is regarded as a separate subdomain and numerical flux is adopted on the common surfaces between neighboring elements. Both methods have their own advantages. The subdomain-level DGTD method not only has practical implicit time stepping approaches for solving electrically small or multi-scale problems but also requires fewer unknowns. However, its effective parallel algorithm is more difficult to implement than the element-level DGTD method, as described below.

Because the number of discretized elements in the element-level DGTD methods are generally much larger than the number of CPUs for a practical problem, fine-grained parallelism can be achieved with the help of mesh partitioning library or software, such as METIS. Moreover, during the time stepping, each element can be moved from one process to another, so it is straightforward and relatively simple to obtain a dynamic load balancing scheme with excellent speedup ratio. Readers interested in this topic can refer to references [20–23] about the parallelism-related progress of the element-level DGTD method. In contrast, there are two difficulties in constructing the parallelization scheme for the subdomain-level DGTD method. *First*, the number of subdomains is comparable to (even if not smaller than) the number of CPUs, which is a typical coarse-grain parallel computing problem instead of a fine-grained one. *Second*, when some elements are transferred between adjacent subdomains, updating the system matrices of the two subdomains and recalculating their lower-upper (LU) decompositions are extremely time-consuming, which hinders the dynamic load balancing. Therefore, efficient parallelism and load balancing schemes for subdomain-level DGTD method (or more generally, domain decomposition-based computational electromagnetic algorithms) have not been well established.

Current load balancing strategies [24, 25] based on domain decomposition are more focused on structural analysis [26] and molecular dynamics problems [27, 28]. They have certain constraints on grid and simulation structure when used in computational electromagnetics. In [29], the final balanced partition is achieved by transferring some elements to adjacent subdomains, which requires that the subdomains have the same grid type, not allowing non-conformal grids. As shown in [30], the binary division method is adopted to reduce the rebalancing time by decreasing the average calculation amount. This method has a high requirement for partition and cannot be used for complex models. In [31], the total time is dynamically reduced to migrate workload and repartition graphics by searching the neighborhood through a traversing tree, but it is seldom used in the subdomain-level DGTD method because dynamic partitioning needs the reassembling of system matrices in each time step, which is inefficient. In [32] and [33], the authors adopt *J Adaptive Unstructured Grid Application Infrastructure* (JAUMIN) framework to accelerate the parallel DGTD method using the components in JAUMIN. However, the technical details of implementation are not available to the readers.

In order to meet the requirement of parallel computing in the subdomain-level DGTD method, the authors of references [34–36] adopt a load balancing strategy combined with some mainstream technologies. But some of these techniques, such as applying multiple types of elements, mixing of low-order and high-order basis functions, and using auxiliary differential equations for a perfectly matched layer (PML), will significantly hinder load balance, that is, simply averaging the number of elements in each process [37–39] does not guarantee load balance. In addition, sometimes overcoming this imbalance will require repeated repartitioning and rebalancing.

To solve this problem, an automatic parallelization scheme of the subdomain-level DGTD method using the message passing interface (MPI) technique is presented in [40]. To estimate the time consumption of subdomains effectively, the time consumption cost databases for different element types with respect to DoFs are obtained by numerical experiments. Note that this database only needs to be generated offline once and can be used in the online calculation procedure. Then, by initially setting the interface location of each subdomain, as well as the type and order of elements within each subdomain, the time consumption of each subdomain can be estimated according to the database above. Multiple repartitioning iterations are automatically performed until the time consumption difference between subdomains is less than a preset threshold. In this method, the system matrix of each subdomain only needs to be assembled and operated once after the final partition is determined.

The arrangement of sections in this chapter is as follows. Section 13.2 reviews the subdomain-level DGTD method and compares it with the element-level DGTD method. Section 13.3 presents the parallel algorithm and automatic load balancing strategy, as well as provides some computational examples simulated by the proposed parallel subdomain-level DGTD method. Section 13.4 shows how the proposed parallel strategy is applied to some large examples, and the effectiveness of the method is verified through the time and speedup ratio recorded in the cases of multi-process configuration. Finally, Section 13.5 draws the conclusions.

13.2 Comparison of Parallel Element- and Subdomain-Level DGTD Methods

13.2.1 Subdomain-Level DGTD

The subdomain-level DGTD method uses the domain decomposition technique. The method divides the whole problem into several nonoverlapping subdomains. The element types in different subdomains can be the same or different, and the finite element mesh generation between the interfaces of subdomains can be consistent or inconsistent. Each subdomain is solved independently within a time step, and the field communicates on the interface between subdomains through numerical flux. The subdomain-level DGTD method is therefore especially suitable for parallel computation, complex geometry, and heterogeneous materials.

13.2.2 Discretized System

The first-order Maxwell's curl equations with electric field intensity $\mathbf{E}$ and magnetic flux density $\mathbf{B}$ are shown below:

$$\varepsilon \frac{\partial \mathbf{E}}{\partial t} = \nabla \times \mu^{-1}\mathbf{B} - \sigma_e\mathbf{E} - \mathbf{J} \tag{13.1}$$

$$\frac{\partial \mathbf{B}}{\partial t} = -\nabla \times \mathbf{E} - \sigma_m\mu^{-1}\mathbf{B} - \mathbf{M} \tag{13.2}$$

where $\mathbf{J}$ and $\mathbf{M}$ are the electric and magnetic current densities; ε, μ, σ_e, and σ_m denote the material's permittivity, permeability, electric conductivity, and magnetic conductivity, respectively.

Denote $\boldsymbol{\Phi}$ and $\boldsymbol{\Psi}$ as the testing functions. After testing, the Galerkin's weak forms of the Maxwell's equations are

$$\int_V \boldsymbol{\Phi}_p^i \cdot \left(\varepsilon \frac{\partial \mathbf{E}^i}{\partial t} + \sigma_e \mathbf{E}^i + \mathbf{J}^i \right) dV = \int_V \nabla \times \boldsymbol{\Phi}_p^i \cdot \mu^{-1} \mathbf{B}^i dV + \int_S \boldsymbol{\Phi}_p^i \cdot (\mathbf{n}^i \times \mu^{-1} \mathbf{B}^t) dS$$

(13.3)

$$\int_V \boldsymbol{\Psi}_p^i \cdot \left(\frac{\partial \mathbf{B}^i}{\partial t} + \sigma_m \mu^{-1} \mathbf{B}^i + \mathbf{M}^i \right) dV = -\int_V \nabla \times \boldsymbol{\Psi}_p^i \cdot \mathbf{E}^i dV - \int_S \boldsymbol{\Psi}_p^i \cdot (\mathbf{n}^i \times \mathbf{E}^t) dS$$

(13.4)

The discretized equations of the i-th subdomain in the subdomain-level DGTD method can be obtained by using the Riemann solver:

$$\mathbf{M}_{ee}^i \frac{d\mathbf{e}^i}{dt} = \mathbf{K}_{eb}^i \mathbf{b}^i + \mathbf{C}_{ee}^i \mathbf{e}^i + \mathbf{j}^i + \sum_{j=1}^N \mathbf{L}_{be}^{ij} \mathbf{e}^j + \sum_{j=1}^N \mathbf{L}_{ee}^{ij} \mathbf{e}^j$$

(13.5)

$$\mathbf{M}_{bb}^i \frac{d\mathbf{b}^i}{dt} = \mathbf{K}_{be}^i \mathbf{e}^i + \mathbf{C}_{bb}^i \mathbf{b}^i + \mathbf{m}^i + \sum_{j=1}^N \mathbf{L}_{eb}^{ij} \mathbf{b}^j + \sum_{j=1}^N \mathbf{L}_{bb}^{ij} \mathbf{b}^j$$

(13.6)

where

$$(\mathbf{M}_{ee})_{kl} = \int_V \varepsilon \boldsymbol{\Phi}_k \cdot \boldsymbol{\Phi}_l dV$$

(13.7)

$$(\mathbf{M}_{bb})_{kl} = \int_V \boldsymbol{\Psi}_k \cdot \boldsymbol{\Psi}_l dV$$

(13.8)

$$(\mathbf{C}_{ee})_{kl} = \int_V \sigma_e \boldsymbol{\Phi}_k \cdot \boldsymbol{\Phi}_l dV$$

(13.9)

$$(\mathbf{C}_{bb})_{kl} = \int_V \boldsymbol{\Psi}_k \cdot \sigma_m \mu^{-1} \boldsymbol{\Psi}_l dV$$

(13.10)

$$(\mathbf{K}_{eb})_{kl} = \int_V \nabla \times \boldsymbol{\Phi}_k \cdot \mu^{-1} \boldsymbol{\Psi}_l dV$$

(13.11)

$$(\mathbf{K}_{be})_{kl} = -\int_V \boldsymbol{\Psi}_k \cdot \nabla \times \boldsymbol{\Phi}_l dV$$

(13.12)

$$(\mathbf{j})_k = -\int_V \boldsymbol{\Phi}_k \cdot \mathbf{J} dV$$

(13.13)

$$(\mathbf{m})_k = -\int_V \boldsymbol{\Psi}_k \cdot \mathbf{M} dV$$

(13.14)

$$\left(\mathbf{L}_{ee}^{ij} \right)_{kl} = \frac{1}{Z^{ij}} \int_S \left(\hat{\mathbf{n}} \times \boldsymbol{\Psi}_k^i \right) \cdot \left(\hat{\mathbf{n}} \times \boldsymbol{\Psi}_l^j \right) dS, \quad i \neq j$$

(13.15)

$$\left(\mathbf{L}_{eb}^{ij} \right)_{kl} = -\frac{Z^j}{Z^{ij}} \int_S \boldsymbol{\Phi}_k^i \cdot \left(\mathbf{n} \times \mu^{-1} \boldsymbol{\Psi}_l^j \right) dS, \quad i \neq j$$

(13.16)

$$\left(\mathbf{L}_{be}^{ij} \right)_{kl} = \sum_{j=1}^N \frac{-Y^i}{Y^{ij}} \int_S \boldsymbol{\Psi}_k^i \cdot \left(\mathbf{n} \times \boldsymbol{\Phi}_l^j \right) dS, \quad i \neq j$$

(13.17)

$$\left(\mathbf{L}_{bb}^{ij}\right)_{kl} = \sum_{j=1}^{N} \frac{1}{Y^{ij}} \int_S \left(\mathbf{n} \times \mathbf{\Psi}_k^i\right) \cdot \left(\mathbf{n} \times \mu^{-1}\mathbf{\Psi}_l^j\right) dS, \quad i \neq j \tag{13.18}$$

$$\left(\mathbf{L}_{ee}^{ij}\right)_{kl} = \frac{1}{Z^{ij}} \int_S \left(\mathbf{n} \times \mathbf{\Phi}_k^i\right) \cdot \left(\mathbf{n} \times \mathbf{\Phi}_l^j\right) dS \tag{13.19}$$

$$\left(\mathbf{L}_{eb}^{ii}\right)_{kl} = -\frac{Z^j}{Z^{ij}} \sum_{j=1}^{N} \int_S \mathbf{\Phi}_k^i \cdot \left(\mathbf{n} \times \mu^{-1}\mathbf{\Psi}_l^i\right) dS \tag{13.20}$$

$$\left(\mathbf{L}_{be}^{ii}\right)_{kl} = \sum_{j=1}^{N} \frac{Y^i}{Y^{ij}} \int_S \mathbf{\Psi}_k^i \cdot \left(\mathbf{n} \times \mathbf{\Phi}_l^i\right) dS \tag{13.21}$$

$$\left(\mathbf{L}_{bb}^{ii}\right)_{kl} = \sum_{j=1}^{N} \frac{1}{Y^{ij}} \int_S \left(\mathbf{n} \times \mathbf{\Psi}_k^i\right) \cdot \left(\mathbf{n} \times \mu^{-1}\mathbf{\Psi}_l^i\right) dS \tag{13.22}$$

and j is the index of the adjacent subdomain. $Z^i = \sqrt{\mu^i}/\sqrt{\varepsilon^i}$ and $Y^i = 1/Z^i$ are the wave impedances and admittance for the i-th subdomain, respectively. We have $Z^{ij} = Z^i + Z^j$ and $Y^{ij} = Y^i + Y^j$.

The interface values $\hat{n} \times \mathbf{E}^t$ and $\hat{n} \times \mathbf{B}^t$ can be obtained by the Riemann solver:

$$(\mathbf{n}^i \times \mathbf{E}^t) = \frac{\mathbf{n}^i \times (Y^i \mathbf{E}^i + Y^j \mathbf{E}^j)}{Y^i + Y^j} - \frac{\mathbf{n}^i \times \mathbf{n}^i \times (\mu^j \mathbf{B}^i - \mu^i \mathbf{B})^j}{\mu^i \mu^j (Y^i + Y^j)} \tag{13.23}$$

$$\left(\mathbf{n}^i \times \frac{\mathbf{B}^t}{\mu}\right) = \frac{\mathbf{n}^i \times (\mu^j Z^i \mathbf{B}^i + \mu^i Z^j \mathbf{B}^j)}{\mu^i \mu^j (Z^i + Z^j)} - \frac{\mathbf{n}^i \times \mathbf{n}^i \times (\mathbf{E}^i - \mathbf{E}^j)}{Z^i + Z^j} \tag{13.24}$$

In this section, the PML is used to truncate the computational domain of the open space problem. The discrete equation with the PML should be modified to

$$\mathbf{M}_{ee}^i \frac{d\tilde{\mathbf{e}}^i}{dt} = \mathbf{K}_{eb}^i \tilde{\mathbf{b}}^i + \mathbf{R}_{ee}^i \tilde{\mathbf{e}}^i + \mathbf{S}_{ee}^i \overline{\mathbf{e}}^i + \mathbf{j}^i + \sum_{j=1}^{N} \mathbf{L}_{eb}^{ij} \tilde{\mathbf{b}}^j + \sum_{j=1}^{N} \mathbf{L}_{ee}^{ij} \tilde{\mathbf{e}}^j \tag{13.25}$$

$$\mathbf{M}_{bb}^i \frac{d\tilde{\mathbf{b}}^i}{dt} = \mathbf{K}_{be}^i \tilde{\mathbf{e}}^i + \mathbf{R}_{bb}^i \tilde{\mathbf{b}}^i + \mathbf{S}_{bb}^i \overline{\mathbf{b}}^i + \mathbf{m}^i + \sum_{j=1}^{N} \mathbf{L}_{be}^{ij} \tilde{\mathbf{e}}^j + \sum_{j=1}^{N} \mathbf{L}_{bb}^{ij} \tilde{\mathbf{b}}^j \tag{13.26}$$

$$\mathbf{M}_{ee}^i \frac{d\overline{\mathbf{e}}^i}{dt} = \mathbf{M}_{ee}^i \tilde{\mathbf{e}}^i + \mathbf{T}_{ee}^i \overline{\mathbf{e}}^i \tag{13.27}$$

$$\mathbf{M}_{bb}^i \frac{d\overline{\mathbf{b}}^i}{dt} = \mathbf{M}_{bb}^i \tilde{\mathbf{b}}^i + \mathbf{T}_{bb}^i \overline{\mathbf{b}}^i \tag{13.28}$$

where

$$(\mathbf{R}_{ee})_{kl} = -\int_V \mathbf{\Phi}_k \cdot \varepsilon \Lambda_1 \mathbf{\Phi}_l \, dV \tag{13.29}$$

$$(\mathbf{S}_{ee})_{kl} = -\int_V \mathbf{\Phi}_k \cdot \varepsilon \Lambda_2 \mathbf{\Phi}_l \, dV \tag{13.30}$$

$$(\mathbf{T}_{ee})_{kl} = -\int_V \mathbf{\Phi}_k \cdot \varepsilon \Lambda_0 \mathbf{\Phi}_l \, dV \tag{13.31}$$

$$(\mathbf{R}_{bb})_{kl} = -\int_V \mathbf{\Psi}_k \cdot \mathbf{\Lambda}_1 \mathbf{\Psi}_l \, dV \qquad (13.32)$$

$$(\mathbf{S}_{bb})_{kl} = -\int_V \mathbf{\Psi}_k \cdot \mathbf{\Lambda}_2 \mathbf{\Psi}_l \, dV \qquad (13.33)$$

$$(\mathbf{T}_{bb})_{kl} = -\int_V \mathbf{\Psi}_k \cdot \mathbf{\Lambda}_0 \mathbf{\Psi}_l \, dV \qquad (13.34)$$

13.2.3 Non-conformal Mesh

The whole computation region is decomposed into several computational subdomains and the interface between two adjacent subdomains can be non-conformal [40, 41]. As shown in Figure 13.1, the overlapping area of these two elements from adjacent subdomains with different types of elements is a planar surface polygon, which is divided into several triangles. And by integrating the functions on these triangles and summing them together, we can obtain the integration value of the function on this surface.

13.2.4 Time Stepping

Using the fourth-order explicit Runge–Kutta method, (13.25) and (13.26) can be described as

$$\mathbf{M}^i \frac{d\tilde{\mathbf{v}}^i}{dt} = \mathbf{L}^{ii}\tilde{\mathbf{v}}^i + \sum_{i=1, j\neq 1}^{N} \mathbf{L}^{ii}\tilde{\mathbf{v}}^j + \mathbf{S}^i\overline{\mathbf{v}}^i \qquad (13.35)$$

$$\mathbf{M}^i \frac{d\overline{\mathbf{v}}^i}{dt} = \mathbf{M}^i\tilde{\mathbf{v}}^i + \mathbf{T}^i\overline{\mathbf{v}}^i \qquad (13.36)$$

where

$$\mathbf{L}^{ii} = \begin{bmatrix} \mathbf{R}^i_{ee} + \mathbf{L}^{ii}_{ee} & \mathbf{K}^i_{eb} + \mathbf{L}^{ii}_{eb} \\ \mathbf{K}^i_{be} + \mathbf{L}^{ii}_{be} & \mathbf{T}^i_{bb} + \mathbf{L}^{ii}_{bb} \end{bmatrix}, \quad \mathbf{L}^{ij} = \begin{bmatrix} \mathbf{L}^{ij}_{ee} & \mathbf{L}^{ij}_{eb} \\ \mathbf{L}^{ij}_{be} & \mathbf{L}^{ij}_{bb} \end{bmatrix}, \quad j \neq i \qquad (13.37)$$

$$\mathbf{S}^i = \begin{bmatrix} \mathbf{S}^i_{ee} & 0 \\ 0 & \mathbf{S}^i_{bb} \end{bmatrix}, \quad \mathbf{T}^i = \begin{bmatrix} \mathbf{T}^i_{ee} & 0 \\ 0 & \mathbf{T}^i_{bb} \end{bmatrix} \qquad (13.38)$$

$$\tilde{\mathbf{v}}^i = \begin{bmatrix} \tilde{\mathbf{e}}^i \\ \tilde{\mathbf{b}}^i \end{bmatrix}, \quad \overline{\mathbf{v}}^i = \begin{bmatrix} \overline{\mathbf{e}}^i \\ \overline{\mathbf{b}}^i \end{bmatrix} \qquad (13.39)$$

Figure 13.1 Treatment of overlapping polygons for interfaces between adjacent subdomains.

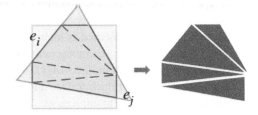

and the coefficients $a_{k,l}$, b_k, and c_k are from the Butcher tableau. For an ordinary physical subdomain, the formula above can be simplified, i.e. we only need to set $\bar{v}$ and $\bar{u}$ to 0. The details can be found in [4].

13.2.5 Element-Level DGTD

The element-level DGTD method can be regarded as an extreme category of the subdomain-level DGTD method. Each element is regarded as a computational subdomain, and all interfaces between adjacent elements adopt numerical flux. It is more flexible in terms of geometric modeling. The discretized system of the element-level DGTD method is similar to that of the subdomain level:

$$\mathbf{M}_{ee}^i \frac{d\mathbf{e}^i}{dt} = \mathbf{K}_{eb}^i \mathbf{b}^i + \mathbf{C}_{ee}^i \mathbf{e}^i + \mathbf{j}^i + \sum_{j=1}^{N} \mathbf{L}_{be}^{ij} \mathbf{e}^j + \sum_{j=1}^{N} \mathbf{L}_{ee}^{ij} \mathbf{e}^j \tag{13.40}$$

$$\mathbf{M}_{bb}^i \frac{d\mathbf{b}^i}{dt} = \mathbf{K}_{be}^i \mathbf{e}^i + \mathbf{C}_{bb}^i \mathbf{b}^i + \mathbf{m}^i + \sum_{j=1}^{N} \mathbf{L}_{eb}^{ij} \mathbf{b}^j + \sum_{j=1}^{N} \mathbf{L}_{bb}^{ij} \mathbf{b}^j \tag{13.41}$$

where i represents the *local* element and j represents the *adjacent* element.

In the element-level DGTD method, after the computation region is meshed into elements, the grid allocates the entire domain into several subdomains, which are processed by blocks, and finally the block allocates each element to the corresponding thread. Compute unified device architecture (CUDA) [42], METIS, and other tools can be used to accelerate the element-level algorithms, making it easy to balance the load of parallel algorithms with fine granularity.

13.2.6 Challenges of Subdomain-Level DGTD Parallelization

According to the above analysis, compared with the element-level DGTD method, the subdomain-level DGTD method has the following characteristics: (i) The number of subdomains is much less than the number of elements. Therefore, the parallel subdomain-level DGTD method is a coarse-grained parallel method, while the parallel element-level DGTD method is a fine-grained parallel method. Usually, the load difference between the most computationally intensive subdomain and the least computationally intensive one in the subdomain-level DGTD method may be large, which makes the load balance of the subdomain-level DGTD method difficult. (ii) The cost of repartitioning is high, and the conversion of different element types between subdomains requires the entire subdomain to be re-meshed and the system matrix to be reassembled. Thus, the cost of repartitioning in the subdomain-level DGTD method is higher than that in the element-level DGTD method. (iii) Repartitioning cannot generate new subdomains. For the parallelization process of the subdomain-level DGTD method, if any subdomain is generated or is deleted, it will affect the interface information of other subdomains, and the algorithm will be difficult to implement.

13.3 Parallelization Scheme for Subdomain-Level DGTD Methods

As known, the efficiency of parallel algorithms depends on the parallel algorithm itself on the one hand and on the load balance between processes on the other hand. This section will describe the specific implementation steps of the automatic load balancing strategy of the subdomain-level DGTD method. As shown in Figure 13.2, the region (the center of the region is a sphere containing a curved structure) is divided into N subdomains. These subdomains may be gridded by different types of elements or even different orders of basis functions, which will affect the system matrices generated in the subdomain-level DGTD method. So even if the DoF of each subdomain is the same, their simulation speed and memory consumption may still be different. For example, basis functions of different orders will lead to different sparsity of the system matrix generated by the DGTD method, so the computational load of each subdomain will be different, even if they have the same DoF. Also, because the PML subdomain requires additional auxiliary differential equations (ADEs) to be solved, the calculation time differs from the physical region even if the element type and the order of basis functions are the same. Therefore, in the subdomain-level DGTD method, the key to parallelization is workload balancing between processes.

Besides the computation load of the subdomains, the communication time is also one of the factors affecting the efficiency of the parallel algorithm. The communication load is generated from the data exchange between adjacent subdomains of different processes, and the main factor affecting the communication time is the DoFs of interprocess interface. Therefore, in addition to the computation load of the processes, the communication between them should also be considered. Therefore, finding a subdomain division method that balances the computation and communication time is the key to improve the efficiency of the

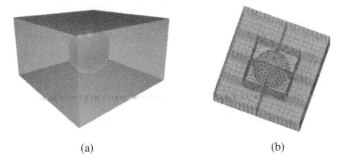

(a)   (b)

Figure 13.2 An example of the domain decomposition with different types of elements. (a) Geometry and (b) the divided subdomains with tetrahedral and hexahedral meshes.

parallel subdomain-level DGTD method. The following is the definition of the symbols used in this chapter:

p: the number of processes;
Th: the maximum load difference acceptable between processes;
N: the number of repartitions;
N_{max}: the preset maximum number of iterations for repartitioning;
w_i: the total workload of the i-th process;
$\overline{w}$: average workload of all processes;
δ_{max}: $\delta_{max} = max_{i,j} |w_i - w_j|$, the maximal workload difference between processes;
T_{cal}^j: calculation time of the j-th process;
T_{com}^j: communication time of the j-th process;
T_{est}^j: estimated calculation time of the j-th process
T_{total}^j: total time consumption of the j-th process;
α_i: time required to solve the first term of matrix **L** in (13.5) and (13.6) $(j \neq i)$;
β_i: the time required for data exchange on the interface of the i-th subdomain
m_1^j: the number of interfaces that do not need to communicate in the j-th process;
m_2^j: the number of interfaces that need to communicate in the j-th process;
m^j: $m^j = m_1^j + m_2^j$, total number of interfaces in the j-th process;

13.3.1 Automatic Load Balancing

There are two difficulties in constructing the parallel scheme of the subdomain-level DGTD method. *First,* the number of subdomains is usually comparable to or even smaller than the number of CPUs, which hinders fine-grained parallelism. *Second,* once an element is transferred from one subdomain to another subdomain, the system matrices of these two subdomains need to be reassembled to continue subsequent calculations, which is very time-consuming and reduces the feasibility of dynamic load balancing. Therefore, efficient parallel and effective load balancing methods for the subdomain-level DGTD method (or more generally, methods based on domain decomposition) have not been well developed. Existing load balancing methods, such as the static domain method based on domain decomposition proposed in [27, 28], have difficulty in achieving good load balancing for the subdomain-level DGTD method.

Therefore, in this section, we will describe an adaptive parallel strategy to solve the problem of load imbalance between processes. The general idea is as follows: (i) Regress the relationship between the DoFs and the calculation time (or communication time) according to the characteristics (such as the mesh element type and/or the interpolation order of the elements) of the subdomains. (ii) Decompose the calculation region into several subdomains and follow step 1 to obtain the calculation time required for each subdomain. (iii) Distribute the load to multiple processes evenly according to the calculation time. The load between the processes will be relatively balanced. (iv) Calculate the communication time between each process and add them to the corresponding calculation time to get the total time of each process. (v) Adjust the subdomain distribution position according to the total time, that is, adjust the load of the subdomain to make the load difference between the processes minimal. (vi) Repeat steps vi and v until the load difference between processes meets the requirements or reaches the maximal number of iterations. Therefore, automatic load balancing is mainly divided into two parts [43]: average process load and reduce inter-process communication time.

13.3.1.1 Preprocessing

As shown in Figure 13.3, the relationship between calculation time and DoFs of hexahedral (physical region and PML region) and tetrahedral (only physical region) elements with different orders was retrieved through numerical experiments. The cubic region is used to generate the time–DoFs curve in Figure 13.3. Since the time consumption only depends on the number of unknowns, not the mesh, this numerical result is representative. Because of the employment of the direct solver, different cases with the same DoFs require almost the same CPU time. For the hexahedral mesh, the slope of time and DoF is close to 1, indicating that the time complexity is close to $O(N)$, while for the tetrahedral mesh, it is close to $O(N^{1.5})$. For the same DoFs, the time efficiency of the hexahedral physical region is the highest, while the time efficiency of the tetrahedral physical region is the lowest.

In this study, we assume that the number of subdomains is larger than the number of the processes. Figure 13.4 shows an adaptive scheme to partition

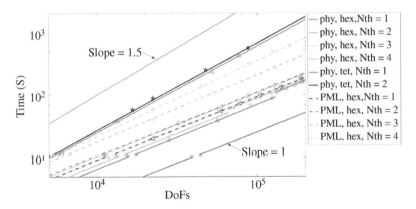

Figure 13.3 Time overhead in 3000 steps with different DoFs (tet: tetrahedral element, hex: hexahedral element, Nth: element order, phy: physical region, PML: PML region). Source: Mi et al. [44]. Reproduced with permission of IEEE.

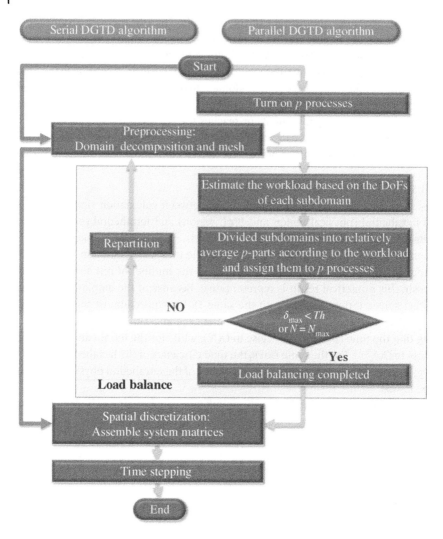

Figure 13.4 Flowchart of the serial subdomain-level DGTD method and the parallel subdomain-level DGTD method with automatic load balancing.

the subdomains. For a particular electromagnetic problem with a required frequency band, initial partitioning can be presented according to its geometric characteristics. Then the order of the basis functions can be selected to meet the accuracy requirement, and the DoFs of each subdomain can be determined. After that, the computation load of each subdomain can be estimated through the curve in Figure 13.3 to provide reference on load distribution for subsequent processes.

13.3.1.2 Repartition of Subdomains

The subdomains are repartitioned to reset the load balancing lines and achieve load balance. Load balancing lines divide the entire computational region into p parts with roughly identical volumes. Since they are virtual lines, they usually pass through the subdomains rather than locating exactly at the boundaries between the adjacent subdomains. However, finally the subdomains' boundaries will adjust to these load balancing lines.

Compared with the serial algorithm, the parallel DGTD method not only has the basic steps, such as preprocessing, spatial discretization, and time stepping, but also requires certain constraints on the subdomain division. Therefore, the load balancing process needs to be added before the spatial discretization.

In each iteration of *repartition*, the subdomain with the largest amount of computational load is selected, and some of its elements will be transferred to its adjacent subdomain across the load balancing line in the adjacent process. The implementation steps are shown below:

1. Calculate the average workload of all processes, $\overline{w} = \frac{1}{p} \sum_{i=1}^{p} w_i$.
2. Select the most computation-intensive subdomain around the load balancing lines in a process with the maximal workload.
3. Determine the moving direction of interface by the workload of the adjacent process.
4. Calculate the density of the elements in this direction.
5. Calculate the moving distance of interfaces in the direction of step 3 according to $\overline{w}$.
6. Repeat steps 1–5 until the maximal workload difference between the processes, δ_{max}, is less than the preset threshold Th or the number of iterations N reaches N_{max}.

Repartitioning has two constraints: (i) The natural interface cannot be modified, which means that the elements inside the physical regions cannot be transferred to the PML regions, and vice versa. (ii) A new interface cannot be created, that is, we cannot increase or decrease the number of subdomains. Figure 13.5 shows the subdomain repartitioning of the cases with four processes in Figure 13.2. In general, the movement is carried out at the boundary of the subdomain with the maximum computational load near the load balancing lines. Different mesh-type elements in adjacent subdomains can be transferred between each other. Tetrahedrons are usually utilized for fine structures and curved surfaces while hexahedrons are selected for the reason of efficiency.

13.3.1.3 Communication Time Reduction

Strips or boxes are the most common strategy to balance the workload on multiple processors [28], effectively reducing the communication load on transmitting

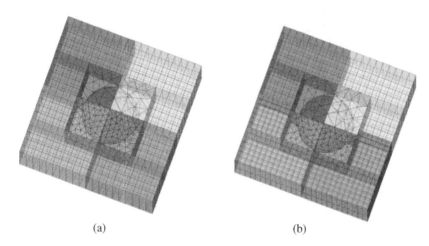

(a)   (b)

Figure 13.5 Repartition of the example in Figure 13.2. (a) Initial mesh and (b) mesh after load balance.

information to adjacent processors. Ignoring the overlap time of communication and the time stepping, the total time for the j-th process is $T^j_{total} = T^j_{cal} + T^j_{com}$. Due to different numbers of interfaces involved in the integration on the interfaces in different subdomains, T^j_{total} will be calculated as follows:

$$T^j_{cal} = T^j_{est} + \sum_{i=1}^{m^j} \alpha_i = T^j_{est} + \sum_{i=1}^{m^j_1} \alpha_i + \sum_{i=m^j_1+1}^{m^j_1+m^j_2} \alpha_i \qquad (13.42)$$

DoFs$_{i(j)}$ represents the DoFs of the $i(j)$-th subdomain. α is related to the matrix **L** determined by the DoFs of the subdomain and its adjacent subdomains, as shown in Table 13.1.

Communication time is affected by m^j_2, which is the number of inter-process interfaces in the j-th process, so the estimated total CPU time of the j-th process

Table 13.1 The time overhead of 3000 time steps for interface integration.

$\alpha(s)$ DoFs$_j$ DoFs$_i$	5000	10,000	20,000	50,000
5 k	0.55	0.59	0.61	0.65
10 k	0.67	0.71	0.72	0.75
20 k	0.86	0.88	0.91	0.95
50 k	1.34	1.35	1.37	1.40

Table 13.2 The relationship between the 3000 communications and DoFs of interface.

DoFs	10	100	500	1000	2000	5000	10,000	20,000
β (ms)	3.67	4.08	5.71	7.57	10.95	18.67	26.39	44.64

can be described as:

$$T^j_{\text{total}} = T^j_{\text{cal}} + \sum_{i=m'_i+1}^{m'_i+m'_i} \beta_i = T^j_{\text{est}} + \sum_{i=1}^{m'_i} \alpha_i + \sum_{i=m'_1+1}^{m'_i+m'_2} (\alpha_i + \beta_i) \tag{13.43}$$

where β_i represents the communication overhead of all the interfaces of the i-th subdomain. Table 13.2 shows the relationship between DoFs of the interfaces and the communication time.

In Figure 13.4, the number of interfaces between processes should be also taken into account when repartitioning the subdomains. As shown in Figure 13.6, the computation load of these four processes in assumption (a) is relatively similar, but the data exchange in assumption (b) is less, so the communication time of (b) is less than (a). If the total time of difference between these four processes in (b) is less than (a), then partition (b) will be used.

13.3.2 Numerical Result

Parallel algorithm has two advantages: remarkable acceleration and the ability of handling larger models. The parallel algorithm using p processes to solve the

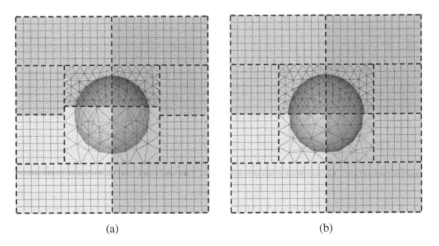

(a)   (b)

Figure 13.6 Two partitions for the same case. (a) Less computational load difference with more communication load configuration. (b) More computational load difference with less communication load configuration.

same problem is ideally p times faster than its serial counterpart. However, this acceleration is practically impossible because of the communication load between processes in parallel algorithms. In order to facilitate the evaluation of the performance of the parallel algorithm procedure, we consider some parameters below [45]:

1. **Speedup ratio (S_p):** T_1 is the execution time of a serial algorithm and T_p is the total execution time of a parallel algorithm using p processes calculating the same case. Then the speedup ratio S_p can be defined as $S_p = T_1/T_p$.
2. **Efficiency (E_p):** $E_p = S_p/p$.
3. **Scalability:** Good scalability means that the efficiency E_p decreases slightly as the number of processes p increases.

All the examples are simulated on a server with two Intel(R) Xeon(R) Silver 4110 CPUs (2.10 GHz/8 cores) and 32 GB of memory using a parallel version of the subdomain-level DGTD method.

13.3.2.1 Metallic Sphere

First, to validate and demonstrate the accuracy of this parallel DGTD method, the scattering property of a metallic sphere is investigated. The incident waveform is a Blackman–Harris window (BHW) pulse with the characteristic frequency of 30 MHz ($\lambda = 10$ m). The radius of this metallic sphere is 8 m (0.8λ). The simulation region is truncated by PML with the thickness of one wavelength. Mesh size of PML is 2.5 m (0.25λ) and the mesh size of vacuum is 1 m (0.1λ) (Figure 13.7).

Figure 13.8 shows the bistatic radar cross section (RCS) of this metallic sphere at 30 MHz. For comparison, the reference by the Mie series method ($N = 40$) is also present.

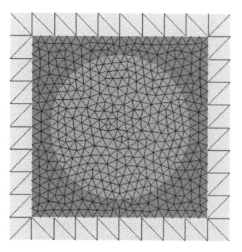

Figure 13.7 The mesh of the metallic sphere on the y–z plane (the coarse mesh grid domain: PML subdomain, the fine mesh grid domain: vacuum and metallic sphere subdomains.

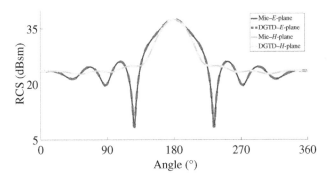

Figure 13.8 The bistatic RCS of the metallic sphere at 30 MHz including E-plane and H-plane.

13.3.2.2 Low-Pass Filter

The simulation of low-pass filter verifies the effectiveness of the parallel subdomain-level DGTD method. The geometry is shown in Figure 13.9a, and the detailed dimensions in the x–z plane and y–z plane are illustrated in Figure 13.9b,c. The negative and positive terminals of the active lumped port are located at (7.8565, 1.5000, 0) mm and (7.8565, 1.5000, 0.7940) mm, respectively, while the passive lumped port is located between (14.4635, 15.0400, 0) mm and (14.4635, 15.0400, 0.7940) mm. The resistance of the ports is 50 Ω. The excitation source is the first derivative of the BHW pulse with a maximal frequency of 20 GHz. The upper and lower layers are perfect electric conductor (PEC), and the dielectric constant of the middle layer is 2.2. The center of the filter is located at (11.1600, 8.2700, 0.5264) mm. Divided into four subdomains with hexahedral mesh, the filter is truncated by PML with the thickness of 7.5 mm in all directions, as shown in Figure 13.10.

To verify the effectiveness of the automatic load balance strategy, we test the performance of parallel subdomain-level DGTD method with multiple processes. Figure 13.11a shows the initial domain decomposition on the x–y plane. As introduced in Section 13.3.1, the initial decomposition boundaries will be adjusted to balance the workload between processes. In order to analyze the scalability, we set the number of process to be 2, 4, and 8, as shown in Figure 13.11b–d. The first consideration when moving the boundary is to reduce the largest subdomain's DoFs. Both the characteristics of the simulation region and the number of processes affect the load balancing. In this filter example, almost the same computational time of processes can be achieved when the number of processes is 2. While in the 8 processes configuration, if the number of subdomains is 12, its efficiency cannot be as good as the configuration with 2 processes. But it still works more efficiently than the static domain decomposition parallel scheme. In the absence

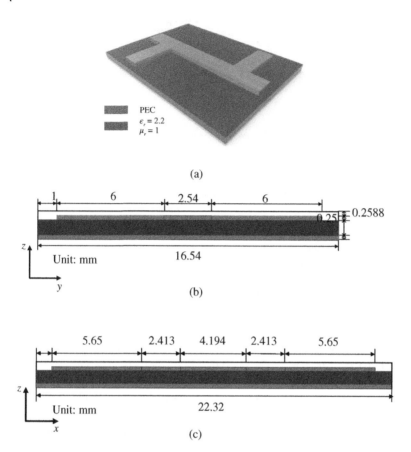

(a)

(b)

(c)

Figure 13.9 Geometry of the low-pass filter. (a) Overview; (b) $y-z$ plane; (c) $x-z$ plane. Source: Mi et al. [44]. Reproduced with permission of IEEE.

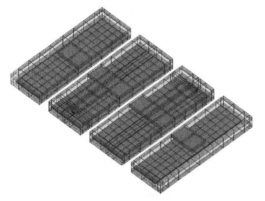

Figure 13.10 Mesh of the low-pass filter with hexahedral elements. Source: Mi et al. [44]. Reproduced with permission of IEEE.

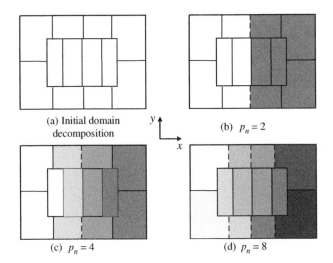

(a) Initial domain decomposition

y

x

(b) $p_n = 2$

(c) $p_n = 4$

(d) $p_n = 8$

Figure 13.11 The domain decomposition of the low-pass filter on the $x-y$ plane. (p_n is the number of processes, and each color represents the subdomains assigned to a certain process. The dotted lines represent the moving boundaries). Source: Mi et al. [44]. Reproduced with permission of IEEE.

of load balancing, the computation load difference is more than 50%, and this automatic load balance strategy can reduce it to less than 15%.

Figure 13.12 presents the total voltage of port 1, the scattering voltages of port 1 and port 2, and the wideband S parameters, which are all in good agreement with the results from the finite-difference time-domain (FDTD) method. Because of the Courant–Friedrichs–Levy (CFL) condition, the time step Δt should not be very large in the explicit time stepping. Figure 13.13 illustrates the advantage of multi-processing in the explicit method.

13.3.2.3 Shielding of a Desktop Case

The third example is a simplified model of a desktop case with the size of $575 \times 216 \times 359$ mm^3, as shown in Figure 13.14. Assume that the two perforated faces of this desktop are infinitely thin PEC with apertures and the remainders are free space. The whole desktop is truncated by PML with a thickness of 0.45 m in all directions. The excitation is a dipole source with a BHW pulse waveform whose characteristic frequency is at 300 MHz. The receivers are set both inside and outside of the desktop case to record the transient electric and magnetic fields.

Divided into six subdomains, the whole desktop is meshed by tetrahedral elements, as shown in Figure 13.14b. After applying the proposed automatic load balancing scheme, the results of the adaptive division of subdomains are presented in Figure 13.15. In this example, the computationally intensive PML regions are

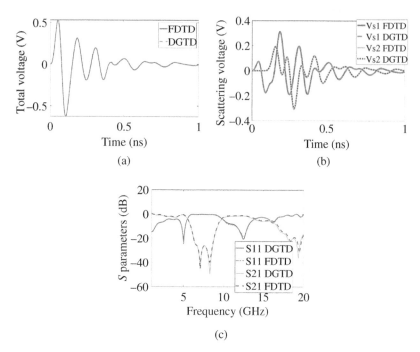

Figure 13.12 Numerical results of the filter case from parallel subdomain-level DGTD method. (a) Total voltage of the active lumped port. (b) Scattering voltages of the active and passive lumped ports. (c) S parameters. Source: Mi et al. [44]. Reproduced with permission of IEEE.

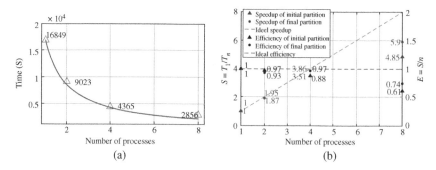

Figure 13.13 The performance of the parallelization. (a) The ideal time overhead (line) and the actual time overhead (triangle markers) calculated through the optimized partition in the subdomain-level DGTD method. (b) Speedup ratio and efficiency for different number of processes. Source: Mi et al. [44]. Reproduced with permission of IEEE.

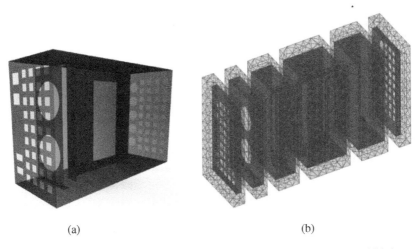

(a)   (b)

Figure 13.14 The overview of the desktop case. (a) The whole structure and (b) the tetrahedral mesh.

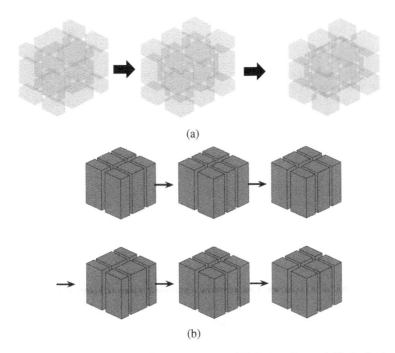

Figure 13.15 Adaptive partitioning process of (a) PML region and (b) physical region. Source: Mi et al. [44]. Reproduced with permission of IEEE.

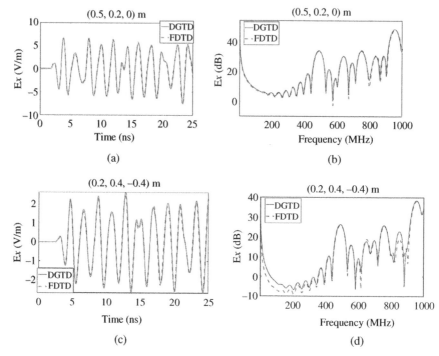

Figure 13.16 The *x* component of E-field signals at the receivers of the desktop case. The FDTD method is used as the reference. Transient E-field signal at (a) point (0.50, 0.20, 0) m and (b) point (0.20, 0.40, −0.40) m. (c) Frequency-domain magnitudes at point (0.50, 0.20, 0) m and (d) point (0.20, 0.40, −0.40) m. Source: Mi et al. [44]. Reproduced with permission of IEEE.

considered first in repartition because of the relative difficulty in changing the partitions of the desktop subdomains.

Figure 13.16 shows the time-domain electric field signals in *x*-direction and the frequency domain amplitudes calculated from fast Fourier transform (FFT). The observation points are located at (0.50, 0.20, 0) m and (0.20, 0.40, −0.40) m, which are in good agreement with the results obtained from the traditional FDTD method.

Figure 13.17 presents the attenuation of the E-field outside the desktop, which shows that this desktop box is like a resonator, using electromagnetic resonance to realize electromagnetic shielding (Figure 13.18).

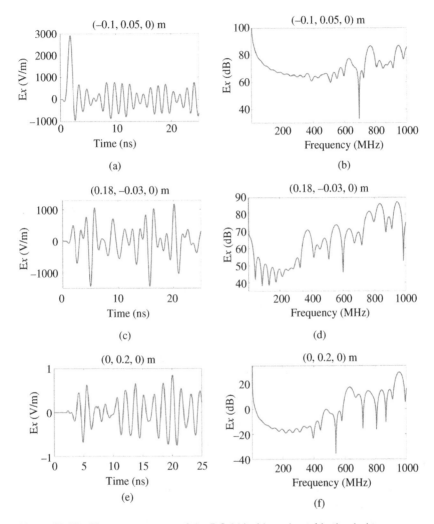

Figure 13.17 The *x* components of the E-field inside and outside the desktop case.
(a) Temporal waveform and (b) frequency-domain magnitude at point (−0.1, 0.05, 0) m.
(c) Temporal waveform and (d) frequency-domain magnitude at point (0.18, −0.03, 0) m.
(e) Temporal waveform and (f) frequency-domain magnitude at point (0, 0.20, 0) m.
Source: Mi et al. [44]. Reproduced with permission of IEEE.

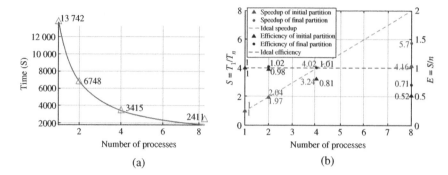

Figure 13.18 The performance of the parallelization for the desktop case. (a) The ideal time overhead (line) and the actual time overhead (triangle markers) calculated through the optimized partition in the subdomain-level DGTD method. (b) Speedup ratio and efficiency for different number of processes. Source: Mi et al. [44]. Reproduced with permission of IEEE.

13.4 Application of the Parallel Subdomain-Level DGTD Method for Large-Scale Cases

The advantage of parallel algorithm is that it can efficiently calculate large problems. The subdomain-level DGTD method with load balancing strategy proposed in this section has to adapt to the trend of increasing complexity and large size of electromagnetic cases. The following two cases are simulated on two nodes of the National Supercomputer Center in Guangzhou. Each node is configured with two Intel Xeon E52692 CPUs (2.200 GHz, 12 cores).

13.4.1 Patch Antenna Array

The first large-scale case is a patch antenna array with 64 units. Figure 13.19 shows the detailed information of the antenna structure. Sixty four lumped ports are located at the center of each unit cell, and the terminals of the only active port are located at $(-140.00, -140.00, 1.00)$ mm and $(-140.00, -140.00, 2.00)$ mm. The characteristic frequency of the excitation BHW pulse is 2.5 GHz ($\lambda = 120$ mm). The relative dielectric constant of the dielectric layer is 3.4. Surrounded by a layer of air as a buffer, the physical region is truncated on all sides by 40 mm PML.

Figure 13.20 presents some representative S parameters. The results are in good agreement with those obtained from the FDTD method.

With the maximal and minimal sizes of the patch antenna being 360 and 0.2 mm, this multi-scale example is divided into 192 subdomains, with 64

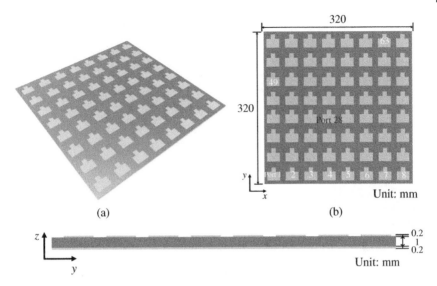

(a)   (b)

Figure 13.19 Geometry of the patch antenna array. (a) Overview; (b) x − y plane; (c) y − z plane. Source: Mi et al. [44]. Reproduced with permission of IEEE.

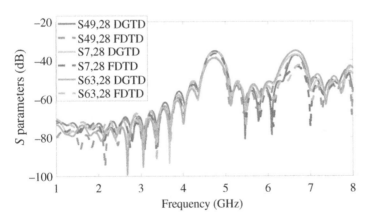

Figure 13.20 Three representative scattering parameters of the patch antenna array. Source: Mi et al. [44]. Reproduced with permission of IEEE.

subdomains of array elements, 64 subdomains of PML, and 64 subdomains of lumped ports. The total DoFs are 7.7 million. Figure 13.21a shows the initial partition of the physical and PML regions. In this example, the computation load of the PML regions is the largest one. Therefore, we first adjust the boundaries

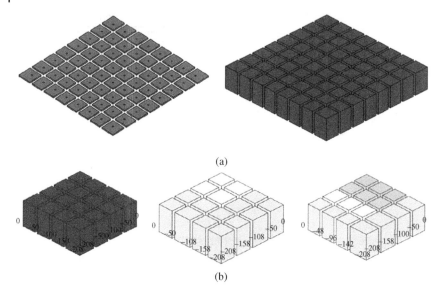

(a)

(b)

Figure 13.21 The partition of the patch antenna array case. (a) Initial partition of the physical region and the PML region. (b) The initial partition, the partition with 16 processes, and the partition with 24 processes for a quarter of the physical region and the PML region. Source: Mi et al. [44]. Reproduced with permission of IEEE.

of the PML regions. Figure 13.22 shows the simulation efficiency and time with different number of processes. As the number of processors increases, the optimized partitions will become more favorable. This tendency is consistent with intuition. When the number of subdomains is large, a very accurate partitioning strategy is required to achieve a better speedup ratio. This is often beyond the capacity of manual manipulation.

13.4.2 RCS of Boeing 737

A Boeing 737 model with the minimal size of about 0.4 cm and many curvilinear structures is analyzed. The geometry is shown in Figure 13.23. The incident plane wave propagates in the $-x$ direction and polarizes in the $+z$ direction. The waveform is a BHW pulse with a characteristic frequency of 100 MHz. The simulation region was truncated by PML with a thickness of the half wavelength of the characteristic frequency in each direction. A hexahedral buffer is inserted between Boeing 737 and the PML region, the receiver is placed in the buffer to record the near field of Boeing 737, and the RCS of the plane is calculated by near-to-far transformation. The parallel subdomain-level DGTD method with load balancing strategy is applied.

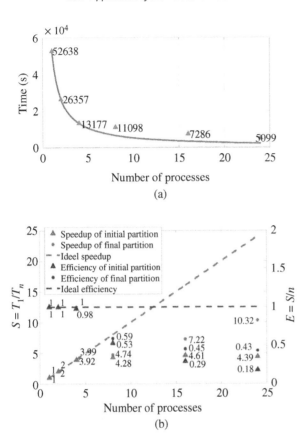

Figure 13.22 The performance of the parallelization for the patch antenna array case. (a) The ideal time overhead (line) and the actual time overhead (triangle markers) calculated through the optimized partition in the subdomain-level DGTD method. (b) Speedup ratio and efficiency for different number of processes. Source: Mi et al. [44]. Reproduced with permission of IEEE.

Figure 13.24 shows the bistatic RCS values of the Boeing 737 model at 100 MHz. The scattered angles first vary from 0° to 360° in φ direction while θ is fixed at 90°. In another configuration, θ varies from 0° to 180° while φ is fixed at 0°. It can be observed that the results of the proposed method are in good agreement with those from the commercial software package FEKO. For this case, the repartition of the computational subdomains containing Boeing 737 is performed first. Figure 13.25 presents the time and efficiency of the solution in a multi-process configuration. Obviously, the proposed parallel method is superior to its serial counterpart.

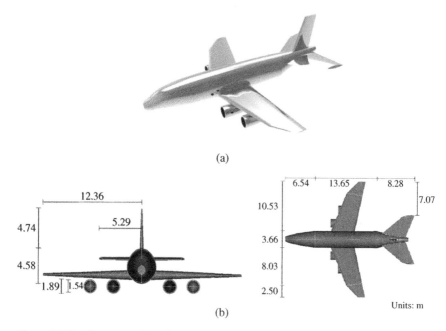

Figure 13.23 Geometry of the Boeing 737 plane. (a) Overview; (b) *y*−*z* plane; (c) *x*−*y* plane.

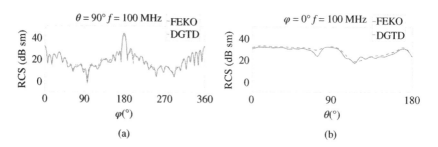

Figure 13.24 RCS of Boeing 737 at 100 MHz in (a) φ direction and (b) θ direction.

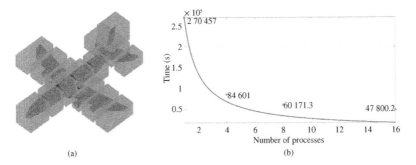

Figure 13.25 The performance of the parallel computation for the Boeing 737 case. (a) The final partition of the physical region. (b) The ideal time (line) and the actual execution time (triangle markers) with the final optimized partition in the subdomain-level DGTD with multiple processes.

13.5 Conclusion

In this chapter, a parallel subdomain-level DGTD method based on MPI library is proposed to simulate complex electromagnetic problems. This method uses a variety of different order elements (tetrahedrons and hexahedrons) to improve efficiency, and domain decomposition makes subdomain-level parallelism possible. First, the relationship between the computational load (CPU time) and the DoFs are extracted from the numerical experiments of hexahedral and tetrahedral meshes with different orders. Then, the automatic iterative load balancing scheme is adopted to optimize the position of the interfaces between adjacent subdomains so as to minimize the CPU time difference between processes, which not only ensures the computational load balancing but also reduces the total data exchange of communication in time stepping. After initial subdivision of the original model, greedy algorithm is used for one-key solution, which is the only step requiring external operation. After several iterations, the computational load of each process will be close. In addition, the whole load balancing scheme only involves the preprocessing part of the code, rather than iteratively modifying the system matrices. Therefore, the implementation of load balancing is relatively straightforward and effective, which can significantly speed up the simulation of large-scale and multi-scale models. The accuracy of the subdomain-level DGTD method is proved by four examples: low-pass filter, shielding of a desktop case, patch antenna array, and scattering of a Boeing 737 plane. Through the acceleration analysis of the above four examples under multi-process, the validity of the parallel algorithm is verified. Therefore, adopting the DGTD method with an easy parallelization scheme can effectively reduce the total execution time, which is very important for solving large-scale and multi-scale electromagnetic problems.

References

1 Zhan, Q., Zhuang, M., and Liu, Q.H. (2018). A compact upwind flux with more physical insight for wave propagation in 3D poroelastic media. *IEEE Transactions on Geoscience and Remote Sensing* 56 (10): 5794–5801.

2 Ren, Q., Tobón, L.E., Sun, Q., and Liu, Q.H. (2015). A new 3D nonspurious discontinuous Galerkin spectral element time domain (DG-SETD) method for Maxwell's equations. *IEEE Transactions on Antennas and Propagation* 63 (6): 2585–2594.

3 Chen, J., Liu, Q.H., Chai, M., and Mix, J.A. (2010). A nonspurious 3-D vector discontinuous Galerkin finite-element time-domain method. *IEEE Microwave and Wireless Components Letters* 20 (1): 1–3.

4 Ren, Q., Sun, Q., Tobón, L. et al. (2016). EB scheme-based hybrid SE-FE DGTD method for multiscale EM simulations. *IEEE Transactions on Antennas and Propagation* 64 (9): 4088–4091.

5 Chen, J., Tobón, L.E., Chai, M. et al. (2011). Efficient implicit–explicit time stepping scheme with domain decomposition for multiscale modeling of layered structures. *IEEE Transactions on Components and Packaging Technologies* 1 (9): 1438–1446.

6 Yan, S. and Jin, J. (2017). A dynamic *p*-adaptive DGTD algorithm for electromagnetic and multiphysics simulations. *IEEE Transactions on Antennas and Propagation* 65 (5): 2446–2459.

7 Li, P. and Jiang, L.J. (2013). A hybrid electromagnetics-circuit simulation method exploiting discontinuous Galerkin finite element time domain method. *IEEE Microwave and Wireless Components Letters* 23 (3): 113–115.

8 Alvarez, J., Angulo, L.D., Bretones, A.R., and Garcia, S.G. (2012). 3-D discontinuous Galerkin time-domain method for anisotropic materials. *IEEE Microwave and Wireless Components Letters* 11: 1182–1185.

9 Reed, W.H. and Hill, T.R. (1973). *Triangular Mesh Methods for the Neutron Transport Equation*. Los Alamos: Los Alamos Scientific Laboratory.

10 Hesthaven, J.S. and Warburton, T. (2002). Nodal high order methods on unstructured grids time-domain solution of Maxwell' equations. *Journal of Computational Physics* 181 (1): 186–221.

11 Chen, G., Zhao, L., Yu, W. et al. (2018). A general scheme for the discontinuous Galerkin time-domain modeling and S-parameter extraction of inhomogeneous waveports. *IEEE Transactions on Microwave Theory and Techniques* 55 (4): 1701–1712.

12 Li, P., Jiang, L.J., and H. Bağci. (2017). Transient analysis of dispersive power-ground plate pairs with arbitrarily shaped antipads by the DGTD method with wave port excitation. *IEEE Transactions on Electromagnetic Compatibility* 59 (1): 172–183.

13 Alvarez, J., Angulo, L.D., Cabello, M.R. et al. (2014). An analysis of the leap-frog discontinuous Galerkin method for Maxwell's equations. *IEEE Transactions on Microwave Theory and Techniques* 62 (2): 197–207.

14 Tian, C., Shi, Y., and Liang, C. (2017). A low-storage discontinuous Galerkin time domain method. *IEEE Microwave and Wireless Components Letters* 27 (1): 1–3.

15 Zhao, L., Peng, X., Li, L., and Li, Z. (2011). A fast waveguide port parameter extraction technique for the DGTD method. *IEEE Transactions on Intelligent Transportation Systems* 12 (1): 2659–2662.

16 Ren, Q., Tobón, L.E., and Liu, Q.H. (2013). A new 2D non-spurious discontinuous Galerkin finite element time domain (DG-FETD) method for Maxwell's equations. *Progress In Electromagnetics Research* 143: 385–404.

17 Gedney, S.D., Young, J.C., Kramer, T.C., and Roden, J.A. (2012). A discontinuous Galerkin finite element time-domain method modeling of dispersive media. *IEEE Transactions on Antennas and Propagation* 60 (4): 1969–1977.

18 Hu, F. and Wang, C. (2013). Higher-order DG-FETD modeling of wideband antennas with resistive loading. *IEEE Antennas and Wireless Propagation Letters* 12: 1025–1028.

19 Wen, P., Ren, Q., Chen, J. et al. Improved memory efficient subdomain level discontinuous Galerkin time domain method for periodic/quasi-periodic structures. *IEEE Transactions on Antennas and Propagation* https://doi.org/10.1109/TAP.2020.2998215.

20 Chen, H., Zhao, L., and Yu, W. (2018). GPU accelerated DGTD method for EM scattering problem from electrically large objects in cross strait quad-reg. *2018 Cross Strait Quad-Regional Radio Science and Wireless Technology Conference,* 21–24 July 2018, IEEE, pp. 1–2.

21 Yan, S. and Jin, J.M. (2016). A GPU accelerated dynamic *p*-adaptation for simulation of EM-plasma interaction. in *2018 IEEE Asia-Pacific Conference on Antennas and Propagation (APCAP),* 05–08 August 2018, IEEE, pp. 2081–2082.

22 G. Chen, L. Zhao, W. Yu, H. Ren and H. Fu. 2018 A novel acceleration method for DGTD algorithm on Sunway TaihuLight, in *2018 IEEE Asia-Pacific Conference on Antennas and Propagation (APCAP),* 05–08 August 2018, IEEE. 153–154.

23 Meng, H. and Jin, J. (2014). Acceleration of the dual-field domain decomposition algorithm using MPI–CUDA on large-scale computing systems. *IEEE Transactions on Antennas and Propagation* 62 (9): 4706–4715.

24 Nikishkov, G.P. and Makinouchi, A. (1996). Efficiency of the domain decomposition method for the parallelization of implicit finite element code, in *Proceedings of 1996 International Conference on Parallel and Distributed Systems,* 03–06 June 1996, IEEE, 49–56.

25 S. Gaurav, P. K. Jimack and M. A. Walkley (2016) A cache-aware approach to domain decomposition for stencil-based codes, in *2016 International Conference on High Performance Computing & Simulation (HPCS)*, 18–22 July 2016, IEEE, 875–885.

26 Giannacopoulos, D.D. (2004). Optimal discretization-based load balancing for parallel adaptive finite-element electromagnetic analysis. *IEEE Transactions on Magnetics* 40 (2): 977–980.

27 Tan, T., Li, K., Tang, Z., and Wang, F. (2011). Parallelization methods for implementation of discharge simulation along resin insulator surfaces, in EmbeddedCom-ScalCom. *Computers & Electrical Engineering* 37 (1): 30–40.

28 N. Sato and J. M. Jezequel 2000. Implementing and evaluating an efficient dynamic load-balancer for distributed molecular dynamics simulation, in *Proceedings 2000. International Workshop on Parallel Processing*, 21–24 August 2000, IEEE, 277–283.

29 Denis, C., Boufflet, J.P., and Breitkopf, P. (2005). A load balancing method for a parallel application based on a domain decomposition. *19th IEEE International Parallel and Distributed Processing Symposium*, 04–08 April 2005, IEEE, pp. 1–10.

30 Berger, M.J. and Bokhari, S.H. (1987). A partitioning strategy for nonuniform problems on multiprocessors. *IEEE Transactions on Computers* 36 (5): 570–580.

31 Li, Y.M., Chen, C.K., Lin, S.S., et al. (2000) An implementation of parallel dynamic load balancing for adaptive computing in VLSI device simulation. *Proceedings 15th International Parallel and Distributed Processing Symposium*, 23–27 April 2001, IEEE, pp. 1–6.

32 Chen, H., Ye, Z., Zhao, L., et al. (2018) Parallel DGTD method for transient electromagnetic problems. in *2018 IEEE International Symposium on Electromagnetic Compatibility and 2018 IEEE Asia-Pacific Symposium on Electromagnetic Compatibility (EMC/APEMC)*, 14–18 May 2018, IEEE, 903–906.

33 Chen, H., Ye, Z., Zhao, L., et al. (2018) Parallel DGTD method based on JAUMIN framework. *2018 IEEE International Conference on Computational Electromagnetics (ICCEM)*, 26–28 March 2018, IEEE, pp. 1–3.

34 W. Yu, L. Zhao and G. Chen. 2016 A novel DGTD method and engineering applications. *2016 International Conference on Electromagnetics in Advanced Applications (ICEAA)*, 19–23 September 2016, IEEE, 324–327.

35 Bernacki, M., Fezoui, L., Lanteri, S., and Piperno, S. (2006). Parallel discontinuous Galerkin unstructured mesh solvers for the calculation of three dimensional wave propagation problems. *Applied Mathematical Modelling* 30 (8): 744–763.

36 Stylianos, D., Zhao, B., and Lee, J. (2013). Non-conformal and parallel discontinuous Galerkin time domain method for Maxwell's equations: EM analysis of IC packages. *Journal of Computational Physics* 238: 48–70.

37 Kiss, I., Gyimothy, S., Badics, Z., and Pavo, J. (2012). Parallel realization of the element-by-element FEM technique by CUDA. *IEEE Transactions on Magnetics* 48 (2): 207–510.

38 S. Dosopoulos and J. Lee. (2010). Discontinuous Galerkin time domain for Maxwell's equations on GPUs, in *2010 URSI International Symposium on Electromagnetic Theory*, 16–19 August 2010, IEEE, 989–919.

39 Mei, G., Zhang, J., Xu, N., and Zhao, K. A sample implementation for parallelizing Divide-and-Conquer algorithms on the GPU. *Heliyon* 4 (1): e00512.

40 Geist, A., Gropp, W., Huss-Lederman, S. et al. (1996). MPI-2: Extending the message-passing interface. In: *Lecture Notes in Computer Science* (ed. L. Bougé, P. Fraigniaud, A. Mignotte and Y. Robert), 127–135. Berlin, Heidelberg: Springer.

41 Bao, H., Kang, L., Campbell, S.D., and Werner, D.H. (2019). PML implementation in a non-conforming mixed-element DGTD method for periodic structure analysis. *IEEE Transactions on Antennas and Propagation* 67 (11): 6979–6988.

42 Sanders, J. and Kandrot, E. (2010). CUDA by example: an introduction to general-purpose GPU programming. *Concurrency and Computation: Practice and Experience* 21: 312.

43 Taylor, V.E., Holmer, B.K., Schwabe, E.J., and Hribar, M.R. (1996). Balancing load versus decreasing communication: exploring the tradeoffs. *Proceedings - IEEE INFOCOM* 1: 585–593.

44 Mi, J., Ren, Q., and Su, D. (2021). Parallel subdomain-level DGTD method with automatic load balancing scheme with tetrahedral and hexahedral elements. *IEEE Transactions on Antennas and Propagation* 69 (4): 2230–2241.

45 Guiffaut, C. and Mahdjoubi, K. (2001). A parallel FDTD algorithm using the MPI library. *IEEE Antennas and Propagation Magazine* 43 (2): 94–103.

14

Alternate Parallelization Strategies for FETD Formulations

Amir Akbari, David S. Abraham, and Dennis D. Giannacopoulos

Department of Electrical & Computer Engineering, McGill University, Montréal, Québec, Canada

14.1 Background and Motivation

Due to the often significant computational burden associated with performing simulations of Maxwell's equations, this chapter will present several possible strategies for accelerating numerical methods based on the finite-element time-domain (FETD) method. In particular, the idea of parallelization and the use of massively parallel architectures, such as graphics processing units (GPUs) and multicore central processing units (CPUs), will be explored as a means of accelerating FETD computations. Though in principle the idea of distributing a workload to multiple workers to be performed in parallel may seem like a straightforward way to achieve good algorithmic speedups, as will be shown, often the underlying algorithms are not ideal for parallel execution and must therefore be adapted or reformulated to yield good results.

A significant amount of the success enjoyed by the FETD method is attributable to the many attractive benefits that the method provides over other similar techniques, such as the finite-difference time-domain (FDTD) method [1]. The most important benefits among these are the FETD method's increased geometric flexibility (alleviating, for example, staircasing errors [2, 3]) and better handling of material discontinuities (removing the need for smoothing techniques [4]). However, as with most aspects of engineering, these additional desirable features as compared to FDTD methods are not free, but rather come at the expense of increased complexity and computational cost.

Indeed, for example, many of the most popular FDTD methods for Maxwell's equations are *explicit* in time, meaning the future field values at a given point in the domain are only a function of past values of nodes in the neighborhood of

Advances in Time-Domain Computational Electromagnetic Methods, First Edition.
Edited by Qiang Ren, Su Yan, and Atef Z. Elsherbeni.

that point. While explicit techniques result in decreased global stability [5], this locality of the update equations means that there is no need to solve a matrix equation at each time step. In this way, while more overall time steps may be required to maintain stability, each individual time step is significantly easier and cheaper to perform. Moreover, due to this explicit and *local* nature of the update equations, each node's update procedure is effectively independent of its neighbours', meaning that parallelization strategies are straightforward to develop and yield good results.

In stark contrast, the FETD method lacks this local feature of the explicit FDTD method, requiring the solution of a global matrix system of equations at each time step. Indeed, even if an explicit temporal discretization (such as the central difference method) is applied in the FETD method, it will still result in a system of equations. Granted that in the explicit case the system of equations will only depend on the mass matrix rather than the mass *and* stiffness matrices (as was the case in Chapter 3[1] for the implicit Newmark-β and Crank–Nicolson methods) [6]. While this results in better conditioning of the explicit matrix, making it easier to solve, the net result is nevertheless the requirement of solving a matrix equation for every time step of the simulation. Thus, FETD methods generally contain greater computational overhead than their FDTD counterparts. Moreover, effective parallelization strategies are much more difficult to develop, as in many cases this equates to developing new techniques for the parallel solution of matrix equations, an immensely important, challenging, and active area of research in its own right [7–9]. Finally, the introduction of nonlinearity, such as with the dielectrics of Chapter 3, only serves to exacerbate the issue. In these cases not only must *nonlinear* systems of equations be solved, but the generation and assembly of the matrices themselves now also constitutes a significant computational burden.

In the following sections, an overview of some of the possible parallelization strategies to tackle these issues will be presented. In particular, a summary of some of the challenges associated with traditional parallelization strategies for FETD methods on CPUs will be discussed, followed by the introduction of a relatively recent highly parallelizable approach known as the *finite-element Gaussian belief propagation* (FGaBP) method, which addresses many of these challenges. Furthermore, in some cases, such as with nonlinear dielectrics, simple and straightforward parallelization strategies will be shown to nevertheless be possible *and* highly effective, especially when combined with FGaBP. Lastly, the application of these techniques to coupled multi-physics problems will also be discussed, and modifications to the base FGaBP will be presented to better address the specific challenges of this application.

---

1 *The FETD method for dispersive and nonlinear media.*

14.2   Challenges in FETD Parallelization

Due to power limitations in single-processor performance, manufacturers are making chips with multiple cores to improve overall computational efficiency by handling more work in parallel. Thus, to obtain efficient performance from evolving parallel computing architectures, the FETD computation must exhibit reasonable parallel scalability on high performance computing (HPC) systems. Traditional finite-element method (FEM) implementations involve two computationally expensive stages, which are the assembly of a global matrix, $\mathbf{A}$, and the solving of the linear system $\mathbf{Ax} = \mathbf{b}$ using (usually) iterative solvers. Owing to their dependence on the sparsity pattern of the stiffness matrix, acceleration of these stages using parallel computing is restricted by memory bandwidth limitations.

To better illustrate this issue, consider the sparse matrix-vector multiplication (SpMV), $\mathbf{Ay}$, which is a fundamental kernel of iterative solvers. To carry out the SpMV operation, the sparse matrix $\mathbf{A}$ needs to be read from memory. Since $\mathbf{A}$ is very sparse, reading its rows or columns is highly dependent on how the matrix is stored in memory and can lead to non-coalesced memory access and very low cache hit rates. Consequently, the CPU's idle time is increased by waiting for data to arrive, causing the CPU to achieve only a small fraction of its peak performance [10]. This issue is even more evident for multicore architectures, since as the number of cores increases, memory access requests from different cores can interfere with each other, causing bottlenecks for the limited memory bandwidth.

Similar concerns exist for FEM parallelization when applied to nonlinear partial differential equations (PDEs). Depending on the type and level of the nonlinearity, nonlinear systems are generally solved using an iterative method that involves approximate linearizations at each iteration, such as Gauss–Seidel, Newton–Raphson, Picard, etc. In consequence, a linear system will need to be solved at each iteration, for which the SpMV scalability limitations can still be a performance bottleneck.

To achieve an efficient SpMV is a challenging task. While researchers have employed various compression techniques to reduce the sparse matrix memory footprint and cache-memory traffic [11], the sparse matrix storage format has a considerable impact on the SpMV performance. Consequently, attempts to optimize SpMV should comprise two stages. In the first stage, a preprocessor analyzes the matrix sparsity pattern and identifies the performance bottlenecks of SpMV for different matrix compression methods on a target architecture. Afterwards, in the second stage, these performance bottlenecks are then addressed by using complex programming techniques such as code transformations, format conversions, parameter tuning, etc.

These optimization stages are computationally expensive and only pay off when the same sparse matrix is used in several SpMV implementations.

Indeed, SpMV optimization efforts have thus far indicated that maintaining reasonable performance for all sparse matrices, even within the same FEM application area, is impossible [10]. In addition, while an extensive body of research exists on optimizing SpMV on multicore architectures [12, 13], it is still unclear how the sparse matrix storage format affects SpMV performance on many-core platforms [11]. As a result of these persistent challenges, in the following section an entirely different approach known as *Gaussian belief propagation* (GaBP) will be introduced that can avoid the need for extensive SpMVs altogether.

14.3 Gaussian Belief Propagation for Solving Linear Systems

The sum-product or belief propagation (BP) algorithm [14] has been successfully used in probability problems for computing either exact or approximate marginal distributions. The BP algorithm is an iterative method based on message-passing updates that infers the marginal distributions of random variables based on their joint distribution. The inference is carried out on probabilistic graphical models, where each node of the graph represents a random variable and the graph edges show the dependency between the variables.

Assuming we have n random variables $\{x_i\}_{i=1}^n$, whose joint distribution $p(x_1, \ldots, x_n)$ is known, the BP algorithm infers the marginal distributions $p_i(x_i)$, based on the probabilistic dependence between the random variables. This inference is carried out on a graphical model of variables $\{x_i\}_{i=1}^n$ and their interdependence. Each node of the graph represents a random variable with the graph edges showing the dependence between the variables. For example, in the graph depicted in Figure 14.1, the undirected edges indicate that there are symmetric probabilistic dependencies between the pairs $\{x_I, x_{II}\}$, $\{x_{II}, x_{III}\}$, $\{x_{III}, x_{IV}\}$, and $\{x_{III}, x_V\}$.

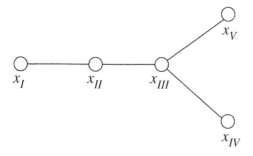

Figure 14.1 A graphical model of random variables shows the probabilistic dependence between them.

In general, an undirected graph $\mathcal{G}$ is defined by a set of nodes $\mathcal{V}$ and a corresponding set of undirected edges $\mathcal{E}$. Assuming the joint distribution $p(\mathbf{x})$ can be represented in a factored form [15]:

$$p(\mathbf{x}) \propto \prod_{(i,j)\in\mathcal{E}} \psi_{ij}(x_i, x_j) \prod_{i\in\mathcal{V}} \phi_i(x_i) \tag{14.1}$$

then such a distribution is called a pairwise Markov random field (MRF). The functions $\psi_{ij}(x_i, x_j) > 0$ are edge potentials and indicate the probabilistic dependencies between nodes i and j. The functions $\phi_i(x_i)$ are the self potentials of each node i, containing information about the marginal distribution $p_i(x_i)$. Note that the self potential $\phi_i(x_i)$ does not equal the marginal distribution $p_i(x_i)$, because of the statistical dependence between node i and the other nodes.

Through its iterations, BP tries to find the marginal distributions. For each node $i \in \mathcal{V}$, let $\mathcal{N}(i) = \{j \in \mathcal{V} \mid (i,j) \in \mathcal{E}\}$ be its set of neighbors. At each iteration t of the BP algorithm, node i sends message $m_{ij}^t(x_j)$ to its neighboring nodes $j \in \mathcal{N}(i)$:

$$m_{ij}^t(x_j) \propto \int_{x_i} \psi_{ij}(x_i, x_j)\phi_i(x_i) \prod_{k\in\mathcal{N}(i)\backslash j} m_{ki}^{t-1}(x_i) \, dx_i \tag{14.2}$$

where $\mathcal{N}(i)\backslash j$ is the neighborhood of i *excluding j*. Intuitively, this message contains statistical information that node i has collected regarding x_j. Eventually these messages will converge, from which the marginals can then be computed as:

$$p_j(x_j) \propto \phi_j(x_j) \prod_{i\in\mathcal{N}(j)} m_{ij}(x_j) \tag{14.3}$$

If the random variables $(x_1, \ldots, x_n)$ are jointly Gaussian, the algorithm is called GaBP, whose update rules can be obtained analytically from Eq. (14.2). The focus of the remainder of this section will now be on how GaBP can be used to solve a linear system of equations.

The application of GaBP to the solving of systems of linear equations was first proposed in [16]. Assuming matrix $\mathbf{A}_{n\times n}$ is non-singular, let the linear system to be solved be expressed as:

$$\mathbf{Ax} = \mathbf{b} \tag{14.4}$$

If matrix $\mathbf{A}$ is symmetric positive definite (SPD), one can define the joint distribution $p(\mathbf{x})$ as:

$$p(\mathbf{x}) \propto \exp\left(-\frac{1}{2}\mathbf{x}^T\mathbf{Ax} + \mathbf{b}^T\mathbf{x}\right) \propto \mathcal{N}(\boldsymbol{\mu} = \mathbf{A}^{-1}\mathbf{b}, \mathbf{A}^{-1}) \tag{14.5}$$

where $\mathcal{N}(\boldsymbol{\mu} = \mathbf{A}^{-1}\mathbf{b}, \mathbf{A}^{-1})$ represents a joint Gaussian distribution of random variable vector $\mathbf{x}$ with mean vector $\boldsymbol{\mu}$ and covariance matrix $\mathbf{A}^{-1}$. Since $\mathbf{A}$ is *symmetric*, an *undirected* graphical model $\mathcal{G}$ based on the non-zero structure of matrix $\mathbf{A}$ can be developed. Each node of $\mathcal{G}$ corresponds to an entry of the random vector $\mathbf{x}$, whereas the set of edges includes one edge for each non-zero entry A_{ij}, for which $j < i$. As matrix $\mathbf{A}$ is SPD, solving the linear system is equivalent to minimizing the quadratic function $q(\mathbf{x}) = \frac{1}{2}\mathbf{x}^T\mathbf{A}\mathbf{x} - \mathbf{b}^T\mathbf{x}$, which in turn is equivalent to maximizing the exponential $p(\mathbf{x}) = e^{-q(\mathbf{x})}$. On the other hand, $p(\mathbf{x})$ is maximized when $\mathbf{x}$ is equal to the marginal mean vector $\boldsymbol{\mu}$. Consequently, by performing GaBP for the joint distribution $p(\mathbf{x})$ and inferring its marginal mean, one can solve the linear system.

Besides the joint distribution $p(\mathbf{x})$ and its graphical model $\mathcal{G}$, this procedure will also require the self potentials ϕ_i and the edge potentials ψ_{ij} of Eq. (14.1). In the special case of GaBP, ϕ_i and ψ_{ij} can be defined as:

$$\phi_i(x_i) = \exp\left(-\frac{1}{2}A_{ii}x_i^2 + b_ix_i\right) \tag{14.6a}$$

$$\psi_{ij}(x_i, x_j) = \exp\left(-\frac{1}{2}x_iA_{ij}x_j\right) \tag{14.6b}$$

Note that the self potentials *also* have a Gaussian form:

$$\phi_i(x_i) \propto \mathcal{N}(\mu_{ii} = b_i/A_{ii},\ P_{ii}^{-1} = A_{ii}^{-1}) \tag{14.7}$$

whose mean and inverse-variance are given by the scalars μ_{ii} and P_{ii}, respectively.

The advantage of GaBP is that its message update rule, i.e. Eq. (14.2), has a closed form. This is not the case for the general BP algorithm, where the integral of equation (14.2) needs to be approximated. Indeed, it can be shown that if the input messages in Eq. (14.2), i.e. $m_{ki}^{t-1}(x_i)$, are Gaussian distributions, then the output messages $m_{ij}^t(x_j)$ are *also* Gaussian. This is due to the fact that the product of Gaussian densities over a common variable is, up to a constant factor, also a Gaussian density. As a result, the update rules for the messages are reduced to the update rules for marginal means and marginal inverse-variances. Assuming the message m_{ij} has a normal distribution $\mathcal{N}(\mu_{ij}, P_{ij}^{-1})$, its mean and inverse-variance can thus be given by:

$$P_{ij} = -A_{ij}^2 P_{i\backslash j}^{-1} \tag{14.8a}$$

$$\mu_{ij} = -P_{ij}^{-1}A_{ij}\mu_{i\backslash j} \tag{14.8b}$$

where

$$P_{i\backslash j} = P_{ii} + \sum_{k\in N(i)\backslash j} P_{ki} \tag{14.9a}$$

$$\mu_{i\backslash j} = P_{i\backslash j}^{-1}\left(P_{ii}\mu_{ii} + \sum_{k\in N(i)\backslash j} P_{ki}\mu_{ki}\right) \tag{14.9b}$$

Here, P_{ki} and μ_{ki} are the inverse-variance and mean of the incoming message $m_{ki}(x_i)$. Thus, as mentioned, all the required messages can be evaluated in closed-form and a solution efficiently found. However, while the GaBP method is effective in its own right, a further modification to GaBP is possible specifically for the FEM. The resulting FEM-based GaBP (FGaBP) algorithm, originally proposed in [17], will now be presented.

14.4 Finite Element Formulation of Gaussian Belief Propagation

Before introducing the finite-element formulation of GaBP, a few concepts from graph theory need to be reviewed. A *bipartite graph* is a graph whose vertices can be divided into two disjoint sets V and U such that every edge connects a vertex in V to one in U. Vertex sets U and V are called the *parts* of the graph. A *factor graph* is a bipartite graph depicting the factorization of a function. For instance, assume $\mathbf{x}$ is a random vector containing four random variables, i.e. $\mathbf{x} = (x_1, \dots, x_4)$. If their joint distribution $p(\mathbf{x})$ can be factorized as:

$$p(\mathbf{x}) = f_{\mathrm{I}}(x_1)f_{\mathrm{II}}(x_1, x_2, x_4)f_{\mathrm{III}}(x_1, x_2)f_{\mathrm{IV}}(x_2, x_3)f_{\mathrm{V}}(x_3, x_4) \tag{14.10}$$

then Figure 14.2 shows the factor graph representing this factorization. Here, the vertices of the factor graph are divided into two parts: the factor nodes $(f_{\mathrm{I}}, \dots, f_{\mathrm{V}})$ shown by squares and the variable nodes (VNs) $(x_1, \dots, x_4)$ shown by circles.

The idea of FGaBP is to reformulate FEM into an inference problem over a factor graph. As shown in Figure 14.3, FGaBP turns the FEM mesh into a factor graph, whose VNs (nodes of unknowns) are represented by circles and factor nodes are represented by squares. Similar to GaBP, the solution at each VN is a random variable with a Gaussian distribution. However, instead of updating the mean and variance of each Gaussian VN directly, two intermediary parameters, α and β, are updated at each FGaBP iteration instead, where α is the reciprocal of the variance and β/α is the mean.

Note that, in accordance with the definition of a bipartite graph, the VNs can only interact with each other through the FNs. By passing messages between each FN and all its connected VNs, the FGaBP algorithm finds the values of α and β at each VN. A message, m_{ai}, is thus sent from factor node a (FN$_a$) to the connected

Figure 14.2 A factor graph representing the factorization of $p(x)$ in (11.10).

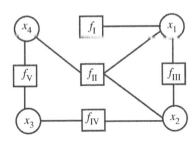

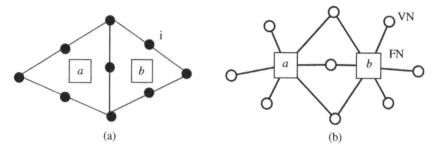

Figure 14.3 A FEM mesh (a) and its corresponding factor graph (b).

variable node i (VN$_i$) and represents the most probable solution value at i, as observed from FN$_a$. In return, VN$_i$ sends a message back to FN$_a$ representing observations from other connected FNs.

By exploiting factor graphs, FGaBP bypasses the assembly stage of FEM. Conventional FEM requires the assembly of a global linear system $\mathbf{Ax} = \mathbf{b}$ by looping through all the elements in the computational domain. A local characteristic matrix, $\mathbf{A}_e$, and a local source vector, $\mathbf{b}_e$, is formed for each element e. The global linear system is then assembled by having each element e contribute $\mathbf{A}_e$ and $\mathbf{b}_e$ to the global matrix $\mathbf{A}$ and the global right-hand side vector $\mathbf{b}$, respectively. In contrast, FGaBP doesn't need to explicitly construct $\mathbf{A}$ and $\mathbf{b}$. Moreover, the computation and exchange of messages back and forth is *highly* parallelizable, since only local data is required, bypassing the memory bandwidth issue described earlier. What follows is a summary of the resulting FGaBP algorithm as originally proposed in [10].

1. $t = 0$: Initialize all messages $\beta^{(0)} = 0$ and $\alpha^{(0)} = 1$.
2. Iterate: $t = 1, 2, \ldots$
   (a) For each VN$_i$, compute messages $\alpha_{ia}^{(t)}$ and $\beta_{ia}^{(t)}$ to each connected FN$_a$ ($a \in \mathcal{N}(i)$) as follows:

$$\alpha_i^{(t)} = \sum_{k \in \mathcal{N}(i)} \alpha_{ki}^{(t-1)}, \quad \alpha_{ia}^{(t)} = \alpha_i^{(t)} - \alpha_{ai}^{(t-1)} \tag{14.11}$$

$$\beta_i^{(t)} = \sum_{k \in \mathcal{N}(i)} \beta_{ki}^{(t-1)}, \quad \beta_{ia}^{(t)} = \beta_i^{(t)} - \beta_{ai}^{(t-1)} \tag{14.12}$$

   where $\mathcal{N}(i)$ is the neighborhood set of node i.
   (b) For each FN$_a$:
       i. Receive messages $\alpha_{ia}^{(t)}$ and $\beta_{ia}^{(t)}$ from all the VNs $i \in \mathcal{N}(a)$.

ii. Assume $\mathcal{A}^{(t)}$ is a diagonal matrix of incoming $\alpha_{ia}^{(t)}$ messages and $\mathcal{B}^{(t)}$ is a vector of incoming $\beta_{ia}^{(t)}$ messages, then define matrix W and vector K as follows:

$$\mathbf{W}^{(t)} = \mathbf{M}_a + \mathcal{A}^{(t)} \tag{14.13}$$

$$\mathbf{K}^{(t)} = \mathbf{B}_a + \mathcal{B}^{(t)} \tag{14.14}$$

where $\mathbf{M}_a$ and $\mathbf{B}_a$ are the characteristic matrix and source vector of element a, respectively.

iii. Partition $\mathbf{W}^{(t)}$ and $\mathbf{K}^{(t)}$ as follows:

$$\mathbf{W}^{(t)} = \begin{bmatrix} W_{\mathcal{L}(i)}^{(t)} & \mathbf{V}^T \\ \mathbf{V} & \overline{\mathbf{W}}^{(t)} \end{bmatrix} \tag{14.15}$$

$$\mathbf{K}^{(t)} = \begin{bmatrix} K_{\mathcal{L}(i)}^{(t)} \\ \overline{\mathbf{K}}^{(t)} \end{bmatrix} \tag{14.16}$$

where $\mathcal{L}(i)$ is the local index corresponding to the global VN i.

iv. Compute and partition $(\mathbf{W}^{(t)})^{-1}$ as follows:

$$(\mathbf{W}^{(t)})^{-1} = \begin{bmatrix} \tilde{W}_{\mathcal{L}(i)} & \tilde{\mathbf{C}}^T \\ \tilde{\mathbf{C}} & \tilde{\mathbf{W}} \end{bmatrix} \tag{14.17}$$

v. Compute and send new FN_a messages $\alpha_{ai}^{(t+1)}$ and $\beta_{ai}^{(t+1)}$ to each VN_i as follows:

$$\alpha_{ai}^{(t+1)} = \frac{1}{\tilde{W}_{\mathcal{L}(i)}} - \alpha_{ia}^{(t)} \tag{14.18}$$

$$\beta_{ai}^{(t+1)} = B_{\mathcal{L}(i)} + \frac{1}{\tilde{W}_{\mathcal{L}(i)}} (\overline{\mathbf{K}}^{(t)})^T \tilde{\mathbf{C}}^T \tag{14.19}$$

3. At message convergence, the mean of the VNs, or solutions, can be obtained by:

$$u_i = \frac{\beta_i}{\alpha_i} \tag{14.20}$$

where again

$$\beta_i = \sum_{k \in N(i)} \beta_{ki}, \quad \alpha_i = \sum_{k \in N(i)} \alpha_{ki}. \tag{14.21}$$

The above steps show how FGaBP can compute the α and β messages, and thus the finite-element solution, *without* constructing global matrices and vectors, based *only* on local data. It should therefore be reiterated that due to the locality of the data, these steps thus benefit immensely from parallelization without the drawbacks of other FETD parallelization strategies.

14.5 Parallelization of Nonlinear Problems

The application of FGaBP to the parallelization of simple *linear* problems has been tested in [10] and has been shown to demonstrate better performance than conventional FEMs. However, the successes of FGaBP need not be confined to the world of linear problems only. In fact, nonlinear problems can also stand to benefit from the FGaBP algorithm.

Indeed, nonlinear problems are often solved using iterative methods, which themselves rely on a series of linearizations, thus resulting in a linear system that needs to be solved at each iteration. In view of this, the FGaBP algorithm can also be of substantial use in accelerating nonlinear solvers. As such, this section introduces techniques for solving nonlinear PDEs numerically in parallel using a few different techniques, with an emphasis on FGaBP. More specifically, two different sources of nonlinearity will be studied: nonlinear media (as discussed in Chapter 3) and multi-physics systems. Both problems are solved via Newton's method (discussed in Section 14.5.1); however, FGaBP is applied in a different manner in each case. With respect to the problem of nonlinear dielectrics, the nonlinear system of equations is solved via the standard Newton's method, however each time the solution of a linear system of equations is required in this process, the original FGaBP algorithm is used. On the other hand, the multi-physics problem further modifies and adapts the FGaBP algorithm to implement a novel parallelization of Newton's method itself.

14.5.1 Newton's Method

Consider the following system of nonlinear equations:

$$\begin{cases} f_1(x_1, x_2, \ldots, x_n) = 0 \\ f_2(x_1, x_2, \ldots, x_n) = 0 \\ \quad\vdots \\ f_n(x_1, x_2, \ldots, x_n) = 0 \end{cases} \tag{14.22}$$

where $f_1, f_2, \ldots, f_n$ are nonlinear functions of $x_1, x_2, \ldots, x_n$. Here, $f_1, f_2, \ldots, f_n$ are called the component residuals, whereas $\mathbf{x} = (x_1, x_2, \ldots, x_n)$ refers to a solution that consists of n components. If the residuals and their derivatives with respect to $\mathbf{x}$ are smooth, Newton's method is a basic algorithm that models the coupling between these components. The problem in (14.22) can equally be formulated in terms of a single residual as:

$$\mathbf{F}(\mathbf{x}) = 0 \tag{14.23}$$

where $\mathbf{F}(\mathbf{x}) = \left(f_1(\mathbf{x}), f_2(\mathbf{x}), \ldots, f_n(\mathbf{x})\right)^T$ is a system of n coupled equations. The key steps of Newton's method are given by Algorithm 14.1. Here, $\Delta\mathbf{x}$ is called the

---

Algorithm 14.1: Newton's method.

---

1 Given initial values $\mathbf{x}^0 = (x_1^0, x_2^0, \ldots, x_n^0)^T$;

2 **for** $k = 1, 2, \ldots$ *(until convergence)* **do**

3 $\quad$ Solve $\mathbf{J}(\mathbf{x}^{k-1})\Delta\mathbf{x} = -\mathbf{F}(\mathbf{x}^{k-1})$;

4 $\quad$ Update $\mathbf{x}^k = \mathbf{x}^{k-1} + \Delta\mathbf{x}$;

5 **end**

---

update vector, $\mathbf{J}(\mathbf{x})$ is the n-by-n Jacobian matrix of $\mathbf{F}(\mathbf{x})$, i.e. $\mathbf{J}_{ij}(\mathbf{x}) = \partial f_i / \partial x_j$, and k is the iteration number. Due to the presence of off-diagonal blocks in the Jacobian, Newton's method is regarded as being *strongly coupled*.

The computation of $\Delta\mathbf{x}$ in Algorithm 14.1 can be very costly for large-scale problems. Hence, inexact or approximative Newton methods, such as the Jacobian-free Newton–Krylov (JFNK) method, are often preferred as they avoid calculating the Jacobian explicitly.

14.5.2 Nonlinear Dielectric Media

In this section, approaches for parallelizing the simulation of problems containing nonlinear dielectric media, as originally put forth in [18], will be discussed. Of course, with the prevalence of nonlinearity in other areas of engineering, the issue of accelerating the solution process for nonlinear systems of equations is not new to nonlinear dielectric media. In fact, many alternatives or modifications to the standard Newton's method have been developed over the years to attempt to reduce the computational burden of nonlinear root finding. In particular, as mentioned in Section 14.5.1, the JFNK method has become quite popular due to its ability to avoid calculating the Jacobian matrix explicitly [19]. Despite the success and popularity of this method, in this section a different approach will be adopted. Rather than approximating the Jacobian as in JFNK, here instead the full non-approximative Newton's method will be employed, and the overhead associated with forming and solving the exact full linear Jacobian system will be directly parallelized. The resulting algorithm will thus exhibit the full accuracy and convergence of Newton's method, but with a significant boost in speed and performance.

14.5.2.1 Elemental Matrix Evaluation

The first bottleneck to be addressed is that which results from the evaluation of the elemental matrices for elements that contain nonlinear dielectrics. As was touched upon in Chapter 3, to evaluate its local $[K^{(e)}]$ and $[J^{(e)}]$ matrices each element must perform numerical integration using Gaussian quadrature, due to the dependence of the permittivity on the electric field (and thus indirectly on the spatial coordinates). However, within the framework of the traditional FEM, each of these elemental matrices is completely independent of the others. In essence, information is only "shared" between elements according to their connectivity when they are assembled into the global system. The numerical evaluation of the elemental matrices thus requires the same computations be executed on different data sets completely independently. This thereby constitutes a so-called *embarrassingly parallel* problem.

This type of problem is ideally suited for implementation on a wide range of parallel systems. Indeed, this is particularly true of architectures based on the single instruction multiple thread (SIMT) execution model, such as GPUs. Under this model, parallel hardware is optimized for execution of the exact same instruction over and over again in parallel, but on different memory locations. Thus, due to the embarrassingly parallel nature of the numerical matrix evaluation, the required parallel algorithm is relatively straightforward to develop and implement. Indeed, if each nonlinear element in the mesh is assigned to a worker or thread, then the Gaussian quadrature calculations can proceed in lockstep with each other, as each worker accesses the geometry and field data for its own element in memory.

Despite the simplicity of this approach, its potential impacts should not be underestimated in the case of nonlinear media. For linear problems, the elemental matrices can often be expressed analytically in closed-form and are only evaluated once, since they are constant. As a result, their numerical evaluation is relatively inexpensive and thus represents a correspondingly much smaller amount of overall computation time. In contrast, the lack of closed-form expressions in the nonlinear case, coupled with the constantly changing and evolving nature of the solution, results in elemental matrix evaluation often dwarfing all other aspects of the solution process. Indeed, Figure 14.4 shows the breakdown of time spent during a typical FETD simulation in which the entire domain is filled with a nonlinear dielectric medium. Clearly, in the nonlinear regime, the elemental matrix evaluations are a significant computational burden across all problem sizes and floating point precisions. The simple parallelization strategy proposed above can thus often lead to substantial performance improvements, as will be demonstrated in the following section. Nevertheless, while the evaluation of the elemental matrices in the nonlinear case is often the rate-determining step, the assembly of the local matrices into their global counterpart and the solution of the linear Jacobian system should not be neglected in terms of added overhead, since they are repeated so often in the nonlinear regime.

14.5.2.2 Matrix Assembly and Solving

Whereas Section 14.5.2.1 was concerned with the evaluation of the individual elemental matrices, here the focus will be on their assembly into a global system of equations and the subsequent solving of that system. Since this is a fundamental underlying process of most finite-element based methods, the parallelization of

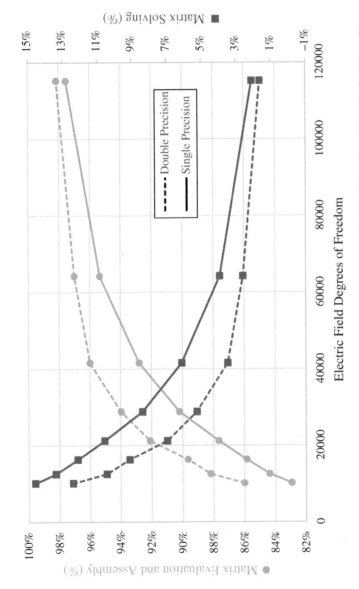

Figure 14.4 Breakdown of FETD computation time for an instantaneous nonlinearity in both single and double floating point precision.

global matrix assembly from local elemental matrices has been widely studied, particularly for frequency-domain or quasi-static analyses where it generally represents a larger fraction of overall computation time [20, 21]. As mentioned earlier, the independent local elements only share information with each other during global matrix assembly. Since this requires connected elements to add their local contributions together to form the global matrix entries, a naive parallelization strategy would inevitably lead to race conditions, in which two elements attempt to write to the same memory location at the same time, yielding undefined behavior. To mitigate this, additional processing layers are often required, such as coloring algorithms [22]. Additionally, as mentioned previously, parallel algorithms for the solution of sparse matrix systems constitute a massive area of study in its own right.

While certainly some combination of these existing techniques and algorithms could be used in conjunction with the elemental evaluation strategy presented in Section 14.5.2.1 to yield good results, Section 14.4 hints at a better alternative. Since the FGaBP algorithm was designed precisely to combine and parallelize the steps of matrix assembly *and* solution, it is an ideal candidate in the present case of nonlinear dielectric media. Indeed, as was alluded to earlier, the impacts of the FGaBP algorithm in this application are far greater than in the linear case since the matrix assembly and solution must occur multiple times within each time step, for each and every time step.

Thus, by combining the FGaBP method with the straightforward elemental evaluation method of Section 14.5.2.1, a simple yet powerful parallelization strategy for nonlinear dielectric media can be obtained. This procedure is outlined in abstract form in Algorithm 14.2.

14.5.3 Multi-Physics Problems

A multi-physics system consists of more than one simultaneously occurring type of physics and their interactions. Specifically, the multi-physics system that is considered in this section is a coupled electrical–thermal problem. For instance, suppose the electrical problem is modeled with Laplace's equation, where the electrical conductivity of the domain is temperature-dependent:

$$\nabla \cdot (\sigma(T)\nabla v) = 0 \qquad (14.24)$$

Here, $\sigma(T)$ is the temperature-dependent electrical conductivity (S/m), and v is the electric potential (V). As for the thermal problem, consider the diffusion equation:

$$\rho\, c_p \frac{\partial T}{\partial t} = Q + \nabla \cdot (d\nabla T) \qquad (14.25)$$

where T is the temperature (K), ρ is the density (kg/m^3), c_p is the specific heat (J/kg/K), d is the thermal conductivity (W/m/K), and Q is the heat source

---

Algorithm 14.2: Suggested parallelization procedure for FETD simulations involving nonlinear dielectric media.

---

1 Define the initial value of the electric field at $t = 0$, $\{e\}^0$;
2 **for** $t = 0, 1, 2, \ldots$ *(until desired end time)* **do**
3   Set the initial guess for $\{e\}^{n+1}_{(0)} = \{e\}^n$;
4   **for** $k = 0, 1, \ldots$ *(until convergence)* **do**
5   Using the current solution estimate $\{e\}^n_{(k)}$, use the method described in Section 14.5.2.1 to evaluate the new mass ($[K^{(e)}]$) and Jacobian ($[J^{(e)}]$) matrices for each element ;
6   Using the elemental data evaluated in the previous step, initiate the FGaBP algorithm to solve for $\{\Delta x\}$ in Algorithm 14.1 ;
7   Once FGaBP has converged, update the solution estimate $\{e\}^{n+1}_{(k+1)} = \{e\}^{n+1}_{(k)} + \{\Delta x\}$
8   **end**
9 **end**

(W/m^3). To make a two-way coupling between Eqs. (14.24) and (14.25), it is assumed that Q is produced by Joule heating, i.e. $Q = \sigma |\nabla v|^2$.

Equations (14.24) and (14.25) thus form a nonlinear multi-physics system. Although the individual electrical and thermal problems are linear, the coupled system is nonlinear due to the temperature dependence of σ in (14.24) and the quadratic relation between Q and magnitude of the gradient of v in (14.25). Such electrical–thermal coupling emerges, for instance, in the radiofrequency ablation of tumors [23].

One possible approach for solving the coupled problem in parallel is to apply FGaBP at each iteration of Newton's method. This method, which was used in Section 14.5.2 (see Algorithm 14.2), solves the linear system $J\Delta x = -F$ arising from Newton's method (Algorithm 14.1) using FGaBP. However, FGaBP needs the Jacobian matrix to be *symmetric*, which is *not* true for the electrical–thermal problem. Fortunately, however, the localized nature of FGaBP computations allows for the definition of *local* nonlinear systems of equations inside each FN instead. In such a case, the number of degrees of freedom (DoF) associated with each FN is the total number of DoF per element for all the physics. Consequently, the number of unknowns for each local nonlinear system is small and independent of the FEM mesh size.

The idea of constructing and solving local nonlinear systems instead of the global $J\Delta x = -F$ is relatively straightforward. In a single-physics problem, each of the FGaBP messages contains information about the mean and variance of the unknown physical quantity, e.g. the electric potential. On the other hand, the different physics involved in a multi-physics problem are interdependent. Thus, the messages should carry information about how the mean and variance of one physics depend on the other physics. This approach thus deals with the multi-physics interactions at the element level. The key property of FGaBP that allows this is that the messages sent by a FN at each FGaBP iteration only depend on the *local data*, i.e. the values in the neighborhood of the FN. As mentioned earlier, this property is the very same that makes FGaBP such an effective parallel algorithm.

To demonstrate this element-level Newton's method, the electrical–thermal problem can again be considered. Assume the theta-scheme [24] is used for the temporal discretization of (14.25), and similar FEM meshes for both (14.24) and (14.25) are used. Then, the following set of coupled discrete equations can be obtained:

$$\left[\mathbf{M}_a^v(T)\right]\{\mathbf{v}\} = 0 \tag{14.26a}$$

$$\left[\mathbf{M}_a^T + \delta t\,\theta\,\mathbf{S}_a\right]\{\mathbf{T}\}^{n+1} = \left[\mathbf{M}_a^T - \delta t\,(1-\theta)\mathbf{S}_a\right]\{\mathbf{T}\}^n$$
$$+ \delta t\,\left\{(1-\theta)\{\mathbf{B}_a(v)\}^n + \theta\{\mathbf{B}_a(v)\}^{n+1}\right\} \tag{14.26b}$$

where $\mathbf{M}_a^v(i,j) = \int \sigma(T)N_iN_j\,dV$, $\mathbf{M}_a^T(i,j) = \rho\,c_p \int N_iN_j\,dV$, $\mathbf{S}_a(i,j) = d\int \vec{\nabla}N_i \cdot \vec{\nabla}N_j\,dV$, and $\mathbf{B}_a(j) = \int \sigma(T)N_j|\nabla v|^2\,dV$. The element characteristic matrix in the electrical problem, $\mathbf{M}_a^v$, and the element source vector in the thermal problem, $\mathbf{B}_a$, are functions of the temperature and potential, respectively.

The integrals $\mathbf{M}_a^v(i,j) = \int \sigma(T)N_iN_j\,dV$ and $\mathbf{B}_a(j) = \int \sigma(T)N_j|\nabla v|^2\,dV$ are computed numerically where $\sigma(T)$ and ∇v are evaluated at quadrature points inside element a. Consequently, $\mathbf{M}_a^v$ and $\mathbf{B}_a$ depend on T_j and v_j, respectively, where $j \in \mathcal{N}(a)$. Returning to the FGaBP equations, (14.11)–(14.21), one can see that the messages α_{ai} and β_{ai} only depend on the local elemental matrix M and vector B that are associated with element a. Therefore, when applied to the coupled electrical–thermal problem, FGaBP messages at each iteration only depend on the temperature and potential values in their neighborhood, leading to immense parallelizability.

For the sake of simplicity, both the α_{ai} and β_{ai} messages sent from FN_a to VN_i will be denoted by m_{ai}, keeping in mind that $m = \{\alpha, \beta\}$. The dependence of m_{ai} on the local data is then expressed as:

$$\begin{cases} m_{ai}^v = f_i(\mathbf{v}_a, \mathbf{T}_a) \\ \\ m_{ai}^T = g_i(\mathbf{v}_a, \mathbf{T}_a) \end{cases} \tag{14.27}$$

Here, m_{ai}^v and m_{ai}^T are messages sent from factor node a to VN $i \in \mathcal{N}(a)$ in the electrical and thermal problems, respectively. The vectors $\mathbf{v}_a$ and $\mathbf{T}_a$ contain potential values and temperature values in the neighborhood of factor node a, $\mathcal{N}(a)$. The functions f_i and g_i are effectively the right-hand sides of (14.18) and (14.19), respectively, describing the dependency of the messages on the potential and temperature values in $\mathcal{N}(a)$. It's important to note that $f_i(g_i)$ is a nonlinear function of $\mathbf{T}_a(\mathbf{v}_a)$.

At each FGaBP iteration, the nonlinear system of (14.27) is solved using Newton's method. A local Jacobian matrix, $\mathbf{J}_a$, is constructed as follows:

$$\mathbf{J}_a = \begin{bmatrix} \mathbf{J}_{v,v} & \mathbf{J}_{v,T} \\ \mathbf{J}_{T,v} & \mathbf{J}_{T,T} \end{bmatrix}_{2n\times 2n} \tag{14.28}$$

where n is the number of nodes per element. The entries of $\mathbf{J}_a$ are analytically computed from the derivatives of f_i and g_i with respect to m_{ai}^v and m_{ai}^T. For instance, the sub-matrix $\mathbf{J}_{v,T}$ is formed as:

$$\mathbf{J}_{v,T}(\mathcal{L}(i), \mathcal{L}(j)) = \frac{\partial(m_{ai}^v - f_i)}{\partial m_{aj}^T} \tag{14.29}$$

where $1 \leq \mathcal{L}(i), \mathcal{L}(j) \leq n$ are the local indices corresponding to the global VNs $i, j \in \mathcal{N}(a)$. After the local Jacobian is found, message updates are calculated as:

$$\left\{ \begin{array}{c} \{\Delta m_{ai}^v\}_{n\times 1} \\ \{\Delta m_{ai}^T\}_{n\times 1} \end{array} \right\}_{2n\times 1} = [\mathbf{J}_a^{-1}] \cdot \left\{ \begin{array}{c} \{f_i - m_{ai}^v\}_{n\times 1} \\ \{g_i - m_{ai}^T\}_{n\times 1} \end{array} \right\}_{2n\times 1} \tag{14.30}$$

The above update rule is performed for each factor node a. The updated messages are then propagated throughout the mesh. After the FGaBP iterations have converged, in the next Newton iteration the local Jacobians and update messages are computed again. This procedure continues until the local update messages are smaller than a predefined threshold, meaning the cell-wise Newton's method has converged for the current time step.

14.6 Implementation on Parallel Hardware

In this section, the algorithms discussed in Sections 14.4 and 14.5 will be implemented on two different kinds of parallel hardware: GPUs and shared memory systems (a distributed memory version of the multi-physics algorithm in Section 14.5.3 has also recently been developed and applied to a thermal-thermal coupled problem in [25]). Due to the nature of these parallel architectures, a haphazard implementation of any particular algorithm can easily result in unwanted serializations, memory bandwidth issues, and underwhelming performance. As a result, the particularities of these architectures will also be discussed, with an emphasis on what measures should be taken to assure good performance in each case. Finally, several example problems will be explored in which the speed, versatility, and performance of the algorithms and hardware discussed in this chapter will be demonstrated.

14.6.1 Graphics Processing Units (GPU)

14.6.1.1 GPU Optimization Strategies

The parallel algorithms presented in the preceding sections should, in principle, provide a significant performance boost when executed on massively parallel hardware, such as a GPU. However, as mentioned, good performance can often only be achieved when the algorithms in question have been carefully implemented to take advantage of the particularities of GPU architecture. Specifically, within the NVIDIA GPU architecture, for example, there are two main points of optimization that can have a significant impact on performance and which will be discussed next.

The first is the interplay and distinction between GPU and CPU memory. In general, each of the multiprocessors on a GPU has access to a large amount of storage known as *global memory*. While this global memory is physically located on the graphics card, it is not on the same chip as the processor itself. Moreover, it is imperative to note that this GPU global memory and the host CPU memory are fundamentally distinct, meaning that the CPU and GPU do not have access to a shared common memory space. A direct consequence of this is that any time information is to be shared between the GPU and CPU must be explicitly transferred from one to the other. Unfortunately, this memory transfer between host CPU memory and global GPU memory must happen over a bus, and is therefore much slower than simply accessing data already stored on the GPU. If data is constantly going back and forth between the CPU and GPU, the resulting transfer overhead can rapidly overshadow any performance increase afforded by the parallelization [26].

Minimizing memory transfers between the host CPU and the GPU is thus an important point of optimization. When implementing the parallel version of the algorithms detailed in the previous sections, it is therefore imperative to ensure that as much of the computation occurs on the GPU as possible. For instance, with nonlinear dielectrics, in addition to the elemental matrix evaluation and the FGaBP algorithm, the evaluation of source terms, updating of the magnetic field (in the case of mixed methods), and updating of any auxiliary variables should all be executed on the GPU within global memory. The result of this optimization strategy is that, in the best-case scenario, memory transfers occur only twice: once at the beginning to transfer problem data to the GPU and once at the end to transfer the solution back to the CPU.

The second memory optimization consideration comes in the form of how NVIDIA GPUs access their global memory. Due to the SIMT design philosophy of the GPU, global memory is designed to be accessed in large contiguous chunks, meaning that all of the data required by a group of executing threads can be fetched in a single memory transaction. In contrast, if each executing thread were to access random scattered locations in global memory, this access pattern would result in serialization and necessitate 32 different memory transactions. Clearly, maximum memory bandwidth will thus result when executing threads require access to sequential memory addresses [26].

As a result of this global memory access pattern, the storage format of the algorithm's various data structures can have a significant impact on memory throughput and performance. For example, consider Figure 14.5 in which three threads, each associated with a different element, are each attempting to access their local matrix data from global memory. If these local matrices are stored in row-major format one after the other, the threads will end up requesting non-consecutive memory locations, requiring multiple memory transactions. In contrast, if the local matrices are interwoven, the resulting memory accesses can be made to be contiguous and served in a single transaction. The result is that for

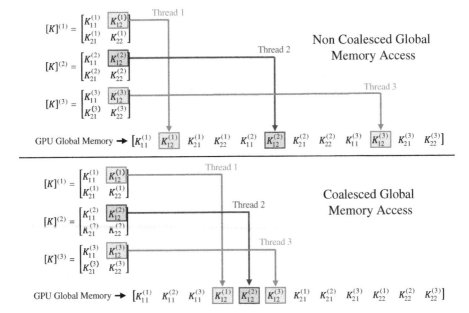

Figure 14.5 Depiction of non-coalesced versus coalesced GPU global memory access.

best performance data should be stored in GPU memory in a way consistent with the required access patterns.

While the two memory considerations detailed above are generally where the most performance can be gained or lost, there are also other smaller optimizations possible that may also increase performance. For example, in addition to global memory, the GPU also has much smaller constant and shared memory available [26]. These differ in their access patterns and scope but if used properly are much faster than global memory. The implementation tested in the next section, for instance, loads values such as the Gaussian Quadrature abscissae and weights, as well as physical parameters, into constant memory as they require little space and are unchanged over the course of the computation.

While having the entire computation execute on the GPU without the need for CPU intervention or memory transfers results in a much faster parallel algorithm, it must be noted that this is counterbalanced by the need to have the entire simulation fit into the GPU memory. For small computations this is inconsequential; however, for very large computations GPU memory is generally less abundant than CPU memory and may become a limiting factor. However, while GPUs offer a very large amount of hardware parallelism, it is important to reiterate that the algorithms described in this chapter are generally architecture independent. Hence, while the implementation details will vary between different hardware, these algorithms could equally be adapted for execution on multicore CPUs and supercomputing clusters.

14.6.1.2 GPU Results

Having discussed the particular implementation strategies and approaches required to obtain good performance on GPUs, here the results of these considerations, and the potential of the GPU, will be demonstrated. To that end, the parallelizations suggested for nonlinear media in Sections 14.5.2.1 and 14.5.2.2 were implemented on an NVIDIA GPU in the CUDA programming language, with special attention given to the implementation details discussed in Section 14.6.1.1.

A test problem was made to simulate the propagation of a spatial soliton in a 2D domain completely filled with an instantaneous nonlinear dielectric (see Chapter 3). Execution time was then compared between a CPU and GPU version of the simulation code. The serial program was executed on a single core of a workstation equipped with an Intel 8700K CPU clocked at 3.7 GHz, supplied with 16 GB of DDR4 RAM operating at 3000 MHz. The GPU code, meanwhile, was compiled in CUDA v9.2 and was executed on an NVIDIA GTX 1070Ti GPU with 2432 cores clocked at 1607 MHz with 8 GB of GDDR5 memory. The workstation hosting the graphics card was the same as that used for the CPU runs, meaning that any non-GPU sections of code were executed on the same processor as in the serial case. Since this model of GPU is capable of both double and single precision arithmetic, two versions of the code were tested.

Figure 14.6 depicts the resulting speedup achieved by the GPU version of the algorithm, obtained by dividing the total execution time on the CPU by the total execution time on the GPU. For the smallest number of DoF tested (about 10^4), the GPU implementation is about three to five times faster than its CPU counterpart (note that the overhead time required to transfer data to and from host and GPU memory has been included in this data). As the number of DoF increases, however, the disparity between the two algorithms grows significantly, as does the difference between single and double precision performance. At the maximum extent tested, equal to roughly 115,000 DoF, the single precision GPU implementation performs an impressive 212 times faster than its CPU counterpart, whereas for double precision it performs 141 times faster. In the case of single precision, this represents a dramatic reduction in computation time from over 1.5 hours down to roughly 30 seconds.

Further analysis of these results reveals, unsurprisingly, that most of the speedup comes from the parallelization of the elemental matrix evaluation. Indeed, with its independent nature and emphasis on simple multiplication and addition, numerical integration benefits immensely from the GPU's FLOPS. With the high proportion of computation time spent on these operations in the CPU version, it is unsurprising, therefore, that the result is a significant boost in speed. This isn't to say that the implementation of FGaBP is inconsequential. In fact, not only does the FGaBP algorithm solve the matrix system up to five times faster than a serial preconditioned conjugate gradient (PCG) solver, but it also completely skips the step of matrix assembly, and allows data to remain on the GPU at all times, saving significantly on memory transfers.

14.6.2 Shared Memory Implementations

In this section, the FGaBP-based Newton's method algorithm will be applied to the coupled electrical–thermal problem of (14.26) in both two-dimensional (2D) and three-dimensional (3D) geometries, using a shared memory implementation. Figure 14.7 depicts the geometry of the coupled problem. The computational domain is an 8.0 cm by 8.0 cm square (8.0 cm by 8.0 cm by 8.0 cm cube in 3D), where a 16 V source voltage is modeled as a Dirichlet boundary condition at the center of the domain and the outer boundaries serve as ground. In the thermal problem, the initial value of the temperature is chosen to be 37 °C with a Neumann boundary condition being applied to the outer boundary. The electrical and thermal parameters in (14.26) are chosen according to Table 14.1. In addition, the electrical conductivity increases with the temperature linearly, with a rate of 2%/°C. The algorithm is implemented in the open-source FEM library deal.II [27].

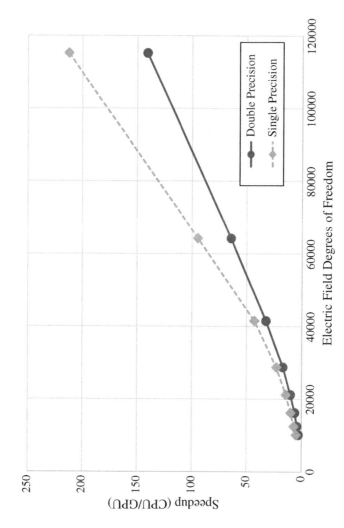

Figure 14.6 GPU over CPU speedup as a function of degrees of freedom for both single and double precision floating point simulations.

Figure 14.7 The geometry of the test case in two dimensions.

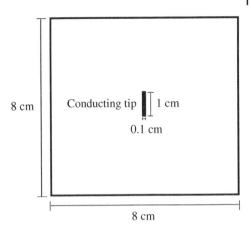

Table 14.1 Numerical values of model parameters.

Parameter	value
σ	0.33 S/m
c_p	3600 J/(K^2, K)
ρ	1060 kg/m^3
d	0.512 W/(K m)
δt	0.5 s
θ	0.5

Figure 14.8 Shared memory model.

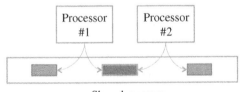

Shared memory

In the shared memory model, concurrent threads interact by reading and writing shared objects in memory (see Figure 14.8). Assume *Thread 1* and *Thread 2* are executed concurrently, sharing the same physical memory. Under these circumstances, a *race condition* describes a situation in which the correctness of a program is governed by the relative timing of events in the concurrent threads, i.e. different results can be obtained depending on when and in what order threads access the same memory location. In the current implementation, a race condition might occur when two adjacent elements, e.g. elements a and b, try to update

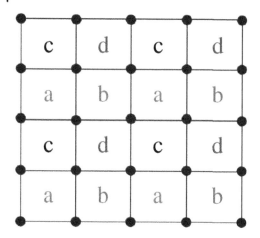

Figure 14.9 Mesh coloring scheme for a structured quadrilateral mesh.

the same global node shared between them. If the two elements belong to different threads, then it must be ensured that the messages m_{aj} and m_{bj} do not try to update node j simultaneously, leading to undefined behavior. One possible solution to avoid race conditions is to schedule the messages based on a so-called *element coloring*. In this scheme, the mesh elements are "colored," or labeled, such that no two adjacent elements have the same color. In this way, the messages in each color group can be computed and safely communicated in parallel, since elements that have the same color do not share any global nodes. An example of a mesh coloring diagram in 2D is illustrated in Figure 14.9 using a quadrilateral mesh.

To accelerate the FGaBP iterations, multiple mesh levels can be used to help pass messages around the domain more effectively, with mesh refinement being conducted in 2D (3D) by splitting each quadrilateral (hexahedral) cell into four (eight) smaller cells successively. Thus, the local messages $m_{ai,v}$ and $m_{ai,T}$ calculated for each element a and node $i \in \mathcal{N}(a)$ are transferred from the coarse mesh (the parent) to the next finest mesh (the child). These messages contain the so-called *beliefs* regarding the temperature and potential values at each VN, as well as the sensitivity of these values with respect to the values in their neighborhood.

The resulting parallel Newton's method with OpenMP directives is provided by Algorithm 14.3. The multigrid version of FGaBP was introduced by El-Kurdi et al. [28]. They defined element-wise belief *residuals* and *corrections* to incorporate the multigrid method into the FGaBP. The belief residuals of each group of child elements are locally restricted into the parent element (line 17 of Algorithm 14.3). On the other hand, the corrections from the parent elements are mapped onto the child elements using interpolation (line 22 of Algorithm 14.3). Since only the potential and temperature values are considered in the prolongation and restriction steps in Algorithm 14.3, these two steps are implemented exactly as proposed by El-Kurdi et al. [28]. The multi-grid scheme is used to transfer information between different refinement levels and is implemented as a V-Cycle, as indicated by lines 14–28. In this way it acts as a preconditioner, in that it reduces the number of iterations required on the finest level. Meanwhile, the FGaBP iterations and the application of the local Jacobian matrices to update each message are executed by Algorithms 14.4 and 14.5, respectively.

First, numerical results from this modified nonlinear FGaBP algorithm are verified against COMSOL Multi-physics software to ensure proper functioning of the parallelized solution process. Accordingly, Figure 14.10 compares the temperature obtained from COMSOL with that obtained by Algorithm 14.3 at a specific location for one minute of ablation. The resulting root-mean-square (RMS) error between these two simulations is only 0.028 °C, corroborating the proper functioning of Algorithm 14.3. The parallel scalability properties of Algorithm 14.3 were also tested using a CPU implementation with multi-threading (OpenMP). To that end, a single node of a high-performance cluster, operated by Compute Canada [29], with 2×20-core Intel Gold 6148 Skylake 2.4 GHz CPUs and 186 GB DRAM was utilized.

Due to the multigrid implementation, the number of iterations is independent of the problem size at the finest level. For example, when the parameters $v_1 = 1$ and $v_2 = 5$ in Algorithm 14.3, the V-cycle required five iterations for all 2D and seven iterations for all 3D runs. However, the number of iterations in the inner FGaBP execution, i.e. line 19 in Algorithm 14.3, remains proportional to the number of unknowns in the coarsest level. In the electrical problem, the starting

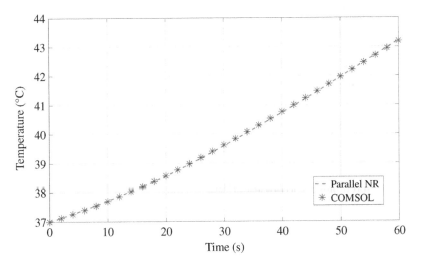

Figure 14.10 Comparison of temperature values over time at a specific point in the 2D domain.

Algorithm 14.3: Parallel Newton's method in each time step.

---

1 Start timer;
2 Import problem geometry and mesh;
3 # pragma omp parallel;
4 **for** *NR iteration m* = 1, 2, … **do**
5   **for** *each level in the mesh hierarchy* **do**
6   # pragma omp for;
7   **for** *cell a* = 1, 2, … **do**
8   **for** *node i* ∈ $\mathcal{N}(a)$ **do**
9   Initialize messages m_{ai}^v and m_{ai}^T;
10   **end**
11   Calculate local vector $\mathbf{B}_a$ and local matrices $\mathbf{M}_a^v$ and $\mathbf{M}_a^T$ inside cell a from (14.26).
12   **end**
13   **end**
14   **for** *cycles* = 1, 2, … *in the V-cycle* **do**
15   **for** *Mesh levels from fine to coarse* **do**
16   Execute v_1 iterations of Algorithm 14.4;
17   Restrict;
18   **end**
19   Execute Algorithm 14.4 on the coarsest level;
20   **for** *Mesh levels from coarse to fine* **do**
21   Execute v_2 iterations of Algorithm 14.4;
22   Prolongate;
23   **end**
24   # pragma omp single;
25   **if** *global tolerance* < *tolerance* **then**
26   break;
27   **end**
28   **end**
29   Execute Algorithm 14.5 on the finest level;
30   # pragma omp single;
31   **if** *global residual* < *NR tolerance* **then**
32   break;
33   **end**
34 **end**
35 End timer;
36 Plot output;

Algorithm 14.4: FGaBP with local Jacobian calculation.

1 **for** *FGaBP iteration t* = 1, 2, ... **do**
2   **for** *color c* = 1, 2, ... **do**
3   # pragma omp for;
4   **for** *cell a in color c* **do**
5   **for** *node i* $\in \mathcal{N}(a)$ **do**
6   Calculate m_{ai}^v and m_{ai}^T from FGaBP update rules using (14.11) to (14.21);
7   Calculate $\mathbf{J}_a$ using (14.28) ;
8   Update v_i and T_i ;
9   Calculate local message tolerance;
10   **end**
11   **end**
12   # pragma omp critical;
13   Update global message tolerance;
14   **end**
15   # pragma omp single;
16   **if** *global tolerance* < *FGaBP tolerance* **then**
17   return(global tolerance);
18   break;
19   **end**
20 **end**

Algorithm 14.5: Parallel message update.

1 **for** *color c* = 1, 2, ... **do**
2   # pragma omp for;
3   **for** *cell a in color c* **do**
4   Update m_{ai}^v and m_{ai}^T using (14.30);
5   Update v_i and T_i;
6   Calculate local NR residual;
7   Calculate local vector $\mathbf{B}_a$ and local matrices $\mathbf{M}_a^v$ and $\mathbf{M}_a^T$ inside cell a using (14.26).
8   **end**
9   # pragma omp critical;
10   Update global NR residual;
11 **end**

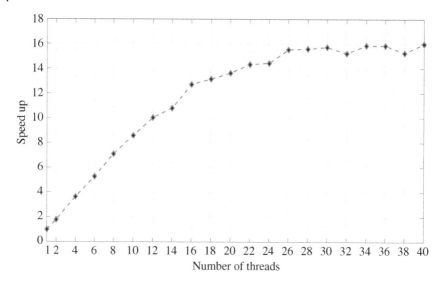

Figure 14.11 Strong scaling of Algorithm 14.3 in terms of speedup with respect to 1 core implementation.

mesh size was 400 cells in 2D (8000 cells in 3D), which took 380 (430) inner iterations to reach a message residual of 10^{-8}. The thermal problem converged faster in both the 2D and 3D implementations. Figure 14.11 shows the shared memory implementation's speedup for 4,173,281 unknowns in 3D.

Algorithm 14.3 was also compared to a parallel implementation of Newton's method provided by the open-source software library PETSc (Portable, Extensible Toolkit for Scientific Computation [30–32]). PETSc has an extensive suite of parallel nonlinear solvers and uses the message passing interface (MPI) standard for communication between parallel tasks. The PETSc implementation was executed on a single cluster node with a total of 40 cores, with one MPI task assigned per core. In addition, the Jacobian matrix was assembled in parallel at each Newton iteration, using deal.II's WorkStream shared-memory model. Regarding the solving stage, PETSc supports a variety of iterative and direct solvers, from which the multifrontal massively parallel sparse direct Solver (MUMPS) [33, 34] was chosen. The average execution time per time step obtained from Algorithm 14.3 and PETSc are shown in Figure 14.12. Algorithm 14.3 was run with 16 threads, whereas PETSc had 16 threads for the assembly and 16 MPI tasks for the solution. The problem sizes varied from 500K to 33M unknowns in 3D. Algorithm 14.3 thus demonstrates faster execution time all while maintaining linear scalability.

Lastly, the convergence plot of Algorithm 14.3 is compared to that of a Gauss–Seidel algorithm in Figure 14.13, where each has been applied to the same

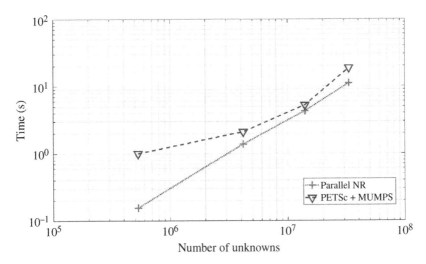

Figure 14.12 Comparison of Algorithm 14.3 versus PETSc execution times on 16 cores.

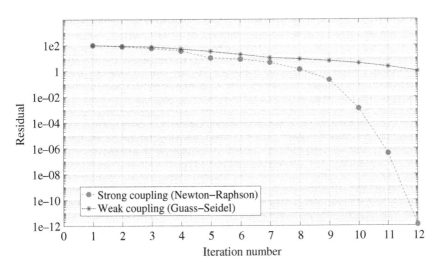

Figure 14.13 The convergence plot of Algorithm 14.3 compared to that of a Gauss-Seidel algorithm.

electrical–thermal problem. In the early iterations of Algorithm 14.3, changes of the residual norm are almost linear; however, the quadratic convergence can be observed in the last three iterations as the algorithm gets sufficiently close to the solution.

14.7 Conclusion

In this chapter, alternative parallelization strategies for FETD formulations have been presented, demonstrating a high degree of parallelism on both GPUs and multicore architectures. In particular, nonlinear PDEs resulting from two different sources, nonlinear dielectric media and multi-physics problems, have been successfully accelerated and solved in parallel. Indeed, the presented nonlinear GPU implementation demonstrated substantial speedup as compared to traditional serial CPU implementations. Moreover, the suggested multicore multi-physics algorithm not only exhibited good speedup and parallelism but also realized faster execution times as compared to the highly optimized open-source alternative PETSc.

References

1 Taflove, A. and Hagness, S.C. (2005). *Computational Electrodynamics: The Finite-Difference Time-Domain Method*, 3e. Boston, MA: Artech House.

2 Cangellaris, A.C. and Wright, D.B. (1991). Analysis of the numerical error caused by the stair-stepped approximation of a conducting boundary in FDTD simulations of electromagnetic phenomena. *IEEE Transactions on Antennas and Propagation* 39 (10): 1518–1525.

3 Akyurtlu, A., Werner, D.H., Veremey, V. et al. (1999). Staircasing errors in FDTD at an air-dielectric interface. *IEEE Microwave and Guided Wave Letters* 9 (11): 444–446.

4 Hwang, K.P. and Cangellaris, A.C. (2001). Effective permittivities for second-order accurate FDTD equations at dielectric interfaces. *IEEE Microwave and Wireless Components Letters* 11 (4): 158–160.

5 Wang, S. and Teixeira, F.L. (2004). Some remarks on the stability of time-domain electromagnetic simulations. *IEEE Transactions on Antennas and Propagation* 52 (3): 895–898.

6 Jin, J. (2014). *The Finite Element Method in Electromagnetics*, 3e. Hoboken, NJ: Wiley-IEEE Press.

7 Rennich, S.C., Stosic, D., and Davis, T.A. (2016). Accelerating sparse Cholesky factorization on GPUs. *Parallel Computing* 59: 140–150.

8 Helfenstein, R. and Koko, J. (2012). Parallel preconditioned conjugate gradient algorithm on GPU. *Journal of Computational and Applied Mathematics* 236 (15): 3584–3590.

9 He, K., Tan, S.X.D., Zhao, H. et al. (2016). Parallel GMRES solver for fast analysis of large linear dynamic systems on GPU platforms. *Integration* 52: 10–22.

10 El-Kurdi, Y., Dehnavi, M.M., Gross, W.J., and Giannacopoulos, D. (2015). Parallel finite element technique using Gaussian belief propagation. *Computer Physics Communications* 193: 38–48.

11 Chen, S., Fang, J., Chen, D. et al. (2018). Optimizing sparse matrix-vector multiplication on emerging many-core architectures. arXiv preprint arXiv:180511938.

12 Liu, W. and Vinter, B. (2015). CSR5: An efficient storage format for cross-platform sparse matrix-vector multiplication. *Proceedings of the 29th ACM on International Conference on Supercomputing*, pp. 339–350.

13 Linte, C.A., Camp, J.J., Holmes, D.R. et al. (2013). Toward online modeling for lesion visualization and monitoring in cardiac ablation therapy. *International Conference on Medical Image Computing and Computer-Assisted Intervention*, Springer, pp. 9–17.

14 Pearl, J. (2014). *Probabilistic Reasoning in Intelligent Systems: Networks of Plausible Inference*. Elsevier.

15 Yedidia, J.S., Freeman, W.T., and Weiss, Y. (2003). Understanding belief propagation and its generalizations. In: *Exploring Artificial Intelligence in the New Millennium*, (ed. Gerhard Lakemeyer and Bernhard Nebel), Morgan Kauffmann Publishing Inc, San Francisco, CA, USA, vol. 8, 236–239.

16 Shental, O., Siegel, P.H., Wolf, J.K. et al. (2008). Gaussian belief propagation solver for systems of linear equations. *2008 IEEE International Symposium on Information Theory*, IEEE, pp. 1863–1867.

17 El-Kurdi, Y. (2015). Parallel finite element processing using Gaussian belief propagation inference on probabilistic graphical models. PhD thesis. McGill University Libraries.

18 Abraham, D.S. and Giannacopoulos, D.D. (2020). A parallel finite-element time-domain method for nonlinear dispersive media. *IEEE Transactions on Magnetics* 56 (2): 1–4.

19 Knoll, D.A. and Keyes, D.E. (2004). Jacobian-free Newton-Krylov methods: a survey of approaches and applications. *Journal of Computational Physics* 193 (2): 357–397.

20 Khalevitsky, Y.V., Burmasheva, N.V., and Konovalov, A.V. (2016). An approach to the parallel assembly of the stiffness matrix in elastoplastic problems. *AIP Conference Proceedings*

21 Scholz, E., Ye, H., Schöps, S., and Clemens, M. (2013). A parallel FEM matrix assembly for electro-quasistatic problems on GPGPU systems. *IEEE Transactions on Magnetics* 49 (5): 1801–1804.

22 Kiran, U., Sharma, D., and Gautam, S.S. (2019). GPU-warp based finite element matrices generation and assembly using coloring method. *Journal of Computational Design and Engineering* 6 (4): 705–718.

23 Audigier, C., Mansi, T., Delingette, H. et al. (2015). Efficient lattice Boltzmann solver for patient-specific radiofrequency ablation of hepatic tumors. *IEEE Transactions on Medical Imaging* 34 (7): 1576–1589.

24 Szabó, T. (2008). On the discretization time-step in the finite element theta-method of the discrete heat equation. *International Conference on Numerical Analysis and Its Applications*, Springer, pp. 564–571.

25 Akbari, A. and Giannacopoulos, D.D. (2021). Efficient solver for a simplified model of the multi-physics heat transfer problem in radiofrequency ablation of hepatic tumors. *IEEE Transactions on Magnetics* 57 (6): 1–4.

26 CUDA C++ Programming Guide (2019). PG-02829-001_v10.2.

27 Bangerth, W., Hartmann, R., and Kanschat, G. (2007). deal. II–a general-purpose object-oriented finite element library. *ACM Transactions on Mathematical Software (TOMS)* 33 (4): 24–es.

28 El-Kurdi, Y., Gross, W.J., and Giannacopoulos, D. (2014). Parallel multigrid acceleration for the finite-element Gaussian belief propagation algorithm. *IEEE Transactions on Magnetics* 50 (2): 581–584.

29 Compute Canada. Available from: www.computecanada.ca (accessed 2021).

30 Balay, S., Abhyankar, S., Adams, M.F. et al. (2020). PETSc Users Manual. Argonne National Laboratory, ANL-95/11 - Revision 3.13. Available from: https://www.mcs.anl.gov/petsc (accessed 1 June 2022).

31 Balay, S., Abhyankar, S., Adams, M.F. et al. (2019). PETSc Web page. https://www.mcs.anl.gov/petsc (accessed 1 June 2022).

32 Balay, S., Gropp, W.D., McInnes, L.C., and Smith, B.F. (1997). Efficient management of parallelism in object oriented numerical software libraries. In: *Modern Software Tools in Scientific Computing* (ed. E. Arge, A.M. Bruaset, and H.P. Langtangen), 163–202. Birkhäuser Press.

33 Amestoy, P.R., Buttari, A., L'Excellent, J.Y., and Mary, T. (2019). Performance and scalability of the block low-rank multifrontal factorization on multicore architectures. *ACM Transactions on Mathematical Software* 45: 2:1–2:26.

34 Amestoy, P.R., Duff, I.S., Koster, J., and L'Excellent, J.Y. (2001). A fully asynchronous multifrontal solver using distributed dynamic scheduling. *SIAM Journal on Matrix Analysis and Applications* 23 (1): 15–41.

Part VI

Multidisciplinary Explorations of Time-Domain Methods

15

The Symplectic FDTD Method for Maxwell and Schrödinger Equations

Zhixiang Huang[1], Guoda Xie[1], Xingang Ren[1], and Wei E.I. Sha[2]

[1]*Key Laboratory of Intelligent Computing and Signal Processing, Ministry of Education, Anhui University, Hefei, China*
[2]*Key Laboratory of Micro-Nano Electronic Devices and Smart Systems of Zhejiang Province, College of Information Science and Electronic Engineering, Zhejiang University, Hangzhou, China*

15.1 Introduction

Classical electromagnetic (EM) theory and its modern extension, known as computational electromagnetism (CEM), are fundamentally important to electrical and electronic engineering, which play a key role in the development and design of cutting-edge technologies [1, 2]. With the growing development of microelectronic and nanoelectronic devices [3–7], the characteristic size of a device is continuously decreasing, and thus the system integration is getting denser. The capacity of the integrated system, e.g. information storage and processing, has improved by a million times. To achieve the integrated system, two common approaches are executed. First, the employment of composite materials toward a densely integrated system. Second, decreasing the characteristic size of each unit in the device from the original micron level to the nanometer level. With these approaches, various nanostructures have been comprehensively studied in electronic systems and are widely applied in ultrahigh-sensitivity sensors [8], plasma circuits [9], high-sensitivity biological detection [10], information storage [11], super-resolution imaging [12], quantum communication [13], etc. These advanced manufacturing electronic systems usually have complex EM system response and require the advancement of traditional CEM in modeling techniques, systematic designs, and practical applications. Actually, the above approaches typically involve the frontier interdisciplinary issues including molecular biology, quantum mechanics (QM), material science, electromagnetism, optics, and computing science. Besides, the integrated and miniaturized structures inevitably involve interactions between EM waves and microscopic particles (such as

Advances in Time-Domain Computational Electromagnetic Methods, First Edition.
Edited by Qiang Ren, Su Yan, and Atef Z. Elsherbeni.
© 2023 The Institute of Electrical and Electronics Engineers, Inc. Published 2023 by John Wiley & Sons, Inc.

microscopic particles: atoms, molecules, and quantum dots, and microscopic-like particles: ultrasmall two-dimensional materials, superconducting artificial atoms, quantum metamaterials, etc.), and the macroscopic average effects described in the constitutive relations of classical EM systems, such as polarization and conduction, will no longer be accurate and applicable. The description of the interactions between electrons, scattering of electrons, and phonons, the influences of external environments (electrode, electric field, and magnetic field), spontaneous emission of atoms and molecules, quantum tunneling, and nonlocal effects urgently requires advanced physical mechanisms. Regarding the interaction between the EM wave and biological molecules/particles, the effect of the terahertz wave on the selective permeability of biological ion channels has been investigated, and the underlying mechanism of the generation, transmission, and coupling of terahertz and other high-frequency EM information in the biological nerve has been explored. Meanwhile, there are also several studies of terahertz technology and methods to enhance brain cognition and regulate brain function, which lay a foundation for realizing the interpretability, readability, and regulation of human intelligence [14–17]. For the ion channels, a quantum-mechanical description of ion motion is essential, and the propagation properties of the terahertz wave are simulated by the macroscopic EM framework. Generally, for emerging nanoscale devices, optoelectronic devices, and the study of high-frequency EM signal in the nervous system, the multiple physical equations and numerical methodology are necessarily developed to explore the interactions between EM and QM systems. Maxwell's equations and time-dependent Schrödinger equation (TDSE) are the fundamental governing equations of EM and QM, respectively [18, 19].

In the past several years, the study of taking quantum effect into EM simulation has attracted a great deal of attention worldwide, which was first proposed by Sui et al. [20]. In addition, the finite-difference time-domain (FDTD) method is used to solve the coupled Maxwell's and Schrödinger equations. The simulation procedure is summarized as follows. Firstly, the electric field components $\mathbf{E}$ and the magnetic field components $\mathbf{H}$ are updated by Maxwell's equations, and then the vector magnetic potential $\mathbf{A}$ and scalar potential φ are calculated according to the physical relations between them. These potential field components are then inserted into the modified Schrödinger equations to solve the wavefunction ψ and the quantum current density $\mathbf{J}_q$. At last, the obtained quantum current density $\mathbf{J}_q$ is regarded as a source for the EM system. Since then, the Maxwell–Schrodinger (M–S) multi-physics problems are widely investigated with similar simulation processes except for the numerical methodology. For instance, Pieratoni et al. combined the FDTD method with the three-dimensional transmission line matrix (TLM) method to solve the coupled M–S equations and studied the interaction between the EM field and the carbon nanostructures [21, 22]. Ahmed et al. adopted a hybrid FDTD method to simulate the electric field distribution of the nanowire [23]. The hybrid FDTD method approach employed the conventional FDTD method to solve the Schrödinger equation, and the locally one-dimensional FDTD method (LOD-FDTD method) is used to solve Maxwell's equations. The proposed hybrid method is efficient due to the extended time stability condition; thus an enlarged time step can be used to solve the coupled M–S system. In order to efficiently couple the EM fields to the Schrödinger equation, the alternative equations based on **A** and φ, instead of E and H based on Maxwell's equations, are introduced, which avoid the conversion process between EM fields (E and H) and potential fields (**A** and φ) [24–26]. Ohnuki et al. adopted the "length gauge" condition and directly substituted the E field components into the modified Schrödinger equation to calculate the wave function, thus avoiding the calculation of **A** and φ in the numerical simulations [27]. Subsequently, Takeuchi et al. proposed a scheme of designing laser pulses for controlling quantum states based on the M–S hybrid simulation [28].

Nevertheless, it should be noted that the time evolution of M–S system always needs more time steps when compared to that of a pure EM system. Therefore, it is necessary to develop a more accurate energy-preservation solution for solving the M–S system. Fortunately, intensive research works suggest that the symplectic algorithm is one of best candidates for long-term simulation, in view of its energy-conserving and highly stable characteristics [29, 30]. The symplectic algorithm is based on the basic principles of Hamiltonian mechanics for preserving Hamiltonian systems. It makes the discretized difference equations preserve the symplectic structure of the original Hamiltonian system. Furthermore, numerous physical systems can be modeled by Hamiltonian differential equations; the time evolutions of the differential equations are symplectic transformations [31, 32].

In this chapter, we introduce the basic theory of the symplectic FDTD method, and then different types of coupled M–S equations without considering the symplectic structure are discussed, including the traditional FDTD method and high-order symplectic FDTD method. At last, we introduce a coupled M–S equation that has the symplectic structure, and the related symplectic FDTD technique is also presented.

15.2 Basic Theory for the Symplectic FDTD Method

Firstly, we introduce the general formulations of the symplectic FDTD (SFDTD) scheme for Maxwell's equations. A function of space and time evaluated at a discrete point in the Cartesian lattice and at a discrete stage in the time step can be notated as:

$$F(x, y, x, t) = F^{n+l/m}(i\Delta_x, j\Delta_y, k\Delta_z, (n + \tau_l)\Delta t_l) \tag{15.1}$$

where Δ_x, Δ_y, and Δ_z are the cell size along each direction, Δ_t is the time step, i, j, k, n, l, and m are integers, $n + l/m$ denotes the l-th stage after the n-th time step, m is the total stage number, and τ_l is the fixed time with respect to the l-th stage. For the spatial direction, the explicit fourth-order-accurate difference expressions in conjugation with the staggered Yee lattice are used to discretize the first-order spatial derivatives, as follows:

$$\left(\frac{\partial F^{n+l/m}}{\partial \Delta}\right)_h = \frac{9}{8} \times \frac{F^{n+l/m}(h+1/2) - F^{n+l/m}(h-1/2)}{\Delta_\delta}$$

$$+ \frac{-1}{8} \times \frac{F^{n+l/m}(h+3/2) - F^{n+l/m}(h-3/2)}{\Delta_\delta} \qquad (15.2)$$

where $\delta = x, y, z$ and $h = i, j, k$.

For the temporal direction, a helicity Hamiltonian [33] for the Maxwell's equations in a homogeneous, lossless, and source-less medium is introduced as:

$$G(\mathbf{H}, \mathbf{E}) = \frac{1}{2}\left(\frac{1}{\varepsilon}\mathbf{H} \cdot \Delta \times \mathbf{H}, \frac{1}{\mu}\mathbf{E} \cdot \Delta \times \mathbf{E}\right) \qquad (15.3)$$

where $\mathbf{E} = (E_x, E_y, E_z)^\mathrm{T}$ is the electric field vector, $\mathbf{H} = (H_x, H_y, H_z)^\mathrm{T}$ is the magnetic field vector, and ε and μ are the permittivity and permeability of the medium, respectively. The symplectic integrator can be generated starting from the canonical Euler–Hamilton equations of the form:

$$\frac{\partial \mathbf{H}}{\partial t} = -\frac{\partial H}{\partial \mathbf{E}}, \frac{\partial \mathbf{E}}{\partial t} = \frac{\partial H}{\partial \mathbf{H}} \qquad (15.4)$$

According to the variational principle, (15.4) can be rewritten as:

$$\frac{\partial}{\partial t}\begin{pmatrix}\mathbf{H} \\ \mathbf{E}\end{pmatrix} = (U + V)\begin{pmatrix}\mathbf{H} \\ \mathbf{E}\end{pmatrix} \qquad (15.5)$$

and

$$U = \begin{pmatrix}\{0\}_{3\times3} & -\mu^{-1}R \\ \{0\}_{3\times3} & \{0\}_{3\times3}\end{pmatrix}, \quad V = \begin{pmatrix}\{0\}_{3\times3} & \{0\}_{3\times3} \\ \varepsilon^{-1}R & \{0\}_{3\times3}\end{pmatrix} \qquad (15.6)$$

$$R = \begin{pmatrix} 0 & -\dfrac{\partial}{\partial z} & \dfrac{\partial}{\partial y} \\[2mm] \dfrac{\partial}{\partial z} & 0 & -\dfrac{\partial}{\partial x} \\[2mm] -\dfrac{\partial}{\partial y} & \dfrac{\partial}{\partial x} & 0 \end{pmatrix} \qquad (15.7)$$

where $\{0\}_{3\times3}$ is the 3×3 null matrix, and R is the 3×3 matrix representing the three-dimensional curl operator. Using the product of elementary symplectic mapping, the exact solution of (15.5) from $t = 0$ to $t = \Delta_t$ can be approximately constructed [34]:

$$\exp(\Delta_t(U + V)) = \prod_{l=1}^{m} \exp(d_l\Delta_t V)\exp(c_l\Delta_t U) + O\left(\Delta_t^{p+1}\right) \qquad (15.8)$$

Table 15.1 The coefficients $c_l = c_{m+1-l}$ $(0 < l < m + 1)$; $d_l = d_{m-l}$ $(0 < l < m)$, $d_m = 0$ in the symplectic algorithms.

Coefficients	2nd-order 2-stage	4th-order 5-stage
c_1	0.5	0.173,996,891,465,41
d_1	1.0	0.623,379,324,513,22
c_2	0.5	−0.120,385,041,214,30
d_2	0.0	−0.123,379,324,513,22
c_3	—	0.892,776,299,497,78
d_3	—	d_2

and c_l and d_l are the coefficients of the symplectic integration algorithm. m is the number of the sub-steps in each time step and p represents the approximation order, generally $m \geq p$. In order to achieve the value of p as high as possible for a given number m, several efforts have been made to obtain the coefficients c_l and d_l. Substantial numerical experiments prove that the coefficients c_l and d_l in [35] perform better than other symplectic integrators. The coefficients for different-order approximations are listed in Table 15.1.

15.2.1 The Basic Update Equations of the SFDTD Method

In order to reduce the truncation error, the electric field component is scaled to the order of amplitude of the magnetic field component.

$$\hat{\mathbf{E}} = \sqrt{\frac{\varepsilon_0}{\mu_0}} \, \mathbf{E} \tag{15.9}$$

After discretization of high-order difference in space and time, the updating equations of the SFDTD method for Maxwell's equations are as follows:

$$
\hat{E}_x^{n+l/m}\left(i+\frac{1}{2},j,k\right) = \hat{E}_x^{n+(l-1)/m}\left(i+\frac{1}{2},j,k\right) + \frac{1}{\varepsilon_r\left(i+\frac{1}{2},j,k\right)}
$$
$$
\times \left\{ \alpha_{y1} \times \left[H_z^{n+l/m}\left(i+\frac{1}{2},j+\frac{1}{2},k\right) - H_z^{n+l/m}\left(i+\frac{1}{2},j-\frac{1}{2},k\right) \right] \right.
$$
$$
- \alpha_{z1} \times \left[H_y^{n+l/m}\left(i+\frac{1}{2},j,k+\frac{1}{2}\right) - H_y^{n+l/m}\left(i+\frac{1}{2},j,k-\frac{1}{2}\right) \right]
$$
$$
+ \alpha_{y2} \times \left[H_z^{n+l/m}\left(i+\frac{1}{2},j+\frac{3}{2},k\right) - H_z^{n+l/m}\left(i+\frac{1}{2},j-\frac{3}{2},k\right) \right]
$$
$$
\left. - \alpha_{z2} \times \left[H_y^{n+l/m}\left(i+\frac{1}{2},j,k+\frac{3}{2}\right) - H_y^{n+l/m}\left(i+\frac{1}{2},j,k-\frac{3}{2}\right) \right] \right\} \tag{15.10}
$$

$$\widehat{E}_y^{n+l/m}\left(i,j+\frac{1}{2},k\right) = \widehat{E}_y^{n+(l-1)/m}\left(i,j+\frac{1}{2},k\right) + \frac{1}{\varepsilon_r\left(i,j+\frac{1}{2},k\right)}$$

$$\times \left\{\alpha_{z1} \times \left[H_x^{n+l/m}\left(i,j+\frac{1}{2},k+\frac{1}{2}\right) - H_x^{n+l/m}\left(i,j+\frac{1}{2},k-\frac{1}{2}\right)\right]\right.$$

$$- \alpha_{x1} \times \left[H_z^{n+l/m}\left(i+\frac{1}{2},j+\frac{1}{2},k\right) - H_z^{n+l/m}\left(i-\frac{1}{2},j+\frac{1}{2},k\right)\right]$$

$$+ \alpha_{z2} \times \left[H_x^{n+l/m}\left(i,j+\frac{1}{2},k+\frac{3}{2}\right) - H_x^{n+l/m}\left(i,j+\frac{1}{2},k-\frac{3}{2}\right)\right]$$

$$\left. - \alpha_{x2} \times \left[H_z^{n+l/m}\left(i+\frac{3}{2},j+\frac{1}{2},k\right) - H_z^{n+l/m}\left(i-\frac{3}{2},j+\frac{1}{2},k\right)\right]\right\}$$

$$(15.11)$$

$$\widehat{E}_z^{n+l/m}\left(i,j,k+\frac{1}{2}\right) = \widehat{E}_z^{n+(l-1)/m}\left(i,j,k+\frac{1}{2}\right) + \frac{1}{\varepsilon_r\left(i,j,k+\frac{1}{2}\right)}$$

$$\times \left\{\alpha_{x1} \times \left[H_y^{n+l/m}\left(i+\frac{1}{2},j,k+\frac{1}{2}\right) - H_y^{n+l/m}\left(i-\frac{1}{2},j,k+\frac{1}{2}\right)\right]\right.$$

$$- \alpha_{y1} \times \left[H_x^{n+l/m}\left(i,j+\frac{1}{2},k+\frac{1}{2}\right) - H_x^{n+l/m}\left(i,j-\frac{1}{2},k+\frac{1}{2}\right)\right]$$

$$+ \alpha_{x2} \times \left[H_y^{n+l/m}\left(i+\frac{3}{2},j,k+\frac{1}{2}\right) - H_y^{n+l/m}\left(i-\frac{3}{2},j,k+\frac{1}{2}\right)\right]$$

$$\left. - \alpha_{y2} \times \left[H_x^{n+l/m}\left(i,j+\frac{3}{2},k+\frac{1}{2}\right) - H_x^{n+l/m}\left(i,j-\frac{3}{2},k+\frac{1}{2}\right)\right]\right\}$$

$$(15.12)$$

$$\alpha_{x1} = \frac{9}{8}d_l \times CFL_x \quad \alpha_{y1} = \frac{9}{8}d_l \times CFL_y \quad \alpha_{z1} = \frac{9}{8}d_l \times CFL_z \qquad (15.13)$$

$$\alpha_{x2} = \frac{-1}{24}d_l \times CFL_x \quad \alpha_{y2} = \frac{-1}{24}d_l \times CFL_y \quad \alpha_{z2} = \frac{-1}{24}d_l \times CFL_z \qquad (15.14)$$

$$CFL_x = \frac{1}{\sqrt{\mu_0\varepsilon_0}}\frac{\Delta_t}{\Delta_x} \quad CFL_y = \frac{1}{\sqrt{\mu_0\varepsilon_0}}\frac{\Delta_t}{\Delta_y} \quad CFL_z = \frac{1}{\sqrt{\mu_0\varepsilon_0}}\frac{\Delta_t}{\Delta_z} \qquad (15.15)$$

where ε_r denotes the averaged relative permittivity at point $\left(i+\frac{1}{2},j,k\right)$, $\left(i,j+\frac{1}{2},k\right)$, and $\left(i,j,k+\frac{1}{2}\right)$. For the cubic lattice case, $\Delta_x = \Delta_y = \Delta_z = \Delta_\delta$, and $CFL_x = CFL_y = CFL_z = CFL_\delta$. $\alpha_{x1} = \alpha_{y1} = \alpha_{z1} = \alpha_{\delta1}$. $\alpha_{x2} = \alpha_{y2} = \alpha_{z2} = \alpha_{\delta2}$.

Similarly, the updating equations for magnetic field **H** along the x, y, and z directions can be written as:

$$H_x^{n+l/m}\left(i,j+\frac{1}{2},k+\frac{1}{2}\right)$$

$$= H_x^{n+(l-1)/m}\left(i,j+\frac{1}{2},k+\frac{1}{2}\right) + \frac{1}{\mu_r\left(i,j+\frac{1}{2},k+\frac{1}{2}\right)}$$

$$\times \left\{\beta_{z1} \times \left[\widehat{E}_y^{n+(l-1)/m}\left(i,j+\frac{1}{2},k+1\right) - \widehat{E}_y^{n+(l-1)/m}\left(i,j+\frac{1}{2},k\right)\right]\right.$$

$$
-\beta_{y1} \times \left[\widehat{E}_z^{n+(l-1)/m} \left(i, j+1, k+\frac{1}{2} \right) - \widehat{E}_z^{n+(l-1)/m} \left(i, j, k+\frac{1}{2} \right) \right]
$$
$$
+\beta_{z2} \times \left[\widehat{E}_y^{n+(l-1)/m} \left(i, j+\frac{1}{2}, k+2 \right) - \widehat{E}_y^{n+(l-1)/m} \left(i, j+\frac{1}{2}, k-1 \right) \right]
$$
$$
-\beta_{y2} \times \left[\widehat{E}_z^{n+(l-1)/m} \left(i, j+2, k+\frac{1}{2} \right) - \widehat{E}_z^{n+(l-1)/m} \left(i, j-1, k+\frac{1}{2} \right) \right] \Big\}
$$

$$(15.16)$$

$$
H_y^{n+l/m} \left(i+\frac{1}{2}, j, k+\frac{1}{2} \right)
$$
$$
= H_y^{n+(l-1)/m} \left(i+\frac{1}{2}, j, k+\frac{1}{2} \right) + \frac{1}{\mu_r \left(i+\frac{1}{2}, j, k+\frac{1}{2} \right)}
$$
$$
\times \Big\{ \beta_{x1} \times \left[\widehat{E}_z^{n+(l-1)/m} \left(i+1, j, k+\frac{1}{2} \right) - \widehat{E}_z^{n+(l-1)/m} \left(i, j, k+\frac{1}{2} \right) \right]
$$
$$
-\beta_{z1} \times \left[\widehat{E}_x^{n+(l-1)/m} \left(i+\frac{1}{2}, j, k+1 \right) - \widehat{E}_x^{n+(l-1)/m} \left(i+\frac{1}{2}, j, k \right) \right]
$$
$$
+\beta_{x2} \times \left[\widehat{E}_z^{n+(l-1)/m} \left(i+2, j, k+\frac{1}{2} \right) - \widehat{E}_z^{n+(l-1)/m} \left(i-1, j, k+\frac{1}{2} \right) \right]
$$
$$
-\beta_{z2} \times \left[\widehat{E}_x^{n+(l-1)/m} \left(i+\frac{1}{2}, j, k+2 \right) - \widehat{E}_x^{n+(l-1)/m} \left(i+\frac{1}{2}, j, k-1 \right) \right] \Big\}
$$

$$(15.17)$$

$$
H_z^{n+l/m} \left(i+\frac{1}{2}, j+\frac{1}{2}, k \right)
$$
$$
= H_z^{n+(l-1)/m} \left(i+\frac{1}{2}, j+\frac{1}{2}, k \right) + \frac{1}{\mu_r \left(i+\frac{1}{2}, j+\frac{1}{2}, k \right)}
$$
$$
\times \Big\{ \beta_{y1} \times \left[\widehat{E}_x^{n+(l-1)/m} \left(i+\frac{1}{2}, j+1, k \right) - \widehat{E}_x^{n+(l-1)/m} \left(i+\frac{1}{2}, j, k \right) \right]
$$
$$
-\beta_{x1} \times \left[\widehat{E}_y^{n+(l-1)/m} \left(i+1, j+\frac{1}{2}, k \right) - \widehat{E}_y^{n+(l-1)/m} \left(i, j+\frac{1}{2}, k \right) \right]
$$
$$
+\beta_{y2} \times \left[\widehat{E}_x^{n+(l-1)/m} \left(i+\frac{1}{2}, j+2, k \right) - \widehat{E}_x^{n+(l-1)/m} \left(i+\frac{1}{2}, j-1, k \right) \right]
$$
$$
-\beta_{x2} \times \left[\widehat{E}_y^{n+(l-1)/m} \left(i+2, j+\frac{1}{2}, k \right) - \widehat{E}_y^{n+(l-1)/m} \left(i-1, j+\frac{1}{2}, k \right) \right] \Big\}
$$

$$(15.18)$$

$$
\beta_{y1} = \frac{9}{8} c_l \times CFL_x \quad \beta_{y1} = \frac{9}{8} c_l \times CFL_y \quad \beta_{z1} = \frac{9}{8} c_l \times CFL_z
$$

$$(15.19)$$

$$
\beta_{x2} = \frac{-1}{24} c_l \times CFL_x \quad \beta_{y2} = \frac{-1}{24} c_l \times CFL_y \quad \beta_{z2} = \frac{-1}{24} c_l \times CFL_z
$$

$$(15.20)$$

where μ_r denotes the averaged relative magnetic permeability at point $\left(i, j+\frac{1}{2}, k+\frac{1}{2} \right)$, $\left(i+\frac{1}{2}, j, k+\frac{1}{2} \right)$, and $\left(i+\frac{1}{2}, j+\frac{1}{2}, k \right)$.

15.2.2 Numerical Results

A large number of numerical cases for analyzing the EM problems have demonstrated high-accuracy and low-dispersion performance [36, 37]. For simplicity, only a one-dimensional propagation problem is introduced bellow. One-dimensional Gaussian pulses propagate over long-term simulation. The field is in the form of a Gaussian pulse $\exp\left[-4\pi\left(\frac{t-T_0}{W}\right)^2\right]$ with the parameters $T_0 = 10^{-8}s$, $W = 1.33 \times 10^{-8}s$. The grid size is chosen as $\Delta_z = 0.1m$, and CFL = 0.5. The waveform of the pulse propagated over a long distance (10,000 grids) is shown in Figure 15.1. The traditional FDTD(2,2) method [38, 39], high-order FDTD(2,4) method [40, 41], and high-order SFDTD(4,4) method have been adopted for demonstration. As shown in Figure 15.1, compared with the traditional FDTD(2,2) method and the high-order FDTD(2,4) method, the high-order symplectic algorithm (SFDTD(4,4)) still maintains good numerical accuracy after long distance propagation or long iteration.

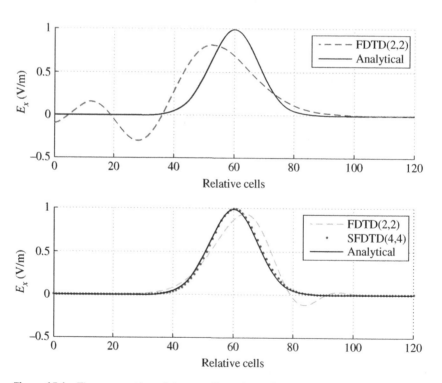

Figure 15.1 The propagation of the one-dimensional Gaussian pulse.

15.3   The Coupled Maxwell–Schrödinger Equations

As introduced in Section 15.1, when the electrons are confined in devices, like quantum dots, they only occupy discrete energy levels, while the classical particles do not have this property but require the QM description. Moreover, at the macroscale, the EM properties of materials are governed by intrinsic parameters such as dielectric constant, magnetic permeability, and electrical conductivity, etc., while at the nanoscale, the EM parameters can no longer reflect the properties of these nanoscale materials, such as quantum and nonlocal effects.

Now, a theoretical model of nanofilms is introduced. Figure 15.2 presents a model of a local region of nanofilms that is uniform in the y–z plane [28, 42]. There are a large number of electrons in the film, and these electrons are constantly moving in the nanometer film, but the distribution of electrons in space remains stable, and this stable charge distribution produces a relatively stable potential field. If the irradiation field intensity is relatively weak, that is, the overall charge distribution inside the film is not changed, then every electron can be considered to move in the same potential field. In this way, the motion behavior of the whole multi-electron system can be described by considering the motion state of one electron, which is a single-electron approximation method. When considering all the electrons in the film, the total current should be $\mathbf{J}_q = N\mathbf{J}_q{}'$, in which the $\mathbf{J}_q{}'$ represents the current deduced by one electron in the film, and N is the total number of electrons, because all electrons will contribute to current of the coupled of the EM field, namely, the current induced by the all the electrons are equals to $\mathbf{J}_q = N\mathbf{J}_q{}'$.

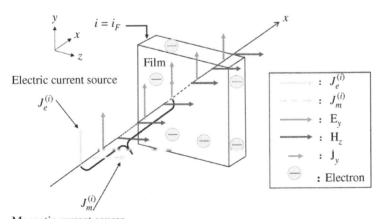

Figure 15.2   Schematic model of nanofilms.

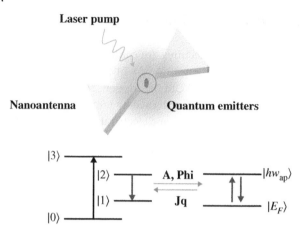

Figure 15.3 A hybrid system of nanoantenna and quantum emitters.

The nanofilm is a thin layer of surface plasmons. The current source excites a uniform plane wave. The electrons in the film are confined in the potential well. Since the electric field is polarized in the x direction, the electrons move along the x direction. In such a theoretical model, the one-dimensional FDTD method can be used to solve the Schrodinger equation and Maxwell's equations.

In addition, as seen in Figure 15.3, due to the resonance effect of the nanometer antenna, there are obvious local field and field gradient enhancements in the antenna gap. Driven by the vector potential **A** and the scalar potential field φ at the gap, the quantum emitter generates quantum current $\mathbf{J}_q$, which transmits near-field energy to the nanoantenna. The scattered field of the nanometer antenna excited by the quantum current "reacts" to the quantum emitter again. Therefore, the mutual coupling is a multi-physical, multi-space–time, and multi-scale coupling between the EM and QM systems. Moreover, the active, nonlinear, and quantum properties of the quantum emitter itself cannot be simply replaced by the classical Hertzian dipole.

Therefore, the design and simulation of new EM devices is a multi-physics problem, which requires the joint solutions of the equations describing both the EM and QM systems. The Schrödinger equation is essential to the hybrid simulation. This is because, in QM, the particle problem can be solved by the time-dependent Schrödinger equation or time-independent Schrödinger equation, which is widely used in atomic physics, nuclear physics, and solid-state physics. The time-independent Schrödinger equation is not related to time; thus, it is suitable for the calculation of steady-state quantum physics problems. For example, the eigenwave function of an intrinsic energy can be calculated according to this method. The time-dependent Schrödinger equation enables to obtain the dynamic process of a quantum system. In order to effectively couple QM theory with traditional EM theory, we mainly discuss the time-dependent Schrödinger equation, because the equation will become nonlinear when coupled to EM fields.

15.3.1 The Coupled System Based on Maxwell's and Schrödinger Equations

For the quantum part, the time-dependent Schrödinger equation is

$$i\hbar\frac{\partial\psi(\mathbf{r},t)}{\partial t} = -\frac{\hbar^2}{2m}\nabla^2\psi(\mathbf{r},t) + V(\mathbf{r})\psi(\mathbf{r},t) \tag{15.21}$$

where ψ is the wave function and is used to describe the quantum state of a particle at position $\mathbf{r}$ and time t, m is the mass of the particle, $-\frac{\hbar^2}{2m}\nabla^2$ is the kinetic energy operator, $V(\mathbf{r})$ is the time-independent potential energy, and $-\frac{\hbar^2}{2m}\nabla^2 + V$ is the Hamiltonian operator.

Equation (15.21) can be rewritten as

$$i\hbar\frac{\partial\psi(\mathbf{r},t)}{\partial t} = \frac{1}{2m}\left(\frac{\hbar}{i}\nabla\right)\left(\frac{\hbar}{i}\nabla\right)\psi(\mathbf{r},t) + V(\mathbf{r})\psi(\mathbf{r},t) \tag{15.22}$$

In the above equation, the first term on the right-hand side represents the kinetic energy, and the second term represents the potential energy. Under static electric field, the scalar potential φ contributes to the potential energy of the equation as follows:

$$i\hbar\frac{\partial\psi(\mathbf{r},t)}{\partial t} = \frac{1}{2m}\left(\frac{\hbar}{i}\nabla\right)\left(\frac{\hbar}{i}\nabla\right)\psi(\mathbf{r},t) + e\varphi(\mathbf{r})\psi(\mathbf{r},t) + V(\mathbf{r})\psi(\mathbf{r},t) \tag{15.23}$$

Under EM radiation, the momentum operators, $\mathbf{p} = \frac{\hbar}{i}\nabla$, need to be modified as $\mathbf{p} = \frac{\hbar}{i}\nabla - e\mathbf{A}$ [43] where e is the elementary charge of the particle. This revision leads to modify the Schrödinger equation as

$$i\hbar\frac{\partial\psi(\mathbf{r},t)}{\partial t} = \frac{1}{2m}\left(\frac{\hbar}{i}\nabla - e\mathbf{A}\right)^2\psi(\mathbf{r},t) + e\varphi\psi(\mathbf{r},t) + V(\mathbf{r})\psi(\mathbf{r},t) \tag{15.24}$$

The above-modified Schrödinger equation fully describes a charged particle under EM radiation. When this particle is excited, it will be in a superposition state. The probability density function for this particle $\psi^2(\mathbf{r}, t)$ needs to obey the following continuity equation, which is analogous to the electric current continuity equation:

$$\frac{\partial}{\partial t}\psi^2(\mathbf{r},t) + \nabla\cdot\mathbf{J}_p = 0 \tag{15.25}$$

where $\mathbf{J}_p$ is the probability current. Using Eqs. (15.23) and (15.24), we can arrive at

$$\mathbf{J}_p = \frac{1}{2}\left\{\left[\left(\frac{-i\hbar\nabla - e\mathbf{A}}{m}\right)\psi(\mathbf{r},t)\right]^*\psi(\mathbf{r},t) + \psi^*(\mathbf{r},t)\left[\left(\frac{-i\hbar\nabla - e\mathbf{A}}{m}\right)\psi(\mathbf{r},t)\right]\right\} \tag{15.26}$$

The particle current that contributes to the $\mathbf{A}$–Φ equations is $\mathbf{J}_q = e\mathbf{J}_p$.

Since the physical quantity of the EM field acting on the wave function in the Schrödinger equation is the magnetic vector potential $\mathbf{A}$ and electric scalar potential φ, it is necessary to build the equations to solve $\mathbf{A}$ and φ. (15.27) is the equation to solve the magnetic vector potential **A** and electric scalar potential φ, and its expression is [42]

$$
\begin{cases}
\nabla \times \mathbf{E} = -\dfrac{\partial \mathbf{B}}{\partial t} \\[2mm]
\nabla \times \mathbf{H} = \dfrac{\partial \mathbf{D}}{\partial t} + \mathbf{J} \\[2mm]
\dfrac{\partial \mathbf{A}}{\partial t} = -\mathbf{E} - \nabla \varphi \\[2mm]
\dfrac{\partial \varphi}{\partial t} = -c_0^2 \nabla \cdot \mathbf{A}
\end{cases}
\tag{15.27}
$$

Equation (15.27) contains two parts. The first part is Maxwell's equations, and **E** and **H** are obtained by solving Maxwell's equations. The second part is about the relationship between the **E** and **A** and φ. **A** and φ are solved under the Lorentz gauge condition.

Combining (15.24), (15.26), and (15.27), the coupled M–S equations can be constructed as follows:

$$
\begin{cases}
\nabla \times \mathbf{E} = -\dfrac{\partial \mathbf{B}}{\partial t} \\[2mm]
\nabla \times \mathbf{H} = \dfrac{\partial \mathbf{D}}{\partial t} + \mathbf{J} \\[2mm]
\dfrac{\partial \mathbf{A}}{\partial t} = -\mathbf{E} - \nabla \varphi \\[2mm]
\dfrac{\partial \varphi}{\partial t} = -c_0^2 \nabla \cdot \mathbf{A} \\[2mm]
i\hbar \dfrac{\partial}{\partial t} \psi(\mathbf{r}, t) = \left[-\dfrac{\hbar^2}{2m_0} \nabla^2 + \dfrac{ie\hbar}{m_0} \mathbf{A} \cdot \nabla + \dfrac{ie\hbar}{2m_0} (\nabla \cdot \mathbf{A}) + \dfrac{e^2}{2m_0} \mathbf{A}^2 + e\varphi + V \right] \psi(\mathbf{r}, t) \\[3mm]
\mathbf{J}(\mathbf{r}, t) = \dfrac{-ie\hbar}{2m} \left[\psi^*(\mathbf{r}, t) \nabla \psi(\mathbf{r}, t) - \psi(\mathbf{r}, t) \nabla \psi^*(\mathbf{r}, t) \right] - \dfrac{e^2}{m} \psi^*(\mathbf{r}, t) \mathbf{A} \psi(\mathbf{r}, t)
\end{cases}
\tag{15.28}
$$

Equation (15.28) represents the coupled M–S equations, and the solution sequence of the equations is shown in Figure 15.4.

The current density **J** obtained by solving the Schrodinger equation is used as the source of the EM system, and **E** and **H** are obtained by solving Maxwell's

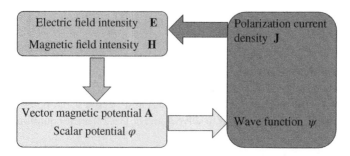

Figure 15.4 The solving flow of quantum–EM coupled equation.

equations. The vector magnetic potential $\mathbf{A}$ and scalar potential φ are solved by the $\mathbf{E}$ field and Lorentz gauge condition. Then, $\mathbf{A}$ and φ are coupled into the Schrodinger equation as the field momentum and potential energy.

15.3.2 The FDTD Method Simulation for the Coupled System

The coupled M–S equations can be solved by using the conventional FDTD method. Such a setting of spatial grid and time steps is based on the differential form of equations (15.28). The grid point configuration in the space and time domain is shown in Figure 15.5. In spatial domain, $\mathbf{E}$, $\mathbf{A}$, and $\mathbf{J}$ are in the center of the edge of the cell, $\mathbf{H}$ is in the center of the cell surface, and φ and ψ are in the corner of the cell. In the time domain, $\mathbf{E}$ and $\mathbf{A}$ are sampled at integer time step nodes. $\mathbf{H}$, φ, and $\mathbf{J}$ are sampled at the semi-integer time step nodes, and ψ is sampled at both integer and semi-integer time step nodes.

Based on Figure 15.5, the updating equations for field components of Eq. (15.28) are obtained by

$$
E_x^n\left(i+\frac{1}{2},j,k\right) = E_x^{n-1}\left(i+\frac{1}{2},j,k\right) - J_x^{n-1/2}\left(i+\frac{1}{2},j,k\right)
$$

$$
+ \frac{\Delta t}{\varepsilon} \left\{
\begin{array}{c}
\dfrac{H_z^{n-1/2}\left(i+\frac{1}{2},j+\frac{1}{2},k\right) - H_z^{n-1/2}\left(i+\frac{1}{2},j-\frac{1}{2},k\right)}{\Delta y} \\[3mm]
- \dfrac{H_y^{n-1/2}\left(i+\frac{1}{2},j,k+\frac{1}{2}\right) - H_z^{n-1/2}\left(i+\frac{1}{2},j,k-\frac{1}{2}\right)}{\Delta z}
\end{array}
\right\}
$$

$$(15.29)$$

Figure 15.5 Space and time configurations of the field components (**E**, **H**, **A**, **J**, φ, ψ, ψ_R, ψ_I) on a Yee grid indexed as (i, j, k), and $n+1/2$ and $n+1$ indicate the points on the time axis.

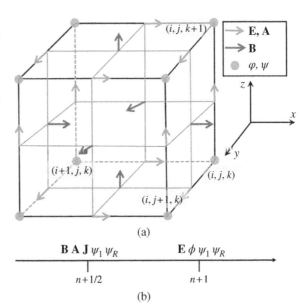

$$
E_y^n \left(i, j + \frac{1}{2}, k \right) = E_y^{n-1} \left(i, j + \frac{1}{2}, k \right) - J_y^{n-1/2} \left(i, j + \frac{1}{2}, k \right)
$$

$$
+ \frac{\Delta t}{\varepsilon} \left\{ \frac{\dfrac{H_x^{n-1/2}\left(i, j + \frac{1}{2}, k + \frac{1}{2} \right) - H_x^{n-1/2}\left(i, j + \frac{1}{2}, k - \frac{1}{2} \right)}{\Delta z}}{- \dfrac{H_z^{n-1/2}\left(i + \frac{1}{2}, j + \frac{1}{2}, k \right) - H_z^{n-1/2}\left(i - \frac{1}{2}, j + \frac{1}{2}, k \right)}{\Delta x}} \right\} \tag{15.30}
$$

$$
E_z^n \left(i, j, k + \frac{1}{2} \right) = E_z^{n-1} \left(i, j, k + \frac{1}{2} \right) - J_z^{n-1/2} \left(i, j, k + \frac{1}{2} \right)
$$

$$
+ \frac{\Delta t}{\varepsilon} \left\{ \frac{\dfrac{H_y^{n-1/2}\left(i + \frac{1}{2}, j, k + \frac{1}{2} \right) - H_y^{n-1/2}\left(i - \frac{1}{2}, j, k + \frac{1}{2} \right)}{\Delta x}}{- \dfrac{H_x^{n-1/2}\left(i, j + \frac{1}{2}, k + \frac{1}{2} \right) - H_x^{n-1/2}\left(i, j - \frac{1}{2}, k + \frac{1}{2} \right)}{\Delta y}} \right\} \tag{15.31}
$$

$$
H_x^{n+1/2} \left(i, j + \frac{1}{2}, k + \frac{1}{2} \right) = H_x^{n-1/2} \left(i, j + \frac{1}{2}, k + \frac{1}{2} \right)
$$

$$
- \frac{\Delta t}{\mu} \left\{ \frac{\dfrac{E_z^n\left(i, j + 1, k + \frac{1}{2} \right) - E_z^n\left(i, j, k + \frac{1}{2} \right)}{\Delta y}}{- \dfrac{E_y^n\left(i, j + \frac{1}{2}, k + 1 \right) - E_y^n\left(i, j + \frac{1}{2}, k \right)}{\Delta z}} \right\} \tag{15.32}
$$

$$
H_y^{n+1/2} \left(i + \frac{1}{2}, j, k + \frac{1}{2} \right) = H_y^{n-1/2} \left(i + \frac{1}{2}, j, k + \frac{1}{2} \right)
$$

$$
- \frac{\Delta t}{\mu} \left\{ \frac{\dfrac{E_x^n\left(i + \frac{1}{2}, j, k + 1 \right) - E_x^n\left(i + \frac{1}{2}, j, k \right)}{\Delta y}}{- \dfrac{E_z^n\left(i + 1, j, k + \frac{1}{2} \right) - E_z^n\left(i, j, k + \frac{1}{2} \right)}{\Delta x}} \right\} \tag{15.33}
$$

$$
H_z^{n+1/2} \left(i + \frac{1}{2}, j + \frac{1}{2}, k \right) = H_z^{n-1/2} \left(i + \frac{1}{2}, j + \frac{1}{2}, k \right)
$$

$$
- \frac{\Delta t}{\mu} \left\{ \frac{\dfrac{E_y^n\left(i + 1, j + \frac{1}{2}, k \right) - E_y^n\left(i, j + \frac{1}{2}, k \right)}{\Delta y}}{- \dfrac{E_x^n\left(i + \frac{1}{2}, j + 1, k \right) - E_x^n\left(i + \frac{1}{2}, j, k \right)}{\Delta y}} \right\} \tag{15.34}
$$

$$A_x^{n+\frac{1}{2}}\left(i+\frac{1}{2},j,k\right) = A_x^{n-\frac{1}{2}}\left(i+\frac{1}{2},j,k\right) - \Delta t E_x^n\left(i+\frac{1}{2},j,k\right)$$

$$- \frac{\Delta t}{\Delta x}\left\{\varphi^n(i+1,j,k) - \varphi^n(i,j,k)\right\} \tag{15.35}$$

$$A_y^{n+\frac{1}{2}}\left(i,j+\frac{1}{2},k\right) = A_y^{n-\frac{1}{2}}\left(i,j+\frac{1}{2},k\right) - \Delta t E_y^n\left(i,j+\frac{1}{2},k\right)$$

$$- \frac{\Delta t}{\Delta y}\left\{\varphi^n(i,j+1,k) - \varphi^n(i,j,k)\right\} \tag{15.36}$$

$$A_z^{n+\frac{1}{2}}\left(i,j,k+\frac{1}{2}\right) = A_z^{n-\frac{1}{2}}\left(i,j,k+\frac{1}{2}\right) - \Delta t E_z^n\left(i,j,k+\frac{1}{2}\right)$$

$$- \frac{\Delta t}{\Delta z}\left\{\varphi^n(i,j,k+1) - \varphi^n(i,j,k)\right\} \tag{15.37}$$

$$\varphi^{n+1}(i,j,k) = \varphi^n(i,j,k) - \frac{\Delta t}{c_0^2}\left\{\frac{1}{\Delta x}\left(A_x^{n+\frac{1}{2}}\left(i+\frac{1}{2},j,k\right) - A_x^{n+\frac{1}{2}}\left(i-\frac{1}{2},j,k\right)\right)\right.$$

$$+ \frac{1}{\Delta y}\left(A_y^{n+\frac{1}{2}}\left(i,j+\frac{1}{2},k\right) - A_y^{n+\frac{1}{2}}\left(i,j-\frac{1}{2},k\right)\right)$$

$$\left. + \frac{1}{\Delta z}\left(A_z^{n+\frac{1}{2}}\left(i,j,k+\frac{1}{2}\right) - A_z^{n+\frac{1}{2}}\left(i,j,k-\frac{1}{2}\right)\right)\right\} \tag{15.38}$$

For the Schrödinger equation, to avoid using complex numbers, one can separate the variable $\psi(\mathbf{r}, t)$ into its real and imaginary parts as

$$\psi(\mathbf{r}, t) = \psi_R(\mathbf{r}, t) + i\psi_I(\mathbf{r}, t) \tag{15.39}$$

Substituting Eq. (15.39) into the Schrödinger equation, we can get the following coupled set of equations [44]:

$$\begin{cases} \dfrac{\partial}{\partial t}\psi_I(\mathbf{r}, t) = \dfrac{\hbar}{2m}\nabla^2\psi_R(\mathbf{r}, t) - \left(\dfrac{e^2}{2m\hbar}\mathbf{A}^2 + \dfrac{e\varphi}{\hbar} + \dfrac{V}{\hbar}\right)\psi_R(\mathbf{r}, t) \\ \qquad + \dfrac{e}{m}\mathbf{A}\cdot\nabla\psi_I(\mathbf{r}, t) + \dfrac{e}{2m}(\nabla\cdot\mathbf{A})\psi_I(\mathbf{r}, t) \\ \\ \dfrac{\partial}{\partial t}\psi_R(\mathbf{r}, t) = -\dfrac{\hbar}{2m}\nabla^2\psi_I(\mathbf{r}, t) + \left(\dfrac{e^2}{2m\hbar}\mathbf{A}^2 + \dfrac{e\varphi}{\hbar} + \dfrac{V}{\hbar}\right)\psi_I(\mathbf{r}, t) \\ \qquad + \dfrac{e}{m}\mathbf{A}\cdot\nabla\psi_R(\mathbf{r}, t) + \dfrac{e}{2m}(\nabla\cdot\mathbf{A})\psi_R(\mathbf{r}, t) \end{cases} \tag{15.40}$$

Similarly, Eq. (15.40) can be solved by the FDTD method (only component A_z is considered in EM system).

$$\psi_R^{n+1}(k) = \psi_R^n(k) - \frac{\hbar \Delta t}{2m} \alpha_k \left\{ \psi_I^{n+\frac{1}{2}}(k) \right\}$$

$$+ \frac{e\Delta t}{2m} \beta_k \left[\frac{1}{2} \left\{ A_z^{n+\frac{1}{2}}\left(i,j,k+\frac{1}{2}\right) + A_z^{n+\frac{1}{2}}\left(i,j,k-\frac{1}{2}\right) \right\} \psi_R^{n+\frac{1}{2}}(k) \right]$$

$$+ \frac{e\Delta t}{4m} \left\{ A_z^{n+\frac{1}{2}}\left(i,j,k+\frac{1}{2}\right) + A_z^{n+\frac{1}{2}}\left(i,j,k-\frac{1}{2}\right) \right\} \beta_k \left\{ \psi_R^{n+\frac{1}{2}}(k) \right\}$$

$$+ \frac{\Delta t}{\hbar} \left[\frac{q^2}{8m} \left\{ A_z^{n+\frac{1}{2}}\left(i,j,k+\frac{1}{2}\right) + A_z^{n+\frac{1}{2}}\left(i,j,k-\frac{1}{2}\right) \right\}^2 \right.$$

$$\left. + e\varphi^{n+\frac{1}{2}}(i,j,k) + V(k) \right] \psi_I^{n+\frac{1}{2}}(k) \tag{15.41}$$

$$\psi_I^{n+1}(k) = \psi_I^n(k) + \frac{\hbar \Delta t}{2m} \alpha_k \left\{ \psi_R^{n+\frac{1}{2}}(k) \right\}$$

$$+ \frac{e\Delta t}{2m} \beta_k \left[\frac{1}{2} \left\{ A_z^{n+\frac{1}{2}}\left(i,j,k+\frac{1}{2}\right) + A_z^{n+\frac{1}{2}}\left(i,j,k-\frac{1}{2}\right) \right\} \psi_I^{n+\frac{1}{2}}(k) \right]$$

$$+ \frac{e\Delta t}{4m} \left\{ A_z^{n+\frac{1}{2}}\left(i,j,k+\frac{1}{2}\right) + A_z^{n+\frac{1}{2}}\left(i,j,k-\frac{1}{2}\right) \right\} \beta_k \left\{ \psi_I^{n+\frac{1}{2}}(k) \right\}$$

$$- \frac{\Delta t}{\hbar} \left[\frac{e^2}{8m} \left\{ A_z^{n+\frac{1}{2}}\left(i,j,k+\frac{1}{2}\right) + A_z^{n+\frac{1}{2}}\left(i,j,k-\frac{1}{2}\right) \right\}^2 \right.$$

$$\left. + e\varphi^{n+\frac{1}{2}}(i,j,k) + V(k) \right] \psi_R^{n+\frac{1}{2}}(k) \tag{15.42}$$

and the operators α and β in these equations are defined to perform the following sixth-order accurate difference formulas and are used to compute the second- and first-order derivatives with respect to the z coordinate, $\partial^2/\partial z^2$ and $\partial/\partial z$, respectively, for an arbitrary function χ [42]:

$$\alpha\{\chi(k)\} = \frac{1}{90\Delta z^2} \{\chi(k+3) - 13.5\chi(k+2) + 135\chi(k+1) - 245\chi(k)$$

$$+ 135\chi(k-1) - 13.5\chi(k-2) + \chi(k-3)\} \tag{15.43}$$

$$\beta\{\chi(k)\} = \frac{1}{60\Delta z} \{\chi(k+3) - 9\chi(k+2) + 45\chi(k+1) - 45\chi(k-1)$$

$$+ 9\chi(k-2) - \chi(k-3)\} \tag{15.44}$$

The quantum current **J** in Eq. (15.28) is evaluated considering the first-order derivative of $\partial/\partial z$ (only component A_z is considered here) as

$$J_z^{n+\frac{1}{2}}(k+1) = \frac{q\hbar}{m}\left[\psi_{\text{real}}^{n+\frac{1}{2}}(k)\beta\left\{\psi_{\text{imag}}^{n+\frac{1}{2}}(k)\right\} - \psi_{\text{imag}}^{n+\frac{1}{2}}(k)\beta\left\{\psi_{\text{real}}^{n+\frac{1}{2}}(k)\right\}\right]$$

$$-\frac{q^2}{2m}\left[\left\{\psi_{\text{real}}^{n+\frac{1}{2}}(k)\right\}^2 + \left\{\psi_{\text{imag}}^{n+\frac{1}{2}}(k)\right\}^2\right]$$

$$\times\left\{A_z^{n+\frac{1}{2}}\left(i_T, j_T, k+\frac{1}{2}\right) + A_z^{n+\frac{1}{2}}\left(i_T, j_T, k-\frac{1}{2}\right)\right\} \quad (15.45)$$

The β operator is defined by Eq. (15.43) and the quantum current $J_z(k+1/2)$ can be calculated by the weighted average of $J_z(k+1)$ and $J_z(k)$.

15.3.3 The SFDTD Simulation for the Coupled System

Since the SFDTD method and traditional FDTD method algorithm are both based on Yee grid subdivision, we can use the SFDTD method to solve the coupled M–S equations. For Maxwell's equations, Eqs. (15.10)–(15.12) and (15.16)–(15.18) can be used to update the **E** and **H** field components, and the updating equations of SFDTD method for **A** and φ are given by

$$A_x^{n+l/m}\left(i+\frac{1}{2},j,k\right) = A_x^{n+(l-1)/m}\left(i+\frac{1}{2},j,k\right) - c_l\Delta t E_x^{n+(l-1)/m}\left(i+\frac{1}{2},j,k\right)$$

$$-\frac{c_l\Delta t}{\Delta x}\{\varphi^{n+(l-1)/m}(i+1,j,k) - \varphi^{n+(l-1)/m}(i,j,k)\} \quad (15.46)$$

$$A_y^{n+l/m}\left(i,j+\frac{1}{2},k\right) = A_y^{n+(l-1)/m}\left(i,j+\frac{1}{2},k\right) - c_l\Delta t E_y^{n+(l-1)/m}\left(i,j+\frac{1}{2},k\right)$$

$$-\frac{c_l\Delta t}{\Delta y}\{\varphi^{n+(l-1)/m}(i,j+1,k) - \varphi^{n+(l-1)/m}(i,j,k)\} \quad (15.47)$$

$$A_z^{n+l/m}\left(i,j,k+\frac{1}{2}\right) = A_z^{n+(l-1)/m}\left(i,j,k+\frac{1}{2}\right) - c_l\Delta t E_z^{n+(l-1)/m}\left(i,j,k+\frac{1}{2}\right)$$

$$-\frac{c_l\Delta t}{\Delta z}\{\varphi^{n+(l-1)/m}(i,j,k+1) - \varphi^{n+(l-1)/m}(i,j,k)\} \quad (15.48)$$

$$\varphi^{n+l/m}(i,j,k) = \varphi^{n+(l-1)/m}(i,j,k)$$

$$-\frac{d_l\Delta t}{c_0^2}\left\{\frac{1}{\Delta x}\left(A_x^{n+(l-1)/m}\left(l+\frac{1}{2},j,k\right) - A_x^{n+(l-1)/m}\left(l-\frac{1}{2},j,k\right)\right)\right.$$

$$+\frac{1}{\Delta y}\left(A_y^{n+(l-1)/m}\left(i,j+\frac{1}{2},k\right) - A_y^{n+(l-1)/m}\left(i,j-\frac{1}{2},k\right)\right)$$

$$+\frac{1}{\Delta z}\left(A_z^{n+(l-1)/m}\left(i,j,k+\frac{1}{2}\right) - A_z^{n+(l-1)/m}\left(i,j,k-\frac{1}{2}\right)\right)\right\}$$

$$(15.49)$$

For the quantum part, as introduced in [26], the modified Schrödinger equation has a symplectic structure, and the symplectic algorithms for Hamiltonian mechanics can be employed. After canonical transformation, the exact solution of the Schrödinger equation $\exp(\Delta_t L)$ is an orthogonal operator. In other words, the time evolution of the Schrödinger equation essentially rotates the normalized wave function with a perfect energy-conserving feature. Hence, utilizing symplectic mapping, the solution of Eq. (15.40) for the quantum system from $t = 0$ to $t = \Delta t$ is established approximately:

$$\exp(\mathbf{tL}) = \prod_{l=1}^{m} \exp(c_l \mathbf{tL_A}) \exp(d_l \mathbf{tL_B}) + O(t^{p+1})$$
$$= \prod_{l=1}^{m} (1 + c_l \mathbf{tL_A})(1 + d_l \mathbf{tL_B}) + O(t^{p+1}) \tag{15.50}$$

and c_l and d_l (listed in Table 15.1) are the coefficients of the symplectic integration algorithm, m is the number of the sub-step in each time step. Finally, the updating equations for the wavefunction can be derived as

$$\psi_R^{n+l/m}(i,j,k) = \psi_R^{n-l/m}(i,j,k) - c_l \Delta t \frac{\hbar}{2m} \nabla^2 \psi_I^{n-l/m}(i,j,k)$$
$$- c_l \Delta t \frac{q}{2m} A^{n+(l-1)/m} \cdot \nabla \psi_I^{n-l/m} - c_l \Delta t \frac{e}{2m}(\nabla \cdot A^{n+(l-1)/m}) \psi_R^{n-l/m}$$
$$+ \frac{\Delta t e^2}{2\hbar m}(A^{n+(l-1)/m})^2 \psi_I^{n-l/m}(i,j,k) - \Delta t \frac{e\varphi - V}{\hbar} \psi_I^{n-l/m}(i,j,k) \tag{15.51}$$

$$\psi_I^{n+l/m}(i,j,k) = \psi_I^{n-l/m}(i,j,k) + d_l \Delta t \frac{\hbar}{2m} \nabla^2 \psi_R^{n+l/m}(i,j,k)$$
$$- d_l \Delta t \frac{e}{m} A^{n+(l-1)/m} \cdot \nabla \psi_I^{n-l/m} - d_l \Delta t \frac{e}{2m}(\nabla \cdot A^{n+(l-1)/m}) \psi_I^{n-l/m}$$
$$- \frac{\Delta t e^2}{2\hbar m}(A^{n+(l-1)/m})^2 \psi_R^{n+l/m}(i,j,k) + \Delta t \frac{e\varphi - V}{\hbar} \psi_R^{n+l/m}(i,j,k) \tag{15.52}$$

$$J_q^{n+l/m} = \frac{q}{2} \left\{ \left[\frac{-i\hbar \nabla - qA^{n+l/m}}{m} \left(\psi_R^{n+l/m} + i \psi_I^{n+l/m} \right) \right]^* \left(\psi_R^{n+l/m} + i \psi_I^{n+l/m} \right) \right.$$
$$\left. + \left(\psi_R^{n+l/m} + i \psi_I^{n+l/m} \right)^* \left[\frac{-i\hbar \nabla - qA^{n+l/m}}{m} \left(\psi_R^{n+l/m} + i \psi_I^{n+l/m} \right) \right] \right\} \tag{15.53}$$

In addition, the first-order and second-order Laplace operators in (15.40) is discretized by fourth-order difference. In view of fourth-order differences, the image theory [45] adopted in Maxwell's equations can be naturally extended to the Schrodinger part. For the rationality of the framework, the image theory will be introduced in next section.

15.3.4 The Coupled System Based on EM Potential and Schrödinger Equations

The M–S coupled equations constructed in the previous section involve six physical quantities, $\mathbf{E}$, $\mathbf{H}$, $\mathbf{A}$, φ, ψ, and $\mathbf{J}$. There are many calculation steps included in the coupled EM–QM system. Particularly, the equations considering the EM field are described by four physical quantities, $\mathbf{E}$, $\mathbf{H}$, $\mathbf{A}$, and φ. The number of equations is increased from four to eight, and the calculation time is also increased. In fact, any groups of $\mathbf{E}$ and $\mathbf{H}$ or $\mathbf{A}$ and φ can describe EM system, so it can simplify the calculation processes by solving wave equation of $\mathbf{A}$ and φ directly. There are many ways to solve the magnetic vector potential $\mathbf{A}$ and the scalar potential φ. One is to solve the two wave equations of $\mathbf{A}$ and φ, and combine the current continuity equation to get the following equations.

$$\begin{cases} \nabla^2 \mathbf{A} - \dfrac{1}{c_0^2} \dfrac{\partial^2 \mathbf{A}}{\partial t^2} = -\mu \mathbf{J} \\[2mm] \nabla^2 \varphi - \dfrac{1}{c_0^2} \dfrac{\partial^2 \varphi}{\partial t^2} = -\dfrac{\rho}{\varepsilon_0} \\[2mm] \nabla \cdot \mathbf{J} = -\dfrac{\partial \rho}{\partial t} \end{cases} \tag{15.54}$$

The first two formulas are the decoupled potential function equations, which are derived from the coupled $\mathbf{A}$ and the φ with the simple Lorentz gauge (Eq. (15.55a)).

$$\nabla \cdot \mathbf{A} = -\mu \varepsilon \frac{\partial \varphi}{\partial t} \tag{15.55a}$$

$$\varepsilon^{-1} \nabla \cdot (\varepsilon \mathbf{A}) = -\mu \varepsilon \frac{\partial \varphi}{\partial t} \tag{15.55b}$$

Equation (15.54) is suitable for the homogeneous medium environment. The generalized Lorenz gauge defined by Eq. (15.55b) can lead to decoupled $\mathbf{A}$–φ formulas for inhomogeneous media [46], i.e.

$$\varepsilon(\mathbf{r})^{-1} \nabla \cdot (\varepsilon(\mathbf{r})\mathbf{A}) = -\mu \varepsilon(\mathbf{r}) \frac{\partial \varphi}{\partial t} \tag{15.56}$$

$$\nabla \cdot \varepsilon(\mathbf{r}) \nabla \varphi - \mu \varepsilon(\mathbf{r})^2 \frac{\partial^2 \varphi}{\partial t^2} = -\rho \tag{15.57}$$

$$-\nabla \times \mu^{-1} \nabla \times \mathbf{A} - \varepsilon(\mathbf{r}) \frac{\partial^2 \mathbf{A}}{\partial t^2} + \varepsilon(\mathbf{r}) \nabla \varepsilon(\mathbf{r})^{-2} \mu^{-1} \nabla \cdot \varepsilon(\mathbf{r})\mathbf{A} = -\mathbf{J} \tag{15.58}$$

Another relatively simple method is to solve the wave equation and combine with Lorentz gauge conditions to obtain the following equations:

$$\begin{cases} \nabla^2 \mathbf{A} - \dfrac{1}{c_0^2} \dfrac{\partial^2 \mathbf{A}}{\partial t^2} = -\mu \mathbf{J} \\[2mm] \nabla \cdot \mathbf{A} + \dfrac{1}{c_0^2} \dfrac{\partial \varphi}{\partial t} = 0 \end{cases} \tag{15.59}$$

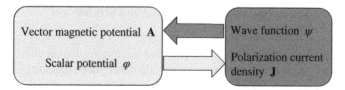

Figure 15.6 The calculation flow of the coupled quantum–EM system.

The introduced **A** and φ formulas provide an alternative way to solve the EM system instead of calculating the **E** and **H** fields.

The solution of the QM system is the same as that in the previous section.

$$\begin{cases} i\hbar\dfrac{\partial}{\partial t}\psi(\boldsymbol{r},t) = \left[-\dfrac{\hbar^2}{2m_0}\nabla^2 + \dfrac{ie\hbar}{m_0}\boldsymbol{A}\cdot\nabla + \dfrac{ie\hbar}{2m_0}(\nabla\cdot\boldsymbol{A}) + \dfrac{e^2}{2m_0}A^2 + e\varphi + V\right]\psi(\boldsymbol{r},t) \\[3mm] \boldsymbol{J}(\boldsymbol{r},t) = \dfrac{-ie\hbar}{2m}\left[\psi^*(\boldsymbol{r},t)\nabla\psi(\boldsymbol{r},t) - \psi(\boldsymbol{r},t)\nabla\psi^*(\boldsymbol{r},t)\right] - \dfrac{e^2}{m}\psi^*(\boldsymbol{r},t)A\psi(\boldsymbol{r},t) \end{cases}$$

$$\tag{15.60}$$

According to above descriptions, the calculation process of the coupled QM–EM system is summarized as shown in Figure 15.6.

The detailed numerical solution processes for **A** and φ with the FDTD method simulation and high-order SFDTD simulation have been introduced in [24, 26]. For the sake of brevity, we will not introduce it here.

15.3.5 Numerical Simulation for the Coupled M–S System

According to the quantum state control theory [42], the control pulse generator is designed to maximize the objective state ψ_1 of the electron from the ground state ψ_0. The control pulse E^s can be generated by solving the equation

$$E^s = -2\frac{E_0}{m_0}Im\langle\psi'|Wez|\psi'\rangle \tag{15.61}$$

where W represents the projection operator of $|\psi_1\rangle\langle\psi_1|$, and $\psi' = \exp\left[-i\frac{e}{\hbar}A_z(z,t)\right]\psi$, which is a unitary transformation for the wave function ψ; and the derivation of Eq. (15.61) is introduced in [42]. At last, applying EM Eqs. (15.54) or (15.57)–(15.58), Schrödinger polarization current density (15.60), and the control pulse E^s (15.61), the proposed scheme provides a configuration of modeling process of the multi-physics problem (as shown in Figure 15.7), in which a uniform plane wave source excites the electrons in a nanotube and then transforms the quantum states of the electron from the ground state ψ_0 to the first excited state ψ_1.

Figure 15.8 shows a procedure illustration of the proposed method for updating the field components in the coupled M–S system. Firstly, the control pulse

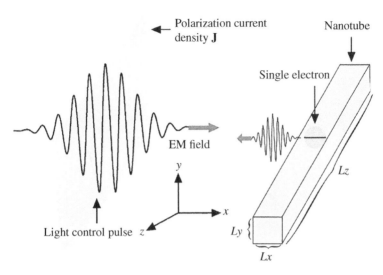

Figure 15.7 A geometric model of a nanotube.

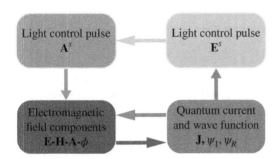

Figure 15.8 A procedure illustration of the proposed method for updating the field components in the coupled M–S system.

E^s (15.61) is obtained by the wave function of the electron, and the polarization current density J_z generated by the electron motion will produce the radiation fields governed by (15.54) or (15.57) and (15.58). And then, the radiation fields will superimpose with the control pulse field F^s as the total external EM field acting on the quantum system governed by (15.51) and (15.52), i.e. one transforms the control pulse E^s to A^s and then superimposes with the A_z for incorporation into a multi-physics simulation, as shown in Figure 15.8.

Finally, according to the program diagram of Figure 15.8, the numerical multi-physics simulation is performed. For the EM system, the computational region contains $50 \times 50 \times 120$ Yee cells with the convolutional perfectly

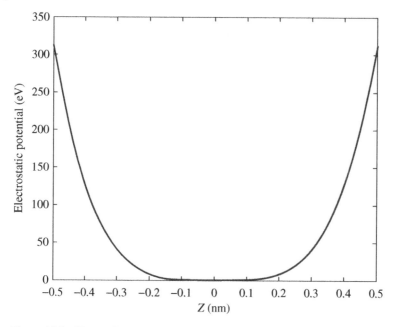

Figure 15.9 The confining potential V along the z direction.

matched layer (CPML) region of 10 layers, and the cell sizes are chosen as $\Delta x = \Delta y = 1$ nm, $\Delta z = 0.01$ nm. For the QM system, as the electron is confined in the quasi-one-dimensional space extending along the z axis like a "tube," its wave function is dependent only on the z coordinate. In other directions, the wave function is set to be uniform and the value is not zero only within one unit Δx and Δy of the Yee cell. Hence the wave function is normalized as

$$\Delta x \Delta y \int_{-\infty}^{\infty} |\psi(z)|^2 dz = 1 \tag{15.62}$$

and the length of the nanotube along z direction is chosen as 1 nm; thus, 100 Yee cells are included in the QM system.

The ground state ψ_0 and the excited state ψ_1 with the confining potential V (Figure 15.9) are shown in Figure 15.10. For numerical convergence, the wave function is initialized to be

$$\psi = \sqrt{0.9999}\psi_0 + \sqrt{0.0001}\psi_1 \tag{15.63}$$

As shown in Figure 15.11, the electron wave packet $|\psi|^2$ initially exists as a single peak, which corresponds to the ground state of the electron, and then the electron wave packet starts fluctuating about the time 2 fs. During the

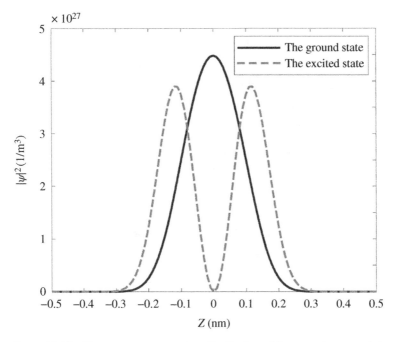

Figure 15.10 The electron wave packet distribution of the ground state and the excited state. The corresponding eigen-energies are 2.05 and 7.35 eV, respectively.

time interval of $2 \sim 10\,\text{fs}$, the electron wave packet oscillates rapidly under the illumination of the pulse and is finally converted into the bimodal form, which corresponds to the first excited state of the electron. Additionally, to quantify the quantum state transition by the proposed scheme, the transition rate factor Ω_k is defined as

$$\Omega_k = \langle \psi' \mid \psi_k \rangle \langle \psi_k \mid \psi' \rangle \tag{15.64}$$

where ψ_k denotes the quantum state of the electron. The resultant variations of the factors are depicted in Figure 15.12. As the simulation time increases, the factor Ω_0 decreases from one to zero and the factor Ω_1 increases from zero to one when the pulse gradually raises ($t = 3 \sim 10\,\text{fs}$). Finally, the value of factor Ω_1 remains stable when the pulse has gone. It can be concluded that the perfect quantum state transition is numerically achieved by using the proposed M–S system.

Finally, the numerical results demonstrate the accuracy of the introduced SFDTD method to solve the coupled M–S equations.

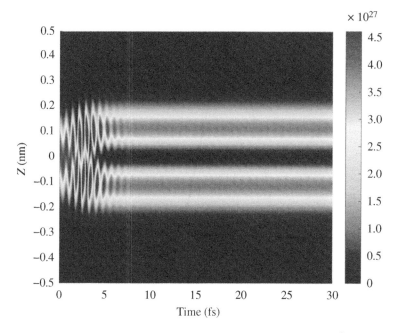

Figure 15.11 The spatiotemporal plots of the probability density $|\psi|^2$.

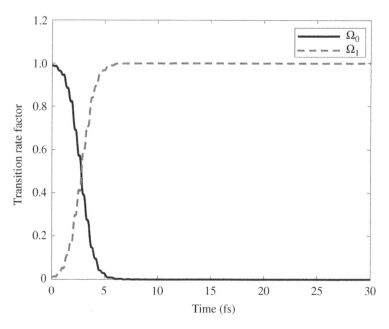

Figure 15.12 The magnitude of the transition rate factor Ω.

15.4 A Unified Symplectic Framework for Maxwell–Schrödinger Equations

In this section, we introduce a method of constructing the symplectic structure for the coupled M–S system and use the high-order SFDTD(4,4) method to discretize the M–S equations. Then, we study the energy-level transition of a particle in a two-level system under the action of an external EM field. Numerical examples show the advantages of the high-order SFDTD(4,4) algorithm in terms of computational accuracy and long time stability. At last, the applications of the presented M–S framework and the SFDTD(4,4) method in lossy media, inhomogeneous media, and free space are also introduced.

15.4.1 Symplectic Structure of the M–S System

For the QM part, an electron with charge e and mass m interacting with external fields can be described by a modified Schrödinger equation:

$$i\hbar \frac{\partial \psi(\mathbf{r}, t)}{\partial t} = \left\{ \frac{1}{2m} [\mathbf{p} - e\mathbf{A}(\mathbf{r}, t)]^2 + e\varphi(\mathbf{r}, t) + V(\mathbf{r}) \right\} \psi(\mathbf{r}, t) \tag{15.65}$$

where $\mathbf{p} = -i\hbar \nabla$ denotes the canonical momentum operator, and the variables $\mathbf{A}(\mathbf{r}, t)$ and $\varphi(\mathbf{r}, t)$ are the magnetic potential and scalar potentials, respectively. $V(\mathbf{r})$ is the electrostatic confinement potential.

For the EM part, the differential Maxwell's equation can be expressed as follows:

$$\nabla \times \mathbf{E} = -\frac{\partial \mathbf{B}}{\partial t} \tag{15.66}$$

$$\nabla \times \mathbf{H} = \frac{\partial \mathbf{D}}{\partial t} + \mathbf{J} \tag{15.67}$$

Since the vector magnetic potential $\mathbf{A}$ and the scalar potential φ can be directly substituted into Eq. (15.65) to solve the wave function ψ of electron e, in order to avoid the numerical conversion between $\mathbf{E}$, $\mathbf{H}$, $\mathbf{A}$, and φ during the iterative process, the identity $\mathbf{B} = \nabla \times \mathbf{A}$ and the auxiliary variable $\mathbf{Y} = -\varepsilon_0 \varepsilon_r \mathbf{E}$ are used and substituted into Eqs. (15.66) and (15.67) to obtain the following results:

$$\frac{\partial \mathbf{A}}{\partial t} = \frac{\mathbf{Y}}{\varepsilon_0 \varepsilon_r} \tag{15.68}$$

$$\frac{\partial \mathbf{Y}}{\partial t} = \frac{\nabla \times \nabla \times \mathbf{A}}{\mu_0} + \mathbf{J} \tag{15.69}$$

According to Eqs. (15.65), (15.68), and (15.69), the Coulomb gauge $\nabla \cdot \mathbf{A} = 0$. The Coulomb gauge splits the EM fields into the transverse (electrodynamic) and longitudinal (electrostatic) parts, and the longitudinal part can be directly inserted into the Schrödinger equation as shown in Eq. (15.65). The Coulomb specification can greatly simplify the complexity of the analysis of quantum optical problems [47], and the QM–EM coupled system can be represented by the total Hamiltonian system:

$$H(\mathbf{A}, \mathbf{Y}, \psi, \psi^*) = H^{em}(\mathbf{A}, \mathbf{Y}) + H^q(\mathbf{A}, \psi, \psi^*) \tag{15.70}$$

and

$$H^{em}(\mathbf{A}, \mathbf{Y}) = \int_\Omega \left(\frac{1}{2\varepsilon_0 \varepsilon_r} |\mathbf{Y}|^2 + \frac{1}{2\mu_0} |\nabla \times \mathbf{A}|^2 \right) d\mathbf{r} \tag{15.71}$$

$$H^{qm}(\mathbf{A}, \psi, \psi^*) = \int_\Omega \left[\psi^* \frac{(\mathbf{p} - e\mathbf{A})^2}{2m} \psi + \psi^* V \psi \right] d\mathbf{r} \tag{15.72}$$

where H^{em} represents the total energy of the EM system, and H^{qm} represents the total kinetic and potential energy in the QM system. Using the variational principle, the Hamiltonian equation of the coupled system can be deduced as following:

$$\frac{\partial \mathbf{A}}{\partial t} = \frac{\partial H}{\partial \mathbf{Y}} = -\frac{\mathbf{Y}}{\varepsilon_0} \tag{15.73}$$

$$\frac{\partial \mathbf{Y}}{\partial t} = -\frac{\partial H}{\partial \mathbf{A}} = -\frac{\nabla \times \nabla \times \mathbf{A}}{\mu_0} + \mathbf{J} \tag{15.74}$$

$$\frac{\partial \psi}{\partial t} = \frac{1}{i\hbar} \frac{\partial H}{\partial \psi^*} = \frac{1}{i\hbar} \left[\frac{(\mathbf{p} - e\mathbf{A})^2}{2m} + V \right] \psi \tag{15.75}$$

$$\frac{\partial \psi^*}{\partial t} = \frac{-1}{i\hbar} \frac{\partial H}{\partial \psi} = -\frac{1}{i\hbar} \left[\frac{(\mathbf{p} + e\mathbf{A})^2}{2m} + V \right] \psi^* \tag{15.76}$$

and the quantum current $\mathbf{J}$ is described by the following form:

$$\mathbf{J} = \frac{e}{2m} [\psi^*(\mathbf{p} - e\mathbf{A})\psi + \psi(-\mathbf{p} - e\mathbf{A})\psi^*] \tag{15.77}$$

The current term $\mathbf{J}$ is generated by the motion of atoms and will radiate EM fields. The generated quantum current will be coupled back to Maxwell's equations.

According to the modified Schrödinger equation and the EM equations based on $\mathbf{A}$ and φ, the Hamiltonian equation of coupled system can be expressed as the following matrix:

$$\begin{bmatrix} 0 & 0 & -1 & 0 \\ 0 & 0 & 0 & -1 \\ 1 & 0 & 0 & 0 \\ 0 & 1 & 0 & 0 \end{bmatrix} \begin{pmatrix} \dot{\mathbf{A}} \\ \dot{\psi} \\ \dot{\mathbf{Y}} \\ \dot{\psi^*} \end{pmatrix} = \begin{pmatrix} \dfrac{\partial H}{\partial \mathbf{A}} \\ \dfrac{1}{i\hbar} \dfrac{\partial H}{\partial \psi} \\ \dfrac{\partial H}{\partial \mathbf{Y}} \\ \dfrac{1}{i\hbar} \dfrac{\partial H}{\partial \psi^*} \end{pmatrix} \tag{15.78}$$

Equation (15.78) is not a standard Hamiltonian equation. The variational principle is applied and the wave function is decomposed into real and imaginary parts:

$$\psi = \frac{1}{\sqrt{2\hbar}} (\psi_R + i\psi_I) \tag{15.79}$$

Then, the Hamiltonian equation of the coupled system can be transformed into the following matrix form:

$$
\begin{bmatrix} 0 & 0 & -1 & 0 \\ 0 & 0 & 0 & -1 \\ 1 & 0 & 0 & 0 \\ 0 & 1 & 0 & 0 \end{bmatrix} \begin{pmatrix} \dot{\mathbf{A}} \\ \dot{\psi}_R \\ \dot{\mathbf{Y}} \\ \dot{\psi}_I \end{pmatrix} = \begin{pmatrix} \dfrac{\partial H}{\partial \mathbf{A}} \\ \dfrac{\partial H}{\partial \psi_R} \\ \dfrac{\partial H}{\partial \mathbf{Y}} \\ \dfrac{\partial H}{\partial \psi_I} \end{pmatrix}
\tag{15.80}
$$

Equation (15.80) is derived as:

$$
\begin{pmatrix} \dot{\mathbf{A}} \\ \dot{\psi}_R \\ \dot{\mathbf{Y}} \\ \dot{\psi}_I \end{pmatrix} = \mathbf{M} \begin{pmatrix} \dfrac{\partial H}{\partial \mathbf{A}} \\ \dfrac{\partial H}{\partial \psi_R} \\ \dfrac{\partial H}{\partial \mathbf{Y}} \\ \dfrac{\partial H}{\partial \psi_I} \end{pmatrix}; \mathbf{M} = \begin{bmatrix} 0 & 0 & 1 & 0 \\ 0 & 0 & 0 & 1 \\ -1 & 0 & 0 & 0 \\ 0 & -1 & 0 & 0 \end{bmatrix} = \begin{bmatrix} \mathbf{0} & \mathbf{I} \\ -\mathbf{I} & \mathbf{0} \end{bmatrix}
\tag{15.81}
$$

It can be seen from Eq. (15.81) that the coupled system based on the above framework is well-posed. In addition, $\mathbf{M}$ is a symplectic matrix, so the coupled system can maintain energy conservation during the long time evolution process [48].

15.4.2 Symplectic Structure Analysis of Subsystems

The symplectic property of M–S coupled system is proved in the previous section, and the symplectic structure of each subsystem is analyzed in this section.

15.4.2.1 The Symplectic Structure of the QM System
According to Eq. (15.81), the time-domain partial differential equations of the real and imaginary parts of the wave function can be obtained:

$$
\frac{\partial \psi_R}{\partial t} = \frac{(x_1 - x_2)}{i\sqrt{2\hbar}} \psi_R + \left(\frac{(x_1 + x_2)}{\sqrt{2\hbar}} + \frac{\sqrt{2V}}{\sqrt{\hbar}} \right) \psi_I
\tag{15.82}
$$

$$
\frac{\partial \psi_I}{\partial t} = -\left(\frac{(x_1 + x_2)}{\sqrt{2\hbar}} + \frac{\sqrt{2V}}{\sqrt{\hbar}} \right) \psi_R + \frac{(x_1 - x_2)}{i\sqrt{2\hbar}} \psi_I
\tag{15.83}
$$

and

$$
x_1 = \frac{(\mathbf{p} - e\mathbf{A})^2}{2m}, x_2 = \frac{(\mathbf{p} + e\mathbf{A})^2}{2m}
\tag{15.84}
$$

For Eqs. (15.82) and (15.83), the differential equations of the real and imaginary parts of the wave function can be written in the form of matrix

$$\frac{\partial}{\partial t}\begin{pmatrix}\psi_R\\\psi_I\end{pmatrix} = (\mathbf{L})\begin{pmatrix}\psi_R\\\psi_I\end{pmatrix} \tag{15.85}$$

It is easy to prove that

$$\mathbf{L}^T = -\mathbf{L} \tag{15.86}$$

$$\mathbf{L}^T\mathbf{E} + \mathbf{E}\mathbf{L} = 0 \tag{15.87}$$

$$\mathbf{E} = \begin{pmatrix}0 & \mathbf{I}\\-\mathbf{I} & 0\end{pmatrix} \tag{15.88}$$

Therefore, the differential equations describing the quantum system have the symplectic structure property and can be solved numerically by the symplectic FDTD method.

15.4.2.2 The Symplectic Structure of the EM System

For Eqs. (15.73) and (15.74), without considering the current term **J**, the time-domain differential equations of **A** and **Y** can be written in the following form:

$$\frac{\partial}{\partial t}\begin{pmatrix}\mathbf{A}\\\mathbf{Y}\end{pmatrix} = \begin{pmatrix}\mathbf{0}_{3\times3} & \varepsilon^{-1}_{3\times3}\\-(\mu^{-1}\nabla\times\nabla\times)_{3\times3} & \mathbf{0}_{3\times3}\end{pmatrix}\begin{pmatrix}\mathbf{A}\\\mathbf{Y}\end{pmatrix} = \mathbf{U}\begin{pmatrix}\mathbf{A}\\\mathbf{Y}\end{pmatrix} \tag{15.89}$$

and

$$\mathbf{U} = \begin{pmatrix}\mathbf{0}_{3\times3} & \mathbf{U}_A\\\mathbf{U}_B & \mathbf{0}_{3\times3}\end{pmatrix} \tag{15.90}$$

$$\mathbf{U}_A = \varepsilon^{-1}_{3\times3} = \begin{pmatrix}1/\varepsilon_0\varepsilon_r & 0 & 0\\0 & 1/\varepsilon_0\varepsilon_r & 0\\0 & 0 & 1/\varepsilon_0\varepsilon_r\end{pmatrix} = \mathbf{U}_A^T \tag{15.91}$$

$$\mathbf{U}_B = -(\mu^{-1}\nabla\times\nabla\times)_{3\times3} = \begin{pmatrix}\left(\frac{\partial^2}{\partial y^2}+\frac{\partial^2}{\partial z^2}\right) & -\frac{\partial^2}{\partial x\partial y} & -\frac{\partial^2}{\partial x\partial z}\\-\frac{\partial^2}{\partial x\partial y} & \left(\frac{\partial^2}{\partial x^2}+\frac{\partial^2}{\partial z^2}\right) & -\frac{\partial^2}{\partial y\partial z}\\-\frac{\partial^2}{\partial x\partial z} & -\frac{\partial^2}{\partial y\partial z} & \left(\frac{\partial^2}{\partial x^2}+\frac{\partial^2}{\partial y^2}\right)\end{pmatrix} = \mathbf{U}_B^T \tag{15.92}$$

The matrix **U** satisfies the following conditions:

$$\mathbf{U}^T\mathbf{J}_{6\times6} + \mathbf{J}_{6\times6}\mathbf{U} = \begin{pmatrix}\mathbf{0}_{3\times3} & \mathbf{U}_B^T\\\mathbf{U}_A^T & \mathbf{0}_{3\times3}\end{pmatrix}\begin{pmatrix}\mathbf{0}_{3\times3} & \mathbf{I}_{3\times3}\\-\mathbf{I}_{3\times3} & \mathbf{0}_{3\times3}\end{pmatrix} + \begin{pmatrix}\mathbf{0}_{3\times3} & \mathbf{I}_{3\times3}\\-\mathbf{I}_{3\times3} & \mathbf{0}_{3\times3}\end{pmatrix}\begin{pmatrix}\mathbf{0}_{3\times3} & \mathbf{U}_A\\\mathbf{U}_B & \mathbf{0}_{3\times3}\end{pmatrix}$$
$$= \begin{pmatrix}-\mathbf{U}_B^T + \mathbf{U}_B & \mathbf{0}_{3\times3}\\\mathbf{0}_{3\times3} & \mathbf{U}_A^T + \mathbf{U}_A\end{pmatrix} = \begin{pmatrix}\mathbf{0}_{3\times3} & \mathbf{0}_{3\times3}\\\mathbf{0}_{3\times3} & \mathbf{0}_{3\times3}\end{pmatrix} \tag{15.93}$$

This means that **U** is an infinitesimal real symplectic matrix, and

$$\mathbf{U} = \begin{pmatrix}\mathbf{0}_{3\times3} & \mathbf{U}_A\\\mathbf{0}_{3\times3} & \mathbf{0}_{3\times3}\end{pmatrix} + \begin{pmatrix}\mathbf{0}_{3\times3} & \mathbf{0}_{3\times3}\\\mathbf{U}_B & \mathbf{0}_{3\times3}\end{pmatrix} = \mathbf{U}_1 + \mathbf{U}_2 \tag{15.94}$$

since $\mathbf{U}_1{}^{\varsigma} = 0$, $\mathbf{U}_2{}^{\varsigma} = 0$, $\varsigma \geq 2$, then

$$\exp(\Delta t \mathbf{U}_1) = I_{6\times6} + \Delta t \mathbf{U}_1 \tag{15.95}$$

$$\exp(\Delta t \mathbf{U}_2) = I_{6\times6} + \Delta t \mathbf{U}_2 \tag{15.96}$$

Therefore, $\exp(\Delta t \mathbf{U}_1)$ and $\exp(\Delta t \mathbf{U}_2)$ can be obtained explicitly. Note that the matrices $\mathbf{U}_1$ and $\mathbf{U}_2$ do not commute:

$$\mathbf{U}_1\mathbf{U}_2 \neq \mathbf{U}_2\mathbf{U}_1 \tag{15.97}$$

Accordingly, the symplectic integrator is applicable to solve the EM subsystem [49].

15.4.3 The SFDTD Simulation for the Coupled M–S System

According to the symplectic analysis of the coupled system and each subsystem in the previous section, the SFDTD algorithm can be used to solve the coupled M–S system [50].

15.4.3.1 The SFDTD Discretized Scheme of M–S Equations
Utilizing symplectic mapping, the solution of Eq. (15.81) for the M–S system from $t = 0$ to $t = \Delta t$ is established approximately:

$$\exp(t\mathbf{F}) = \prod_{l=1}^{m} \exp(c_l t \mathbf{F}_1) \exp(d_l t \mathbf{F}_2) + O(t^{p+1})$$

$$= \prod_{l=1}^{m} (1 + c_l t \mathbf{F}_1)(1 + d_l t \mathbf{F}_2) + O(t^{p+1}) \tag{15.98}$$

where $\mathbf{F} = \mathbf{L}$ or $\mathbf{U}$. The coefficients for different-order approximations are listed in Table 15.1. Additionally, the components of the M–S equations at discrete points in time and space can be described as

$$G(i, j, k, t) = G^{n+l/m}(i\Delta_x, j\Delta_y, k\Delta_z, (n + l/m)\Delta t) \tag{15.99}$$

where Δ_x, Δ_y, and Δ_z are the cell sizes along three different directions.

In addition, a combination of the high-order time algorithm and the high-order space algorithm will deliver satisfactory results with high accuracy and low numerical dispersion errors [29]. Based on this, the explicit fourth-order collocated difference is adopted for discretizing the differential operators in the space domain.

From the above descriptions, the qth-order accurate collocated difference expressions, which are applied to discretize the second-order differential operators, are of the following forms:

$$\frac{\partial^2 G^{n+l/m}(r)}{\partial \varsigma^2} = \frac{1}{\Delta_\varsigma^2} \sum_{d=-\frac{q}{2}}^{d=\frac{q}{2}} \beta_d G^{n+l/m}(r + d) + O\left(\Delta_\varsigma^{q+1}\right) \tag{15.100}$$

Table 15.2 The coefficients for different-order accurate collocated differences.

Order	β_{-2}	β_{-1}	β_0	β_1	β_2
2	—	1	-2	1	—
4	-1/12	4/3	-5/2	4/3	-1/12

Where $\zeta = x, y, z, r = i, j, k$, and β_d listed in Table 15.2 denotes the difference coefficients at different grid points for the second- and fourth-order differences.

The SFDTD(4,4) update equations can be given by (taking the y-direction as an example)

$$A_y^{n+\frac{l}{m}}\left(i,j+\frac{1}{2},k\right) = A_y^{n+\frac{l-1}{m}}\left(i,j+\frac{1}{2},k\right) + \frac{c_1\Delta t}{\varepsilon_0\varepsilon_r}Y_y^{n+\frac{l-1}{m}}\left(i,j+\frac{1}{2},k\right)$$

(15.101)

$$
\begin{aligned}
&Y_y^{n+\frac{l}{m}}\left(i,j+\frac{1}{2},k\right) \\
&= Y_y^{n+\frac{l-1}{m}}\left(i,j+\frac{1}{2},k\right) + J_y^{n+\frac{l-1}{m}}\left(i,j+\frac{1}{2},k\right) \\
&+ {}_{x1}\left[y^{n+\frac{l-1}{m}}\left(i+1,j+\frac{1}{2},k\right) - 2y^{n+\frac{l-1}{m}}\left(i,j+\frac{1}{2},k\right) + y^{n+\frac{l-1}{m}}\left(i-1,j+\frac{1}{2},k\right)\right] \\
&+ {}_{x2}\left[y^{n+\frac{l-1}{m}}\left(i+2,j+\frac{1}{2},k\right) - 2y^{n+\frac{l-1}{m}}\left(i,j+\frac{1}{2},k\right) + y^{n+\frac{l-1}{m}}\left(i-2,j+\frac{1}{2},k\right)\right] \\
&+ {}_{z1}\left[y^{n+\frac{l-1}{m}}\left(i,j+\frac{1}{2},k+1\right) - 2y^{n+\frac{l-1}{m}}\left(i,j+\frac{1}{2},k\right) + y^{n+\frac{l-1}{m}}\left(i,j+\frac{1}{2},k-1\right)\right] \\
&+ {}_{z2}\left[y^{n+\frac{l-1}{m}}\left(i,j+\frac{1}{2},k+2\right) - 2y^{n+\frac{l-1}{m}}\left(i,j+\frac{1}{2},k\right) + y^{n+\frac{l-1}{m}}\left(i,j+\frac{1}{2},k-2\right)\right]
\end{aligned}
$$

(15.102)

$\frac{4}{3}d_l \times \kappa_x$, $\alpha_{z1} = \frac{4}{3}d_l \times \kappa_z$, $\alpha_{x2} = -\frac{1}{12}d_l \times \kappa_x$, and $\alpha_{z2} = -\frac{1}{12}d_l \times \kappa_z$. For the cubic grid, we have $\Delta x = \Delta y = \Delta z = \Delta$ and $\kappa_x = \kappa_y = \kappa_z = \kappa_\zeta$. The constant κ_ζ is defined as a type of Courant–Friedrichs–Levy (CFL) coefficient in the FDTD method approach.

In theory, the system in Eq. (15.81) constitutes a complete coupled system framework, and the field component of ψ_R, ψ_I, $\mathbf{A}$, and $\mathbf{Y}$ can be calculated to be self-consistent. However, the problem of scale mismatch exists between the macroscopic EM system and the microscopic QM system. If the whole simulation area is discretized by fine quantum mesh, it is bound to bring great burden to the numerical calculation. In order to avoid the spatial multi-scale problem between the QM and EM systems, the eigenstate extension technique and the two-level approximation model are used to replace the space–time dependent Schrödinger equation. The wave function $\psi(\mathbf{r}, t)$ can be expressed as

$$\psi(\mathbf{r}, t) = C_g(t)e^{-iE_g t/\hbar}\psi_g(\mathbf{r}) + C_e(t)e^{-iE_e t/\hbar}\psi_e(\mathbf{r}) \tag{15.103}$$

where $C_g(t)$ and $C_e(t)$ denote the expansion coefficients and satisfy the condition

$$|C_g(t)|^2 + |C_e(t)|^2 = 1 \tag{15.104}$$

The exponential terms of $e^{-iE_g t/\hbar}$ and $e^{-iE_e t/\hbar}$ characterize the time evolution of the eigenstates, and $E_g = \hbar\omega_g$, $E_e = \hbar\omega_e$. These two dominant atomic states are denoted by $\psi_g(\mathbf{r})$ (for ground state) and $\psi_e(\mathbf{r})$ (for excited state). The QM system is nothing but a two-level system, and the two states can be expressed by

$$\psi_g(\mathbf{r}_{s=x,y,x}) = \left(\frac{1}{a\sqrt{\pi}}\right)^{3/2} e^{-(x^2+y^2+z^2)/2a^2} \tag{15.105}$$

$$\psi_e(\mathbf{r}_{s=x,y,x}) = \left(\frac{1}{a\sqrt{\pi}}\right)^{3/2} \left(\frac{s}{a}\right)\sqrt{2}e^{-(x^2+y^2+z^2)/2a^2} \tag{15.106}$$

and $a = \sqrt{\hbar/m\omega}$.

Define the inner product form as

$$\langle\psi_i(\mathbf{r}) \mid \psi_j(\mathbf{r})\rangle = \int_\Omega d\mathbf{r}\psi_i^*(\mathbf{r})\cdot\psi_j^*(\mathbf{r}) \tag{15.107}$$

According to the orthogonality and parity of the eigenmode, the following conditions need to be satisfied (the integral of an odd function over the whole space is equal to zero):

$$\langle\psi_i(\mathbf{r}) \mid \psi_j(\mathbf{r})\rangle = \begin{cases} 1 & i = j \\ 0 & i \neq j \end{cases} \tag{15.108}$$

$$\langle\psi_i(\mathbf{r}) \mid \mathbf{p} \mid \psi_i(\mathbf{r})\rangle = 0 \tag{15.109}$$

$\mathbf{p}$ is the momentum operator.

At last, two sets of equations about $C_g(t)$ and $C_e(t)$ are deduced as

$$i\hbar\frac{dC_g(t)}{dt} = -\frac{e\mathbf{A}}{m}\langle\psi_g|\mathbf{p}|\psi_e\rangle C_e(t)e^{-i\omega_0 t} + \frac{e^2\mathbf{A}^2}{2m}C_g(t) \tag{15.110}$$

$$i\hbar\frac{dC_e(t)}{dt} = -\frac{e\mathbf{A}}{m}\langle\psi_e|\mathbf{p}|\psi_g\rangle C_g(t)e^{i\omega_0 t} + \frac{e^2\mathbf{A}^2}{2m}C_e(t) \tag{15.111}$$

and the current term $\mathbf{J}$ is given as

$$\langle\mathbf{J}\rangle = -\frac{e^2\mathbf{A}}{m}\left(|C_g(t)|^2 + |C_e(t)|^2\right)$$
$$+ \frac{e}{m}\left[C_g^*(t)C_e(t)e^{-i\omega_0 t}\langle\psi_g|\mathbf{p}|\psi_e\rangle + C_e^*(t)C_g(t)e^{i\omega_0 t}\langle\psi_e|\mathbf{p}|\psi_g\rangle\right] \tag{15.112}$$

where ω_0 is the transition frequency, which is equal to $\omega_e - \omega_g$.

According to the SFDTD scheme, the multi-stage time stepping iteration formulas of (15.110), (15.111), and (15.112) are discretized as follows:

$$C_g^{n+\frac{l}{m}} = C_g^{n+\frac{l-1}{m}} - \frac{eA\Delta t}{i\hbar m}\langle\psi_g|\mathbf{p}|\psi_e\rangle C_e^{n+\frac{l-1}{m}} e^{-i\omega_0\left(n+\frac{l-1}{m}\right)\Delta t} + \frac{e^2A^2\Delta t}{2i\hbar m}C_g^{n+\frac{l-1}{m}}$$

$$(15.113)$$

$$C_e^{n+\frac{l}{m}} = C_e^{n+\frac{l-1}{m}} - \frac{eA\Delta t}{i\hbar m}\langle\psi_e|\mathbf{p}|\psi_g\rangle C_g^{n+\frac{l-1}{m}} e^{i\omega_0\left(n+\frac{l-1}{m}\right)\Delta t} + \frac{e^2A^2\Delta t}{2i\hbar m}C_e^{n+\frac{l-1}{m}}$$

$$(15.114)$$

$$J_y^{n+\frac{l}{m}} = -\frac{e^2A}{m}\left(\left|C_g^{n+\frac{l-1}{m}}\right|^2 + \left|C_e^{n+\frac{l-1}{m}}\right|^2\right)$$
$$+ \frac{e}{m}\left[\left(C_g^{n+\frac{l-1}{m}}\right)^* C_e^{n+\frac{l-1}{m}} e^{-i_0\left(n+\frac{l-1}{m}\right)t}\langle\psi_g|\mathbf{p}|\psi_e\rangle\right.$$
$$\left. + \left(C_e^{n+\frac{l-1}{m}}\right)^* C_g^{n+\frac{l-1}{m}} e^{i_0\left(n+\frac{l-1}{m}\right)t}\langle\psi_e|\mathbf{p}|\psi_g\rangle\right]$$

$$(15.115)$$

Finally, Eqs. (15.101), (15.102), (15.113), (15.114), and (15.115) constitute a high-order SFDTD numerical solution for solving M–S coupled equations.

Figure 15.13 shows the simulation process of M–S coupled system based on the high-order SFDTD method. It can be seen that a complete numerical iteration

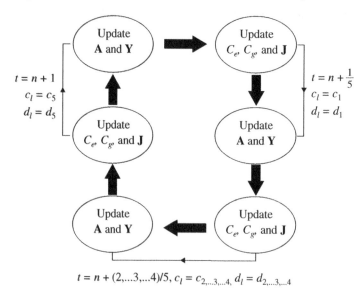

Figure 15.13 The simulation procedures of the symplectic algorithm for the coupled M–S system.

for the coupled system includes a multi-stage numerical solution process. Each stage of numerical solution involves the calculation process of the field components from the microscopic quantum system (the macroscopic EM system) to the macroscopic EM system (the microscopic quantum system). For example, $\mathbf{A}$ and $\mathbf{Y}$ are calculated by Eqs. (15.101) and (15.102), then $\mathbf{A}$ is used to calculate $C_g(t)$ and $C_e(t)$ (Eqs. (15.113) and (15.114)), and the quantum current $\mathbf{J}$ is calculated by the value of $\mathbf{A}$, $C_g(t)$, and $C_e(t)$ (Eq. (15.115)). Finally, the quantum current $\mathbf{J}$ is regarded as an excitation source to the EM system.

15.4.3.2 The Boundary Condition for the M–S System

The analysis of the QM–EM multi-physical problems often involves the closed domain and open domain problem, for example, the simulation of the energy-level transition of a particle in a resonant cavity or free space under an external EM field. For the cavity problem, the Dirichlet boundary is needed, and for the free space problem, the absorbing boundary conditions such as perfect matched layer (PML) need to be implemented to truncate the calculation area.

Dirichlet Boundary Condition
For the coupled M–S system introduced in this section, only $\mathbf{A}$ and $\mathbf{Y}$ contain spatial position information, so Dirichlet boundary conditions are only applied to EM systems. It is easy to set up the Dirichlet boundary conditions for the spatial second-order difference. For simplicity, taking a one-dimensional resonator as an example (as shown in Figure 15.14), $\mathbf{A}(0) = \mathbf{A}(L) = 0$ and $\mathbf{Y}(0) = \mathbf{Y}(L) = 0$ can be set directly at both ends of $\mathbf{A}$ and $\mathbf{Y}$, where L is the length of the resonator. However, for the spatial fourth-order difference, the image theory method should be used for accurate modeling. As shown in Figure 15.14, for the left end of $\mathbf{A}$, let $\mathbf{A}(0) = 0$, $\hat{\mathbf{A}}(1) = -\mathbf{A}(1)$, and $\hat{\mathbf{A}}(2) = -\mathbf{A}(2)$. $\hat{\mathbf{A}}(1)$ and $\hat{\mathbf{A}}(2)$ are the mirror points of $\mathbf{A}(1)$ and $\mathbf{A}(2)$, respectively. Similarly, for the right side of $\mathbf{A}$, $\hat{\mathbf{A}}(L) = 0$, $\hat{\mathbf{A}}(L-1) = -\hat{\mathbf{A}}(L-1)$, and $\hat{\mathbf{A}}(L-2) = -\hat{\mathbf{A}}(L-2)$. $\hat{\mathbf{A}}(L-1)$ and $\hat{\mathbf{A}}(L-2)$ are the mirror points of a $\mathbf{A}(L-1)$ and $\mathbf{A}(L-2)$, respectively. Similarly, for $\mathbf{Y}$ components, $\mathbf{Y}(0)$ 0, $\hat{\mathbf{Y}}(1) = \mathbf{Y}(1)$, and $\hat{\mathbf{Y}}(2) = \mathbf{Y}(2)$; $\mathbf{Y}(L) = 0$, $\hat{\mathbf{Y}}(L-1) = \mathbf{Y}(L-1)$, and $\hat{\mathbf{Y}}(L-2) = \mathbf{Y}(L-2)$.

It should be noted that, for the settings of Dirichlet boundary conditions, if there is lack of special handling of high-order difference, such as the mirror principle, the calculation accuracy of the numerical simulation results is worse than that of the ordinary second-order difference.

$\hat{\mathbf{A}}(2) \qquad \hat{\mathbf{A}}(1) \qquad \mathbf{A}(0) \qquad \mathbf{A}(1) \qquad \mathbf{A}(2)$

$\mathbf{A}(L-2) \qquad \mathbf{A}(L-1) \qquad \mathbf{A}(0) \qquad \hat{\mathbf{A}}(L-1) \qquad \hat{\mathbf{A}}(L-2)$

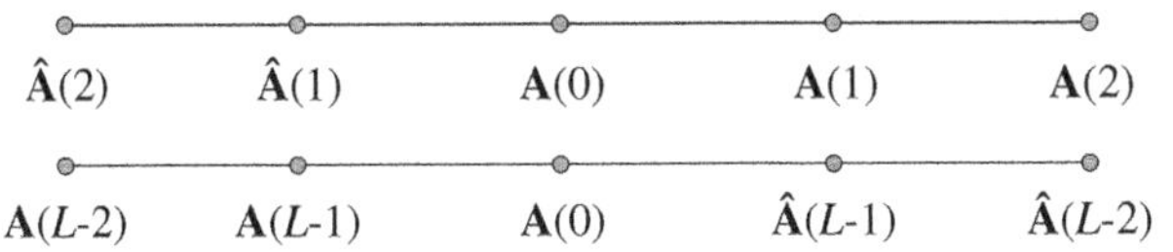

Figure 15.14 Boundary treatment of **A** field by the image theory.

Absorbing Boundary Conditions

For the above M–S system, the absorbing boundary conditions are needed to simulate the infinite space domain. Rewriting Eq. (15.74) (ignoring current term **J**):

$$\frac{\partial \mathbf{Y}}{\partial t} = -\frac{\nabla \times \nabla \times \mathbf{A}}{\mu_0} \tag{15.116}$$

It can be seen from formula (15.116) that the calculation of the **Y** component involves the second-order spatial partial derivatives; it is difficult to apply PML technology in the SFDTD method. In order to reduce the application difficulty of PML technology, $\mathbf{B} = \nabla \times \mathbf{A}$ is applied and substituted into Eq. (15.116); then we get

$$j\omega \mathbf{Y} = \nabla_s \times \mathbf{B} \tag{15.117}$$

and

$$\nabla_s = \frac{1}{S_x}\frac{\partial}{\partial x}e_x + \frac{1}{S_y}\frac{\partial}{\partial y}e_y + \frac{1}{S_z}\frac{\partial}{\partial z}e_z \tag{15.118}$$

At last, using the recursive convolution method in [51], the CPML absorbing boundary conditions can be constructed to realize the numerical simulation of open domain problems.

15.5 Numerical Simulation

In order to verify the correctness and effectiveness of the M–S model and corresponding high-order SFDTD(4,4) method in analyzing and simulating multi-physical problems, an analytical quantum EM model is studied, which provides a benchmark for verifying the correctness of the proposed algorithm.

As shown in Figure 15.15a, the particle is placed in the metal resonant cavity. When the particle is irradiated by the EM wave, it periodically absorbs photons and re-emits photons (as shown in Figure 15.15b), giving rise to the Rabi oscillation. If the cavity is ideal without any material loss and radiation loss, the EM wave will act on the particles again, and the Rabi oscillation will be cyclic. Once the expansion

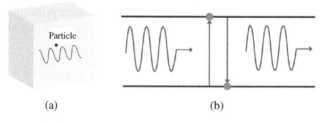

(a) (b)

Figure 15.15 (a) The particle is placed in a metal resonator. (b) The particle evolves between the ground state and the excited state when interacting with the external EM field, resulting in Rabi oscillation.

coefficients $C_g(t)$ and $C_e(t)$ in Eqs. (15.113) and (15.114) are numerically obtained, the population inversion can be calculated by the expression:

$$W(t) = |C_e(t)|^2 - |C_g(t)|^2 \tag{15.119}$$

This coupled quantum–EM model can be calculated and analyzed by using the analytical method (Rabi model). The formulas of the analytical method are as follows:

$$i\hbar \frac{dC_g(t)}{dt} = -e\mathbf{E}_0 \cdot \langle \psi_g | \mathbf{r} | \psi_e \rangle C_e(t) \cos(t) e^{-i\omega_0 t} \tag{15.120}$$

$$i\hbar \frac{dC_e(t)}{dt} = -e\mathbf{E}_0 \cdot \langle \psi_e | \mathbf{r} | \psi_g \rangle C_g(t) \cos(t) e^{i\omega_0 t} \tag{15.121}$$

where ω denotes the EM frequency, ω_0 is the transition frequency ($\omega_0 = \omega_e - \omega_g$).

The size of the nanocavity is $L_x = L_y = L_z = 40\,\text{nm}$ and the grid size $\Delta x = \Delta y = \Delta z = 1\,\text{nm}$. The resonating cavity is excited at its fundamental mode (TE_{101}), with the initial solution:

$$Y_y\Big|_{t=0} = -\varepsilon_r\varepsilon_0 E_0 \sin\left(\frac{\pi}{L_x}x\right) \sin\left(\frac{\pi}{L_z}z\right) \cos(\omega t)\Big|_{t=0} \tag{15.122}$$

where

$$\omega = \sqrt{\left(\frac{\pi}{L_x}\right)^2 + \left(\frac{\pi}{L_x}\right)^2} \tag{15.123}$$

and the relative permittivity ε_r equals to 1 in the vacuum. The atom is placed at the center of the structure and is in a superposed state with $C_g = 1/\sqrt{2}$ and $C_e = 1/\sqrt{2}$.

Based on the coupled M–S system introduced above and the established QM–EM model, we use the high-order SFDTD(4,4) method, the traditional FDTD(2,2) method, and the analytical method (Rabi model) to calculate the population inversions of the particle in different EM environments and verify the correctness and effectiveness of the high-order SFDTD(4,4) method.

15.5.1 Effect of the EM Field Strength

Firstly, the case of quantum transition frequency matching the resonant frequency of the cavity is considered ($\Delta = \omega - \omega_0 = 0$, ω denotes the EM frequency, ω_0 is transition frequency). We calculate the population inversions of the particle with different strengths of the $\mathbf{E}_0$ field (weak field $\Omega = 0.02\omega$ and strong field $\Omega = 0.2\omega$), and

$$\Omega = ((-e\mathbf{E}_0 \cdot \langle \psi_e | \mathbf{r} | \psi_g \rangle)/\hbar) \tag{15.124}$$

As shown in Figures 15.16a and 15.17a, the results obtained by the SFDTD(4,4), FDTD(2,2) and Rabi analytical models are consistent in case of strong and weak

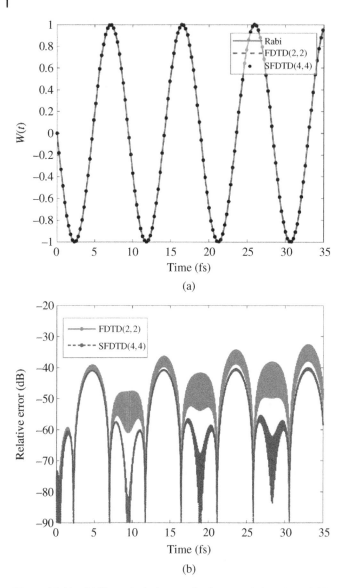

Figure 15.16 (a) The calculation results for a weak field ($\Omega = 0.02\omega$) in the exact resonance condition. (b) The relative errors of the FDTD(2,2) approach and SFDTD(4,4) algorithm for a weak field ($\Omega = 0.02\omega$) in the exact resonance condition.

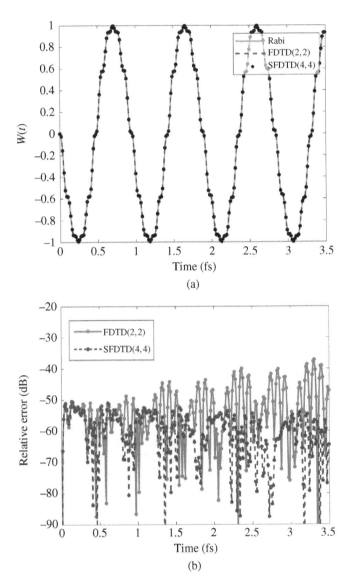

Figure 15.17 (a) The calculation results for a strong field ($\Omega = 0.2\omega$) in the exact resonance condition. (b) The relative errors of the FDTD(2,2) approach and SFDTD(4,4) algorithm for a strong field ($\Omega = 0.02\omega$) in the exact resonance condition.

fields, which verifies the correctness of the introduced theoretical framework and SFDTD method in solving M–S equations.

To further verify the numerical performance of the SFDTD method, Eq. (15.125) is used to quantify the relative calculation error:

$$EE_{rr} = 20\log_{10}\frac{|EE(t) - EE_{ref}(t)|}{\max |EE_{ref}(t)|} \tag{15.125}$$

where $EE(t)$ represents the results calculated by two numerical algorithms (SFDTD(4,4) and FDTD(2,2) methods), and $EE_{ref}(t)$ represents the results calculated by the analytical method. The denominator $\max|EE_{ref}(t)|$ represents the maximum modulus of the result calculated by the analytical method.

As shown in Figures 15.16b and 15.17b, compared with the traditional FDTD(2,2) method, the relative calculation error of SFDTD(4,4) algorithm is significantly smaller. In addition, it can be seen that the relative calculation error of the traditional FDTD(2,2) method increases with time evolution in both strong and weak field cases. However, in both cases, the relative error of SFDTD(4,4) algorithm remains stable, which demonstrates that the SFDTD(4,4) algorithm has high accuracy and stability in solving M–S equations.

15.5.2 Effect of Detuning

Secondly, the effect of the detuning factor $\Delta \neq 0$ is considered. We calculate the population inversion of the particle with different conditions of detuning (small detuning $\Delta = 0.05\omega$, large detuning $\Delta = 0.3\omega$). As shown in Figures 15.18a and 15.19a, the results of the SFDTD(4,4), FDTD(2,2), and Rabi model are consistent with the two detuning conditions, which verifies the correctness of the introduced symplectic framework and SFDTD method in solving QM–EM problems again. At the same time, the relative calculation errors of the SFDTD(4,4) and FDTD(2,2) methods in two cases (small detuning $\Delta = 0.05\omega$, large detuning $\Delta = 0.3\omega$) are calculated. It can be seen from Figures 15.18b and 15.19b that the SFDTD(4,4) algorithm has a smaller calculation error than the FDTD(2,2) method.

It is worth noting that for the case of detuning, we find that the calculation formulas of Rabi analytical method (Eqs. (15.120) and (15.121)) must be modified as following:

$$i\hbar\frac{dC_g(t)}{dt} = -eE_0 \cdot (R \cdot \langle\psi_g|\mathbf{r}|\psi_e\rangle)C_e(t)\cos(t)e^{-i\omega_0 t} \tag{15.126}$$

$$i\hbar\frac{dC_e(t)}{dt} = -eE_0 \cdot (R \cdot \langle\psi_e|\mathbf{r}|\psi_g\rangle)C_g(t)\cos(t)e^{i\omega_0 t} \tag{15.127}$$

This is because the analytical Rabi model is based on $\mathbf{r}\cdot\mathbf{E}$ Hamiltonian, while the numerical method introduced in this section is based on $\mathbf{p}\cdot\mathbf{A}$ Hamiltonian. It is pointed out that the difference between the two Hamiltonian operators is related

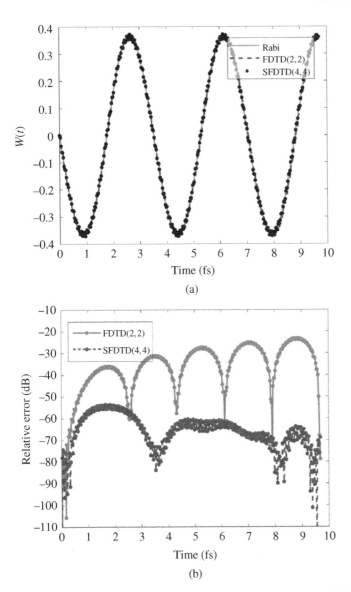

Figure 15.10 (a) The calculation results for a small detuning $\Delta = 0.05\omega$ condition. (b) The relative errors of the FDTD(2,2) approach and SFDTD(4,4) algorithm for a small detuning $\Delta = 0.05\omega$ condition.

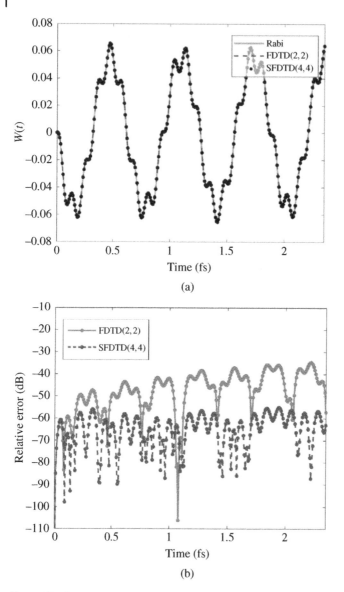

Figure 15.19 (a) The calculation results for a large detuning $\Delta = 0.3\omega$ condition. (b) The relative errors of the FDTD(2,2) approach and SFDTD(4,4) algorithm for a large detuning $\Delta = 0.3\omega$ condition.

to the matching relationship between the atomic transition frequency (ω_0) and the applied EM field frequency (ω) [52], that is $R = \omega_0/\omega$. When two frequencies are equal ($\omega_0 = \omega$, $R = 1$), there is no difference between the two Hamiltonian operators. However, when two frequencies are not equal ($\omega_0 \neq \omega$, $R \neq 1$), they are different.

To further verify the high accuracy and stability of the high-order SFDTD(4,4) method, the results of FDTD(2,2) method with different grid sizes and time steps are compared with those of the SFDTD(4,4) method under the condition of coarse grids. Figure 15.20 shows the relative calculation errors of the three methods (FDTD(2,2) ($\Delta = 1$ nm), FDTD(2,2)-DG (dense grid) ($\Delta = 0.5$ nm), SFDTD(4,4) ($\Delta = 1$ nm)). The results show that the accuracy of the FDTD(2,2) method can be improved by increasing the grid resolution and reducing the time step size, but its relative error is still higher than that of the SFDTD(4,4) method. Furthermore, it can be found that the relative error of the FDTD(2,2) method with different mesh sizes and time step sizes increases with the increase of simulation time. Therefore, for the traditional FDTD (2,2) method, the high grid resolution in the time and space domain can only slow down the growth rate of the accumulated error but cannot eliminate the error accumulation.

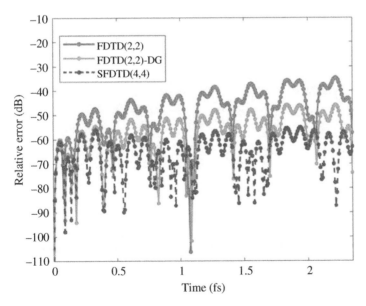

Figure 15.20 The relative errors of the FDTD(2,2), FDTD(2,2)-DG, and SFDTD(4, 4) approaches for a large detuning $\Delta = 0.3\omega$ condition. The FDTD(2,2) and SFDTD(4,4) use coarse grids of 1 nm. The FDTD(2,2)-DG uses fine grids of 0.5 nm.

15.5.3 Application of the Coupled M–S System in a Complex Environment

The correctness and effectiveness of the introduced M–S system and the SFDTD (4,4) method are verified by the simple and ideal QM–EM models. However, the actual simulation of the QM–EM model often involves lossy medium, inhomogeneous medium, and free space. In the following, the multi-physical simulations in complex environments are introduced.

15.5.3.1 Lossy Medium Environment

Assuming that the cavity model shown in Figure 15.15a is filled with lossy medium, we analyze the population inversions of the particle with or without quantum current $\mathbf{J}_q$ feedback for the cases of small loss medium case ($\sigma = 0.0001$) and large loss medium case ($\sigma = 0.02$). The ohmic loss can be introduced by adding another conduction current $\mathbf{J}_c = \sigma \mathbf{Y}/\varepsilon_0$ to Eq. (15.69).

It can be seen from Figure 15.21 that when the dielectric loss is small, the EM wave attenuation is slow, and the Rabi oscillation will be cyclic. However, when the loss becomes larger, the EM field decays rapidly, and the periodic Rabi oscillation is no longer maintained, as shown in Figure 15.22. In addition, in order to unveil the influence of the quantum current $\mathbf{J}_q$ on the coupled system, the population inversions of the particle are calculated in the following two cases

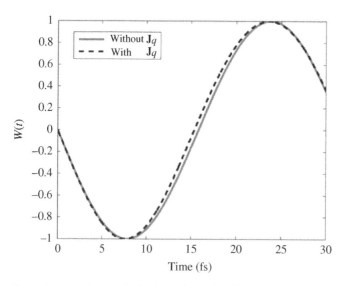

Figure 15.21 The population inversions of particles with or without quantum current $\mathbf{J}_q$ feedback for a small loss medium case ($\sigma = 0.0001$).

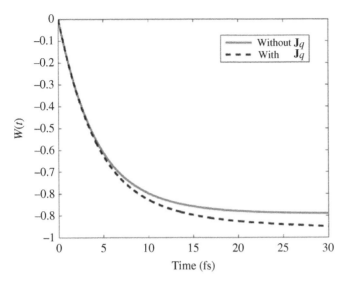

Figure 15.22 The population inversions of particles with or without quantum current J_q feedback for a small loss medium case ($\sigma = 0.02$).

(without J_q and with J_q feedback). It can be found from Figures 15.21 and 15.22 that when the loss in the environment is small, the effect of quantum current J_q on the coupled system is relatively small. After the increment of loss, the effect is more significant. Because the actual coupling system will have a certain degree of loss, it is necessary to study the "active property" of quantum current in a lossy environment.

15.5.3.2 Inhomogeneous Media Environment
In addition to the loss in the coupling system, the inhomogeneous medium environment also exists in the actual EM simulation requirements. Therefore, we study the effect of the EM field on the energy-level transition of particles in the inhomogeneous medium environment. Assume that a lossless dielectric sphere ($\varepsilon_r = 3$) with a radius of 8 nm is placed in the center of the resonant cavity shown in Figure 15.15a. First, the fundamental mode of the cavity is calculated by the SFDTD method, and the transition frequency (ω_0) of the particles is matched with the fundamental mode frequency (ω) of the cavity. Figure 15.23 shows the population inversions of the particles at two positions. The EM field will decay fast as far away from the center of the sphere, the population inversion at the offset point shows a lower Rabi frequency, which is consistent with the actual situation.

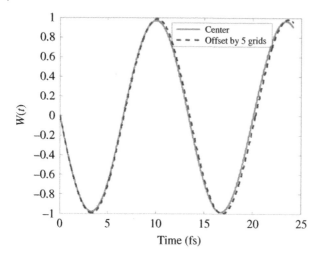

Figure 15.23 Changes of population inversion at different positions.

15.5.3.3 Free Space Environment

At last, the population inversions of the particle in free space under the external EM field is studied and compared with that of the perfect electric conductor (PEC) boundary condition. In free space, when the pulse passes through the particle, its population inversion remains unchanged (black dotted line). However, if the PEC

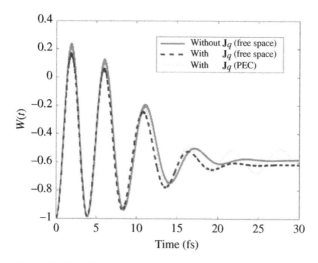

Figure 15.24 Changes of population inversion with different factors.

boundary condition is used, the population inversion of the particle oscillates periodically (pink dotted line). This is because the EM wave radiated by the particle will be reflected back and absorbed again, thus forming periodic Rabi oscillation repeatedly. In addition, if the way of "no feedback" approximation is adopted, that is, the quantum current and radiation field produced by the particles are ignored, the result (red solid line in Figure 15.24) is obviously different from that of the self-consistent coupling solution (black dotted line), which also proves the existence of radiation displacement and attenuation.

15.6   Conclusion

In this chapter, we first introduced the basic theory about the SFDTD method, which outperforms in solving the EM system and provides an excellent choice to solve the coupled M–S system. For the M–S system, some commonly used M–S models and corresponding numerical methods, including the FDTD and SFDTD methods, are introduced to handle the multi-physical problems. The numerical results demonstrate the correctness and accuracy of the numerical methods. Then, a coupled M–S system with symplectic structure, which has the non-dissipative property is introduced. It is proved that each subsystem of the coupled system satisfies symplectic-structure property, and then the SFDTD(4,4) method is used to discretize the M–S equations. To verify the correctness and effectiveness of the SFDTD(4,4) method, the Rabi oscillation of microscopic particles in a two-level system is studied. The results of different types of numerical examples fully demonstrate the advantages of the SFDTD(4,4) algorithm in terms of numerical accuracy and long time stability. Finally, SFDTD(4,4) is successfully applied to the numerical calculation of multi-physical problems, involving lossy media, inhomogeneous medium, and free space, which enables us to consider the actual QM–EM environment. The content of this chapter is systematic, which lays a theoretical foundation and provides technical support for the subsequent, more complex simulation of the EM–QM system, high-precision simulation software development, and other related work.

Acknowledgments

The authors acknowledge the support of the NSFC (U20A20164, 62171001, 61975177 61971001,61871001, 61701003), Open Fund for Discipline Construction, Institute of Physical Science and Information Technology, Anhui University (No. 2019AH001) and NSFC of Anhui Province (Nos. KJ2018A0015, K120436040).

Author Biography

Zhixiang Huang (M'16-SM'18) was born in Anhui, China, in 1979. He received the BS and PhD degrees from Anhui University, Hefei, China, in 2002 and 2007, respectively.

From September 2010 to September 2011, he was a Visiting Scholar with Iowa State University, USA. From August 2013 to October 2013, he was a Visiting Professor with the University of Hong Kong. From February 2014 to February 2015, he was a Visiting Professor with the Beijing National Laboratory from September 2010 to September 2011, he was a Visiting Scholar with Iowa State University, USA. From August 2013 to October 2013, he was a Visiting Professor with the University of Hong Kong. From February 2014 to February 2015, he was a Visiting Professor with the Beijing National Laboratory Prof. Huang was the recipient of the Second Prize of Science and Technology from the Anhui Province Government, China, in 2015. He was also the recipient of the National Science Foundation for Outstanding Young Scholar of China in 2017. He is a member of the OSA.

Guoda Xie was born in Anhui, China, in 1991. He received the MS degree from the School of Electronics and Information Engineering, Anhui University, China, in 2017, and is currently a post-doctoral at the School of Electronics and Information Engineering, Anhui University, China. His research interests include computational electromagnetics, unconditionally stable finite difference time domain algorithm, multiphysics modeling with time domain numerical methods.

Xingang Ren received his PhD degree from the University of Hong Kong, Hong Kong in 2016. He is now an Associate Professor in School of Electronics and Information Engineering, Anhui University. His research interests cover computational electromagnetics, nano-optics, plasmonic structures, and optoelectronic devices.

Wei E.I. Sha (M'09-SM'17) received the BS and PhD degrees in Electronic Engineering at Anhui University, Hefei, China, in 2003 and 2008, respectively. From July 2008 to July 2017, he was a Postdoctoral Research Fellow and then a Research Assistant Professor in the Department of Electrical and Electronic Engineering at the University of Hong Kong, Hong Kong. From March 2018 to March 2019, he worked at University College London as a Marie Skłodowska-Curie Individual Fellow. From October 2017, he joined the College of Information Science & Electronic Engineering at Zhejiang University, Hangzhou, China, where he is currently a tenure-tracked Assistant Professor. His research interests include theoretical and computational research in electromagnetics and optics, focusing on the multiphysics and interdisciplinary research. His research involves fundamental and applied aspects in computational and applied electromagnetics, nonlinear and quantum electromagnetics, micro- and nano-optics, optoelectronic device simulation, and multiphysics modeling. Dr. Sha has authored or coauthored 151 refereed journal papers, 130 conference publications (including 33 invited talks), five book chapters, and two books. His Google Scholar citation is 6050 with h-index of 37. He is a senior member of IEEE and a member of OSA. He served as Reviewers for 60 technical journals and Technical Program Committee Members of 9 IEEE conferences. He also served as Associate Editors of Progress In Electromagnetics Research, IEEE Access and IEEE Open Journal of Antennas and Propagation, and Guest Editors of IEEE Journal on Multiscale and Multiphysics Computational Techniques and The Applied Computational Electromagnetics Society Journal. In 2015, he was awarded Second Prize of Science and Technology from Anhui Province Government, China. In 2007, he was awarded the Thousand Talents Program for Distinguished Young Scholars of China. Dr. Sha also received 5 Best Student Paper Prizes and one Young Scientist Award with his students.

References

1 Jin, J.-M. (2010). *Theory and Computational of Electromagnetic Fields* Hoboken, NJ: Wiley.
2 Chew, W.C., Jin, J.M., Michielssen, E., and Song, J.M. (2001). *Fast and Efficient Algorithms in Computational Electromagnetics*. Boston: Artech House.
3 Schwierz, F. (2010). Graphene transistors. *Nature Nanotechnology* 5: 487–496.
4 Hanson, G.W. (2005). Fundamental transmitting properties of carbon nanotube antennas. *IEEE Transactions on Antennas and Propagation* 53 (11): 3426–3435.

5 Rouhi, N., Jain, D., and Burke, P.J. (2010). Nanoscale devices for large-scale applications. *IEEE Microwave Magazine* 11 (7): 72–80.

6 Russer, P., Fichtner, N., Lugli, P., and Porod, W. (2010). Nanoelectronics-based integrated antennas. *IEEE Microwave Magazine* 11 (7): 58–71.

7 Mueller, T., Xia, F., and Avouris, P. (2010). Graphene photodetectors for high-speed optical communications. *Nature Photonics* 4 (5): 297–301.

8 Dintinger, J., Klein, S., Bustos, F. et al. (2005). Strong coupling between surface plasmon-polaritons and organic molecules in subwavelength hole arrays. *Physical Review B* 71: 035424.

9 Ozbay, E. (2006). Plasmonics: merging photonics and electronics at nanoscale dimensions. *Science* 311: 189–193.

10 Nakagawa, K., Tajiri, A., Tamura, K. et al. (2013). Thermally assisted magnetic recording applying optical near field with ultra short-time heating. *Journal of the Magnetics Society of Japan* 37 (3): 119–122.

11 Ding, X., Yu He, Z.C., Duan, N.G. et al. (2016). On-demand single photons with high extraction efficiency and near-unity indistinguishability from a resonantly driven quantum dot in a micropillar. *Physical Review Letters* 116: 020401.

12 Shnaiderman, R., Wissmeyer, G., Ulgen, O., and Mustafa, Q. (2020). A submicrometre silicon-on-insulator resonator for ultrasound detection. *Nature* 585: 372–378.

13 Jing, B., Wang, X., Yu, Y. et al. (2019). Entanglement of three quantum memories via interference of three single photons. *Nature Photonics* 13: 210–213.

14 Li, Y.M., Chang, C., Zhu, Z. et al. (2021). Terahertz wave enhances permeability of the voltage-gated calcium channel. *Journal of the American Chemical Society* 143 (11): 4311–4318.

15 Liu, G.Z., Chang, C., Qiao, Z. et al. (2018). Myelin sheath as a dielectric waveguide for signal propagation in the mid-infrared to terahertz spectral range. *Advanced Functional Materials* 1807862.

16 Xiang, Z.X., Chang, C., Zhu, Z. et al. (2020). A primary model of THz and far-infrared signal generation and conduction in neuron systems based on the hypothesis of the ordered phase of water molecules on the neuron surface I: signal characteristics. *Science Bulletin* 65 (4): 308–317.

17 Wu, K.J., Chang, C., Zhu, Z. et al. (2020). Terahertz wave accelerates DNA unwinding: a molecular dynamics simulation study. *Journal of Physical Chemistry Letters* 11: 7002–7008.

18 Yankwich, P.E. (1960). Introduction to quantum mechanics. *Journal of the American Chemical Society* 82 (14): 3803–3803.

19 Rae, A.I.M. (1996). The picture book of quantum mechanics. *Physics Today* 49 (1): 65–66.

20 Sui, W., Yang, J., Yun, X.H., and Wang, C. (2007). Including quantum effects in electromagnetic system-an FDTD solution to Maxwell-Schrödinger equations. *International Microwave Symposium* 1979–1982.

21 Pierantoni, L., Mencarelli, D., and Rozzi, T. (2008). A new 3-D transmission line matrix scheme for the combined Schrödinger-Maxwell problem in the electronic/electromagnetic characterization of nanodevices. *IEEE Transactions on Microwave Theory and Techniques* 56 (3): 654–662.

22 Pierantoni, L., Mencarelli, D., and Rozzi, T. (2009). Boundary immittance operators for the Schrödinger-Maxwell problem of carrier dynamics in nanodevices. *IEEE Transactions on Microwave Theory and Techniques* 57 (5): 1147–1155.

23 Ahmed, I. et al. (2010). A hybrid approach for solving coupled Maxwell and Schrödinger equations arising in the simulation of nano-devices. *IEEE Antennas and Wireless Propagation Letters* 9: 914–917.

24 Ryu, C.J., Liu, A.Y., Sha, W.E.I., and Chew, W.C. (2016). Finite-difference time-domain simulation of the Maxwell-Schrödinger system. *IEEE Journal on Multiscale and Multiphysics Computational Techniques* 1: 40–47.

25 Chen, A.Q., Zeng, H., Chen, S.T., and Chen, R.S. (2019). A d'Alembert-Schrödinger hybrid simulation for laser-induced multi-quantum states transition in a three-dimensional artificial atom. *Optics Letters* 44 (17): 4399–4402.

26 Xiang, C., Kong, F., Li, K., and Liu, M. (2017). A high-order symplectic FDTD scheme for the Maxwell-Schrödinger system. *IEEE Journal of Quantum Electronics* 54 (99): 1–7.

27 Ohnuki, S., Takeuchi, T., Sako, T. et al. (2013). Coupled analysis of Maxwell–Schrödinger equations by using the length gauge: harmonic model of a nanoplate subjected to a 2D electromagnetic field. *Journal of Numerical Modelling: Electronic Networks, Devices and Fields* 26 (6): 533–544.

28 Takeuchi, T., Ohnuki, S., and Sako, T. (2014). Hybrid simulation of Maxwell-Schrodinger equations for multi-physics problems characterized by anharmonic electrostatic potential (Invited Paper). *Progress In Electromagnetics Research* 148: 73–82.

29 Sha, W., Huang, Z., Chen, X.M., and Wu, X. (2008). Survey on symplectic finite-difference time-domain schemes for Maxwell's equations. *IEEE Transactions on Antennas and Propagation* 56 (2): 493–500 Jan.

30 Sheu, T.W.H. and Lin, L. (2015). Dispersion relation equation preserving FDTD method for nonlinear cubic Schrödinger equation. *Journal of Computational Physics* 299 (4): 1–21 Oct.

31 Chen, Q., Qin, H., Liu, J. et al. (2017). Canonical symplectic structure and structure-preserving geometric algorithms for Schrödinger–Maxwell systems. *Journal of Computational Physics* 349: 441–452.

32 Chen, Z., You, X., Shi, W., and Liu, Z. (2012). Symmetric and symplectic ERKN for oscillatory Hamiltonian systems. *Computer Physics Communications* 183 (1): 86–98.

33 Anderson, N. and Arthurs, A.M. (1983). Helicity and variational principles for Maxwell's equations. *International Journal of Electronics* 54: 861–864.

34 Yoshida, H. (1990). Construction of higher order symplectic integrators. *Physics Letters A* 150: 262–268.

35 Huang, Z., Xu, J., Sun, B. et al. (2015). A new solution of Schrödinger equation based on symplectic algorithm. *Computers & Mathematics with Applications* 69 (11): 1303–1312, Jun.

36 Sha, W., Huang, Z.X., Wu, X.L., and Chen, M.S. (2007). Application of the symplectic finite-difference time-domain scheme to electromagnetic simulation. *Journal of Computational Physics* 225: 33–50.

37 Sheu, T.W.H., Liang, L.Y., and Li, J.H. (2013). Development of an explicit symplectic scheme that optimizes the dispersion- relation equation of the Maxwell's equations. *Communications in Computational Physics* 4 (13): 1107–1133.

38 Yee, K.S. (1966). Numerical solution of initial boundary value problems involving Maxwell's equations in isotropic media. *IEEE Transactions on Antennas and Propagation* 14: 302–307.

39 Taflove, A. and Hagness, S.C. (2005). *Computational Electrodynamics the Finite Difference Time-Domain Method*, 3e. Artech House Boston London.

40 Georgakopoulos, S.V., Birtcher, C.R., Balanis, C.A., and Renaut, R.A. (2002). Higher-order finite-difference schemes for electromagnetic radiation, scattering, and penetration, part I: theory. *IEEE Antennas and Propagation Magazine* 44 (1): 134–142.

41 Georgakopoulos, S.V., Birtcher, C.R., Balanis, C.A., and Renaut, R.A. (2002). Higher-order finite-difference schemes for electromagnetic radiation, scattering, and penetration, part 2: applications. *IEEE Antennas and Propagation Magazine* 44 (2): 92–101.

42 Takeuchi, T., Ohnuki, S., and Sako, T. (2015). Maxwell-Schrodinger hybrid simulation for optically controlling quantum states: a scheme for designing control pulses. *Physical Review A* 91 (3): 033401.

43 Feynman, R., Leighton, R.B., and Sands, M.L. (1965). *The Feynman Lectures on Physics*. Reading, MA: Addison-Wesley Publishing Co., Inc.

44 Sullivan, D.M. and Citrin, D.S. (2005). Determining quantum eigenfunctions in three-dimensional nanoscale structures. *Journal of Applied Physics* 97 (10): 104305.

45 Cao, Q., Chen, Y., and Mittra, R. (2002). Multiple image technique (MIT) and anisotropic perfectly matched layer (APML) in implementation of MRTD scheme for boundary truncations of microwave structures. *IEEE Transactions on Microwave Theory and Techniques* 50 (6): 1578–1589, Jun.

46 Chew, W.C. (2014). Vector potential electromagnetics with generalized gauge for inhomogeneous media: formulation. *Progress In Electromagnetics Research* 149: 69–84.

47 Kira, M. and Koch, S.W. (2012). *Semiconductor Quantum Optics*. Cambridge University Press.

48 Chen, Y.P., Sha, W.E.I., Jiang, L. et al. (2017). A unified Hamiltonian solution to Maxwell-Schrödinger equations for modeling electromagnetic field-particle interaction. *Computer Physics Communications* 215: 63–70.

49 Wang, H., Wu, B., Huang, Z., and Wu, X. (2014). A symplectic FDTD algorithm for the simulation of lossy dispersive materials. *Computer Physics Communications* 185 (3): 862–872.

50 Xie, G.D., Huang, Z.X., Fang, M., and Sha, W.E.I. (2019). Simulating Maxwell-Schrödinger equations by high-order symplectic FDTD algorithm. *IEEE Journal on Multiscale and Multiphysics Computational Techniques* 4 (1): 143–151.

51 Roden, J.A. and Gedney, S.D. (2000). Convolution PML (CPML): an efficient FDTD implementation of the CFS-PML for arbitrary media. *Microwave and Optical Technology Letters* 27 (5): 334–339, Dec.

52 Marlon, O.S. and Suhail Zubairy, M. (1997). *Quantum Optics*. Cambridge: Cambridge University Press.

16

Cylindrical FDTD Formulation for Low Frequency Applications

Abdullah Algarni[1], Atef Z. Elsherbeni[2], and Mohammed Hadi[2]

[1]*Department of Electrical Engineering, King Fahd University of Petroleum and Minerals, Dhahran, Saudi Arabia*
[2]*Department of Electrical Engineering, Colorado School of Mines, Golden, CO, USA*

Electromagnetic numerical simulations are very efficient tools for accelerating research and development as well as to reduce the costs of destructive testing and general product design and retooling. They can also be used in field operations as operator-assisting toolboxes for adaptive product control and logging analysis. There are multiple applications, for example, electric generators/motors and oil and gas well logging, where the numerical solver is most efficient when the underlying Maxwell's equations are formulated for cylindrical coordinates and specifically for low frequency applications. This chapter present a detailed finite difference time-domain method formulation based on cylindrical coordinates (CFDTD) with emphasis on low frequency applications. Emphasis will be based on formulating the proper absorbing boundary conditions (ABCs) and the integration of linear circuit elements models for practical simulations.

Being a numerical solver, CFDTD relies on a suite of modeling tools that need to be optimized in order to ensure overall simulation accuracy and reliability. This goal gets more challenging when the frequency bandwidth of interest is low enough to place the modeled structure in close electrical proximity to the axis of rotation. That happens when at least part of the structure's radial distance is within a fraction of a wavelength. Under such circumstances, the underlying cylindrical coordinates-based equations (whether continuous or discrete) allow wave solutions as a superposition of cylindrical harmonics, usually represented by mode numbers. Therefore, the discrete CFDTD solver needs to faithfully reproduce cylindrical harmonics in the near field region, which is challenging considering their heavily evanescent nature, especially at higher rotational modes.

Advances in Time-Domain Computational Electromagnetic Methods, First Edition.
Edited by Qiang Ren, Su Yan, and Atef Z. Elsherbeni.

Even the simplest of structure-wave interactions, unless they are perfectly symmetric around the axis of rotation, require large numbers of high rotational-mode cylindrical harmonics to properly articulate. This objective can be achieved, albeit at a large computational cost, if the solution space is gridded with a dense enough discrete mesh.

To alleviate this extreme computational cost, the simulation space needs to be truncated in the radial direction, by an appropriate set of ABCs, beyond the physical space containing the modeled structure and axis of rotation. ABCs are also needed to truncate the resulting cylindrical-shaped discrete space beyond the extreme axial reaches of the modeled structure. The first task of this chapter details an efficient implementation of the perfectly-matched-layer (PML) as the chosen ABC tool [1] in cylindrical coordinates. This PML will be formulated to effectively absorb outgoing cylindrical waves regardless of their evanescence content. It will also be designed such that it can function effectively with outgoing waves containing large number of rotational modes.

Many electromagnetics applications with cylindrical structures require lumped circuit elements. The linear circuit elements may be active such as voltage sources and current sources or passive elements such as resistors, capacitors, and inductors or combinations. These active and passive elements are important components in many electromagnetic simulations. The second task of this chapter is to develop formulation to integrate circuit elements in the CFDTD formulation following the procedure presented in [2] for Cartesian coordinates system.

16.1 Cylindrical Finite-Difference Time-Domain Method

In this section, we will provide the detailed formulation for the CFDTD method, followed by new derivation of the corresponding convolutional perfectly-matched-layer (CPML) absorbing boundary [3]. The performance of this derived CPML boundary will be verified by numerical examples spanning various parameters with emphasis on cylindrical simulation domains that are much smaller than the wavelength. This is usually the case for low frequency applications.

The FDTD method is one of the methods that uses the grid-based differential time-domain equations. Maxwell's equations as presented in Eqs. (16.1a)–(16.1d) are usually discretized using central-difference approximations to the space and time partial derivatives,

$$\nabla \times \vec{H} = \frac{\partial \vec{D}}{\partial t} + \vec{J} \tag{16.1a}$$

$$\nabla \times \vec{E} = -\frac{\partial \vec{B}}{\partial t} - \vec{M} \tag{16.1b}$$

$$\nabla \cdot \vec{D} = \rho_e \tag{16.1c}$$

$$\nabla \cdot \vec{B} = \rho_m \tag{16.1d}$$

where $\vec{E}$ is the electric field intensity vector in V/m, $\vec{D}$ is the electric displacement vector in C/m^2, $\vec{H}$ is the magnetic field intensity vector in A/m, $\vec{B}$ is the magnetic flux density vector in Weber/m^2, $\vec{J}$ is the electric current density vector in A/m^2, $\vec{M}$ is the magnetic current density vector in V/m^2, ρ_e is the electric charge density in C/m^3, and ρ_m is the virtual magnetic charge density in Weber/m^3. The electric displacement vector and the magnetic flux density vector can be written for the linear, isotropic and non-dispersive material as:

$$\vec{D} = \varepsilon \vec{E} \tag{16.2a}$$

$$\vec{B} = \mu \vec{H} \tag{16.2b}$$

where ε is the permittivity, and μ is the permeability of the material. The electric current density and the magnetic current density are the sum of the conduction current densities and the impressed current densities as $\vec{J} = \vec{J}_C + \vec{J}_i$ and $\vec{M} = \vec{M}_C + \vec{M}_i$ where $\vec{J}_C = \sigma^e \vec{E}$ and $\vec{M}_C = \sigma^m \vec{H}$. Here σ^e and σ^m are the material electric conductivity in S/m and the magnetic conductivity in Ω/m, respectively. Then Faraday and Ampere equations can be rewritten as:

$$\nabla \times \vec{H} = \varepsilon \frac{\partial \vec{E}}{\partial t} + \sigma^e \vec{E} + \vec{J}_i \tag{16.3a}$$

$$\nabla \times \vec{E} = -\mu \frac{\partial \vec{H}}{\partial t} - \sigma^m \vec{H} - \vec{M}_i \tag{16.3b}$$

The two vector equations in (16.3a)–(16.3b) can be split into six scalar equations in a cylindrical coordinates (ρ, ϕ, z) as follows:

$$\frac{\partial E_\rho}{\partial t} = \frac{1}{\varepsilon_\rho} \left(\frac{1}{\rho} \frac{\partial H_z}{\partial \phi} - \frac{\partial H_\phi}{\partial z} - \sigma_\rho^e E_\rho - J_{i\rho} \right) \tag{16.4a}$$

$$\frac{\partial E_\phi}{\partial t} = \frac{1}{\varepsilon_\phi} \left(\frac{\partial H_\rho}{\partial z} - \frac{\partial H_z}{\partial \rho} - \sigma_\phi^e E_\phi - J_{i\phi} \right) \tag{16.4b}$$

$$\frac{\partial E_z}{\partial t} = \frac{1}{\varepsilon_z} \left(\frac{1}{\rho} \frac{\partial}{\partial \rho}(\rho H_\phi) - \frac{1}{\rho} \frac{\partial H_\rho}{\partial \phi} - \sigma_z^e E_z - J_{iz} \right) \tag{16.4c}$$

$$\frac{\partial H_\rho}{\partial t} = \frac{1}{\mu_\rho} \left(\frac{\partial E_\phi}{\partial z} - \frac{1}{\rho} \frac{\partial E_z}{\partial \phi} - \sigma_\rho^m H_\rho - M_{i\rho} \right) \tag{16.4d}$$

$$\frac{\partial H_\phi}{\partial t} = \frac{1}{\mu_\phi} \left(\frac{\partial E_z}{\partial \rho} - \frac{\partial E_\rho}{\partial z} - \sigma_\phi^m H_\phi - M_{i\phi} \right) \tag{16.4e}$$

$$\frac{\partial H_z}{\partial t} = \frac{1}{\mu_z} \left(\frac{1}{\rho} \frac{\partial E_\rho}{\partial \phi} - \frac{1}{\rho} \frac{\partial}{\partial \rho}(\rho E_\phi) - \sigma_z^m H_z - M_{iz} \right) \tag{16.4f}$$

For CFDTD, time and space derivatives are approximated by central finite differences of the form [2]:

$$\frac{\partial f(u)}{\partial u} \approx \frac{f(u + \Delta u) - f(u - \Delta u)}{2\Delta u} \tag{16.5}$$

where Δu is the sampling period along the axis of interest. This difference approximation is implemented within the space containing the simulated cylindrical structure after such a space is divided into a grid of CFDTD cells with dimensions $(\Delta\rho, \rho\Delta\phi, \Delta z)$. For example, Figure 16.1 shows such a structure which is divided into a total of $N_\rho \times N_\phi \times N_z$ CFDTD cells.

Figure 16.2 shows a typical CFDTD cell along with discrete indexing convention for the various electromagnetic field components. Each surface of this cell has

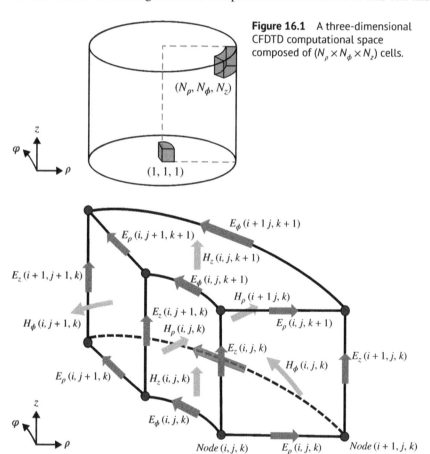

Figure 16.1 A three-dimensional CFDTD computational space composed of $(N_\rho \times N_\phi \times N_z)$ cells.

Figure 16.2 Staggered arrangement of electromagnetic field components for a typical CFDTD cell.

four tangential electric field components. A fifth magnetic field component exists at the surface's center and is normal to that surface. This type of staggered electromagnetic field component placements allows convenient conversion of Maxwell's equations from the continuous to the discrete domain whether using their differential or integral forms. The top surface, for example, has two E_ρ components which are spaced $\rho\Delta\phi$ apart, with an H_z component centered between both along ϕ. Similarly, the H_z is centered between two E_ϕ components which are spaced $\Delta\rho$ apart. The i, j, k indices correspond to storage array indices for each field components, along ρ, ϕ, z, respectively. They can be mapped into cylindrical coordinate positions using the following translations for each field component which are particular to how the indexing is laid out in Figure 16.2:

$$E_\rho(i, \ j, \ k) \Rightarrow ((i - 0.5)\Delta\rho, \ (j - 1)\Delta\phi, \ (k - 1)\Delta z)$$
$$E_\phi(i, \ j, \ k) \Rightarrow ((i - 1)\Delta\rho, \ (j - 0.5)\Delta\phi, \ (k - 1)\Delta z)$$
$$E_z(i, \ j, \ k) \Rightarrow ((i - 1)\Delta\rho, \ (j - 1)\Delta\phi, \ (k - 0.5)\Delta z)$$
$$H_\rho(i, \ j, \ k) \Rightarrow ((i - 1)\Delta\rho, \ (j - 0.5)\Delta\phi, \ (k - 0.5)\Delta z)$$
$$H_\phi(i, \ j, \ k) \Rightarrow ((i - 0.5)\Delta\rho, \ (j - 1)\Delta\phi, \ (k - 0.5)\Delta z)$$
$$H_z(i, \ j, \ k) \Rightarrow ((i - 0.5)\Delta\rho, \ (j - 0.5)\Delta\phi, \ (k - 1)\Delta z)$$

As an exception, the first layer of cells surrounding the cylindrical coordinates z-axis have their inner curved surfaces collapsing into a line, resulting in wedge-shaped cells instead of cylindrical shell-shaped cells. Figure 16.3 shows

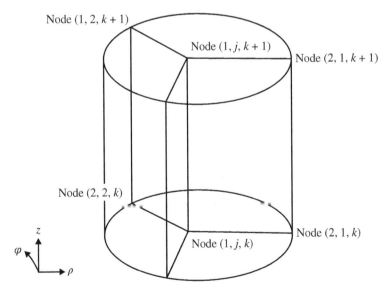

Figure 16.3 The first layer of CFDTD cells around the z-axis are wedge-shaped.

a simplistic representation where the entire ϕ span is divided into only three CFDTD cells. The j counter along the z-axis is meaningless for this case.

The discrete electromagnetic field components of Figure 16.2 are also staggered along time. During the simulation time-marching, all electric field components are updated synchronously and repetitively every Δt time increment. All magnetic field components are also updated synchronously and repetitively every Δt time increment, except they will be interlaced among the electric field updates in a leap-frog manner. Thus, if all electric field components are updated at $t = n\Delta t$, with n being the time counter, the next update for all magnetic field components occurs at $t = \left(n + \frac{1}{2} \right) \Delta t$, and so on. To convert Eq. (16.4c), into a discrete CFDTD update equation, for example, we expand all present derivatives around each E_z component position index (i, j, k) and time index $n + \frac{1}{2}$, keeping in mind that for E_z, $\rho = (i-1)\Delta\rho$, $\phi = (j-1)\Delta\phi$, and $z = (k-0.5)\Delta z$:

$$
\frac{E_z^{n+1}(i,j,k) - E_z^n(i,j,k)}{\Delta t}
$$

$$
= \frac{1}{\Delta\rho\Delta\rho(i-1)\varepsilon_z(i,j,k)} \left(\Delta\rho(i-0.5)H_\phi^{n+\frac{1}{2}}(i,j,k) - \Delta\rho(i-1.5)H_\phi^{n+\frac{1}{2}}(i-1,j,k) \right)
$$

$$
- \frac{1}{\Delta\phi\Delta\rho(i-1)\varepsilon_z(i,j,k)} \left(H_\rho^{n+\frac{1}{2}}(i,j,k) - H_\rho^{n+\frac{1}{2}}(i,j-1,k) \right)
$$

$$
- \frac{\sigma_z^e(i,j,k)}{\varepsilon_z(i,j,k)} E_z^{n+\frac{1}{2}}(i,j,k) - \frac{1}{\varepsilon_z(i,j,k)} J_{iz}^{n+\frac{1}{2}}(i,j,k) \tag{16.6}
$$

By taking the average time of the electric field in the right-hand side and rearranging the terms such that the highest time-indexed term is kept at the left-hand side, we get

$$
\frac{E_z^{n+1}(i,j,k) - E_z^n(i,j,k)}{\Delta t}
$$

$$
= \frac{1}{\Delta\rho\Delta\rho(i-1)\varepsilon_z(i,j,k)} \left(\Delta\rho(i-0.5)H_\phi^{n+\frac{1}{2}}(i,j,k) - \Delta\rho(i-1.5)H_\phi^{n+\frac{1}{2}}(i-1,j,k) \right)
$$

$$
- \frac{1}{\Delta\phi\Delta\rho(i-1)\varepsilon_z(i,j,k)} \left(H_\rho^{n+\frac{1}{2}}(i,j,k) - H_\rho^{n+\frac{1}{2}}(i,j-1,k) \right)
$$

$$
- \frac{\sigma_z^e(i,j,k)}{\varepsilon_z(i,j,k)} \left(\frac{E_z^n(i,j,k) + E_z^{n+1}(i,j,k)}{2} \right) - \frac{1}{\varepsilon_z(i,j,k)} J_{iz}^{n+\frac{1}{2}}(i,j,k) \tag{16.7}
$$

which after some manipulation results in

$$
E_z^{n+1}(i,j,k) = \frac{2\varepsilon_z(i,j,k) - \Delta t\sigma_z^e(i,j,k)}{2\varepsilon_z(i,j,k) + \Delta t\sigma_z^e(i,j,k)} E_z^n(i,j,k)
$$

$$
+ \frac{2\Delta t}{2\varepsilon_z(i,j,k) + \Delta t\sigma_z^e(i,j,k)} \frac{(i-0.5)}{(i-1)} H_\phi^{n+\frac{1}{2}}(i,j,k)
$$

$$- \frac{2\Delta t}{2\varepsilon_z(i,j,k) + \Delta t \sigma_z^e(i,j,k)} \frac{(i-1.5)}{(i-1)} H_\phi^{n+\frac{1}{2}}(i-1,j,k)$$

$$- \frac{2\Delta t}{2\varepsilon_z(i,j,k) + \Delta t \sigma_z^e(i,j,k)} \frac{1}{\Delta \rho \Delta \phi (i-1)}$$

$$\times \left(H_\rho^{n+\frac{1}{2}}(i,j,k) - H_\rho^{n+\frac{1}{2}}(i,j-1,k) \right)$$

$$- \frac{2\Delta t}{2\varepsilon_z(i,j,k) + \Delta t \sigma_z^e(i,j,k)} J_{iz}^{n+\frac{1}{2}}(i,j,k) \tag{16.8}$$

This $E_z^{n+1}(i,j,k)$ update equation can be re-written with respective coefficient terms, following the formulations in [2] as

$$E_z^{n+1}(i,j,k) = C_{eze}(i,j,k) \times E_z^n(i,j,k)$$

$$+ \left(C_{ezh\phi1}(i,j,k) \times H_\phi^{n+\frac{1}{2}}(i,j,k) - C_{ezh\phi2}(i,j,k) \times H_\phi^{n+\frac{1}{2}}(i-1,j,k) \right)$$

$$+ C_{ezh\rho}(i,j,k) \times \left(H_\rho^{n+\frac{1}{2}}(i,j,k) - H_\rho^{n+\frac{1}{2}}(i,j-1,k) \right)$$

$$+ C_{ezj}(i,j,k) \times J_{iz}^{n+\frac{1}{2}}(i,j,k) \tag{16.9}$$

where

$$C_{eze}(i,j,k) = \frac{2\varepsilon_z(i,j,k) - \Delta t \sigma_z^e(i,j,k)}{2\varepsilon_z(i,j,k) + \Delta t \sigma_z^e(i,j,k)}$$

$$C_{ezh\phi1}(i,j,k) = \frac{2\Delta t}{2\varepsilon_z(i,j,k) + \Delta t \sigma_z^e(i,j,k)} \frac{(i-0.5)}{(i-1)\Delta \rho}$$

$$C_{ezh\phi2}(i,j,k) = \frac{2\Delta t}{2\varepsilon_z(i,j,k) + \Delta t \sigma_z^e(i,j,k)} \frac{(i-1.5)}{(i-1)\Delta \rho}$$

$$C_{ezh\rho}(i,j,k) = -\frac{2\Delta t}{2\varepsilon_z(i,j,k) + \Delta t \sigma_z^e(i,j,k)} \frac{1}{\Delta \rho \Delta \phi (i-1)}$$

$$C_{ezj}(i,j,k) = -\frac{2\Delta t}{2\varepsilon_z(i,j,k) + \Delta t \sigma_z^e(i,j,k)}$$

Similarly, the H_z update equation can be derived from (16.4f) with finite differences centered around position index (i,j,k) and time index n,

$$\frac{H_z^{n+\frac{1}{2}}(i,j,k) - H_z^{n-\frac{1}{2}}(i,j,k)}{\Delta t}$$

$$= \frac{1}{\Delta \phi \Delta \rho \left(i - \frac{1}{2} \right) \mu_z(i,j,k)} \left(E_\rho^n(i,j+1,k) - E_\rho^n(i,j,k) \right)$$

$$-\frac{1}{\Delta\rho\Delta\rho\left(i-\frac{1}{2}\right)\mu_z(i,j,k)}\left(\Delta\rho(i)E_\phi^n(i,j+1,k)-\Delta\rho(i-1)E_\phi^n(i,j,k)\right)$$

$$-\frac{\sigma_z^m(i,j,k)}{\mu_z(i,j,k)}H_z^n(i,j,k)-\frac{1}{\mu_z(i,j,k)}M_{iz}^n(i,j,k) \tag{16.10}$$

By taking the average time of the magnetic field in the right-hand side and rearranging the terms such that the future term is kept at the left-hand side, we get

$$\frac{2\mu_z(i,j,k)+\Delta t\sigma_z^m(i,j,k)}{2\Delta t}H_z^{n+\frac{1}{2}}(i,j,k)$$

$$=\frac{2\mu_z(i,j,k)-\Delta t\sigma_z^m(i,j,k)}{2\Delta t}H_z^{n-\frac{1}{2}}(i,j,k)$$

$$+\frac{1}{\Delta\rho\Delta\phi(i-0.5)}\left(E_\rho^n(i,j+1,k)-E_\rho^n(i,j,k)\right)$$

$$-\left(\frac{(i)}{(i-0.5)\Delta\rho}E_\phi^n(i+1,j,k)-\frac{(i-1)}{(i-0.5)\Delta\rho}E_\phi^n(i,j,k)\right)$$

$$-M_{iz}^n(i,j,k) \tag{16.11}$$

which eventually reduces to

$$H_z^{n+\frac{1}{2}}(i,j,k)=C_{hzh}(i,j,k)\times H_z^{n-\frac{1}{2}}(i,j,k)$$

$$+C_{hze\rho}(i,j,k)\times\left(E_\rho^n(i,j+1,k)-E_\rho^n(i,j,k)\right)$$

$$+\left(C_{hze\phi 1}(i,j,k)\times E_\phi^n(i+1,j,k)-C_{hze\phi 2}(i,j,k)\times E_\phi^n(i,j,k)\right)$$

$$+C_{hzm}(i,j,k)\times M_{iz}^n(i,j,k) \tag{16.12}$$

where

$$C_{hzh}(i,j,k)=\frac{2\mu_z(i,j,k)-\Delta t\sigma_z^m(i,j,k)}{2\mu_z(i,j,k)+\Delta t\sigma_z^m(i,j,k)}$$

$$C_{hze\rho}(i,j,k)=\frac{2\Delta t}{2\mu_z(i,j,k)+\Delta t\sigma_z^m(i,j,k)}\frac{1}{\Delta\rho\Delta\phi(i-0.5)}$$

$$C_{hze\phi 1}(i,j,k)=-\frac{2\Delta t}{2\mu_z(i,j,k)+\Delta t\sigma_z^m(i,j,k)}\frac{(i)}{(i-0.5)\Delta\rho}$$

$$C_{hze\phi 2}(i,j,k)=-\frac{2\Delta t}{2\mu_z(i,j,k)+\Delta t\sigma_z^m(i,j,k)}\frac{(i-1)}{(i-0.5)\Delta\rho}$$

$$C_{hzm}(i,j,k)=-\frac{2\Delta t}{2\mu_z(i,j,k)+\Delta t\sigma_z^m(i,j,k)}$$

The remaining updating field equations with respective coefficient terms can be derived as

$$E_\rho^{n+1}(i,j,k)=C_{e\rho e}(i,j,k)\times E_z^n(i,j,k)$$

$$+C_{e\rho hz}(i,j,k)\times\left(H_z^{n+\frac{1}{2}}(i,j,k)-H_z^{n+\frac{1}{2}}(i,j-1,k)\right)$$

$$+ C_{e\rho h\phi}(i,j,k) \times \left(H_\phi^{n+\frac{1}{2}}(i,j,k) - H_\phi^{n+\frac{1}{2}}(i,j,k-1) \right)$$

$$+ C_{e\rho j}(i,j,k) \times J_{i\rho}^{n+\frac{1}{2}}(i,j,k) \tag{16.13}$$

where

$$C_{e\rho e}(i,j,k) = \frac{2\varepsilon_\rho(i,j,k) - \Delta t \sigma_\rho^e(i,j,k)}{2\varepsilon_\rho(i,j,k) + \Delta t \sigma_\rho^e(i,j,k)}$$

$$C_{e\rho hz}(i,j,k) = \frac{2\Delta t}{2\varepsilon_\rho(i,j,k) + \Delta t \sigma_\rho^e(i,j,k)} \frac{1}{(i-0.5)\Delta\rho\Delta\phi}$$

$$C_{e\rho h\phi}(i,j,k) = -\frac{2\Delta t}{2\varepsilon_\rho(i,j,k) + \Delta t \sigma_\rho^e(i,j,k)} \frac{1}{\Delta z}$$

$$C_{e\rho j}(i,j,k) = -\frac{2\Delta t}{2\varepsilon_\rho(i,j,k) + \Delta t \sigma_\rho^e(i,j,k)}$$

$$E_\phi^{n+1}(i,j,k) = C_{e\phi e}(i,j,k) \times E_\phi^n(i,j,k)$$

$$+ C_{e\phi h\rho}(i,j,k) \times \left(H_\rho^{n+\frac{1}{2}}(i,j,k) - H_\rho^{n+\frac{1}{2}}(i,j,k-1) \right)$$

$$+ C_{e\phi hz}(i,j,k) \times \left(H_z^{n+\frac{1}{2}}(i,j,k) - H_z^{n+\frac{1}{2}}(i-1,j,k) \right)$$

$$+ C_{e\phi j}(i,j,k) \times J_{i\phi}^{n+\frac{1}{2}}(i,j,k) \tag{16.14}$$

where

$$C_{e\phi e}(i,j,k) = \frac{2\varepsilon_\phi(i,j,k) - \Delta t \sigma_\phi^e(i,j,k)}{2\varepsilon_\phi(i,j,k) + \Delta t \sigma_\phi^e(i,j,k)}$$

$$C_{e\phi h\rho}(i,j,k) = -\frac{2\Delta t}{2\varepsilon_\phi(i,j,k) + \Delta t \sigma_\phi^e(i,j,k)} \frac{1}{\Delta z}$$

$$C_{e\phi hz}(i,j,k) = -\frac{2\Delta t}{2\varepsilon_\phi(i,j,k) + \Delta t \sigma_\phi^e(i,j,k)} \frac{1}{\Delta\rho}$$

$$C_{e\phi j}(i,j,k) = -\frac{2\Delta t}{2\varepsilon_\phi(i,j,k) + \Delta t \sigma_\phi^e(i,j,k)}$$

$$H_\rho^{n+\frac{1}{2}}(i,j,k) = C_{h\rho h}(i,j,k) \times H_z^{n-\frac{1}{2}}(i,j,k)$$

$$+ C_{h\rho ez}(i,j,k) \times \left(E_z^n(i,j+1,k) - E_z^n(i,j,k) \right)$$

$$+ C_{h\rho e\phi}(i,j,k) \times \left(E_\phi^n(i,j,k+1) - E_\phi^n(i,j,k) \right)$$

$$+ C_{hzm}(i,j,k) \times M_{iz}^n(i,j,k) \tag{16.15}$$

where

$$C_{h\rho h}(i,j,k) = \frac{2\mu_\rho(i,j,k) - \Delta t \sigma_\rho^m(i,j,k)}{2\mu_\rho(i,j,k) + \Delta t \sigma_\rho^m(i,j,k)}$$

$$C_{h\rho e\phi}(i,j,k) = \frac{2\Delta t}{2\mu_\rho(i,j,k) + \Delta t \sigma_\rho^m(i,j,k)} \frac{1}{\Delta z}$$

$$C_{h\rho ez}(i,j,k) = -\frac{2\Delta t}{2\mu_\rho(i,j,k) + \Delta t \sigma_\rho^m(i,j,k)} \frac{1}{(i-1)\Delta\rho\Delta\phi}$$

$$C_{h\rho m}(i,j,k) = -\frac{2\Delta t}{2\mu_\rho(i,j,k) + \Delta t \sigma_\rho^m(i,j,k)}$$

$$H_\phi^{n+\frac{1}{2}}(i,j,k) = C_{h\phi h}(i,j,k) \times H_\phi^{n-\frac{1}{2}}(i,j,k)$$
$$+ C_{h\phi ez}(i,j,k) \times \left(E_z^n(i+1,j,k) - E_z^n(i,j,k) \right)$$
$$+ C_{h\phi e\rho}(i,j,k) \times \left(E_\rho^n(i,j,k+1) - E_\rho^n(i,j,k) \right)$$
$$+ C_{h\phi m}(i,j,k) \times M_{i\phi}^n(i,j,k) \tag{16.16}$$

where

$$C_{h\phi h}(i,j,k) = \frac{2\mu_\phi(i,j,k) - \Delta t \sigma_\phi^m(i,j,k)}{2\mu_\phi(i,j,k) + \Delta t \sigma_\phi^m(i,j,k)}$$

$$C_{h\phi ez}(i,j,k) = -\frac{2\Delta t}{2\mu_\phi(i,j,k) + \Delta t \sigma_\phi^m(i,j,k)} \frac{1}{\Delta\rho}$$

$$C_{h\phi e\rho}(i,j,k) = \frac{2\Delta t}{2\mu_\phi(i,j,k) + \Delta t \sigma_\phi^m(i,j,k)} \frac{1}{\Delta z}$$

$$C_{h\phi m}(i,j,k) = -\frac{2\Delta t}{2\mu_\phi(i,j,k) + \Delta t \sigma_\phi^m(i,j,k)}$$

One can observe that the E_z at z-axis has no variation along ϕ. There are also no E_ϕ and H_ρ in the Yell cell adjacent to the z-axis, as shown in Figure 16.4. Equation (16.4c), therefore, cannot be used to get the update equation for E_z there. Instead, the modified Ampere's law [3] can be used to determine the E_z update equation. Figure 16.5 shows the contour path which helps derive the update equation for $E_z(1,j,k)$.

The modified Ampere's law

$$\oint \vec{H}_\phi \cdot \vec{dl} = \int \left(\varepsilon_z \frac{\partial \vec{E}_z}{\partial t} + \sigma_z^e \vec{E}_z + \vec{J}_{iz} \right) \cdot \vec{ds} \tag{16.17}$$

can be rewritten in FDTD discrete form as

$$\frac{\Delta\rho\Delta\phi}{2} \sum_{j=1}^{N_\phi} H_\phi^{n+\frac{1}{2}}(1,j,k) = \left(\begin{matrix} \varepsilon_z(1,j,k) \times \frac{E_z^{n+1}(1,j,k)-E_z^n(1,j,k)}{\Delta t} \\ + \sigma_z^e(1,j,k) \times E_z^{n+\frac{1}{2}}(1,j,k) + J_{iz}^{n+\frac{1}{2}}(1,j,k) \end{matrix} \right) \frac{\pi\Delta\rho^2}{4} \tag{16.18}$$

Figure 16.4 Field components at the vicinity of the z-axis.

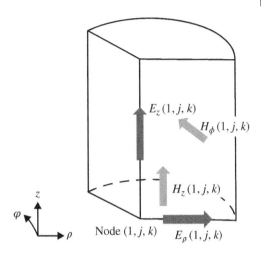

Figure 16.5 Contour path for updating $E_z(1,j,k)$.

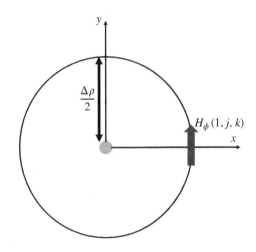

The $E_z^{n+1}(1,j,k)$ update equation is then produced by re-arranging the above equation:

$$E_z^{n+1}(1,j,k) = C_{eze}(1,j,k) \times E_z^n(1,j,k)$$

$$+ \, C_{ezh\phi}(1,j,k) \times \sum_{j=1}^{N_\phi} H_\phi^{n+\frac{1}{2}}(1,j,k)$$

$$+ \, C_{ezj}(1,j,k) \times J_{iz}^{n+\frac{1}{2}}(1,j,k) \tag{16.19}$$

where

$$C_{eze}(1,j,k) = \frac{2\varepsilon_z(1,j,k) - \Delta t \sigma_z^e(1,j,k)}{2\varepsilon_z(1,j,k) + \Delta t \sigma_z^e(1,j,k)}$$

$$C_{ezh\phi}(1,j,k) = \frac{2\Delta t}{2\varepsilon_z(1,j,k) + \Delta t \sigma_z^e(1,j,k)} \frac{2\Delta\phi}{\pi\Delta\rho}$$

$$C_{ezj}(1,j,k) = -\frac{2\Delta t}{2\varepsilon_z(1,j,k) + \Delta t \sigma_z^e(1,j,k)}$$

Each H_ϕ component from the above equation will share $E_z(1,j,k)$ in its respective update equation. Therefore, their special update equations can be written as

$$H_\phi^{n+\frac{1}{2}}(1,j,k) = \frac{2\mu_\phi(1,j,k) - \Delta t \sigma_\phi^m(1,j,k)}{2\mu_\phi(1,j,k) + \Delta t \sigma_\phi^m(1,j,k)} H_\phi^{n-\frac{1}{2}}(1,j,k)$$

$$+ \frac{2\Delta t}{2\mu_\phi(1,j,k) + \Delta t \sigma_\phi^m(1,j,k)} \frac{1}{\Delta\rho} \left(E_z^n(2,j,k) - E_z^n(1,j,k) \right)$$

$$- \frac{2\Delta t}{2\mu_\phi(1,j,k) + \Delta t \sigma_\phi^m(1,j,k)} \frac{1}{\Delta z} \left(E_\rho^n(1,j,k+1) - E_\rho^n(1,j,k) \right)$$

$$- \frac{2\Delta t}{2\mu_\phi(1,j,k) + \Delta t \sigma_\phi^m(1,j,k)} M_{i\phi}^{n+\frac{1}{2}}(1,j,k) \tag{16.20}$$

The Courant–Friedrichs–Lewy (CFL) condition is used in this CFDTD formulation to determine Δt which is limited by the smallest dimensioned CFDTD cell [2]. The corresponding numerical stability condition of Δt in CFDTD [4–7] is then given by:

$$\Delta t \leq \frac{1}{c\sqrt{\left(\frac{1}{\Delta\rho}\right)^2 + \left(\frac{1}{\Delta z}\right)^2 + \left(\frac{1}{\rho_{min}\Delta\phi}\right)^2}} \tag{16.21}$$

where c is the speed of light and $\rho_{min} = 0.5\Delta\rho$ is the shortest distance of field components from the z-axis within the CFDTD grid.

The CFDTD algorithm as detailed so far is usable for numerical analysis where the radial and axial dimensions are terminated by perfectly conducting enclosures, or at least highly lossy materials. Otherwise, ABCs need to be implemented. The most common ABC implementation in this type of discrete formulation is the CPML method [3].

16.2 Convolutional PML in Cylindrical Coordinates

When modeling unbounded structures with CFDTD, a CPML absorber can be used to truncate the computational domain. It is an encompassing finite-thickness special layer of CFDTD cells where special update equations are implemented.

They are designed and optimized to maximize wave attenuation with minimal wave reflection. Since the radial variable ρ appears explicitly in CFDTD equations, special attention needs to be paid when choosing the appropriate PML variant. The complex frequency-shifted PML [8] is one of the variant of CPML since it allows absorption of evanescent waves. Such waves are encountered when the PML layers are placed in close proximity to the z-axis due to the radial nature of cylindrical waves.

16.2.1 Coordinate Stretching Approach

CPML formulation utilizes complex stretching variables along either the radial or axial dimensions, whichever is normal to the PML layer. The stretching in the z-direction will not be included here as it can be easily determined using the same procedure for CPML in Cartesian FDTD formulation [2]. The stretching complex variable in the ρ direction is given by

$$s_\rho(\rho, \omega) = \kappa_\rho(\rho) + \frac{\sigma_\rho(\rho)}{j\omega\varepsilon_0}. \tag{16.22}$$

where the parameters $\sigma_\rho(\rho)$ and $\kappa_\rho(\rho)$ are given by

$$\sigma_\rho(\rho) = \begin{cases} \sigma_{\max}\left(\dfrac{\rho - \rho_0}{\rho_{pml}}\right)^{n_\sigma}, & \rho > \rho_0 \\ 0, & \rho \leq \rho_0 \end{cases} \tag{16.23a}$$

$$\kappa_\rho(\rho) = \begin{cases} 1 + (\kappa_{\max} - 1)\left(\dfrac{\rho - \rho_0}{\rho_{pml}}\right)^{n_\kappa}, & \rho > \rho_0 \\ 1, & \rho \leq \rho_0 \end{cases} \tag{16.23b}$$

The parameters n_σ, n_κ, $\sigma_{\max}$, $\kappa_{\max}$ are specific to the absorption profile and should be optimized for each problem configuration. The stretched radial dimension within the CPML region is mapped to [9, 10]

$$\bar{\rho}(\rho, \omega) = \rho_0 + \int_{\rho_0}^{\rho} s_\rho(\rho', \omega)d\rho' \tag{16.24}$$

where ρ_0 is the innermost CPML interface's radial distance from the z-axis. The associated derivative is determined from the above relation as

$$\frac{\partial}{\partial\bar{\rho}} = \frac{1}{s_\rho(\rho, \omega)}\frac{\partial}{\partial\rho} \tag{16.25}$$

In introducing these stretching variables and derivatives into Maxwell's equations in cylindrical coordinates, care should be exercised by limiting this stretching mechanism to equations where $\partial/\partial\rho$ is present. Therefore, the ρ variable in Eqs. (16.4a) and (16.4d) should not be stretched since there is no

ρ derivative there. Most of the CFDTD literature skims on this important point which would render the CPML implementation inefficient except for rotationally invariant and strictly radial waves where Eqs. (16.4a) and (16.4d) vanish. To illustrate how the remaining equations in (16.4b), (16.4c), (16.4e), and (16.4f) are modified for the CPML region, we start by applying Ampere's law on a path transverse to the E_ϕ field component, that is

$$\int j\omega\varepsilon_0 E_\phi ds = \oint \overline{H} \cdot d\overline{l} \tag{16.26}$$

leading to

$$\int j\omega\varepsilon_0 E_\phi s_\rho \partial\rho\partial z = \oint \overline{H} \cdot (\overline{a}_\rho s_\rho \partial\rho + \overline{a}_z \partial z) \tag{16.27}$$

The integrals in (16.27) can be converted to finite sums with the help of Figure 16.6 and reformulated into finite differences. The corresponding derivative limits will then produce the corresponding CPML differential equation:

$$j\omega\varepsilon_0 E_\phi = \frac{\partial H_\rho}{\partial z} - \frac{1}{s_\rho}\frac{\partial H_z}{\partial \rho} \tag{16.28}$$

A similar treatment can be applied to a transverse path surround the field component E_z with the help of Figure 16.7:

$$\int j\omega\varepsilon_0 E_z s_\rho \overline{\rho} \partial\rho\partial\phi = \oint \overline{H} \cdot (\overline{a}_\rho s_\rho \partial\rho + \overline{a}_\phi \overline{\rho}\partial\phi) \tag{16.29}$$

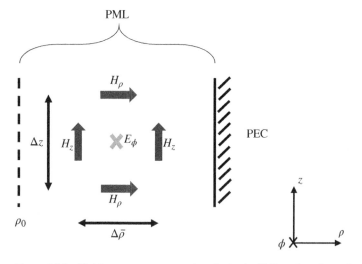

Figure 16.6 Field components to update $E\varphi$ in the PML region along the ρ-axis.

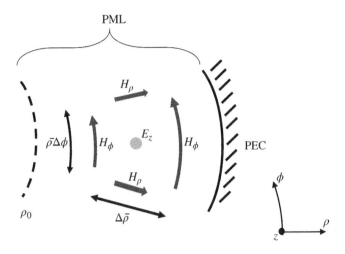

Figure 16.7 Field components to update E_z in the PML region along the ρ-axis.

which results in

$$j\omega\varepsilon_0 E_z = \frac{1}{s_\rho\bar{\rho}}\frac{\partial(\bar{\rho}H_\phi)}{\partial\rho} - \frac{1}{\bar{\rho}}\frac{\partial H_\rho}{\partial\phi} \tag{16.30}$$

The complete set of the stretched Maxwell's equation for the radial CPML region, after Faraday's law is similarly applied to the magnetic field components, is given by:

$$j\omega\varepsilon_0 E_\rho = \frac{1}{\rho}\frac{\partial H_z}{\partial\phi} - \frac{\partial H_\phi}{\partial z} \tag{16.31a}$$

$$j\omega\varepsilon_0 s E_\phi = \frac{\partial(sH_\rho)}{\partial z} - \frac{\partial H_z}{\partial\rho} \tag{16.31b}$$

$$j\omega\varepsilon_0 s\bar{\rho}E_z = \frac{\partial(\bar{\rho}H_\phi)}{\partial\rho} - \frac{\partial(sH_\rho)}{\partial\phi} \tag{16.31c}$$

$$j\omega\mu_0 H_\rho = \frac{\partial E_\phi}{\partial z} - \frac{1}{\rho}\frac{\partial E_z}{\partial\phi} \tag{16.31d}$$

$$j\omega\mu_0 s H_\phi = \frac{\partial E_z}{\partial\rho} - \frac{\partial(sE_\rho)}{\partial z} \tag{16.31e}$$

$$j\omega\mu_0 s\bar{\rho}H_z = \frac{\partial(sE_\rho)}{\partial\phi} - \frac{\partial(\bar{\rho}E_\phi)}{\partial\rho} \tag{16.31f}$$

To facilitate converting these equations to CFDTD update equations, certain products will be combined into intermediate variables, i.e. $sE_z = E_{zs}$, $sH_z = H_{zs}$,

$sH_\rho = H_{\rho s}$, $sE_\rho = E_{\rho s}$, $\bar\rho E_\phi = \bar E_\phi$, and $\bar\rho H_\phi = \bar H_\phi$, then the Maxwell's equation for the radial CPML region are

$$j\omega\epsilon_0 E_\rho = \frac{1}{\rho}\frac{\partial H_z}{\partial\phi} - \frac{\partial H_\phi}{\partial z} \tag{16.32a}$$

$$j\omega\epsilon_0 sE_\phi = \frac{\partial H_{\rho s}}{\partial z} - \frac{\partial H_z}{\partial\rho} \tag{16.32b}$$

$$j\omega\epsilon_0\bar\rho E_{zs} = \frac{\partial \bar H_\phi}{\partial\rho} - \frac{\partial H_{\rho s}}{\partial\phi} \tag{16.32c}$$

$$j\omega\mu_0 H_\rho = \frac{\partial E_\phi}{\partial z} - \frac{1}{\rho}\frac{\partial E_z}{\partial\phi} \tag{16.32d}$$

$$j\omega\mu_0 sH_\phi = \frac{\partial E_z}{\partial\rho} - \frac{\partial E_{\rho s}}{\partial z} \tag{16.32e}$$

$$j\omega\mu_0\bar\rho H_{zs} = \frac{\partial E_{\rho s}}{\partial\phi} - \frac{\partial \bar E_\phi}{\partial\rho} \tag{16.32f}$$

For the above equations any outgoing cylindrical wave will split into a radial and a transverse components. The transverse projection (along z) will continue propagating uninterrupted, until they encounter the CPML layers terminating the axial dimension. The radial wave on the other hand, will be attenuated, or absorbed. The efficiency of this absorption will depend on a balanced choice of the various CPML parameters. Within the discrete CFDTD grid, this efficiency will also depend on the spatial and temporal steps chosen for the simulation.

16.2.2 CPML Update Equations

To derive the CFDTD update equations within the CPML region, we start by explicitly deriving the stretched radial variable in terms of the chosen absorption profile in Eqs. (16.23a)–(16.23b) as

$$\bar\rho(\rho,\omega) = \int_0^\rho s(\rho',\omega)\,d\rho' = \int_0^\rho \left(\kappa_\rho(\rho') + \frac{\sigma_\rho(\rho')}{j\omega\epsilon_0}\right)d\rho' = b(\rho) + \frac{d(\rho)}{j\omega\epsilon_0} \tag{16.33}$$

where

$$b(\rho) = \begin{cases} \rho + (\kappa_{max}-1)\dfrac{\rho_{pml}}{n_\kappa+1}\left(\dfrac{\rho-\rho_0}{\rho_{pml}}\right)^{n_\kappa+1}, & \rho \ge \rho_0 \\[2mm] \rho, & \rho < \rho_0 \end{cases}$$

$$d(\rho) = \begin{cases} \sigma_{max} \dfrac{\rho_{pml}}{n_\sigma + 1} \left(\dfrac{\rho - \rho_0}{\rho_{pml}} \right)^{n_\sigma + 1}, & \rho \geq \rho_0 \\ 0, & \rho < \rho_0 \end{cases}$$

Next, we need to find the inverse Laplace transform for each term in frequency-domain Eqs. (16.32a)–(16.32f). For the terms in the right-hand side of Eqs. (16.32a)–(16.32f), there are three sets of terms which need to be inverted into their time domain forms. The first set is in the form of the variable "X_s" which represents either $E_{\rho s}$ or $H_{\rho s}$:

$$\chi_s(\omega) = s(\omega)\chi(\omega) = \left(\kappa + \frac{\sigma}{j\omega\varepsilon_0} \right) \chi(\omega) = \left(\kappa + \frac{\sigma}{\varepsilon_0} \frac{1}{j\omega} \right) \chi(\omega) \tag{16.34}$$

Taking the inverse Fourier transform, we get the time-domain equivalent

$$X_s(t) = \left(\kappa\delta(t) + \frac{\sigma}{\varepsilon_0} u(t) \right) * X(t) = \kappa X(t) + \frac{\sigma}{\varepsilon_0} \int_0^t X(t - t')dt' \tag{16.35}$$

The convolution integral is then converted to a discrete sum to match CFDTD's time-stepping such that:

$$X_s^n(i,j,k) = \kappa(i,j,k) \times X^n(i,j,k) + \frac{\sigma(i,j,k)}{\varepsilon_0} \sum_{m=0}^n X^{n-m}(i,j,k) \int_{m\Delta t}^{(m+1)\Delta t} dt'$$

$$= \kappa(i,j,k) \times X^n(i,j,k) + \frac{\sigma(i,j,k)}{\varepsilon_0} \sum_{m=0}^n X^{n-m}(i,j,k)[t]_{t=m\Delta t}^{t=(m+1)\Delta t}$$

$$= \kappa(i,j,k) \times X^n(i,j,k) + \underbrace{\frac{\sigma(i,j,k)\Delta t}{\varepsilon_0}}_{\psi(i,j,k)} \underbrace{\sum_{m=0}^n X^{n-m}(i,j,k)}_{Q^n(i,j,k)}. \tag{16.36}$$

The discrete convolution sum can be computed recursively during the CFDTD time-marching, which allows replacing the above equation with

$$Q^n(i,j,k) = \psi(i,j,k) \times X^n(i,j,k) + Q^{n-1}(i,j,k)$$
$$X_s^n(i,j,k) = \kappa(i,j,k) \times X^n(i,j,k) + Q^n(i,j,k) \tag{16.37}$$

Similarly, for the second set of terms in frequency domain equations (16.32a)–(16.32f), "$\overline{X}$," which represents either $\overline{F}_\psi$ or $\overline{H}_\psi$, we have

$$\overline{\chi}(\omega) = \overline{\rho}(\omega)\chi(\omega) = \left(b + \frac{d}{j\omega\varepsilon_0} \right) \chi(\omega) = \left(b + \frac{d}{\varepsilon_0} \frac{1}{j\omega} \right) \chi(\omega) \tag{16.38}$$

After taking the inverse Fourier transform, we get

$$\overline{X}(t) = \left(b\delta(t) + \frac{d}{\varepsilon_0} u(t) \right) * X(t) = bX(t) + \frac{d}{\varepsilon_0} \int_0^t X(t - t')dt' \tag{16.39}$$

and the corresponding discrete form is given by

$$\overline{X}^n(i,j,k) = b(i,j,k) \times X^n(i,j,k) + \frac{d}{\varepsilon_0} \sum_{m=0}^{n} X^{n-m}(i,j,k) \int_{m\Delta t}^{(m+1)\Delta t} dt'$$

$$= b(i,j,k) \times X^n(i,j,k) + \frac{d}{\varepsilon_0} \sum_{m=0}^{n} X^{n-m}(i,j,k)[t]_{t=m\Delta t}^{t=(m+1)\Delta t}$$

$$= b(i,j,k) \times X^n(i,j,k) + \underbrace{\frac{d\Delta t}{\varepsilon_0}}_{\Omega(i,j,k)} \sum_{m=0}^{n} X^{n-m}(i,j,k)$$

$$= b(i,j,k) \times X^n(i,j,k) + \Omega(i,j,k) \times \underbrace{X^{n-m}(i,j,k)}_{Q^n(i,j,k)} \qquad (16.40)$$

which can be computed recursively as

$$Q^n(i,j,k) = X^n(i,j,k) + Q^{n-1}(i,j,k)$$

$$\overline{X}^n(i,j,k) = b(i,j,k) \times X^n(i,j,k) + \Omega(i,j,k) \times Q^n(i,j,k) \qquad (16.41)$$

Finally, the third set of terms in the right-hand side of Eqs. (16.32a)–(16.32f) involves the need to compute X field components from the updated X_s components where "X" represents either E_z or H_z.

$$\chi(\omega) = \frac{1}{s(\omega)} \chi_s(\omega) = \left(\kappa + \frac{\sigma}{j\omega\varepsilon_0} \right)^{-1} \chi_s(\omega) = \frac{1}{\kappa} \left(1 - \frac{\sigma}{\kappa\varepsilon_0} \frac{1}{\frac{\sigma}{\kappa\varepsilon_0} + j\omega} \right) \chi_s(\omega)$$

$$(16.42)$$

The inverse Fourier transform treatment produces the time-domain expression

$$X(t) = \frac{1}{\kappa} \left(\delta(t) - \frac{\sigma}{\kappa\varepsilon_0} e^{-\frac{\sigma}{\kappa\varepsilon_0}t} u(t) \right) * X_s(t)$$

$$= \frac{1}{\kappa} \left(X_s(t) - \frac{\sigma}{\kappa\varepsilon_0} \int_0^t X_s(t-t') e^{-\frac{\sigma}{\kappa\varepsilon_0}t'} dt' \right) \qquad (16.43)$$

and the subsequent discrete form is given by

$$X^n(i,j,k) = \frac{1}{\kappa(i,j,k)}$$

$$\times \left(X_s^n(i,j,k) - \frac{\sigma(i,j,k)}{\kappa(i,j,k) \times \varepsilon_0} \sum_{m=0}^{n} X_s^{n-m}(i,j,k) \int_{m\Delta t}^{(m+1)\Delta t} e^{-\frac{\sigma(i,j,k)}{\kappa(i,j,k)\times\varepsilon_0}t'} dt' \right)$$

$$= \frac{1}{\kappa(i,j,k)}$$

$$\times \left(X_s^n(i,j,k) - \underbrace{\left(1 - e^{-\frac{\sigma(i,j,k)}{\kappa(i,j,k)\times\varepsilon_0}\Delta t}\right)}_{\Lambda(i,j,k)} \underbrace{\sum_{m=0}^{n} X_s^{n-m}(i,j,k)\, e^{-\frac{\sigma(i,j,k)}{\kappa(i,j,k)\times\varepsilon_0}m\Delta t}}_{Q^n(i,j,k)} \right) \quad (16.44)$$

The involved sum can also be computed recursively during CFDTD time-marching:

$$Q^n(i,j,k) = X_s^n(i,j,k) + Q^{n-1}(i,j,k)e^{-\frac{\sigma(i,j,k)}{\kappa(i,j,k)\times\varepsilon_0}\Delta t}$$

$$X^n(i,j,k) = \frac{1}{\kappa(i,j,k)} \times \left(X_s^n(i,j,k) - \Lambda(i,j,k) \times Q^n(i,j,k) \right) \quad (16.45)$$

As for the left-hand side of Eqs. (16.32a)–(16.32f), there are two sets of terms which need to be converted into their time domain forms. The first set regards

$$j\omega s(\omega)\chi(\omega) = j\omega\left(\kappa + \frac{\sigma}{j\omega\varepsilon_0}\right)\chi(\omega) \quad (16.46)$$

where "X" represents either E_ϕ or H_ϕ. Taking the inverse Fourier transform, we get

$$F^{-1}(\cdot) = \kappa\frac{\partial X(t)}{\partial t} + \frac{\sigma}{\varepsilon_0}(\delta(t)) * X(t) = \kappa\frac{\partial X(t)}{\partial t} + \frac{\sigma}{\varepsilon_0}X(t) \quad (16.47)$$

The corresponding discrete form is

$$F^{-1^n}(i,j,k) = \kappa(i,j,k) \times \frac{X^{n+\frac{1}{2}}(i,j,k) - X^{n-\frac{1}{2}}(i,j,k)}{\Delta t} + \frac{\sigma(i,j,k)}{\varepsilon_0}$$

$$\times \frac{X^{n+\frac{1}{2}}(i,j,k) + X^{n-\frac{1}{2}}(i,j,k)}{2} \quad (16.48)$$

which can be regrouped into

$$F^{-1^n}(i,j,k) = \left(\frac{\kappa(i,j,k)}{\Delta t} + \frac{\sigma(i,j,k)}{2\varepsilon_0}\right) X^{n+\frac{1}{2}}(i,j,k)$$

$$- \left(\frac{\kappa(i,j,k)}{\Delta t} - \frac{\sigma(i,j,k)}{2\varepsilon_0}\right) X^{n-\frac{1}{2}}(i,j,k) \quad (16.49)$$

The second set of terms in the left-hand side of Eqs. (16.32a)–(16.32f) involves $j\omega\bar{\rho}(\omega)\chi(\omega)$ where "X" represents either E_{zs} or H_{zs}.

$$j\omega\bar{\rho}(\omega)\chi(\omega) = j\omega\left(b + \frac{d}{j\omega\varepsilon_0}\right)\chi(\omega) \quad (16.50)$$

Taking the inverse Fourier transform produces

$$F^{-1}(\cdot) = b\frac{\partial X(t)}{\partial t} + \frac{d}{\varepsilon_0}(\delta(t)) * X(t) = b\frac{\partial X(t)}{\partial t} + \frac{d}{\varepsilon_0}X(t) \tag{16.51}$$

with the corresponding discrete form

$$F^{-1"}(i,j,k) = b(i,j,k) \times \frac{X^{n+\frac{1}{2}}(i,j,k) - X^{n-\frac{1}{2}}(i,j,k)}{\Delta t}$$
$$+ \frac{d(i,j,k)}{\varepsilon_0} \times \frac{X^{n+\frac{1}{2}}(i,j,k) + X^{n-\frac{1}{2}}(i,j,k)}{2} \tag{16.52}$$

This last equation can also be regrouped to a more useful version as

$$F^{-1"}(i,j,k) = \left(\frac{b(i,j,k)}{\Delta t} + \frac{d(i,j,k)}{2\varepsilon_0}\right) X^{n+\frac{1}{2}}(i,j,k)$$
$$- \left(\frac{b(i,j,k)}{\Delta t} - \frac{d(i,j,k)}{2\varepsilon_0}\right) X^{n-\frac{1}{2}}(i,j,k) \tag{16.53}$$

All the above treatments can now be assembled and sequenced into CPML update equations ready for direct coding. Going through all the six electromagnetic field components, we start by updating E_ρ^{n+1} with

$$E_\rho^{n+1}(i,j,k) = E_\rho^n(i,j,k)$$
$$+ \frac{1}{\Delta\phi\Delta\rho(i-0.5)\varepsilon_0}\left(H_z^{n+\frac{1}{2}}(i,j,k) - H_z^{n+\frac{1}{2}}(i,j-1,k)\right)$$
$$- \frac{1}{\Delta z\varepsilon_0}\left(H_\phi^{n+\frac{1}{2}}(i,j,k) - H_\phi^{n+\frac{1}{2}}(i,j,k-1)\right) \tag{16.54}$$

followed by updating $E_{\rho s}^{n+1}$

$$E_{\rho s}^{n+1}(i,j,k) = \kappa(i,j,k) \times E_\rho^{n+1}(i,j,k) + \psi(i,j,k) \times Q_{E\rho s}^{n+1}(i,j,k)$$
$$Q_{E\rho s}^{n+1}(i,j,k) = E_\rho^{n+1}(i,j,k) + Q_{E\rho s}^n(i,j,k) \tag{16.55}$$

where $\psi(i,j,k) = \frac{\sigma(i,j,k)\Delta t}{\varepsilon_0}$.

For E_ϕ^{n+1} we have

$$E_\phi^{n+1}(i,j,k) = \frac{\left(\frac{\kappa(i,j,k)}{\Delta t} - \frac{\sigma(i,j,k)}{2\varepsilon_0}\right)}{\left(\frac{\kappa(i,j,k)}{\Delta t} + \frac{\sigma(i,j,k)}{2\varepsilon_0}\right)}E_\phi^n(i,j,k)$$
$$+ \frac{1}{\Delta z\varepsilon_0\left(\frac{\kappa(i,j,k)}{\Delta t} + \frac{\sigma(i,j,k)}{2\varepsilon_0}\right)}\left(H_{\rho s}^{n+\frac{1}{2}}(i,j,k) - H_{\rho s}^{n+\frac{1}{2}}(i,j,k-1)\right)$$
$$- \frac{1}{\Delta\rho\varepsilon_0\left(\frac{\kappa(i,j,k)}{\Delta t} + \frac{\sigma(i,j,k)}{2\varepsilon_0}\right)}\left(H_z^{n+\frac{1}{2}}(i,j,k) - H_z^{n+\frac{1}{2}}(i-1,j,k)\right) \tag{16.56}$$

followed by updating $\overline{E}_\phi^{n+1}$

$$\overline{E}_\phi^{n+1}(i,j,k) = b(i,j,k) \times E_\phi^{n+1}(i,j,k) + \Omega(i,j,k) \times Q_{\overline{E}\phi}^{n+1}(i,j,k)$$

$$Q_{\overline{E}\phi}^{n+1}(i,j,k) = E_\phi^{n+1}(i,j,k) + Q_{\overline{E}\phi}^n(i,j,k) \tag{16.57}$$

where $\Omega(i,j,k) = \frac{d(i,j,k)\Delta t}{\varepsilon_0}$.

For E_{zs}^{n+1} we have

$$E_{zs}^{n+1}(i,j,k) = \frac{\left(\frac{b(i,j,k)}{\Delta t} - \frac{R(i,j,k)}{2}\right)}{\left(\frac{b(i,j,k)}{\Delta t} + \frac{R(i,j,k)}{2}\right)} E_{zs}^n(i,j,k)$$

$$+ \frac{1}{\Delta\rho\varepsilon_0 \left(\frac{b(i,j,k)}{\Delta t} + \frac{R(i,j,k)}{2}\right)} \left(\overline{H}_\phi^{n+\frac{1}{2}}(i,j,k) - \overline{H}_\phi^{n+\frac{1}{2}}(i-1,j,k)\right)$$

$$- \frac{1}{\Delta\phi\varepsilon_0 \left(\frac{b(i,j,k)}{\Delta t} + \frac{R(i,j,k)}{2}\right)} \left(H_{\rho s}^{n+\frac{1}{2}}(i,j,k) - H_{\rho s}^{n+\frac{1}{2}}(i,j-1,k)\right) \tag{16.58}$$

followed by updating E_z^{n+1}

$$E_z^{n+1}(i,j,k) = \frac{1}{\kappa(i,j,k)} \times \left[E_{zs}^{n+1}(i,j,k) - \Lambda(i,j,k) \times Q_{Ez}^{n+1}(i,j,k)\right]$$

$$Q_{Ez}^{n+1}(i,j,k) = E_{zs}^{n+1}(i,j,k) + Q_{Ez}^n(i,j,k) \times e^{-\frac{\sigma(i,j,k)\Delta t}{\kappa(i,j,k)\varepsilon_0}} \tag{16.59}$$

where

$$\Lambda(i,j,k) = 1 - e^{-\frac{\sigma(i,j,k)\Delta t}{\kappa(i,j,k)\varepsilon_0}}$$

Once all the above quantities are updated, the corresponding ones for the magnetic fields are updated in a similar manner. For $H_\rho^{n+\frac{1}{2}}$ we have

$$H_\rho^{n+\frac{1}{2}}(i,j,k) = H_\rho^{n-\frac{1}{2}}(i,j,k)$$

$$+ \frac{1}{\Delta z\mu_0} \left(E_\phi^n(i,j,k+1) - E_\phi^n(i,j,k)\right)$$

$$- \frac{1}{\Delta\phi\Delta\rho(i-1)\mu_0} \left(E_z^n(i,j+1,k) - E_z^n(i,j,k)\right) \tag{16.60}$$

followed by updating $H_{\rho s}^{n+\frac{1}{2}}$

$$H_{\rho s}^{n+\frac{1}{2}}(i,j,k) = \kappa(i,j,k) \times H_\rho^{n+\frac{1}{2}}(i,j,k) + \psi(i,j,k) \times Q_{H\rho s}^{n+\frac{1}{2}}(i,j,k)$$

$$Q_{H\rho s}^{n+\frac{1}{2}}(i,j,k) = H_\rho^{n+\frac{1}{2}}(i,j,k) + Q_{H\rho s}^{n-\frac{1}{2}}(i,j,k) \tag{16.61}$$

For $H_\phi^{n+\frac{1}{2}}$ we have

$$H_\phi^{n+\frac{1}{2}}(i,j,k) = \frac{\left(\frac{\kappa(i,j,k)}{\Delta t} - \frac{\sigma(i,j,k)}{2\varepsilon_0}\right)}{\left(\frac{\kappa(i,j,k)}{\Delta t} + \frac{\sigma(i,j,k)}{2\varepsilon_0}\right)} H_\phi^{n-\frac{1}{2}}(i,j,k)$$

$$+ \frac{1}{\Delta\rho\mu_0\left(\frac{\kappa(i,j,k)}{\Delta t} + \frac{\sigma(i,j,k)}{2\varepsilon_0}\right)}\left(E_z^n(i+1,j,k) - E_z^n(i,j,k)\right)$$

$$- \frac{1}{\Delta z\mu_0\left(\frac{\kappa(i,j,k)}{\Delta t} + \frac{\sigma(i,j,k)}{2\varepsilon_0}\right)}\left(E_{\rho s}^n(i,j,k+1) - E_{\rho s}^n(i,j,k)\right) \quad (16.62)$$

followed by updating $\overline{H}_\phi^{n+\frac{1}{2}}$

$$\overline{H}_\phi^{n+\frac{1}{2}}(i,j,k) = b(i,j,k) \times H_\phi^{n+\frac{1}{2}}(i,j,k) + \Omega(i,j,k) \times Q_{\overline{H}\phi}^{n+\frac{1}{2}}(i,j,k)$$

$$Q_{\overline{H}\phi}^{n+\frac{1}{2}}(i,j,k) = H_\phi^{n+\frac{1}{2}}(i,j,k) + Q_{\overline{H}\phi}^{n-\frac{1}{2}}(i,j,k) \quad (16.63)$$

For $H_{zs}^{n+\frac{1}{2}}$ we have

$$H_{zs}^{n+\frac{1}{2}}(i,j,k) = \frac{\left(\frac{b(i,j,k)}{\Delta t} - \frac{R(i,j,k)}{2}\right)}{\left(\frac{b(i,j,k)}{\Delta t} + \frac{R(i,j,k)}{2}\right)} H_{zs}^{n-\frac{1}{2}}(i,j,k)$$

$$+ \frac{1}{\Delta\phi\mu_0\left(\frac{b(i,j,k)}{\Delta t} + \frac{R(i,j,k)}{2}\right)}\left(E_{\rho s}^n(i,j+1,k) - E_{\rho s}^n(i,j,k)\right)$$

$$- \frac{1}{\Delta\rho\mu_0\left(\frac{b(i,j,k)}{\Delta t} + \frac{R(i,j,k)}{2}\right)}\left(\overline{E}_\phi^n(i+1,j,k) - \overline{E}_\phi^n(i,j,k)\right) \quad (16.64)$$

followed by updating $H_z^{n+\frac{1}{2}}$

$$H_z^{n+\frac{1}{2}}(i,j,k) = \frac{1}{\kappa(i,j,k)} \times \left[H_{zs}^{n+\frac{1}{2}}(i,j,k) - \Lambda(i,j,k) \times Q_{Hz}^{n+\frac{1}{2}}(i,j,k)\right]$$

$$Q_{Hz}^{n+\frac{1}{2}}(i,j,k) = H_{zs}^{n+\frac{1}{2}}(i,j,k) + Q_{Hz}^{n-\frac{1}{2}}(i,j,k) \times e^{-\frac{\sigma(i,j,k)\Delta t}{\kappa(i,j,k)\varepsilon_0}} \quad (16.65)$$

Depending on where the CPML region's inner interface is located along the grid, one of the update equations needs to be adjusted. Assuming this interface is aligned with the cylindrical surface that hosts E_z, E_ϕ, and H_ρ field components, the updating equation for E_z needs to be adjusted. To update $\overline{E}_z^{n+1}(i,j,k)$ at the interface, $\overline{H}_\phi^{n+\frac{1}{2}}(i,j,k)$ and $\overline{H}_\phi^{n+\frac{1}{2}}(i-1,j,k)$ are required. However, $\overline{H}_\phi^{n+\frac{1}{2}}(i-1,j,k)$ component is located outside the CPML region. Therefore, after updating $H_\phi^{n+\frac{1}{2}}(i-1,j,k)$, we need to compute $\overline{H}_\phi^{n+\frac{1}{2}}(i-1,j,k)$ by multiplying

$H_\phi^{n+\frac{1}{2}}(i-1,j,k)$ by $\rho(i-1,j,k)$ to get $\overline{H}_\phi^{n+\frac{1}{2}}(i-1,j,k)$. This results in the following slightly modified updating equation:

$$
E_{zs}^{n+1}(i,j,k) = \frac{\left(\frac{b(i,j,k)}{\Delta t} - \frac{R(i,j,k)}{2}\right)}{\left(\frac{b(i,j,k)}{\Delta t} + \frac{R(i,j,k)}{2}\right)} E_{zs}^n(i,j,k) + \frac{1}{\Delta\rho\varepsilon_0\left(\frac{b(i,j,k)}{\Delta t} + \frac{R(i,j,k)}{2}\right)}
$$
$$
\times \left(\overline{H}_\phi^{n+\frac{1}{2}}(i,j,k) - (i-1.5)\Delta\rho \times H_\phi^{n+\frac{1}{2}}(i-1,j,k)\right)
$$
$$
- \frac{1}{\Delta\phi\varepsilon_0\left(\frac{b(i,j,k)}{\Delta t} + \frac{R(i,j,k)}{2}\right)} \left(H_{\rho s}^{n+\frac{1}{2}}(i,j,k) - H_{\rho s}^{n+\frac{1}{2}}(i,j-1,k)\right)
$$

$$(16.66)$$

16.2.3 Numerical Examples

In this section, spurious reflections from the CPML interface will be analyzed for different rotational mode orders. We will consider radial computational domain sizes of 0.1λ, 0.25λ, and λ, excluding the CPML region, where λ is the wavelength. The reflection coefficient from the vacuum-PML interface can be calculated using the difference of the sampled time domain data between the CPML truncated simulation and a reference simulation where the radial dimension is extended far enough to allow data collection before eventual boundary reflections approach the measuring probe. The expression for reflection coefficient used here is

$$
\Gamma = 20 \times \log_{10}\left(\frac{\max\left|E_z^{pml} - E_z^{ref}\right|}{\max\left|E_z^{ref}\right|}\right)
$$

$$(16.67)$$

A cylindrical shell source around the z-axis will be used for the simulation to avoid any reflections from the axial terminations while allowing full control of rotational mode excitations. The CPML test configuration is shown in Figure 16.8 which illustrates both the source orientation and the measurement probe at the CPML interface.

The source waveform used here is the cosine-modulated Gaussian [2] function given as

$$
\text{waveform }(m,t) = \cos(m\phi) \times \cos(\omega(t - t_0)) \times e^{-\left(\frac{t-t_0}{\tau}\right)^2}
$$

$$(16.68)$$

where $t_0 = 4.5\tau$, $\tau = nc \times \Delta S$, nc is the number of CFDTD cells per wavelength, ΔS is the largest spatial step in the domain, and m represents the rotational mode order. Figure 16.9 shows a sampling of this waveform, in both time and frequency domains, where the central frequency is chosen as 1 GHz.

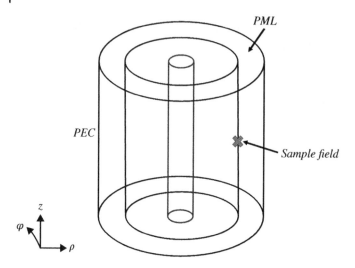

Figure 16.8 Three-dimensional CFDTD domain for testing radial CPML implementations.

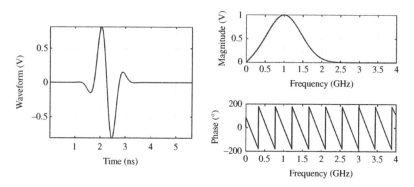

Figure 16.9 A cosine-modulated Gaussian waveform with a central frequency of 1 GHz.

This waveform is added to the updating equation for E_z along the cylindrical shell source shown in Figure 16.8:

$$E_z^{n+1}(i,j,k) = C_{eze}(i,j,k) \times E_z^n(i,j,k)$$

$$+ \left(C_{ezh\phi1}(i,j,k) \times H_\phi^{n+\frac{1}{2}}(i,j,k) - C_{ezh\phi2}(i,j,k) \times H_\phi^{n+\frac{1}{2}}(i-1,j,k) \right)$$

$$+ C_{ezh\rho}(i,j,k) \times \left(H_\rho^{n+\frac{1}{2}}(i,j,k) - H_\rho^{n+\frac{1}{2}}(i,j-1,k) \right)$$

$$+ C_{ezj}(i,j,k) \times \text{waveform}(m,n) \tag{16.69}$$

The remaining CPML parameters are

- Number of cells per wavelength, $nc = 20$ (grid density)
- $\Delta\rho = \Delta z = \lambda/nc = 0.015$ m at 1 GHz, $\Delta\varphi = 7.2°$, and $\Delta t = 3.1293$ ps
- Reference domain size in the ρ-direction is 25λ
- Domain size in the z-direction is 4 cells
- The number of CPML layers is NL $= \lambda/2 = nc/2 = 10$
- CPML region has the following parameters: $n_\sigma = n_\kappa = 3$, $\kappa_{max} = 1$, and σ_{max} variable
- Number of time step is $N = 2000$

Figure 16.10 demonstrates the effect of varying σ_{max} on the CPML performance when placed at a challenging distance of 0.1λ from the z-axis. As can be seen, the CPML performs best at low rotational modes. Figure 16.11 shows that increasing

Figure 16.10 CPML reflections for the first four rotational modes when the CPML region is located 0.1λ away from the z-axis with $nc = 20$.

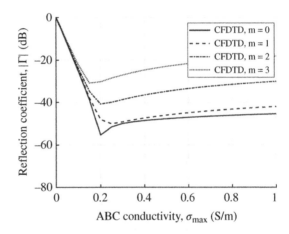

Figure 16.11 CPML reflections for the first four rotational modes when the CPML region is located 0.1λ away from the z-axis when the grid density is increased to $nc = 160$.

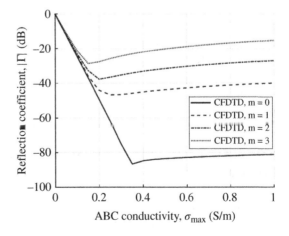

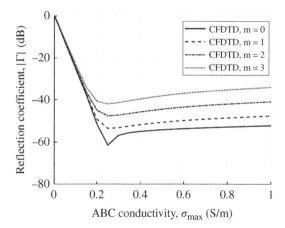

Figure 16.12 CPML reflections for the first four rotational modes when the CPML region is located 0.25λ away from the z-axis.

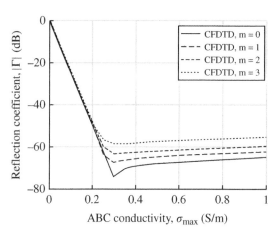

Figure 16.13 CPML reflections for the first four rotational modes when the CPML region is located λ away from the z-axis.

the CFDTD grid density will improve CPML performance for the $m = 0$, though not for higher rotational modes.

Figures 16.12 and 16.13 demonstrate progressively improved CPML performance as the CPML region is further pushed away from the z-axis to 0.25λ and λ, respectively. At distances beyond one wavelength, CPML performance at all rotational modes converges and approaches that of Cartesian FDTD CPML.

16.3 Cylindrical FDTD Formulation for Circuit Elements

For accurate simulation of many practical applications, in cylindrical coordinates, it is usually required to include circuit elements. This section will present the

Figure 16.14 A voltage source placed between nodes (i, j, k) and $(i, j, k+1)$ (a) with and (b) without an internal impedance.

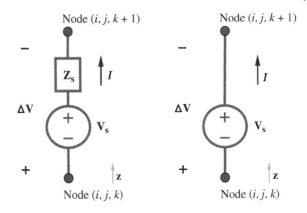

CFDTD formulation and the corresponding updating equations for several passive and active circuit elements. Only elements where the associated voltage and current follows a linear relationship will be considered.

16.3.1 Voltage Source with Internal Impedance

The first circuit element to be considered is a voltage source with its internal impedance as shown in Figure 16.14a that is referred to as a soft source. On the contrary, a voltage source without internal impedance is called a hard source as shown in Figure 16.14b but this will not be used in the simulations here as it introduces artificial reflections.

The voltage V_s is a time-varying function selected such that its Fourier transform covers all frequencies of interest.

The internal impedance in Figure 16.14a can be reduced to a resistor as shown in Figure 16.15. Then by applying Kirchhoff's voltage law (KVL) between nodes (i, j, k) and $(i, j, k+1)$, we can obtain

$$I = \frac{\Delta V + V_s}{R_v} \tag{16.70}$$

where ΔV is the potential difference between nodes (i, j, k) and $(i, j, k+1)$ which can be represented as:

$$\Delta V = \Delta z \times E_z^{n+\frac{1}{2}}(i, j, k)$$

$$= \frac{\Delta z}{2} \times \left(E_z^{n+1}(i, j, k) + E_z^n(i, j, k) \right) \tag{16.71}$$

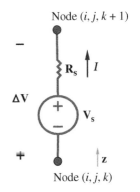

Figure 16.15 A voltage source placed between nodes (i, j, k) and $(i, j, k+1)$ with internal resistor.

The current density J_i can be expressed in term of the current I along the z-axis by

$$I = \int_S \vec{J}_i \cdot d\vec{s} \tag{16.72}$$

where S is the cross-sectional area normal to the current direction. The development of the FDTD updating equation of the voltage source placed between nodes (i, j, k) and $(i, j, k+1)$ continues by modifying the corresponding E_z field component updating equation. Two cases are considered, i.e. when the voltage source is placed along the z-axis or along the edge of an arbitrary cell in the simulation domain. In the first case, the relationship between the current density and the current can be expressed as

$$I = A_{z0} \times J_{iz}^{n+\frac{1}{2}}(1, j, k) \tag{16.73}$$

where A_{z0} is the area of the surrounding magnetic field as shown in Figure 16.16 and it is given by

$$A_{z0} = \frac{\pi \Delta \rho^2}{4} \tag{16.74}$$

For the second case, the above relation can be expressed as

$$I = A_z(i, j, k) \times J_{iz}^{n+\frac{1}{2}}(i, j, k) \tag{16.75}$$

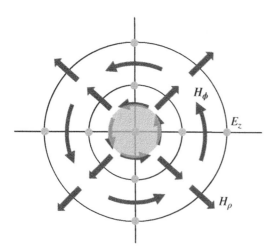

Figure 16.16 Top view of the cylindrical FDTD space where the area of the surrounding magnetic field of $E_z(1, j, k)$ is illustrated (the dark area).

Figure 16.17 Top view of the cylindrical FDTD space where the area of the surrounding magnetic field of $E_z(i, j, k)$ is illustrated (the dark area).

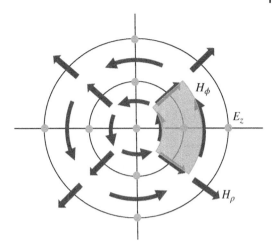

where A_z now depends on the voltage source position. An example is shown in Figure 16.17 and it is given by

$$A_z(i,j,k) = (i-1)\Delta\rho_2\Delta\phi \tag{16.76}$$

By substituting Eqs. (16.71), (16.73), and (16.74) into Eq. (16.70) and rearranging the terms, we get

$$J_{iz}^{n+\frac{1}{2}}(1,j,k) = \frac{\Delta z}{2A_{z0}R_s}\left(E_{iz}^{n+1}(1,j,k) + E_{iz}^n(1,j,k)\right) + \frac{1}{A_{z0}R_s}V_s \tag{16.77}$$

and if the voltage source is placed anywhere in the simulation domain other than the z-axis, we have

$$J_{iz}^{n+\frac{1}{2}}(i,j,k) = \frac{\Delta z}{2A_z(i,j,k)R_s}\left(E_{iz}^{n+1}(i,j,k) + E_{iz}^n(i,j,k)\right) + \frac{1}{A_z(i,j,k)R_s}V_s \tag{16.78}$$

Now, the current density in Eqs. (16.77) and (16.78) can be substituted in Eqs. (16.19) and (16.9), respectively. After some manipulations, the updating equation for $E_z(1, j, k)$ with a voltage source and internal resistor along the z-axis is given by

$$E_z^{n+1}(1,j,k) = C_{eze}(1,j,k) \times E_z^n(1,j,k)$$
$$+ C_{ezh\phi}(1,j,k) \times \sum_{j=1}^{N_\phi} H_\phi^{n+\frac{1}{2}}(1,j,k)$$
$$+ C_{ezs}(1,j,k) \times V_s \tag{16.79}$$

where

$$C_{eze}(1,j,k) = \frac{2\varepsilon_z(1,j,k) - \Delta t\sigma_z^e(1,j,k) - a_{Ez0}}{2\varepsilon_z(1,j,k) + \Delta t\sigma_z^e(1,j,k) + a_{Ez0}}$$

$$C_{ezh\phi}(1,j,k) = \frac{2\Delta t}{2\varepsilon_z(1,j,k) + \Delta t\sigma_{z0}^e(1,j,k) + a_{Ez0}} \frac{2\Delta\phi}{\pi\Delta\rho}$$

$$C_{ezs}(1,j,k) = -\frac{2\Delta t}{2\varepsilon_{z0}(1,j,k) + \Delta t\sigma_{z0}^e(1,j,k) + a_{Ez0}} \frac{1}{A_{z0}R_S}$$

and $a_{Ez0} = \frac{\Delta z \Delta t}{A_{z0}R_S}$.

For an arbitrary positioned voltage source, the updating equation for $E_z(i, j, k)$ is given by

$$E_z^{n+1}(i,j,k) = C_{eze}(i,j,k) \times E_z^n(i,j,k)$$

$$+ \left(C_{ezh\phi1}(i,j,k) \times H_\phi^{n+\frac{1}{2}}(i,j,k) - C_{ezh\phi2}(i,j,k) \times H_\phi^{n+\frac{1}{2}}(i-1,j,k) \right)$$

$$+ C_{ezh\rho}(i,j,k) \times \left(H_\rho^{n+\frac{1}{2}}(i,j,k) - H_\rho^{n+\frac{1}{2}}(i,j-1,k) \right)$$

$$+ C_{ezs}(i,j,k) \times V_S \tag{16.80}$$

where

$$C_{eze}(i,j,k) = \frac{2\varepsilon_z(i,j,k) - \Delta t\sigma_z^e(i,j,k) - a_{Ez}}{2\varepsilon_z(i,j,k) + \Delta t\sigma_z^e(i,j,k) + a_{Ez}}$$

$$C_{ezh\phi1}(i,j,k) = \frac{2\Delta t}{2\varepsilon_z(i,j,k) + \Delta t\sigma_z^e(i,j,k) + a_{Ez}} \frac{(i-0.5)}{(i-1)\Delta\rho}$$

$$C_{ezh\phi2}(i,j,k) = \frac{2\Delta t}{2\varepsilon_z(i,j,k) + \Delta t\sigma_z^e(i,j,k) + a_{Ez}} \frac{(i-1.5)}{(i-1)\Delta\rho}$$

$$C_{ezh\rho}(i,j,k) = -\frac{2\Delta t}{2\varepsilon_z(i,j,k) + \Delta t\sigma_z^e(i,j,k) + a_{Ez}} \frac{1}{\Delta\rho\Delta\phi(i-1)}$$

$$C_{ezs}(i,j,k) = -\frac{2\Delta t}{2\varepsilon_z(i,j,k) + \Delta t\sigma_z^e(i,j,k) + a_{Ez}} \frac{1}{A_z(i,j,k)R_S}$$

and $a_{Ez} = \frac{\Delta z \Delta t}{A_z(i,j,k)R_S}$.

By using the same methodology, the updating equation for $E_\phi(i, j, k)$ with a voltage source and internal resistor along the φ direction is given by

$$E_\phi^{n+1}(i,j,k) = C_{e\phi e}(i,j,k) \times E_\phi^n(i,j,k)$$

$$+ C_{e\phi h\rho}(i,j,k) \times \left(H_\rho^{n+\frac{1}{2}}(i,j,k) - H_\rho^{n+\frac{1}{2}}(i,j,k-1) \right)$$

$$+ C_{e\phi hz}(i,j,k) \times \left(H_z^{n+\frac{1}{2}}(i,j,k) - H_z^{n+\frac{1}{2}}(i-1,j,k) \right)$$

$$+ C_{e\phi s}(i,j,k) \times V_S \tag{16.81}$$

where

$$C_{e\phi e}(i,j,k) = \frac{2\varepsilon_\phi(i,j,k) - \Delta t \sigma_\phi^e(i,j,k) - a_{E\phi}}{2\varepsilon_\phi(i,j,k) + \Delta t \sigma_\phi^e(i,j,k) + a_{E\phi}}$$

$$C_{e\phi h\rho}(i,j,k) = -\frac{2\Delta t}{2\varepsilon_\phi(i,j,k) + \Delta t \sigma_\phi^e(i,j,k) + a_{E\phi}} \frac{1}{\Delta z}$$

$$C_{e\phi hz}(i,j,k) = -\frac{2\Delta t}{2\varepsilon_\phi(i,j,k) + \Delta t \sigma_\phi^e(i,j,k) + a_{E\phi}} \frac{1}{\Delta\rho}$$

$$C_{e\phi s}(i,j,k) = -\frac{2\Delta t}{2\varepsilon_\phi(i,j,k) + \Delta t \sigma_\phi^e(i,j,k) + a_{E\phi}} \frac{1}{A_\phi R_S}$$

and $a_{E\phi} = \frac{i\Delta\rho\Delta\phi\Delta t}{A_\phi R_S}$, $A_\phi = \Delta\rho\Delta z$.

Also, the updating equation for $E_\rho(i,j,k)$ with a voltage source and internal resistor along the ρ direction is given by

$$E_\rho^{n+1}(i,j,k) = C_{e\rho e}(i,j,k) \times E_z^n(i,j,k)$$

$$+ C_{e\rho hz}(i,j,k) \times \left(H_z^{n+\frac{1}{2}}(i,j,k) - H_z^{n+\frac{1}{2}}(i,j-1,k) \right)$$

$$+ C_{e\rho h\phi}(i,j,k) \times \left(H_\phi^{n+\frac{1}{2}}(i,j,k) - H_\phi^{n+\frac{1}{2}}(i,j,k-1) \right)$$

$$+ C_{e\rho s}(i,j,k) \times V_s \qquad (16.82)$$

where

$$C_{e\rho e}(i,j,k) = \frac{2\varepsilon_\rho(i,j,k) - \Delta t \sigma_\rho^e(i,j,k) - a_{E\rho}}{2\varepsilon_\rho(i,j,k) + \Delta t \sigma_\rho^e(i,j,k) + a_{E\rho}}$$

$$C_{e\rho hz}(i,j,k) = \frac{2\Delta t}{2\varepsilon_\rho(i,j,k) + \Delta t \sigma_\rho^e(i,j,k) + a_{E\rho}} \frac{1}{(i-0.5)\Delta\rho\Delta\phi}$$

$$C_{e\rho h\phi}(i,j,k) = -\frac{2\Delta t}{2\varepsilon_\rho(i,j,k) + \Delta t \sigma_\rho^e(i,j,k) + a_{E\rho}} \frac{1}{\Delta z}$$

$$C_{e\rho s}(i,j,k) = -\frac{2\Delta t}{2\varepsilon_\rho(i,j,k) + \Delta t \sigma_\rho^e(i,j,k) + a_{E\rho}} \frac{1}{A_\rho(i,j,k)R_S}$$

and $a_{E\rho} = \frac{\Delta\rho\Delta t}{A_\rho(i,j,k)R_S}$, $A_\rho(i,j,k) = (i-0.5)\Delta\rho\Delta\phi\Delta z$.

The internal impedance in Figure 16.14a can be represented by a resistor and a capacitor as shown in Figure 16.18.

By applying KVL between nodes (i,j,k) and $(i,j,k+1)$ as in Eq. (16.70), and when the voltage source is a long the z-axis, the current density in frequency domain is given by

$$J_{iz}(\omega) = \frac{\Delta z}{A_z} \frac{j\omega C_s}{1 + j\omega C_s R_S} E_z + \frac{1}{A_z} \frac{j\omega C_s}{1 + j\omega C_s R_S} V_s \quad (16.83)$$

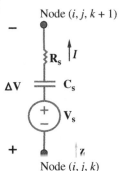

Figure 16.18 A voltage source placed between nodes (i, j, k) and $(i, j, k+1)$ with internal resistor and capacitor.

The inverse Fourier transform of (16.83) in discrete form is given by

$$
\begin{aligned}
J_{iz}^{n+\frac{1}{2}}(i,j,k) = {} & \frac{\Delta z}{A_z(i,j,k)} \left(E_z^n(i,j,k) + \left(e^{-\frac{\Delta t}{C_s R_s}} - 1 \right) \times Q_{z1}^{n+\frac{1}{2}}(i,j,k) \right) \\
& + \frac{1}{A_z(i,j,k)} \left(V_s(n) + \left(e^{-\frac{\Delta t}{C_s R_s}} - 1 \right) \times Q_{z2}^{n+\frac{1}{2}}(i,j,k) \right)
\end{aligned}
\tag{16.84}
$$

where

$$
Q_{z1}^{n+\frac{1}{2}}(i,j,k) = E_z^n(i,j,k) + Q_{z1}^{n-\frac{1}{2}}(i,j,k) \times e^{-\frac{\Delta t}{C_s R_s}}
\tag{16.85}
$$

$$
Q_{z2}^{n+\frac{1}{2}}(i,j,k) = V_s(n) + Q_{z2}^{n-\frac{1}{2}}(i,j,k) \times e^{-\frac{\Delta t}{C_s R_s}}
\tag{16.86}
$$

The internal impedance in Figure 16.14a can be represented by a resistor and a inductor as shown in Figure 16.19.

By applying KVL between nodes (i,j,k) and $(i,j,k+1)$, we get

$$
\Delta V + V_s - IR_s - V_L = 0
\tag{16.87}
$$

where V_L is the voltage across the inductor and the relation between this voltage and the current through the inductor is given by

$$
V_L = L_s \frac{dI}{dt}
\tag{16.88}
$$

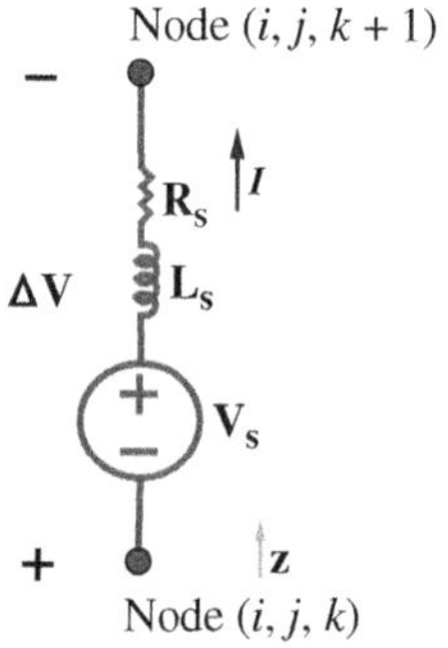

In discrete form, Eq. (16.88) can be reduced to

$$
\Delta z E_z^n(i,j,k) + V_s(n) - \frac{A_z R_s}{2} \left(J_{iz}^{n+\frac{1}{2}}(i,j,k) + J_{iz}^{n-\frac{1}{2}}(i,j,k) \right)
$$
$$
- \frac{A_z L_s}{\Delta t} \left(J_{iz}^{n+\frac{1}{2}}(i,j,k) - J_{iz}^{n-\frac{1}{2}}(i,j,k) \right) = 0
\tag{16.89}
$$

After rearranging the term, we get

$$
\begin{aligned}
J_{iz}^{n+\frac{1}{2}}(i,j,k) = {} & \left(\frac{2L_s - \Delta t R_s}{2L_s + \Delta t R_s} \right) \times J_{iz}^{n-\frac{1}{2}}(i,j,k) \\
& + \Delta z \left(\frac{2\Delta t}{2A_z(i,j,k)L_s + \Delta t A_z(i,j,k)R_s} \right) \times E_z^n(i,j,k) \\
& + \left(\frac{2\Delta t}{2A_z(i,j,k)L_s + \Delta t A_z(i,j,k)R_s} \right) \times V_s(n)
\end{aligned}
\tag{16.90}
$$

Figure 16.19 The voltage source places between nodes (i,j,k) and $(i,j,k+1)$ with internal resistor and inductor.

16.3.2 Cylindrical FDTD Updating Equation for a Load Impedance

16.3.2.1 Resistor

Figure 16.20 represent a resistor along the z-direction between the nodes (i, j, k) and $(i, j, k + 1)$.

The updating equation for $J_{iz}^{n+\frac{1}{2}}(i, j, k)$ for such a resistor can be derived from the soft voltage source Eq. (16.78) by letting $V_s \to 0$. The resulting update equation for $E_z(i, j, k)$ is then given by

$$
\begin{aligned}
E_z^{n+1}(i, j, k) = \ & C_{eze}(i, j, k) \times E_z^n(i, j, k) \\
& + \Big(C_{ezh\phi1}(i, j, k) \times H_\phi^{n+\frac{1}{2}}(i, j, k) \\
& \quad - C_{ezh\phi2}(i, j, k) \times H_\phi^{n+\frac{1}{2}}(i - 1, j, k) \Big) \\
& + C_{ezh\rho}(i, j, k) \\
& \quad \times \Big(H_\rho^{n+\frac{1}{2}}(i, j, k) - H_\rho^{n+\frac{1}{2}}(i, j - 1, k) \Big) \quad (16.91)
\end{aligned}
$$

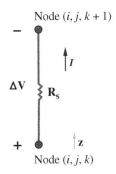

Node $(i, j, k + 1)$

ΔV R_s

Node (i, j, k)

Figure 16.20
A resistor placed between nodes (i, j, k) and $(i, j, k + 1)$.

where

$$
C_{eze}(i, j, k) = \frac{2\varepsilon_z(i, j, k) - \Delta t \sigma_z^e(i, j, k) - a_{Ez}}{2\varepsilon_z(i, j, k) + \Delta t \sigma_z^e(i, j, k) + a_{Ez}}
$$

$$
C_{ezh\phi1}(i, j, k) = \frac{2\Delta t}{2\varepsilon_z(i, j, k) + \Delta t \sigma_z^e(i, j, k) + a_{Ez}} \frac{(i - 0.5)}{(i - 1)\Delta\rho}
$$

$$
C_{ezh\phi2}(i, j, k) = \frac{2\Delta t}{2\varepsilon_z(i, j, k) + \Delta t \sigma_z^e(i, j, k) + a_{Ez}} \frac{(i - 1.5)}{(i - 1)\Delta\rho}
$$

$$
\begin{aligned}
C_{ezh\rho}(i, j, k) = \ & -\frac{2\Delta t}{2\varepsilon_z(i, j, k) + \Delta t \sigma_z^e(i, j, k) + a_{Ez}} \\
& \times \frac{1}{\Delta\rho\Delta\phi(i - 1)}
\end{aligned}
$$

and $a_{Ez} = \frac{\Delta z \Delta t}{A_z(i,j,k) R_S}$.

Node $(i, j, k + 1)$

ΔV C_s

Node (i, j, k)

Figure 16.21
A capacitor placed between nodes (i, j, k) and $(i, j, k + 1)$.

16.3.2.2 Capacitor

Figure 16.21 represents a capacitor along the z-direction between nodes (i, j, k) and $(i, j, k + 1)$ where the updating equation for $J_{iz}^{n+\frac{1}{2}}(i, j, k)$ with a capacitor is given by

$$
J_{iz}^{n+\frac{1}{2}}(i, j, k) = \frac{C_s \Delta z}{\Delta t A_z(i, j, k)} \Big(E_{iz}^{n+1}(i, j, k) - E_{iz}^n(i, j, k) \Big) \quad (16.92)
$$

and the corresponding updating equation for $E_z(i,j,k)$ becomes

$$E_z^{n+1}(i,j,k) = C_{eze}(i,j,k) \times E_z^n(i,j,k)$$
$$+ \left(C_{ezh\phi1}(i,j,k) \times H_\phi^{n+\frac{1}{2}}(i,j,k) - C_{ezh\phi2}(i,j,k) \times H_\phi^{n+\frac{1}{2}}(i-1,j,k) \right)$$
$$+ C_{ezh\rho}(i,j,k) \times \left(H_\rho^{n+\frac{1}{2}}(i,j,k) - H_\rho^{n+\frac{1}{2}}(i,j-1,k) \right) \tag{16.93}$$

where

$$C_{eze}(i,j,k) = \frac{2\varepsilon_z(i,j,k) - \Delta t\sigma_z^e(i,j,k) + b_{Ez}}{2\varepsilon_z(i,j,k) + \Delta t\sigma_z^e(i,j,k) + b_{Ez}}$$

$$C_{ezh\phi1}(i,j,k) = \frac{2\Delta t}{2\varepsilon_z(i,j,k) + \Delta t\sigma_z^e(i,j,k) + b_{Ez}} \frac{(i-0.5)}{(i-1)\Delta\rho}$$

$$C_{ezh\phi2}(i,j,k) = \frac{2\Delta t}{2\varepsilon_z(i,j,k) + \Delta t\sigma_z^e(i,j,k) + b_{Ez}} \frac{(i-1.5)}{(i-1)\Delta\rho}$$

$$C_{ezh\rho}(i,j,k) = -\frac{2\Delta t}{2\varepsilon_z(i,j,k) + \Delta t\sigma_z^e(i,j,k) + b_{Ez}} \frac{1}{\Delta\rho\Delta\phi(i-1)}$$

and $b_{Ez} = \frac{2C_s\Delta z}{A_z(i,j,k)}$.

16.3.2.3 Inductor
Figure 16.22 represents an inductor along the z-direction between nodes (i,j,k) and $(i,j,k+1)$ where the updating equation for $J_{iz}^{n+\frac{1}{2}}(i,j,k)$ with an inductor is given by

$$J_{iz}^{n+\frac{1}{2}}(i,j,k) = J_{iz}^{n-\frac{1}{2}}(i,j,k) + \frac{\Delta t\Delta z}{L_sA_z(i,j,k)}$$
$$\times E_z^n(i,j,k) \tag{16.94}$$

and the corresponding updating equation for $E_z(i,j,k)$ becomes

$$E_z^{n+1}(i,j,k) = C_{eze}(i,j,k) \times E_z^n(i,j,k)$$
$$+ \left(C_{ezh\phi1}(i,j,k) \times H_\phi^{n+\frac{1}{2}}(i,j,k) - C_{ezh\phi2}(i,j,k) \times H_\phi^{n+\frac{1}{2}}(i-1,j,k) \right)$$
$$+ C_{ezh\rho}(i,j,k) \times \left(H_\rho^{n+\frac{1}{2}}(i,j,k) - H_\rho^{n+\frac{1}{2}}(i,j-1,k) \right)$$
$$+ C_{ezj}(i,j,k) \times J_{iz}^{n+\frac{1}{2}}(i,j,k) \tag{16.95}$$

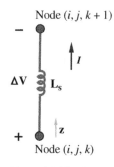

Node $(i,j,k+1)$

ΔV L_s

I

z

Node (i,j,k)

Figure 16.22 An inductor placed between nodes (i,j,k) and $(i,j,k+1)$.

where

$$C_{eze}(i,j,k) = \frac{2\varepsilon_z(i,j,k) - \Delta t \sigma_z^e(i,j,k)}{2\varepsilon_z(i,j,k) + \Delta t \sigma_z^e(i,j,k)}$$

$$C_{ezh\phi 1}(i,j,k) = \frac{2\Delta t}{2\varepsilon_z(i,j,k) + \Delta t \sigma_z^e(i,j,k)} \frac{(i-0.5)}{(i-1)\Delta\rho}$$

$$C_{ezh\phi 2}(i,j,k) = \frac{2\Delta t}{2\varepsilon_z(i,j,k) + \Delta t \sigma_z^e(i,j,k)} \frac{(i-1.5)}{(i-1)\Delta\rho}$$

$$C_{ezh\rho}(i,j,k) = -\frac{2\Delta t}{2\varepsilon_z(i,j,k) + \Delta t \sigma_z^e(i,j,k)} \frac{1}{\Delta\rho\Delta\phi(i-1)}$$

$$C_{ezj}(i,j,k) = -\frac{2\Delta t}{2\varepsilon_z(i,j,k) + \Delta t \sigma_z^e(i,j,k)}$$

Notice that $J_{iz}^{n+\frac{1}{2}}(i,j,k)$ should be calculated before $E_{iz}^{n+1}(i,j,k)$.

16.3.3 Simulation Examples

In this section, different circuit configurations composed of a voltage source with internal impedance connected to a load impedance will be simulated for the verification of the developed equations. The first configuration has a circuit with a voltage source connected to a load impedance. In the first example of this category, all circuit elements will be placed along the z-direction as shown in Figure 16.23 where the voltage source with an internal impedance is placed between nodes (i,j,k) and $(i,j,k+1)$. The load impedance is placed between nodes $(i+1,j,k)$ and $(i+1,j,k+1)$. The circuit elements will be connected by wires coinciding with a grid boundary. Along these boundaries the conductivity $\sigma_\rho^e(i,j,k)$ and $\sigma_\rho^e(i,j,k+1)$ is to be set large enough (usually in the order of 10^{20} S/m) to act as a perfect electric conductor.

The voltage source time waveform is a sinusoidal function with a $50\,\Omega$ impedance and the impedance at the load side is also a 50 Ω resistor. The voltage source is located parallel to the z-axis at node (i, j, k). The sampled voltage in CFDTD in the z direction for node $(i+1, j, k)$ is calculated as

$$V_L(n) = \Delta z E_z^n(i+1,j,k) \tag{16.96}$$

The results of the sampled voltage based on the CFDTD implementation and the analytical solution from circuit theory are shown in Figure 16.24. Good agreement is clearly shown between the two results which verified the validity of the developed CFDTD equation and their implementation.

The previous example is repeated but now the source impedance is replaced by a combination of a $50\,\Omega$ resistor and a 1 nF capacitor. The voltage source sinusoidal waveform is also replaced by a unit step function. The results for this case

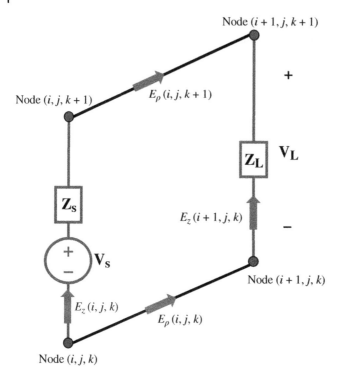

Figure 16.23 A simple circuit diagram in FDTD cell configuration where the elements are placed parallel to the *z*-direction.

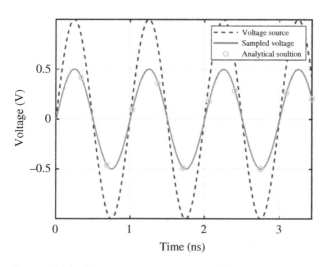

Figure 16.24 Sampled voltage across $Z_L = R_L$ from the CFDTD simulation.

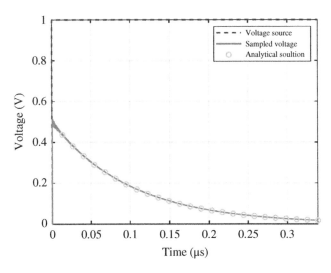

Figure 16.25 A voltage source with $Z_S = R_S + Z_C$ connected to a load impedance $Z_L = R_L$.

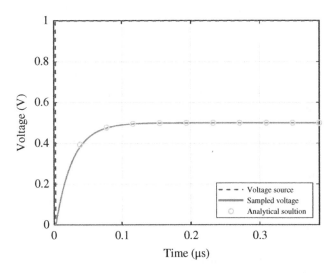

Figure 16.26 Sampled voltage across $Z_l = R_l$ from the CFDTD simulation

are shown in Figure 16.25 where the sampled voltage across the load is in good agreement with the corresponding analytical solution.

Then the source impedance is replaced by a 50 Ω resistor and 2.5 μH inductor. The results of this case are shown in Figure 16.26 where the sampled voltage again compared with the analytical solution and good agreement is observed.

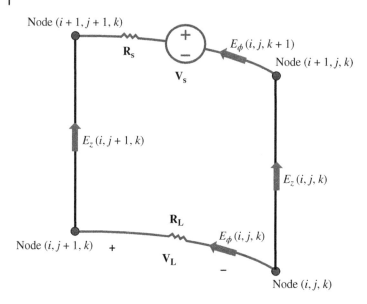

Figure 16.27 A circuit configuration where the elements are placed in the ϕ direction.

In the following example, the circuit elements will be placed along the ϕ direction as shown in Figure 16.27 where the voltage source and an internal resistor will be placed between nodes $(i, j, k + 1)$ and $(i, j + 1, k + 1)$. The resistive load will be placed between nodes (i, j, k) and $(i, j + 1, k)$. The connecting wires will be generated by forcing the conductivity $\sigma_z^e(i, j, k)$ and $\sigma_z^e(i, j + 1, k)$ to be large enough to act as a good conductor.

A sinusoidal waveform will be assigned to the voltage source with a 50 Ω internal resistance connected to a resistive load of 50 Ω. The sampled FDTD voltage in the ϕ direction will be calculated as

$$V_L(n) = (i - 1)\Delta\rho\Delta\varphi E_\phi^n(i + 1, j, k) \tag{16.97}$$

The result of this example is shown in Figure 16.28 where the sampled voltage agrees with the analytical solution.

In the example below, the circuit elements will be placed along the ρ direction as shown in Figure 16.29 where the voltage source with an internal resistor will be placed between nodes $(i, j, k + 1)$ and $(i + 1, j, k + 1)$. The resistive load will be placed between nodes (i, j, k) and $(i + 1, j, k)$.

The voltage source waveform is a sinusoidal function, the impedance at the source and at the load sides is a 50 Ω resistance. The sampled voltage across the load in the ρ direction will be calculated as

$$V_L(n) = \Delta\rho E_\rho^n(i + 1, j, k) \tag{16.98}$$

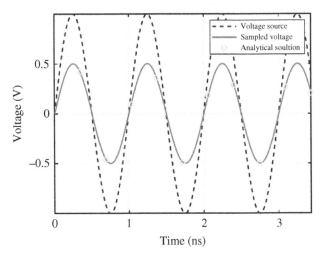

Figure 16.28 The voltage across the resistive load R_L in the circuit in Figure 16.27.

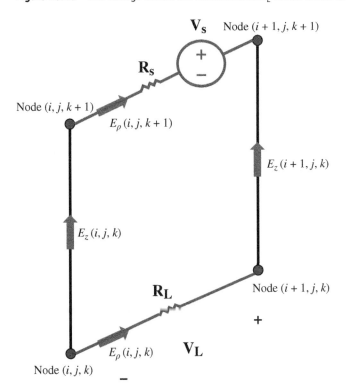

Figure 16.29 A circuit configuration in CFDTD where the elements are placed in the ρ direction.

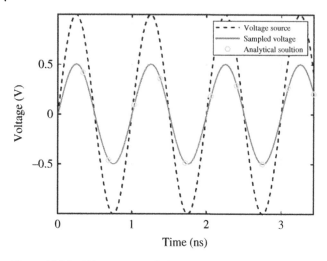

Figure 16.30 Voltage across R_L for the elements placed in the ρ direction.

The result of this example is shown in Figure 16.30 with good agreement between the sampled voltage from the CFDTD simulation and the analytical voltage based on circuit theory.

The second circuit configuration to be examined in CFDTD is shown in Figure 16.31 where the two parallel loads are connected to the source. In this example a voltage source placed along the z-axis as shown in Figure 16.32 with two loads sharing the same source. The voltage source with an internal impedance will be placed between nodes $(1, j, k)$ and $(1, j, k + 1)$ with the first load impedance between nodes $(2, j, k)$ and $(2, j, k+1)$. The second load impedance is located between nodes $(2, j+1, k)$ and $(2, j+1, k+1)$.

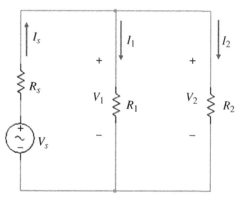

Figure 16.31 Circuit configuration of a source connected to two parallel loads.

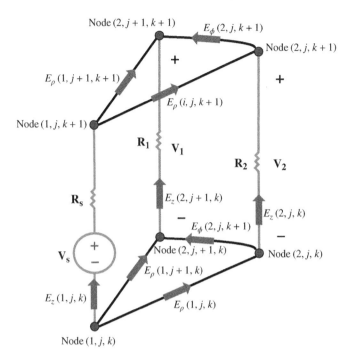

Figure 16.32 A FDTD circuit diagram for the circuit in Figure 16.31.

The sinusoidal waveform is considered for the voltage source with 50 Ω internal impedance. The load will be two 100 Ω resistors in parallel. The result of this example is shown in Figure 16.33 with good agreement between the sampled CFDTD voltage across the load compared with the corresponding analytical solution.

The third circuit configuration to be tested in CFDTD is shown in Figure 16.34.

The elements will be placed parallel to the z-direction as shown in Figure 16.35. The voltage sources and their impedances will be places between nodes $(2, 1, 2)$, $(2, 1, 3)$, and between nodes $(2, 2, 2)$, $(2, 2, 3)$. The impedances at the load side will be placed between nodes $(3, 1, 2)$, $(3, 1, 3)$, and between nodes $(3, 2, 2)$, $(3, 2, 3)$.

The sinusoidal waveform is considered for the two voltage sources with 50 Ω internal impedance. The load will be two 50 Ω resistors in parallel. The result of this study is shown in Figure 16.36 with good agreement between the CFDTD sampled voltage across the two loads compared to that of the analytical solutions.

Similar test has been conducted for the third study for the parallel sources and resistors placed in the ρ and ϕ directions and good agreements were observed.

A final test configuration is for a circuit composed of a voltage source with internal impedance consisting of an inductor and a resistor. The voltage sources is

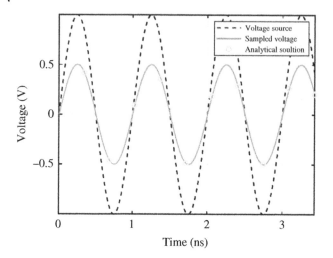

Figure 16.33 Voltage across the two parallel loads in Figure 16.32.

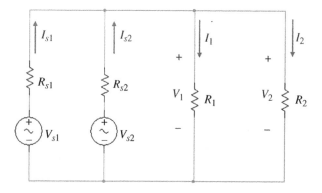

Figure 16.34 Circuit configuration of parallel sources and parallel resistors.

connected to a load impedance which is purely resistive. This circuit configuration verifies the CFDTD formulation when a solution for wide range of frequencies is desired. The circuit has been employed with $Z_s = R_s + X_L$ and $Z_L = R_L$ where $R_s = R_L = 50\ \Omega$ and $L = 50$ pH. The following parameters are used in the CFDTD simulation:

- Domain sizes: $N_\rho = 11, N_\phi = 9, N_z = 10$.
- The step size: $\Delta\rho = \Delta z = 0.1873$ mm, $\Delta\phi = 40°$.
- The time step size and the number of timesteps:

$$\Delta t = 1.956 \times 10^{-13}\ \text{s}, N = 20000$$

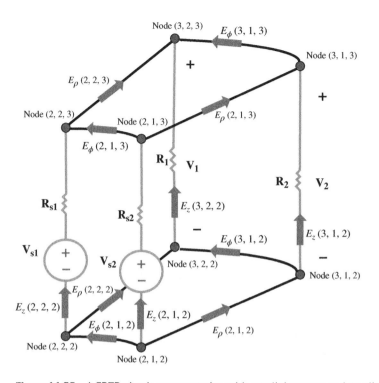

Figure 16.35 A FDTD circuit representation with parallel sources and parallel resistors.

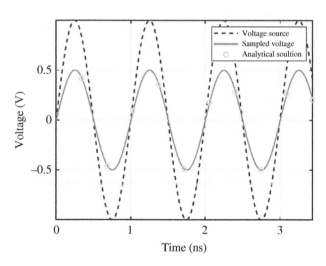

Figure 16.36 Voltage across the two loads in Figure 16.35.

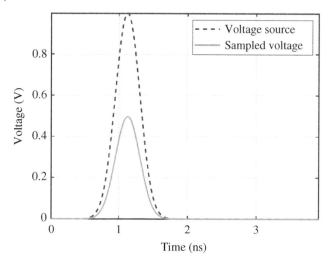

Figure 16.37 Time domain waveform and sampled voltage across the voltage source.

The voltage across the load side will be used to compare the theoretical result with the result from the CFDTD simulation. Theoretically, the voltage across the load side in is given by:

$$V_L = \left(\frac{R_L}{R_L + R_s + j\omega L} \right) V_s \tag{16.99}$$

In CFDTD, the source waveform to be used will determine the valid range of the covered frequencies. Thus, the input voltage V_s and the sampled voltage across the load side in time domain are shown Figure 16.37. The transformation to frequency domain for the voltage source and sampled voltage are performed to compare with the results from the theoretical solution.

The sampled voltage across the load form FDTD and the theoretical voltage in Eq. (16.99) are compared in Figure 16.38 as a function of frequency. The absolute error between sampled voltage and the theoretical results are shown in Figure 16.39.

The results obtained from the above test cases gives us confidence in using the developed updating equations for passive and active circuit element in cylindrical FDTD simulations.

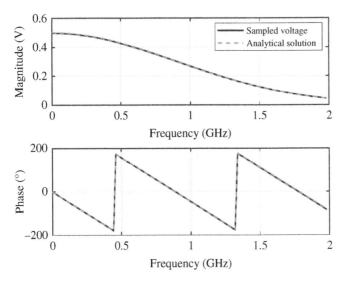

Figure 16.38 Frequency domain sampled voltage and analytical solution across the load.

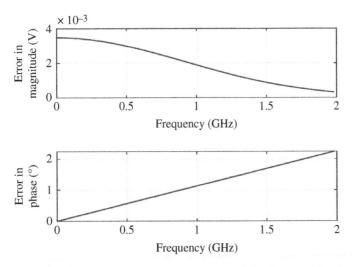

Figure 16.39 The absolute error between sampled voltage and the theoretical results.

16.4 Concluding Remarks

This chapter showcased a detailed presentation of CFDTD, a computational electromagnetics tool for analyzing cylindrical structures. The bulk of the presentation was concerned with proper implementation of the convolutional perfectly-matched-layer ABCs in cylindrical coordinates for FDTD simulations. Key point in this analysis is using a mathematically sound approach of implementing the necessary variable stretching which allows usable wave absorption for multiple rotational modes, provided the CPML region is placed at least a quarter-wavelength away from the z-axis. The investigated CPML region placements correspond to relatively low frequency simulations where modeled structures fall well below a single wavelength in overall size. Furthermore, the development of the formulation of integrating circuit elements in the CFDTD formulation is presented. Very good agreement is observed for simulations of different configurations of circuit elements. The formulations presented in this chapter provides confidence in using CFDTD method for low frequency applications.

References

1 Berenger, J.P. (1994). A perfectly matched layer for absorbing of electromagnetic waves. *Journal of Computational Physics* 114 (2): 185–200.

2 Elsherbeni, A.Z. and Demir, V. (2016). *The Finite-Difference Time-Domain Method for Electromagnetics with MATLAB Simulations*, ACES Series on Computational Electromagnetics and Engineering, 2e. Edison, NJ: SciTech Publishing, an Imprint of IET.

3 Alan Roden, J. and Gedney, S.D. (2000). Convolution PML (CPML): an efficient FDTD implementation of the CFS–PML for arbitrary media. *Microwave and Optical Technology Letters* 27 (5): 334–339.

4 Hadi, M.F., Elsherbeni, A.Z., Piket-May, M.J., and Mahmoud, S.F. (2017). Radial waves based dispersion analysis of the body-of-revolution FDTD method. *IEEE Transactions on Antennas and Propagation* 65 (2): 721–729.

5 Hadi, M.F. and Elsherbeni, A.Z. (2019). *Axial PML Performance Near the Axis of Rotation in Cylindrical FDTD*. Miami, USA: ACES.

6 Taflove, A. and Hagness, S. (2005). *Computational Electrodynamics: The Finite-Difference Time-Domain Method*, 3e. Boston, MA: Artech House.

7 Yu, W., Li, W., Elsherbeni, A., and Rahmat-Samii, Y. (2015). *Advanced Computational Electromagnetic Methods and Applications*. Boston, MA: Artech House.

8 Kuzuoglu, M. and Mittra, R. (1996). Frequency dependence of the constitutive parameters of causal perfectly matched anisotropic absorbers. *IEEE Microwave Guided Wave Letters* 6 (12): 447–449.

9 Teixeira, F.L. and Chew, W.C. (1997). PML-FDTD in cylindrical and spherical grids. *IEEE Microwave and Guided Wave Letters* 7 (9): 285–287.

10 Hadi, M.F. and Elsherbeni, A.Z. (July 2019). Oblique incidence PML reflection analysis for cylindrical FDTD. In: *2019 IEEE International Symposium on Antennas and Propagation and USNC-URSI Radio Science Meeting*, 1655–1656. IEEE.

Index

Advances in Time-Domain Computational Electromagnetic Methods, First Edition.
Edited by Qiang Ren, Su Yan, and Atef Z. Elsherbeni.
© 2023 The Institute of Electrical and Electronics Engineers, Inc. Published 2023 by John Wiley & Sons, Inc.

IEEE PRESS SERIES ON ELECTROMAGNETIC WAVE THEORY

Printed and bound by CPI Group (UK) Ltd, Croydon, CR0 4YY

16/04/2025

14658346-0004